Eleventh Edition

The Marriage and Family Experience

Intimate Relationships in a Changing Society

Bryan Strong
Formerly of University of California, Santa Cruz

Christine DeVault
Cabrillo College

Theodore F. Cohen
Ohio Wesleyan University

WADSWORTH
CENGAGE Learning™

Australia • Brazil • Japan • Korea • Mexico • Singapore • Spain • United Kingdom • United States

The Marriage and Family Experience: Intimate Relationships in a Changing Society, Eleventh Edition
Bryan Strong, Christine DeVault, and Theodore F. Cohen

Publisher: Linda Schreiber-Ganster

Sociology Editor: Erin Mitchell

Developmental Editor: Kristin Makarewycz

Assistant Editor: Rachael Krapf

Editorial Assistant: Pamela Simon

Media Editor: Melanie Cregger

Marketing Manager: Andrew Keay

Marketing Assistant: Jillian Myers

Marketing Communications Manager: Laura Localio

Content Project Manager: Cheri Palmer

Creative Director: Rob Hugel

Art Director: Caryl Gorska

Print Buyer: Judy Inouye

Rights Acquisitions Account Manager, Text: Roberta Broyer

Rights Acquisitions Account Manager, Image: Robyn Young

Production Service: S4Carlisle Publishing Services

Text Designer: Ellen Pettengell

Photo Researcher: Sarah Evertson

Copy Editor: Bruce Owens

Illustrator: S4Carlisle Publishing Services

Cover Designer: RHDG

Cover Image: Creatas/Photolibrary

Compositor: S4Carlisle Publishing Services

For product information and technology assistance, contact us at **Cengage Learning Customer & Sales Support, 1-800-354-9706.**

For permission to use material from this text or product, submit all requests online at **www.cengage.com/permissions**
Further permissions questions can be emailed to **permissionrequest@cengage.com**

Library of Congress Control Number: 2009942041

Student Edition:

ISBN-13: 978-0-534-62425-5

ISBN-10: 0-534-62425-1

Loose-leaf Edition:

ISBN-13: 978-0-8400-3221-8

ISBN-10: 0-8400-3221-8

Wadsworth
20 Davis Drive
Belmont, CA 94002-3098
USA

Cengage Learning is a leading provider of customized learning solutions with office locations around the globe, including Singapore, the United Kingdom, Australia, Mexico, Brazil, and Japan. Locate your local office at **www.cengage.com/global.**

Cengage Learning products are represented in Canada by Nelson Education, Ltd.

To learn more about Wadsworth, visit **www.cengage.com/wadsworth**

Purchase any of our products at your local college store or at our preferred online store **www.CengageBrain.com.**

Printed in Canada
1 2 3 4 5 6 7 13 12 11 10

I dedicate this edition to my parents, Kalman and Eleanor Cohen, whose sixty-year-long marriage has been a wonderful example of how commitment and compromise, along with love and trust, can carry a couple through the many challenges of raising a family and sharing a lifetime together.

—Ted Cohen

brief contents

CONTENTS

contents

boxes

 **Real Families**

 **Public Policies, Private Lives**

preface

It has been thirty years since Bryan Strong wrote the first edition of *The Marriage and Family Experience*. Over three decades and through many editions, it has appealed to teachers and students in a number of different types of institutions and across a range of academic and applied disciplines. From the first edition through the current, eleventh edition, the book has addressed the rich diversity of family experience, the dynamic nature of "the family" and of individual families, and the ways in which family life is affected by the wider social, cultural, economic and political contexts within which we live. The book recognizes that in whatever form(s) we experience them, our families shape who we are and who we become, provide us with our most intimate and loving relationships, and need to be valued and supported.

My personal involvement with *The Marriage and Family Experience* has a shorter history. This is the fourth time I have had the opportunity to revise and update this text. Each time I have incorporated the latest available research and official statistics on subjects such as sexuality (sexual orientation and expression), marriage, cohabitation, childbirth, child care, divorce, remarriage, blended families, adoption, abuse, the division of housework, and connections between paid work and family life. There are more than 600 new references in this edition, drawn mainly though not exclusively from research in sociology, psychology, and family studies. I have also tried to feature many real life examples, sometimes drawn straight from recent news stories or narrative accounts, to give readers a better appreciation for how the more academic content applies to real life.

During my involvement with *The Marriage and Family Experience,* I have been reminded on a personal level, of the range of family experiences people have and of the dynamic quality of family life. When I first began working on the eighth edition of this book,

I was fortunate to be in a stable marriage of more than twenty years. I had no reason to imagine ever being single again or remarrying. My wife and I were parents of two young teenagers who were the center of our too hectic life together. I was a husband and father, two roles that I valued above all others and which I juggled along with my career as a sociologist and teacher. In the years since, I have been a fulltime caregiver for my wife, a widower after her passing, a single parent, a remarried husband and a stepfather. My son and daughter are now in their twenties, and I have two stepsons and a stepdaughter. My wife and I continue to struggle with the challenges of blending families, and continue to experience and mourn the absence of our late spouses. I am fortunate to still have my parents in my life, but I watch anxiously as they age and face a variety of health challenges. My experiences of marriage, fatherhood, caregiving, widowerhood, single parenting, remarriage, and stepfatherhood heighten my sensitivity to and appreciation of the roles and relationships covered in this book. They also are constant reminders to me of how within a single lifetime or across a society, none of us can necessarily anticipate or fully control the directions our families may take.

New to This Edition

Returning users will notice a number of noteworthy changes from the prior edition, designed to make the text more usable within the time constraints of a semester calendar, more comprehensive in its coverage of nonmarital lifestyles, and as up-to-date as is possible. There is a new boxed feature—*Public Policies, Private Lives*—that focuses on legal issues and public policies that affect the lives of our families. There is also a new chapter (Chapter 9), *Unmarried Lives: Singlehood and Cohabitation,* addressing two of the more visible trends and dramatic changes in contemporary family life. Although prior editions covered these issues in the chapter *Singlehood, Pairing, and Cohabitation,* in the 11th edition they are addressed at much greater length and in substantially more detail.

Knowing first hand how difficult it was trying to cover all fifteen text chapters in a typical semester, I have shortened this edition by one chapter, by merging the two chapters on parenthood from the last edition into a single chapter (Chapter 10), *Becoming and Being Parents.* It retains and updates the most important material regarding pregnancy, child birth, infant mortality, and adoption, along with coverage of being parents and raising children.

The sequence of chapters has been altered so that the coverage of singlehood and cohabitation follows the chapter on marriage. Material on dating (including new material on "hooking up") has been moved into Chapter 5, "Intimacy, Friendship, and Love." Material on theories and patterns of mate selection and partner choice are now in the chapter on marriage (Chapter 8, "Marriage in Societal and Individual Perspective"). Data and discussion on child care has been moved into Chapter 11, "Marriage, Work and Economics."

Content Changes by Chapter

Chapter 1, *The Meaning of Marriage and the Family.* In an effort to ensure that the text remains current and fresh, I have changed the chapter opening examples of controversial and contested family issues. The examples used in the new edition include the Nebraska Safe Haven law, the issue of polygamy among fundamentalist Mormons, the attempt by a California legislator to ban spanking, and the case of Thomas Beattie, who came to be called (and who called himself) the first pregnant man. These fit the chapter's continued emphasis on different and competing viewpoints about the meaning of family and the interpretation of changing family patterns. The chapter also includes an updated discussion of same gender marriage laws. There are two new boxed features—an *Issues and Insights* box entitled "Technological Togetherness" which examines the use of technology to allow family members to "dine together" despite being separated by hundreds of miles, and a *Real Families* box on "The Care Families Give" which looks at the caregiving roles played by families of the mentally ill.

Chapter 2, *Studying Marriages and Families.* There is a newly expanded discussion of research ethics and secondary analysis, a new section on applied family research and a new discussion of cross national survey data on dating violence in the *Exploring Diversity* box on "Researching Dating Violence Cross-Culturally." In addition, the theories addressed are now grouped into macro and micro level theories. Finally, new examples are included to illustrate certain theories (e.g., symbolic interaction and the idea of quality time; social exchange theory and involuntarily celibate marriages) or methods of data collection.

Chapter 3, *Variations in American Family Life.* Along with much new data on economic, racial and ethnic differences in social and family characteristics, the discussion of poverty has been significantly enlarged and updated, with particular reference to the working poor, and with sustained attention to the impact of the recession on family life. Two new features—one a *Real Families* box entitled "In Times of Trouble" on the familial impact of the recession and the other a *Public Policies, Private Lives* box on "A Multiracial First Family" on the Obama's multiracial family history—keep the text relevant to the world with which students are most familiar.

Chapter 4, *Gender and Family.* This chapter contains new material on gender inequality and patriarchy, both cross culturally and within the contemporary United States; new material on transgendered people; expanded and updated discussion of the media and gender socialization; and discussions of survey data from the Pew Foundation on marital power and decision making, and on the division of housework. Data on the wage gap between women and men has been updated. Features examining girls and violence (the *Issues and Insights* box called "Girls and Violence" addresses the YouTube gang assault of a teenage girl by other teenage girls) and gender diversity (the *Real Families* box "Third Genders and Pregnant Men") have been added.

Chapter 5, *Intimacy, Friendship, and Love.* This chapter has been revised to include coverage on dating in a section "How and Where We Find Love." Additional new material addresses "hooking up" among college students, relationships that are formed online, and same-gender breakups. The latter is part of an attempt to update material on aspects of same gender relationships and to extend coverage of same gender relationships more widely throughout the book.

Chapter 6, *Understanding Sex and Sexualities.* There is expanded coverage of sexual socialization, especially on the roles played by parents and media, as well as a new section on the roles siblings play in our learning about sex. A new *Popular Culture* feature "Sex, Teens, and Television" examines the impact of television depiction of teenagers and sex on teen sexual behavior. There are also new discussions cautioning about problems in gathering data on sexuality and sexual expression, and addressing efforts to count the gay, lesbian and bisexual populations. Despite the cautions, the chapter includes updated information drawn from surveys about sexuality. There is also a discussion of celibate marriages, and a second new *Public Policies, Private Lives* feature on "Sexting and the Law."

Chapter 7, *Communication, Power, and Conflict*. There is enlarged coverage of nonverbal communication, including material on cross cultural differences. Other new material, such as the *Popular Culture* box on "Staying Connected with Technology," discusses telecommunication and text messaging as used by family members to communicate throughout the day. There is more material on the demand-withdraw pattern of communication and conflict management, a new discussion that compares areas of conflict experienced by cohabiting and married couples, and expanded discussion of consequences of conflict for couples and children. There is another new *Public Policies, Private Lives* feature entitled "Can We Learn How to Manage and Avoid Conflict?", addressing the question of whether and how couples can be taught how to more effectively manage conflict.

Chapter 8, *Marriages in Societal and Individual Perspective*. The chapter now includes the discussion of theories and patterns of spousal choice, including more and more up-to-date discussion of religious homogamy. There is updated material on changing patterns in age at marriage and in who we can marry. There is a review of the recent history and current legal status of same gender marriage. The chapter also includes discussions of attitudes toward marriage in the United States compared with attitudes in other cultures, and the effects of marriage on kinship and other social ties. Although the chapter still includes material looking at marriage across the life span, there is less emphasis on family life cycle stages. The focus instead remains more directly on marriage. A new *Popular Culture* feature explores what we can learn about marriage from reality programs such as "Wife Swap" and "Trading Spouses."

Chapter 9, *Unmarried Lives: Singlehood and Cohabitation*. As described above, this new chapter looks in detail at the trends in singlehood and cohabitation, tracing demographic patterns in both lifestyles. In the new material on singlehood there is an up-to-date profile of the unmarried population, a discussion of different types of singles, and greater attention to ways in which singles often have been stereotyped, stigmatized and discriminated against. Using Bella DePaulo's concepts of singlism and matrimania, the chapter addresses some mistreatment that the unmarried may suffer. The material on cohabitation examines factors associated with the increase and with the demographic distribution of cohabitation. Consideration of the advantages and disadvantages of cohabitation in general is included. There are also discussions of: cohabitation prior to remarriage and its effects; domestic partnerships and common law marriages; how cohabitation compares to marriage; the factors that create the "cohabitation effect" on marriages; and the phenomenon of serial cohabitation. Another substantial section considers same gender cohabitation and ways in which gay men and lesbians choose and redesign their families. There are new features: the *Issues and Insights* box "Living Apart Together," and a *Public Policies, Private Lives* box on legal considerations pertaining to cohabitation. The chapter ends with a brief discussion of friends as family.

Chapter 10, *Becoming and Being Parents*. As described above, this is a newly revised chapter, blending together material from two chapters (10 and 11) from the prior edition. It combines material on trends in pregnancy, birth, child-free choices, and adoption, with material on expectations of parents, parental roles and experiences, responsibility for children and child care, caring for adult children, and children caring for aging parents. Special attention is paid to trends in teen and unmarried parenthood; how women experience and evaluate their childbirth experiences; types of adoption (including adoptions that dissolve), post-partum depression; how parenthood affects marriage and mental health; the effects of parents marital status, sexuality age and ethnicity on their parenting experiences; parenting in later life and grandparenthood. A *Popular Culture* feature "Calling Nanny 911" and a *Public Policies, Private Lives* feature explores "When Adoptions Dissolve" are also new to this edition.

Chapter 11, *Marriage, Work, and Economics*. The chapter includes new material on time at work, both within the United States and comparatively. Data from the American Time Use Survey on actual expenditure of time and from recent Pew and Gallup surveys on work-family time preferences are introduced and discussed. In addition, the chapter contains new material on the distribution of housework and childcare is included, along with expanded and updated discussions of shift work, unemployment, the wage gap and the availability and cost of child care. There are two new features, the *Real Families* box on the impact of recession induced unemployment and the *Public Policies, Private Lives* box on "The Family and Medical Leave Act." The latter is part of an expanded discussion of workplace policies.

Chapter 12, *Intimate Violence and Sexual Abuse.* The chapter has much new material. Data on the prevalence of intimate partner violence, child maltreatment, sibling and elder abuse, and child sexual abuse have been updated. There is more discussion of policies for dealing with intimate partner violence, including programs with abusers. Material on factors associated with the range of forms of family violence and abuse has been enlarged and updated. There is also an expanded discussion of the costs, economic and otherwise, of family violence and abuse. New boxes include the new *Issues and Insights* feature focusing on dating abuse via text messaging, the *Public Policies, Private Lives* feature "Nixzmary's Law" discussing legal attempts to address child abuse, and the *Issues and Insights* box on the relationship between divorce and the risk of victimization.

Chapter 13, *Coming Apart: Separation and Divorce.* The data on divorce, custody, child support and alimony have been updated and coverage of these issues has been enlarged. The discussion of factors associated with divorce has also been revised to reflect how personal, familial, societal, demographic, and life course factors are currently associated with divorce patterns. There is now more coverage of non-custodial parents, including issues of visitation and non-payment of child support. The new features include the *Exploring Diversity* box on divorce in India, the *Popular Culture* box discussing the use (or misuse) of blogs and web-postings by divorcing spouses, and the dilemma faced by married same-gender couples who seek a divorce in the *Public Policies, Private Lives* box on "How Can You Get a Divorce if They Don't Recognize Your Marriage?"

Chapter 14, *New Beginnings: Single-Parent Families, Remarriages, and Blended Families.* There are updated discussions of trends in single-parenting and remarriage, and of the economic status and diversity of living arrangements of single parents. In addition there is more attention paid to stepfamilies, especially to the effects of stepfamily life on children. There are new features, including the *Public Policies, Private*

Lives box on the lack of legal policies and guidelines surrounding stepparenting and the *Real Families* box on the complexity of relationships that ensue "When Families Blend."

Pedagogy

What Do You Think? Self-quiz chapter openers let students assess their existing knowledge of what will be discussed in the chapter. We have found these quizzes engage the student, drawing them into the material and stimulating greater interaction with the course.

Chapter Outlines. Each chapter contains an outline at the beginning of the chapter to allow students to organize their learning.

Public Policies, Private Lives. These eleven boxed features are new additions to the 11th edition. They focus on legal issues and public policies that affect how we think about and/or experience family life. Among them are features on sexting, adoptions that dissolve, the multiracial Obama family, the lack of legal policies about stepfamilies, and the dilemma faced by same gender couples who entered marriage but because of where they've moved to can't divorce. There are also features on what legal marriage consists of, the Family and Medical Leave Act, and legal advice for cohabiting couples.

Exploring Diversity. These boxes let students see family circumstances from the vantage point of other cultures, other eras, or within different lifestyles in the contemporary United States. Several are new to this edition and they address divorce in India, the phenomenon of posthumous marriage, and recent cross-national survey data on dating violence.

Issues and Insights. These boxes focus on current and high-interest topics. These address such issues as girls and violence, "living apart together," and the uses and abuses of technology in families and relationships. One feature in Chapter 1, *Technological Togetherness,* describes an innovation called the "virtual family dinner," designed to allow families to feel as though they are eating together even when separated by great distance. A second technology feature, "CALL ME!!! Where ARE U? ☺" looks at the use of texting to control and emotionally abuse dating partners.

Popular Culture. These features discuss the ways family issues are portrayed through various forms of popular culture. Topics new to this edition include features on television programs *Wife Swap* and *Nanny*

911, a feature on sex, teens and television, a feature on the use of blogging by divorcing spouses, and a feature on celebrating and studying singlehood.

Real Families. These features give "up close," sometimes first person, accounts of issues raised in the text as they are experienced by people in their everyday lives. In this edition there are new boxes on blending families, co-parenting by gay men and lesbians, family caregivers for mentally ill family members, and a feature on third genders and pregnant men. There are also two Real Families features inspired by the U.S. recession as it impacts families.

Matter of Fact items provide statistics and quick data relating to the concepts in the chapter.

Critical Thinking thought questions bring students closer to the material by encouraging them to consider their own ideas and beliefs.

Each chapter also has a *Chapter Summary* and a list of *Key Terms*, all of which are designed to maximize students' learning outcomes. The Chapter Summary reviews the main ideas of the chapter, making review easier and more effective. The Key Terms are boldfaced within the chapter and listed at the end, along with the page number where the term was introduced. Both Chapter Summaries and Key Terms assist students in test preparation.

Glossary. There is a comprehensive glossary of key terms included at the back of the textbook.

Appendices. There are appendixes on sexual structure and the sexual response cycle, fetal development, and managing money.

Supplements and Resources

The Marriage and Family Experience, Eleventh Edition, is accompanied by a wide array of supplements prepared for both the instructor and student. Some new resources have been created specifically to accompany the Eleventh Edition, and all of the continuing supplements have been thoroughly revised and updated.

Supplements for the Instructor

Online Activities for Courses on the Family

This collection of ideas for classroom and online activities and lectures is made up of contributions from Marriage and Family instructors around the country. Free to adopters of *The Marriage and Family Experience,* this supplement will provide you with some creative ideas to enhance your lectures.

Instructor's Edition of The Marriage and Family Experience, Eleventh Edition

The Instructor's Edition contains a visual preface, a walk-through of the text that provides an overview of its key features, themes, and supplements.

Instructor's Resource Manual with Test Bank

This manual will help the instructor to organize the course and to captivate students' attention. The manual includes a chapter focus statement, key learning objectives, lecture outlines, in-class discussion questions, class activities, student handouts, extensive lists of reading and online resources, and suggested Internet sites and activities. The Test Bank portion includes approximately 40 to 50 multiple choice, 20 true/false, 10 short answer, and 5 to 10 essay questions, all with answers and page references, for each chapter of the text.

PowerLecture™ with ExamView®

PowerLecture instructor resources are a collection of book-specific lecture and class tools on either CD or DVD. The fastest and easiest way to build powerful, customized media-rich lectures, PowerLecture assets include chapter-specific PowerPoint presentations, images, animations and video, instructor manuals, test banks, useful web links and more. PowerLecture media-teaching tools are an effective way to enhance the educational experience. **ExamView** is also available within PowerLecture, allowing you to create, deliver, and customize tests and study guides (both print and online) in minutes. See assessments onscreen exactly as they will print or display online. Build tests of up to 250 questions using up to 12 question types and enter an unlimited number of new questions or edit existing questions. PowerLecture also includes the text's Instructor's Resource Manual and Test Bank as Word documents.

WebTutor™ on Blackboard® and WebCT®

Jumpstart your course with customizable, rich, text-specific content within your Course Management System. Simply load a content cartridge into your course management system to easily blend, add, edit, reorganize, or delete content, all of which is specific to Strong's *The Marriage and Family Experience,* Eleventh Edition, and includes media resources, quizzing, web links, interactive games and exercises.

Wadsworth Sociology Video Library

This large selection of thought-provoking films, including many from the Films for the Humanities collection, is available to adopters who meet adoption criteria. Please contact your local sales representative for more information.

ABC®Video: Marriage and Family, Volumes 1 & 2

Launch your lectures with riveting footage from ABC, a leading news television network. ABC Videos allow you to integrate the newsgathering and programming power of ABC into the classroom to show students the relevance of course topics to their everyday lives. Organized by topics covered in a typical course, these videos are divided into short segments—perfect for introducing key concepts. The wide selection includes thought-provoking clips such as "David Reimer: Raised as a Girl," "All-Female Fire Department," and "Effect of Holding Hands."

Families and Society: Classic and Contemporary Readings

This reader is designed to promote a sociological understanding of families, at the same time demonstrating the diversity and complexity of contemporary family life. The different sections of the reader are designed to "map" onto most textbooks and course syllabi relating to sociology of the family. Edited by Scott L. Coltrane, University of California, Riverside.

The Marriage and Families Activities Workbook

This workbook of interactive self-assessments was written by Ron J. Hammond and Barbara Bearnson, both of Utah Valley State College. They present questions such as: What are your risks of divorce? Do you have healthy dating practices? What is your cultural and ancestral heritage, and how does it affect your family relationships? The answers to these and many more questions are found in this workbook of nearly a hundred interactive self-assessment quizzes designed for students studying marriage and family. These self-awareness instruments, all based on known social science research studies, can be used as in-class activities or homework assignments.

Supplements for the Student

Study Guide

For each chapter of the text, this student study tool contains a chapter focus statement, key learning objectives, key terms, chapter outlines, assignments, Internet activities and websites, and practice tests containing 20 multiple choice and 15 true/false with answers and page references, and 5 short answer questions with page references.

Relationship Skills Exercises

This supplement by Lue K. Turner is full of assessments and questionnaires that will make students think more reflectively on important topics related to marriage, such as finances and intimacy. Assignments can be done in-class or at home, alone or with a partner.

Online Resources

Book Companion Website

www.cengage.com/sociology/strong

Students can gain an even better understanding of the material by using the additional study resources at the book companion website. For example, students can prepare for quizzes and exams with online resources—including tutorial quizzes, a glossary, interactive flash cards, crossword puzzles, self assessments, virtual explorations, and more.

Acknowledgments

Many people deserve to be recognized for the roles they played in the revision and production of this edition. I thank the following reviewers, whose comments and reactions were encouraging and whose suggestions were helpful: Beth Williams, University of South Carolina Aiken; Nita Jackson, Butler Community College; Kris Koehne, University of Tennessee; Rachel Hagewen, University of Nebraska; Von Bakanic, College of Charleston; Pam Monaco, Widener University; Katrina Akande, University of Kentucky.

This book remains the product of many hands. Bryan Strong and Christine DeVault created a wonderful book from which to teach or study families and relationships. I hope that once again I have retained their emphasis on the meaning and importance of families. I am gratified to continue their efforts.

A number of people at Cengage Learning deserve thanks. Senior Publisher Linda Schreiber-Ganster and Sociology Editor Chris Caldeira showed considerable faith in and continued support for this book. Through the last two editions, Chris brought enthusiasm and energy to the revision process, offering many fresh ideas and wise suggestions. I hope that I have done at least some of them justice. My developmental editor, Kristin Makarewycz, was everything an author could wish for in an editor. She offered encouragement, expressed enthusiasm, reminded me of deadlines (and helped me meet them) and helped immensely with her careful editing, thoughtful suggestions and wise commentary as she read through the drafts of each chapter. This book has been made stronger by her guidance and the processes of writing and revising have been made smoother and more gratifying because of her involvement.

Rachael Krapf helped tremendously in readying the manuscript for production and putting together the strong ancillary package that accompanies the Eleventh Edition; Melanie Cregger, the Media Editor, is responsible for the state-of-the-art technology and website supporting the text; while Andrew Keay, marketing manager, and Laura Localio, marketing communications manager, have skillfully introduced the Eleventh Edition to adopters and prospective adopters. I want to extend my thanks to Cheri Palmer, the senior production project manager at Cengage, who oversaw the complex production process with great skill. Lynn Steines, Bruce Owens, and S4Carlisle Publishing Services were tremendously helpful and highly competent in the copyediting and production phases. The text looks and reads better because of their involvement. My appreciation again goes to Sarah Evertson, photo editor and researcher, for finding such good examples of what were occasionally vaguely requested subjects.

Once again, I wish to express deep appreciation to my colleagues and friends at Ohio Wesleyan University for the support they provided me. A sabbatical semester at Ohio Wesleyan afforded me the opportunity to invest the time and energy I have in this edition. My current and former Ohio Wesleyan colleagues Jan Smith, Mary Howard, Akbar Mahdi, Jim Peoples, John Durst, and Pam Laucher, along with the many enthusiastic and curious students I have had in classes, make me realize how very fortunate I have been in my academic career.

I want again to express my appreciation to my family. My parents, Kalman and Eleanor Cohen, and sisters, Laura Cohen and Lisa Merrill, are always supportive. My now adult children, Dan and Allison, make me incredibly proud. No matter where their lives take them, they will remain in the center of my heart. They are wonderful legacies to their beautiful mother. By opening their hearts to me and letting me share in and be part of their lives, my stepchildren, Daniel, Molly, and Brett have greatly enriched my life.

I want to express special thanks to my wife, Julie, for her willingness to blend her life and family with mine. Everyday I feel blessed to be her partner, to draw upon her wisdom, strength, and encouragement, and to be able to count upon her love and support. Even as she and I look forward to our future together, she allows our past lives—and her Karl and my Sue—to remain very much a part of the life she and I share with each other and with our children.

1

© Punchstock / Getty Images

The Meaning of Marriage and the Family

What Do YOU Think? Are the following statements TRUE or FALSE?
You may be surprised by the answers (see answer key on the following page).

T	F	**1** Most American families are traditional nuclear families in which the husband works and the wife stays at home caring for the children.
T	F	**2** Families are easy to define and count.
T	F	**3** No U.S. state prohibits interracial marriage.
T	F	**4** All cultures traditionally divide at least some work into male and female tasks.
T	F	**5** Nowhere in the United States can same-gender couples legally marry.
T	F	**6** There is widespread agreement about the nature and causes of change in American family patterns.
T	F	**7** Most cultures throughout the world prefer monogamy—the practice of having only one husband or wife.
T	F	**8** Married people tend to live longer than single men.
T	F	**9** Most people who divorce eventually marry again.
T	F	**10** Nuclear families, single-parent families, and stepfamilies are equally valid family forms.

A course in marriage and the family is unlike almost any other course you are likely to take. At the start of the term—before you purchase any books, before you attend any lectures, and before you take any notes—you may believe you already know a lot about families. Indeed, each of us acquires much firsthand experience of family living before being formally instructed about what families are or what they do. Furthermore, each of us comes to this subject with some pretty strong ideas and opinions about families: what they're like, how they should live, and what they need. Our personal beliefs and values shape what we think we know as much as our experience in our families influences our thinking about what family life is like. But if pressed, how would we describe American family life? Are our families "healthy" and stable? Is marriage important for the well-being of adults and children? Are today's fathers and mothers sharing responsibility for raising their children? How many spouses cheat on each other? What happens to children when parents divorce? Do stepfamilies differ from biological families? How common are abuse and violence in families? Questions such as these will be considered throughout this book; they encourage us to think about what we know about families and where our knowledge comes from.

In this chapter, we examine how marriage and family are defined by individuals and society, paying particular attention to the discrepancies between the realities of family life as uncovered by social scientists and the impressions we have formed elsewhere. We then look at the functions that marriages and families fulfill and examine extended families and kinship. We close by introducing the themes that will be pursued through the remaining chapters.

Personal Experience, Social Controversy, and Wishful Thinking

As we begin to study family patterns and issues, we need to understand that our attitudes and beliefs about families may affect and distort our efforts. In contemplating the wider issues about families that are the substance of this book, it is likely that we will consider our own households and family experiences. How we respond to the issues and information presented over the 14 chapters that follow may be influenced by what we have experienced in and come to believe about families.

Experience versus Expertise

For some of us, those experiences have been largely loving ones, and the relationships have remained stable. For others, family life has been characterized by conflict and bitterness, separations and reconfigurations. Most people experience both sides of family life, the love and the conflict, whether their families remained intact or not.

The temptation to draw conclusions about families from personal experiences of particular families is understandable. Thinking that experience translates to expertise, we may find ourselves tempted to generalize from what we experience to what we assume others must also encounter in family life. The dangers of doing that are clear; although the knowledge we have about our own families is vividly real, it is also both highly subjective and narrowly limited. We "see" things, in part, as we want to see them. Likewise, we overlook some things because we don't want to accept them. Our family members are likely to have different perceptions and attach different meanings to even those same relationships. Thus, the understanding we have of our families is very likely a distorted one.

Furthermore, no other family is exactly like our family. We don't all live in the same places, and we don't all possess the same financial resources, draw from the same cultural backgrounds, and build on the same sets of experiences. These make our families somewhat unique. No matter how well we might think we know our families, they are poor sources of more general knowledge about the wider marital or family issues that are the focus of this book.

Ongoing Social Controversy

Learning about marriage and family relationships is challenging for another reason. Few areas of social life are more controversial than family matters. Just consider the following recent news stories. What

Answer Key to What Do YOU Think?

1 False, see p. 12;   **2** False, see p. 11;   **3** True, see p. 8;   **4** True, see p. 13;   **5** False, see p. 9;   **6** False, see p. 20;   **7** False, see p. 9;   **8** True, see p. 12;   **9** True, see p. 10;   **10** True, see p. 11.

underlying issues can you identify? What is your position on such issues?

- Over the past few years, Americans became more aware of the existence of polygamy among some fundamentalist Mormon groups in the southwestern United States, most notably the Fundamentalist Church of Jesus Christ of Latter-Day Saints and the Apostolic United Brethren. In 2008, amidst allegations of bigamy, rape, underage marriage, child abuse and sexual assault, the state attorney general authorized that more than 450 children of polygamous families at the Fundamentalist Church of Jesus Christ of Latter-Day Saints' Yearning for Zion ranch in Eldorado, Texas, should be removed from their parents' custody and taken by authorities. After state courts determined that there was insufficient evidence of abuse, custody was returned to parents. The controversy surrounding this case was substantial and led to proposed senate legislation calling for a national task force on polygamy (Kovach 2008).

The controversies surrounding polygamists of the Fundamentalist Church of Jesus Christ of Latter-day Saints' Yearning for Zion ranch in Eldorado Texas illustrate one of the more recent instances of controversy over marriage and the well-being of children.

- Decisions to marry or divorce are typically decisions based on falling in or out of love. Sometimes, though, as was the case for Bo and Dena McLain of Milford, Ohio, such decisions are also heavily influenced by much more practical and mundane motives, such as the need to attain or retain health insurance. The McLains married so that Dena could be added to Bo's health insurance and thus meet the requirement for insurance imposed by her nursing school. Aside from whatever else brought the McLains together, their decision to marry reflects some of the privileges found in marriage. According to a poll by the Kaiser Family Foundation, 7% of adults indicated that, in the past year, a member of their household had married to gain access to health care coverage (Sack 2008).
- In July 2008, the state of Nebraska passed a "safe haven" law, allowing distressed parents to leave a child at a licensed hospital, surrendering custody,

without being charged with abandonment. Such laws exist in all 50 states, so, in that sense, Nebraska was merely keeping pace with the other 49 states. Typically, such laws cover only very young children. In 42 of the 50 states, the law only covered infants one month of age or younger. However, Nebraska's law was unique; it carried no age limit on the "child" that one could surrender without facing charges of abandonment. In other words, parents could, without penalty, abandon children up to 18 years of age. Between July and November 21, 2008, when Nebraska's governor signed Legislative Bill-1, restricting the safe haven protection to children 30 days old or younger, there were 27 separate uses of Nebraska's safe haven provision, involving 36 children. These children ranged in age from one to 17, but only five children were under 10 years of age, and only two were under the age of five. Although most were from Nebraska, there were also cases of parents traveling from Florida, California, Indiana, Michigan, Iowa, and Georgia to abandon their children in Nebraska.

- Susan Beatie was born on June 29, 2008, to parents Thomas and Nancy. What makes Susan's birth so unusual is that she was born to 34-year-old Thomas, making him the "first fully legal male and husband to get pregnant and give birth to a child"

(Beattie 2008, 6). Thomas was born female and for 24 of his 34 years lived as Tracy. He writes that throughout his life, "for as long as I can remember, and certainly before I fully knew what this meant, I wanted to live my life as a man" (Beattie 2008, 6). With the aid of testosterone to stimulate muscle development and grow facial hair, exercise, and surgery to remove the breasts, Tracy became Thomas. Choosing not to have his female reproductive organs removed, Thomas retained the ability to conceive, carry, and bear a child. Using donor sperm, Thomas became pregnant in the fall of 2007, four years into his marriage to Nancy.

- In January 2007, California State Assemblywoman Sally Lieber introduced legislation (California AB 755) to make California the first state in the United States to make spanking a child under three years of age a crime. The proposed legislation would have made spanking a misdemeanor, punishable by up to a $1,000 fine and a year in jail. Successful passage of the bill would have put California in the same company as 15 other, mostly European, countries that have laws banning corporal punishment. By the end of February, Lieber had withdrawn her legislation after being inundated with calls and e-mails from proponents of spanking and realizing the bill could not pass. In 2008, Lieber introduced a new bill (California AB 2943) that would make it a crime to spank a child with "an implement, including, but not limited to, a stick, a rod, a switch, an electrical cord, an extension cord, a belt, a broom or a shoe." Again, there was considerable negative reaction to the legislation, which many critics felt limited parents' right to decide how to discipline their children.

The preceding cases raise interesting questions without clear answers. For example, how much should the state restrict people's marriage choices? How do policies that privilege married couples influence decisions to enter or exit a marriage? At what point should the protection of children take precedence over the privacy of family life? Is biology or identity more definitive of one's gender? What are parents' responsibilities to their children? Is parenthood revocable? What constitutes appropriate and effective parental discipline of children? As a society, we are often divided, sometimes strongly and bitterly, on such family issues. That we are so deeply invested in certain values regarding family life makes a course about families a different kind of learning experience than if you were studying material to which you, yourself, were less connected. Ideally, as a result, you will find yourself more engaged, even provoked, to think about and question things you take for granted. At minimum, you will be exposed to information that can help you more objectively understand the realities behind the more vocal debates.

What Is Marriage? What Is Family?

To accurately understand marriage and family, it is important to define these terms. Before reading any further, think about what the words *marriage* and *family* mean to you. As simple and straightforward as this may seem, as you attempt to systematically define these words, you may be surprised at the complexity involved.

© Fox Searchlight/Courtesy Everett Collection

Families are often the focus of popular movies. In 2006, "Little Miss Sunshine," a comedy focusing on a highly dysfunctional three-generational family, was an Academy Award nominee for Best Picture.

Defining Marriage

Close to three out of every five adults in the United States, age 18 and older, are married. As shown in Figure 1.1, among males, 59.9% are currently married, and 71.4% have at least experienced marriage (i.e., are married, divorced, or widowed). Although a smaller percentage of females is currently married (56.7%), 78% of females 18 and older are or have been married (U.S. Census Bureau 2008, *Current Population Reports, P20-537*).

A **marriage** is a legally recognized union between two people, generally a man and a woman, in which they are united sexually, cooperate economically, and may give birth to, adopt, or rear children. The union is assumed to be permanent (although it may be dissolved by separation or divorce). As simple as such a definition may make marriage seem, it differs among cultures and has changed considerably in our society.

With one exception, the Na of China, marriage has been a universal institution throughout recorded history (Peoples and Bailey 2006). Despite the universality of marriage, widely varying rules across time and cultures dictate whom one can, should, or must marry; how many spouses one may have at any given time; and where married couples can and should live—including whether husbands and wives are to live together or apart, whether resources are shared between spouses or remain the individual property of each, and whether children are seen as the responsibility of both partners or not (Coontz 2005). Among non-Western cultures, who may marry whom and at what age varies greatly from our society. In some

areas of India, Africa, and Asia, for example, children as young as six years may marry other children (and sometimes adults), although they may not live together until they are older. In many cultures, marriages are arranged by families who choose their children's partners. In many such societies, the "choice" partner is a first cousin. And in one region of China as well as in certain parts of Africa (e.g., the Nuer of Sudan), marriages are sometimes arranged between unmarried young men and women who are dead.

Considerable cultural variation exists in what societies identify as the essential characteristics that define couples as married. In many societies, marriage entails an elaborate ceremony, witnessed and legitimated by others, which then bestows a set of expectations, obligations, rights, and privileges on the newly married. Far from this relatively familiar construction of marriage, historian Stephanie Coontz notes that in some "small-scale societies," the act of eating alone together defines a couple as married. In such instances, as found among the Vanatinai of the South Pacific, for example, dining together alone has more social significance than sleeping together (Coontz 2005). Anthropological study of Sri Lanka revealed that when a woman cooked a meal for a man, this indicated that the two were married. Likewise, if a woman stopped cooking for a man, their marriage might be considered a thing of the past.

Although cultural and historical variation abounds, the following seem to be shared among all arrangements defined as marriages (Coontz 2005):

- Marriage typically establishes rights and obligations connected to gender, sexuality, relationships with kin and in-laws, and legitimacy of children.
- Marriage establishes specific roles within the wider community and society. It specifies the rights and duties of husbands and wives, as well as of their respective families, to each other and makes such duties and responsibilities enforceable by the wider society.
- Marriage allows the orderly transfer of wealth and property from one generation to the next.
- Additionally, as anthropologists James Peoples and Garrick Bailey (2006) note, marriage assigns the responsibility for caring for and socializing children to the spouses or their relatives.

Many Americans believe that marriage is divinely instituted; others assert that it is a civil institution involving the state. The belief in the divine institution of marriage is common to religions such as

Figure 1.1 Marital Status of U.S. Population, 18 and Older

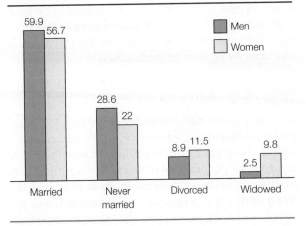

SOURCE: U.S. Census Bureau, Current Population Reports P20-537, Table 55.

Exploring Diversity: Ghost or Spirit Marriage

Looking across cultures, there are many marriage customs that may strike most Americans as unusual. Few, if any, can rival the custom of a marriage where one or both spouses are deceased. There are a number of versions of so-called ghost, spirit, or posthumous marriages that can be found among some African countries, in France, and in parts of rural China. In Sudan, among the Nuer, a dead groom can be replaced by a male relative (e.g., a brother) who takes his place at the wedding. Despite his being deceased, he—not the living substitute—is considered the husband. Any children born subsequently will be considered children of the deceased man who is recognized socially as the father. In this way, a man who died before leaving an heir can have his family line continue. Among the Iraqw of Tanzania, the "ghost" groom could be the imagined son of a woman who never had a son.

In France, in 1959, Parliament drafted a law that legalized "postmortem matrimony" under certain circumstances. These included proof of the couple's intention to marry before one of the partners died and permission from the deceased's family. After a request is submitted to the president, it is passed to a justice minister and ultimately to the prosecutor who has jurisdiction over the locality in which the marriage is to occur. It is then the prosecutor's responsibility to determine whether the conditions have been met and the marriage is to be approved. In a 2004 *New York Times* article, a French attorney estimated that about 20 such marriages occur each year, most of which are kept quiet. Such posthumous marriages are largely for sentimental reasons. In fact, French law prevents spouses from any inheritance. Nonetheless, the marriages are retroactive to the eve of the groom's demise. They allow the woman to "carry her husband's name and identify herself as a widow" on official documents. If the woman is pregnant at the time of the man's death, the children are considered legitimate heirs to his estate (Smith 2004).

In some parts of rural China, parents of a son who died before marrying may "procure the body of a (dead) woman, hold a 'wedding,' and then bury the couple together," in keeping with the Chinese tradition of deceased spouses sharing a grave (Economist 2007). As reported in the *New York Times*, the custom of "minghun" (afterlife marriage) follows from the Chinese practice of ancestor worship, which holds that people continue to exist after death and that the living are obligated to tend to their wants—or risk the consequences. Traditional Chinese beliefs also hold that an unmarried life is incomplete, which is why some parents worry that an unmarried dead son may be an unhappy one (Yardley 2006).

Parents whose daughters had died might sell their daughter's body for economic reasons but also are motivated by the desire to give their daughters a place in Chinese society. As stated by sociologist Guo Yuhua, "China is a paternal clan culture. . . . A woman does not belong to her parents. She must marry and have children of her own before she has a place among her husband's lineage. A woman who dies unmarried has no place in this world" (Yardley 2006).

Christianity, Judaism, and Islam and to many tribal religions throughout the world. But the Christian church only slowly became involved in weddings; early Christianity was at best ambivalent about marriage, despite being opposed to divorce (Coontz 2005). Over time, as the church increased its power, it extended control over marriage. Traditionally, marriages had been arranged between families (the father "gave away" his daughter in exchange for goods or services); by the tenth century, marriages were valid only if they were performed by priests. By the thirteenth century, the ceremony was required to take place in a church. As states competed with organized religion for power, governments began to regulate marriage.

In the United States today, for a marriage to be legal it must be validated through government-issued marriage licenses, regardless of whether the ceremony is officiated by legal or religious officials.

Who May Marry?

Who may marry has changed over the past 150 years in the United States. Laws once prohibited enslaved African Americans from marrying because they were regarded as property. Marriages between members of different races were illegal in more than half the states until 1967, when the U.S. Supreme Court declared, in *Loving vs. Virginia*, that such prohibitions were unconstitutional. Each state enacts its own laws regulating marriage, leading to some discrepancies from state to state. For example, in some states, first cousins may marry; other states prohibit such marriages as incestuous.

The greatest current controversy regarding legal marriage continues to be over the question of same-sex marriage. As of late 2009, five states—Massachusetts, Connecticut, Iowa, Vermont, and New Hampshire—allow same-sex couples to legally marry, with all the rights, benefits, and privileges marriage entails. Two other states, California and Maine, legalized same sex marriage in 2008 and 2009, only to have voters pass legislation to alter the state constitutions and/or explicitly restrict legal marriage to heterosexuals. Once again, at least for now, legal marriage is now available to gay men or lesbians in either California or Maine. We more fully explore such legal aspects of marriage (such as the age at which one can marry, whom one may marry, and so on) in Chapter 9.

© Phill Snel/Getty Images

Same sex marriage is now legal in the United States but as of late 2009 only in Massachusetts, Connecticut, Iowa, Vermont, and New Hampshire.

Forms of Marriage

In Western cultures such as the United States, the only legal form of marriage is **monogamy,** the practice of having only one spouse at one time. Thus, the fundamentalist Mormon polygamists depicted earlier were in violation of state marriage laws. Monogamy is also the only form of marriage recognized in *all* cultures. Interestingly, and possibly surprisingly, it is not always the *preferred* form of marriage. One survey of world cultures indicated that only 24% of the known cultures perceive monogamy as the ideal form of marriage (Murdock 1967). The preferred marital arrangement worldwide is **polygamy,** specifically polygyny—the practice of having two or more wives. One study of 850 non-Western societies found that 84% of the cultures studied (representing, nevertheless, a minority of the world's population) practiced or accepted **polygyny,** the practice of having two or more wives (Gould and Gould 1989). Anthropologist James Peoples notes that where polygyny is allowed, it tends

to be the preferred form of marriage. He points out that, even today, polygyny is allowed in many modern Middle Eastern societies and is commonly practiced by tribal societies in Africa and Southeast Asia (Peoples and Bailey 2006). Conversely, **polyandry,** the practice of having two or more husbands, is actually quite rare: where it does occur, it often coexists with poverty, a scarcity of land or property, and an imbalanced ratio of men to women.

Even within polygynous societies, monogamy is the *most widely practiced* form of marriage. In such societies, plural marriages are in the minority, primarily for simple economic reasons: they are a sign of status that relatively few people can afford and require wealth that few men possess. As we think about polygyny, we may imagine high levels of jealousy and conflict among wives. Indeed, problems of jealousy may and do arise in plural marriages—the Fula in Africa, for example, call the second wife "the jealous one." Based on data from 69 polygynous societies (56% of which were in Africa), Jankowiak, Sudakov, and Wilreker suggest that co-wife conflict and competition for access to the husband is common, but there are also circumstances that reduce conflict (e.g., when the wives are sisters, when one is fertile and one barren or postmenopausal). For

Lieutenant Laurel Hester, 49 years old and a longtime officer for the Ocean County, New Jersey, county prosecutor's office, was dying of lung cancer. For months, as her disease spread, Hester fought against county officials, seeking the right to have her pension benefits pass to her same-sex partner, Stacie Andree. But pension benefits for those in the Police and Fire Retirement System could be passed only to spouses (Bonafide 2006), and, at the time, New Jersey state law did not allow same-sex couples to marry. Just weeks prior to her death, county officials relented. Facing considerable public outcry, Ocean County freeholders voted to allow the $13,000 benefit to be transferred to Andree. New Jersey has since legalized civil unions, thus granting same-sex couples the same rights as married couples, minus the right to marry.

Heterosexuals rarely stop to think about the privileges that sexual orientation offers. One such privilege is the right to marry. Those couples who do marry receive many more rights and protections than couples who don't marry. For heterosexuals, marriage versus cohabitation is a matter of choice. Heterosexual couples who choose cohabitation may do so because they prefer the more informal arrangement. They, too, will lack the protections and privileges that accompany marriage, but they elect to cohabit anyway. For many same-sex couples, the historical *inability* to marry has cost them many protections, a dozen of which are listed below. It is the lack of these rights and protections that state courts in the United States (Vermont, Connecticut, New Jersey, and Massachusetts, for example) have found unconstitutional.

- Right to enter into a premarital agreement
- Income tax deductions, credits, rates exemption, and estimates
- Legal status with partner's children
- Partner medical decisions
- Right to support from spouse
- Right to inherit property
- Award of child custody in divorce proceedings
- Control, division, acquisition, and disposition of community property
- Division of property after dissolution of marriage
- Payment of worker's compensation benefits after death
- Public assistance from the Department of Human Services
- Right to support after divorce

There are also potential personal and emotional benefits related to the right to marry. Knowing that the wider society recognizes, accepts, or respects a relationship may cause feelings of greater self-validation and comfort within the relationship. On the other hand, knowing that people do not respect, accept, or recognize a commitment may cause additional emotional suffering and personal anguish for the partners involved. Opposition to same-sex marriage is rarely based on issues such as legal rights. Opponents most often question the moral acceptability of gay or lesbian relationships, often referring to religious grounds for their rejection of gay marriage. Morality is harder to address objectively than the question of legal rights. Clearly, those who can and do marry receive substantial privileges and protections that those who don't or can't must live without.

SOURCE: Parents, Families, and Friends of Lesbians and Gays, "What Is Marriage, Anyway?" www.pflag.org/education/marriage.html.

---

both the men and the women involved, polygyny brings higher status.

~~Even though conflict and competition among co-wives is often found in polygynous societies, the level is probably less than would result if our monogamous society was to suddenly allow people multiple spouses. In~~ part because of our culture's traditional roots in Christianity, polygamy has been illegal in the United States since a U.S. Supreme Court decision in 1879. Polygamy was prohibited because it was considered a potential threat to public order (Tracy 2002). As a result, polygamy was looked on as strange or exotic. However, it may not seem so strange if we look at actual American marital practices. Considering the high divorce and remarriage rates in this country, monogamy may no longer be the best way of describing our marriage forms. For many, our marriage system might more accurately be called **serial monogamy** or **modified polygamy**, a practice in which one person may have several spouses over his or her lifetime although no more than one at any given time.

Defining Family

As contemporary Americans, we live in a society composed of many kinds of families—married couples, stepfamilies, single-parent families, multigenerational families, cohabiting adults, child-free families, families headed by gay men or by lesbians, and so on. With such variety, how can we define family? What are the criteria for identifying these groups as families?

The U.S. Census Bureau attempts to count and characterize American families. The Census Bureau defines a **family** as "a group of two or more persons related by birth, marriage, or adoption and residing together in a household" (U.S. Census Bureau 2001). A distinction is made between a family and a **household**. A household consists of "one or more people—everyone living in a housing unit makes up a household" (Fields 2003). Single people who live alone, roommates, lodgers, and live-in domestic service employees are all counted among members of households, as are family groups. *Family households* are those in which at least two members are related by birth, marriage, or adoption (Fields 2003). Thus, the U.S. Census reports on characteristics of the nation's households *and* families (Figure 1.2). Of the 116,011,000 households in the United States in 2007, 78,425,000, or 68%, were family households. Among family households, 75% (58,945,000)

consisted of married couples, either with or without children (U.S. Census Bureau 2007, Current Population Survey, table H2).

In individuals' perceptions of their own life experiences, *family* has a less precise definition. For example, when we asked our students whom they included as family members, their lists included such expected relatives as mother, father, sibling, and spouse. Most of those designated as family members are individuals related by descent, marriage, remarriage, or adoption, but some are **affiliated kin**—unrelated individuals who feel and are treated as if they were relatives, such as the following:

best friend	lover	priest
boyfriend	minister	rabbi
girlfriend	neighbor	teacher
godchild	pet	

Furthermore, being related by blood or through marriage is not always sufficient to be counted as a family member or kin. Emotional closeness may be more important than biology or law in defining family.

There are also ethnic differences as to what constitutes family. Among Latinos, for example, *compadres* (or godparents) are considered family members. Similarly, among some Japanese Americans, the *ie* (pronounced "ee-eh") is the traditional family. The *ie* consists of living members of the extended family (such as grandparents, aunts, uncles, and cousins) as well as deceased and yet-to-be-born family members (Kikumura and Kitano 1988). Among many traditional Native American tribes, the **clan**, a group of related families, is regarded as the fundamental family unit (Yellowbird and Snipp 1994).

A major reason we may have difficulty defining *family* is that when we think of families, we tend to think of the **nuclear family**, consisting of mother, father, and children. The term "nuclear family" is roughly 60 years old, coined by anthropologist

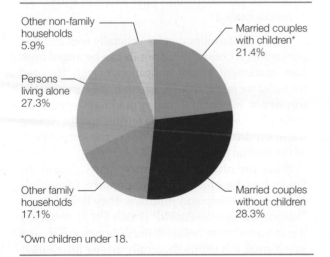

Figure 1.2 Household Composition, 2007

- Other non-family households 5.9%
- Married couples with children* 21.4%
- Persons living alone 27.3%
- Other family households 17.1%
- Married couples without children 28.3%

*Own children under 18.

SOURCE: U.S. Census Bureau, American Community Survey, Selected Social Characteristics in the U.S., 2007.

Critical Thinking

Think about all the people you consider your family. What criteria—biological, legal, affectional—did you use? Did you exclude any biological or legal family? If so, whom and why?

Robert Murdock in 1949 (Levin 1993). In addition, what most Americans envision when they consider the **traditional family** is a mostly middle-class version of the nuclear family in which women's primary roles are wife and mother and men's primary roles are husband and breadwinner. As we discuss in Chapter 3, such traditional families exist more in our imaginations than they ever did in reality.

Because we believe that the nuclear or traditional family is the real family, we compare all other family forms against these models. To include these diverse forms, the definition of family needs to be expanded beyond the boundaries of the "official" census definition. A more contemporary and inclusive definition describes family as "two or more persons related by birth, marriage, adoption, *or choice* [emphasis added]. Families are further defined by socio-emotional ties and enduring responsibilities, particularly in terms of one or more members' dependence on others for support and nurturance" (Allen, Demo, and Fine 2000). Such a definition more accurately and completely reflects the diversity of contemporary American family experience.

Functions of Marriages and Families

Whether it is a mother–father–child nuclear family, a married couple with no children, a single-parent family, a stepfamily, a dual-worker family, or a cohabiting family, the family generally performs four important societal functions: (1) it provides a source of intimate relationships, (2) it acts as a unit of economic cooperation and consumption, (3) it may produce and socialize children, and (4) it assigns social roles and status to individuals. Although these are the basic functions that families are "supposed" to fulfill, families do not have to fulfill them all (as in families without children), nor do they always fulfill them well (as in abusive families).

Intimate Relationships

Intimacy is a primary human need. Human companionship strongly influences rates of illnesses, such as cancer or tuberculosis, as well as suicide, accidents, and mental illness. Thus, it is no surprise that studies consistently show married couples and adults living with others to be generally healthier and have lower mortality rates than divorced, separated, and never-married individuals. Although some of this difference results from what is known as the *selection* factor—wherein healthier people are more likely to marry or live with someone—both marriage and cohabitation yield benefits to health and well-being. Chapter 9 considers in more detail whether it is the selection into marriage of healthier individuals or the protective benefits of marriage that accounts for the health benefits of marriage.

Family Ties

Marriage and the family usually furnish emotional security and support. This has probably been true from the earliest times. Thousands of years ago, in the Judeo-Christian Bible, the book of Ecclesiastes (4:9–12) emphasized the importance of companionship:

> Two are better than one, because they have a good reward for their toil. For if they fall, one will lift up his fellow; but woe to him who is alone when he falls and has not another to lift him up. Again if two lie together, they are warm; but how can one be warm alone? And though a man might prevail against one, two will withstand him. A three-fold cord is not quickly broken.

It is in our families that we generally seek and find our strongest bonds. These bonds can be forged from love, attachment, loyalty, obligation, or guilt. The need for intimate relationships, whether they are satisfactory or not, may hold unhappy marriages together indefinitely. Loneliness may be a terrible specter. Among the newly divorced, it may be one of the worst aspects of the marital breakup.

Since the nineteenth century, marriage and the family have become even more important as sources of companionship and intimacy. They have become "havens in a heartless world" (Lasch 1977). As society has become more industrialized, bureaucratic, and impersonal, it is within the family that we increasingly seek and expect to find intimacy and companionship.

In the larger world around us, we are generally seen in terms of some formal status. A professor may see us primarily as students, a used-car salesperson relates to us as potential buyers, and a politician views us as voters. Only among our intimates are we seen on a personal level, as Maria or Matt. Before marriage, our friends are our intimates. After marriage, our spouses are expected to be the ones with whom we are *most* intimate. With our spouses we disclose ourselves most completely, share our hopes, rear our children, and hope to grow old together.

A major function of marriages and families is to provide us with intimacy and social support, thus protecting us from loneliness and isolation.

Economic Cooperation

The family is a unit of economic cooperation and interdependence. Traditionally, heterosexual families divide responsibilities along gender lines—that is, between males and females (Ferree 1991; Fox and Murry 2000). Although a division of labor by gender is characteristic of virtually all cultures, the work that males and females perform varies from culture to culture. Among the Nambikwara in Africa, for example, the fathers take care of the babies and clean them when they soil themselves; the chief's concubines, secondary wives in polygamous societies, prefer hunting over domestic activities. In American society, since the late nineteenth century until recently, men were expected to work away from home, whereas women were to remain at home caring for the children and house.

Such tasks are assigned by culture, not biology. Only a man's ability to impregnate and a woman's ability to give birth and produce milk are biologically determined. We commonly think of the family as a consuming unit, but it also continues to be an important producing unit. The husband is not paid for building a shelf or bathing the children; the wife is not paid for fixing the leaky faucet or cooking.

Although children contribute to the household economy by helping around the house, they generally are not paid (beyond an "allowance") for such things as cooking, cleaning their rooms, or watching their younger brothers or sisters. Yet they are all engaged in productive, sometimes essential, labor (Dodson and Dickert 2004).

As a service unit, the family is dominated by women. Because women's work at home is unpaid, the productive contributions of homemakers have been overlooked. Yet if women were paid wages for their labor as mothers and homemakers according to the wage scale for chauffeurs, physicians, babysitters, cooks, therapists, and so on, many women would make more for their work in the home than most men do for their jobs outside the home. One recent economic estimate of a typical full-time homemaker's work (including caring for children) placed the yearly value at more than $130,000 (Sahadi 2006). Because family power is partly a function of who earns the money, a stay-at-home partner may have less power because their financial contribution to the family is invisible since there is no paycheck.

Reproduction and Socialization

The family makes society possible by producing (or adopting) and rearing children to replace the older members of society as they die off. Traditionally, reproduction has been a unique function of the married family. But single-parent and cohabiting families also perform reproductive and socialization functions. Technological advances in assisted reproductive techniques such as artificial insemination and in vitro fertilization have separated reproduction from sexual intercourse and now allow for the participation of others (e.g., sperm or egg donors, surrogate mothers, and so on) in the reproductive process.

Real Families: The Care Families Give

In talking with sociologist David Karp, 37-year-old Angie reflected on her relationship with her mentally ill brother. Her words convey her effort to determine where her obligations to care for her family members begin and end:

> It's kind of hard to put . . . into words. I mean I love my parents dearly and I love my brother, and I think you're raised to know what's right and wrong. And I do feel like your family comes first. But by the same token, how much . . . is realistic for a sibling to give up? Are you supposed to give up your life . . . your career . . . your hopes? . . . Just where do you draw the line? Do you do what's right for your family and just do it un-selfishly? It's a hard thing. It's easy to say, "Yeah, I'd do anything for my family" until you really have to, until you are faced with it. (Karp 2001, 130)

Karp interviewed 60 people with family members who were suffering from diagnosed mental illness. He spoke with parents dealing with a child's mental illness as well as "children of emotionally sick parents, spouses with a mentally ill partner, and siblings of those suffering from depression, manic depression, or schizophrenia" (14). His 60 interviewees each presented a story that is somewhat distinctive. Yet his sociological approach sought to detail "the consistencies and uniformities" that surfaced (24). He raises the following provocative questions—"What do we owe each other?" "What are the moral boundaries of family relationships?" and "To what extent are we bound to care for each other?" (30)—and speaks of "the extraordinary power of love" displayed by his interviewees:

Even when an ill person treated them with anger and disdain, denied that they were sick, completely disrupted the coherence of everyday life, and did things that were incomprehensible, distressing beyond mea-sure, socially repugnant, or downright dangerous, love kept caregivers caring. (16)

Karp's interviews revealed a sort of hierarchy of caring, such that

- we have greater obligations to members of our immedi-ate families than those in our wider kin group.
- the marital and parental relationships carry the stron-gest caregiving obligations.
- before turning to their adult children, spouses should provide care for each other.
- sibling relationships, such as Angie's relationship with her late brother Paul, carry less obligation than one's responsibilities to one's spouse, children or parents.
- as children marry and/or have families of their own, the claim for support and care that parents can make are diminished.
- the "most powerful and enduring of all obligations" is owed to one's ill children (148).

Karp suggests that families have been "abandoned" by American society, left on their own without social sup-ports to solve any problems individual members may face. Still, his interviews with caregiving spouses, parents, children and siblings reveal an "extraordinary reservoir of love, caring and connection that holds families together, even at a time when family life is so meagerly supported" (263).

Depending on their contraceptive choices, couples can engage in sexual intercourse with relatively high confidence that they will not become parents. Innova-tions in reproductive technology permit many other-wise infertile couples to give birth. Such techniques have also made it possible for lesbian couples and single women without partners to become parents.

The family traditionally has been responsible for **socialization**—the shaping of individual behavior to conform to social or cultural norms. Children are helpless and dependent for years following birth. They must learn how to walk and talk, how to take care of themselves, how to act, how to love, and how to touch and be touched. Teaching children how to fit

into their particular culture is one of the family's most important tasks.

This socialization function, however, often includes agents and caregivers outside the family. The involve-ment of nonfamily in the socialization of children need not indicate a lack of parental commitment to their children or a lack of concern for the quality of care received by their children. Increasing numbers of dual-earner households and employed single mothers have resulted in the placement of many infants, tod-dlers, and small children under the care of nonfamily members, thus broadening the role of others (such as neighbors, friends, or paid caregivers) and reducing the family's role in child rearing. Additionally, since the rise of compulsory education in the nineteenth century, the state has become responsible for a large part of the socialization of children older than age five.

Assignment of Social Roles and Status

We fulfill various social roles as family members, and these roles provide us with much of our identities. During our lifetimes, most of us will belong to two families: the family of orientation and the family of procreation. The **family of orientation** (sometimes called the *family of origin*) is the family in which we grow up, the family that orients us to the world. The family of orientation may change over time if the marital status of our parents changes. Originally, it may be an intact nuclear family or a single-parent family; later it may become a stepfamily. We may even speak of *binuclear families* to reflect the experience of children whose parents separate and divorce. With parents maintaining two separate households and one or both possibly remarrying, children of divorce are members in two different, parentally based nuclear families (Ahrons 1995, 2004).

The common term for the family formed through marriage and childbearing is **family of procreation**. Because many families have stepchildren, adopted children, or no children, we can use a more recent term—**family of cohabitation**—to refer to the family formed through living or cohabiting with another person, whether we are married or unmarried. Most Americans will form families of cohabitation sometime in their lives. Much of our identity is formed in the crucibles of families of orientation, procreation, and cohabitation. In a family of orientation, we are given the roles of son or daughter, brother or sister, stepson or stepdaughter. We internalize these roles until they become a part of our being. In each of these roles, we are expected to act in certain ways. For

Much childhood socialization occurs in nonfamily settings such as preschools or day-care centers.

example, children obey their parents, and siblings help one another. Sometimes our feelings fit the expectations of our roles; other times they do not. We may not wish to follow our parents' suggestions or loan money to an unemployed sister and yet feel compelled to do so because of the role expectations we face.

Our family roles as offspring and siblings are most important when we are living in a family of orientation. After we leave home, these roles gradually diminish in everyday significance, although they continue throughout our lives. In relation to our parents, we never cease being children; in relation to our siblings, we never cease being brothers and sisters. The roles simply change as we grow older.

As we leave a family of orientation, we usually are also leaving adolescence and entering adulthood. Being an adult in our society is defined in part by entering new family roles—those of husband or wife, partner, or father or mother. These roles formed in a family of procreation take priority over the roles we had in a family of orientation. In our nuclear family system, when we marry, we transfer our primary loyalties from our parents and siblings to our partners. Later, if we have children, we form additional bonds with them. When we assume the role of spouse or bonded partner, we assume an entirely new social identity linked with responsibility, work, and parenting. In earlier times, such roles were considered lifelong. Because of divorce or separation, however, these roles today may last for considerably less time.

The status or place we are given in society is acquired largely through our families. Our families place us in a certain socioeconomic class, such as blue collar (working class), middle class, or upper class. We learn the ways of our class through identifying with our families. As shown in Chapter 3, different classes experience the world differently. These differences include the ability to satisfy our needs and wants but may extend to how we see men's and women's roles, how we value education, and how we bear and rear our children (Edin and Kefalas 2005; Lareau 2003). Our families also give us our ethnic identities as African American, Latino, Jewish, Irish American, Asian American, Italian American, and so forth. Families also provide us with a religious tradition as Protestant, Catholic, Jewish, Greek Orthodox, Islamic, Hindu, or Buddhist—as well as agnostic, atheist, or New Age. These identities help form our cultural values and expectations. These values and expectations may then influence the kinds of choices we make as partners, spouses, or parents.

Why Live in Families?

As we look at the different functions of the family, we can see that, at least theoretically, most of them can be fulfilled outside the family. For example, artificial insemination permits a woman to be impregnated by a sperm donor, and embryonic transplants allow one woman to carry another's embryo. Children can be raised communally, cared for by foster families or child care workers, or sent to boarding schools. Most of our domestic needs can be satisfied by microwaving prepared foods or going to restaurants, sending our clothes to the laundry, and hiring help to clean our bathrooms, cook our meals, and wash the mountains of dishes accumulating (or growing new life forms) in the kitchen. Friends can provide us with emotional intimacy, therapists can listen to our problems, and sexual partners can be found outside marriage. With the limitations and stresses of family life, why bother living in families?

Sociologist William Goode (1982) suggests that there are several advantages to living in families:

- *Families offer continuity as a result of emotional attachments, rights, and obligations.* Once we choose a partner or have children, we do not have to search continually for new partners or family members who can better perform a family task or function such as cooking, painting the kitchen, providing companionship, or bringing home a paycheck. We expect our family members—whether partner, child, parent, or sibling—to participate in family tasks over their lifetimes. If at one time we need to give more emotional support or attention to a partner or child than we receive, we expect the other person to reciprocate at another time. We count on our family members to be there for us in multiple ways. We rarely have the same extensive expectations of friends.

- *Families offer close proximity.* We do not need to travel across town or the country for conversation or help. With families, we do not even need to leave the house; a husband or wife, parent or child, or brother or sister is often at hand (or underfoot). This close proximity facilitates cooperation and communication.

- *Families offer intimate awareness of others.* Few people know us as well as our family members because they have seen us in the most intimate circumstances throughout much of our lives. They have seen us at our best and our worst, when we are kind or selfish, and when we show understanding or intolerance. This familiarity and close contact teach us to make adjustments in living with others. As we do so, we expand our knowledge of ourselves and others.

- *Families provide many economic benefits.* They offer us economies of scale. Various activities, such as laundry, cooking, shopping, and cleaning, can be done almost as easily and with less expense for several people as for one. As an economic unit, a family can cooperate to achieve what an individual could not. It is easier for a working couple to purchase a house than an individual, for example, because the couple can pool their resources. Because most domestic tasks do not take great skill (a corporate lawyer can mop the floor as easily as anyone else), most family members can learn to do them. As a result, members do not need to go outside the family to hire experts. For many family tasks—from embracing a partner to bandaging a child's small cut or playing peekaboo with a baby—there are no experts to compete with family members.

These are only some of the theoretical advantages families offer to their members. Not all families perform all these tasks or perform them equally well. But families, based on mutual ties of feeling and obligation, offer us greater potential for fulfilling our needs than do organizations based on profit (such as corporations) or compulsion (such as governments).

Extended Families and Kinship

Society "created" the family to undertake the task of making us human. According to some anthropologists, the nuclear family of man, woman, and child is universal, either in its basic form or as the building block for other family forms (Murdock 1967). Other anthropologists disagree that the father is necessary, arguing that the basic family unit is the mother and child *dyad*, or pair (Collier, Rosaldo, and Yanagisako 1982). The use of artificial insemination and new reproductive technologies, as well as the rise of female-headed single-parent families, are cited in support of the mother–child model.

Extended Families

The **extended family**, as already described, consists not only of the cohabiting couple and their children but also of other relatives, especially in-laws, grandparents, aunts and uncles, and cousins. In most non-European countries, the extended family is often regarded as the basic family unit. For many Americans, especially those with strong ethnic identification and those in certain groups (discussed in Chapter 3), the extended family takes on great importance. Sometimes, however, we fail to recognize the existence of extended families because we assume the nuclear family model as our definition of family. Thus, when someone asks us to name our family members, if we are unmarried, many of us will probably name our parents, brothers, and sisters. If we are married, we will probably name our spouses and, if we have any, our children. Only if questioned further will some bother to include grandparents, aunts or uncles, cousins, or even friends or neighbors who are "like family." We may not name all our blood relatives, but we will probably name the ones with whom we feel emotionally close, as shown earlier in the chapter.

Although most households in the United States are nuclear in structure, the number of multigenerational households continues to increase. According to a study by the American Association of Retired Persons (AARP), there were approximately 6.2 million multi-generational households in the United States in 2008, up from 5 million in 2000. The study further estimates that 5.3% of all households are multigenerational households, an increase from 4.8% in 2000 (Green 2009). The U.S. Census Bureau estimates that roughly 1.3 million children live in households with grandparents and no parents (www.census.gov 2007).

In the United States today, extended family households are somewhat more common among immigrants and where economic necessity dictates. They can be found in greater proportion in states where there are large concentrations of certain ethnic populations. For example, in Hawaii, which has a large Asian population, more than 8% of households are multigenerational. Among families in California, where there is a large Hispanic population, close to 6% of households fall under this arrangement (Max 2004).

But even in the absence of multigenerational *households*, many Americans maintain what have been called *modified extended families*, in which care and support are shared among extended family members even though they don't share a residence. Think about your own family. What, if any, role or roles have your grandparents played in your life? Did they babysit for you when you were younger? Did you visit them regularly? Talk on the phone? Exchange gifts? The point is that, even in the absence of sharing a household, grandparents and other extended kin may be important figures in your life and, hence, broaden and enrich your family experiences beyond the nuclear households in which you may live or have lived.

Journalist Tamar Lewin suggests that "in many families, grandparents are the secret ingredients that make the difference between a life of struggle and one of relative ease." They may provide assistance that allows their grandchildren to go to camp, get braces for their teeth, go on vacation, and get music lessons or necessary tutoring, all of which enrich their grandchildren's lives beyond what parents alone could manage. Sociologist Vern Bengston has 20 years of data that he has gathered from his undergraduates about how they finance their college educations. Bengston contends that among his own students, grandparents are now the third most frequently mentioned source, behind parents and scholarships but ahead of both jobs and loans. And the importance of grandparents includes but goes well beyond those instances in which they either share the households of or provide child care for their young grandchildren. We should note that there are also many instances in which

The maintenance of wider kin ties has been enhanced by technological innovations from bicycles, through telephones and automobiles, and most recently through advances made in information technology. Just as wheels allowed family members to visit kin some distance away and telecommunications allowed people to speak with relatives nearly anywhere in the world, more recent innovations may make it possible to have family get-togethers and even eat family dinners "together" despite being separated by great distance.

Imagine the following family: Mom, widowed and in her mid-seventies, has a home in a small town in Ohio. Her son lives with his family in southern California, and her daughter lives in North Carolina. The son and daughter worry about their mother, who as she ages becomes somewhat more forgetful. Although they speak with her regularly on the phone, her children and grandchildren are too far away to visit more than once a year. Modern technology may soon offer families in situations such as these a means to feel more closely connected and to stay in more intimate contact. In fact, they may be able to "get together" for dinner on a frequent and regular basis.

Through developments in videoconferencing, the high-tech consulting company Accenture is close to marketing the "Virtual Family Dinner." With this system, when our hypothetical mother in Ohio sits down to eat her dinner, the system would detect this and notify her son in California and daughter in North Carolina. Her children could then go into their own kitchens, where, with the assistance of small cameras, microphones, speakers, and small screens, they could see and hear each other. Although there is wide use of similar devices for business related videoconferencing, the intention of the Virtual Family Dinner is to allow people with no technological savvy to use it almost effortlessly.

Using computers, broadband, and television sets, "motion detection sensors near the dinner table determine when someone is about to sit down for a meal. The information is then relayed to a remote server, which checks the status of all family members on the list to see if they are available. Family members who are available press a button on their system and automatically connect to their elderly relative" (Kawamoto 2007). Fully automated, the system requires nothing of the elderly relative, who, instead, is automatically connected with others on their contact list.

Innovations such as this have obvious appeal for those who wish to bridge physical distance and create a sense of togetherness that transcends geographic boundaries. This could be used to connect families such as the fictional one described above as well as spouses separated for work assignments or military deployment, parents and their college-age children, or even noncustodial divorced parents and their children. In fact, this kind of "technological togetherness" has the potential to "reextend" families and allow for the maintenance of ever closer contact between kin living in geographically dispersed households.

adults help their elderly parents. In the previously mentioned AARP survey, 25% of "baby boomers" expected to have their parents move in with them at some point in time (Green 2009). In either direction, such assistance and support remind us that extended families are important sources of aid and support for one another.

Kinship Systems

The **kinship system** is the social organization of the family. It is based on the reciprocal rights and obligations of the different family members, such as those between parents and children, grandparents and grandchildren, and mothers-in-law and sons-in-law.

Conjugal and Consanguineous Relationships

Family relationships are generally created in two ways: through marriage and through birth. Family relationships created through marriage are known as **conjugal relationships**. (The word *conjugal* is derived from the Latin *conjungere*, meaning "to join together.") In-laws, such as mothers-in-law, fathers-in-law, sons-in-law, and daughters-in-law, are created by law—that is, through marriage. **Consanguineous relationships** are created through biological (blood) ties—that is, through birth. (The word *consanguineous* is derived from the Latin *com-*, "joint," and *sanguineous*, "of blood.") Relationships between adopted children and parents, though not related by blood, might be considered "fictive consanguineous" relationships in that they are culturally treated as having the same kinds of ties and obligations.

Families of orientation, procreation, and cohabitation provide us with some of the most important roles we will assume in life. The nuclear family roles (such as parent, child, husband, wife, and sibling) combine with extended family roles (such as grandparent, aunt, uncle, cousin, and in-law) to form the kinship system.

Kin Rights and Obligations

In some societies, mostly non-Western or nonindustrialized cultures, kinship obligations may be more extensive than they are for most Americans in the twenty-first century. In cultures that emphasize wider kin groups, close emotional ties between a husband and a wife are viewed as a threat to the extended family.

In a marriage form found in Canton, China, women do not live with their husbands until at least three years after marriage, as their primary obligation remains with their own extended families. Under the traditional marriage system among the Nayar of India, men had a number of clearly defined obligations toward the children of their sisters and female cousins, although they had few obligations toward their own children (Gough 1968).

In American society, the basic kinship system consists of parents and children, but it may include other relatives as well, especially grandparents. Each person in this system has certain rights and obligations as a result of his or her position in the family structure. Furthermore, a person may occupy several positions at the same time. For example, an 18-year-old woman may simultaneously be a daughter, a sister, a cousin, an aunt, and a granddaughter. Each role entails different rights and obligations. As a daughter, the young woman may have to defer to certain decisions of her parents; as a sister, to share her bedroom; as a cousin, to attend a wedding; and as a granddaughter, to visit her grandparents during the holidays.

In American culture, the nuclear family has many norms regulating behavior, such as parental support of children and sexual fidelity between spouses, but the rights and obligations of relatives outside the basic kinship system are less strong and less clearly articulated. Because neither culturally binding nor legally enforceable norms exist regarding the extended family, some researchers suggest that such kinship ties have become voluntary. We are free to define our kinship relations much as we wish. Like friendship, these relations may be allowed to wane (Goetting 1990).

Despite the increasingly voluntary nature of kin relations, our kin create a rich social network for us. Adult children and their parents often live close to one another, make regular visits, and/or help one another with child care, housework, maintenance, repairs, loans, and gifts. The relations among siblings also are often strong throughout the life cycle (Lee, Mancini, and Maxwell 1990). In fact, as vividly illustrated by sociologist Karen Hansen's research on "networks of care," kin are frequently essential supports in the ever more complicated tasks associated with raising children in dual-earner households or single-parent households. Although they are invisible when we focus so intensively on nuclear families, to effectively raise children may require the help of "'other mothers,' aunties, grandmothers and child-care workers (as well as) . . . uncles, grandfathers and male friends" (Hansen 2005, 215). Where kin are unavailable or where certain family members are either uncooperative or deemed to be unsuitable, these networks might expand to include neighbors, friends, and paid caregivers.

Multiple Viewpoints of Families

As we noted earlier, marriage and family issues inspire much debate. For instance, those who believe that families of male providers, female homemakers, and their dependent children living together, 'til death do they part, are what families *should be* would not be encouraged by the continued high rates of divorce, increases in cohabitation, or the declining rates of marriage or full-time motherhood. Those on the "other" side who claim that there are basic inequities within the traditional family, especially regarding the status of women, will not mourn the diminishing numbers of breadwinner–housewife families. Similarly, the question of gay marriage will divide those who believe that marriage *must* be a relationship between a man and a woman from those who believe that we *must* recognize and support all kinds of families and provide equal marriage rights to all people.

There are numerous sources of such different viewpoints. One such source is religion, as the following example nicely illustrates. In October 2005, PBS (the public broadcasting network) conducted a poll of American attitudes and opinions on a host of family issues. Sampling 1,130 American adults for the program *Religion and Ethics Newsweekly*, the pollsters asked about a number of family issues. The survey garnered interesting results. Consider a few:

- 80% of respondents agreed that it is better for children if their parents are married.
- 71% believe that "God's plan for marriage is one man, one woman, for life."
- 49% agree that it is okay for a couple to live together without intending to marry.
- 52% agree that divorce is the best solution for a couple who cannot work out their marriage problems.
- 55% agree that "Love makes a family . . . and it doesn't matter if parents are gay or straight, married or single."

Interestingly, within each of these items there were big differences in attitudes based on respondents' religious backgrounds. Looking again at these same items and comparing respondents of different religious backgrounds, the data clearly indicate that big differences exist between those of more traditional or conservative Christian backgrounds and "mainline Protestants" or liberal Catholics (no other religious groups were included in the sample). The most liberal attitudes were expressed by those who identified themselves as having no religious preference (or as atheists or agnostics). Such differences are often obscured when we look at overall attitudes of Americans or even at attitude differences between Protestants and Catholics (see Table 1.1).

Divisiveness such as this is neither new nor unique to the United States. In the early twentieth century, we witnessed considerable pessimism about whether families would survive the changing and liberalizing culture of sexuality, the increasing numbers of women delaying marriage for educational or occupational reasons, and the declining birthrate and increases in divorce. In considering the same sorts of changes, others advocated that these trends were positive signs of families adapting to changes in the wider society (Mintz and Kellogg 1988).

In recent years, many other countries have faced similar cultural clashes over trends and changes in family life. In Spain, for example, there is a dispute pitting the Spanish socialist government against the Catholic Church, as governmental initiatives to legalize same-sex marriage and make abortion and divorce easier or quicker have met with strong and vocal opposition from the church. Whereas some in the Spanish Socialist Party or among its allies such as the United Left Party believe that Spain has not gone far enough in recognizing and embracing change, organizations aligned with the Church, such as the Institute of Family Policy, consider the climate in Spain "family phobic" (Fuchs 2004).

Ultimately, the ways we view families depend on what we conceive of as families. Such disagreements reflect both different definitions of family and different values regarding particular kinds of families. Often the product of personal experience as much as of religious background, personal values reflect what we want families to be like and, thus, what we come to believe about the kinds of issues that are raised throughout this book.

Half Full versus Half Empty

With so much "noise" in the wider society around what family life is and should be like, how families are changing, and whether those changes are good or bad, you may find it difficult to know what conclusions to

Table 1.1 Religious Differences in Attitudes toward Family Issues: Results from PBS "Faith and Family" Survey, October 2005

Item	Total (%)	Evangelical Christian (%)	Mainline Protestant (%)	Traditional Catholic (%)	Liberal Catholic (%)	No Preference/ Atheist/Agnostic (%)
Better for children if parents are married	80	86	82	88	75	58
God's plan for marriage . . .	71	92	62	91	60	31
Divorce is usually best	52	48	61	46	63	50
All right to live together	49	21	57	38	72	78
Love is what makes a family	55	33	62	41	77	80

draw about family issues. Given the lack of societal consensus, it is easy to become confused or be misled about what American families are really like. To some, contemporary family life is weaker because of cultural and social changes and is now, to some extent, endangered (Popenoe 1993; Wilson 2002). More optimistic interpretations of changing family patterns celebrate the increased domestic diversity of numerous family types and the richer range of choices now available to Americans (Coontz 1997; Stacey 1993). Like the proverbial glass, some see the family as "half empty," whereas others see it as "half full." What makes the "half full, half empty" metaphor so apt is that even when looking at the same phenomenon or the same trend, some interpret it as evidence of the troubled state U.S. families are in, and others see today's families as different or changing. So, for instance, although the rates of divorce and marriage, the numbers of children in nonparental childcare, or the extent of increase in cohabitation can, like the volume of liquid in a partially filled glass, be objectively measured, the meaning of those measures can vary widely, depending on perspective.

Conservative, Liberal, and Centrist Perspectives

In the wider, societal discourse about families, we can identify opposing ideological positions on the well-being of families (Glenn 2000). The two extremes, which sociologist Norval Glenn calls conservative and liberal, are like the half empty–half full disagreement, a difference between pessimistic and optimistic viewpoints. Conservatives are fairly pessimistic about the state of today's families. To **conservatives**, cultural values have shifted from individual self-sacrifice toward personal self-fulfillment. This shift in values is seen as an important factor in some major changes in family life that occurred beginning with the last three or four decades of the twentieth century (especially higher divorce rates, more cohabitation, and more births outside marriage).

Furthermore, conservatives believe that as a result of such changes, today's families are weaker and less able to meet the needs of children, adults, or the wider society (Glenn 2000). Conservatives therefore recommend social policies to reverse or reduce the extent of such changes (recommendations to repeal no-fault divorce and the introduction of covenant marriage are two examples we examine later).

Compared with conservatives, liberals are more optimistic about the status and future of family life in the United States. **Liberals** tend to believe that the changes in family patterns are just that—changes, not signs of familial decline (Benokraitis 2000). The liberal position also portrays these changing family patterns as products of and adaptations to wider social and economic changes rather than a shift in cultural values (Benokraitis 2000; Glenn 2000). Such changes in family experience create a wider range of contemporary household and family types and require greater tolerance of such diversity. Placing great emphasis on economic issues, liberal family policies are often tied to the economic well-being of families, such as the increasing numbers of employed mothers and two-earner households.

According to Glenn, there is yet a third position in the discourse about families. **Centrists** share aspects of both conservative and liberal positions. Like conservatives, they believe that some familial changes have had negative consequences. Like liberals, they identify wider social changes (e.g., economic or demographic) as major determinants of the changes in family life, but they assert greater emphasis than liberals do on the importance of cultural values. They note that too many people are too absorbed in their careers or too quick to surrender in the face of marital difficulties (Benokraitis 2000; Glenn 2000).

The assumptions within and the differences between these positions are more important than they might first appear to be. The perceptions we have of what accounts for the current status of family life or the directions in which it is heading influence what we believe families *need*. These, in turn, influence social policies regarding family life. As Nijole Benokraitis (2000) states, "Conservatives, centrists, liberals, and feminists who lobby for a variety of family-related 'remedies' affect our family lives on a daily basis" (19).

Disagreement among Family Scientists

It should be noted that social scientists are similarly divided in how they perceive contemporary families. In other words, changing family patterns and trends in marriage, divorce, parenting, and child care are explained and interpreted differently even by the experts who study them. For example, consider the following statements about the effects of divorce on children, each of which comes from published research rather than mere opinion. Nonetheless, it is hard to reconcile the alternate viewpoints.

The first statement comes from Constance Ahrons, a noted scholar on family relationships and author of numerous articles and books on divorce. A member of

Popular Culture: Cartoon Controversy: Are *SpongeBob SquarePants* and *The Simpsons* Threats to Family Values?

Whether or not you have ever been a regular viewer, many of you are likely familiar with the cartoon *SpongeBob SquarePants*, one of the most popular cartoons in recent memory. The Simpsons, of course, are the dysfunctional family in the show of the same name. They have entertained us for 20 years as the longest-running show on television. Both of these animated programs have been targets of more conservative critics who see them as threats to traditional family values.

In SpongeBob's case, he is among a number of characters—including Kermit the Frog and Winnie the Pooh—singing the disco-era hit "We Are Family" in a video produced by the We Are Family Foundation. The video was designed to be used in elementary schools around the country to teach tolerance, cooperation, and appreciation of diversity (www.wearefamilyfoundation.org).

In January 2005, the video and organization that produced it became the target of Dr. James Dobson, the 70-year-old founder and board chairperson of Focus on the Family, a nonprofit evangelical Christian organization Dobson started in 1977. To Dobson, the "We Are Family" video was an attempt by a gay-supporting organization to convince children to accept homosexuality, although no mention of homosexuality can be found in the video. He claimed that the We Are Family Foundation's efforts to use the video to teach "tolerance" and recognize "diversity" extended to teaching children that homosexuality is an acceptable lifestyle. Dobson (2005) contends that "tolerance and diversity . . . are almost always buzzwords for homosexual advocacy."

In that same year, *The Simpsons* aired an episode, "There's Something about Marrying," that became one of the more controversial episodes of a frequently controversial cartoon. It was the first episode in the then 16-year history of the popular program to carry a viewer

Everett Collection

Popular cartoon characters such as those in SpongeBob SquarePants and The Simpsons have been criticized by conservative critics in the culture wars about families and family values.

discretion warning acknowledging sexual content. In the episode, the town of Springfield legalizes same-sex marriage in the hope of attracting tourism to the

the Council on Contemporary Families, Ahrons offers these mostly encouraging words:

> The good news about divorce is that the vast majority of children develop into reasonably competent individuals, functioning within a normal range. . . . Overall, the findings thus far clearly indicate that it is not the divorce per se, but the quality of the

relationship between divorced parents that has an important long term impact on adult children's lives. Good or "good enough" divorces (characterized by parents who are able to minimize their conflict and continue to share parenting, even if minimally) maintain the bonds of family and extended kinship ties.

town. When numerous gay and lesbian couples come to Springfield to marry, the minister (Reverend Lovejoy) refuses to marry them, claiming that marriage should be between only a man and a woman. When he learns that he can collect $200 for every marriage he performs, Homer Simpson joins and becomes an ordained minister of the online e-Piscopal Church, ultimately presiding over a number of weddings. Also in this episode, Marge Simpson's sister, Patty Bouvier, comes out of the closet and announces her desire to marry a woman. The marriage does not occur, but the story line depicts different and changing reactions to same-sex marriage, culminating in Marge's acceptance of her sister's sexuality.

The episode was praised by many, including the Gay and Lesbian Alliance Against Discrimination, but was also harshly criticized by such conservative groups as the Parents Television Council and the American Family Association, which accused the show and Hollywood more generally of a "blatant, pro-homosexual bias" (Rettig 2005).

These two examples illustrate how different perspectives on families and intimate relationships can lead to different interpretations of popular cultural images.

Popular cartoon characters such as those in *SpongeBob SquarePants* and *The Simpsons* have been criticized by conservative critics in the culture wars about families and family values.

The resulting controversies involving animated images and popular cartoon characters pitted those with more conservative views about families against those with more liberal views, creating what one author called a "stink beneath the ink" (Baiocchi 2006). They serve as reminders not only that family issues are differently defined and interpreted but also that these differences are often expressed in highly divisive and heated ways.

Now, contrast Ahrons's comments with the following from David Popenoe, also a well-known sociologist, author, and/or editor of numerous books about contemporary American families. Popenoe, a Rutgers University sociologist and codirector of the National Marriage Project, provides a different perspective:

Divorce increases the risk of interpersonal problems in children. There is evidence, both from small,

qualitative studies and from large-scale, long-term empirical studies that many of these problems are long lasting. In fact, they may even become worse in adulthood. . . .

Based on the findings of this study, except in the minority of high-conflict marriages, it is better for the children if their parents stay together and work out their problems than if they divorce.

Finally, consider these comments from Elizabeth Marquardt, author of the book *Between Two Worlds: The Inner Lives of Children of Divorce* (Marquardt 2006), based on a telephone survey she conducted with sociologist Norval Glenn of 1,500 young adults from divorced and intact families.

> What we found in . . . this first ever study, of the young adults who grew up in divorced families, a nationally representative study, was that there is no such thing as a good divorce. A good divorce is better than a bad divorce, but it is not good. . . . [I]t turns out that the way the parents divorce, whether or not they're good at divorce, matters less than the divorce itself. The divorce itself is the main problem for children of divorce, not whether their parents fight. (cnn.com 2008)

Although there may be ways to accommodate these contrary points of view, clearly they reflect different overall perspectives about marriage, divorce, and the well-being of children. Thus, it is important to realize that, just as the wider society and culture are fraught with conflicting opinions and values about marriage and family relationships, the academic disciplines that study family life frequently suffer from a similar lack of consensus.

As we set off on our exploration of marriage and family issues, it is important to realize that many of the topics we cover are part of similar ongoing debates about families. As you try to make sense of the material we introduce throughout this book, we require you not to take a particular viewpoint but rather to keep in mind that multiple interpretations are possible. Where different interpretations are particularly glaring (as in the many issues surrounding divorce), we present them and allow you to decide which better fits the evidence presented.

The Major Themes of This Text

Throughout the many chapters and pages that follow, as we examine in detail intimate relationships, marriage, and family in the United States, we will introduce a range of theories, provide much data, and look at a number of family issues and relationships in ways you may never have considered before. As we do so, we will visit and revisit the following points.

Families Are Dynamic

As we will see shortly (in Chapter 3) and throughout the text, the family is a dynamic social institution that has undergone considerable change in its structure and functions. Similarly, values and beliefs about families have changed over time. We are more accepting of divorce, employed mothers, and cohabitation. We expect men to be more involved in hands-on child care. We place more importance on individual happiness than on self-sacrifice for family.

In Chapter 3, we explore some of the major changes that have occurred in how Americans experience families. Then, throughout the text, as we address topics such as marriage, divorce, cohabitation, raising children, and managing employment and family, we ask, In what ways have things changed, and why? What consequences and implications result from these changes? Because familial change is often differently perceived and interpreted (see the final theme in this section), we also present different possible interpretations of the meaning of change. Are families merely changing, or are they declining?

Throughout much of the text, we also look at how individual family experience changes over time. From the formation of love relationships; to the entry into marriage or intimate partnerships; to the bearing, raising, and aging of children; and to the aging and death of spouses, families are ever changing.

Families Are Diverse

Not all families experience things the same way. Beginning with Chapter 3, we look closely at a variety of factors that create differences in family experience. We consider, especially, the following major sources of patterned variation in family experience: race, ethnicity, gender, social class, sexuality, and lifestyle choice.

"Race" and Ethnicity

There were more than 240 different native cultures that lived in what is now the United States when the colonists first arrived (Mintz and Kellogg 1988). Since then, American society has housed immigrant groups from the world over who bring with them some of the customs, beliefs, and traditions of their native lands, including those about families. Thus, we can speak of African American families, Latino families, Asian families, Native American families, European families, Middle Eastern families, and so on. In Chapter 3, we provide a brief sketch of the major characteristics of the family experiences of each of these racial or ethnic groups. As we proceed from there, we compare and contrast, where relevant and possible, major differences in family experiences across racial and ethnic lines.

Social Class

Different social classes (categories of individuals and families that share similar economic positions in the wider society) have different experiences of family life. Because of both the material and the symbolic (including cultural and psychological) dimensions of social class, our chances of marrying, our experiences of marriage and parenthood, our ties with kin, our experience of juggling work and family, and our likelihood of experiencing violence or divorce all vary. And this is but a partial list of major areas of family experience that differ among social classes.

Gender

Although gender roles have changed considerably over time, gender differences still surface and loom large in each area of marriage and family on which we touch. Love and friendship, sexual freedom and expression, marriage responsibilities and gratifications, involvement with children, experience of abuse, consequences of divorce and becoming a single parent, and chances for remarriage all differ between women and men. Throughout the book, we identify where women and men have different experiences of relationships and family and consider possible causes and consequences of these differences.

Sexuality and Lifestyle Variation

A striking difference between twenty-first-century families and early American families is the diversity of family lifestyles that people choose or experience. There is no family form that encompasses most people's aspirations or experiences. Statistically, the dual-earner household is the most common form of family household with children, but there is considerable variation among dual-earner households and between such households as traditional or single-parent families.

Increasingly, people choose to cohabit and many same-sex cohabitors continue to press for legal marriage rights. Increasing numbers of couples choose not to have children, while increasing numbers of others choose expensive procedures to assist their efforts and enable them to bear and rear children. This diversity of family types and lifestyles will not soon abate. In the chapters that follow, specific attention is directed at singles (with and without children), cohabitation, childless or child-free couples, and role-reversed households. In addition, we examine sexual orientation and, where data are available, compare and contrast how experiences of such things as intimacy, sexual expression, parenting, abuse, and separation differ among heterosexuals, gay men, and lesbians.

Outside Influences on Family Experience

This book takes a mostly sociological approach to relationships, marriage, and families in that we repeatedly stress the outside forces that shape family experiences. The family is one of the core social institutions of society, along with the economy, religion, the state, education, and health care. As such, the shape and substance of family life is heavily affected by the needs of the wider society in which it is located. In addition, other social institutions influence how we experience our families.

Similarly, cultural influences in the wider society, such as the values and beliefs about what families are or should be like and the norms (or social rules) that distinguish acceptable from unacceptable behavior, guide how we choose to live in relationships and families. Thus, although each of us as an individual makes a series of decisions about the kinds of family lives we want, the choices we make are products of the societies in which we live.

In addition, options available to each of us may not reflect what we would freely choose if we faced no constraints on our choices. So, for example, parents who might prefer to stay at home with their children might find such a choice impractical to impossible because economic necessity forces them to work outside the home. Working parents may find the time they spend with their children more a reflection of the demands of their jobs and the inflexibility of their workplaces than of their own personal preferences, just as some at-home parents might prefer to be employed but find that their children's needs, the cost and availability of quality child care, the jobs available to them, and the demands and benefits contained in those jobs push them to stay home.

Even the decision to marry requires a pool of potential and suitable spouses from which to choose and the preferred marital choice to be accepted within the society in which we live. We cannot marry if there are no "marriageable options available" (as may be the case in many inner-city, low-income areas) or if our choice of spouse is not legally allowed (as in gay or lesbian relationships).

Our familial lives reflect decisions we face, the choices we make, and the opportunities and/or constraints we confront. In the wider discourse about families, we tend to encounter mostly individualistic explanations for what people experience, focusing sometimes exclusively on personal choices. Throughout this text, we examine the wider environments within which our family choices are made and the ways in which some of us are given more opportunities whereas others face limited choices.

The Interdependence of Families and the Wider Society

Following the prior theme, we indicate throughout the book how societal support is essential for family well-being. Equally true, healthy, well-functioning families are essential to societal well-being. To function effectively, if not optimally, families need outside assistance and support. Better child care, more flexible work environments, economic assistance for the neediest families, protection from violent or abusive partners or parents, and a more effective system for collecting child support are just some examples we consider of where families clearly have needs for greater societal or institutional support.

In turn, the health and stability of our society depend largely on strong and stable families. When families fail, individuals must turn to society for assistance; social institutions must be designed to fill the voids left by failing families, and the pathologies created by weak family structures make society a less livable place. There are enormous costs that result from neglecting the needs of America's families and children.

Family is the irreplaceable means by which most of the social skills, personality characteristics, and values of individual members of society are formed. Hope, purpose, and general attitudes of commitment, perseverance, and well-being are nurtured in the family.

Indeed, even the rudimentary maintenance and survival care provided by families is no small contribution to the well-being of a community.

Some of the services provided by families are such a basic part of our existence that we tend to overlook them. These include such essentials as the provision of food and shelter—a place to sleep, rest, and play—as well as caretaking, including supervision of health and hygiene, transportation, and the accountability of family members involving their activities and whereabouts. Without families, communities would have to provide extensive dormitories and many personal-care workers with different levels of training and responsibility to perform the many activities in which families are engaged.

On a more emotional level, without families individuals must look elsewhere to satisfy basic needs for intimacy and support. We marry or form marriagelike cohabiting relationships, have children, and maintain contact with other kin (adult siblings, aging parents, and extended kin) because such relationships retain importance as bases for our identities and sources of social and emotional sustenance. We bring to these relationships high affective expectations. When our intimacy needs are not met (in marriage or long-term cohabitation), we terminate those relationships and seek others that will provide them. We believe, however, that those needs are best met in families.

As you now begin studying marriage and the family, it is hoped that you will see that such study is both abstract and personal. It is abstract insofar as you will learn about the general structure, processes, and meanings associated with marriage and the family, especially within the United States. In the chapters that follow, the things that you learn should also help you better understand your own family, how it compares to other families, and why families are the way they are. In other words, as we address family more generally, in some ways it is *your* present, *your* past, and *your* future that you are studying. By providing a wider sociological context to marriage, family, and intimate relationships, we show you how and where your experiences fit and why.

Summary

- Marriage is a legally recognized union between a man and a woman in which they are united sexually, cooperate economically, and may give birth to, adopt, or rear children.

- Marriage differs among cultures and has changed historically in our own society.

- In Western cultures, the preferred form of marriage is *monogamy*, in which there are only two spouses. *Polygyny*, the practice of having two or more wives, is common throughout many cultures in the world.

- Legal marriage provides a number of rights and protections to spouses that couples who live together lack.

- The current legal definitions of marriage are changing in the United States and in many other countries. The greatest change relates to same-sex marriage.

- Defining the term *family* is complex. Most definitions of family include individuals related by descent, marriage, remarriage, or adoption; some also include affiliated kin.

- Four important family functions are (1) the provision of intimacy, (2) the formation of a cooperative economic unit, (3) reproduction and socialization, and (4) the assignment of social roles and status, which are acquired both in a *family of orientation* (in which we grow up) and in a *family of cohabitation* (which we form by marrying or living together).

- Advantages to living in families include (1) continuity of emotional attachments, (2) close proximity, (3) familiarity with family members, and (4) economic benefits.

- The *extended family* consists of grandparents, aunts, uncles, cousins, and in-laws. It may be formed *conjugally* (through marriage), creating in-laws or stepkin, or *consanguineously* (by birth) through blood relationships.

- The *kinship system* is the social organization of the family. It includes our nuclear and extended families. Kin can be *affiliated*, as when a nonrelated person is considered "kin," or a relative may fulfill a different kin role, such as a grandmother taking the role of a child's mother.

Key Terms

affiliated kin 11	household 11
centrists 21	kinship system 18
clan 11	liberals 21
conjugal relationships 18	marriage 7
consanguineous relationships 18	modified polygamy 10
conservatives 21	monogamy 9
extended family 17	nuclear family 11
family 11	polyandry 9
family of cohabitation 15	polygamy 9
family of orientation 15	polygyny 9
family of procreation 15	serial monogamy 10
	socialization 14
	traditional family 12

RESOURCES ON THE WEB

Book Companion Website

www.cengage.com/sociology/strong

Prepare for quizzes and exams with online resources—including tutorial quizzes, a glossary, interactive flash cards, crossword puzzles, self-assessments, virtual explorations, and more.

2

Studying Marriages and Families

What Do YOU Think? Are the following statements TRUE or FALSE?
You may be surprised by the answers (see answer key on the following page).

T	F	**1** To answer questions about families, we need to rely most on our "common sense."
T	F	**2** "Everyone should get married" is an example of an objective statement.
T	F	**3** Many researchers believe that both love and conflict are normal features of families.
T	F	**4** Stereotypes about families, ethnic groups, and gays and lesbians are easy to change.
T	F	**5** We tend to exaggerate how much other people's families are like our own.
T	F	**6** Family researchers formulate generalizations derived from carefully collected data.
T	F	**7** Every method of collecting data on families is limited in some way.
T	F	**8** A belief that our own ethnic group, nation, or culture is innately superior to another is an example of an ethnocentric fallacy.
T	F	**9** According to some scholars, in marital relationships we tend to weigh the costs against the benefits of the relationship.
T	F	**10** It is impossible to observe family behavior.

A word of warning: The subjects covered in this book come up often and unexpectedly in everyday experience. You may be reading the paper or watching television and come upon some news about research on the effects of divorce or day care on children. You might be having lunch with friends or dinner with your parents, and before you know it, someone may make claims about what marriages need or lack or how some kinds of families are better or stronger than others. The following hypothetical situation is not an altogether uncommon or unrealistic one.

Imagine you are out having coffee with a close friend. You sense that she is troubled, and she confides that she is worried about her relationship with her boyfriend of two years now that they are separated by nearly 600 miles while at different colleges. You feel for your friend, think hard about her predicament, and, wanting to be supportive, smile reassuringly. She shares the following: "I don't know, I guess I sometimes think I'm worrying too much. After all, how many times have I heard, 'absence makes the heart grow fonder'? Everyone knows that, right?" As you open your mouth to respond, she continues: "But then I guess I do think too much. Don't they say, 'Out of sight, out of mind'? I wonder if I should just prepare myself, even start looking for someone new. Now." In obvious distress and confusion, she looks to you for advice.

How would you answer your friend? "Absence makes the heart grow fonder" and "out of sight out of mind" both can't be true. Moreover, how can "everyone know" one thing even though "they" say the opposite? Surely, there must be a way to resolve such a contradiction.

In this chapter, we examine how family researchers attempt to explore issues such as the hypothetical one posed here. In that sense, this chapter differs from the other thirteen chapters. Instead of presenting material about different aspects of the marriage and family experience, it explains and illustrates how we learn the information about relationships and families found in the rest of the book. However, it will enable you to understand better and appreciate more how much our knowledge and understanding of families is enriched by the theories and research procedures we introduce. In learning *how information is obtained and interpreted*, we set the stage for the in-depth exploration of family issues in the chapters that follow.

How Do We Know?

As sociologist Earl Babbie (2007) suggests, social research is one way we can learn about things. However, most of what we "know" about the social world we have "learned" elsewhere through other less systematic means (Babbie 2007; Neuman 2006). In the previous chapter, we noted the dangers inherent in generalizing from personal experiences. We all do this. If you or someone you know had an unfavorable experience with a long-distance relationship, you probably favor the "out of sight, out of mind" response more than the optimistic one your friend is hoping to hear.

The opening scenario illustrates the difficulty involved in relying on what are often called common sense–based explanations or predictions (Neuman 2006). Commonsense understanding of family life may be derived from "tradition," what everyone knows because it has always been that way or been thought to be that way; from "authority figures," whose expertise we trust and whose knowledge we accept; or from various media sources. Cumulatively, these alternatives to research-based information are typically poor sources of accurate and reliable knowledge about social and family life. Often, what we consider and accept as common sense is fraught with the kinds of contradictions depicted previously (or, for example, "birds of a feather flock together" but "opposites attract"). Even in the absence of contradiction, many commonsense beliefs are simply untrue. Thus, if we "really want to know" about how families work or what people in different kinds of family situations or relationships experience, we would be better informed by seeking and acquiring more trustworthy information.

How Popular Culture Misrepresents Family Life

Because so much of the day-to-day stuff of family life (e.g., caring for children, arguing, dividing chores, and engaging in sexual behavior) takes place in private and

---

Answer Key to What Do YOU Think?

1 False, see p. 30; **2** False, see p. 34; **3** True, see p. 41; **4** False, see p. 35; **5** True, see p. 35; **6** True, see p. 35; **7** True, see p. 51; **8** True, see p. 35; **9** True, see p. 45; **10** False, see p. 53.

behind closed doors, we do not have access to what really goes on. But we are privy to those behaviors on television and in movies and magazines. Thus, those depictions can influence what we *assume* happens in real families. The various mass media are so pervasive that they become invisible, almost like the air we breathe. Yet they affect us. Popular culture, in all its forms, is a key source of both information and misinformation about families. Popular culture conveys images, ideas, beliefs, values, myths, and stereotypes about every aspect of life and society, including the family.

Cumulatively, television, popular music, the Internet, magazines, newspapers, and movies help shape our attitudes and beliefs about the world in which we live. The U.S. Census Bureau estimates that, on average, each of us will spend almost 3,600 hours a year using one of these media (U.S. Census Bureau 2009 *Statistical Abstract*, table 1089). Because television has an especially powerful effect on our values and beliefs, we ought to look more closely at its content.

Matter of Fact As of 2008, more than 98% of U.S. households had television sets. According to Nielsen Media Research, during the third quarter of 2008, the average person watched television four hours and 45 minutes per day, an increase of five hours from the same period in 2007 (Semuels 2008). The average household was tuned in for eight hours and 18 minutes per day, the highest reported level since Nielsen Media Research began measuring television viewing in the 1950s. The group with the greatest number of hours of viewing per week was those 65 and older, who watched approximately six hours of television (including "time-shifted television") per day (Nielsen's Three Screen Report" 2008).

Prime-time television, in both dramas and situation comedies, unrealistically depicts married life. It understates the unique issues faced by various ethnic families, inaccurately depicts single-parent family life, inaccurately portrays the relative sexual activity levels of marrieds and singles, and portrays conflict as something easily resolved within 20 or so minutes, often with humor. It also is not inclusive in portraying the family experiences of nontraditional families.

The combined portrayal of family life on daytime television that results from soap operas and talk shows

is unrealistic and highly negative. Those who have scrutinized daytime soap operas note the extremely high rates of conflict, betrayal, infidelity, and divorce that afflict soap opera families; the stereotyping of women as starry-eyed romantics or scheming manipulators; and the distorted and exaggerated portrayal of sex and sexual infidelity (Benokraitis and Feagin 1995; Lindsey 1997). Daytime talk shows are even worse. From *Jerry Springer* and *Maury* to *Dr. Phil*, talk shows contribute to the idea that American family life is deeply dysfunctional, that parents are anything from "irresponsible fools" to "in-your-face monsters" (Hewlett and West 1998), that spouses and partners routinely cheat on each other and often strike each other, and that teenagers are recklessly out of control. Half-naked fisticuffs on *Jerry Springer* and contested allegations and paternity tests on *Maury* are especially distasteful distortions of families and relationships.

Ron P. Jaffe/© CBS/Courtesy Everett Collection

Danny Feld/© ABC/Courtesy Everett Collection

Television programs, such as the popular *How I Met Your Mom* or *Desperate Housewives*, can influence our beliefs and attitudes about marriage and family. What messages and expectations do these programs convey?

The various radio or television talk shows, columns, articles, and advice books have several things in common. First, their primary purpose is to sell books or periodicals or to raise program ratings. They must capture the attention of viewers, listeners, or readers. In contrast, the primary purpose of scholarly research is the pursuit of knowledge.

Second, the media must entertain while disseminating information about relationships and families. Thus, the information and advice must be simplified. Complex explanations and analyses must be avoided because they would interfere with the entertainment purpose. Furthermore, the media rely on high-interest or shocking material to attract readers or viewers. Consequently, we are more likely to read or view stories about finding the perfect mate or protecting our children from strangers than stories about new research methods or the process of gender stereotyping.

Third, the advice and information genre focuses on how-to information or morality. The how-to material advises us on how to improve our relationships, sex lives, child-rearing abilities, and so on. Advice and normative judgments (evaluations based on norms) are often mixed together. Advice columnists act as moral arbiters, much as do ministers, priests, rabbis, and other religious leaders.

To reinforce their authority, the media also incorporate statistics, which are key features of social science research. But not all statistics are of equal value. Unless the research from which the statistics are drawn is sound, statistics have little usefulness. They may

Courtesy of Everett Collection

Dr. Philip McGraw is a licensed clinical psychologist who in addition to his television program has authored six *New York Times* No. 1 best-selling books.

misrepresent family reality by poor sampling techniques, inappropriately worded questions, or unwarranted generalizations.

As you will see throughout this book, although families are not without their share of serious problems, daily family life is as poorly represented by daytime television as by prime-time programming.

(Un)reality Television

The newest genre of television programming is what has been termed *reality television*. Operating without scripts or professional actors, reality television typically puts "ordinary people" into situations or locations that require them to meet various challenges. Much of what is considered reality television has nothing to do with relationships, marriages,

or families. However, there have been a number of reality programs that focus directly on relationships and family life, including the following current or canceled shows: *Who Wants to Marry a Millionaire?, Trading Spouses, The Hills, The Bachelor, Supernanny, Jon and Kate Plus Eight,* and *Renovate My Family*. Note that these represent just a fraction of the reality genre. Whether these shows match and/or marry people or showcase aspects of families or relationships, they hardly represent what their genre claims as its name. By highlighting extreme cases or introducing artificial circumstances and/or competitive goals, these shows are no more representative of familial reality than the daytime talk shows. It would be dangerous to draw

With the media awash in advice and information about relationships, marriage, and family, how can we evaluate what is presented to us? Here are some guidelines:

- *Be skeptical.* Remember: Much of what you read or see is meant to entertain you. Consider the reliability of the sources and the representativeness of the people interviewed.
- *Search for biases, stereotypes, and lack of objectivity.* Is conflicting information omitted? Are diverse family forms and experiences represented?
- *Look for moralizing.* Many times, what passes as fact is disguised moral judgment. What are the underlying values of the article or program?
- *Go to the original source or sources.* The media simplify, so find out for yourself what the studies said and how well or poorly they were done.
- *Seek additional information.* The whole story is probably not told, so seek out additional information, especially information in scholarly books and journals, reference books, or college textbooks.

AP Images/Harpo Productions Inc., George Burns

Oprah Winfrey is among a number of television and radio talk show hosts who often focus on family issues, although she approaches them with more seriousness than most others.

Throughout this book, you will be exposed to a variety of information or data about families. This information may or may not reflect your experiences, but its value is that it will enable you to learn about how other people experience family life. This knowledge and the results of different kinds of responses to family situations enable a more informed understanding of families in general and of yourself as an individual.

generalizations from shows such as *Supernanny* and conclude that "kids today" are disrespectful or out of control. Although you may consider yourself too sophisticated to make such a generalization, millions of others watch programs such as these. Are all of them equally sophisticated?

Mass Media's Depiction of Families

Although you may suspect that most people who watch television comedies, dramas, talk shows, or reality programs realize that they are not the best sources for accurate information about families, there is one genre of the media where such realization may be absent. A veritable industry exists to support what's called "the advice and information genre." It produces self-help and child-rearing books, advice columns, radio and television shows, and numerous articles in magazines and newspapers. In fact, as it transmits information, it also conveys values and norms—cultural rules and standards—about relationships, marriage, and family, often intended as entertainment.

In newspapers in the past, this genre was represented by such popular "advice columnists" as Abigail Van Buren (whose real name was Pauline Esther Friedman and whose column "Dear Abby" is now written by her daughter, Jeanne Phillips), Dan Savage (whose sex-advice column "Savage Love" is widely syndicated in numerous newspapers), and the late Ann Landers (Abby's twin sister, Esther Pauline Friedman, whose daughter, Margo Howard, now writes her own advice column, "Dear Margo").

Newer, Web-based columnists, such as Alison Blackman Dunham and her late twin sister Jessica Blackman Freedman, the self-proclaimed "Advice Sisters" (or "Ann and Abby for the new millennium"), helped carry this genre to the Internet. Allison Blackman Dunham continues to give relationship advice online at her "Great Relationships" Web site (http://advicesisters.net). Radio therapists, such as Dr. Joy Browne and Dr. Laura Schlessinger, have daily callers seeking advice or information about relationships, family crises, and so on, and the program *Loveline* with a number of different hosts has been offering listeners medical and relationship advice since 1983.

On television, Dr. Philip McGraw's *Dr. Phil* has become a ratings success. McGraw, a psychologist of some 25 years, was featured often on *The Oprah Winfrey Show* before landing his own talk show in 2002. His shows cover a range of personal and family issues. In a recent two-week period, for example, episodes included "Time to Grow Up!," "Fighting Fair," "Honeymoon Hangover," "For Better or Worse," "Child Caught in the Middle," and "What's Wrong with Men?" Dr. Phil is also the author of a number of best-selling books, including *Family First: Your Step-by-Step Plan for Creating a Phenomenal Family* and *Relationship Rescue: Seven Steps for Reconnecting with Your Partner*, and has a Web site from which visitors can obtain a variety of suggestions for how to deal with the kinds of relationship and personal issues featured on his show or in his books.

Of all media genres, the advice and information genre can be especially misleading. After all, it offers advice to and depicts the experiences of "real people." It uses experts, cites studies, and may sometimes sound scientific. However, as the previous Popular Culture feature suggests, there are good reasons to be skeptical of the information such sources provide or the advice they offer.

By now, between the previous chapter and the present one, we hope you are aware of the limitations of both personal experience and popular culture as sources of reliable information about familial reality. That doesn't mean that we can't learn from our own experiences or become more aware of various aspects of family life because of how they are portrayed or presented in the mass media. However, if one's goal is to discover what family life looks and feels like for most people or for people unlike ourselves, we need to turn to more trustworthy sources of information.

Researching the Family

Before we examine the specific theories and research techniques used by family researchers, it is important to emphasize that the attitudes of the researcher (or you as you read research) are important. To obtain valid research information, we need to keep in mind the rules of critical thinking. The term *critical thinking* is another way of saying "clear and unbiased thinking."

The Importance of Objectivity

We all have perspectives, values, and beliefs regarding marriage, family, and relationships. These can create blinders that keep us from accurately understanding the research information. We need instead to develop a sense of **objectivity** in our approach to information—to suspend the beliefs, biases, or prejudices we have about a subject until we understand what is being said (Kitson et al. 1996). We can then take that information and relate it to the information and attitudes we already have. Out of this process, a new and enlarged perspective may emerge.

One area in which we may need to be alert to maintaining an objective approach is that of family lifestyle. The values we have about what makes a successful family can cause us to decide ahead of time that certain family lifestyles are "abnormal" because they differ from our experience or preference. We may refer to single-parent families as "broken" or say that adoptive parents are "not the real parents."

A clue that can sometimes help us "hear" ourselves and detect whether we are making value judgments or **objective statements** is as follows: A **value judgment** usually includes words that mean "should" and imply that our way is the correct way. An example is, "Everyone should get married." This text presents information based on scientifically measured findings—for example, concluding that "about 90% of Americans marry."

Opinions, biases, and stereotypes are ways of thinking that lack objectivity. An **opinion** is based on our experiences or ways of thinking. A **bias** is a strong opinion that may create barriers to hearing anything contrary to our opinion. A **stereotype** is a set of simplistic, rigidly held, and overgeneralized beliefs about the personal characteristics of a group of people. Stereotypes are fairly resistant to change. Furthermore, stereotypes are often negative. Common stereotypes related to marriages and families include the following:

- Nuclear families are best.
- Stepfamilies are unhappy.
- Lesbians and gay men cannot be good parents.
- Women are instinctively nurturing.
- People who divorce are selfish.

We all have opinions and biases; most of us, to varying degrees, think stereotypically. But the commitment to objectivity requires us to become aware of these opinions, biases, and stereotypes and to put them aside in the pursuit of knowledge.

Fallacies are errors in reasoning. These mistakes come as the result of errors in our basic presuppositions. Two common types of fallacies that especially affect our understanding of families are egocentric fallacies and ethnocentric fallacies. The **egocentric fallacy** is the mistaken belief that everyone has the same experiences and values that we have and therefore should think as we do. The **ethnocentric fallacy** is the belief that our ethnic group, nation, or culture is innately superior to others. In the next chapter, when we consider the differences and strengths of families from different ethnic and economic backgrounds, you need to keep both of these fallacies from distorting your understanding.

From the day of your birth, you have been forming impressions about human relationships and developing ways of behaving based on these impressions. Hence, you might feel a sense of "been there, done that" as you read about an aspect of personal development or family life. However, your study of the information in this book will provide you the opportunity to reconsider your present attitudes and past experiences and relate them to the experiences of others. As you do this, you will be able to use the logic and problem-solving skills of critical thinking so that you can effectively apply that which is relevant to your life.

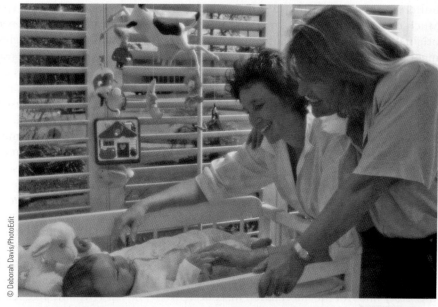

© Deborah Davis/PhotoEdit

Despite the fact that many people believe stereotypes of gay men and lesbians and have biases against same-sex couples as parents, research reveals that heterosexuals, lesbians, and gay men can all be equally effective and loving parents.

The Scientific Method

Family researchers come from a variety of academic disciplines—from sociology, psychology, and social work to communication and family studies (sometimes known as "family and consumer sciences"). Although these disciplines may differ in terms of the specific questions they ask or the objectives of their research, they are unified in their pursuit of accurate and reliable information about families through the use of social scientific theories and research techniques. Scholarly research about the family brings together information and formulates generalizations about certain areas of experience. These generalizations help us predict what happens when certain conditions or actions occur.

Family science researchers use the **scientific method**—well-established procedures used to collect and analyze information about family experiences. This rigorous approach allows other people to know the source of the information and to be confident of the accuracy of the findings. Much of the research family scientists do is shared in specialized journals (e.g., the *Journal of Marriage and the Family* and the *Journal of Family Issues*), on Web sites, or in book form. By communicating their results through such channels, other researchers can build on, refine, or further test research findings. Much of the information contained in this book originally appeared in scholarly journals or government reports.

Concepts and Theories

One of the most important differences between the knowledge about marriage and family derived from family research and that acquired elsewhere is that family research is influenced or guided by **theories**—sets of general principles or concepts used to explain a phenomenon and to make predictions that may be tested and verified experimentally. Although researchers collect and use a variety of kinds of data on marriages and families, these data alone do not automatically convey the meaning or importance of the information gathered. Concepts and theories supply the "story line" for the information we collect.

Concepts are abstract ideas that we use to represent the reality in which we are interested. We use concepts to focus our research and organize our data. Many examples of concepts—for example, nuclear families, monogamy, and socialization—were introduced in the previous chapter. Family research involves the processes of **conceptualization**, the specification and definition of concepts used by the researcher, and of **operationalization**, the identification and/or development of research strategies to observe or measure concepts. For example, to study the relationship between social class and child-rearing strategies, we need to define and specify how we are going to identify and measure a person's social class position and child-rearing strategies.

In **deductive research**, concepts are turned into **variables**, concepts that can vary in some meaningful way. Marital status is an example of a variable used by family researchers. We may be married, divorced, widowed, or never married. As researchers explore the causes and/or consequences of marital status, they may formulate **hypotheses**, or predictions, about the relationships between marital status and other variables. We might hypothesize that race or social class influences whether someone is married or not. In such an example, race is an **independent variable** and marital status the **dependent variable** in that race is thought to influence the likelihood of becoming or staying married. Marital status, on the other hand, may be a causal or independent variable in a hypothesized relationship between being married and life expectancy or satisfaction. Finally, marital status might be hypothesized as an **intervening variable**, affected by the independent variable, race, and in turn affecting the dependent variable, life expectancy. In that instance, the hypothesis suggests that race differences in marital status account for race differences in life expectancy (Figure 2.1).

Rarely do researchers construct theories with only two or three variables. They may hypothesize multiple independent and intervening variables and seek to identify those having the greatest effect on the dependent variable (Neuman 2006). In Figure 2.1, panel (d)

Figure 2.1 The four examples illustrate different ways researchers might hypothesize about the causes or effects of the variable marital status

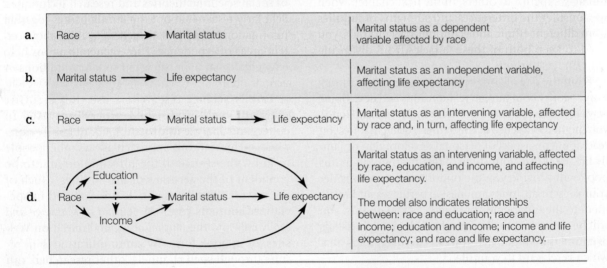

is an illustration of this. Race is hypothesized to have direct and indirect effects on marital status. Race is alleged to have effects on both income and education, which in turn are hypothesized to affect marital status. And, finally, race, income, and marital status are all hypothesized to have effects on life expectancy.

Inductive research is not hypothesis-testing research. Instead, it begins with a topic of interest to the researcher (e.g., what happens when couples reverse traditional roles) and perhaps some vague concepts. As researchers gather their data, typically in the form of field observations or interviews, they refine their concepts, seek to identify recurring patterns out of which they can make generalizations, and, perhaps, end by building a theory (or asserting some hypotheses) based on the data collected. Theory that emerges in this inductive fashion is often referred to as grounded theory in that it is *grounded* or "rooted in observations of specific, concrete details" (Neuman 2006).

Theoretical Perspectives on Families

On a more abstract level of theory, we can identify major theoretical frameworks or perspectives that guide much of the research about families. These perspectives (sometimes also called *paradigms*) are sets of concepts and assumptions about how families work and/or how they fit into society. Theoretical frameworks guide the kinds of questions we raise, the types of predictions we make, where we look to find answers, and how we construct explanations (Babbie 2007).

In this section, we discuss several of the most influential theories sociologists and psychologists use to study families, including ecological, symbolic interaction, social exchange, developmental, structural functional, conflict, and family systems theory. We also look at the influence of feminist perspectives on family studies. As you examine them, notice how the choice of a theoretical perspective influences the way data are interpreted. Furthermore, as you read this book, ask yourself how different theoretical perspectives would lead to different conclusions about the same material.

Macro-Level Theories

The following four theories are what we call macro-level theories in that they focus on the family as a social institution (see Table 2.1). A social institution is

Table 2.1 Macro- and Micro-Level Theories That Guide Research on Families

Macro-Level Theories	Micro-Level Theories
Family ecology theory	Symbolic interactionism
Structural functionalism	Social exchange theory
Conflict theory	Developmental theory
Feminist theory	Family systems theory

the organized pattern of statuses and structures, roles, and rules by which a society attempts to meet certain of its most basic needs. For example, the family refers to groups and relationships through which a society regulates reproduction, socializes children, and provides us with emotional support. The economy is the ways a society organizes the production and distribution of essential goods and services. Other institutions include religion, politics, and education. Macro-level theories examine how the family as a social institution is affected by the wider society, including the interconnections and influences between families and other social institutions. Sometimes these perspectives are also applied to analyzing the internal structure of relationships within families. Still, when they do, they connect that structure to wider social, economic, and political influences.

Family Ecology Theory

The emphasis of family ecology theory is on how families are influenced by and in turn influence the wider environment. Ecological theory was introduced in the late nineteenth century by plant and human ecologists. German biologist Ernst Haeckel first used the term *ecology* (from the German word *oekologie*, or "place of residence") and placed conceptual emphasis on environmental influences. This focus was soon picked up by Ellen Swallows Richards, the founder and first president of the American Home Economics Association (now known as the American Association of Family and Consumer Sciences), who believed that scientists needed to focus on home and family, "for upon the welfare of the home depends the welfare of the commonwealth" (quoted in White and Klein 2002).

The core concepts in ecological theory include environment and adaptation. Initially used to refer to the adaptation of plant and animal species to their physical environments, these concepts were later extended to humans and their physical, social, cultural,

For many of us, even years later, the awful images are still vividly with us and recalled anytime we hear the name "Katrina." Hurricane Katrina was a true tragedy, a category 5 hurricane, that struck the Gulf coast states of Mississippi, Louisiana, Alabama, and Florida in August 2005. With winds that occasionally reached 170 miles an hour, Katrina devastated the region. Hardest hit was New Orleans, where 80% of the city was submerged underwater.

The human cost of Katrina was enormous. More than 1,300 people in five states died from Hurricane Katrina. Suddenly, thousands of people lost spouses, parents, siblings and children. In addition, many thousands of families, for at least a time, were left in limbo, without news about the whereabouts of missing loved ones or knowledge of whether they had survived. Months after Hurricane Katrina, more than 3,000 of the nearly 11,500 people reported missing were still missing (Associated Press, January 19, 2006). What must families feel in such situations?

Pauline Boss has spent more than 30 years studying families dealing with either physically missing or psychologically missing members. Beginning in the early 1970s by looking at *psychological father absence* (fathers who were present but distant), Boss broadened her interest to include situations in which any family member might be said to be "there but not there." She labeled such circumstances **ambiguous loss**, "a situation of unclear loss resulting from not knowing whether a loved one is dead or alive, absent or present" (Boss 2004, 554). She suggested that such loss is the most stressful because it remains unresolved and creates lasting confusion "about who is in or out of a particular family" (553). There is no death

certificate, no funeral, and no opportunity to honor the deceased or bury remains. It prevents family members from reaching psychological closure. As a result, families are immobilized, roles are confused, and tasks remain undone.

Boss considers two situations of ambiguous loss. First is the ambiguous loss of "there but not there," of "physical presence and psychological absence" mentioned previously and applicable in unexpected situations, such as when a family member suffers from dementia (including Alzheimer's disease), autism, depression, or addictions, and in more common situations, such as preoccupation with work, obsessive involvement with the Internet, or divorce followed by remarriage. In the second form of ambiguous loss, members remain psychologically present despite physical absence. This "not there (physically) but there (psychologically)" version of ambiguous loss can be found in tragic situations of war (for families of soldiers missing in action), among families of incarcerated inmates, in families where a member deserts, and in such events as occurred on 9/11 or in the Gulf coast states, especially if no body is recovered. Even more common versions of "not there but there" can occur after divorce or adoption, work relocations, and children leaving home and the "nest" emptying (Boss 2004).

Not all situations of ambiguous loss result in the same outcomes or suffering. Some families manage to redraw otherwise ambiguous boundaries (such as when an aunt or uncle steps in and is viewed as a parent). It appears as if some people have a higher tolerance for ambiguity and therefore may be more resilient in instances such as Hurricane Katrina. Other individuals may suffer emotional or psychological wounds after such tragedies. Although surviving family members may also experience posttraumatic stress disorder (PTSD), ambiguous loss is not the same as PTSD. Treatment of PTSD focuses on individuals, not families as whole systems. In addition, PTSD is a pathology, a psychological illness. Ambiguous loss is a situation of stress that can lead to individual suffering but needs to be understood *on the familial level* (Boss 2004).

More than 8,000 of those reported missing in the Gulf coast after Katrina have been found or their bodies identified. Still, 3,200 or more families struggle to find closure and come to terms with what the storm took from them. Using concepts such as *ambiguous loss* enables us to better understand what they suffer from and why. Such understanding won't alter their suffering or reduce the pain of their losses, but it may make it possible to be more effective in any efforts to help them move on.

Disasters such as 2005's Hurricane Katrina often throw families into extreme situations of ambiguous loss.

and economic environments (White and Klein 2002). As applied to family issues, the family ecology perspective asks, How is family life affected by the environments in which families live? We use the plural *environments* to reflect the multiple environments that families encounter.

In Urie Brofenbrenner's ecologically based theory of human development, for example, the environment to which individuals adapt as they develop consists of four levels: (1) *microsystem*, (2) *mesosystem*, (3) *exosystem*, and (4) *macrosystem*. Cumulatively, these levels make up the environments in which we live. The *microsystem* contains the most immediate influences with which individuals have frequent contact (e.g., our families, peers, schools, and neighborhoods). Each of these exerts influence over how we develop. The *mesosystem* consists of the interconnections between microsystems (e.g., school experiences and home experiences) and ways they influence each other. The *exosystem* consists of settings in which the individual does not actively participate but that nonetheless affect his or her development (e.g., one's parents' work experiences, salaries, and schedules). Finally, the *macrosystem* operates at the broadest level, encompassing the laws, customs, attitudes, and belief systems of the wider society, all of which influence individual development and experience (Rice and Dolgin 2002).

In constructing an ecological framework to better understand marriage relationships, sociologist Ted Huston illustrated how marital and intimate unions are "embedded in a social context" (Huston 2000). This social context includes the *macroenvironment*—the wider society, culture, and physical environment in which a couple lives—and their particular *ecological niche*—the behavior settings in which they function on a daily basis (e.g., a poor, urban neighborhood as opposed to a small town or suburb). Also included in the social context is the marriage relationship itself, especially as it is affected by a larger network of relationships. The final key element in Huston's ecological approach contains the physical, psychological, and social attributes of each spouse, including attitudes and beliefs about their relationship and each other. Each of these environments influences and is influenced by the others. His ecological approach suggests that in order to fully understand marriage, we must explore the interconnections among these three elements.

Ecological theory has been and continues to be a popular perspective among family researchers. In recent years, it has been used broadly in the study of a wide range of family issues. For example, in recent years, it has been applied to such topics as how children of incarcerated mothers maintain family relationships (Poehlman et al. 2008), the effects of economic restructuring on rural women and their families (Ames et al. 2006), the mothering behavior of women involved in street-level prostitution (Dalla 2004), and changes in the quality of lesbian relationships as couples become parents (Goldberg and Sayer 2006). It has been especially prominent in studying low-income families and in exploring the multiple influences on children's and adolescents' behavior and school performance (e.g., Bowen et al. 2008).

Regardless of the specific topic, ecological approaches examine how family experience is affected by the broader social environments. In many ways, much of what we examine in subsequent chapters shares at least this level of ecological focus. We cannot understand what happens within families without considering the wider cultural, social, and economic environments within which family life takes place.

Critique

There have been a variety of criticisms of ecological theory (White and Klein 2002). It is not always entirely clear which system best accounts for the behavior we attempt to explain (e.g., microsystem, mesosystem, exosystem, or macrosystem), how exactly that outcome results, or how the different systems influence each other. A second criticism contends that the theory has been more effectively applied to individual or familial development and growth. Families are prone to degeneration or decline as much as they are to development and growth, yet ecological theory often fails to address this. Finally, some suggest that the theory may not apply as well to a range of diverse, especially nontraditional, families.

Structural Functionalism Theory

Structural functionalism theory explains how society works, how families work, and how families relate to the larger society and to their own members. The theory is used largely in sociology and anthropology, disciplines that focus on the study of society rather than of individuals. When structural functionalists study the family, they look at three aspects: (1) what functions the family as an institution serves for society (discussed in Chapter 1), (2) what functional requirements family members perform for the family, and (3) what needs the family meets for its individual members.

Structural functionalism is deeply influenced by biology. It treats society as if it were a living organism, like a person, animal, or tree. The theory sometimes uses the analogy of a tree in describing society. In a tree, there are many substructures or parts, such as the trunk, branches, roots, and leaves. Each structure has a function. The roots gather nutrients and water from the soil, the leaves absorb sunlight, and so on. Society is like a tree insofar as it has different structures that perform functions for its survival. These structures are called **subsystems**, each of which is necessary for the survival of the wider societal system.

The subsystems are the major institutions, such as the family, religion, government, and the economy. Each of these structures performs functions that help maintain society, just as the different parts of a tree serve a function in maintaining the tree. Religion gives spiritual support, government ensures order, and the economy produces goods. The family provides new members for society through procreation and socializes its members so that they fit into society. In theory, all institutions work in harmony for the good of society and one another.

The Family as a System

Families may also be regarded as systems. When looking within "the family," structural functionalists examine how the family organizes itself for survival and what functions the family as a whole performs for its members. For the family to survive, its members must perform certain functions that are traditionally divided along gender lines. Men and women have different tasks: Men work outside the home to provide an income, whereas women perform household tasks and child rearing.

According to structural functionalists, the family molds the kind of personalities it needs to carry out its functions. It encourages different personality traits for men and women to ensure its survival. Men develop instrumental traits, and women develop expressive traits. *Instrumental traits* encourage competitiveness, coolness, self-confidence, and rationality—qualities that will help a person succeed in the outside world. *Expressive traits* encourage warmth, emotionality, nurturing, and sensitivity—qualities appropriate for someone caring for a family and a home.

Such a division of labor and differentiation of temperaments is seen as efficient because it allows each spouse to specialize, thus minimizing competition, creating interdependence, and reducing ambiguity or uncertainty over such things as who should work outside the home or whose outside employment is more important. For these reasons, such role allocation may be deemed functional.

Critique

Although structural functionalism has been an important theoretical approach to the family, it has declined in significance in recent decades for several reasons.

First, how do we know which family functions are vital? The family, for example, is supposed to socialize children, but much socialization has been taken over by the schools, peer groups, and the media. Is this "functional"? Does it diminish the importance of families?

Second, structural functionalism looks at the family abstractly, from a distance far removed from the daily lives and struggles of men, women, and children. It views the family in terms of functions and roles. Family interactions, the lifeblood of family life, are absent, making structural functionalism of little relevance to real families in the real world.

Third, it is not always clear what function a particular structure serves. For example, what is the function of the traditional division of labor along gender lines—efficiency, survival, or the subordination of women? If interdependence, specialization, and clarity of role responsibilities make breadwinner–homemaker households most "functional," those same objectives could be met by household arrangements wherein men stay home, rear kids, and tend house and women earn incomes. In some relationships, these role reversals might be even more functional. But structural functionalism has a conservative bias against change. Aspects that reflect stability are called functional, and those that encourage instability (or change) are called dysfunctional. By that logic, traditional roles would be described as functional, and nontraditional roles would be seen as dysfunctional.

Conflict Theory

The basic premise of conflict theory sets it apart from structural functionalism. Where structural functionalists assert that existing structures benefit society, conflict theorists ask, "Who benefits?" **Conflict theory** holds that life involves discord and competition.

Conflict theory is sometimes applied to the conflict and inequality that exists within families between members over such things as the division of responsibilities, allocation of resources, and levels of commitment.

Conflict theorists see society not as basically cooperative but as divided, with individuals and groups in conflict with one another. Given the scarcity of society's resources, such as wealth, income, prestige, or power, groups are in competition for those resources. As a consequence of such competition, inequality results. Conflict theory directs attention to how the wider gender, racial, and economic inequalities that result influence families.

Sources of Conflict in Families

We can also analyze marriages and families in terms of internal conflicts and power struggles. Conflict theorists agree that love and affection are important elements in marriages and families, but they believe that conflict and power are also fundamental. Marriages and families are composed of individuals with different personalities, ideas, values, tastes, and goals. Each person is not always in harmony with every other person in the family.

Conflict theorists do not believe that conflict is bad; instead, they think that it is a natural part of family life. Families always have disagreements, from small ones, such as what movie to see, to major ones, such as how to rear children. Families differ in the number of underlying conflicts of interest, the degree of underlying hostility, and the nature and extent of the expression of conflict. Conflict can take the form of competing goals or differences in role expectations and responsibilities. For example, an employed mother may want to divide housework 50–50, whereas her husband insists that household chores are not his responsibility.

Sources of Power

When intrafamily conflict occurs, who wins? Family members have different resources and amounts of power. There are four important sources of power: legitimacy, money, physical coercion, and love. When arguments arise in a family, a man may want his way "because I'm the head of the household" or a parent may argue "because I'm your mother." These appeals are based on legitimacy—that is, the belief that the person is entitled to prevail by right. Money is also a potentially powerful resource within marriages and families, much as it is in the wider society. "As long as you live in this house . . ." is a parental directive based on the power of the purse. Because men have tended to earn more than women, they have had greater economic power, which, in part, can translate into marital power. More than money is involved, however, since when wives outearn husbands, they don't reap the same benefits or enjoy the same amount of power. Physical coercion, threats, or force make up another important source of power. Spanking children and physical abuse of spouses are examples of this. Finally, there is the power of love. Love can be used to coerce someone emotionally, as in "If you really loved me, you'd do what I ask."

Critique

A number of difficulties arise in conflict theory. First, conflict theory derives from politics and economics, in which self-interest, egotism, and competition are dominant elements. Yet is such a harsh judgment of human nature justified? People's behavior is also characterized by self-sacrifice and cooperation.

Love is an important quality in relationships. Conflict theorists do not often talk about the power of love or bonding, yet the presence of love and bonding may distinguish the family from all other groups in society. We often will make sacrifices for the sake of those we love. We will defer our wishes to another's desires; we may even sacrifice our lives for a loved one. Second, conflict theorists assume that differences lead to conflict. Differences can also be accepted, tolerated, or appreciated. Differences do not necessarily imply conflict. Third, conflict in families is not easily measured or evaluated. Families live much of their lives privately, and outsiders are not always aware of whatever conflict exists or how pervasive it is. In addition, much overt conflict is avoided because it is regulated through family and societal rules. Most children obey their parents, and most spouses, although they may argue heatedly, do not employ violence.

Feminist Perspectives

Thanks to the feminist movement of the 1960s and 1970s, new questions and ways of thinking about the meaning and characteristics of families arose. By 2002, nearly 25% of the articles published in the two main scholarly journals about families (*Journal of Marriage and Family* and *Journal of Family Issues*) were assessed as having feminist and/or gender content (Wills and Risman 2006). Although there are some notable distinctions between different kinds of feminist perspectives, they share certain concerns: a focus on women's family experiences, family diversity, and gender-based imbalances of power (Humble et al. 2006).

Feminists critically examine the ways in which family experience is shaped by **gender**—the social aspects of being female or male. This is the orienting focus that unifies most feminist writing, research, and advocacy. Feminists maintain that family and gender roles have been constructed by society and do not derive from biological or absolute conditions. They further tend to believe that family and gender roles have been created by men to maintain power over women. Basically, the goals of the feminist perspectives are to work to accomplish changes and create conditions in society that remove barriers to opportunity and oppressive conditions and are, instead, "good for women" (Thompson and Walker 1995).

Although that characterization applies to the multiple feminist perspectives, there are differences of emphasis between them (Budig 2004). For example, *liberal feminism* emphasizes the acquisition and protection of equal legal rights and equal economic opportunities as essential for women to achieve a greater quality of life. *Radical feminism* tends to depict the male-dominated family as a major source of the oppression of women through either men's control of women's household labor or their sexuality and reproduction. *Socialist feminism* asserts that women's status is a product of both capitalism and patriarchy, which exploit women's roles in both production and reproduction, whereas *Marxist feminism* focuses on how capitalism exploits women, who—especially as wives and mothers—provide essential yet unpaid care (Budig 2004). Sociologist Michelle Budig asserts that there are also profamily versions of radical and socialist feminism, both of which contend that it is women's family roles and "exclusion from politics and economics" that lead women to develop "superior female virtues," such as cooperation, caring, and protecting.

Gender and Family: Concepts Created by Society

Who or what constitutes a family cannot be taken for granted. The "traditional family" is no longer the predominant family lifestyle. Today's families have great diversity. What we think family should be is influenced by our own values and family experiences. Research demonstrates that couples actually may construct gender roles in the ongoing interactions that make up their marriages (Zvonkovic et al. 1996).

Are there any basic biological or social conditions that require the existence of a particular form of family? Some feminists would emphatically say no. Some object to efforts to study the family because to do so accepts as "natural" the inequalities built into the traditional concept of family life. Feminists urge an extended view of family to include all kinds of sexually interdependent adult relationships regardless of the legal, residential, or parental status of the partnership. For example, families may be formed of committed relationships between lesbian or gay individuals, with children obtained through adoption, from previous marriages, or through artificial insemination.

Feminist Agenda

Feminists have an action orientation alongside their analytical one as they strive to raise society's level of awareness regarding the oppression of women. Furthermore, some feminists make the point that all groups defined on the basis of age, class, race, ethnicity, disability, or sexual orientation are oppressed; they extend their concern for greater sensitivity to all disadvantaged groups (Allen and Baber 1992). Feminists assume that the experiences of individuals are influenced by the social system in which they live.

Therefore, the experiences of each individual must be analyzed to form the basis for political action and social change. The feminist agenda is to attend to the social context as it affects personal experience and to work to translate personal experience into community action and social critique.

Feminists share a belief in the need to challenge and change the system that exploits and devalues women. They are aware of the dangers of speaking out but feel that their integrity will be threatened if they fail to do so. Some feminists have described themselves as having "double vision"—the ability to be successful in the existing social system and simultaneously work to change oppressive practices and institutions.

Men as Gendered Beings

Inspired and influenced by the writing and research of feminist scholars, many social scientists now focus on how men's experiences are shaped by cultural ideas about masculinity and by their efforts to either live up to or challenge those ideas (Kimmel and Messner 2007). Instead of assuming that gender only matters to or includes women, this perspective looks at men as men, or as "gendered beings," whose experiences are shaped by the same kinds of forces that shape women's lives (Kimmel and Messner 2007). We now have a greatly enlarged and still growing body of literature about men as husbands, fathers, sexual partners, ex-spouses, abusers, and so on (e.g., see Cohen 1987; Coltrane 1996; Daly 1993; Gerson 1993; LaRossa 1988; Marsiglio 1998; Johnson 1996). Throughout this book, we explore how gender shapes women's and men's experiences of the family issues we examine.

Critique

The feminist perspective is not a unified theory; rather, it represents thinking across the feminist movement. It includes a variety of viewpoints that have, however, an integrating focus relating to the inequity of power between men and women in society and especially in family life (MacDermid et al. 1992).

Some family scholars who conceptualize family life and work as a "calling" have taken issue with feminists' focus on power and economics as a description of family. This has created a moral dialogue concerning the place of family life and work in "the good society" (Ahlander and Bahr 1995; Sanchez 1996). Feminists today recognize considerable diversity within their ranks, and the ideas of feminist theorists and other family theorists often overlap.

Micro-Level Theories

The four **micro-level theories** discussed next look at families from a different angle. They emphasize what happens within families, looking at everyday behavior, interaction between family members, patterns of communication, and so on. Rather than attempting to analyze "the family," they are useful for examining what happens between individuals in "families" that accounts for the relationships we form and maintain. Whereas macro-level theories focus on the family at the institutional level, addressing how it is related to other institutions and varies from one society to another and/or one historical period to another, micro-level theories focus their attention at how members of families interact, communicate, make decisions, divide responsibilities, and raise children—in other words, how they experience family life.

Symbolic Interaction Theory

Symbolic interaction theory looks at how people interact with one another. An **interaction** is a reciprocal act, the everyday words and actions that take place between people. For an interaction to occur, there must be at least two people who both act and respond to each other. When you ask your sister to pass the potatoes and she does it, an interaction takes place. Even if she intentionally ignores you or tells you to "get the potatoes yourself," an interaction occurs (even if it is not a positive one). Such interactions are conducted through symbols, words, or gestures that stand for something else.

When we interact with others, we do more than simply react to others; we interpret or define the meaning of their words, gestures, and actions. If your sister did not respond to your request for the potatoes, what did her nonresponse mean or symbolize? Hostility? Rudeness? A hearing problem? We interpret the meaning of her action and, *on the basis of that interpretation,* act accordingly. If we interpret the nonresponse as not hearing, we may repeat the request. If we believe that it symbolizes hostility or rudeness, we may become angry.

The notion of "meaning" is central to symbolic interaction theory. We interpret or attach meanings to interactions, situations, roles, relationships and other individuals whenever we encounter them. Moreover, for interaction to occur, these meanings must, to some degree, be shared. Symbolic interactionists study how we arrive at or construct such shared meanings and how they affect relationships and the roles we play in them.

Family as the Unity of Interacting Personalities

In the 1920s, sociologist Ernest Burgess (1926) defined the family as a "unity of interacting personalities." This definition has been central to symbolic interaction theory and in the development of marriage and family studies. Marriages and families consist of individuals who interact with one another over time. Such interactions and relationships define the nature of a family: a loving family, a dysfunctional family, a conflict-ridden family, an emotionally distant family, a high-achieving family, and so on.

In marital and family relationships, our interactions are partly structured by **social roles**—established patterns of behavior that exist independently of a person, such as the role of wife or husband existing independently of any particular husband or wife. Each member in a marriage or family has one or more roles, such as husband, wife, mother, father, child, or sibling. These roles help give us cues as to how we are supposed to act. When we marry, for example, these roles help us "become" wives and husbands; when we have children, they help us "become" mothers and fathers.

Symbolic interactionists study how one's sense of self is maintained in the process of acquiring these roles. We are, after all, more than simply the roles we fulfill. There is a core self independent of our being a husband or wife, father or mother, or son or daughter. Symbolic interactionists ask how we fulfill our roles and continue to be ourselves and, at the same time, how our roles contribute to our sense of self. Our identities as humans emerge from the interplay between our unique selves and our social roles.

Only in the most rudimentary sense are families created by society. According to symbolic interactionists, families are "created" by their members. Each family has its own unique personality and dynamics created by its members' interactions. Classifying families by structure, such as nuclear family, stepfamily, and single-parent family, misses the point of families. Structures are significant only insofar as they affect family dynamics. It is what goes on inside families— the construction, communication, and interpretation of shared meanings—that is important.

A recent study exploring how parents differently define the idea of "quality time" with their children nicely illustrates the symbolic-interactionist emphasis on meanings and definitions of social and familial reality. Sociologist Karrie Ann Snyder analyzed interview transcripts of 110 married, mostly dual-earner, middle- and upper-middle-class couples with teenage children and identified the different meanings that parents attach to the concept of "quality time" with their children. Assessing what parents said about the purpose of spending quality time, the activities associated with quality time, their main role or responsibility during quality time, and how much time parents thought was necessary to qualify as "quality" time, Snyder (2007) identified three categories of parents:

- *Structured planning parents* perceived that they had an overall shortage of time with their children and believed that only through efforts to specifically set aside time from their normal, hectic family and work lives for special and carefully scheduled family activities could they experience "quality time" with their teenagers.
- *Child-centered parents* also felt a deficit in the amount of time spent with their children but defined quality time in terms of having intimate, heart-to-heart conversations with their teenagers about their (children's) needs and interests *whenever and wherever such conversations occurred.*
- *Time-available parents* felt that all time spent at home with their families, good or bad, was quality time. They emphasized the amount of time together more than what occurred during that time and felt that both structured planning parents and child-centered parents use their claims of "quality time" to ease their guilt for not spending enough time with their families. Time-available parents were somewhat skeptical of the culturally understood version of quality time.

Snyder (2007) suggests that parents' definitions of quality time may differ between mothers and fathers in the same households as well as vary across different families. She further suggests that the versions of quality time that parents construct and convey may be strategic, designed to fit their parenting behaviors, especially their efforts to juggle their job demands and time with their families. In this way, they can "make sense of and explain their actions to outsiders and themselves by drawing on socially acceptable words and images to talk about their behaviors" (322).

In symbolic-interactionist terms, we arrive at definitions of parenting responsibility and family needs that then guide our behavior. But, at the same time, we employ a definition that serves to justify our behavior and allows us to maintain a positive sense of self. We then communicate that definition to others in the hope of influencing how they will define and assess our parenting behavior.

In general, whenever we assess our relationships, when we feel that our partner does (or does not) understand us, that we communicate well (or poorly), or that our relationship can (or cannot) withstand the difficulties created by long distance, we are illustrating dynamics that are at the heart of symbolic interaction research. The quality and stability of our relationships will depend on how we characterize such features of our relationships. Accurate or not, these subjective interpretations influence how we feel about, think about, and act within our intimate and family relationships.

© discpicture/Alamy

How family members interact with one another is partly determined by how they define their roles and by the meanings they attach to such behaviors as housework and childcare.

Critique

Although symbolic interaction theory focuses on the daily workings of the family, it suffers from several drawbacks. First, the theory tends to minimize the role of power in relationships. A partner with greater power may well have a greater chance to influence how the other person comes to define social reality. If a conflict exists, it may reveal more than differences in meaning, and it may take more than simply communicating to resolve it. If one partner strongly wants to pursue a career in Los Angeles and the other just as strongly wants to pursue a career in Boston, no amount of communication and role adjustment may be sufficient to resolve the conflict. The partner with the greater power in the relationship may prevail.

Second, symbolic interaction does not fully account for the psychological aspects of human life, especially personality and temperament. It sees us more in terms of our roles, thus neglecting the self that exists independently of our roles and limiting our uniqueness as humans.

Perhaps most important, the theory does not place marriage or family within a larger social context. It thereby disregards or minimizes the forces working on families from the outside, such as economic or legal discrimination against minorities and women.

Social Exchange Theory

According to **social exchange theory**, we measure our actions and relationships on a cost–benefit basis, seeking to maximize rewards and minimize costs by employing our resources to gain the most favorable outcome. An outcome is basically figured by the equation *Reward − Cost = Outcome*. However, one's assessment of the outcome also takes into account what he or she expected or believes one should get from their relationship and what one believes he or she could get from available alternatives.

How Exchange Works

At first glance, exchange theory may be the least attractive theory we use to study marriage and the family. It seems more appropriate for accountants than for lovers. But all of us use a cost–benefit analysis to some degree to measure our actions and relationships.

One reason many of us do not recognize our use of this interpersonal accounting is that we do much of it unconsciously. If a friend is unhappy with a partner, you may ask, "What are you getting out of this relationship? Is it worth it?" Your friend will start listing pluses and minuses: When the emotional costs outweigh the benefits of the relationship, your friend will probably end it. This weighing of costs and benefits is social exchange theory at work.

We may tend to think of rewards and costs as tangible objects, like money. In personal relationships, however, resources, rewards, and costs are more likely to be things such as love, companionship, status, power, fear, and loneliness. As people enter into relationships, they have certain resources—either tangible or intangible—that others consider valuable, such as intelligence, warmth, good looks, or high social status. People consciously or unconsciously use their various resources to obtain what they want, as when they "turn on" the charm.

Equity

A corollary to exchange is **equity**: exchanges that occur between people have to be fair, to be balanced. We are always exchanging favors: You do the dishes tonight, and I'll take care of the kids. Often we do not even articulate these exchanges; we have a general sense that they will be reciprocated. If, in the end, we feel that the exchange was not fair, we are likely to be resentful and angry. Some researchers suggest that people are most happy when they get what they feel they deserve in a relationship. Oddly, both partners (the one feeling deprived and the one feeling "overbenefited") may well feel uneasy in an inequitable relationship: When partners recognize that they are in an inequitable relationship, they try to restore equity in one of three ways:

- They attempt to restore actual equity in the relationship.
- They attempt to restore psychological equity by trying to convince themselves and others that an obviously inequitable relationship is actually equitable.
- They decide to end the relationship.

Society regards marriage as a permanent commitment. Because marriages are expected to endure, exchanges take on a long-term character. Instead of being calculated on a day-to-day basis, outcomes are judged over time.

An important ingredient in these exchanges is whether the relationship is fundamentally cooperative or competitive. In cooperative exchanges, both husbands and wives try to maximize their "joint profit" (Scanzoni 1979). These exchanges are characterized by mutual trust and commitment. Thus, a husband might choose to work part time and care for the couple's infant so that his wife may pursue her education. In a competitive relationship, however, each is trying to maximize individual profit. If both spouses want the freedom to go out whenever or with whomever they wish, despite opposition from the other, the relationship is likely to be unstable.

Applying Exchange Theory: Involuntarily Celibate Relationships

Exchange theory has been applied to a number of areas of marriage and family, including mate selection or partner choice, transition to parenthood, caregiving by aging spouses and adult children, and decisions to divorce. A recent application of exchange theory applied it to long-term marital or cohabiting relationships where couples engage in little to no sexual activity. Denise Donnelly and Elisabeth Burgess (2008) studied 352 people who were involuntarily celibate, desiring but not having had sex for at least six months prior to being interviewed. The social exchange perspective was applied specifically to a subset of 77 people (51% males) who were either married or partners in cohabiting relationships of one year or more. Although they asked other questions as well (e.g., How do relationships become involuntarily celibate?), here we consider their analysis regarding why partners in involuntarily celibate relationships stay in their relationships:

> [They] weigh the rewards and costs of staying in their relationships, keeping in mind past experiences and available alternatives, investments of resources such as time, money and psychic energy, and social prescriptions regarding partnerships, marriages, and families. (Donnelly and Burgess 2008, 529)

Although the assessment of rewards and costs differ depending on gender as well as on one's personal attitudes and beliefs, couples were more likely to stay together when the following occurred:

- They were happy with other, nonsexual aspects of the relationship.
- They perceived a lack of alternatives, believing that the probability of finding another partner is not worth the risks associated with leaving their current partner. Women were more likely to come to such a conclusion.
- They perceived that, for the sake of their families, they had a responsibility to stay in their relationships.

- They emphasized the commitment that marriage entails and expressed awareness of social expectations.

Summarizing their results, Donnelly and Burgess contend that "people stay in sexless relationships when they perceive both the benefits of staying and the costs of leaving as high" (530). It isn't until they assess the sexual situation—along with other aspects of the relationship—as "unbearable" that they consider leaving.

Critique

Social exchange theory assumes that we are all rational, calculating individuals, weighing the costs and rewards of our relationships and making cost–benefit comparisons of all alternatives. In reality, sometimes we are rational, and sometimes we are not. Sometimes we act altruistically without expecting any reward. This may be most true of love relationships and parent–child interactions. Social exchange theory also has difficulty ascertaining the value of costs, rewards, and resources, as such values may vary considerably from person to person or situation to situation.

Family Development Theory

Of all the theories discussed here, **family development theory** is the only one exclusively directed at families (White and Klein 2002). It emphasizes the patterned changes that occur in families through stages and across time. In its earliest formulations, family development theory borrowed from theories of individual development and identified a set number of stages that all families pass through as they are formed: growth with the birth of children, change during the raising of children, and contract as children leave and spouses die. Such stages created the *family life cycle*. Eventually, other concepts were introduced to replace the idea of a family life cycle. Roy Rodgers (1973) and Joan Aldous (1978, 1996) proposed the notion of the *family career*, which was said to consist of subcareers like the marital or the parental career, which themselves were affected by an educational or occupational career. Most recently, the idea of the *family life course* has been used to examine the dynamic nature of family experience.

The family life course consists of "all the events and periods of time (stages) between events traversed by a family" (White and Klein 2002). Because all these concepts emphasize the change and development of families over time, they are complementary and overlapping.

Family development theory looks at the changes in the family that typically commence in the formation of the premarital relationship, proceed through marriage, and continue through subsequent sequential stages. The specification of stages may be based on family economics, family size, or developmental tasks that families encounter as they move from one stage to the next. The stages are identified by the primary or orienting event characterizing a period of the family history. An eight-stage family life cycle might consist of the following: (1) beginning family, (2) childbearing family, (3) family with preschool children, (4) family with schoolchildren, (5) family with adolescents, (6) family as launching center, (7) family in middle years, and (8) aging family.

As we grow, each of us responds to certain universal developmental challenges (Person 1993). For example, all people encounter **normative age-graded influences**, biological or social influences that are clearly correlated with age, such as the biological processes of physical maturation, puberty, and menopause, or typical events, such as the beginning of school, death of parents, and the advent of retirement, that are linked to age. **Normative history-graded influences** come from historical facts that are common to a particular generation, such as the political and economic influences of wars and economic depressions, and that are similar for individuals in a particular age-group (Santrock 1995).

The life cycle model gives us insights into the complexities of family life and the different tasks that families perform. This model describes the interacting influences of changing roles and circumstances through time and how such changes produce corresponding changes in family responsibilities and needs. Planning that uses the developmental model alerts the family to seek resources appropriate to the upcoming needs and to be aware of vulnerabilities associated with each family stage (Higgins, Duxbury, and Lee 1994).

There are a variety of developmental theories that examine the stages involved in specific family phenomena, such as "falling in love," choosing a spouse, or experiencing divorce. Instead of attempting to depict all stages families might encounter, these theories look at the unfolding of specific aspects of family life across stages. You will find such approaches in a number of later chapters.

Critique

An important criticism sometimes made of family development theory is that it assumes the sequential processes of intact, nuclear families. It further assumes that all families go through the same process of change across the same stages. Thus, the theory downplays both the diversity of family experience and the experiences of those who divorce, remain childless, or bear children but never marry (Winton 1995). For example, stepfamilies experience different stages and tasks (Ahrons and Rogers 1987). Nevertheless, the universality of the family life cycle may transcend the individuality of the family form. Single-parent and two-parent families go through many of the same development tasks and transitions. They may differ, however, in the timing and length of those transitions.

A related criticism points out that gender, sexual orientation, race, ethnicity, and social class all create variations in how we experience family dynamics. The very sequence of stages may reflect a middle- to upper-class-family reality. Many lower- and working-class families do not have lengthy periods of early childless marriage. The transitions to marriage and parenthood may be encountered simultaneously or in reverse of what the stages specify. In neglecting these sorts of variations, the developmental model can appear overly simplistic.

Family Systems Theory

Family systems theory combines two of the previous sociological theories, structural functionalism and symbolic interaction, to form a more psychological—even therapeutic—theory. Mark Kassop (1987) notes that family systems theory creates a bridge between sociology and family therapy.

Structure and Patterns of Interaction

Like functionalist theory, family systems theory views the family as a structure of related parts or subsystems: the spousal subsystem, the parent–child subsystem, the parental subsystem (spouses relating to each other as coparents), the sibling subsystem, and the personal subsystem (the individual and his or her relationships). Each part carries out certain functions. One of the important tasks of these subsystems is maintaining their boundaries. For the family to function well, the subsystems must be kept separate (Minuchin 1981). Husbands and wives, for example, should try to prevent their conflicts from spilling over into the parent–child subsystem. Sometimes a parent will turn to the child for the affection that he or she ordinarily receives from a spouse. When the boundaries of the separate subsystems blur, as in incest, the family becomes dysfunctional.

As in symbolic interaction theory, interaction is important in systems theory. A family system consists of more than simply its members. It also consists of the pattern of interactions of family members: their communication, roles, beliefs, and rules. Marriage is more than a husband and wife; it is also their pattern of interactions, how they act in relation to each other over time (Lederer and Jackson 1968). Each partner influences and in turn is influenced by the other partner. And each interaction is determined in part by the previous interactions. This emphasis on the pattern of interactions within the family is a distinctive feature of the systems approach.

Any change in the family—such as a child leaving the family, family members forming new alliances, and hostility distancing the mother from the father—can create disequilibrium, which often manifests itself in emotional turmoil and stress. The family may try to restore the old equilibrium by forcing its "errant" member to return to his or her former position, or it may adapt and create a new equilibrium with its members in changed relations to one another.

Analyzing Family Dynamics

In looking at the family as a system, researchers and therapists believe the following:

- *Interactions must be studied in the context of the family system.* Each action affects every other person in the family. The family exerts a powerful influence on our behaviors and feelings, just as we influence the behaviors and feelings of other family members.
- *The family has a structure that can be seen only in its interactions.* Each family has certain preferred patterns of acting that ordinarily work in response to day-to-day demands. These patterns become strongly ingrained "habits" of interactions that make change difficult.
- *The family is a purposeful system; it has a goal.* In most instances, the family's goal is to remain intact as a family, achieving **homeostasis**, or stability. This makes change difficult, for change threatens the old patterns and habits to which the family has become accustomed.
- *Despite resistance to change, each family system is transformed over time.* A well-functioning family constantly changes and adapts to maintain itself in response to its members and the environment. The family changes as partners age and as children are born, grow older, and leave home. The family system adapts to stresses to maintain family continuity while making restructuring possible.

Although it has been applied to a variety of family dynamics, systems theory has been particularly influential in studying *family communication* (White and Klein 2002). As applied by systems theorists, interaction and communication between spouses are the kinds of systems wherein a husband's (next) action or communication toward his wife depends on her prior message to him. But through research in family communications, we recognize that marital communication is more complex than a simple quid pro quo or reciprocity expectation, such as, "If she is nasty, he is nasty." Well-known marriage researcher John Gottman has explored marital communication patterns that differentiate distressed from nondistressed couples. He identifies the importance of nonverbal communication over that of verbal messages spouses send (White and Klein 2002).

Critique

It is difficult for researchers to agree on exactly what family systems theory is. Many of the basic concepts are still in dispute, even among the theory's adherents, and the theory is sometimes accused of being so abstract that it loses any real meaning (White and Klein 2002).

Family systems theory originated in clinical settings in which psychiatrists, clinical psychologists, and therapists tried to explain the dynamics of dysfunctional families. Although its use has spread beyond clinicians, its greatest success is still in the analysis and treatment of dysfunctional families. However, the basic question is whether its insights apply to healthy families as well as to dysfunctional ones. Do healthy families, for example, seek homeostasis as their goal, or do they seek individual and family well-being?

Applying Theories to Family Experiences

Although the preceding theories were illustrated with numerous examples, it is worthwhile to stress that each will raise different questions and suggest different explanations for family phenomena. For instance, returning for a moment to the chapter-opening scenario, each of the theories would pose different questions about long-distance relationships, given their major assumptions. This would be equally true for any other research topic, such as how couples divide housework and childcare, why family violence and abuse occur, or how unemployment

affects marital stability. Table 2.2 gives an example of the kind of question each might raise about long-distance relationships.

Keep in mind that whether theory guides research or is the product of research, theories alone will not give us the kind of understanding we want about relationships and families. For that, we need to collect information, or data, from or about real people and real families.

Conducting Research on Families

In gathering their data, researchers use a variety of techniques. Some researchers ask the same set of questions of great numbers of people. They collect information from people of different ages, sexes, living situations, and ethnic backgrounds. This is known as "representative sampling." In this way, researchers can discover whether age or other background characteristics influence people's responses. This approach to research is called **quantitative research** because it deals with large quantities of information that is analyzed and presented statistically. Quantitative family research often uses sophisticated statistical techniques to assess the relationships between variables. Survey research and, to a lesser extent, experimental research (discussed in the following sections) are examples of quantitative research.

Other researchers study smaller groups or sometimes individuals in a more in-depth fashion. They may place observers in family situations, conduct intensive interviews, do case studies involving information provided by several people, or analyze letters, diaries, or other records of people whose experiences represent special aspects of family life. This form of research is known as **qualitative research** because it is concerned with a detailed understanding of the object of study.

In addition to using information provided specifically by people participating in a research project, researchers use information from public sources. This research is called **secondary data analysis**. It involves reanalyzing data originally collected for another purpose. Examples might include analyzing U.S. Census data and official statistics, such as state marriage, birth, and divorce records. Secondary data analysis also includes content analysis of various communication media, such as newspapers, magazines, letters, and television programs.

Table 2.2 Applying Theories to Long-Distance Relationships

Theory Assumptions about Families Applied to Long-Distance Relationships		
Ecological	Families are influenced by and must adapt to environments.	How do the characteristics of each partner's different living environments affect their abilities to maintain their commitments to the relationship?
Symbolic Interaction	Family life acquires meaning for family members and depends on the meanings they attach.	What meaning do couples attach to being separated? How does this alter their perceptions of the relationship?
Social Exchange	Individuals seek to maximize rewards, minimize costs, and achieve equitable relationships.	How do both partners define the costs and rewards associated with their relationship?
Family Development	Families undergo predictable changes over time and across stages.	What are the stages or phases that couples encounter as they adjust to being separated?
Structural Functionalism	The institution of the family contributes to the maintenance of society. On a familial level, roles and relationships within the family contribute to its continued well-being.	How does physical separation function to maintain or threaten the stability of the relationship?
Conflict	Family life is shaped by social inequality. Within families, as within all groups, members compete for scarce resources (e.g., attention, time, power, and space).	To what extent does one partner benefit more from being apart?
Family Systems	Families are systems that function and must be understood on that level.	How does being physically separated make it difficult for the couple to communicate effectively or maintain the equilibrium of the relationship?
Feminist	Gender affects our experiences of and within families. Gender inequality shapes how women and men experience families. Families perpetuate gender difference.	How are women and men differently affected by separation (i.e., do women carry more of the burden of managing and maintaining the relationship)?

Ethics in Family Research

Family science researchers conduct their investigations using **ethical guidelines** agreed on by professional researchers. These guidelines protect the privacy and safety of people who provide information in the research. For example, any research conducted with college students requires the investigator to present the plan and method of the research to a review committee. This ensures that participants' involvement is voluntary, that their privacy is protected, and that they will be free of harm.

To protect the privacy of participants, researchers promise them either anonymity or confidentiality. **Anonymity** insists that no one, including the researcher, can connect particular responses to the individuals who provided them. Much questionnaire research is of this kind, providing that no identifying information is found on the questionnaires. According to the rules of **confidentiality**, the researcher knows the identities of participants and can connect what was said to who said it but promises not to reveal such information publicly.

To protect the safety of research participants, researchers design their studies with the intent to minimize any possible and controllable harm that might come from participation. In social research on relationships and families, such harm is not typically physical harm but rather embarrassment or discomfort. Much of what family researchers study is ordinarily kept private. Talking about personal matters with an interviewer or even answering a series of survey questions may create unintended anxiety on the part of the participants. At best, researchers carefully design their studies to minimize the likelihood and reduce the extent of such reactions. Unfortunately, they cannot always be completely prevented (Babbie 2007).

Research ethics also require researchers to conduct their studies and report their findings in ways that assure readers of the accuracy, originality, and trustworthiness of their reports. Falsifying data, misrepresenting patterns of findings, and plagiarizing the research of others are all unethical.

What researchers know about marriage and the family comes from four basic research methods: survey research, clinical research, observational research, and experimental research. No single method is best for studying marriage and the family. Each technique has its strengths and weaknesses, and each may provide important and useful information that another method may not (Cowan and Cowan 1990).

Survey Research

The **survey research** method, using questionnaires or interviews, is the most popular data-gathering technique in marriage and family studies. Surveys may be conducted in person, over the telephone, or by written questionnaires. Typically, the purpose of survey research is to gather information from a smaller, representative group of people and to infer conclusions valid for a larger population. Questionnaires offer anonymity, may be completed fairly quickly, and are relatively inexpensive to administer.

Quantitative questionnaire research is an invaluable resource for gathering data that can be generalized to the wider population. Because researchers who use such techniques typically draw or use *probability-based random samples*, they can estimate the likelihood that their sample data can be safely inferred to the population in which they are interested. Furthermore, preestablished response categories or existing scales or indexes used by all respondents allow more comparability across a particular sample and between the sample data and related research.

Questionnaires usually do not allow in-depth responses, however; a person must respond with a short answer, a *yes* or *no*, or a choice on a scale of, for example, 1 to 10, from *strongly agree* to *strongly disagree*, from *very important* to *unimportant*, and so on.

Surveys are often used to look at how daily housework, such as cooking, is divided between marriage partners.

Unfortunately, marriage and family issues are often too complicated for questionnaires to explore in depth.

Interview techniques avoid some of this shortcoming of questionnaires because interviewers are able to probe in greater depth and follow paths suggested by the interviewee. They are also typically better able to capture the particular meanings or the depth of feeling that people attach to their family experiences.

Consider the following example from sociologist Sharon Hays's interview study of 38 mothers of two- to four-year olds. In describing how priorities are restructured when a woman becomes a mother, one of Hays's informants offered this comment on the life changes associated with being a mother:

> I think the reason people are given children is to realize how selfish you have been your whole life—you are just totally centered on yourself and what you want. And suddenly here's this helpless thing that needs you constantly. And I kind of think that's why you're given children, so you kinda think, okay, so my youth was spent for myself. Now, you're an adult, they come first. . . . Whatever they need, they come first.

Narrative data of this kind convey much about the experience of parenthood, including a depth of feeling and degree of nuance that quantitative questionnaire data cannot. By having respondents circle or check the appropriate preestablished response categories to a researcher's questions, we may never identify what that response truly means to the respondent or how it fits within the wider context of her or his life. However, interviewers are less able to determine how commonly such experiences or attitudes are found. Interviewers may also occasionally allow their own preconceptions to influence the ways in which they frame their questions and to bias their interpretation of responses.

There are problems associated with survey research, whether done by questionnaires or interviews. First, how representative is the sample (the chosen group) that volunteered to take the survey? In the case of a probability-based sample, this is less of a concern. However, self-selection (volunteering to participate) also tends to bias a sample. Second, how well do people understand their own behavior? Third, are people underreporting undesirable or unacceptable behavior? They may be reluctant to admit that they have extramarital affairs or how often they yell at their children, for example. If for any reason people are unable or unwilling to answer questions honestly, the survey technique will produce incomplete or inaccurate data. Nevertheless, surveys are well suited for determining the incidence of certain behaviors or for discovering traits and trends.

Much of the research that family scientists conduct and use—on topics as far reaching as the division of housework and child care, the frequency of and satisfaction with sex, or the effect of divorce on children or adults—is derived from interview or questionnaire data. Surveys are more commonly used by sociologists than by psychologists because they tend to deal on a more general or societal level rather than on a personal or small-group level. But surveys are less able to measure well how people interact with one another or what they actually do. For researchers and therapists interested in studying the dynamic flow of relationships, surveys are not as useful as clinical, experimental, and observational studies.

Secondary Analysis of Existing Survey Data

Secondary analysis is one of the techniques used most frequently by family researchers. Because of the various costs associated with conducting surveys on large, nationally representative samples, researchers often turn to one of the many available survey data sets, such as the General Social Survey (GSS) conducted by the National Opinion Research Center at the University of Chicago. The GSS includes many social science variables of interest to family researchers. Family researchers also often use data issued by the U.S. Census Bureau, the National Center for Health Statistics, the Bureau of Labor Statistics, and other government agencies that include many descriptive details about the U.S. population, including characteristics of families and households; birth, marriage, and divorce rates; and employment and time use.

Additional examples of available survey data of particular value to family researchers include the National Survey of Families and Households (NSFH), the National Health and Social Life Survey (NHSLS), the National Longitudinal Study of Adolescent Health (Add Health), and the Fragile Families and Child Wellbeing Study. The NSFH has provided much information about a range of family behaviors, including the division of housework, the frequency of sexual activity, and the relationships between parents and their adult children. The NHSLS is based on a representative sample of 3,432 Americans, aged 18 to 59, and contains much useful data about sexual behavior. Add Health is the largest longitudinal study (following a sample of respondents over a period of time) ever of more than 90,000 adolescents, from 80 high schools and more than 50 middle schools. They were initially studied in 1994 and then again in 1996, 2001–2002, and 2007–2008. The study examines the "social, economic, psychological and physical well-being" and includes data on family, friendships, peer groups and romantic relationships (www.cpc.unc.edu/projects/addhealth).

The Fragile Family and Child Wellbeing Study follows almost 5,000 children born between 1998 and 2000, three-quarters of whom were born to unmarried parents (i.e., "fragile families" who are at "greater risk of breaking up"). In addition to studying children's cognitive and emotional development, physical health, and home environment, the project contains interview data with parents on their attitudes, relationships, parenting behavior, health, economic situation, and employment status (www.fragilefamilies.princeton.edu/about.asp). At least a dozen different articles in the 2008 issues of the *Journal of Marriage and the Family* were based on data from the Fragile Families study.

The major difficulty associated with secondary data analysis is that the material collected in the original survey may "come close to" but not be exactly what you wanted to examine. Perhaps you would have worded the items differently to capture the essence of what you are interested in. Likewise, perhaps you would have asked additional questions to further or more deeply explore your topical interest (Babbie 2007). This disadvantage, although real, does not negate the enormous benefits associated with secondary analysis, the greatest of which is the availability of and easy access to broad sets of data on large, representative samples from which a host of family-related issues can be quickly and inexpensively examined.

Clinical Research

Clinical research involves in-depth examination of a person or a small group of people who come to a psychiatrist, psychologist, or social worker with psychological or relationship problems. The **case-study method**, consisting of a series of individual interviews, is the most traditional approach of all clinical research; with few exceptions, it was the sole method of clinical investigation through the first half of the twentieth century (Runyan 1982).

Clinical researchers gather a variety of additional kinds of data, including direct, firsthand observation, or analysis of records. Rather than a specific technique of data collection, clinical research is distinguished by its examination of individuals and families that have sought some kind of professional help. The advantage of clinical approaches is that they offer long-term, in-depth study of various aspects of marriage and family life. The primary disadvantage is that we cannot necessarily make inferences about the general population from them. People who enter psychotherapy are not a representative sample. They may be more motivated to solve their problems or have more intense problems than the general population (Kitson et al. 1996).

One of the more widely cited and celebrated clinical studies is Judith Wallerstein's longitudinal study of 60 families who sought help from her divorce clinic. Wallerstein has published three books—*Surviving the Breakup: How Children and Parents Cope with Divorce*, *Second Chances: Men, Women, and Children a Decade after Divorce*, and *The Unexpected Legacy of Divorce: The 25 Year Landmark Study*—following the experiences of most of the children in these families (she has retained 93 of the original 131 children whom she first interviewed in 1971) at 5, 10, and 25 years after divorce (Wallerstein 1980, 1989, 2000). All three books are sensitively written and richly convey the multitude of short- and long-term effects of divorce in the lives of her sample. Her critics have questioned whether findings based on such a clinically drawn sample (60 families from Marin County, California, who sought help as they underwent divorce) apply to divorced families more generally (Coontz 1998).

Clinical studies, however, have been fruitful in developing insight into family processes. Such studies have been instrumental in the development of family systems theory, discussed earlier in this chapter. By analyzing individuals and families in therapy, psychiatrists, psychologists, and therapists such as R. D. Laing, Salvador Minuchin, and Virginia Satir have been able to understand how families create roles, patterns, and rules that family members follow without being aware of them.

Observational Research

Observational research and experimental studies (discussed in the next section) account for a small minority of published research articles. In **observational research**, scholars attempt to study behavior systematically through direct observation while remaining as unobtrusive as possible. To measure power in a relationship, for example, an observer researcher may sit in a home and videotape exchanges between a husband and a wife or bring couples into a lab situation.

Some observational research involves family members being given structured activities to carry out. These activities involve interaction between family members (e.g., cohabiting couples) that can be observed and analyzed. They may include problem-solving tasks, putting together puzzles or games, responding to a contrived family dilemma, or managing conflict. Different tasks are intended to elicit different types of family interaction, which then provide the researchers with opportunities to observe behaviors of interest. For example, researchers have used observational techniques designed to elicit conflict between couples. How couples manage the conflict, what kind of negative affect they express, and how they interact are all observed and recorded.

A notable example of this kind of observational research followed 100 couples over a 13-year period from their premarital period through the period of greatest risk of divorce. Multiple sources of data were used, including the videotaping of couple interactions. The research revealed that premarital interaction patterns can help predict relationship outcomes some ten or more years later (Clements, Stanley, and Markman 2004).

An example of more "naturalistic" observation, wherein people are unaware that they are being watched, is sociologist Paul Amato's study of who takes care of children in such public settings as parks, shopping malls, and restaurants in San Diego, California, and Lincoln, Nebraska. Amato and his assistants compiled some 2,500 observations of children with their male or female caregivers, enabling them to determine that more than 40% of the young children were cared for by males, with boys more likely than girls to be cared for by a man (Amato 1989).

The most obvious disadvantage of observational research is that individuals may hide or suppress unacceptable ways of dealing with decisions, such as threats of violence, when the observer is present. Individuals within families, as well as families as groups, are concerned with appearances and the impressions they make.

Researchers may bring their own biases into what they see and how they interpret what they see. Some of that is unavoidable and could affect any social research. But observational research is the most subjective of the techniques. It is the hardest in which to remain objective, and, as a consequence, research is often hard to replicate.

A third problem that observational researchers encounter involves the essentially private nature of most family relationships and experiences. Because we experience most of our family life "behind closed doors," researchers typically cannot see what goes on "inside," without being granted access. For more public family behavior (e.g., caring for children in public places), observational data can be effectively used. However, some critics may point out the ethical dilemma of observing people when they don't know they are being studied. This puts researchers who use naturalistic observation in the position of having to justify their research strategy as necessary despite it falling outside the standard of "voluntary participation."

Critical Thinking

Because a major limitation of strictly observational data is identifying meanings that people attach to their behavior or attributing motives for why people are doing what they are observed doing, researchers often combine observational data with other sorts of data in a process known as **triangulation**.

Experimental Research

In **experimental research**, researchers isolate a single factor under controlled circumstances to determine its influence. Researchers are able to control their experiments by using *variables*, aspects or factors that can be manipulated in experiments. Recall the earlier discussion of types of variables, especially independent and dependent variables. In experiments, independent variables are factors manipulated or changed by the experimenter; dependent variables are factors affected by changes in the independent variable.

Because it controls variables, experimental research differs from the previous methods we have examined. Clinical studies, surveys, and observational research are correlational in nature. Correlational studies

There are aspects of family life that can be easily observed, such as care for children in public.

What goes on at home, behind closed doors, may not be easily accessible to observational researchers.

measure two or more naturally occurring variables to determine their relationship to one another. Because correlational studies do not manipulate the variables, they cannot tell us which variable causes the others to change. But because experimental studies manipulate the independent variables, researchers can reasonably determine which variables affect the other variables.

Experimental findings can be powerful because such research gives investigators control over many factors and enables them to isolate variables. For example, psychologists Michael Wohl and April McGrath designed a study to test the notion that "time heals all wounds." Specifically, they were interested in the effect of "temporal distance" (i.e., the passage of time) on the willingness people have to forgive someone who committed an interpersonal transgression against them. Using three different experimental designs and

Exploring Diversity: Researching Dating Violence Cross-Culturally

Family researchers use surveys to examine all sorts of family issues, but few aspects of intimate and family relationships are as disturbing as the issues of violence and abuse. They represent the worst of family relationships and the opposite of what we believe such relationships ought to be like. We will look in some detail in a later chapter (Chapter 12) at many of the issues that surround family and intimate partner violence. For now, we look at comparative survey data on one form of intimate partner violence—dating violence—because such data are representative of the kind of data that family researchers can obtain through survey instruments.

The findings reported below are from the International Dating Violence Study, a remarkable effort by a consortium of researchers to examine partner violence among samples of college students at 68 different colleges and universities in 32 different countries (including two in Africa, seven in Asia, 13 in Europe, four in Latin America, two in North America, two in the Middle East, plus Australia and New Zealand). A total of 13,701 college students were surveyed (71% female), most of whom were enrolled in sociology, psychology, family studies, or criminology classes. The surveys explored a number of interesting issues, including respondents' relevant attitudes (e.g., toward intimate partner violence and corporal punishment) and experiences (had they ever assaulted a partner, injured a partner, been spanked as a child, and so on). Researchers used the revised Conflict Tactics Scales (CTS2), which measures both more "minor" and more severe assaults. As described by sociologist and principal investigator Murray Straus (2008),

> The CTS2 items to measure "minor" assault are: (1) pushed or shoved, (2) grabbed, (3) slapped, (4) threw something at partner and (5) twisted arm or hair. The items in the "severe" assault scale are: (1) punched or hit a partner, (2) kicked, (3) choked, (4) slammed against a wall, (5) beat up, (6) burned or scalded and (7) used a knife or gun on partner.

Considering only students who had been in a relationship of at least a month, the following findings are based on reports of 4,239 respondents. If a respondent reported that one or more of these acts had occurred (either as perpetrator or victim) in the prior 12 months, an assault was recorded. Among the key findings that Straus reports, the following stand out:

- The level of self-reported assaults against a dating partner ranged from a low of 16.6% (Portugal) to a high of 77.1% (Iran) with a median of 31.2%. The rate in the United States was 30%.
- The percentage of students reporting severely assaulting a partner ranged from 1.7% (Sweden) to 23.2% (Taiwan), with a median of 10.8%. The U.S. rate was 11.0%.
- In every one of the 32 national settings, the largest category of partner violence was "bidirectional" (i.e., both are violent). In the United States, nearly 70% of the self-reported partner violence was bidirectional.
- When looking at "severe violence," again the largest category is "both violent" (56.6% in the United States).
- In only four of the 32 national settings was the percentage of unilateral male-only violence greater than the percentage of unilateral female-only violence (Iran, Tanzania, Greece, and Brazil).
- The second-largest category of both overall and severe violence was "female only," wherein the female was the only partner to use violence.

Straus is careful to caution against overgeneralizing from these findings. He points out that because the sample is a *convenience sample* of *college students*, the findings cannot be extrapolated either to the general populations of those countries (college students are not representative of national populations) or to college students in the settings included (convenience samples of students in a handful of social science courses do not allow us to generalize to all college students). Still, the data are consistent with much additional research in indicating high levels of violence among dating partners.

These results document internationally what has been known for a long time—that physical assaults against partners in dating and marital relationships are by far the most prevalent type of violent crime (Straus, 2004). In Chapter 12, we explore the issues addressed here in much greater detail.

both real and hypothetical transgressions, they were able to determine that the more time that passed since a transgression, the more likely individuals were to forgive their transgressor. In addition, one's subjective sense of time affects one's ability to forgive. They note that "the more a person feels temporally removed from the transgression, regardless of elapsed clock time, forgiveness becomes a more likely response" (Wohl and McGrath 2007,1032).

The obvious problem with such experimental studies is that we may well respond differently to people in real life than we do in controlled situations, especially in paper-and-pencil situations.

Experimental situations are often faint shadows of the complex and varied situations we experience in the real world. Thus, even when researchers can design and carry out a carefully controlled experiment, "the artificial setting . . . and its accompanying controls still can limit the researcher's ability to generalize . . . to naturalistic settings" (i.e., real life) (Small 2004, 324).

Applied Family Research

Despite our earlier lengthy consideration of theories, a good amount of family research is what we call **applied research**. The focus of such research is more practical than theoretical. It tends to be less concerned with formulating theories, generating concepts, or testing hypotheses. Data are gathered in an effort to solve problems, evaluate policies or programs, or estimate the outcome of some proposed future change in policy. For example, an applied researcher might study the effectiveness of a new mandatory arrest policy in reducing domestic violence incidents or the success of abstinence-only education in reducing teen pregnancy rates, the likely outcomes of new welfare policies on poor single mothers and their children, or whether and how repealing no-fault divorce laws might influence divorce rates.

Data from applied research tends to be of less interest to academic family researchers who are engaged in what is called basic or pure research than to policymakers, program directors, and heads of agencies. Such individuals may not even publish the findings of their research in journals or books but rather disseminate them in organizational reports read by only a small number of others. Although they may recognize the need for careful attention to detail in data collection and analysis, the practical needs that motivate them also push applied researchers to make "trade-offs" and to "compromise scientific rigor to get quick, usable results" (Newman 2004,11).

How to Think about Research

The last chapter discussed research into the effects of divorce on children and demonstrated that researchers occasionally arrive at very different findings and draw even contrary conclusions. You may have been left wondering how research results can be so different. This will occur frequently throughout the next 12 chapters. Differences in sampling and methodological techniques help explain why studies of the same phenomenon may arrive at different conclusions. They also help explain a common misperception many of us hold regarding scientific studies. Many of us believe that because studies arrive at different conclusions, *none* are valid. What conflicting studies may show us, however, is that researchers are constantly exploring issues from different perspectives and with different techniques as they attempt to arrive at a consensus.

In addition, researchers may discover errors or problems in sampling or methodology that lead to new and different conclusions. They seek to improve sampling and methodologies to elaborate on or disprove earlier studies. In fact, the very word *research* is derived from the prefix *re-*, meaning "over again," and *search*, meaning "to examine closely." And that is the scientific endeavor: searching and re-searching for knowledge.

It is also important to recognize that there are always going to be exceptions to any identified pattern of findings that family scientists identify. This is important for two reasons. First, don't assume that the patterns reported in this book *will happen* in your life. Your experience may constitute an exception to the more general pattern. All human beings don't behave the same way even when faced with very similar circumstances or constraints. Second, and equally important, don't dismiss findings reported here because they don't fit your experiences or those of people you may know. Instead, try to account for why your experience departs from the more generally observed social regularities.

By using critical thinking skills and by understanding something about the methods and theories used by family researchers, we are in a position to more effectively evaluate the information we receive about families. We are also better able to step outside our personal experience, go beyond what we've always been told, and begin to view marriage and family from a sounder and broader perspective. In Chapters 3 and 4, we take such steps and explicitly examine the factors and forces that create differences in family experience.

Summary

- We need to be alert to maintain *objectivity* in our consideration of different forms of family lifestyle. *Opinions*, *biases*, and *stereotypes* are ways of thinking that lack objectivity.

- *Fallacies* are errors in reasoning. Two common types of fallacies are *egocentric fallacies* and *ethnocentric fallacies*—the belief that all people are or should be the same as we are or that our way of living is superior to all others.

- *Theories* attempt to provide frames of reference for the interpretation of data. Macrotheories of marriage and families include family ecology, structural functionalism, conflict theory, and feminist theories. Microtheories include symbolic interaction, social exchange, family development, and family systems.

- Theories are built from concepts—abstract ideas about reality. Conceptualization is the process of identifying and defining the concepts we are studying, and operationalization is the development of research strategies to observe our concepts.

- Deductive research tests hypotheses—statements in which we turn our concepts into variables and specify how variables are related to each other.

- Inductive research does not test hypotheses. It begins with a more general interest, and, as data are collected, concepts are specified in more detail, leading to the development of hypotheses and to grounded theory.

- *Family ecology theory* examines how families are influenced by and, in return, influence the wider environments in which they function.

- *Symbolic interaction theory* examines how people interact and how we interpret or define others' actions through the symbols they communicate (their words, gestures, and actions).

- *Social exchange theory* suggests that we measure our actions and relationships on a cost–benefit basis. People seek to maximize their rewards and minimize their costs to gain the most favorable outcome.

- *Family development theory* looks at the changes in the family, beginning with marriage and proceeding through seven sequential stages reflecting the interacting influences of changing roles and circumstances through time.

- *Family systems theory* approaches the family in terms of its structure and pattern of interactions.

- *Structural functionalism theory* looks at society and families as though they were organisms containing different structures, each of which has a function.

- *Conflict theory* assumes that individuals in marriages and families are in conflict with one another. Power is often used to resolve the conflict. Four important sources of power are legitimacy, money, physical coercion, and love.

- *Feminist perspectives* provide an orienting focus for considering gender differences relating to family and social issues. The attention to gender led to men's studies, a field in which scholars examine how masculinity and male socialization shape men's experiences, including their family lives.

- Family researchers apply the *scientific method*—well-established procedures used to collect information.

- Professional family researchers follow ethical principles to protect participants from having their identities revealed and to minimize the discomfort the subjects experience from their participation in the research.

- Research data come from surveys, clinical studies, and direct observation, in which naturally occurring variables are measured against one another. Data are also obtained from experimental research.

- *Survey research* uses questionnaires and interviews. They are more useful for dealing with societal or general issues than for personal or small-group issues.

- Frequently, researchers conduct secondary analyses on already existing data. This allows researchers to examine large representative samples at little cost of time or resources.

- *Clinical research* involves in-depth examinations of individuals or small groups that have entered a clinical setting for the treatment of psychological or relationship problems.

- In *observational research*, interpersonal behavior is examined by an unobtrusive researcher, either in a natural setting, such as the home, or in a controlled setting, such as a laboratory.

- In *experimental research*, the researcher manipulates variables. Such studies are of limited use in marriage and family research because of the difficulty of controlling behavior and duplicating real-life conditions.

- To overcome limitations with any particular method, researchers often engage in triangulation—the use of multiple methods and/or multiple sources of data.

- Family researchers strive to identify and account for patterns of behavior. There will be exceptions to all patterns. Exceptions do not negate the importance or validity of research conclusions.

Key Terms

RESOURCES ON THE WEB

Book Companion Website

www.cengage.com/sociology/strong

Prepare for quizzes and exams with online resources—including tutorial quizzes, a glossary, interactive flash cards, crossword puzzles, self-assessments, virtual explorations, and more.

3

Variations in American Family Life

What Do YOU Think? Are the following statements TRUE or FALSE?
You may be surprised by the answers (see answer key on the following page).

T	F	
T	F	**1** Compared with contemporary families, colonial family life was considered more private.
T	F	**2** Industrialization transformed the role families played in society as well as the roles women and men played in families.
T	F	**3** Slavery destroyed the African American family system.
T	F	**4** Compared with what came both before and after, families of the 1950s were unusually stable.
T	F	**5** Within upper-class families, husbands and wives are relatively equal in their household roles and authority.
T	F	**6** Lower-class families are the most likely to be single-parent families.
T	F	**7** Family relationships can suffer as a result of either downward or upward mobility.
T	F	**8** Compared with Caucasian families, relationships between African American husbands and wives are more traditional.
T	F	**9** Asian American or Latino families show much variation within each group, depending on the country from which they came, why they left, and when they arrived in the United States.
T	F	**10** European ethnic groups are as different from one another as they are from African Americans, Latinos, Asian Americans, or Native Americans.

One thing you can almost always count on is that sometime during the term or semester, whether in class or in conversation, someone will make the oft-heard statement, "Well, all families are different." There is a lot of truth to that sentiment. For example, your family is not like your best friend's family in every way. Furthermore, assuming your best friend is someone a lot like you (which, as you've probably noticed, is common among people who become best friends), the differences between your families likely understate how richly variable family experience actually is.

It is true that in some ways family researchers are interested in discovering, describing, and explaining patterns in family experience. Of course, *every family is different*, but over the next few chapters we look closely at some patterned variations that separate and diversify family experiences. Although there are a number of factors that we could include as sources of such variation, the current chapter is concerned with the following four: time, social class, race, and ethnicity. In subsequent chapters, we also look at how gender and sexual orientation shape people's experiences of relationships and families. Then, throughout the remainder of the book, we draw comparisons and make contrasts among different types of households and families—singles, cohabiting and married couples, parents and nonparents, single-parent households and two-parent households, dual earners, male-breadwinner–female-homemaker households and "role reversers," first marriages and remarriages, and step relationships in blended families and blood relationships in birth families. Therefore, the task we start here won't end until you finish this book.

In this chapter, we begin by detailing the historical development of the kinds of families that predominate in the United States today, noting key transformations and the forces that created them. This accomplishes two things: it gives you a better sense of where today's American families have come from, and it enables you to see how different family life has been across generations, even within the same families. We then shift our attention to some major economic, ethnic, and racial variations that diversify contemporary American families.

American Families across Time

In Chapter 1, we noted that American marriages and families are dynamic and must be understood as the products of wider cultural, demographic, and technological developments (Mintz and Kellogg 1988). Although more attention tends to be paid to changes that have occurred in the post–World War II era, those changes represent only some of the more recent instances of more than 300 years of change that make up the history of American family life from the colonial period on into the twenty-first century. Armed with this brief history, we can recognize and make connections between changes in society and changes in families. In addition, we will be better positioned to assess the meaning of some of the more dramatic changes that have occurred recently in American family life. Finally, on a more personal level, you can better understand your own genealogies and family histories by recognizing the shifting stage on which they were played out.

The Colonial Era (1607–1776)

The colonial era is marked by differences among cultures, family roles, customs, and traditions. These families were the original crucible from which our contemporary families were formed.

Native American Families

The greatest diversity in American family life may have existed during our country's earliest years, when it was populated by 2 million Native Americans, representing more than 240 groups with distinct family and kinship patterns. Many groups were **patrilineal**: rights and property flowed from the father. Others, such as the Zuni and Hopi in the Southwest and the Iroquois in the Northeast, were **matrilineal**: rights and property descended from the mother.

Native American families tended to share certain characteristics, although it is easy to overgeneralize. Most families were small. There was a high child mortality rate, and mothers breastfed their infants; during breastfeeding, mothers abstained from sexual intercourse. Children were often born in special birth huts. As they grew older, the young were rarely physically disciplined. Instead, they were taught by example, praised when they were good and publicly shamed when they were bad.

Children began working at an early stage. Their play, such as hunting or playing with dolls, was modeled on adult activities. Ceremonies and rituals marked transitions into adulthood. Girls underwent puberty ceremonies at first menstruation. For boys, events such as growing the first tooth and killing the first large animal when hunting signified stages of growing up. A vision quest often marked the transition to manhood.

Critical Thinking

How far back can you trace your family's history? What would you like to know about it? What values, traits, or memories do you wish to pass on to your descendants?

Marriage took place early for girls, usually between 12 and 15 years; for boys, it took place between 15 and 20 years. Some tribes arranged marriages; others permitted young men and women to choose their partners. Most groups were monogamous, although some allowed two wives. Some tribes permitted men to have sexual relations outside of marriage when their wives were pregnant or breastfeeding.

Colonial Families

From earliest colonial times, America has been an ethnically diverse country. In the houses of Boston, the mansions and slave quarters of Charleston, the mansions of New Orleans, the haciendas of Santa Fe, and the Hopi dwellings of Oraibi (the oldest continuously inhabited place in the United States, dating back to A.D. 1150), American families have provided emotional and economic support for their members.

The Family. Colonial America was initially settled by waves of explorers, soldiers, traders, pilgrims, servants, prisoners, farmers, and slaves. In 1565, in St. Augustine, Florida, the Spanish established the first permanent European settlement in what is now the United States. In 1607, America's first permanent English colony was founded in Jamestown, Virginia. Some 13 years later, Plymouth colony was founded in what is now Plymouth, Massachusetts. But the members of these first groups came as single men—as explorers, soldiers, and exploiters.

In 1620, the leaders of the Jamestown colony, hoping to promote greater stability, began importing English women to be sold in marriage. The European colonists who came to America attempted to replicate their familiar family system. This system, strongly influenced by Christianity, emphasized **patriarchy** (rule by father or eldest male), the subordination of women, sexual restraint, and family-centered production. In the North, the colonial family was basically a multifunctional social and economic institution, the primary unit for producing most goods and caring for the needs of its members. The family planted and harvested food, made clothes, provided shelter, and cared for the necessities of life. As a social unit, the family reared children and cared for the sick, infirm, and aged. Its responsibilities included teaching reading, writing, and arithmetic because there were few schools. The family was also responsible for religious instruction: it was to join in prayer, read scripture, and teach the principles of religion.

Unlike in New England, the plantation system that came to dominate the southern colonies did not give the same priority to family life. Hunting, entertaining, and politics provided the greatest pleasure. The elite plantation owners continued to idealize gentry ways until the Civil War destroyed the slave system on which they based their wealth.

Marital Choice. Romantic love was not a factor in choosing a partner; one practical seventeenth-century marriage manual advised women that "this boiling affection is seldom worth anything" (Fraser 1984). Because marriage had profound economic and social consequences, parents often selected their children's mates. In the seventeenth century, 8 of the 13 colonies had laws *requiring parental approval* and imposed sanctions as harsh as imprisonment or whipping on men who "insinuated" themselves into a woman's affections without her parents' approval (Coontz 2005). Even in instances without such restrictions, in which individuals were "free to choose," children rarely went against their parents' wishes. If parents disapproved, their children typically gave up out of fear of the social and financial consequences of defying their parents (Coontz 2005). Love was not irrelevant but came after marriage. It was a person's duty to love his or her spouse. The inability to desire and love a marriage partner was considered a defect of character.

Although the Plymouth colonists prohibited premarital intercourse, they were not entirely successful. **Bundling**, the New England custom in which a young man and woman spent the night in bed together, separated by a wooden bundling board, provided a courting couple with privacy; it did not, however, encourage restraint. An estimated one-third of all marriages in the eighteenth century took place with the bride pregnant (Smith and Hindus 1975).

Family Life. The colonial family was strictly patriarchal, and such paternal authority was reinforced by both the church and the community (Mintz 2004). Steven Mintz (2004) describes the range of fathers' influence. Fathers were

> responsible for leading their households in daily prayers and scripture reading, catechizing their children and servants, and teaching house-hold members to read so that they might study the Bible. . . . Child-rearing manuals were thus addressed to men, not their wives. They had an obligation to help their sons find a vocation or calling, and a legal right to consent to their children's marriage. Massachusetts Bay Colony and Connecticut underscored the importance of paternal authority by making it a capital offense (i.e., punishable by death) for youths sixteen or older to curse or strike their father. (13)

The authority of the husband/father rested in his control of land and property. In a society dependent on agriculture and farming, such as colonial America, land was the most precious resource. The manner in which the father decided to dispose of his land affected his relationships with his children. In many cases, children were given land adjacent to the father's farm, but the title did not pass into their hands until the father died. This power gave fathers control over their children's marital choices and kept them geographically close.

This strongly rooted patriarchy called for wives to submit to their husbands. The wife was not an equal but was a helpmate. This subordination was reinforced by traditional religious doctrine. Like her children, the colonial wife was economically dependent on her husband. On marriage, she transferred to her husband many rights she had held as a single woman, such as the right to inherit or sell property, to conduct business, and to attend court.

For women, marriage marked the beginning of a constant cycle of childbearing and child rearing. On average, colonial women had six children and were consistently bearing children until around age 40. In addition to their maternal responsibilities, colonial women were expected to do a wide range of chores from cooking and cleaning to spinning, sewing, gardening, keeping chickens, and even brewing beer (Mintz and Kellogg 1988).

Childhood and Adolescence. The colonial conception of childhood was radically different from ours. First, children were believed to be evil by nature. The community accepted the traditional Christian doctrine that children were conceived and born in sin.

Second, childhood did not represent a period of life radically different from adulthood. Such a conception is distinctly modern (Aries 1962; Meckel 1984). In colonial times, a child was regarded as a small adult. From the time children were six or seven years old, they began to be part of the adult world, participating in adult work and play.

Third, children between the ages 7 and 12 were often "bound out" or "fostered" as apprentices or domestic servants (Mintz 2004). They lived in the home of a relative or stranger where they learned a trade or skill, were educated, and were properly disciplined. **Adolescence**—the separate life stage between childhood and adulthood—did not exist. They went from a shorter childhood (than what we are accustomed to) to adulthood (Mintz 2004; Mintz and Kellogg 1988). Thus, our contemporary notions of a rebellious life stage filled with inner conflicts, youthful indiscretions, and developmental crises do not fit well with the historical record of Plymouth Colony (Demos 1970; Mintz 2004).

African American Families

In 1619, a Dutch man-of-war docked at Jamestown in need of supplies. Within its cargo were 20 Africans who had been captured from a Portuguese slaver. The captain quickly sold his captives as indentured servants. By 1664, when the British gained what had been Dutch governed New Amsterdam, 40% of the colony's population consisted of African slaves. During the seventeenth century and much of the eighteenth, enslaved Africans and their descendants faced difficulty forming and maintaining families. It was hard for men, who often outnumbered women 60% to 40% or worse, to find wives. Enslaved African Americans were more successful in continuing the traditional African emphasis on the extended family, in which aunts, uncles, cousins, and grandparents played important roles. Although slaves were legally prohibited from marrying, they created their own marriages.

Childhood experience was often bitter and harsh. It was common for children to be separated from their parents because of a sale, a repayment of a debt, or a plantation owner's decision to transfer slaves from one property to another (Mintz 2004). Despite the hardships placed on them, enslaved Africans and African Americans developed strong emotional bonds and family ties. Slave culture discouraged casual sexual relationships and placed a high value on marital stability. On the large plantations, most enslaved people lived in two-parent families with their children. To maintain family identity, parents named their children after themselves or other relatives or gave them African names. In the harsh slave system, the family provided strong support against the daily indignities of servitude. As time went on, the developing African American family blended West African and English family traditions (McAdoo 1996).

Strong family ties endured in enslaved African American families. The extended family, important in West African cultures, continued to be a source of support and stability.

Nineteenth-Century Marriages and Families

In the nineteenth century, the traditional colonial family form gradually vanished and was replaced by the modern family. In this transition, families became more egalitarian, less patriarchal, and more affection based.

Industrialization and the Shattering of the Old Family

In the nineteenth century, the industrialization of the United States transformed the face of America. It also transformed American families from self-sufficient farm families to wage-earning, increasingly urban families. As factories began producing gigantic harvesters, combines, and tractors, significantly fewer farmworkers were needed. Looking for employment, workers migrated to the cities, where they found employment in the ever expanding factories and businesses. Because goods were now bought rather than made in the home, the family began its shift from being primarily a production unit to being more of a consumer- and service-oriented unit. With this shift, a radically new **division of labor** arose in the family. Men began working outside the home in factories or offices for wages they then used to purchase the family's necessities and other goods. Men became identified as the family's sole provider or breadwinner. Their work was given higher status than women's work because it was paid in wages. Men's work began to be increasingly identified as "real" work, distinct from the unpaid domestic work done by women.

Marriage and Families Transformed

Without its central importance as a work unit and less and less the source of other important societal functions (e.g., education, religious worship, protection, and recreation), the family became the focus and abode of feelings. The emotional support and well-being of adults and the care and nurturing of the young became the two most important family responsibilities.

The Power of Love. This new affectionate foundation of marriage brought love to the foreground as the basis of marriage and represented the triumph of individual preference over family, social, or group considerations. Stephanie Coontz (2005) reports that "by the middle of the nineteenth century there was near unanimity in the middle and upper classes throughout western Europe and North America that the love-based marriage, in which the wife stayed home and was protected and supported by her husband, was a recipe for heaven on earth" (162).

Women now had a new degree of power: they were able to choose whom they would marry. Women could rule out undesirable partners during courtship; they could choose mates with whom they believed they would be compatible. Mutual esteem, friendship, and confidence became guiding ideals. Without love, marriages were considered empty shells. Although such middle- and upper-class ideas of marital love and intimacy were only gradually adopted by working-class men and women, the working class more readily accepted middle-class ideas about gender roles.

Changing Roles for Women. The two most important family roles for middle-class women in the nineteenth century were that of housewife and mother. As there was a growing emphasis on domesticity in family life, the role of housewife increased in significance and status. Home was the center of life, and the housewife was responsible for making family life a source of fulfillment for everyone. For many women, especially middle class, this "doctrine of separate spheres" was wholeheartedly accepted and enthusiastically embraced (Coontz 2005). But even many among the working and lower classes ultimately came to recognize "good economic sense" in having a wife at home. With what women could contribute to their families through their performance of such tasks as making clothing, preparing food, growing vegetables, and so on, families benefited more than they would from the contribution of women's wages, estimated at only about one-third of the wages earned by men. Although many working-class and poor women did work outside the home for large stretches in their lives, "the ideology of male breadwinning and female homemaking became entrenched in working-class aspirations" (Coontz 2005, 175).

Women also increasingly focused their identities on motherhood. The nineteenth century witnessed the most dramatic decline in fertility in American history. Between 1800 and 1900, fertility dropped by 50%.

Where at the beginning of the nineteenth century American mothers typically gave birth to between 7 and 10 children, beginning "in her early twenties and (giving birth) every two years or so until menopause," by 1900 the average number of births had fallen to just three (Mintz 2004).

Women reduced their childbearing by insisting that they, not men, control the frequency of intercourse. Child rearing rather than childbearing became one of the most important aspects of a woman's life. Having fewer children and having them in the early years of marriage allowed more time to concentrate on mothering and opened the door to greater participation in the world outside the family. This outside participation manifested itself in women's heavy involvement in abolition, prohibition, and women's emancipation movements.

Childhood and Adolescence. A strong emphasis was placed on children as part of the new conception of the family. A belief in childhood innocence replaced the idea of childhood corruption. A new sentimentality surrounded the child, who was now viewed as born in total innocence. Protecting children from experiencing or even knowing about the evils of the world became a major part of child rearing.

The latter part of the nineteenth century also witnessed changes in the lives of young people that eventuated in the "beginning" of adolescence. In contrast to colonial youths, who participated in the adult world of work and other activities, in the late nineteenth and early twentieth centuries, young people were kept economically dependent and separate from adult activities, remained in school until their mid-teens, and often felt apprehensive when they entered the adult world. This apprehension sometimes led to the emotional conflicts associated with adolescent identity crises. In a 1904 book, psychologist G. Stanley Hall described the "emotional upheaval and fluctuating emotions, . . . contradictory tendencies toward hyperactivity and inertness, selfishness and altruism, bravado and a sense of worthlessness," that characterized this stage of development (Mintz 2004, 196).

Education also changed as schools, rather than families, became responsible for teaching reading, writing, and arithmetic as well as educating students about ideas and values. Conflicts between the traditional beliefs of the family and those of the impersonal school were inevitable. At school, the child's peer group increased in importance.

The African American Family: Slavery and Freedom

Although there were large numbers of free African Americans—100,000 in the North and Midwest and 150,000 in the South—most of what we know about the African American family before the Civil War is limited to the slave family.

The Slave Family. By the nineteenth century, the slave family had already lost much of its African heritage. Under slavery, the African American family lacked two key factors that helped give free African American and Caucasian families stability: autonomy and economic importance. Slave marriages were not recognized as legal. Final authority rested with the owner in all decisions about the lives of slaves. The separation of families was a common occurrence, spreading grief and despair among thousands of slaves. Furthermore, slave families worked for their masters, not themselves. It was impossible for the slave husband/father to become the provider for his family. The slave women worked in the fields beside the men. When an enslaved woman was pregnant, her owner determined her care during pregnancy and her relation to her infant after birth.

Slave children endured deep and lasting deprivation. Often shoeless, sometimes without underwear or adequate clothing, hungry, underfed and undernourished, and forced into hard physical labor as young as age five or six, slave children suffered considerably. Rates of illness and death in infancy and childhood were high. Furthermore, family life was fragile and often disrupted. Steven Mintz reports that separation of children from parents, especially fathers, was so common that at least half of all enslaved children experienced life separate from their father because he died, lived on another plantation, or was a white man who declined to acknowledge that they were his children. By their late teens, either temporary or permanent separation from their parents was something virtually all slave children had suffered (Mintz 2004).

Still, it is important to reiterate that slavery did not destroy all aspects of slave families. Despite the intense oppression and hardship to which they were subjected, many slaves displayed resilience and survived by relying on their families and by adapting their family system to the conditions of their lives (Mintz and Kellogg 1988). This included, for example, relying on extended kinship networks and, where necessary, on unrelated adults to serve as surrogates for parents absent because of the forced breakup of families.

Furthermore, enslavement did not forever destroy the African American family system. In no way does saying this diminish the horrors of slavery. Instead, it acknowledges the resilience of those who survived enslavement, and it illustrates how family systems may be pivotal sources of support and key mechanisms of surviving even the most extraordinary distress.

After Freedom. In 1865, with the ratification of the Thirteenth Amendment to the U.S. Constitution, slavery was outlawed. When freedom came, the formerly enslaved African American families had strong emotional ties and traditions forged from slavery and from their West African heritage (Guttman 1976; Lantz 1980). Because they were now legally able to marry, thousands of former slaves formally renewed their vows. The first year or so after freedom was marked by what was called "the traveling time," in which African Americans traveled up and down the South looking for lost family members who had been sold. Relatively few families were reunited, although many continued the search well into the 1880s.

African American families remained poor, tied to the land, and segregated. Despite poverty and continued exploitation, the southern African American family usually consisted of both parents and their children. Extended kin continued to be important.

Immigration: The Great Transformation

In the nineteenth and early twentieth centuries, great waves of immigration swept over America. Between 1820 and 1920, 38 million immigrants came to the United States. Historians commonly divided them into "old" immigrants and "new" immigrants.

The Old and the New Immigrants. The old immigrants, who came between 1830 and 1890, were mostly from western and northern Europe. During this period, Chinese also immigrated in large numbers to the West Coast. The new immigrants, who came from eastern and southern Europe, began to arrive in great numbers between 1890 and 1914 (when World War I virtually stopped all immigration).

Japanese also immigrated to the West Coast and Hawaii during this time. Today, Americans can trace their roots to numerous ethnic groups.

As the United States expanded its frontiers, surviving Native Americans were incorporated. The United States acquired its first Latino population when it

Except for Native Americans, most of us have ancestors who came to America—voluntarily or involuntarily. Between 1820 and 1920, more than 38 million immigrants came to the United States.

annexed Texas, California, New Mexico, and part of Arizona after its victory over Mexico in 1848.

The Immigrant Experience. Most immigrants were uprooted; they left only when life in the old country became intolerable. The decision to leave their homeland was never easy. It was a choice between life and death and meant leaving behind ancient ties.

Most immigrants arrived in America without skills. Although most came from small villages, they soon found themselves in the concrete cities of America. Again, families were key ingredients in overcoming and surviving extreme hardship. Because families and friends kept in close contact even when separated by vast oceans, immigrants seldom left their native countries without knowing where they were going—to the ethnic neighborhoods of New York, Chicago, Boston, San Francisco, Vancouver, and other cities. There they spoke their own tongues, practiced their own religions, and ate their customary foods. In these cities,

immigrants created great economic wealth for America by providing cheap labor to fuel growing industries.

In America, kinship groups were central to the immigrants' experience and survival. Passage money was sent to their relatives at home, information was exchanged about where to live and find work, families sought solace by clustering together in ethnic neighborhoods, and informal networks exchanged information about employment locally and in other areas.

The family economy, critical to immigrant survival, was based on cooperation among family members. For most immigrant families, as for African American families, the middle-class idealization of motherhood and childhood was a far cry from reality. Because of low industrial wages, many immigrant families could survive only by pooling their resources and sending mothers to work and even sending their children to work in the mines, mills, and factories.

Most groups experienced hostility. Crime and immorality were attributed to the newly arrived ethnic groups; ethnic slurs became part of everyday parlance. Strong activist groups arose to prohibit immigration and promote "Americanism." Literacy tests required immigrants to be able to read at least 30 words in English. In the early 1920s, severe quotas were enacted that slowed immigration to a trickle.

Critical Thinking

As you read through these historical perspectives, what are your feelings about such struggles and triumphs? How does knowledge of your family history affect you, your values, and your behavior?

Twentieth-Century Marriages and Families

By the beginning of the twentieth century, the functions of American middle-class families had been dramatically altered from earlier times. Families had lost many of their traditional economic, educational, and welfare functions. Food and goods were produced outside the family, children were educated in public schools, and the poor, aged, and infirm were increasingly cared for by public agencies and hospitals. The primary focus of the family was becoming even more centered on meeting the emotional needs of its members. In time, cultural emphasis would shift from self-sacrificing familism (i.e., making sacrifices in one's own pursuit of happiness or satisfaction for the well-being of one's family) to more self-centered individualism (wherein one's family's well-being was less important than one's own), and individuals' sense of their connections and obligations to their families would be greatly transformed.

The New Companionate Family

Beginning in the 1920s, a new ideal family form was beginning to emerge that rejected the "old" family based on male authority and sexual repression. This new family form was based on what's called the **companionate marriage**.

There were four major features of this companionate family (Mintz and Kellogg 1988): (1) men and women were to share household decision making and tasks; (2) marriages were expected to provide romance, sexual fulfillment, and emotional growth; (3) wives were no longer expected to be guardians of virtue and sexual restraint; and (4) children were no longer to be protected from the world but were to be given greater freedom to explore and experience the world; they were to be treated more democratically and encouraged to express their feelings.

Through the Depression and World Wars

The history of twentieth-century family life cannot be told without considering how profoundly family roles and relationships were affected by the Great Depression and two world wars. Although many different connections could be drawn, two seem particularly significant: changes in the relationship between the family and the wider society and changes in women's and men's roles in and outside of the family.

Linking Public and Private Life. The economic crisis during the Depression was staggering in its scope. Unemployment jumped from less than 3 million in 1929 to more than 12 million in 1932, and the rate of unemployment rose from 3.2% to 23.6%.

Over that same span of time, average family income dropped 40% (Mintz and Kellogg 1988). To cope with this economic disaster, families turned inward, modifying their spending, increasing the numbers of wage earners to include women and children, and pooling their incomes. Often it was a broadened "inward" to which they turned because people often took in relatives or relied on kinship ties for economic assistance (Mintz and Kellogg 1988).

Ultimately, these more personal, intrafamilial efforts proved insufficient. President Franklin Roosevelt's New Deal social programs attempted to respond to the social and economic despair that more localized efforts were unable to alleviate. Farm relief, rural electrification, Social Security, and a variety of social welfare provisions were all implemented in the hope of doing what local communities and individual families could not. Such federal initiatives reflected a dramatic ideological shift wherein government now bore responsibility for the lives and well-being of families (Mintz and Kellogg 1988).

Precipitated by the mass entrance into the workforce of millions of previously unemployed women, including many with young children, there was a clear need and opportunity for public resources to be committed to child care. Unfortunately, the federal government's response was slow and inadequate given the sudden and dramatic increase in need and demand (Filene 1986; Mintz 2004; Mintz and Kellogg 1988). Most mothers who entered the labor force had to rely on neighbors and grandparents to provide child care. When such supports were unavailable, many had no choice but to turn their children into "latchkey" kids fending for themselves (Mintz 2004). Unlike some of our European allies who invested more heavily in policies and services to accommodate employed mothers (Mintz and Kellogg 1988), it took the federal government two years to "appropriate funds to build and staff day-care centers, and the funds were sufficient for only one-tenth of the children who needed them" (Filene 1986, 164). Despite having engineered a propaganda campaign to entice women into jobs vacated by the 16 million men who entered the service, the government remained ambivalent about welcoming mothers of young children into those positions. However inadequate or slow government efforts were, they were still greater than what followed for most of the rest of the century.

Gender Crises: The Great Depression and World Wars. Both the Depression and the two world wars (especially World War II) reveal much about the gender foundation on which twentieth-century families rested. During the Depression, it was men whose gender identities and family statuses were threatened by their lost status as providers. During each world war, women were the ones who faced challenges that required them to abandon their gender socialization and step into roles and situations that fell outside their traditional familial roles. In each instance, the familial gender roles and identities had to be altered to match extraordinary circumstances.

During Word War II women were urged to enter the labor force and especially to enter nontraditional occupations left vacant by the deployment of men overseas. The images here illustrate the kinds of messages women received and the kinds of jobs they helped fill.

What is especially striking about men's reactions to their job loss is their internalization of fault for what was a society-wide economic crisis. Given how widespread unemployment was, one might think that men would at least take some comfort in knowing that the predicaments they faced were not of their own making. Yet they had so deeply internalized their sense of themselves as providers that their identities, family statuses, and sense of manhood were all invested in wage earning and providing. When unable to provide, many men were deeply shaken. Some were even driven to the point of emotional breakdown or suicide by their sense of economic failure (Filene 1986).

For many families, survival depended on the efforts of wives or the combination of women's earnings, children's earnings, assistance from kin, or some kind of public assistance. For those who depended at least somewhat on women's earnings, there were other gender consequences of running the household. Sometimes, men were pressed by their wives to contribute domestically in the women's "absence." Although some did, many others resisted (Filene 1986). Sometimes women displayed ambivalence about the meaning of male unemployment and male housework. Whereas 80% of the women who were surveyed in 1939 by the *Ladies' Home Journal* thought an

unemployed husband should do the domestic work in the absence of his employed wife, 60% reported they would lose respect for men whose wives outearned them (Filene 1986).

If the Depression revealed male anxiety about their familial roles as providers, we see in women's experiences during World Wars I and II that gender crises were not limited to men. During both wars, the absence of millions of men meant that women were pressed to step into their vacant shoes and participate in wartime production. During World War I, 1.5 million women entered the wartime labor force, many in jobs previously held largely by men (Filene 1986). During World War II, the number of employed women rose even more dramatically. Between 1941 and 1945, the numbers of employed women increased by more than 6 million to a wartime high of 19 million (Degler 1980; Lindsey 1997). Furthermore, "nearly half of all American women held a job at some time during the war" (Mintz and Kellogg 1988). Whereas single women had long worked and poor or minority women had worked even after marriage, the biggest change in women's labor force participation during World War II was among married, middle-class women. Thus, despite the widely held cultural emphasis on the special nurturing role of women and the belief that the home was a woman's "proper place," American society needed women to take over for the absent men. Once enticed into nontraditional female employment, women received both material and non-material benefits that were hard for many to surrender once the war ended and men returned.

Materially, women in traditionally male occupations received higher wages than they had in their past, more sex-segregated work experiences. As important, they also found a sense of gratification and enhanced self-esteem that were often missing from the jobs they were more accustomed to. However reluctant they may have been to take on such work, many were clearly more than a little ambivalent to leave it. To assist women in their departures from these jobs, profamily rhetoric and a new ideology extolling the value and importance of women's roles as mothers and caregivers were broadly conveyed by a variety of sources (e.g., popular media, social workers, and educators).

Families of the 1950s

In the long history of American family life, no other decade has come to symbolize so much about that history despite actually representing relatively little of it (Coontz 1997; Mintz and Kellogg 1988). In many ways, the 1950s appear to be a period of unmatched family stability, during which marriage and family seemed to be central to American lives. It was a time of youthful marriages, unusually high marriage and birthrates, stable and uncharacteristically low divorce rates, and economic growth. With a prosperous economy, many couples were able to buy homes on the income of only one wage earner—the husband. This was the period during which the breadwinner–homemaker family, with its traditional gender and marital roles, reached its peak. Man's place was in the world, and woman's place was in the home. Women were expected to place motherhood first and to sacrifice their opportunities for outside advancement to ensure the success of their husbands and the well-being of their children.

Given the meaning often invested in this era, it is important to understand that the 1950s were atypical. Compared with both what came before and what followed, families of the 1950s were unique. This is important: it means that anyone who uses this decade as a baseline against which to compare more recent trends in such family characteristics as birth, marriage, or divorce rates starts with a faulty assumption about how representative it is of American family history. Looking at those same trends with a longer view reveals that the changes that followed the 1950s were more

consistent with some familial patterns evident in the nineteenth and earlier part of the twentieth centuries (Mintz and Kellogg 1988). For example, during the 1950s, the divorce rate increased less than in any other decade of the twentieth century. Similarly, after more than 100 years of declining birthrates and shrinking family sizes, during the 1950s "women of childbearing age bore more children, spaced . . . closer together, and had them earlier and faster" than had previous generations (Mintz and Kellogg 1988, 179).

It can't be emphasized enough how much familial experience of the 1950s was created, sustained by, and depended on the unprecedented economic growth and prosperity of the postwar economy (Coontz 1997). The combination of suburbanization and economic prosperity, supplemented by governmental assistance to veterans, allowed many married couples to achieve the middle-class family dream of home ownership while raising their children under the loving attention of full-time caregiving mothers. We must be careful, though, not to oversimplify the family experience of the 1950s. Americans did not all benefit equally from the economic prosperity and opportunity of the decade. Thus, overgeneralizations would leave out the experiences of poor and working-class families and racial minorities for whom neither full-time mothering nor home ownership were common (Coontz 1997). In addition, many, especially women, found that the ideal lifestyle of the period left them longing for something more (Friedan 1963).

When we look at family changes that occurred in subsequent decades, we need to recognize that economic factors, again, were among the most important determinants of some more dramatic departures from the 1950s model. This especially pertains to the emergence of the dual-earner household. As Stephanie Coontz (1997, 47) points out, "By the mid-1970s, maintaining the prescribed family lifestyle meant for many couples giving up the prescribed family form. They married later, postponed children, and curbed their fertility; the wives went out to work." They did this not in rejection of the family lifestyle of the 1950s but in the pursuit of central features of that lifestyle, such as home ownership, no longer as attainable on the earnings of one wage earner.

Aspects of Contemporary Families

The remaining 11 chapters of this book look closely at families of the latter decades of the twentieth century and the first decade of the current century. The characteristics displayed by these families did not emerge suddenly but were established over years. Beginning with the latter years of the 1950s and escalating through and then beyond the 1960s and 1970s, some striking family trends surfaced. These trends persisted through and beyond the end of the twentieth century, leaving marriages and families reshaped and the meaning and experience of family life significantly altered.

Tables 3.1, 3.2, and 3.3 depict a variety of such changes. As Table 3.1 shows, beginning in the 1960s, Americans increasingly delayed marriage. By the mid-1990s, the median age at marriage for both genders was higher than it had been in more than a century. After further increases in the first years of this century, it remains four to five years older for men and women today than in 1960.

Table 3.1 Median Age at First Marriage, 1960–2008

Year	Males (age)	Females (age)
1960	22.8	20.3
1970	23.2	20.8
1980	24.7	22.0
1990	26.1	23.9
2000	26.8	25.1
2008	27.4	25.6

SOURCE: Table MS-2, www.census.gov/population/socdemo/hh-fam/ms2.xls.

Where Are We Now?

Careful inspection of the trends in Tables 3.2 and 3.3 shows that for much of the time period that is portrayed, an increasing divorce rate joined the falling marriage rates and that a dramatic increase in cohabitation occurred alongside a similarly marked increase in births outside marriage and in single-parent families. But family trends are neither always linear nor always unambiguous. They may rise, unexpectedly stabilize, or even reverse direction. This is evident in the divorce rates as seen in Table 3.2. Similarly, the birthrate dropped before rising again and ultimately falling even lower. Thus, it appears that, even in the short term, the only constant in family life is change (Mintz and Kellogg 1988).

To some, many of the changes depicted in Tables 3.2 and 3.3 might appear to signal family decline, implying especially a diminishing importance or appeal of marriage, leaving the future of the family as most Americans perceived it and marriage, in particular, seemingly in some doubt (Popenoe 1993). However, one could also look at these same trends more optimistically, taking a more liberal position that change, in and of itself, is not a bad thing. In fact, with these changes come more choices for people about the kinds of families or lifestyles they wish to create and experience (Coontz 1997; Mintz and Kellogg 1988).

Certainly, today's families do reflect considerable diversity of structure. In painting a picture of today's families, we would include many categories: breadwinner–homemaker families with children, two-earner couples with children, single-parent households

Table 3.2 Trends in Marriages, Divorces, and Births: 1970–2008

	1970	1980	1990	2000	2008*
Marriages	2,159,000	2,390,000	2,443,000	2,329,000	2,153,000
Marriage rate	10.6%	10.6%	9.8%	8.5%	7.1%
Divorces	708,000	1,189,000	1,182,000	1,135,000	NA
Divorce rate	2.2%	3.5%	5.2%	4.7%	3.7%
Births	3,731,000	3,612,000	4,158,000	4,063,000	4,309,000
Birthrate**	18.4%	15.9%	16.7%	14.5%	14.2%

* 2008 data are for the 12-month period from June 2007 to June 2008.
NA means data not available.
** Rate per 1,000 people.

SOURCES: Munson and Sutton (2005); Tejada-Vera and Sutton (2009); U.S. Census Bureau (2002).

Table 3.3 Couples and Children: 1970–2008

	1970	1980	1990	2000	2008
Married couples	44,728,000	49,112,0000	52,317,000	55,311,000	60,129,000
Unmarried cohabiting couples	523,000	1.6 million	2.9 million	4.5 million	6.8 million
Children living with two parents	59,681,000	47,543,000	46,820,000	49,688,000	51,785,000
Children living with one parent	8,426,000	12,349,000	15,842,000	19,227,000	19,501,000
Births to unmarried mothers	399,000	666,000	1,165,000	1,308,000	1,642,000
As percent of all births	11%	18%	28%	33%	39%

NOTE: Birth data are from 2006.

SOURCES: U.S. Census Bureau, Current Population Survey Reports, America's Families and Living Arrangements, 2008; National Center for Health Statistics, Births: Final Data, National Vital Statistics Reports, vol. 57 #7, 2009.

with children, marriages without children, cohabiting couples with or without children, blended families, role-reversed marriages, and gay and lesbian couples with or without children. These are the families that will concern us through the remainder of this book.

Of the many trends indicated in the data, there are four that merit special comment:

- *Cohabitation*. As commonly used, **cohabitation** refers to unmarried couples sharing living quarters and intimate and sexual relationships. Few trends rival cohabitation in terms of the extent of change. In 1970, cohabitation was a relatively uncommon and socially questionable lifestyle. Many cohabitors hid the lifestyle from their families and, sometimes, from their friends. Today, cohabitation is widely and more openly practiced by heterosexual and same-sex couples, by young people, and by middle-age or even elderly couples and by people of all races and classes. Despite sharing the characteristics of living together in sexually and emotionally intimate relationships, couples who cohabit do so for many different reasons. As we shall see in Chapter 8, some cohabitors live together as an alternative to marriage, others as a test of their compatibility and suitability for marriage, and still others for convenience. Many same-sex cohabitors live together because they are unable to legally marry. Cohabitation and its effect on marriage are to be considered.
- *Marriage*. Although between 75% and 80% of the U.S. population is predicted to marry at least once in their lifetimes, marriage has undergone significant changes in recent decades. The expectations people bring, the kinds of relationships they attempt to construct, the roles they play as spouses,

and the levels of satisfaction they express all have changed. As of mid-2009, the legalization of same-sex marriage in Massachusetts, Connecticut, Iowa, Vermont, New Hampshire, and Maine and the ongoing fight for same-sex marriage rights in other states have led to more public discourse about marriage in legal, social, and religious contexts. Overall, marriage rates have declined, especially among lower-income nonwhites. However, even among the population least likely to marry, marriage remains a highly valued life goal. At the "other end," marriage rates for college-educated women have actually increased. Thus, the status of marriage is more complicated than meets the eye.

- *Divorce, remarriage, and blended families*. After decades of escalating divorce rates, more recently divorce rates have declined and stabilized. Yet divorce remains a traumatic disruption in the lives of millions of American couples and children each year. The effects of divorce are not the same on women and men, adults and children, or children of varying ages. Furthermore, the social acceptability of divorce and the legal processes that surround divorce remain controversial public issues. The prevalence of divorce is also associated with both remarriages and blended families. Around half of men and women over 25 who have ever divorced are currently remarried. Twelve percent of men and 13% of women age 15 and older have been married twice. These subsequent marriages face stresses beyond those of first marriages, especially when children are involved. As a result, they fail at a rate higher than first marriages. A 2007 U.S. Census Report estimated that nearly 10% of U.S. children lived with a biological parent and a stepparent or adoptive parent. Five and a half million, 7.6% of all children, lived with a stepparent, and another 1.5 million, or 2.1%, lived with at least one adoptive parent (Kreider 2007). **Blended families**, or families containing either a stepparent, a step-sibling, or half sibling, are even more common. Seventeen percent of all children younger than 18 years old (12 million children) were estimated to live in blended families (Kreider 2007). Relationships in blended families are more complicated than those in "intact" families in part because of divided loyalties to one's "real" (i.e., original) family.

- *Unmarried motherhood and single-parent families.* The increase in births outside marriage, though not as dramatic as the increase in cohabitation, is nonetheless among the most striking changes in family patterns. Among some racial and ethnic groups, African Americans, American Indian and Alaska Natives, and Puerto Ricans, for example, the percent of all births that are to unmarried mothers exceeds 60%. Along with divorce, unmarried childbearing has helped drive the increase in single-parent–headed households. In fact, the proportion of single mothers who never married (as opposed to having been married and divorced) has become increasingly large. By 2007, almost 30% of U.S. children lived in households headed by either an unmarried mother or an unmarried father, though the vast majority of the more than 11 million children in single-parent households, 85%, lived with their mothers without their fathers. Female-headed households typically face economic hardships, and more children in single-parent versus two-parent households experience a variety of social, psychological, and educational difficulties.

Factors Promoting Change

Trends in marriages and families are driven by a number of different forces in society. In looking over the major changes to American families, we can identify four especially important factors that initiated these changes: (1) economic changes, (2) technological innovations, (3) demographics, and (4) gender roles and opportunities for women.

Economic Changes

As noted earlier, over time, the family has moved from being an economically productive unit to a consuming, service-oriented unit. Where families once met most needs of their members—including providing food, clothing, household goods, and occasionally surplus crops that it bartered or marketed—most of today's families must purchase what they need.

Economic factors have been responsible for major changes in the familial roles played by women and men. Inflation, economic hardship, and an expanding economy led to married women entering the labor force in unprecedented numbers. More than half of married women with preschool-aged children are typically employed outside the home (Figure 3.1). As a result, the dual-earner marriage and the employed mother have become common features of contemporary families. As women have increased their participation in the paid labor force, other familial changes have occurred. For instance, women are less economically dependent on either men or marriage. This provides them greater legitimacy in attempts to exercise marital power. It has also increased the tension around the division of household chores and raised anxiety and uncertainty over who will care for the children. The current recession in the U.S. economy is affecting families at all economic levels, albeit in different ways and to greater or lesser extent.

Technological Innovations

The family has been affected by most major innovations in technology—from automobiles, telephones, cell phones, televisions, and microwaves to personal computers and the Internet. These devices were designed or invented not to transform families

Figure 3.1 Percentage of Married Women Employed outside the Home Who Have Children 6 Years Old or Younger

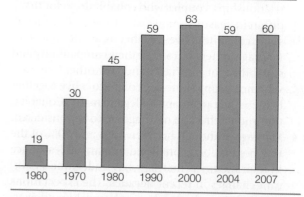

SOURCES: U.S. Census Bureau 2006: Table 586. U.S. Department of Labor, Bureau of Labor Statistics, "Employment Characteristics of Families in 2007," Table 4.

but to improve transportation, enhance communication, expand choices and quality of entertainment, and maximize efficiency. Nevertheless, they have had major repercussions in how family life is experienced.

For example, older devices, such as automobiles and telephones, as well as more recent innovations, such as personal computers and cell telephones, have aided families in maintaining contact across greater distances, thus allowing extended families to sustain closer relations and nuclear family members to stay available to one another through school- and job-related travel or relocation. The proliferation of automobiles also changed the residential and relationship experiences of many Americans. It became possible for people to live greater distances from where they work—thus contributing to the suburbanization of America—and to experience premarital relationships away from more watchful adult supervision.

Televisions and, more recently, the Internet have altered the recreation and socialization activities in which families engage, with both beneficial and negative consequences. As important as the entertainment function of both television and the Internet are, they also operate as additional socialization agents beyond parents and other relatives. What we watch on television or view and read on the Internet helps shape our values and beliefs about the world around us. As shown in a subsequent chapter, the Internet has also greatly expanded our options for meeting friends, potential partners, and spouses. Finally, a host of innovations in communications, including cell phones, e-mail, instant messaging, and texting, have altered the ways in which parents monitor children and family members remain in contact with one another.

The range of domestic appliances—from washing machines and dishwashers to microwaves—has altered how the tasks of housework are done. Although we might be tempted to conclude that such devices free people from some time- and labor-intensive burdens associated with maintaining homes, historical research has shown that this is not automatically so. For instance, as technology made it possible to more easily wash clothes, the standards for cleanliness increased. In the case of microwaves, the time needed for tasks associated with meal preparation has been reduced, freeing people to spend more time in other activities (not necessarily as families and often away from their families—at work, for example).

Finally, revolutions in contraception and biomedical technology have reshaped the meaning and experience of sexuality and parenthood. Much of what we call the "sexual revolution" in the 1970s and beyond was fueled partly by safer and more reliable methods of preventing pregnancy, such as the birth control pill. Regarding parenthood, people who in the past would have been unable to become parents have the opportunity to enjoy childbearing and child rearing as a result of **assisted reproductive technologies**—including medical advances such as in vitro fertilization as well as surrogate motherhood and sperm donation. Such developments have thus altered the meaning of parenthood, as multiple individuals may be involved in any single conception, pregnancy, and eventual birth. Sperm and/or egg donors, surrogate mothers, and the parent or parents who nurture and raise the child all can claim in some way to have reared the child in question. Such changes have complicated the social and legal meanings of parenthood, as they have opened the possibility of parenthood to previously infertile couples or same-sex couples.

Demographics

The family has undergone dramatic demographic changes in areas that include family size, life expectancy, divorce, and death. Three important changes have emerged:

- *Increased longevity.* As people live longer, they are experiencing aspects of family life that few experienced before. In colonial times, because of a relatively short life expectancy, husbands and wives could anticipate a marriage lasting 25 years. Today, couples can remain married 50 or 60 years. Today's couples can anticipate living more years together after their children are grown; they can also look forward to grandparenthood or great-grandparenthood. Since men tend to marry women younger than themselves and on average die younger than women do, American women can anticipate a prolonged period of widowhood.

- *Increased divorce rate.* Even with the more recent decrease, the long-term increase in the divorce rate, beginning in the late nineteenth century (even before 1900, the United States had the highest divorce rate in the world), led to the rise of single-parent families and stepfamilies. In this way, it has dramatically altered the experience of both childhood and parenthood and has altered our expectations of married life.

Popular Culture: Can We See Ourselves in *Zits*?
Comic Strips and Changes in Family Life

Every day, millions of people in hundreds of countries open thousands of newspapers and pause to read comic strips. The list of more popular strips will have familiar names on it for just about everyone: *Calvin & Hobbes*, *Cathy*, *Doonesbury*, *Dilbert*, *Sally Forth*, and *Peanuts*, the last of which is so popular that it has run repeated comics for years since the retirement and later death of creator Charles Schulz.

Many comic strips focus on family life, typically featuring a couple and their young to adolescent children. Such is the case with the popular award-winning strip *Zits*, written by Jerry Scott and illustrated by Jim Borgman.

Zits looks at the life of 15-year-old Jeremy Duncan, "high school freshman with, thank God, four good friends, but other than that a seriously boring life in a seriously boring town made livable only by the knowledge that someday in the far-off future at least this will all be over" (www.kingfeatures.com/features/comics/zits/about.htm). His parents—mother Connie, a frustrated novelist, and father Walt, an orthodontist—struggle to find ways to communicate with Jeremy, a brooding "handful" of adolescent hormones and moods.

Comics like *Zits* allow us to laugh at some familiar, occasionally exaggerated situations and conversations, but they do much more. They offer us a window through which to explore the wider cultural attitudes and values about family life. Sociologist Ralph LaRossa, one of the leading experts on changes in parenthood and especially fatherhood through the twentieth century, has studied the portrayal of gender and parental roles across six decades (1940–1999) of popular comic strips (LaRossa et al. 2000, 2001). Some comics that they examined included

Blondie, Cathy, Dennis the Menace, For Better or Worse, Hi and Lois, Garfield, and *Ziggy* (LaRossa et al. 2000). They looked to see how prominently the Mother's Day or Father's Day theme was represented, what activities fathers and mothers were portrayed doing, whether the fathers and mothers were portrayed as incompetent, whether they were mocked or made to look foolish, and whether father and mother characters were engaged in nurturant behaviors such as expressing affection toward, caring for, comforting, listening to, teaching, or praising a child or children. The data revealed fluctuating portrayals of fathers as nurturant or competent. Looking across the six decades (in five-year increments) revealed a U-shaped curve; in the late 1940s and early 1950s, there were high percentages of nurturant fathers unmatched until the 1990s. However, an increase in father nurturance can be seen beginning in the 1980s. This may be surprising to students who think that only late in the twentieth century did nurturing qualities become valued or expected of fathers. At least comic strip fathers of the late 1940s and early 1950s were often nurturing and supportive toward their children. Nurturant portrayals of mothers "spiked" in the late 1950s and late 1970s and were consistently high from the mid-1980s to the end of the century. LaRossa and colleagues note that, although cartoonists seem to have tried to acknowledge the new ideology of nurturant fathers, they did not do it "at the expense" of mothers. "Indeed, if anything, they seemed to pay homage to fatherhood and motherhood at the end of the millennium" (LaRossa et al. 2000, 385).

In chapter 10, we examine more explicitly the "culture" and "conduct" of fatherhood and motherhood. For now,

• *Decreased fertility rate.* As women bear fewer children, they have fewer years of child-rearing responsibility. With fewer children, partners are able to devote more time to each other and expend greater energy on each child. Children from smaller families benefit in a variety of ways from the greater levels of parental attention, although they may lack the advantages of having multiple siblings, such as learning to share, to care for, and to negotiate with others (Downey and Condron 2004). From the adults' perspective, smaller families afford women greater opportunity for entering the workforce and enjoying the enhanced economic status that follows.

Gender Roles and Opportunities for Women

Changes in gender roles are the fourth force contributing to alterations in American marriages and families. The history summarized earlier indicated some major changes that took place in women's and men's responsibilities and opportunities. These gender shifts then directly or indirectly led to changes in both the ideology surrounding and the reality confronting families.

The emphasis on child rearing and housework as women's proper duties lasted until World War II, when, as we saw, there was a massive influx of women into factories and stores to replace the men fighting

we can use the research on comic strips to reiterate and illustrate that cultural changes have occurred in our ideas about families, in this case our expectations of fathers. We can also better appreciate how much Walt Duncan and Darryl MacPherson fit the wider context of involved, if exasperated, comic strip fathers.

overseas. This initiated a trend in which women increasingly entered the labor force, became less economically dependent on men, and gained greater power in marriage.

The feminist movement of the 1960s and 1970s led many women to reexamine their assumptions about women's roles. Betty Friedan's *The Feminine Mystique* challenged head-on the traditional assumption that women found their greatest fulfillment in being mothers and housewives. The women's movement emerged to challenge the female roles of housewife, helpmate, and mother, appealing to some women as it alienated others.

More recently, the dual-earner marriage made the traditional division of roles an important and open question for women. Today, contemporary women have dramatically different expectations of male–female roles in marriage, child rearing, housework, and the workplace than did their mothers and grandmothers. Changes in marriage, birth, and divorce rates and in the ages at which people enter marriage have all been affected by women's enlarged economic roles.

We also have witnessed changes in what men expect and are expected to do in marriage and parenthood. Although it may still be assumed that men will be "good providers," that is no longer enough. Married men face greater pressure to share housework and participate in child care. Although they have been slow to increase the amount of housework they do, there has been more acceptance of the idea that greater father involvement benefits both children and fathers. New standards and expectations of paternal behavior and more participation by fathers in raising children help explain the ongoing changes—from how dual-earner households function to why we are more accepting of fathers staying home to care for their young children.

The U.S. Census Bureau estimates that there are 1 million fathers (and 6.6 million mothers) of children age 15 or younger who were out of the labor force and home full time for all of 2003. For 159,000 (nearly 16%) of the fathers, the "primary reason" they were home was to care for home and family. Among at-home mothers, 5.9 million of the 6.6 million were home to care for home and family (U.S. Census Bureau 2008). Furthermore, 2.9 million preschoolers are cared for by their fathers while their mothers are at work. This amounts to 25% of the 11.3 million preschool-age children of employed mothers. Only grandparents provided more noninstitutionalized care than fathers. Finally, there are approximately 2.5 million single fathers, representing 19% of all single parents (U.S. Census Bureau 2008b).

Gender issues are so central to family life that they are the subject of the entire next chapter. It is not an exaggeration to say that we cannot truly understand the family without recognizing the gender roles and differences on which it rests.

Cultural Changes

We can, in conclusion, point to a shift in American values from an emphasis on obligation and self-sacrifice to individualism and self-gratification (Amato et al. 2007; Bellah et al. 1985; Coontz 1997; Mintz and Kellogg 1998). The once strong sense of **familism**, in which individual self-interest was expected to be subordinated to family well-being, has given way to more open and widespread individualism in which even marriages and/or families can be sacrificed for individual happiness and personal fulfillment.

This shift in values has had consequences for how people choose between alternative lifestyle paths. For example, complex decisions—about whether and how much to work, whether to stay married or to divorce, and how much time and attention to devote to children or to spouses—are increasingly weighed and made against a backdrop of pursuing self-gratification and individual happiness. We have embraced the idea that what makes individuals happy, in the long run, is what is best for them and their families. Such shifts in values, alone, have not changed families, but they have contributed to the choices people make, out of which new family forms predominate (Coontz 1997).

How Contemporary Families Differ from One Another

The preceding discussion traced some ways families have changed throughout history and why. In that sense, it has led us to family life of today. But today's families differ from one another, a topic we now explore. We look first at economic factors that differentiate families and then at cultural characteristics, social class, and race and ethnicity.

Economic Variations in Family Life

A **social class** is a category of people who share a common economic position in the stratified (i.e., unequal) society in which they live. We typically identify classes using economic indicators such as ownership of property or wealth, amount of income earned, the level of prestige accorded to work, and so forth. Social class has both a structural and a cultural dimension. Structurally, social class reflects the occupations we hold (or depend on), the income and power they give us, and the opportunities they present or deny us. The cultural dimension of social class refers to any class-specific values, attitudes, beliefs, and motivations that distinguish classes from one another. Cultural aspects of social class are somewhat controversial, especially when applied to supposed "cultures of poverty"—an argument holding that poor people become trapped in poverty because of the values they hold and the behaviors in which they engage (Harrington 1962; Lewis 1966). What is unclear regarding "cultures of class" is how much difference there is in the values and beliefs of different classes and whether such differences cause or follow the more structural dimensions that separate one class from another.

To an extent, there is also a psychological aspect to social class, an internalization of economic status in the self-images we form and the self-esteem we possess. These may also be seen as consequences of other aspects of class position, such as the self-identity that results from the prestige accorded to work or the respect paid to accomplishments. Like the structural and cultural components of social class, these are brought home and affect our experiences in our families.

The effect of social class is far reaching and deep. In an article about how social class affects marriage, *New York Times* reporter Tamara Lewin quotes one of her sources, Della Mae Justice, of Piketown, Kentucky. Justice grew up in the coal-mining world of Appalachia, in a house without indoor plumbing. Having put herself through college and later law school, she is now solidly and unambiguously middle class. Justice says, "I think class is everything, I really do. When you're poor and from a low socioeconomic group, you don't have a lot of choices in life. To me, being from an upper class is all about confidence. It's knowing you have choices, knowing you have connections" (Lewin 2005).

Clearly, many facets of our lives (often referred to by sociologists as **life chances**) are affected by our **socioeconomic status**, including our health and well-being, safety, longevity, religiosity, and politics. A host of family experiences also vary up and down the socioeconomic ladder. For instance, class variations can be found in such family characteristics as age at marriage, age at parenthood, timing of marriage and parenthood, division of household labor, ideologies of gender, socialization of children, meanings attached to sexuality and intimacy, and likelihood of violence or divorce.

Conceptualizations of social class vary in both how class is defined and how many classes are identified and counted in American society. For example, in Marxian formulations of social class, it is a person's relationship to the means of production—as owner or worker—that defines class position. The capitalist class owns the factories and machinery and employs workers in the processes of production. The workers own nothing but their own ability to work, for which they are paid a wage. In other models, people are grouped into classes because of similar incomes, amounts of wealth, degrees of occupational status, and years of education. Whether we claim that the United States has two (owners and workers), three (upper, middle, and lower), four (upper, middle, working, and lower), six (upper-upper, lower-upper, upper-middle, lower-middle, upper-lower, and lower-lower), or more classes, the important point about the concept of social class is that life is differently experienced by individuals across the range of identified classes and similarly experienced by people within any one of the class categories.

Based on some common sociological models, we can describe six or seven social classes in the United States: the upper-upper class, or "elites"; the lower-upper class, or "everyday rich"; the upper-middle class; the "middle-middle class"; the working class; and the lower class. The latter can be further subdivided into the "working poor" and the underclass.

Upper Class(es)

Roughly 1% of the U.S. population occupies an "upper-class" position. At the very top of this group are those referred to as "upper-upper class" or the "ruling class" or "elite." They own a third of all assets, and their wealth is worth more than the "bottom" 90% of the population (Henslin 2006). Their "extraordinary wealth" often takes them into the hundreds of millions if not billions of dollars (Curry, Jiobu, and Schwirian 2002). They own stocks and bonds, prosperous businesses, and commercial real estate, allowing them to earn great amounts of income whether they are employed or not (Gilbert 2008). Many among the upper-upper class were born into their fortunes, and others joined the "superrich" through the information technology boom or in entertainment or sports (Kimmel 2009). Sociologist Dennis Gilbert suggests that an income of $2 million a year is "typical" for this upper 1%.

The rest of the upper class live on yearly incomes ranging from hundreds of thousands to a million dollars, own substantial amounts of wealth, and enjoy much prestige. The members of this "lower-upper class" might be considered the "everyday rich" (Kimmel 2009). They are professionals and corporate executives and managers who tend to have degrees from prestigious universities (Kimmel 2009). They live well in private homes in exclusive communities and enjoy considerable privilege. In some formulations of class, a distinction is drawn between the elite and the lower-upper class in terms of "old" versus "new" money (Langman 1988; Steinmetz, Clavan, and Stein 1992). In other words, some separate the upper class based on how they achieved and how long they have enjoyed their affluence.

Middle Classes

In some analyses, the middle class is considered the largest class, representing between 45% and 50% of the population (Gilbert 2008; Henslin 2006). Often, the middle class is subdivided into two groupings: the upper-middle class and the middle-middle class.

The **upper-middle class** consists of highly paid professionals (e.g., lawyers, doctors, accountants, and engineers) who have annual incomes that may reach into the hundreds of thousands of dollars. They are typically college educated, although they may not have attended the same elite colleges as the upper-upper class, and often possess advanced degrees (Henslin 2006). Women and men of the upper-middle class have incomes that allow them luxuries such as home ownership, vacations, and college educations for their children. Roughly 15% of the population is upper-middle class (Gilbert 2008; Henslin 2006).

The **middle class** comprises a larger portion of the population, typically estimated at around 30% (Gilbert 2008) to 35% (Henslin 2006). Although it is impossible to specify an exact income threshold that separates the middle (or middle-middle class) from the upper-middle class, sociologist Michael Kimmel (2009) suggests that the upper-middle class has household incomes above $80,000, while the middle-middle class is between $40,000 and $80,000. Of course, in areas of the country with higher costs of living, such incomes will not bring you the comforts associated with being middle class. The middle class is comprised of white-collar service workers, technicians, educators, salespeople, and nurses. Additionally, many "high-demand" service personnel (e.g., police or firefighters) and some blue-collar workers do reach the income threshold as indicated above. Generally, those in the middle class possess less wealth, live on less income, and have less education (or less prestigious degrees) and social standing than their professional and managerial counterparts (e.g., physicians, attorneys, and managers) in the upper-middle class. They own or rent more modest homes and purchase more affordable automobiles than their upper-middle-class counterparts, and they hope, but with less certainty, to send their children to college. Importantly, they do not have the same sorts of safety nets protecting them from sliding "down the ladder" with periods of increased and prolonged unemployment as is occurring at present in the United States.

Working Class

About 30% of the U.S. population can be considered **working class**, though the line separating them from the middle class is not always clear (Gilbert 2008). Members of this class tend to work in low-level white-collar or blue-collar occupations (as more likely unskilled or semiskilled laborers) and typically have high school or vocational educations, though some have attended college. Factory workers, clerical workers, retail salespeople, and custodians are all examples of the working class. Typically, they have jobs that offer little opportunity to move up and generate low levels of job satisfaction. Household incomes in the working class range from $20,000 to $40,000, depending on whether they are two-earner or one-earner households (Kimmel 2009). Members of the working class live somewhat precariously, with little savings and few liquid assets should illness or job loss occur. They also have difficulty buying their own homes or sending their children to college (Curry et al. 2002). Significantly, as we shall see, the extended family is of great importance to many in the working class.

Lower Class: The Working Poor and the Underclass

According to the U.S. Census Bureau, in 2007 the official poverty rate in the United States was 12.5%, unchanged from 2006. This translates to some 37.3 million Americans. It is further estimated that 18% of all children (age 18 and younger) live in poverty (www.census.gov/hhes/www/poverty/poverty07/pov07hi.html).

Some researchers contend that a more accurate estimate of how many Americans are poor would be closer to 20% of the population (Seccombe 2000). As originally established, the *poverty line* (or "poverty threshold") was determined by calculating the annual costs of a "minimal food budget" multiplied by three since 1960s survey data estimated that families spent one-third of their budgets on food (Seccombe 2000). The 2008 threshold for a family of four with two children was drawn at $21,910. Families whose incomes are just *$1 above* the threshold for their size are neither officially classified nor counted as poor.

The Department of Health and Human Services utilizes a different set of poverty guidelines to reflect who is eligible for governmental assistance. The 2009 poverty guidelines are illustrated in Table 3.4.

Poverty is consistently associated with marital and family stress, increased divorce rates, low birth weight and infant deaths, poor health, depression, lowered life expectancy, and feelings of hopelessness and despair. Poverty is a major contributing factor to family dissolution. Poor families are characterized by irregular employment or chronic underemployment. Individuals work at unskilled jobs that pay minimum wage and offer little security or opportunity for advancement. Although many lower-class individuals rent substandard housing, homelessness is a problem

Table 3.4 Poverty Guidelines, 2009

The 2009 Poverty Guidelines for the 48 Contiguous States and the District of Columbia	
Persons in Family	**Poverty Guideline**
1	$10,830
2	14,570
3	18,310
4	22,050
5	25,790
6	29,530
7	33,270
8	37,010

For families with more than eight persons, add $3,740 for each additional person.

among poor families. Karen Seccombe (2000) effectively describes the problems: "Poverty affects one's total existence. It can impede adults' and children's social, emotional, biological, and intellectual growth and development" (1096). She further notes that over a year, most poor families experience one or more of the following: "eviction, utilities disconnected, telephone disconnected, housing with upkeep problems, crowded housing, no refrigerator, no stove, or no telephone" (Seccombe 2000, 1096).

Despite stereotypes of the poor being African Americans and Latinos, most poor families—and most of those who receive assistance—are Caucasian. However, African Americans and Hispanics or Latinos are more likely to experience poverty than are Caucasians (U.S. Census Bureau 2008). Those living in poverty, like their upper- and middle-class counterparts, can be subdivided.

Working Poor. Since 1979, there have been large increases in the proportion of the population who, despite paid employment, live in poverty. The label *working poor* refers to people who spent at least 27 weeks in the labor force but whose incomes fell below the poverty threshold (U.S. Bureau of Labor Statistics 2005). Factors such as low wages, occupational segregation, and the dramatic rise in single-parent families account for why having a job and an income may not be enough to keep people out of poverty.

The working poor represents 10% to 15% of the population. They work at low-skilled and low-paid occupations, paying them approximately minimum wage and thus putting them very close to the "poverty line" established by the U.S. government. They may rely on food stamps, food pantries, and welfare even when employed and on unemployment assistance should they lose their jobs. The vulnerability of the working poor cannot be overstated. They live from paycheck to paycheck, week to week, with no opportunity to save money. Their jobs (e.g., delivering pizza, cleaning homes, migrant farmwork, day labor, low-pay factory work) don't typically offer benefits such as health insurance or sick pay, making them very vulnerable to illness or accidents (Henslin 2006; Kimmel 2009). Many single mothers and their children fall into this class, but even a family headed by two minimum-wage-earning spouses can find themselves among the working poor.

Based on a report from the Bureau of Labor Statistics (2008), we can make the following statements about the working poor:

- Of the nation's poor, 20% can be classified as "working poor." This amounts to 7.4 million people and 4.0 million families. In 6% of U.S. families, "despite having at least one family member in the labor force for half the year or more, families fell below the poverty level."
- Certain categories of people are more vulnerable to being among the working poor. Younger workers, people who fail to finish high school, and people who work part time are more likely to be among the working poor than are older workers, college graduates, and full-time workers.
- Racially, the percentages of African Americans and Hispanics who are among the working poor are more than twice the percentage of whites at that same class level. Asian Americans are less likely than whites to be among the working poor. However, because of differences in population sizes, whites represent 70% of the working-poor population.
- Women are more likely to be in the working poor than are men (5.8% to 4.5%).
- As to family characteristics, at 7.5%, married couples are less likely than either female-headed (22%) or single-male-headed households (12%) to be in the working poor.
- Families with children have a higher working-poor rate than families without children (9.4% vs. just under 2%). More than one of every five (23%) single-female-headed families are working poor compared to 13.5% of single-male-headed

families and 8.4% of families of married couples. Looking at families with children under 18 years of age, 23% of female-headed, 11% of male-headed, and 5% of married-couple-headed households were among the working poor.

Although their family members may be working or looking for work, these families cannot earn enough to raise themselves out of poverty. An individual working full time at minimum wage simply does not earn enough to support a family of three. Thus, this kind of poverty results from problems in the economic structure—low wages, job insecurity or instability, or lack of available jobs.

The Underclass. Finally, at the very bottom, we find what researchers call the "underclass." The underclass tends to be concentrated in the inner cities, in substandard housing if not homeless. They lack education and job skills and have inadequate nutrition and no health care (Kimmel 2009). When employed, they typically do low-paying, menial work but are often dependent on government assistance to survive. Some engage in criminal activities. Their lives are bleak at best.

Disproportionately African Americans and Latinos, the underclass represents a deeply disturbing counterpoint to wider cultural values and beliefs that are definitive features of American life. Their lifestyles and circumstances challenge cherished images of wealth, opportunity, and economic mobility. Their behaviors, actions, and problems are often responses to lack of opportunity, urban neglect, and inadequate housing and schooling. With the flight of manufacturing, few job opportunities exist in the inner cities; the jobs that do exist are usually service jobs that fail to pay their workers sufficient wages to allow them to rise above poverty. Schools are substandard. The infant death rate approaches that of Third World countries. The housing projects are infested with crime and drug abuse, turning them into kingdoms of despair. Estimating the size of the underclass is difficult, and estimates range from around 5% of the population (Henslin 2006) to 12% (Gilbert 2008).

Spells of Poverty. Most of those who fall below the poverty threshold tend to be there for spells of time rather than permanently. In typical times, for most poor people, poverty lasts less than a year (Henslin 2006). About a quarter of the American population may require some form of assistance at one time during their lives because of changes in families caused by divorce, unemployment, illness, disability, or death. About half of our children are vulnerable to poverty spells at least once during their childhood. Many families who receive assistance are in the early stages of recovery from an economic crisis caused by the death, separation, divorce, or disability of the family's major wage earner. Many who accept government assistance return to self-sufficiency within a year or two. Most children in these families do not experience poverty after they leave home.

Two major factors are related to the beginning and ending of spells of poverty: changes in income and changes in family composition. Many poverty spells begin with a decline in earnings of the head of the household, such as a job loss or a cut in work hours. Other causes include a decline in earnings of other family members, the transition to single parenting, the birth of a child to a single mother, and the move of a youth to his or her own household.

Poor Women and Children. The **feminization of poverty** is a painful fact that has resulted primarily from high rates of divorce, increasing numbers of unmarried women with children, and women's lack of economic resources. When women with children divorce, their income and standard of living fall, occasionally dramatically. By family type, 28% of single-mother families are below the poverty line (U.S. Census Bureau 2007).

In 2007, 13.2 million children, 18% of children under 18, were poor. The rate is higher among younger children; 21% of children under age 6, living in families, were poor. Sixteen percent of children age 6 or older live in impoverished families (Fass and Cauthen 2008). Like their parents, they move in and out of spells of poverty, depending on major changes in family structure, employment status of family members, or the disability status of the family head (Duncan and Rodgers 1988). These variables affect ethnic groups differently and account for differences in child poverty rates. African Americans, for example, have significantly higher unemployment rates and numbers of never-married single mothers than do other groups. As a result, their child poverty rates are markedly higher. More than a third (34%) of African American children are poor, as are nearly 30% of Hispanic children. In contrast, 13% of Asian children and 10% of Caucasian children lived in poverty in 2007 (Fass and Cauthen 2008). Being poor puts the most ordinary needs—from health care to housing—out of reach.

These statistics may understate the extent of economic hardship faced by children in the United States. Many researchers recommend using a standard other than the official poverty threshold to gauge the breadth of childhood economic vulnerability. With evidence that families typically need incomes that amount to twice the official poverty threshold to make ends meet, if we use 2008 levels, a family of four would need $42,400 to avoid being characterized as low income. In 2007, more than 28 million children, the equivalent of 39% of U.S. children could be classified as low income (Fass and Cauthen 2008).

Class and Family Life

Working within this framework, we can note some ways in which family life is differently experienced by each of the four classes. Although there are a number of family characteristics we could consider (including divorce, domestic violence, and the division of labor), we look briefly at class-based differences in marriage relationships, parent–child relationships, and ties between nuclear and extended families.

Marriage Relationships

Within upper-class families, we tend to find sharply sex-segregated marriages in which women are subordinated to their husbands. Upper-class women often function as supports for their husbands' successful economic and political activities, thus illustrating the **two-person career** (Papanek 1973).

Although their supportive activities may be essential to the husbands' success, such wives are neither paid nor widely recognized for their efforts. Rather than having their own careers, they often volunteer within charitable organizations or their communities. They are free to pursue such activities because they have many servants—from cooks to chauffeurs to nannies—who do the domestic work and some child care or supervision.

Middle-class marriages tend to be *ideologically* more egalitarian and are often two-career marriages. In fact, middle-class lifestyles increasingly *require* two incomes. This creates both benefits and costs for middle-class women. The benefits include having more say in family decision making and greater legitimacy in asking for help with domestic and child-rearing tasks. The costs may include the failure to receive the help they request. Because most employed wives earn less than their husbands, the strength of their role in family decision making may still be less than that of their

husbands. Middle-class marriages are "ideologically" more egalitarian because middle-class couples more highly value and more readily accept the ideal of marriage as a sharing, communicating relationship in which spouses function as "best friends."

Once more explicitly traditional, working-class marriages are becoming more like their middle-class counterparts. Whereas such marriages in the past were clearly more traditional in both rhetoric and division of responsibilities, in recent years they have moved toward a model of sharing both roles and responsibilities (Komarovsky 1962; Rubin 1976, 1994). The sharply segregated, traditional marriage roles evident even just two decades ago have given way to two-earner households, increasingly driven by the need for two incomes.

Especially among those working-class couples who work "opposite" shifts, higher levels of sharing domestic and child care responsibilities are often found, as is greater male involvement in home life (Rubin 1994). The reality of being the only parent home forces men to take on tasks that otherwise might be done by wives. In such instances, it is necessity more than ideology that influences men's participation. Such involvement may be more valued in the circles in which middle-class men live and work but be more of a practicality or necessity for working-class men. Thus, working-class men may understate, and middle-class men may exaggerate, what and how much they do, making class comparisons that much more difficult to draw.

Marriage is least common and least stable among the lower classes. Men are often absent from day-to-day family life. Resulting from the combination of high divorce rates and widespread nonmarital childbearing, 28% of single mothers and their families are poor, five and a half times the rate of poverty among married couple families with children. Furthermore, although they represent less than a fifth of all families in the United States, they are more than half of the 7.6 million poor families (Denavas-Walt, Proctor, and Smith, 2008). The cultural association of men's wage earning with fulfillment of their family responsibilities subjects lower-class men to harsher experiences within families. They are less likely to marry. If married, they are less likely to remain married, and when married, they derive fewer of the benefits that supposedly accrue in marriage.

When marriages cross class lines, other problems can arise. People may find themselves feeling out of step, as if they are in a world where there are different, perhaps dramatically different, assumptions about

Family experiences are affected by such variables as social class and ethnicity.

how to discipline and raise children, where to go and what to do on vacation, and how to save or spend money (Lewin 2005). It is more difficult to measure than interracial marriage or religious intermarriage, but using education as an indicator of class, there appear to be fewer cross-class marriages than in the past. Most of those marriages that do cross class lines are now between women with more education marrying men with less. This combination does not bode especially well for the future stability of the marriages (Lewin 2005).

Parents and Children

The relationships between parents and children vary across social lines, but most research has focused on the middle and working classes (Kohn 1990; Lareau 2003). Among the upper class, some hands-on child rearing may be done by nannies or au pairs. Certainly, mothers are involved, and relationships between parents and children are loving, but parental involvement in economic and civic activities may sharply curtail time with children. For upper-class parents, an important objective is to see that children acquire the appropriate understanding of their social standing and that they cultivate the right connections with others like themselves. They may attend private and exclusive boarding schools and later join appropriate clubs and organizations. Their eventual choice of a spouse receives especially close parental scrutiny.

A considerable amount of research indicates that working- and middle-class parents socialize their children differently and have different objectives for child rearing (Hays 1996; Kohn 1990; Lareau 2003; Rubin 1994). Although all parents want to raise happy and caring children, middle-class parents tend to empha-

size autonomy and self-discipline, and working-class parents tend to stress compliance (Hays 1996; Kohn 1990). Whereas both middle- and working-class mothers value education, working-class mothers saw and therefore stressed education as essential for their children's later life chances. Middle-class mothers took for granted that their children would receive good-quality educations and emphasized, instead, the importance of building children's self-esteem. And although both classes of mothers acknowledged using spanking to discipline their children, middle-class mothers spanked more selectively and favored other methods of discipline (e.g., "time-out") (Hays 1996).

One of the more recent and fascinating class comparisons is Annette Lareau's (2003) *Unequal Childhoods*. Lareau contends that "social class does indeed create distinctive parenting styles . . . that parents differ by class in the ways they define their own roles in their children's lives as well as in how they perceive the nature of childhood" (748). Lareau introduces the concepts of *concerted cultivation* and *accomplishment of natural growth* to represent class-based differences in philosophy of childrearing.

Middle-class families engage in concerted cultivation. Parents enroll their children in numerous extracurricular activities—from athletics to art and music—that come to dominate their children's lives as well as the life of the whole family. Through these activities, however, children partake in and enjoy a wider range of outside activities and interact with a range of adults in authoritative positions, giving them experiences and expertise that can serve them well later. Because of the way household life tends to center around children's schedules and activities, the other members of these middle-class families (parents and siblings) are forced to endure a frenzied pace and a shortage of family time (see the Real Families feature in this chapter).

Working-class parents, lacking the material resources to enroll their children in such activities, tend to focus less on developing their children and more on letting them grow and develop naturally, play freely in unsupervised settings, and spend time with relatives and in the neighborhood. For a sense of how these two approaches may have been experienced by children, see the Real Families feature in this chapter.

Lower-class families are the most likely of all families to be single-parent families. Single parents, in general, may suffer stresses and experience difficulties that parents in two-parent households do not, but this situation is exacerbated for low-income single parents (McLanahan and Booth 1989). Parent–child relationships suffer from a variety of characteristics of lower-class life: unsteady, low-pay employment; substandard housing; and uncertainty about obtaining even the most basic necessities (food, clothing, and so forth). All of these can affect the quality of parent–child relationships and the ability of parents to supervise and control what happens to and with their children.

Extended Family Ties

Links between nuclear family households and extended kin vary in kind and meaning across social class. By some measures, the least closely connected group may be the middle class, which, because of the geographic mobility that accompanies their economic status, may find themselves the most physically removed from their kin. As Matthijs Kalmijn (2004) observed among the upper-middle-class families he studied in the Netherlands, they live almost three times as far from their siblings and more than three times as far from their parents and their (adult) children as does the lowest-educated class. Similar class differences can be observed in the United States.

Middle-class families do visit kin or phone regularly and are available to exchange aid when needed. Still, the emphasis is on the conjugal family of spouses and children. Closer connections may be found among both the working and the upper classes, although the reasons differ. In the case of working-class families, there are often both the opportunity and the need for extensive familial involvement. Opportunity results from lesser levels of geographic mobility, which results in closer proximity and allows more continuous contact to result. The need for involvement is created by the pooling of resources and exchange of

services (e.g., child care) that often result between adults and their parents or among adult siblings. Intergenerational upward mobility may lessen the reliance on extended families (see discussion later in this chapter).

Upper-class families, especially among the "old-money" upper class, highly value the importance of family name and ancestry. They tend to maintain strong and active kinship groups that exert influence in the mate selection processes of members and monitor the behavior of members. Inheritance of wealth gives the kin group more than symbolic importance in their ability to influence behavior of individual members.

Among the lower class, kin ties—both real and fictive—may be essential resources in determining economic and social survival. Grandparents, aunts, and uncles may fill in for or replace absent parents, and multigenerational households (for example, children living with their mothers and grandmothers) are fairly common. **Fictive kin ties** refer to the extension of kinship-like status to neighbors and friends, thus symbolizing both an intensity of commitment and a willingness to help one another meet needs of daily life (Liebow 1967; Stack 1974).

The Dynamic Nature of Social Class

Like other aspects of family life, social class position is not set in stone. Individuals may experience **social mobility**, movement up *or* down the social class ladder. Either kind of social mobility can affect family relationships, especially, although not exclusively, intergenerational relationships (Kalmijn 2004; Newman 1988; Sennett and Cobb 1972). For example, children who see their parents "fall from grace" through job loss and dwindling assets look differently at those parents. Fathers who once seemed heroic may become the source of concern and even resentment as their job loss threatens the lifestyle of the family on which children depend (Newman 1988). Children who in adulthood climb upward occasionally find their relationships with their parents suffering as a result. As they are exposed to new values and ideas that differ from those held by their parents, generational tension and social distance may follow. Furthermore, as they move into a new social circle, parents (as well as less mobile siblings) may appear to fit less well with their new life circumstances. The more they strive to fit into new circles and circumstances accompanying their increased social standing, the less well they may fit comfortably within their ongoing family relationships.

Louise and Don Tallinger are proud parents of three boys, 10-year-old Garrett, 7-year-old Spencer, and 4-year-old Sam. They are also busy professionals; Louise is a personnel consultant, and Don is a fund-raising consultant. Between them, they earn $175,000, making them comfortably upper-middle class. They travel a lot for work; Don is out of town an average of three days a week, and Louise, four or five times a month, flies out of state early in the morning and returns sometime after dinner. Don often doesn't return home from work until 9:30 P.M. This middle-class family of five is one of the families studied by Annette Lareau (2003) in her fascinating class comparison *Unequal Childhoods*.

With 10-year-old Garrett's involvement in baseball, soccer, swim team, piano, and saxophone lessons, his schedule dictates the pace and routines of the household. Lareau (2003) describes what life is like for Don and Louise:

> Rush home, rifle through the mail, prepare snacks, change out of . . . work clothes, make sure the children are appropriately dressed and have the proper equipment for the upcoming activity, find the car keys, put the dog outside, load the children and equipment into the car, lock the door and drive off. (42)

The Tallingers epitomize the middle-class child-rearing strategy that Lareau called "concerted cultivation." This lifestyle, dedicated as it is to each child's individual development and enrichment, is exhausting just to read about. It may be familiar to you as an extreme example of your experiences; Lareau argues that among the middle class, it is not uncommon.

The Tallinger brothers hardly go long stretches without some scheduled activity, most of which require adult-supplied transportation, adult supervision, and adult planning and scheduling. Rarely can they count on playing outside all day like Tyrec Taylor, a working-class child from Lareau's study. Like other children in working- and lower-class families, Tyrec's daily life is less structured, more flexible and free flowing, more oriented around kin and neighborhood peers, and more suited to developing such skills as negotiation, mediation, and conflict resolution.

Of course, working- and lower-class children like Tyrec could not participate in all the activities that the Tallinger boys do. By the Tallingers' estimate, Garrett's activities alone cost more than $4,000 a year. Despite this obvious advantage, Garrett feels disadvantaged because he cannot attend the private school he once attended.

Although Lareau is careful to illustrate what middle-class children like the Tallinger boys miss out on—free play, closer connections to relatives, more time for themselves away from adult supervision and control, comfort with and ability to amuse themselves, and less fatigue—she also illustrates the many benefits they receive beyond involvement in activities that they enjoy. The Tallingers believe that all the activities that their boys participate in teach them to work as part of a team, to perform on a public stage and in front of adults, to compete, to grow familiar with the many performance-based assessments that will come at them through school and work experiences, and to prioritize. The children travel to tournaments, eat in restaurants, and stay in hotels; they may fly to summer camps or special programs out of state or overseas. Indeed, Lareau suggests that children like the Tallinger boys may travel more than working-class and poor *adults*. These experiences, combined with what the Tallingers teach the boys at home, promote skills that enhance their chances of staying or even moving higher up in the middle class.

In lifestyles such as this one, family life is organized and ruled by large calendars that detail the children's sports, play activities, music, and scouting events. It then falls on the parents to see that their children arrive at these activities, often directly from one to another. As Lareau (2003) somberly puts, "At times, middle-class homes seem to be little more than holding places for the occupants during the brief periods when they are between activities" (64).

Aside from the difficulty fitting parents and siblings into a new social standing, practical considerations, imposed by a job, may create obstacles preventing individuals from maintaining closer relationships. As is true elsewhere, ascending the ladder to a higher rung may require geographic relocation. Such jobs may also impose greater demands on the individual's time. To these constraints of time and distance one can add that as someone establishes new friendships and participates in leisure activities, further reductions in opportunity and availability may result (Kalmijn 2004).

Marital relationships, too, may be altered by either downward or upward mobility. Research indicates that some men who lose their jobs and "slide downward" react to their economic misfortune by abusing their spouses, turning to alcohol or other substances, withdrawing emotionally, or leaving the home (Newman 1988; Rubin 1994). Changes in the marriage are not entirely of men's doing; after an initial period of sympathy and support, wives may grow impatient with their husbands' unemployment or alter their positive views of the husbands' dedication as a worker or job seeker. In addition, as couples are forced to scale back their accustomed lifestyle, tensions may rise, and resentment and distance may grow.

Upward mobility may transform marriage relationships as well. Some women face situations where, after sacrificing to help launch their husbands' careers by supporting them through school, they are left by those same husbands once they have achieved their career goals. With their own increasing economic opportunity, some women find that marriage becomes less desirable because of the constraints it continues to impose on their career development.

Social mobility is differently experienced by Americans of different races. A report by economist Julia Isaacs (2007) for the Brookings Institution compared adult children's incomes in the late 1990s and early 2000s to their parents' incomes in the 1960s. She detailed the following differences between whites and African Americans:

- Close to two-thirds of African American and white children exceed the incomes of their parents after controlling for inflation.
- Looking specifically at middle-income families, African Americans are much less likely to exceed their parents—experience upward intergenerational mobility—than are whites. Sixty-eight percent of whites from the middle quintile (20%) of the income distribution versus 31% of blacks have family incomes greater than their parents.
- Nearly half (47%) of middle-income African American children end up falling to the bottom of the income distribution at some point in their lives. Only 16% of white children experience this same sort of fall.
- While just under a third of white children from families in the bottom 20% of the income distribution stay at that same level, more than half (54%) of African American children from the bottom 20% stay there.
- In general, one's place in the income distribution is heavily influenced by where one starts out: 42%

of children born to parents in the bottom fifth stay there while 39% of children from the uppermost 20% stay there (Isaacs 2007, 2008).

Structural Downward Mobility: U.S. Families Face the Recession

The examples above depict consequences of individual upward or downward mobility. Similarly, families can see their circumstances dramatically altered by **structural mobility**, which refers to when large segments of a population experience upward or downward mobility resulting from changes in the society and economy. The recent and current state of the U.S. economy in 2009–2010 has made downward mobility a harsh and painful reality for large numbers of U.S. families across the social class spectrum. Widespread job loss, increases in home foreclosures, crises on Wall Street, and the failures of many big employers have caused many families to suffer a variety of familial consequences along with their economic losses. With increased debt, lower incomes, and dwindling savings, many families are pushed beyond their limit.

For some couples, the consequence has been a reallocation of household roles and responsibilities in two-earner families as husbands who suffer job losses become dependent on their wives incomes. This may generate more and more frequent tense conversation and negotiation, as couples must make hard financial decisions and alter their accustomed lifestyles. For others, the consequences have been more visibly dramatic, including increased risk of suicide, domestic violence, or homelessness. Looking at domestic violence, the National Domestic Violence Hotline (2009) reported that calls to the hotline had increased 21% between the third quarter of 2007 and 2008. In a study of some 8,000 callers, 54% indicated that their financial situation had worsened, and 64% indicated that there had been an increase in abusive behavior (National Domestic Violence Hotline 2009). The impact of the recession on divorce is less certain. While the mounting economic pressures certainly strain many marriages to or beyond the breaking point, the wider economy may force them to stay married even if they had intended to divorce.

Childbearing and child rearing have also been affected. Couples who were planning to have or adopt children are having to consider delaying their entry into parenthood or downsizing the number of children they hoped to have. The effects on children are perhaps more chilling. It is estimated that the

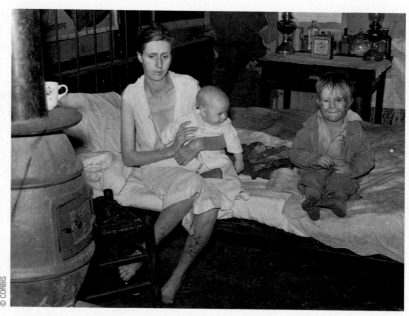

Families are affected by wider economic conditions, as was evident during the Great Depression and is also the case during the current economic recession.

economic crisis may push an additional 3 million children into poverty, costing the United States a yearly loss of $35 billion over the lifetimes of these children. More than half of children who fall, with their families, into poverty will remain impoverished long after the recession ends. An estimated 25% will remain there through half of their remaining childhoods. In adulthood, those who spend most of their childhoods in poverty will earn, on average, 39% less than the median income, while those who spend between one-quarter to one-half of their childhoods in poverty are projected to suffer about half that level of reduced earnings (Linden 2008). Even young adults who graduate from college during the recession are vulnerable. Faced with reduced career prospects, many of them may get stuck in lower-prestige jobs even after the economy recovers (Shellenbarger 2009). See the following Real Families feature on the effects of the recession on two families: one working class and one upper-middle class.

Racial and Ethnic Diversity

The United States is a richly diverse society. This is not news to you; Americans take pride in the multicultural mix of groups, whether conceptualized as "melting" together into one large pot or, like a salad bowl, retaining their uniqueness even when tossed together. As we begin to look at the racial and ethnic variations in family experience, we need to first note the multiplicity of different groups that make up the U.S. population. To capture and count this diversity, the U.S. Census Bureau asked the following question:

> What is this person's ancestry or ethnic origin? (For example, Italian, Jamaican, African American, Cambodian, Cape Verdeian, Norwegian, Dominican, French Canadian, Haitian, Korean, Lebanese, Polish, Nigerian, Mexican, Taiwanese, Ukrainian, and so on.)

The census also contains items about a person's race and whether he or she is of Hispanic origin. These different attempts to get at the diversity of the population have yielded somewhat inconsistent data. For example, although African American is among the options for people to select as their "ancestry," 12 million fewer people chose African American as their ancestry than answered that their race was African American.

Defining Race, Ethnicity, and Minority Groups

Before we begin to look more closely at ethnic or racial diversity in family experience, we need to define several important terms. A **race** or **racial group** is a group of people, such as whites, blacks, and Asians, classified

Karen Marshall and her two children are certainly not alone. But that may be of little consolation to the Marshalls, who in 2008 went from their two-bedroom apartment in Bakersfield, California, to living out of Ms. Marshall's car after she lost her job. Their days are spent at a Bakersfield homeless shelter. Acknowledging a "total sense of loss of control," Marshall worries about her children and how to keep them clothed, fed, and safe. The shelter they turned to reported in February 2009 that they had seen a 38% increase in homeless clients (Turnto23 .com, February 24, 2009).

In 2008–2009, homeless shelters around the country reported substantial increases in the numbers of homeless families seeking shelter in their facilities. New York City saw a 40% increase in July to November 2008 compared to the same period in 2007. Similar jumps from 2007 to 2008 were reported in Connecticut, Massachusetts, Minnesota, and Kentucky. Two national surveys show the problem to be widespread. A 22-city survey by the U.S. Conference of Mayors reported that 16 of the 22 had experienced increases in the numbers of homeless families with children. Another survey, by First Focus and the National Association for the Education of Homeless Children and Youth, found that 20% of the surveyed school districts reported increases in the numbers of homeless students in the fall of 2008 compared to all of the 2007–2008 school year.

Even when the impact is less dramatic, it is nonetheless real for families. At the end of 2007, Scott Berry lost his job as a technology analyst for a Manhattan firm. Berry, too, knows that he is not alone, as he has noticed many middle-aged men just like himself as he walks through downtown Darien, Connecticut (Tyre 2009). Scott and Tracey, parents of a seven- and eight-year-old, had planned to have Tracey stay home with the children. The Berrys had paid off their mortgage and set up college funds for their children and carried little debt, so they were somewhat more protected from the harshest of impacts. Still, the life-style they enjoyed and the plans they had made had to be reconfigured. They had to replace a full-time nanny with an au pair and give up weeklong vacations for weekend get-aways. They also had to rethink their plans regarding retire-ment and where they would be able to send their children to college. With ample savings, though, they are still able to enroll their children in an assortment of organized activities and retain membership in a country club (Tyre 2009).

Clearly, the Marshalls and Berrys are not experiencing the same effects of the struggling economy. The Berrys are able to "roll with it a little bit," making sacrifices and changing their roles and responsibilities. The Marshalls, with less of a cushion with which to buffer their fall, have had much more extreme misfortune. We spoke of homelessness earlier in our discussion of the underclass. Certainly, the Marshalls don't seem to belong in that discussion at all. Yet for circumstances beyond their mak-ing or control, the family of three finds itself in need of some of the same assistance. Neither the Berrys nor the Marshalls, though, are experiencing changes of their own choosing. Like hundreds of thousands of families throughout the United States and Canada, they illustrate the extent to which the private world of family relation-ships is affected by larger structural and economic realities.

---

according to their **phenotype**—their anatomical and physical characteristics. Racial groups share common phenotypical characteristics, such as skin color and fa-cial structure. The concept of race is often misused and misunderstood to the extent that some sociologists refer to it as a myth (Henslin 2006). This refers spe-cifically to any assumption of racial purity or homo-geneity within racial groupings (in skin color, facial features, and so on) and any belief that certain racial groups as superior or inferior in comparison to one another. Socially, however, we perceive or identify our-selves and others within racial classifications, and we are treated and act toward others on the basis of race. As a consequence, race becomes a highly significant and very real factor in shaping our life experiences. Although its biological importance may be doubtful, its social significance remains great.

An **ethnic group** is a set of people distinct from other groups because of cultural characteristics. Such things as language, religion, and customs are shared within and allow us to differentiate between ethnic groups. These cultural characteristics are transmitted from one generation to another and may then shape how each person thinks and acts—both inside and outside families.

Either a racial or an ethnic group can be considered a **minority group**, depending on social experience, not numerical size. Minority groups are so designated because of their status (position in the social hierarchy), which places them at an economic, social, and political disadvantage (Taylor 1994b). Thus, African Americans are simultaneously an ethnic, a racial, and a minority group in the United States (as well as an *ancestry category* as shown previously). The term *African American*, used interchangeably with and, increasingly, instead of *black*, reflects the growing awareness of the importance of ethnicity (culture) in contrast to race (skin color) (Smith 1992; but see Taylor 1994b).

As we will soon see, ethnic and/or racial differences are often difficult to untangle from social class differences. It may be that some differences in family patterns reflect cultural background factors or distinctive values. However, it is equally plausible that ethnic or racial differences in family patterns, such as in nonmarital childbearing or relations with extended kin, reflect the different socioeconomic circumstances under which different groups live (Sarkisian, Gerena, and Gerstel, 2007; Wildsmith and Raley 2006).

The best estimates of the racial composition of the U.S. population come from data gathered by the U.S. Census Bureau's decennial census of the population. Complying with the constitutional mandate, in each year ending in a zero a census of the population is undertaken. As we were completing this edition, the 2010 census was just getting started. As a result, much of the data in the discussions that follow are census estimates, drawn instead from the Population Estimates Program or the 2007 American Community Survey.

Census data reveal that more than a third of the U.S. population are ethnic minorities: 14% are African American, 15% are Hispanic, 5% are Asian/Pacific Islander, and 2% are Native American. By 2050, more than half of the U.S. population is expected to be minorities: 50% Caucasian, 30% Hispanic, 15% African American, 9% Asian, and 2% Native American (www.census.gov/Press-Release/www/releases/archives/population/012496.html).

Until the latter decades of the twentieth century, most research about American marriages and families tended to focus on white, middle-class families. The nuclear family was the norm against which all other families, including single-parent and stepfamilies, were evaluated and often viewed as pathological because they differed from the traditional norm. A similar distortion unfortunately influenced our understanding of African American, Latino, Asian American, and Native American families. Instead of recognizing the strengths of diverse ethnic family systems, misguided researchers viewed these families as "tangles

of pathology" for failing to meet the model of the traditional nuclear family (Moynihan 1965). Part of this distortion resulted from the long-term scarcity of studies on families from African American, Latino, Asian American, Native American, and other ethnic groups. Furthermore, many earlier studies focused on weaknesses rather than strengths, giving the impression that all families from a particular ethnic group were riddled with problems.

For example, the "culture of poverty" approach depicts African American families as being deeply enmeshed in illegitimacy, poverty, and welfare as a result of their slave heritage. As one scholar notes, the culture of poverty approach "views black families from a white middle-class vantage point and results in a pejorative analysis of black family life" (Demos 1990). This approach ignores most families that are intact or middle class. It also fails to see African American family strengths, such as strong kinship bonds, role flexibility, love of children, commitment to education, and care for the elderly.

America is a pluralistic society. Thus, it is important that students and researchers alike reexamine diversity among our different ethnic groups as possible sources of strength rather than pathology (DeGenova 1997). For instance, cultures may vary widely in how the best interests of the child are defined (Murphy-Berman, Levesque, and Berman 1996). Differences may not necessarily be problems but solutions to problems; they may be signs of adaptation rather than weakness (Adams 1985). As two family scholars pointed out, "Whether a phenomenon is viewed as a problem or a solution may not be objective reality at all but may be determined by the observer's values" (Dilworth-Anderson and McAdoo 1988).

As we embark on our discussion of race and ethnicity in family life, it is important to try to avoid ethnocentric fallacies (a term introduced in Chapter 2)—beliefs that your ethnic group, nation,

Matter of Fact According to the U.S. Census Bureau, nearly 52 million Americans, more than 20%, speak languages other than English at home. Of those, nearly 35 million speak Spanish. More than 8 million people, 3% of the population, speak Asian and Pacific Island languages, the most common being Chinese (SOURCE: U.S. Census Bureau, 2007 American Community Survey).

Public Policies, Private Lives: A Multiracial First Family

As is well known, the 2008 presidential election was an unprecedented one. A woman and a man of mixed race were the last two candidates standing in the quest for the Democratic nomination for president, and a woman ran on the Republican ticket for vice president. On November 4, the United States elected Barack Obama as president, making him the first nonwhite person to reach that political peak. Barack and Michelle (Robinson) Obama are a source of inspiration to millions of Americans who previously felt locked out of the highest reaches of American society. They are also products of an increasingly common type of family—a multiracial family.

In the Obama–Robinson family histories, one finds a multitude of races, religions, languages, and classes. Their wider families contain members who are "black and white and Asian, Christian, Muslim and Jewish. They speak English; Indonesian; French; Cantonese; German; Hebrew; African languages including Swahili, Luo and Igbo; and even a few phrases of Gullah, the Creole dialect of the South Carolina Lowcountry. Very few are wealthy, and some . . . are quite poor" (Kantor 2009).

How many such families are there in the United States? Until the 2000 census, we would have had a much harder time estimating the size or growth of the *multiracial* population. Beginning with the 2000 census, Americans were given the chance to identify themselves with more than one racial category; close to 7 million people did. Between 2000 and 2006, demographers estimate that the multiracial population increased 25%, while the overall population increased only 7%. By 2050, the Census Bureau estimates that there will be more than 16 million people who identify themselves as being of two or more races (Population Division, U.S. Census Bureau, Table 4, "Projections of the Population by Sex, Race and Hispanic Origin for the United States: 2010–2050," NP2008-T4).

Multiracial families are products of racial intermarriage. Until 1967, it was illegal in 16 states of the United States for people to marry across racial lines. Ruling in *Loving v. Virginia*, the U.S. Supreme Court decreed that "the decision to marry, or not to marry, a person of another race resides with the individual and cannot be infringed by the State" (*Loving v. Virginia*, 388 U.S. 1 1967).

Today, it is not only legal everywhere in the United States but also meets with increasing acceptance, at least as measured by public opinion data. When asked whether they would accept a child or grandchild marrying someone of a different race, 66% of whites, 79% of Hispanics, and 86% of African Americans said they would. Where more than a third of whites surveyed in the 1970s said that they favored laws prohibiting black–white intermarriages, by the 2000s only 10% said that they favored such legal proscriptions (Lee and Edmonston 2005). Even if we accept that many people may be unwilling to admit opposing intermarriage, such data are indications of greater acceptance of and comfort with interracial relationships.

Interracial marriages and multiracial families have great social significance. By their presence, they help challenge the social and cultural boundaries and reduce the social distance that otherwise separate racial groups. As the population of biracial and multiracial individuals grows, the meaning and measurement of race are likely to change as well. Even the seemingly straightforward task of counting populations will be more difficult and less clear as more people possess multiracial or biracial identities (Lee and Edmonston 2005). Although the Obamas are not an "interracial couple," President Obama's parents were. Today, the wider Obama–Robinson family is an example of the increasing number of multicultural and multiracial families.

or culture is innately superior to others. In the following sections, we consider briefly some distinctive characteristics and strengths of families from various ethnic and cultural groups.

Ethnic Groups in the United States

The following brief profiles highlight some of the major features of family life for the major ethnic and racial minorities in the United States. As shown,

family patterns vary both between and within these groups, with economic factors helping to account for the extent and nature of the differences.

African American Families

According to the Census Bureau, in 2007 there were nearly 39 million African Americans in the United States, making African Americans 13% of the U.S. population. If we include those who consider themselves biracial, in this case black combined with one or more other races, the 2008 total reached just under 41 million people, or 14% of the population of the United States (Bernstein and Edwards 2008).

Compared with the total U.S. population, African Americans are younger and less likely to be married (see Figure 3.2). Although they are no more likely to be divorced or widowed, a much greater percentage of blacks than whites have never married (46.8% vs. 26.4%) (U.S. Census Bureau, 2007 American Community Survey). Blacks are much more likely to bear children outside of marriage and more likely to live in single-parent, mostly mother-headed, families (U.S. Census Bureau, 2009 *Statistical Abstract of the United States*, Tables 85, 68). Although African Americans are as likely as the general population to live in family households, their households differ from the family households in the general population. In 2008, 30% of black households and 50% of households in the wider population were headed by married couples (Current Population Survey, 2008 Social and Economic Supplement, Table H1, 2009). Because of high rates of divorce and of births to unmarried women, in 2008 only 39% of African American children lived in households with two parents. Meanwhile, 53% lived with either single mothers (50%) or single fathers (3%) (Current Population Survey, 2008 Social and Economic Supplement, Table C9).

Considering families rather than households, data from 2008 reveal that 45% of African American families were married-couple families, 31% were single mothers and their children, and 3% were single fathers and their children. The equivalent percentages for whites show nearly 80% of white families headed by married couples with or without children, 8% by unmarried women with children, and 2% by single fathers and their children. African American families are also more likely to consist of grandparents living with their grandchildren or adult children living with their parents than are white, Asian, or Hispanic families (Current Population Survey, 2008 Social and Economic Supplement, Table FG10).

In addition, we can note the following:

- Compared to the general population, African Americans are less likely to have completed college (19% vs. 29%). Among blacks, women are slightly more likely than men to have completed college (19% vs. 17.5%).
- Overall, African Americans are less likely than the general population to be employed. In the fall of 2008, unemployment began increasing rapidly, dramatically, and unpredictably. As of March 2009, the Bureau of Labor Statistics reported a national unemployment rate of 8.1. Among African Americans, the unemployment rate was 13.4. This statistic alone is potentially misleading for two reasons. First, it does not include those who were no longer looking for work (i.e., "discouraged workers"). Second, there was a large gap in unemployment between African American women and men. Black women had an unemployment rate of 9.9, whereas black men had seen unemployment rise to 14.9 (Bureau of Labor Statistics, *Employment Situation Summary: March 6, 2009*).
- Median household income for African Americans in 2007 was $33,916. This was the lowest among racial groups in the United States, 68% of the median for all U.S. households and 62% of the median income for non-Hispanic whites. The median incomes for African American men and women employed full time year-round were $36,068 and $31,009, respectively. Black men's median earnings were 80% of the median for men overall. Black women's earnings were much closer (88%) to earnings among all women. The gender wage gap among blacks is narrower than it is among the general population; black women's median income is 86% of the median among black men. In the general population, the gender wage gap was 78% in 2007 (McKinnon and Bennett 2005).

Figure 3.2 Marital Status of the Population, 18 and Over, 2007

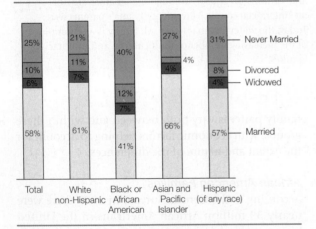

NOTE: Percentages may not add up to 100%.

SOURCE: U.S. Census Bureau, *2009 Statistical Abstract of the United States*, Table 55.

- Data from 2007 reveal that 25% of African Americans live below the poverty line. The percentage of impoverished blacks is nearly three times the percentage of poor whites (8%) and twice the percentage of the general population who are poor (12.5%). Among families, 21% of African American families and 32% of families with children under age five were below poverty. Looking at female-headed families with children under age five, nearly half (47.8%) of African American families were in poverty as of 2007.
- Related to the previous point, nationwide 18% of children under 18 live in poverty. The poverty rate for black children is nearly twice as high at 34%, and black children have three times the rate of poverty as white children (10%) (Fass and Cauthen 2008).

There are several noteworthy features of African American families, some of which are related to the aforementioned economic facts. First, African American families, in contrast to Caucasian families, have a long history of being dual-earner families as a result of economic need. As a consequence, employed women have played important roles in the African American family. They also have more egalitarian family roles. Black men have more positive attitudes toward working wives, take on a slightly larger share of household labor, and spend more time on domestic tasks and child care activities (McLoyd et al. 2000). Second, marital relations more often show signs of greater distress than is true of the general population. Some evidence indicates a greater likelihood of spousal violence and lower levels of reported marital happiness among African American marriages (McLoyd et al. 2000). Third, kinship bonds—whether blood related or affiliated kin—are especially important because they provide economic assistance and emotional support in times of need (Taylor 1994c; Taylor et al. 1991). Fourth, African Americans have a strong tradition of familism (emphasis on family and family loyalty), with an important role played by intergenerational ties. Fifth, the African American community values children highly. Finally, African Americans are much more likely than Caucasians to live in **extended households**, households that contain several different generations. Black children are more likely than other children to live in a household with a grandparent. In half of these households, the grandparent has assumed responsibility for her grandchild(ren) (typically, this grandparent is a grandmother) (U.S. Census Bureau 2008).

Many of these characteristics are often associated with harsher economic circumstances and thus may not be features inherent in African American families. When divorce rates are adjusted according to socioeconomic status, racial differences are minimal (Table 3.5). Poor African Americans have divorce rates more similar to poor Caucasians, and middle-class African Americans have divorce rates more similar to middle-class Caucasians. Thus, understanding socioeconomic status, especially poverty, is critical in examining African American life (Julian, McKenry, and McKelvey 1994; Wilkinson 1997).

As the preceding data reveal, African Americans and their families are at a clear economic disadvantage relative to the wider population. On the whole, compared to Caucasians, they have nearly twice the unemployment rate, three times the poverty rate, and two-thirds the median income. These economic indicators point out the potential difficulty of comparing black and white family characteristics. Combined with the tendency of upper-status African American families (i.e., middle and upper-middle class) to be as stable as Caucasian families of comparable status, these economic indicators suggest that much of what we may assume to be race differences are confounded

Table 3.5 Race, Ethnicity, and Socioeconomic Status

	Total	Whites	African Americans	Latinos	Asians
Median family income, 2007	$50,740	$55,096	$40,300	$42,074	$77,046
Percent unemployed, February 2009*	8.1	7.3	13.4	10.9	6.9
Percent of families in poverty, 2007	9.5	6.0	21.3	18.5	8.0
Percent of children in poverty, 2007	14.9	10.6	34.6	27.5	11.6

* Bureau of Labor Statistics, Economic News Release, "The Employment Situation: February 2009."

SOURCE: U.S. Census Bureau 2008, Selected Population Profile in the United States, 2007 American Community Survey.

by economic differences or may be social class differences *masquerading as race ones*.

This more economic argument pertains especially well to an understanding of race differences in marriage rates, divorce rates, and the numbers of single-mother-headed families. The most widely applied argument is that blacks "marital prospects" have shifted dramatically, especially among the poor (Aponte, Beal, and Jiles 1999). Wilson's notion of the "male marriage-able pool index" emphasizes the importance of male employment to their "marriageability" (Wilson 1987). Downward shifts in male employment patterns would then account for some of the decline in marriage rates and the increase in single-mother-headed families. Meanwhile, middle-class African Americans tend to have high rates of intact marriages.

Not only are African Americans unlikely to devalue marriage, they may actually more highly value marriage than do other groups. Asked about the importance of marriage when a man and woman have decided to spend their lives together, African Americans were more likely than either whites or Hispanics to say "very important" or "somewhat important" (77% vs. 69% vs. 71%). African Americans are also most likely to claim that divorce should be avoided except in an "extreme situation" (Pew Research Center 2007). Thus, despite being the least likely to be married, African Americans may be most likely to value marriage.

Despite the benefits of linking class and race in our efforts to understand family diversity, we cannot simply interpret all race differences as economic in nature. Don't forget that a major feature of race in American society is that it determines much treatment we receive from others. Thus, the opportunities we are offered or refused and whether others insult, avoid, or think less of us are all affected by race. The interpretation of race differences as only (or even largely) class differences unfortunately minimizes or ignores such expressions of racism and discrimination and fails to acknowledge patterns that may have cultural origins to them—such as greater emphasis on extended family ties or gender equality.

Hispanic Families

Hispanics (or Latinos) are now the largest and fastest-growing ethnic group in the United States. As of July 2007, there were an estimated 45.5 million Hispanics, representing 15% of the U.S. population. Furthermore, it is projected that by 2050, at least 30% of the population will be of Hispanic origin. These increases result from both immigration and relatively high birthrates. Currently, 64% of Hispanics are of Mexican descent, 9% are Puerto Rican, and another 3.4% are Cuban. The remaining 24% includes 8% from Central American countries and 5% from South American countries.

Compared to the population overall, the population of non-Hispanic whites, and the other ethnic or racial minorities discussed in this chapter, as a whole, Hispanics have the youngest median age: 27.6. Socioeconomically, Hispanics are less well off than the general population. The median family income for Hispanics in 2006 was $40,074—lower than the income of Asians ($72,305) or whites ($65,180) but higher than the median incomes for African American ($38,385) or American Indian ($38,800) families. Whereas the overall poverty rate was 12.5% in 2007, 21.6% of Hispanics were poor. Roughly 20% of Hispanic families were below poverty, and—at 29%—a greater proportion of

Children are highly valued in Latino culture, and the family is understood to be the basic source of emotional support for children.

Hispanic children live in poverty than either Asian (13%) or white children (10%). Still higher child poverty rates are found among African Americans (Fass and Cauthen 2008).

Familial characteristics tell a similar story. Generally, Hispanics are less likely than either Asians or non-Hispanic whites to be married. They are more likely to be divorced than Asian Americans but less likely than either African Americans or non-Hispanic whites to be divorced. Nearly a third (31%) of Hispanic families with children under 18 are headed by single parents; this is a greater percentage than found among whites (23%) or Asians (12%) but a smaller percentage than found among African Americans (58%).

It is important to note that, in addition to these sorts of differences between Hispanics and non-Hispanics, there is considerable diversity *among* Hispanics in terms of ethnic heritage (such as Mexican, Cuban, Puerto Rican, or Central or South Americans), socioeconomic status, and family characteristics. For example, Hispanics of Cuban descent have the highest socioeconomic status, as indicated by incomes, poverty rates, home ownership, unemployment rates, and educational attainment. Puerto Ricans, Mexicans, and Central Americans tend to have more similar economic characteristics, with Puerto Rican families being slightly more likely to be poor. These differences, although real, are not as distinctive as the ways in which Cubans differ from the other Hispanic groups. Cuban Americans are much more likely to have high school or college degrees and own their own homes and less likely to be unemployed or fall below the poverty line than are other Hispanic groups.

Among Hispanics, there are interesting connections between the aforementioned familial and economic characteristics. The most affluent Hispanic group, Cubans, are among the most likely to have their families be married couple families and the least likely to have their families be single-parent families. Research indicates that the percentage of children born to either teenage or unmarried mothers varies similarly. The lowest percentage of births to either unmarried or teenage mothers is among Cubans and the highest among Puerto Ricans. This diversity is partly but not entirely because of economics. It is further accentuated by the varying proportions of U.S.-born and foreign-born Latinos in each group. Finally, keep in mind that characterizations of Mexican, Puerto Rican, Cuban, or any other Latino family types must avoid overgeneralization. None of these groups have a singular family system. More specifically, in Mexico, Cuba, or Puerto Rico, there is much diversity, some of which results from socioeconomics, some from rural versus urban living, some from religion, and so on (Aponte et al. 1999).

Traditional Mexican and Puerto Rican families can be characterized by a few distinctive cultural traits: high value placed on children and childbearing, devotion to family (i.e., *familism*) and male dominance (i.e., *machismo*). Children are especially important. Birth and fertility rates are still higher among Hispanics than among the general U.S. population, non-Hispanic whites, African Americans, or Asian Americans. Among Hispanics, birth and fertility rates range; they are highest among Mexican Americans and lowest among Cuban Americans (*National Vital Statistics Reports*, vol. 57 no. 7, 2009). Furthermore, Hispanics are significantly more likely than whites or African Americans to identify the purpose of marriage as having children and much less likely to see marriage as for mutual happiness and fulfillment (Pew Research Center 2007).

La familia is based on the nuclear family, but it also includes the extended family of grandparents, aunts, uncles, and cousins. All tend to live close by, often on the same block or in the same neighborhood. There is close kin cooperation and mutual assistance, especially in times of need, when the family bands together. Family unity and interdependence, sometimes extended to include fictive kin (e.g., Cuban *compadres* and *comadres*—godparents), reflect the importance of extended kin ties.

One indicator of the importance Hispanics place on family ties can be seen in answers to a national survey conducted by the Pew Hispanic Center. When asked about the importance of family over friends, 90% of the 2,000 Hispanics surveyed "strongly agreed" (78%) or "agreed somewhat" (12%) that relatives are more important than friends. Additionally, 82% agreed strongly (68%) or somewhat (14%) that it is better for children to live in their parents home until they get married. Yet even the greater integration of Hispanic extended families (living nearby, staying in touch, and providing assistance) cannot be entirely disconnected from social class, and—among Mexican Americans, for example—may account for their greater tendency to live with or close to and provide household and practical help to extended kin (Sarkisian et al. 2007).

Male dominance, as suggested, although often exaggerated in the misuse or misunderstanding of *machismo*, is part of traditional Latino family systems but has declined, especially among dual-earner couples. Migration and mobility disrupt traditional Latino family forms and lead to change. This change can be seen as part of a wider process of "convergence," in which distinctive ethnic traits diminish over time (Aponte et al. 1999). Although there has been a tradition of male dominance, current day-to-day living patterns suggest that noteworthy change has occurred. Women have gained power and influence in the family as they have increased their participation in paid employment. When wives are coproviders, Hispanic men spend more time on household tasks (Aponte et al. 1999; McLoyd et al. 2000).

Asian American Families

Asian Americans are the second-fastest-growing minority in the United States after Hispanics. In 2007, the more than 15 million Asian Americans and Pacific Islanders represented 5% of the U.S. population. Of this population, 13.4 million people identified themselves as "Asian alone," while an additional 1.8 million reported themselves as "Asian in combination" with some other racial group (Populations Division, U.S. Census Bureau, 2008).

Similar to Hispanics, Asian Americans are richly diverse comprising Chinese, Filipino, Asian Indians, Japanese, Vietnamese, Cambodian, Thai, Hmong (an Asian ethnic minority who live mostly in China and Southeast Asian countries such as Vietnam, Laos, and Thailand), and other groups. The six largest Asian American groups—Chinese Americans, Filipino Americans, Asian Indians, Korean Americans, Vietnamese Americans, and Japanese Americans—each number a million people or more. Cumulatively, three groups make up 60% of the Asian American population: Chinese at 23%, Filipino Americans at 19%, and Asian Indians at 18% (U.S. Census Bureau, *Facts for Features: Asian Pacific-American Heritage Month*, 2009).

General comparisons show that in many key ways, Asian Americans are unlike all other racial or ethnic minorities. In fact, on a variety of social, educational, and economic indicators, Asian Americans stand out from the general population and the population of non-Hispanic whites for the levels of success they enjoy.

As of 2007, Asians were very slightly more likely to have graduated from high school than the general population (among those 25 and older, 86% of Asians and 85% of the general population were high school graduates) but were much more likely to obtain higher levels of education. Half of all Asian Americans 25 and older had bachelor's degrees or higher in 2007. For the general population 28% had similar levels of educational attainment. Additionally, 20% of Asian Americans had graduate or professional degrees compared to 10% of the general population (Crissey 2009).

When examining income and comparing between groups, one can use household or family income. Remember, in the calculating of population data such as income, households can consist of individuals living alone, groups of unrelated people, or families. Families are defined as groups of people who are related by blood, marriage, or adoption and share a household. Thus, family incomes are higher. With median household incomes of nearly $67,000 and median family incomes of over $77,000, Asian Americans have higher incomes than either the general population or non-Hispanic whites. They are also less likely than the societal population to be unemployed or fall beneath the poverty level. Given these economic characteristics, it is perhaps unsurprising that Asians are also more likely to be married, less likely to be divorced, and less likely to bear children outside of marriage, as these familial characteristics often vary by income and education.

As with Hispanics, there is noteworthy educational, economic, and familial variation between different Asian American groups. Where Asian Indians and Chinese, Filipino, Korean, and Japanese Americans tend to have higher incomes and greater levels of education than the overall population, certain other groups, such as Hmong, Laotian, and Cambodian Americans, have much lower levels of income and education and higher rates of poverty and unemployment.

Familial characteristics also differ between Asian populations. In marital status, for example, although Asians were overall more likely than the general population to be married and much less likely to bear children outside of marriage, this was not uniformly true. With more than a third of all U.S. births in 2007 occurring outside of marriage, among Asians as a whole the level was less than a third (10%) of the national level. However, among Cambodians, the percentage was 45%. Among most Asian groups, more than half the adults were in intact marriages. In 2007, less than half of Cambodians and Hmong were married.

Clearly, much diversity can be observed within Asian American families based on where they're from, time of arrival in the United States, and reasons for coming to this country (e.g., political vs. economic). More recent immigrants retain more culturally distinct characteristics, such as family structure and values, than do older groups, such as Chinese Americans and Japanese Americans. Asian American families tend to be slightly larger than the average U.S. family (averaging 3.5 compared to 3.2 members), although there is wide variation between older and more recent immigrants. Among the more assimilated Japanese, average family size is just under three members. Among more recent Asian immigrants (e.g., Cambodians, Laotians, Vietnamese, and Hmong), families average closer to or over four members. The greater family size reflects the presence of extended kin.

Values that continue to be important to Asian Americans in general include a strong sense of importance of family over the individual, self-control to achieve societal goals, and appreciation of cultural heritage. Chinese Americans tend to exercise strong parental control while encouraging their children to develop a sense of independence and strong motivation for achievement (Ishii-Kuntz 1997; Lin and Fu 1990).

Migration and assimilation alter many traditional Asian family patterns. For example, among Japanese families, there are considerable differences among the *Issei* (immigrant generation), the *Nisei* (first-generation American born), and the *Sansei* and subsequent generations on such family characteristics as the relative importance of marriage over extended kin ties, the role of love in the choice of a spouse, and the relationship between the genders (Kitano and Kitano 1998). Similarly, we can draw distinctions between traditional Vietnamese families and American-born Vietnamese. Attitudes toward marriage and family, changes in familial gender roles, increased prevalence of divorce, and single-parent households all separate the generations. We can also see marked change between parents' and children's attitudes about individualism and self-fulfillment versus family obligation and self-sacrifice (Tran 1998).

The most dramatic change affecting Chinese Americans has been their sheer increase in numbers over the past 40 years. The Chinese American population increased from 431,000 in 1970 to over 3 million in 2007. More recent immigrants tend to be from Taiwan or Hong Kong rather than mainland China (Glenn and Yap 1994). Because of the large numbers of new immigrants, it is important to distinguish between American-born and foreign-born Chinese Americans; little research is available concerning the latter. Contemporary American-born Chinese families continue to emphasize familism, although filial piety and strict obedience to parental authority have become less strong. Chinese Americans tend to be better educated, have higher incomes, and have lower rates of unemployment than the general population. Their sexual values and attitudes toward gender roles tend to be more conservative. Chinese American women are expected to be employed and to contribute to the household income. More than 1.2 million speak Chinese at home.

Native American Families

More than 4 million Americans identify themselves as being of native descent, that is, as American Indian or Alaska Native. This includes 2.4 million Americans who identify themselves as American Indian or Alaska Native alone and an additional 1.6 million who identify themselves as American Indian/Alaska Native as well as one or more other races. Cumulatively, this population represents 1.5% of the population of the United States. It is also a population that has increased more than the overall U.S. population (2009 Statistical Abstract of the United States, Table 6).

Those who continue to be deeply involved with their own traditional culture give themselves a tribal identity, such as Dine (Navajo), Lakota, or Cherokee, rather than a more general identity as Indians or Native Americans (Kawamoto and Cheshire 1997). The largest tribal groups include the Cherokee, Navajo, Latin American Indian, Choctaw, Sioux, and Chippewa. Those who are more acculturated, such as urban dwellers, tend to give themselves an ethnic identity as Native Americans or Indians. Most Americans of native descent consider themselves members of a tribal group rather than an ethnic group.

Based on data from the 2007 Population Profile compiled by the U.S. Census Bureau, we can make the following points regarding the American Indian and Alaska Native populations (U.S. Census Bureau 2007):

- Approximately 68% of American Indian and Alaska Native households were family households, and 35% were family households with children under 18 years of age.
- The American Indian and Alaska Native populations were less likely to complete high school and college and more likely to drop out of school before completing high school than were the general population. Where 85% of the general population, age 25 and older, had completed high school, 76% of the American Indian and Alaska Native populations had. Of the American Indian and Alaska Native populations, 12.7% had completed college, less than half the percentage of the general population (27.5%) (U.S. Census Bureau, 2007 American Community Survey, SO201).
- Although similar percentages of American Indian and Alaska Native populations were in the labor force as the general population, with similar levels of female employment and employment of mothers with children under age six, the American Indian and Alaska Native populations were twice as likely

to be unemployed and twice as likely to be below poverty and had substantially lower median household and family incomes (U.S. Census Bureau, 2007 American Community Survey, SO201).

Although there is considerable variation among different tribal groups and hence no single type of American Indian or Alaska Native family, three aspects of Native American families are important. First, extended families are significant. These extended families may be different from what the larger society regards as an extended family (Wall 1993). They often revolve around complex kinship networks based on clan membership rather than birth, marriage, or adoption. Concepts of kin relationships may also differ. A child's "grandmother" may be an aunt or great-aunt in a European-based conceptualization of kin (Yellowbird and Snipp 1994).

Second, increasingly large numbers of Native Americans are marrying non–Native Americans. Among married Native Americans, more than half have non-Indian spouses. With such high rates of intermarriage, a key question is whether Native Americans can sustain their ethnic identity. Michael Yellowbird and Matthew Snipp (1994, 236) wonder if "Indians, through their spousal choices, may accomplish what disease, Western civilization, and decades of federal Indian policy failed to achieve" (236).

Third, family characteristics are affected by the economic status of American Indians and Alaska Natives. Given the higher levels of unemployment and poverty and lower overall earnings and educational attainment, once again social class may be confounding our attempt to look at patterns of family living.

Families of Middle Eastern Background

People of Middle Eastern ethnic backgrounds living in the United States are among the fastest-growing ethnic minorities in the country. Estimates of the population vary, depending on such issues as what countries are included and whether we count only naturalized citizens or include all immigrants—both legal and illegal and temporary (e.g., students and guest workers) and permanent (Brittingham and de la Cruz 2005; Camarota 2002). Furthermore, the census provides more detailed analysis of people of Arab ancestries than people whose ancestry is Middle Eastern. Thus, estimates from the census tend to undercount the overall population of Middle Eastern background.

As defined by Steven Camarota of the Center for Immigration Studies, "Middle Eastern" includes people whose backgrounds can be traced to one of the following: Pakistan, Bangladesh, Afghanistan, Turkey, the Levant, the Arabian Peninsula, and Arab North Africa. In terms of specific countries, the designation "Middle Eastern" encompasses Afghanistan, Bangladesh, Pakistan, Iran, Iraq, Israel, Jordan, Kuwait, Lebanon, Syria, Turkey, Oman, Qatar, Bahrain, Saudi Arabia, the United Arab Emirates, Yemen, Algeria, Egypt, Libya, Morocco, Sudan, Tunisia, Western Sahara, and Mauritania (Camarota 2002).

The Middle Eastern immigrant population is relatively recent and very diverse. The population includes non-Arab countries such as Israel, Iran, Turkey, and Pakistan, representing half of the top eight Middle Eastern countries of origin in 2000. Further complicating counts, among those Middle Eastern immigrants from Arab countries, we find many non-Arabs. Similarly, many immigrants from non-Arab countries, such as Israel, are Arabs.

The U.S. Census Bureau's American Community Survey, counting the more narrowly defined Arab population, estimates that in 2007, 1.5 million people claimed some Arab ancestry, either alone or in combination. Meanwhile, the Center for Immigration Studies states that among Middle Eastern immigrants in 2000, 40% were of Arab background. The center further estimates that within a decade (i.e., by 2010), the number is likely to be 2.5 million or more. Putting aside the question of counts, the U.S. Census 2007 American Community Survey provides the following profile:

- In comparison to the general population, the Arab population is disproportionately male. Males made up approximately 54% of the 2007 Arab American population, compared to 49% of the total U.S. population. Furthermore, 31% of the Arab population consisted of men ages 20 to 49.
- The Arab American population is more likely than the total population to be married and less likely to be widowed, separated, or divorced. Where 50% of the total U.S. adult population is married, 55% of the Arab population is married. As was true of Asians and Latinos, much variation exists among Arab ethnicities. Moroccans are the least likely to be married (53.4%) and Jordanians the most likely (67%). Nearly one out of five adults in the U.S. population is separated, widowed, or divorced; among Arab Americans, 13% fall into those categories (Brittingham and de la Cruz 2005).
- Compared to the total population, a greater proportion of Arab households consisted of married couples, with or without children, in 2007. Married

couples made up 56% of Arab households, compared to 50% of all U.S. households. Among Palestinians and Jordanians, the percentage of married couples reached 70%. Meanwhile, where more than 1 in 10 (12%) U.S. households were headed by a woman with no husband present, only about one in five (6%) of Arab American households were headed by a woman (Brittingham and de la Cruz 2005).

- Arab Americans tend to be more highly educated and employed than the total population. Forty-five percent of Arab Americans 25 and older have at least college educations, compared to 28% overall. Unemployment rates were nearly identical for Arabs (6.9) and the overall population (6.3). However, Arab women were less likely to be employed (48%) than were women overall (55%).
- As a group, Arab Americans enjoyed higher median individual, household, and family incomes but also higher poverty rates than the wider population.
- As was true among Hispanics and Asians, there is much social and economic diversity among Arab populations. Generally, Egyptian, Lebanese and Syrian Americans have higher median incomes and levels of education than do Palestinians and Morrocans.
- Although the Middle East is approximately 98% Muslim, immigrants to the United States from the region historically were not. In the past, most were Christian. This changed in the 1990s, and estimates are that the vast majority of the Middle Eastern immigrant population is Muslim (Camarota 2002). The fact of their Muslim faith may influence certain family patterns, although, as happens to other ethnic groups, assimilation operates against strict adherence to even religiously reinforced customs. Such is the case, for example, with both dating and mate selection. Sharply sex-segregated customs surrounding dating and a preference toward arranged marriage are characteristic of Muslim family life, yet both undergo considerable challenge from sons and especially daughters who are exposed to and may come to value Western notions of love, marriage, and family life (Zaidi and Shuraydi 2002). As Arshia Zaidi and Muhammad Shuraydi report from their examination of Pakistanis, "Families, depending on their educational, religious, economic, and social backgrounds, are coping with these changes by modifying the traditional authoritarian structure of the family system and their attitudes" (497).
- Educational, employment, and income characteristics of certain non-Arab Middle Eastern Americans,

such as Israeli or Iranian Americans, show them to be more highly educated, more affluent, less likely to be unemployed, and less likely to be poor than the general population. They are also more likely to be married and less likely to be divorced than the general population. Among Iranians, the percentage of births to unmarried women was less than a third of the percentage in the wider U.S. population (data for Israeli Americans were not available).
- Other groups, such as Afghanis and Pakistanis, have lower incomes and higher poverty rates than the wider population. Still, they are more likely to be married and less likely to be divorced than the wider population.

European Ethnic Families

The sense of ethnicity among Americans of European descent varies considerably. Much of this variation is generational, as among immigrants and their children, ethnic identification may be much greater than it is among subsequent generations. In fact, across and despite ethnic differences, many families experience similar pressures toward and processes of assimilation and acculturation. Historically, members of European ethnic groups sought to assimilate—to adopt the attitudes, beliefs, and values of the dominant culture, especially since assimilation seemed necessary in order to achieve upward mobility. Most white ethnic groups have assimilated to a considerable degree—they have learned English, moved from their ethnic neighborhoods, and married outside their group, but many continue to be bound emotionally to their ethnic roots. These roots are psychologically important, giving them a sense of community and a shared history. This common culture is manifested in shared rituals, feast days, and saint's days, such as St. Patrick's Day.

Except for some West Coast enclaves, such as Little Italy in San Francisco, white ethnicity is strongest in the East and Midwest. The Irish neighborhoods of Boston, the Polish areas of Chicago, and the Jewish sections of Brooklyn, for example, have strong ethnic identities. Common languages and dialects are spoken in the homes, stores, and parks. Traditional holidays are celebrated; the foods are prepared from recipes passed down through generations. Elders speak of the old country and their villages—even if it was their parents or grandparents who immigrated.

As is true of some non-European ethnic groups, as children grow up and move from their neighborhoods, their ethnic identity often becomes weaker in terms of language and marriage to others within their group—but they may retain some elements of ethnic pride. Their ethnicity is what Herbert Gans (1979) calls *symbolic ethnicity*—an ethnic identity that's used only when the individual chooses. Symbolic ethnicity has little effect on day-to-day life. It is not linked to neighborhoods, accents, the use of a foreign language, or working life. Others cannot easily identify the person's ethnicity; he or she "looks" American. Nevertheless, for many Americans, ethnicity has emotional significance. A person is Irish, Jewish, Italian, or German, for example—not only an American.

European ethnic groups differ from one another in many ways. However, a major study of contemporary American ethnic groups (Lieberson and Waters 1988) found that European ethnic groups are more similar to one another than they are to African Americans, Latinos, Asian Americans, and Native Americans. The researchers concluded that a European–non-European distinction remains a central division in our society. There are several reasons for this. First, most European ethnic groups no longer have **minority status**—that is, unequal access to economic and political power. Some scholars suggest that what separates ethnic groups into distinctive lifestyles is their social placement. As groups become more similar in their access to opportunities, their family lifestyles may "converge" toward a common pattern, one that includes smaller families, increased divorce, less interdependent ties with extended families, and less male dominance (Aponte et al. 1999). Second, because most European ethnic groups are not physically distinguishable from other white Americans, they are not discriminated against racially.

This chapter has covered much ground. As we have now seen, in a host of ways, American families are diverse. They vary across time and, within any given period, between racial, ethnic, and socioeconomic groups. Family diversity is reflected throughout subsequent, more specialized chapters as relevant variations by race, class, or ethnicity are discussed. Thus, our goal of understanding American families will be made more complete and representative.

Acknowledging the diversity that exists across families has personal consequences as well. It ought to make us a bit more cautious in generalizing from our particular set of family experiences to what others "must also experience." In addition, in noting how historical, economic, and cultural factors shape our families, we link our personal experiences to broader societal forces. In that way, we are better able to apply "sociological imaginations" to family experiences, identifying how our private and personal family worlds are largely products of when, where, and how we live (Mills 1959). Simply put, if we come of age during a period of great economic upheaval, we may put off marrying, bearing children, or divorcing because of the opportunities and constraints we face. Similarly, the kinds of family experiences we are able to have are limited or enhanced by the economic resources at our disposal, regardless of what we might otherwise choose to do.

Despite the extent to which the factors discussed in this and the next chapter may limit your opportunities or narrow your range of choices, remember that you do and will make choices about what kind of family you wish to create. You decide whether to marry, whether to bear children, how to rear your children, whether to stay married, and so on. A major goal of this book is to equip you with a foundation of accurate information about family issues from which you can make sound choices more effectively.

Summary

- In the early years of colonization, there were 2 million Native Americans in what is now called the United States. Many families were *patrilineal*; rights and property flowed from the father. Other tribal groups were *matrilineal*. Most families were small.

- In colonial America, marriages were arranged. Marriage was an economic institution, and the marriage relationship was *patriarchal*.

- African American families began in the United States in the early seventeenth century. They continued the African tradition that emphasized kin relations. Most

  slaves lived in two-parent families that valued marital stability.

- In the nineteenth century, industrialization revolutionized the family's structure and transformed gender roles, childhood was sentimentalized, and adolescence was invented. Marriage came to be based more on emotional bonds.

- Enslaved families were broken up by slaveholders, and marriage between slaves was not legally recognized. African American families formed solid bonds nevertheless.

- Beginning in the twentieth century, *companionate marriage* became the marital ideal. In this ideal, household decision making and tasks were to be shared, marriages were expected to be romantic; both husbands and wives were expected to be sexually active, and children were to be treated more democratically.

- During the Great Depression and the two world wars, women and men were faced with challenges to their traditional gender roles and identities. The Depression also initiated an enlarged role for the government in protecting poor families.

- The decade of the 1950s was an exception to more general trends of rising divorce and nontraditional gender roles. Prosperity was unusually high; suburbanization led to increased isolation, and middle-class families were able to live on the wages of male breadwinners.

- The terms *ethnic group*, *racial group*, and *minority group* are conceptually distinct. An ethnic group is a group of people distinct from other groups because of cultural characteristics. A *racial group* is a group of people, such as whites, blacks, or Asians, classified according to phenotype as well as anatomical and physical characteristics. A *minority group* is a group whose status (position in the social hierarchy) places its members at an economic, social, and political disadvantage.

- Family life differs between African Americans, Hispanics, Asians, and Native Americans. *Socioeconomic status* is an important element in understanding diversity between and within ethnic and racial groups.

- In recent years, increasing numbers of people from Middle Eastern countries have come to the United States. Overall, people of Middle Eastern background are economically better off than the general population, more highly educated, and more likely to live in married-couple-headed households, though there is much social, economic, and familial diversity within the Middle Eastern population.

- For many Americans of European descent, ethnicity is symbolic and has little effect on day-to-day life. Most members of European ethnic groups are physically indistinguishable from other white Americans and no longer have minority status.

Key Terms

adolescence 64

assisted reproductive technologies 75

blended families 74

bundling 64

cohabitation 73

companionate marriage 69

division of labor 65

ethnic group 89

extended households 93

familism 78

feminization of poverty 82

fictive kin ties 85

life chances 79

matrilineal 62

middle class 80

minority group 89

minority status 100

patriarchy 63

patrilineal 62

phenotype 89

race 88

racial group 88

social class 78

social mobility 85

socioeconomic status 79

structural mobility 87

two-person career 83

upper-middle class 80

working class 80

RESOURCES ON THE WEB

Book Companion Website

www.cengage.com/sociology/strong

Prepare for quizzes and exams with online resources—including tutorial quizzes, a glossary, interactive flash cards, crossword puzzles, self-assessments, virtual explorations, and more.

4

© Juice Images/Photolibrary

Gender and Family

What Do YOU Think? Are the following statements TRUE or FALSE?
You may be surprised by the answers (see answer key on the following page).

T	F	
T	F	**1** Gender differences reflect the instinctive nature of males and females.
T	F	**2** Gendered roles are influenced by ethnicity.
T	F	**3** The only universal feature of gender is that all societies sort people into only two categories.
T	F	**4** Parents are not always aware that they treat their sons and daughters differently.
T	F	**5** During adolescence, the influence of peers on gender greatly diminishes.
T	F	**6** Both boys and girls suffer from gender-related problems in school.
T	F	**7** For African Americans, the traditional female gender role includes both employment and motherhood.
T	F	**8** Research shows it is possible for women and men to establish work or family roles that are counter to their socialization.
T	F	**9** Compared with traditional roles, contemporary male gender roles place more emphasis on the expectation that men will be actively involved with their children.
T	F	**10** Men's and women's movements have consistently stressed the importance of family.

As each of us moves through life, including into or out of family roles and relationships, the road we take—how it twists or turns—and the places it takes us are shaped, in part, by gender. Think about your own life. Did you ever stop to consider how similar or different your life might be if you hadn't been born into the gender you presently occupy? Would you be the kind of person you are? Have the same roles and relationships within your families or have the same friends? Would your goals be the same as they are now? Would you envision and later experience the same familial future? Asking ourselves such questions reminds us that much of what we do, who we are, what we expect, and what happens to us is influenced by gender. In this chapter we examine how deeply interconnected family experience is with gender. It is no exaggeration to say that we cannot fully understand either without taking the other into account.

The traditional view of gender depicts male and female, masculinity and femininity, and men and women as polar opposites. Familiar gender stereotypes fit this same pattern of polar differences. We believe that if men are aggressive, women must be passive; if men are **instrumental traits** (task oriented), women must be **expressive traits** (emotion oriented); if men are rational, women must be irrational; and if men want sex, women want love (Duncombe and Marsden 1993; Kimmel 2008; Lips 1997).

As shown in this chapter, these *perceptions* of male–female differences are much greater than actual differences (Hare-Mustin and Marecek 1990b). Males and females may be accustomed to thinking that they are as different as night and day, but, in fact, the genders are not as different as is commonly assumed. In fact, aside from physical differences, there are greater differences *among* men or among women than there are *between* women and men. At the same time, our family experience *is* highly "gendered" (i.e., differently experienced for women and men). For example, women and men tend to have different ideas and experiences of love. In marriage, roles and responsibilities differ greatly between the genders. As parents, caregiving activities tend to be divided by gender. The difficulties juggling family responsibilities and paid employment are not typically the same for both genders. Women and men seek divorces for different reasons, play different roles in the divorce process, and suffer different outcomes when their marriages end. The causes, context, and consequences of intimate violence between spouses or partners differ between women and men. In short, as we examine both in this chapter and throughout the remaining ones, gender affects nearly all aspects of families and intimate relationships.

The present chapter examines how and why so many of our experiences differ. We consider some gender and socialization theories to illustrate how much our families influence how we learn to act masculine and feminine. We then explore how gender influences some areas of family experience. Finally, we discuss efforts to change gender roles on both an individual and a collective level.

Studying Gender

Before we commence, we need to note some important points about gender. *First, your gender is not the same as your sex*, even though in everyday usage the terms are often treated as though they were interchangeable. As social scientists use these terms, **sex** refers to the biological aspect of being male or female. As such, it includes chromosomal, hormonal, and anatomical characteristics that differentiate most females from most males. One might say that **gender** refers to *everything else*—such as how one sees oneself, the expectations one forms of others based on their being females or males, the roles one plays in households and families, and the opportunities one has in the wider world of education, work, and politics—about being a female or male. This leads to our *second* point, that is, *gender is multidimensional*. In other words, there are psychological, social, cultural, political, and economic dimensions to gender. Looking over that list points to a *third*, critical point about gender: *at one and the same time, gender is both highly personal and eminently political*. For each of us, gender is an essential element of our personal identity and an important social status in our relationships with others. In the wider society, gender is a basis for social inequality. Just as one's gender helps account for much of one's tastes, style of interpersonal communication, how one presents oneself to others, and how one sees oneself,

---

Answer Key to What Do YOU Think?

1 False, see p. 110; **2** True, see p. 117; **3** False, see p. 105; **4** True, see p. 115; **5** False, see p. 119; **6** True, see p. 118; **7** True, see p. 125; **8** True, see p. 121; **9** True, see p. 122; **10** False, see p. 133.

it also limits or expands opportunities in life and gives shape to the distribution of such societal resources as income, power, respect, and safety throughout society. In other words, gender is important on *both* a micro—that is, personal and interpersonal—and macro—societal—level. *Fourth*, on both individual and societal levels, *gender is dynamic and highly variable*. On an individual level, this means that our understanding of what it means to be a male or female may change as we age and varies some from person to person and/or from one group to another. Thus, what a child or teenager thinks about gender will likely differ from and become broader in adulthood. Similarly, what parents believe a female or male ought to do or not do or be or not be will likely differ from what friends or classmates believe.

Critical Thinking

Have your own ideas about gender changed over time? To what, if any, extent can you identify differences between your own assumptions about gender and those of your parents?

On a more macro or societal level, this means that how gender is understood and the expected characteristics and attributes for males and females vary from one society to another, as do the opportunities afforded to and the extent of inequality between women and men in the distribution of desired roles, responsibilities, and resources. Within a society, they may also vary over time or from one (class, racial, or ethnic) group to another. Even the most basic assumptions we have about gender—that there are two and only two genders and that your gender is unchangeable—are not shared universally.

In many societies, one can identify more than two gender categories. Among a number of Native American cultures—including the Navajo, Sioux, Zuni, Mohave, and Lakota as well as some Asian societies, such as India—we find instances of more than two genders (Peoples and Bailey 2006). For example, the Hjira of India are socially recognized as *neither* male nor female (Nanda 1990). Even in societies that identify only two gender categories, we find numerous examples where people can "cross" from one gender to the other (Peoples and Bailey 2006).

There are other, less extreme cultural variations in conceptualizations of gender. Among the Arapesh of New Guinea, both males and females possess what we consider feminine traits. Men and women alike tend to be passive, cooperative, peaceful, and nurturing. The father is said to "bear a child," as well as the mother; only the father's continual care can make a child grow healthily both in the womb and in childhood. Only 80 miles away, the Mundugumor live in remarkable contrast to the peaceful Arapesh. Anthropologist Margaret Mead (1975) offered this description:

Both [Mundugumor] men and women are expected to be violent, competitive, aggressively sexed, jealous and ready to see and avenge insult, delighting in display, in action, in fighting.

Given the great differences between the two cultures regarding gender, Mead concluded,

Many, if not all, of the personality traits which we have called masculine or feminine are as lightly linked to sex as are the clothing, the manners, and the form of headdress that a society at a given period assigned to either sex.

Considerable cultural and historical variation in the characteristics and attributes expected of males and females—in other words, masculinity and femininity—can also be found. In fact, although a society may embrace one version of masculinity or femininity as more appropriate, in every society there are actually *multiple masculinities* and *femininities*, out of which emerges a version that is more widely accepted and expected (Kimmel 2006). These **hegemonic models of gender** are then held up as the standards for all women and men to emulate. Such standards are, themselves, dynamic and culturally variable, changing over time, and—within a given time and place—being challenged for cultural dominance by those who advocate other versions of masculinity or femininity (Connell 1995; Gilmore 1990; Kimmel 2006).

Gender and Inequality

Most societies, both past and present, have been characterized by inequality between females and males, or **gender stratification**. The vast majority can be considered **patriarchies**, societies in which males dominate political and economic institutions and exercise power in interpersonal relationships. Although there are societies that have been identified as more **egalitarian** (where women and men possess similar amounts of power and neither dominates the economic or political institutions), true **matriarchies**—societies wherein women rule over men and where men are denied the right to political office or are excluded from participating in the most significant religious rituals—have not been documented (Peoples and Bailey 2006).

There are many examples across a range of societies of practices that illustrate rather extreme male dominance and female subordination. For example, consider the following:

- In India, there is a disturbing and widespread practice of controlling the sex of one's offspring via selective abortion of female fetuses. Decried as a "national shame" by the prime minister, the practice has led to a decline in the sex ratio of the number of girls for every 1,000 boys from 962 in 1981 to 927 in 2001 (Gentleman 2008).
- Although they were banned in 2004, honor killings claim hundreds of Pakistani women each year. The practice came under considerable public scrutiny in 2008 after news of the deaths of five women who had been buried alive. Among them were three young women who had defied their families by planning to marry men of their own choosing. The three women were kidnapped by men from their village, beaten, shot, and, while still alive, "covered with earth and stones" (Masood 2008).
- In Yemen, 10-year-old Nojuud Mohammed Ali went to a courthouse seeking a divorce from her husband, Faez Ali Thamer, age 30. Forced to marry at age eight, Nojuud wanted freedom from her marriage, claiming that her husband beat and raped her. Pulled out of the second grade by her father and then married off to her husband, Nojuud's case became an international news story. Although she is only one of thousands of young girls forced into such circumstances—a 2006 study estimated that 52% of rural Yemeni girls are married by age 15— she drew considerable acclaim for her successful determination to escape her marriage.
- In June 2007, Badour Ahmad Shaker, a 12-year-old Egyptian girl, underwent a procedure known as female circumcision, or female genital mutilation. Undertaken largely to control female sexuality, the procedure "involves partial or total removal of the external female genitalia," provides no health benefits, and carries considerable risk for the girls and young women who undergo such operations. Yet the World Health Organization estimates that somewhere between 100 million and 140 million girls and young women worldwide have been "circumcised," including more than 90 million females age 10 and older in Africa, where, every

© Tom Carter/PhotoEdit

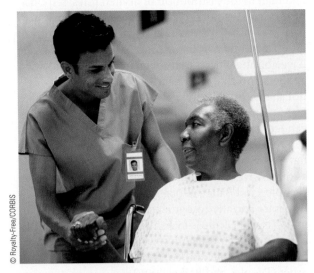

© Royalty-Free/CORBIS

Generally, the only limit on the jobs that women or men hold is social custom, not biology or individual ability. Even sex-segregated jobs such as nursing and firefighting can be performed by either gender.

year, approximately 3 million girls are at risk of undergoing these procedures (World Health Organization, www.who.int/mediacentre/factsheets/fs241/en).

In addition to examples such as these, one can examine political and economic institutions and find numerous other indicators of gender inequality. Doing so allows us to recognize elements of male dominance in the United States, where political office has been dominated by men, especially at the national level. As of November 2008, 17% of the U.S. Senate (17 senators) and the House of Representatives (74 out of 435 members of the House of Representatives) were female. Never has a woman been elected president or vice president of the United States.

Economically, although women continue to make gains, men in the United States have enjoyed long-standing advantage in the workplace. They tend to earn greater amounts of income, experience more upward mobility, exercise more authority and enjoy more autonomy in their jobs, and are given more prestige for the work they do (Padavic and Reskin 2002). Given the inequality in wages, women are more likely than men to be poor, and female-headed households are especially prone to poverty (see Figure 4.1). Such gender inequalities are not the same across all racial and ethnic groups, as is evident from data on the wage gap, where white females fare better than either African American or Hispanic males (see Table 4.1).

Within male-dominated societies, families tend to be male dominated as well. In many societies, men have the right to have more than one spouse, descent is traced through males, property is passed from fathers to sons, and authority is exercised by men. In the United States, the division of responsibilities may be the most visible familial advantage that men

Table 4.1 2008 Median Usual Weekly Earnings, Full-Time Wage and Salary Workers

	Male	Female	Female as % of Male
All races	$798	$638	80%
White, non-Hispanic	$825	$654	79%
African American	$620	$554	89%
Asian American	$966	$753	78%
Hispanic	$559	$501	90%

SOURCE: U.S. Department of Labor, Bureau of Labor Statistics, www.bls.gov/news.release/pdf/wkyeng.pdf.

hold. Women bear the vast bulk of responsibility for domestic work and child care. Even women who are employed full time come home to what sociologist Arlie Hochschild called a "**second shift**" of housework and child care at the end of their paid workday. Whether one uses absolute (i.e., numbers of hours per day or week) or relative measures (percent of total housework that is done by women and men) and whether one takes men's or women's estimates of how they divide housework, women do considerably more (Bianchi, Robinson, and Milkie 2006; Lee and Waite 2005). How much more varies some from study to study, but equal sharing of housework is unusual regardless of how one chooses to measure it. We return to the division of housework and child care later in this chapter when we look at contemporary gender roles in marriages and families. Especially for employed women with children, having more of the responsibility for household and children translates into having more to think about, more to worry about, and less free time. It is likely a contributing factor in accounting for why women feel more rushed than men do and why, over the past three decades, women's levels of happiness have declined relative to men's happiness (Stevenson and Wolfers 2007).

Using decision making as a measure of power or dominance among heterosexual cohabitants and married couples tells a different story. Survey data by the Pew Research Center indicates that women have more decision-making input

Figure 4.1 Percentage in Poverty, 2007

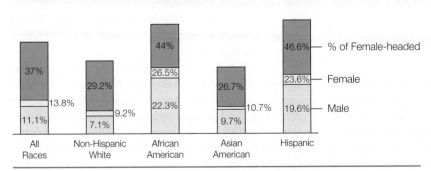

SOURCE: U.S. Census Bureau, Current Population Survey, 2008 Annual Supplement.

Figure 4.2 Who Makes the Decisions at Home?

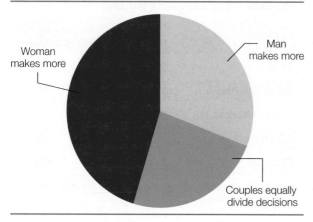

Woman makes more

Man makes more

Couples equally divide decisions

SOURCE: Morin and Cohn 2008.

than do men (Morin and Cohn 2008). Looking at decisions of four different kinds—choosing what to do on the weekend, buying major purchases for the home, deciding what to watch on television, and managing household finances—in 43% of couples, women make more decisions than men, while in only 26% did respondents indicate that men make more of the decisions (see Figure 4.2). Furthermore, in 43% of the couples, men have final say in *none* of the four areas.

Decision-making data of this kind are not ideal. They are based on what people say, without opportunity for behavioral confirmation. These data were obtained from only one partner reporting about his or her relationship. It is probable that the reports from their spouses or partners would differ some. Furthermore, in some ways, decision making can be another responsibility that one carries in the household rather than an indication of how much power one has. Nevertheless, data such as these prevent us from claiming that U.S. men dominate families via their role in decision making.

Gender Identities, Gendered Roles, and Gender Stereotypes

On a more personal level, each of us possesses a **gender identity**, or a sense of ourselves as of a certain gender. Gender identity may be the most basic element of one's identity; whatever else one is or becomes, most people identify themselves as a female or a male, usually—but not always—in accord with their biological sex characteristics. We say "most"

and "usually" because some individuals identify themselves as *both* male and female or as alternately male and female, living and acting sometimes as one and other times as the other, while still other people develop a more stable sense of themselves as males despite being biologically females or as females despite being biologically males. Typically, when individuals develop gender identities that are opposite their biological sex characteristics, they are considered **transsexuals** or, increasingly, **transgendered**. Some may opt for reconstructive surgery to bring their biology into line with the identity they have developed, seeking "to become—physically, socially and legally—the sex they have always been psychologically. If they succeed in doing so, they typically consider themselves simply as members of their new sex, rejecting any significance to how they arrived there" (Coombs 1997). Others reject such surgeries and simply live in the gender with which they identify. Such is the case with Thomas Beattie, hailed in the press as the "first man to give birth."

Transgendered is the broadest category and includes those who alter their social but not necessarily their physical characteristics, presenting themselves as of the opposite sex. Those who are transgendered include **transvestites**, or cross-dressers, who wear clothing of the opposite sex, as well as people who are **intersexed** and are "born with reproductive or sexual anatomy that doesn't seem to fit the typical definitions of female or male" (www.isna.org/faq/what_is_intersex).

Estimates of the prevalence of transgenderism are hard to come by. The American Psychological Association estimates that one in 10,000 men and one in 30,000 women are transsexuals. They also estimate that between 2% and 3% of males engage in cross-dressing behavior at least occasionally (American Psychological Association, Task Force on Gender Identity, Gender Variance, and Intersex Conditions, 2009). *Intersexuality*, making up a variety of different conditions, is estimated to occur in between one in 1,500 to one in 2,000 births (Intersex Society of America, "How Common Is Intersex?" www.isna.org/faq/frequency).

There are also some individuals who perceive themselves as *neither* male nor female (terms include "genderqueer," "androgyne," "third gendered," "bigendered," and "gender-benders"). Each of these terms refers to individuals who stand outside and apart from societal conventions regarding gender, neither identifying with nor acting within societal conventions for males or females. As one can see, gender identity is

Real Families: Third Genders and Pregnant Men

Tafi Toleafoa is a student at the University of Alaska, Anchorage. Of Samoan origin, Tafi, who lives in Anchorage with her parents, siblings, and other relatives, creates confusion for many of those she encounters. Puzzled by her long hair and "plump, glossy lips," along with her deep voice and "tall, substantial body," people don't know how to categorize her (O'Malley 2007). When pressed to tell whether she is a boy or a girl, her answer is neither of those. She is *fa'afafine*, a biological male who had been raised in Samoa as a female. Translated as "like a woman," *fa'afafine* is a label for a son who prefers the dress and work associated with women. She lives as a female. *Fa'afafine* and the less common *fa'atama* (or "like a man") occupy a "third gender" category, neither male nor female but thought to possess the best qualities of both males and females (O'Malley 2007).

Thomas Beattie, introduced in Chapter 1, is hailed as the first man to have become pregnant and, subsequently, to give birth. Retaining the female reproductive organs with which he had been born but identifying himself as a male and living as a married man, Thomas's story raises a different question than Tafi's. What makes one a man or a woman? If one applies strictly biological criteria in answering this question, Thomas (formerly Tracy) isn't *really* the first man to become pregnant and bear a child. If, however, one uses gender as the definitive answer, indeed Thomas was the first man to experience pregnancy and childbirth. He is not the last. Ruben Noe Coronado, born Estefania, is a female-to-male transsexual in Spain who interrupted his sex change to get pregnant via artificial insemination. At the time this is being written, he is pregnant with twins. On giving birth, Ruben plans to resume and complete the remaining stages of his sex change. Along with his partner, Esperanza—already a mother of two—he plans to raise his twins as their father. Meanwhile, Thomas Beattie is expecting his second child.

However much more common such cases become, if at all, they raise fundamental questions about the meaning of gender, the number of genders, and the permanence of gender.

Some individuals identify themselves as of both genders or neither gender.

far more varied than a simple dichotomy of male and female.

The differing expectations to which males and females are held are what we mean by the idea of **gendered roles**. We use this term instead of the more commonly used idea of "gender roles" because while it is clear that males and females are socialized toward different statuses and as women and men perform different roles in a variety of social institutions, groups, and situations, it is less useful to talk of "*a* male role" or "*a* female role." In fact, there are multiple roles that are assigned to and performed differently by females than by males or that are performed by females more often than or instead of males and vice versa. It makes more sense to say that such roles are *gendered* than to consider all of them part of a singular female or male role in society.

A **gender stereotype** is a rigidly held and oversimplified belief that all males and females, as a result of their sex, possess distinct psychological and behavioral traits. Stereotypes tend to be false not only for the group but also for any individual member of the group. Even if the generalization is statistically valid in describing a group average, such as males are taller

© Jeffrey Braverman/Getty Images/Stone

than females, we cannot necessarily predict whether Jason will be taller than Tanya. When we discuss gendered roles, it is important not to confuse stereotypes with reality.

Historically, most studies of gender in the United States focused on the Caucasian middle class. This made it difficult to know whether and how ideas and expectations about gender. may have differed among African Americans, Latinos, Asian Americans, and other ethnic groups. Students and researchers must be just as careful not to project onto other groups concepts or expectations characteristic of their own groups. Too often such projections can lead to distortions or moral judgments.

Believing in Gender Differences

Culturally, we tend to view the genders as *dichotomous and opposite*: females and males are seen as polar opposites, with males possessing exclusively instrumental traits and females possessing exclusively expressive ones. Psychologist Sandra Bem (1993) calls this assumption *gender polarization* and suggests that it is the fundamental assumption that we tend to make about gender. Our entire society is organized around such supposedly polar differences (Kimmel 2008).

Traditional views of masculinity and femininity as opposites have several implications. First, if a person differs from the male or female stereotype, he or she is seen as being more like the other gender. If a woman is sexually assertive, for example, she not only is less feminine but also is believed to be more masculine. Similarly, if a man is nurturing, he not only is less masculine but also is seen as more feminine. Second, because males and females are perceived as opposites, they cannot share the same traits or qualities. A "real man" possesses exclusively masculine traits and behaviors, and a "real woman" possesses exclusively feminine traits and behaviors. A man is assertive, and a woman is receptive; in reality, both men and women are often both assertive and receptive. Third, because males and females are viewed as opposites, they are believed to have little in common with each other. Men and women can't understand each other, nor can they expect to do so. Difficulties in their relationships are attributed to their "oppositeness."

The fundamental problem with the view of men and women as opposites is that it is erroneous. As men and women, we are significantly more alike than we are different, yet our culture encourages us to look for differences and, when we find them, to exaggerate their degree and significance. It has taught us to ignore the most important fact about males and females: that we are both human. As humans, we are significantly more alike biologically and psychologically than we are different (Hyde 2004). As men and women, we share similar respiratory, circulatory, neurological, skeletal, and muscular systems. (Even the penis and the clitoris evolve from the same undifferentiated embryonic structure.) Hormonally, both men and women produce androgens and estrogen (but in different amounts). Where men and women biologically differ most significantly is in terms of their reproductive functions: men impregnate, whereas women menstruate, gestate, and nurse. Beyond these reproductive differences, biological differences are not great. In terms of social behavior, studies suggest that men are more aggressive both physically and verbally than women; the gender difference, however, is not large. Most differences can be traced to culturally specific expectations, male–female status, and gender stereotyping. We are capable of displaying both masculine and feminine qualities as is suggested by the notion of **androgyny**.

Although we are more similar than different in our attributes and abilities, large and meaningful differences do exist in the *statuses* (or positions in various groups and organizations) we occupy and the privileges and responsibilities these carry. Although either gender may have the *ability* to nurture children, support families, clean, or cook, these tasks are assumed to be more appropriate for one gender than the other. Although women and men may possess the *ability* to do many kinds of jobs, as noted earlier, the labor force is gender segregated into jobs that are disproportionately male or female. Men's jobs typically carry more prestige, earn higher salaries, and offer more opportunity for advancement than do women's jobs. We often refer to these differences as "gaps." The "wage gap" refers to the difference between what men tend to earn and what women tend to earn (see Table 4.1). We can also speak of "prestige gaps" or "mobility gaps." Jobs that tend to be among the most highly respected jobs (typically, jobs such as physician, attorney, and engineer) tend to be held disproportionately by men. Jobs held largely by women (such as types of clerical work, elementary and preschool teaching, household service, and nursing) are often undervalued (and, not surprisingly, underpaid). Gaps such as these illustrate the persisting gender stratification in the United

States. In much the same way that we conceptualize the upper and middle classes as being "above" the working and lower classes, we can consider the genders to be stratified.

Unfortunately, believing that the genders are fundamentally different reinforces the idea that individuals should *not* have the attributes culturally identified or associated with the other gender. This inhibits many people from displaying the full range of human qualities, as females may suppress their instrumental traits (perceived as their "masculine side") and, to an even greater extent, males their expressive traits (perceived as their "feminine side").

When we initially meet a person, we unconsciously note whether the individual is male or female and respond accordingly, expecting the person to fit our expectations for his or her gender. This procedure, known as **gender attribution**, is the "methodical procedures through which (we) come to identify others as unambiguously male or female" (Speer 2005). But what happens if we cannot immediately classify a person as male or female? Many of us feel uncomfortable because we don't know how to act if we don't know the gender. This is true even if gender is irrelevant, as in a bank transaction, walking past someone on the street, or answering a query about the time. Our need to classify people as male or female is an indication of how much our ideas about gender influence everyday interaction.

Gender Theory

Beginning in the 1980s, gender theory emerged as an important model explaining inequality. It has proven to be an especially popular approach, especially with feminist scholars. It begins with two assumptions: (1) that male–female relationships are characterized by power issues and (2) that society is constructed in such a way that males dominate females. Gender theorists argue that on every level, male–female relationships—whether personal, familial, or societal—reflect and encourage male dominance, putting females at a disadvantage. Male dominance is neither natural nor inevitable, however. Instead, it is created by social institutions, such as religious groups, government, and the family (Acker 1993; Ferree 1991). The question is, How is male–female inequality created?

Social Construction of Gender
According to this theory, gender is a **social construct**, an idea or concept created by society through the use of social power. **Gender theory** asserts that society may

be best understood by how it is organized according to gender and that social relationships are based on the *socially perceived* differences between females and males that are used to justify unequal power relationships (Scott 1986; White 1993). Imagine, for example, a dual-earner couple whose infant daughter wakes up sick. Both parents work outside the home, but one will have to take the day off to take the child to the doctor. In the mother–father parenting relationship, which parent is most likely to stay home? In most cases, the mother does because women are socially perceived to be nurturing and it's the woman's "responsibility" as a mother.

Gender theory focuses on (1) how specific behaviors (such as nurturing or aggression) or roles (such as child rearer, truck driver, or administrative assistant) are defined as male or female; (2) how labor is divided into man's work and woman's work, both at home and in the workplace; and (3) how different institutions bestow advantages on men (such as male-only clergy in many religious denominations or women receiving less pay than men for the same work).

Central to the creation of gender inequality are the belief that men and women are fundamentally different and the fact that the differences between the genders—in personalities, abilities, skills, and traits—are unequally valued: reason and aggression (defined as male traits) are considered more valuable than sensitivity and compliance (defined as female traits). Making men and women appear to be opposite and of unequal value requires the suppression of natural similarities by the use of social power. The exercise of social power might take the form of greater societal value being placed on looks than on achievement for women, of sexual harassment of women in the workplace or university, of patronizing attitudes toward women, and so on.

"Doing Gender"
Some gender scholars emphasize the situational and performative nature of gender: how it is reproduced or constructed in everyday social encounters. They argue that more than what we *are*, gender is something that we *do* (Risman 1998; West and Zimmerman 1987). As Greer Fox and professor of family and child development Velma Murry (2000, 1164) explain it, "Men and women not only vary in their degree of masculinity or femininity but have to be constantly persuaded or reminded to be masculine and feminine. That is, men and women have to 'do' gender rather than 'be' a gender."

We "do gender" whenever we take into account the gendered expectations in social situations and act accordingly. We don't so much perform an internalized role as tailor our behaviors to convey our suitability as a woman or a man in the particular situation in which we find ourselves (West and Zimmerman 1987). To fail to conform to the expectations for someone of our gender in a given situation exposes us to potential criticism, ridicule, or rejection as an incompetent or immoral man or woman (Risman 1998). Much of our day-to-day behavior—such as the way we walk, talk, sit, and dress—is a means of communicating our gender to others. But in living up to or within those social expectations, we help create and sustain the idea of gender difference. According to Michael Kimmel (2000, 104), "Successfully being a man or a woman simply means convincing others that you are what you appear to be."

Although we see the social construction or "doing" of gender in all kinds of social settings, the family is a particularly gendered domain (Risman 1998). There are cultural expectations about how wage earning, housework, child care, and sexual intimacy should be allocated and performed between women and men. Thus, much of the experience that people have in their families is understandable as both an exercise in and a consequence of how they and others "do gender."

Gender as Social Structure

Another key idea shared by many gender theorists is the notion that gender is, itself, a *social structure* that constrains behavior by the opportunities it offers or denies us (Connell 1987; Lorber 1994; Risman 1987, 1998). The consequences of the different opportunities afforded women and men can be seen at the *individual* level in the development of gendered selves, at the *interactional* level in the cultural expectations and situational meanings that shape how we "do gender," and at the *institutional* level in such things as sex-segregated jobs, a wage gap, and other economic and institutional realities that differentiate women's and men's experiences (Risman 1998). Although we may more often focus on individuals making choices that reflect their internalization of gender expectations, situations and institutions also shape behavior.

Gender Socialization

To most family scientists, gender differences are products of differently socializing males and females. Even if they recognize the potential effect of some biological characteristic or physical process in contributing to gendered behavior, psychologists and sociologists emphasize the importance of socialization. In doing so, they may take different approaches to accounting for how we are socialized to be females or males. Two prominent theories used to explain how we learn what is expected of us are social learning theory and cognitive development theory.

Socialization through Social Learning Theory

Many theorists see gender like any other socially acquired role. They stress that we have to be socialized to act according to the expectations attached to our status as female or male. The emphasis on socialization has been considerable, although consensus on the process of socialization has not. In other words, there is considerable agreement that we undergo gender socialization, but there are different theories of how such socialization proceeds. **Social learning theory** is derived from behaviorist psychology and its emphasis on observable events and their consequences rather than internal feelings and drives. According to behaviorists, we learn attitudes and behaviors as a result of social interactions with others (hence the term *social learning*).

The cornerstone of social learning theory is the belief that consequences control behavior. Acts regularly followed by a reward are likely to occur again; acts regularly punished are less likely to recur. Girls are rewarded for playing with dolls ("What a nice mommy!"), but boys are not ("What a sissy!").

This behaviorist approach has been modified recently to include **cognition**—that is, mental processes (such as evaluation and reflection) that intervene between stimulus and response. The cognitive processes involved in social learning include our ability to use language, anticipate consequences, and make observations. These cognitive processes are important in learning gender roles. By using language, we can tell our daughter that we like it when she does well in school and that we don't like it when she hits someone. A person's ability to anticipate consequences affects behavior. A boy does not need to wear lace stockings in public to know that such dressing will lead to negative consequences. Finally, children observe what others do. A girl may learn that she "shouldn't" play video games by seeing that the players in video arcades are mostly boys.

We also learn gender roles by imitation, according to social learning theory. Learning through imitation is called **modeling**. Most of us are not even aware

Playing "dress up" is one way children model the characteristics and behaviors of adults. It is part of the process of learning what is and isn't appropriate for someone of their gender.

of the many subtle behaviors that make up gender roles—the ways in which men and women use different mannerisms and gestures, speak differently, and so on. We don't "teach" these behaviors by reinforcement. Children tend to model friendly, warm, and nurturing adults; they also tend to imitate adults who are powerful in their eyes—that is, adults who control access to food, toys, or privileges. Initially, the most powerful models that children have are their parents. Reflecting on your own family, you might examine the division of labor in your household. How is housework divided? How is unpaid household work valued in comparison with employment in the workplace?

As children grow older and their social world expands, so do the number of people who may act as their role models: siblings, friends, teachers, media figures, and so on. Children sift through the various demands and expectations associated with the different models to create their unique selves.

Cognitive Development Theory

In contrast to social learning theory, **cognitive development theory** focuses on the child's active interpretation of the messages he or she receives from the environment. Whereas social learning theory assumes that children and adults learn in fundamentally the same way, cognitive development theory stresses that we learn differently, depending on our age. Swiss psychologist Jean Piaget (1896–1980) showed that children's abilities to reason and understand change as they grow older.

Lawrence Kohlberg (1969) took Piaget's findings and applied them to how children assimilate gender-role information at different ages. At age two, children can correctly identify themselves and others as boys or girls, but they tend to base this identification on superficial features, such as hair and clothing. Girls have long hair and wear dresses; boys have short hair and never wear dresses. Some children even believe that they can change their sex by changing their clothes or hair length. They don't identify sex in terms of genitalia, as older children and adults do. No amount of reinforcement will alter their views because their ideas are limited by their developmental stage.

When children are six or seven years old and capable of grasping the idea that basic characteristics do not change, they begin to understand that gender is permanent. A woman can be a woman even if she has short hair and wears pants. Interestingly enough, although children can understand the permanence of sex, they tend to insist on rigid adherence to gender-role stereotypes. Even though boys can play with dolls, children of both sexes believe they shouldn't because "dolls are for girls." Researchers speculate that children exaggerate gender roles to make the roles "cognitively clear."

According to social learning theory, children learn appropriate gender-role behavior through reinforcement and modeling. But according to cognitive development theory, once children learn that gender is permanent, they independently strive to act like "proper" girls or boys. They do this on their own because of an internal need for congruence, the agreement between what they know and how they act. Also, children find that performing the appropriate gender-role activities is rewarding. Models and reinforcement help show them how well they are doing, but the primary motivation is internal.

Learning Gender Roles and Playing Gendered Roles

Although biological factors, such as hormones, clearly are involved in the development of male and female differences, the extent of biological influences is not well understood. Moreover, it is difficult to analyze the relationship between biology and behavior because learning—or gender socialization—begins at birth. In this section, we explore gender-role learning from infancy through adulthood, emphasizing the influence of our families in the construction of our ideas about gender. It should be stressed that gender socialization is *lifelong*; it begins at birth or even before we are born and continues throughout our lives. Along the way, numerous **agents of socialization** contribute to our development as males and females. First and, perhaps, foremost is the family.

Childhood and Adolescence

Parents are major influences in our gender development. As early as the first day after birth, parents tend to describe their daughters as soft, fine featured, and small and their sons as hard, large featured, big, and attentive. Fathers tend to stereotype their sons more extremely than mothers do, but fathers and mothers are reinforced by others in relating differently to

female and male infants (Kimmel 2008). Although it is impossible for strangers to know the gender of a diapered baby, once they learn the baby's gender, they respond "accordingly"; in other words, they describe and react to babies based on social and cultural ideas about gender.

In the United States, after the first few months of our lives, infant girls and boys are treated and talked to differently (Clearfield and Nelson 2006; Lippa 2005). In fact, as psychologists Melissa Clearfield and Naree Nelson report, research, including their own, repeatedly has revealed differences in the amount and nature of parental verbal interaction and play with male and female infants and toddlers. They point out that these differences are evident well before children are able to speak and even before they become independently mobile (Clearfield and Nelson 2006). Mothers of two-year-olds ask daughters more questions than they ask sons, including asking more interpretive questions to daughters (e.g., "You like that, don't you"), whereas they issue more directives to their sons (e.g., "Come here") (Clearfield and Nelson 2006; Lippa 2005).

Girls are usually held more gently and treated more tenderly than boys, who are initially the recipients of more holding, kissing, rocking, and touching. By six months of age, it is girls who are held, talked to, and soothed more than boys and who receive more parental engagement. Clearfield and Nelson (2006) suggest that one implication of such differences is that infant daughters are being encouraged to seek help, whereas infant sons are encouraged to be independent.

As they grow, boys are ordinarily subjected to rougher forms of play, and both girls and boys begin to absorb expectations about appropriate behavior. In many ways, socialization into masculinity is more rigid, and gender-inappropriate behavior is less tolerated among boys than among girls. Boys who behave in any way that appears effeminate are often stigmatized as "sissies" or "fags," whereas girls who behave in more masculine ways may receive the more ambiguous label of "tomboy." The connotations of the terms are quite different.

Such gender-role socialization occurs throughout our lives. By middle childhood, although conforming to gender-role behavior and attitudes becomes increasingly important, there is still considerable flexibility (Absi-Semaan, Crombie, and Freeman 1993). It is not until late childhood and adolescence that conformity becomes most characteristic. The primary agents forming our gender roles are parents. Eventually, teachers, peers, and the media also play important roles.

Parents as Socialization Agents

During infancy and early childhood, a child's most important source of learning is the primary caretaker—often both parents but also often just the mother, father, grandmother, or someone else. Most parents may not be aware of how much their words and actions contribute to their children's gender-role socialization, nor are they aware that they treat their sons and daughters differently because of their gender. Although parents may recognize that they respond differently to sons than to daughters, they usually have a ready explanation—the "natural" differences in the temperament and behavior of girls and boys. Parents may also believe that they adjust their responses to each particular child's personality. In an everyday living situation that involves changing diapers, feeding babies, stopping fights, and providing entertainment, it may be difficult for harassed parents to recognize that their own actions may be largely responsible for the very differences that they attribute to nature or personality.

The role of nature cannot be ignored completely, however. Temperamental characteristics may be present at birth. Also, many parents who have conscientiously tried to raise their children in a nonsexist way have been frustrated to find their toddler sons shooting each other with carrots or their daughters primping in front of the mirror. Indeed, it is increasingly likely that some gender differences are influenced by hormones and/or chromosomes. At the same time, it is undeniable that children are socialized differently based on their gender.

How Parents Shape Gender Differences

Childhood gender socialization occurs in many ways, with parents the strongest influencers. In the clothing, toys, and books they buy for their children; the activities in which they involve them; the environments in which they raise them; and the behavioral expectations they convey to them, parents give shape to gender differences. Children's literature, for example, traditionally depicts girls as passive and dependent, whereas boys are instrumental and assertive (Kortenhaus and Demarest 1993).

Females are socialized with more attention placed on the importance of physical appearance. Women's studies professor Lori Baker-Sperry and sociologist Liz Grauerholz examined one particular genre of children's literature—classic fairy tales—and determined that such tales tended to emphasize the importance of physical appearance for female characters, especially

Generally, daughters are given more responsibilities than are sons.

young female characters, and associated a character's "goodness" with her beauty and attractiveness. The tales that have been most popular, republished in books and/or remade into films, are the tales that most stress the importance of female beauty (Baker-Sperry and Grauerholz 2003). In the more than 4,000 children's books published annually, females are rarely portrayed as brave or independent and are typically presented in supporting roles (Renzetti and Curran 2003).

Children's toys and clothing also reflect and reinforce gender differences. Boys prefer such toys as blocks, trucks and trains, guns and swords, and tool sets. Girls prefer to play with dolls and domestic toys and to dress up. At least some, if not most, of the preference appears to be the result of parental preferences, peer influence, and positive reinforcement (Lippa 2005).

Parents construct the physical environments in which they raise their children in very gender differentiated ways. More than 30 years ago, research established that parents furnished and decorated their children's rooms in ways that reflect traditional notions and expectations about gender. Psychologists Erin Sutfin, Megan Fulcher, Ryan Bowles, and Charlotte Patterson (2007) found that the "great majority" of children in their study were being raised similarly in gender-differentiated settings (i.e., bedrooms). Interestingly, in comparing heterosexual parents to lesbian parents, they reported that lesbian mothers were less likely to create highly stereotypical environments for their children. This is in keeping with their finding that lesbian mothers had more liberal attitudes and egalitarian beliefs than expressed by heterosexual parents.

In general, children are socialized by their parents through four subtle processes: manipulation, channeling, verbal appellation, and activity exposure (Oakley 1985):

- *Manipulation.* From an early age, parents treat daughters more gently (e.g., telling them how beautiful they are or advising them that nice girls do not fight) and sons more roughly (telling them how strong they are or advising them not to cry). Eventually, children incorporate such views as integral parts of their personalities. Differences in girls' and boys' behaviors may result from parents expecting their children to behave differently.
- *Channeling.* Children are directed toward specific objects and activities and away from others. Toys, for example, are differentiated by gender and are

marketed with gender themes. Parents purchase different toys for their daughters and sons, who—influenced by advertising, the reinforcement by their parents, and the enthusiasm of their peers—are attracted to gendered toys (Kimmel 2008).
- *Verbal appellation.* Parents use different words with boys and with girls to describe the same behavior. A boy who pushes others may be described as "active," whereas a girl who does the same may be called "aggressive."
- *Activity exposure.* Both genders are usually exposed to feminine activities early in life, but boys are discouraged from imitating their mothers, whereas girls are encouraged to be "mother's little helpers." Chores are categorized by gender (Dodson and Dickert 2004; Gager, Cooney, and Call 1999). Boys' domestic chores take them outside the house, whereas girls' chores keep them in it—another rehearsal for traditional adult life.

Critical Thinking

Can you recall any occasions when you, a sibling, or a friend wanted to play with a toy that was more commonly associated with the opposite gender? How did others react? Do you think the reaction would be different today?

Although it is generally accepted that parents socialize their children differently according to gender, there are differences between fathers and mothers. Fathers pressure their children more to behave in gender-appropriate ways. Fathers set higher standards of achievement for their sons than for their daughters, play more interactive games with their sons, and encourage them to explore their environments (Renzetti and Curran 2003). Fathers emphasize the interpersonal aspects of their relationships with their daughters and encourage closer parent–child proximity. Mothers also reinforce the interpersonal aspect of their parent–daughter relationships, typically engaging in more "emotion talk" with their daughters than with their sons, and—unsurprisingly—as early as first grade, girls are more adept at monitoring emotion and social behavior (Renzetti and Curran 2003).

Both parents of teenagers and the teenagers themselves believe that parents treat boys and girls differently. It is not clear, however, whether parents are reacting to existing differences or creating them. It is probably both, although by that age, gender

is spent playing video games. Of course, the exact mix of these varies by age, race, sex, and personal interest.

With such heavy exposure, the content of media messages becomes even more significant. In all its forms, the mass media depict females and males quite differently. We can safely assert that the media typically have stereotyped, objectified, trivialized, or ignored women. Researchers suggest that gender and sexual stereotypes characterize television (Ward and Friedman 2006), advertising, popular music lyrics (Bretthauer, Zimmerman, and Banning 2007), music videos (Bell, Lawton, and Dittmar 2007), and video games (Burgess, Stermer, and Burgess 2007; Miller and Summers 2007).

Much of television programming promotes or condones negative stereotypes about gender, ethnicity, age, and gay men and lesbians. Women are underrepresented and narrowly depicted. The women depicted on television represent women less than the men depicted represent men. A 2003 study of gender and age of characters revealed that female characters continue to be younger than male characters. The largest percentage of female characters was in the 20- to 29-year-old range, and the largest age range among male characters was 30 to 39. Within age-groups, such as adolescence, females are underrepresented and stereotyped (Walsh and Ward 2008). As summarized by researchers Jennifer Walsh and Monique Ward, young female characters "groomed, whined, shopped, and did chores," while males worked, fought, and rebelled (Walsh and Ward 2008, 137). Across media, heavy emphasis is placed on female appearance, especially thinness, and on females as sex objects. Male characters are shown as more aggressive and constructive than female characters. They solve problems, take action, and rescue others from danger, often by resorting to violence.

Do such gendered portrayals matter? Research indicates that they do. For example, after viewing select music videos, daytime talk shows, and prime-time programs, individuals are more likely to endorse stereotypical gender roles, emphasize female thinness, and accept myths about women and sexual assault (Kahlor and Morrison 2007). Even exposure to images of Barbie dolls can affect attitudes and self-esteem of young girls. Adolescent girls who viewed just three music videos featuring thin and glamorous women showed an increase in their own level of dissatisfaction with their bodies (Bell et al. 2007).

Continued Gender Development in Adulthood

Although more attention is directed at early experiences and socialization in childhood and adolescence, gender development doesn't stop there. Many life experiences that we have in adulthood alter our ideas about and actions as males and females. Again, families loom large in reshaping our gendered ideas and behaviors.

Research offers examples in which adult life experiences transform how we act as males or females. This is especially evident in the work of sociologists like Kathleen Gerson and Barbara Risman (Gerson 1985, 1993; Risman 1986, 1987, 1998). Their more structural analyses have shown how adult life experiences both inside and outside of families have the potential to restructure our identities, redefine our role responsibilities, and take us in directions quite different from those suggested by our early gender socialization. The life one leads is often different from what one was raised to lead or expected to lead (Gerson 1993; Risman 1987, 1998).

Focusing on adulthood is important because it reveals the gaps that often exist between earlier gender socialization and adult experiences. To some scholars, this diminishes the importance of socialization and discredits theories that deterministically link early socialization to later life outcomes (Gerson 1993). In some ways, those theories may be no better than *biological determinism*, in which we are limited to those behaviors that our genetic or hormonal characteristics allow. They simply substitute socialization for biology (Risman 1989).

Socialization is important, especially in affecting our expectations and offering us role models for lives we might live. But life is more circuitous than linear. Unanticipated twists and turns often take us in directions we neither expected nor intended. Research on women's and men's career and family experiences bear this out. For example, Kathleen Gerson's research on women's and men's career and family choices reveals that many people develop commitments to either careers or parenting that were unexpected and stem from their experiences in jobs and relationships (Gerson 1985, 1993). Some women and men who anticipate "traditional" adult outcomes move in nontraditional directions based on the levels of fulfillment and opportunity at work, the experiences and aspirations of their partners, and their experiences with children. Similarly, men and women who aspire to nontraditional outcomes (career attachment for women and involved fatherhood for men) may "reluctantly" abandon those directions as a result of firsthand experiences at home and work.

Barbara Risman's research on single custodial fathers pointed to similar adult development. Men who reluctantly found themselves as lone, custodial parents developed nurturing abilities that their socialization had not included. More important than how they were raised was how they interacted with their children as well as the lack of a female in their lives to whom caregiving tasks could be assigned. Thus, these single fathers "mothered" their children in ways that were more like women's relationships with children than what one would predict (Risman 1986). Importantly, socialization contributes to but neither guarantees us nor restricts us to any particular family outcome.

In addition, as one moves through adulthood, one has new experiences or encounters different sources of gender-role learning, both of which can alter one's sense of their gendered roles. Marriage and parenthood can alter one's understanding of what being a female or male means, as can experiences in the workplace and/or attending college.

College

Within the college setting, many young adults learn to think critically, to exchange ideas, and to discover the bases for their actions. There, many young adults first encounter alternatives to traditional gender roles, either in their personal relationships or in their courses. Research indicates that college liberalizes people's attitudes about a host of issues, including ideas about gender, attitudes toward women's employment opportunities, and family roles and responsibilities. This is especially so for students who major in the humanities, social sciences, or arts; live on campus; have jobs while attending school; socialize with others of different racial and ethnic backgrounds; and take women's studies courses. Both women's and men's attitudes tend to become more liberal, though in general, at both entry and exit from school, women's attitudes are more liberal than men's (Bryant 2003).

Marriage and Parenthood

People's experiences in marriage, parenthood, and the workplace can also lead to changes in gender attitudes, beliefs, and behavior. Marriage is an important source of gender-role learning because it creates the roles of husband and wife, which are not merely the same as male and female. For many individuals, no one

is more important than a partner in shaping gender-role behaviors through interaction. Our partners have expectations of how one should act as a husband or wife, and these expectations are important in shaping behavior. Such expectations may require significant adjustments to one's prior expectations of what a male or female does in marriage.

Husbands tend to believe in innate gender roles more than wives do. This should not be especially surprising because men tend to be more traditional and less egalitarian about gender roles. Husbands stand to gain more in marriage by believing that women are "naturally" better at cooking, cleaning, shopping, and caring for children.

For most men and women, life is dramatically transformed by parenthood. Still, motherhood alters life more significantly and visibly for women than fatherhood does for men. For some men, fatherhood may still mean little more than providing financially for their children. One is unlikely to find many who would associate motherhood only with providing. As parents, mothers do more, are expected to do more, and are expected to juggle what they do with paid employment. As a consequence, fatherhood does not typically create the same degree of work–family conflict that motherhood does. Those fathers who strive to be a fully or nearly equal coparent will, however, discover the ways in which the demands of parenthood clash with demands of the workplace. Whereas traditionally a man's work role allowed him to fulfill much of his perceived parental obligation, we now expect more out of fathers.

Not only have our expectations shifted toward more nurturing versions of fatherhood, but where traditional fatherhood was tied to marriage and often to a breadwinner–homemaker marital structure, today nearly 40% of all current births occur outside of marriage, and nearly half of all current marriages end in divorce. In most two-parent family households with children, both parents are employed. What, then, is the father's role for a man who is not married to his child's mother, who is divorced and does not have custody, or who shares wage earning with his spouse? What are his role obligations as a single father as distinguished from those of married fathers? For many men, the answers to such questions are not altogether clear and must be actively constructed in response to circumstances.

Women today have somewhat greater latitude as wives and mothers. It is now both accepted and expected that women will work outside the home, at least until they become mothers, and, more than likely, continue or return to paid employment sometime after they have children. Even with increases in the numbers of women who remain childless, women may be expected to become mothers and be subjected to social pressure toward motherhood. Once children are born, roles tend to become more traditional, even in previously nontraditional marriages. Often, the wife remains at home, at least for a time, and the husband continues full-time work outside the home. The woman must then balance her roles as wife and mother against her needs and those of her family.

The Workplace

Experiences in the workplace can also lead to changes in one's attitudes and expectations. Both men and women are psychologically affected by their occupations. Work that encourages self-direction, for example, makes people more active, flexible, open, and democratic; restrictive jobs tend to lower self-esteem and make people more rigid and less tolerant. If we accept that sex-segregated female occupations are often of lower status with little room for self-direction, we can understand why some women are not as achievement oriented as men. With different opportunities for promotion, men and women may express different attitudes toward achievement. Women may downplay their desire for promotion, suggesting that promotions would interfere with their family responsibilities. But this really may be related to a need to protect themselves from frustration because many women are in jobs where promotion to management positions is unlikely.

Household work can affect women psychologically in many of the same ways that paid work affects them in female-dominated occupations, such as clerical and service jobs (Schooler 1987). Women in both situations feel greater levels of frustration because of the repetitive nature of the work, time pressures, and being held responsible for things outside their control. Such circumstances do not encourage self-esteem, creativity, or a desire to achieve. Moreover, housework *on top of paid work* is a reality with which many women but much fewer men contend.

Gendered Family Experiences

Characterizing contemporary gendered roles is difficult because they are in such a state of flux. Within the past generation, there has been a significant shift from traditional toward more egalitarian gender roles (Brewster and Padavic 2000; Lang and Risman 2007). Women have changed more than men, but men are changing. These changes seem to affect all classes, although not to the same extent or at the same pace. Middle-class women and men have tended to be more liberal and egalitarian than working-class or upper-class women and men. In practice, working-class women and men may behave in more equal ways, more out of necessity than because of an egalitarian ideology. Their beliefs may remain more traditional than their behavior suggests.

Also, there are some notable differences across religious groups, with those from more conservative religious groups, such as Mormons, and conservative and evangelical Protestants continuing to adhere more strongly to beliefs in more traditional gendered family roles (Ammons and Edgell 2007). Such beliefs don't always get expressed in choices and/or behaviors (Ammons and Edgell 2007). Religion seems to make even more of a difference in one's attitudes toward premarital sex, divorce, or abortion than in one's advocacy for or against families with traditional gender roles (Pearce and Thornton 2007).

Contemporary gender-role attitudes have changed partly as a response to the steady increase in women's participation in the labor force. Although this increase was especially evident in the 1970s and 1980s, as was the move toward more egalitarian attitudes, it continued in the 1990s through to the present. As illustrated above, college-educated women and men, especially, are considerably less likely to hold traditional ideas about gender, work, and family roles (Brewster and Padavic 2000; Bryant 2003).

Within the family, attitudes toward gender roles have become more liberal, moving in the direction of more sharing of housework and child care. Researchers estimate that from the 1960s to the twenty-first century, men's share of housework had doubled from about 15% to about 30%. Moreover, this trend has been observed in industrialized countries throughout the world. A 20-country comparison showed an overall increase of men's share of housework and child care from less than a fifth of all housework in the 1960s to more than a third by 2003 (Sullivan and Coltrane 2008). This trend still places women at a disadvantage, especially by making them responsible for more of the housekeeping and child care activities, but it also gives one reason to assume that movement in the direction of greater sharing will continue.

Popular Culture: Video Gender: Gender, Music Videos, and Video Games

Over the past 25 years, recreation and entertainment, especially for young people, increasingly encompass video images and technologies. Popular music was revolutionized by the "invention" of the music video and by the inception of MTV, which premiered almost 30 years ago in 1981. An estimated 350 million households worldwide tune in to MTV, and three-fourths of all 16- to 24 year-old females and males watch MTV (Kaiser Family Foundation 2003).

Meanwhile, the video game industry has revolutionized "play" for millions of Americans, especially males. Billions of dollars and countless hours are spent on arcade, handheld, or home video games. Together these media have also altered the experience while reinforcing more traditional content of gender socialization.

Video Games

Considered the "fastest growing" of any form of entertainment, video games generated record sales of over $21 billion in total sales in 2008 (of which $11 billion was spent on game software), a 19% increase over 2007 (Hefflinger 2009). Of all media, these games also convey "the most uniform and unsubtle" gender stereotypes (Dill 2007).

Although the average age of video game players is now 35 (Entertainment Software Association 2009) concern is perhaps greatest regarding the quantity and quality of exposure of younger populations. Research indicates that 97% of teens as compared to 53% of people over 18 play video games (Lenhart, Jones, and McGill 2008). In terms of gender, among adults over age 18, 55% of men and 50% of women play video games. Among teens, 99% of males and 94% of females play, with younger teen boys being most likely to play and older female teens the least likely. Among parents surveyed by the Pew Foundation's Internet and American Life project (Lenhart 2008), 18% percent of parents of boys say that games have had "negative effects" on their sons, with parents of 12- to 14-year-old boys being most likely to claim negative effects. Only 7% of parents of girls believe that gaming has had negative effects on their daughters. Perhaps they have good reason. One-third of all teenage gamers from the same Pew study admit playing games rated M (for mature audiences) or AO (adults only). Of these teens, 79% are boys. Also, Douglas Gentile and J. Ronald Gentile (2005) found that 8.5% of a national sample of 8- to 18-year-olds could be considered to show signs of "pathological play," including the possibility of "addiction" to games. For each indicator of pathological play (e.g., skipping household chores to play, using video game play to escape one's problems, skipping homework, and/or doing poorly on tests because of video game play), males were more likely

Further reason to suspect more movement in the direction of more sharing can be found in some national public opinion data. A report by the Pew Research Center (2007) reveals the extent to which "sharing household chores" has come to be seen as a vital ingredient in achieving marital success. More than 60% of respondents identified such sharing as "very important," while only 7% considered it "not very important." This placed sharing housework third, behind faithfulness (93%) and a happy sexual relationship (70%), in importance but made it more important than adequate income, good housing, shared religious beliefs, shared tastes or interests, children, and agreement on politics. There was minimal difference between the percent of women (61%) and men (64%) or between fathers (66%) and mothers (62%) in stressing the importance of sharing housework. Of course, how people answer in a survey may be neither an indicator of their true feelings nor a predictor of their behavior.

Women's Roles in Families and Work

Traditionally, women's lives centered around marriage and motherhood. When a woman left adolescence, she was expected to either get a job, go to college, or marry and have children. Although a traditional woman might work before marriage, she was not expected to defer marriage for work goals and soon after marriage was expected to be "expecting." Within the household, she was expected to subordinate herself to her husband. Often this subordination was sanctioned by religious teachings. Circumstances permitting, for women with more traditional values this may still

than females to acknowledge such problems (Gentile 2008).

Aggression and violence are major components of many games, but one doesn't even need to play to be exposed to these themes. Popular magazines about games and gaming depict more than 80% of the male characters as violent, and hypermasculine male characters are common. Female characters, on the other hand, are depicted as sex objects, "busty," and scantily clad (Dill, Brown, and Collins 2008; Miller and Summers 2007). Female characters are under-represented and are rarely the central figures in the stories. Those few who are (e.g., the Lara Craft character in *Tomb Raiders*) tend to be portrayed more as sexual objects.

Although most male violence in games is directed at other male characters, some researchers describe a disturbing trend of glorified male on female violence. This is especially evident in the *Grand Theft Auto* series, which an estimated 75% of teenage males in the United States have played (Dill et al. 2008). Overall, violent themes, aggressive action, and sexual stereotypes are common.

Music Videos

Even with innovations such as the iPod revolutionizing the experience of listening to music, for nearly 30 years music has come to be "seen" as well as heard. Visual images are as important as the music and the lyrics; indeed, the images may even be more important than the music.

Studies of gender stereotypes and ratios of males and females portrayed in music videos have consistently found males featured more extensively and portrayed more broadly than females. Of the "characters" featured (including performers, dancers, and any characters in more storytelling videos), females are frequently portrayed as sex objects in how they act and what they wear (Walsh and Ward 2008). There is considerable evidence detailing effects of music videos on female body image and self-acceptance and on one's attitudes, beliefs, and behavior regarding gender and sexuality (Bell et al. 2007).

Cumulatively, video games and music videos have become more important parts of the gender socialization process. Their themes—male as aggressive and violent, females as sex objects and victims—fit both with each other and with other popular media content (e.g., television, film, cartoons, and advertising). Clearly, no single game or video will determine a person's attitudes toward women or propensity toward violence. Collectively, however, such images help shape and reinforce traditional gender attitudes and make aggressive outcomes more likely.

reflect their intentions and preferences. However, there are class and racial differences in women's traditional work and family roles.

This description of the traditional Caucasian female gender role pertains more to middle-class women than their less privileged counterparts and to middle-class ideals more than the reality experienced by all women. It does not extend to African American women. This may be attributed to a combination of the African heritage, slavery (which subjugated women to the same labor and hardships as men), and economic discrimination that pushed women into the labor force. One study (Leon 1993) found that African American women appear more instrumental than either Caucasian or Latina women; they also have more flexible gender and family roles. African American men are generally more supportive than Caucasian or Latino men of egalitarian gender roles.

In traditional Latina gender roles, the notion of *marianismo* has been the cultural counterpart to *machismo*. Drawn from the Catholic ideal of the Virgin Mary, *marianismo* stresses women's roles as self-sacrificing mothers suffering for their children and subordinating themselves to males (McLoyd et al. 2000b; Vasquez-Nuttall et al. 1987). But this subordination is based more on respect for the male's role as provider than on subservience (Becerra 1988). It also appears to be waning. Latina women are increasingly adopting values incompatible with a belief in male dominance and female subordination. They also display higher levels of marital satisfaction and less depression when their husbands share more of the domestic work (McLoyd et al. 2000b). Wives have greater equality if they are employed; they also have more rights in the family if they are educated (Baca Zinn 1994).

Latina gender roles, unlike those of Caucasians, are strongly affected by age roles in which the young subordinate themselves to the old. In this dual arrangement, notes Rosina Becerra (1988), "females are viewed as submissive, naive, and somewhat childlike. Elders are viewed as wise, knowledgeable, and deserving of respect." As a result of this intersection of gender and age roles, older women are treated with greater deference than younger women.

Even though the traditional roles for white women have typically been those of wife and mother, increasingly over the past few decades these roles have been joined by the additional role of employed worker or professional. It is now generally expected that most women will be employed at various times in their lives. It is also typically economically necessary. This raises the issue of how to manage the competing demands of paid employment and caring for family, a situation that forces women more than men to "juggle" responsibilities. How they respond to such circumstances depends in part on their relative commitment to work and family, what sociologist Mary Blair-Loy (2003) calls their *schemas of devotion* or what Arlie Hochschild (1989) called their *gender ideologies*. Family-committed women most often attempt to reduce the conflict between work and family roles by giving family roles precedence. As a result, they tend to work outside the home in greatest numbers before motherhood and after divorce, when single mothers generally become responsible for supporting their families. After marriage, most women are employed even after the arrival of the first child. Regardless of whether a woman is working full time, she almost always continues to remain the one with primary responsibility for housework and child care. The end result is that women more than men are likely to feel pinched for time, face a time bind, and feel rushed (Blair-Loy 2003; Hochschild 1997; Mattingly and Sayer 2006; Sayer 2005).

Cultural expectations impose high standards of devotion and labor-intensive self-sacrifice on women who become mothers, what is described as the **intensive mothering ideology**—the belief that children need full-time, unconditional attention from mothers to develop into healthy, well-adjusted people. This puts all mothers in a demanding position, but it creates a particularly difficult dilemma for mothers who also choose or need to work outside the home (Hays 1996). It leads increasing numbers of women

to question whether they should have children and, if they do, how much of their time and attention their children need.

Women from ethnic and minority groups, however, are less likely than Caucasians to view motherhood as an impediment. African American women and Latinas tend to place greater value on motherhood than Caucasians. For African Americans, tradition has generally combined work and motherhood; the two are not viewed as necessarily antithetical (Basow 1993). For Latinas, the cultural and religious emphasis on family, the higher status conferred on motherhood, and their own familial attitudes have contributed to high birthrates (Jorgensen and Adams 1988).

Although husbands were once the final authority, wives have greatly increased their power in decision making. Between the late 1970s and mid-1990s, data on attitudes about marriage indicated increasing acceptance of more gender equal relationships, where women contribute to decision making, share wage earning, and have assistance with family caregiving. After a slight drop in such egalitarian attitudes in the late 1990s, research indicates that they are rising again, especially regarding the acceptance of women's employment and, as we saw earlier, men's involvement in domestic work (Lang and Risman 2007).

In fact, some couples develop and act on an ideology of sharing and fairness, valuing and pursuing such relationship characteristics as equality and equity (Risman and Johnson-Sumerford 1997; Schwartz 1994). Although these "peer" and **postgender relationships** (i.e., relationships lived outside the constraints of gender expectations) are not yet the norm, they reflect the most concerted efforts to establish greater equality in marriage. Research also indicates that couples who share housework and wage earning are less likely to divorce than couples in which men earn all the income and women do all the housework (Cooke 2008).

Men's Roles in Families and Work

Central features of the traditional male role—whether among Caucasians, African Americans, Latinos, or Asian Americans—stressed male dominance and men as breadwinners. Males are generally regarded as being more power oriented than females and more competitive, even at a young age (Gneezy and Rustichini 2002; Stewart and McDermott 2004). Statistically, men demonstrate higher degrees of aggression, especially violent aggression (such as assault, homicide, and rape). They are more often arrested for violent crimes and are more common victims of violent offenses other than rape and sexual assault (Kimmel and Messner 2010). Aggressiveness is one of a handful of attributes on which consistent moderate to large differences between the genders have been documented (Hyde 2005). Although aggressive traits are thought to be useful in the corporate world, politics, and the military, such characteristics are rarely helpful to a man in fulfilling marital and family roles requiring understanding, cooperation, communication, and nurturing.

Traditionally, across ethnic and racial lines, male roles centered on providing, and the centrality of men's work identity affected their family roles as husbands and fathers. Men's identity as providers took precedence over all other family functions, such as nurturing and caring for children, doing housework, preparing meals, and being intimate. Because of this focus, men who retain traditional attitudes may become confused if their spouses expect greater sharing or intimacy; they believe that they are good husbands when they are good providers. When circumstances render them unable to provide, the blow to their self-identities can be quite powerful (Rubin 1994).

However, because race, ethnicity, and economic status often overlap, certain categories of men face more difficulty meeting the expectations of the traditional provider role. Because African Americans and Latinos often fare less well economically, men often are left unable to lay claim to the household status and power that traditional masculine roles promise.

Occasionally, characterizations of Latino families have exaggerated the extent of male dominance, as suggested by the notion of *machismo*. Although such a notion may have been somewhat more accurate in depicting gender ideologies of rural Mexico and the Caribbean in the first half of the twentieth century, it is inaccurately applied to contemporary Latino families (McLoyd et al. 2000b). Both African American and Latino men have more positive attitudes toward employed wives. Ethnic differences in traditional notions of masculinity and men's roles are more evident among older and less educated African Americans and among Mexican Americans not born in the United States (McLoyd et al. 2000b). The additional attention to race, ethnicity, and class alongside gender is at the heart of an approach to studying gender called **intersectionality** (Stewart and McDermott 2004).

Critical Thinking

Have you observed ways in which gender roles and ideas differ across lines of ethnicity, race, class, or sexual orientation?

Because the key assumption about male gender roles has been the centrality of work and economic success, many earlier researchers failed to look closely at how men interacted within their families. Increasingly, over the past 25 years, we have witnessed a dramatic increase in the popular and scholarly attention paid to men's family lives. Researchers began to ask some of the same questions about men's lives that were previously asked about women's, looking at whether and how men juggle paid work and family, and whether they maintain sufficient involvement in each (Coltrane 1996; Daly 1996; Gerson 1993). Although we may not yet treat employed fathers with the same concern we bring to working mothers, we have made strides in examining how men experience conflicts between work and parenting.

As contemporary male gender roles allow increasing expressiveness, men are encouraged to nurture their children.

In addition, research indicates that men consider their family role to be much broader than that of family breadwinner (Cohen 1993; Coltrane 1996; Gerson 1993). Sociologists Oriel Sullivan and Scott Coltrane note that national public opinion data further indicate increasing acceptance of gender equality and along with that the expectation that spouses will share both wage earning and family responsibilities. Additionally, men appear to be more involved in caring for their children than ever before, having tripled their time spent in child care from 1965 to 2003 (Sullivan and Coltrane 2008). Later chapters look at men's experiences of marriage, parenting, and the division of household labor.

Still, alongside enlarged emphasis on men's more nurturing qualities, men continue to be expected to work and to support or help support their families. Although women's financial contributions may be no less essential in maintaining their family's standard of living or even remaining out of poverty, women are not judged as successful wives and mothers based on whether they succeed at paid employment. As a result, men have less role freedom than women to choose whether to work. Those men who aspire toward equal involvement in child care may discover that the complexity of juggling work and family is not restricted to women. Men who attempt the same juggling act often experience similar role strain and role overload (Gerson 1993; Hansen 2005).

In addition, many men continue to have greater difficulty expressing their feelings than do women (Real 1997). Men tend to cry less and show love, happiness, and sadness less. When men do express their feelings, they are more forceful, domineering, and boastful; women, in contrast, tend to express their feelings more gently and quietly. When a woman asks a man what he feels, a common response is "I don't know" or "Nothing." Such men have lost touch with their inner lives because they have repressed feelings that they have learned are inappropriate.

Continued Constraints of Contemporary Gendered Roles

Both women and men often reinforce traditional gender-role stereotypes. For example, both genders react more negatively to men displaying so-called female traits (e.g., crying easily or needing security) than to women displaying so-called male traits (e.g., assertiveness or worldliness), and both define male gender-role stereotypes more rigidly than they do female stereotypes. Men, however, do not define women as rigidly as women do men. And both men and women describe the ideal female in more neutral terms (Hort, Leinbach, and Fagot 1990).

Despite the limitations that traditional roles may place on us, changing them is not easy. Gendered roles are closely linked to self-evaluation. Our sense of adequacy often depends on our gender performance as defined by parents and peers in childhood ("You're a good boy" or "You're a good girl"). Because gendered roles often seem to be an intrinsic part of our personality and temperament, we may defend these roles as being natural, even if they are destructive to a relationship or to ourselves. To threaten an individual's gender role is to threaten his or her gender identity as male or female because people do not generally make the distinction between gendered roles and gender identity. Such threats are an important psychological mechanism that keeps people in traditional roles.

Even though substantially more flexibility is offered to men and women today, contemporary gendered roles and expectations continue to limit our potential. Indeed, there is evidence that some stereotypes about gender traits are still very much alive. Asked in a national survey to identify whether certain traits were truer characteristics of men, women, or both men and women, respondents answered in ways that reinforced gender stereotypes. Both women and men associated arrogant, stubborn, and decisive with men more than with women. Traits identified as truer of women than of men included compassionate, emotional, creative, intelligent, honest, and manipulative. Interestingly, two traits—ambitious and hardworking—were said by men to be truer of men than women, whereas women said the reverse (Pew Research Center Reports, "Men or Women: Who's the Better Leader?" 2008).

The situation of contemporary women in dual-earner households imposes its own constraints on women's lives. Because they continue to shoulder the bulk of responsibility for housework and child care *on top of full-time jobs*, they often experience fatigue, stress, resentment, and a lack of leisure and, as shown in Figure 4.3, are much more likely than their male counterparts to feel rushed (Hochschild 1989; Pew Research Center Reports, "Who's Feeling Rushed?" February 28, 2006).

Especially for women who try to be "supermoms," the volume and complexity of work and family can force them to cut back on their aspirations or compromise their expectations for marriage and motherhood (Hochschild 1989, 1997). It can also lead them to be harsh critics of their own performance as mothers.

Figure 4.3 Feeling Rushed: Percentage of Americans Who "Always Feel Rushed"

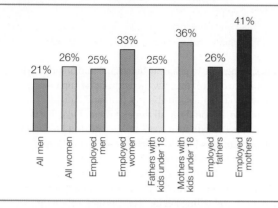

SOURCE: Pew Research Center: "Who's Feeling Rushed? (Hint: Ask a Working Mom)," 2006.

Mothers who are employed full time are less likely than at-home mothers or mothers employed part time to rate highly their performance as parents. "On a self rating scale of 0–10, 28% of full-time working moms give themselves a 9 or a 10" (Morin and Taylor 2008, http://pewsocialtrends.org/pubs/709/politics-gender-parenthood). At-home mothers (43%) and mothers employed part time (41%) were much more likely to rate their performances that highly.

In addition to self-criticism, public opinion data reflect more opposition to than support for full-time employed mothers, with nearly twice as many people seeing "the trend toward more mothers of young children working outside the home" a bad thing (41%) than those who see it as a good thing (22%) (Morin and Taylor 2008). Significantly, though, despite the ongoing stresses, women who "juggle" employment and motherhood are less distressed and more fulfilled than full-time homemakers (Crosby 1991).

The social structure also reinforces traditional gender norms and behaviors and makes some changes more difficult. Some religious groups, for example, strongly support traditional gender roles. The Catholic Church, conservative Protestantism, Orthodox Judaism, and fundamentalist Islam, for example, view traditional roles as being divinely ordained. Accordingly, to violate these norms is to violate God's will. The workplace also helps enforce traditional gender roles. The wage disparity between men and women is a case in point. Such a significant difference in income makes it "rational" for many couples for the man's work role to take precedence over the woman's work role. If someone needs to remain at home to care for the children or an elderly relative, it makes "economic sense" for a heterosexual woman to quit her job because her male partner probably earns more money.

Gender Movements and the Family

Gender issues have been the source of much collective action and the focus of a number of social movements that press for change. These movements include the range of perspectives within the contemporary women's movement but also various "men's movements" that, although less visible, have organized to change aspects of men's lives. We look briefly here at some of the ways these movements have framed and acted on family matters.

Women's Movements

A complete history of American feminism is beyond the scope of this book. In the eighteenth, nineteenth, and twentieth centuries, women organized around issues such as economic justice, abolition of slavery, temperance, and women's suffrage. In their antislavery activity during the nineteenth century, many women were sensitized to the extent of their own oppression and disadvantage, which helped energize their pursuit of voting rights (Lindsey 1997; Renzetti and Curran 1999). After gaining the right to vote with the passage of the Nineteenth Amendment in 1919, many women withdrew from active feminist involvement because they thought they had reached equality with men (Renzetti and Curran 1999).

During the 1960s, feminism resurfaced dramatically. Catalyzed by the publication of Betty Friedan's *The Feminine Mystique*, many women began to look critically at the sources of their "problems with no names," and the family was seen as a major culprit. In addition, wage inequality was made a public issue through President John F. Kennedy's Commission on the Status of Women in 1961 and the passage of the Equal Pay Act of 1963. Then, in 1966, the National Organization for Women (NOW) was established. Over the past 40-plus years, this liberal, reform-oriented feminist organization has grown to include more than half a million members in its more than 500 chapters throughout the United States. It is the largest, although not the only, organized plank of the women's movement, and its philosophy represents one of a number of "feminisms" (Lindsey 1997; Renzetti and Curran 2003). Like NOW, the Feminist Majority Foundation has, since its founding in 1987 and through its education, research and

Real Families: Making Gender Matter Less

When asked by sociologist John Durst to reflect on her situation, then 32-year-old Karen Wilson described having what she considered an almost ideal life. "Oh, how much time do you have. . . . Do you have like three hours? I love it! I love it! God, there's just so many things about it." Karen is a success in her career in sales and promotions for a communications company. She has a husband she loves and two children she adores, a 3-year-old daughter and a 4-month-old son. As the full-time breadwinner for her family (husband Kevin and two young children) she describes how she and Kevin reached the arrangement they have and what she most enjoys about it:

> You know what, we talked about this before we got married. . . . I planned on working after I had kids. [I told Kevin] . . . "if you want to stay home, great, but I can't stand it!" And he said, "Yeah, we can do that." Then, BAM! This opportunity came along for me to make more money than he was and I said, "You wanna do it? You wanna live the dream?"

Pushed to identify what she sees as the biggest positives of her lifestyle, she enthusiastically replied,

I like being able to get away between 8 and 5 and to have a lot more control over my life without having to worry about two other responsibilities (son and daughter) and Kevin, too. I should say all three of them. I like that. I like being able to turn it off and just go, but I like coming back and having my daughter's little face pressed against the window (waiting), Kevin standing there with a beer in his hand, the dog running around me, it's really nice to come home to. . . . I love bringing home the paycheck and telling Kevin, "Here, honey, split it up. . . ." I love that. I love contributing; I just think it's the ultimate.

I love not having all the responsibilities he has. I hated cooking. I hated the dishes, the laundry—I felt like it was the least rewarding job anybody could have because you never get any pats on the back. I like having a title and being able to say, "This is what I do. I'm contributing to my family.". . . And I like that Kevin's just so calm and relaxed and really laid back. The kids keep him moving constantly yet at the end of the day he's still relaxed enough to talk to me. I think it's been really wonderful.

---

action programs, and policy development, striven to "empower women economically, socially and politically" (http://feminist.org/welcome/index.html). Both organizations have among their concerns women's reproductive rights and health, safe and affordable contraception, and the protection of women's abortion rights. NOW also focuses on family issues such as paid parental leave and work–family balance, mothers' and caregivers' rights, and support for equal marriage rights.

As we saw in Chapter 2, there are multiple feminist theoretical positions. So, too, are there multiple feminist political ideologies. Contemporary feminist positions range across a spectrum including liberal, socialist, radical, lesbian, multiracial, and postmodern feminism (Lorber 1998; Renzetti and Curran 1999). Each has a specific emphasis on issues and advocates different strategies to improve women's lives.

Judith Lorber sorts these various feminist perspectives into three broader categories: **Gender-reform feminism** is geared toward giving women the same rights and opportunities that men enjoy; **gender-resistant feminism** advocates more radical, separatist strategies for women out of the belief that their subordination is too embedded in the existing social system; and **gender-rebellion feminism** tends to emphasize overlapping and interrelated inequalities of gender, sexual orientation, race, and class (Lorber 1998; Renzetti and Curran 1999).

Given this diversity of opinion, it is difficult to characterize *one* "feminist" position on families. Furthermore, such attempts occasionally exaggerate or simplify complex positions. In her critique of American feminism, for example, economist Sylvia Hewlett (1986) notes that neither liberal feminism (what she called "equal rights" feminism) nor radical feminist positions have recognized the commitments that women feel toward their families and the consequences of those commitments. By stressing equal rights and full equality with men, liberal feminism

Research on intimate relationships, marriage, and family consistently reveals the importance of gender in dividing up domestic responsibilities and shaping personal and familial experiences. Typically, women perform two to three times as much housework as men, and employed wives experience greater stress and enjoy less leisure than their husbands (Coltrane 2000). The consistency with which such inequalities are reported may give the impression of inevitability, that they are somehow unavoidable parts of marriage and parenthood, but couples such as the Wilsons offer a more hopeful scenario to those who might wish to someday depart from the norm, whether to create more equal partnerships or, more dramatically, reverse roles.

Sociologists Barbara Risman and Danette Johnson-Sumerford interviewed their own sample of 15 "post-gender" couples who explicitly reject conventional conceptions of gender, opting instead for more gender-neutral relationships. That is, they carefully and intentionally share responsibility for paid work and share responsibility as caregivers for their children.

At a minimum, they "changed how gender works in their families." Furthermore, "in the negotiation of marital roles and responsibilities, they have moved beyond using gender as their guidepost" (Risman and Johnson-Sumerford 1998, 24).

Regardless of the route couples took to arrive at their postgender family arrangements, they used criteria other than gender to organize their daily activities. They have rejected the ideas that "wifehood involves a script of domestic service or that breadwinning is an aspect of successful masculinity" (Risman and Johnson-Sumerford 1998, 38). Such couples are still rare, and their lifestyles may require high levels of female income and professional autonomy if women are to be able to move beyond male dominance or privilege.

The significance of couples like the Wilsons or Risman and Johnson-Sumerford's postgender couples is that they reveal a wider range of possible marital outcomes than most literature reports. There is no inevitable inequality that engulfs married couples. Equality and fairness take work and persistence but are possible for those who seek them.

may have downplayed the unique responsibilities women tend to carry within families and not recognized that women may need different supports than those needed by men. Some of the more radical feminist positions articulated in the 1960s and 1970s may have been fairly antimarriage or antimotherhood, as either or both have at times been seen as relationships that oppress women and keep them from achieving their full capabilities.

Hewlett (1986) compares both approaches to a movement more characteristic of European feminist activity: **social feminism**—the belief that workplace and family supports are essential if women are to experience a high quality of life. Feminist critics of Hewlett rightly point out that the greatest activism on behalf of public support for families has and continues to come most strongly from women; thus, her characterization is said to be unfair. Although they have emphasized issues such as abortion rights, reproductive freedom, ending bigotry against gays and lesbians, and ending violence against women, American feminists have been active at the forefront of pushing for parental leave, child care, and so forth.

Various attempts have been made to document how many women in the United States consider themselves feminists. Such estimates have varied, though most place the figure at 30% or less (McCabe 2005; Reid and Purcell 2004; Roy, Weibust, and Miller 2007). Data from large national surveys, such as the General Social Survey, suggest that around a quarter of female respondents consider themselves feminists, with better-educated women, white women, and lower-income women being more likely and more religiously active women less likely to identify as feminists (Peltola, Milkie, and Presser 2004). Roy et al. (2007) found that the willingness to identify oneself as a feminist was affected by stereotypes about feminists and could be altered by exposing young women to positive descriptions of feminism and feminists. Complicating the situation, Alyssa Zucker (2004) found that some women express feminist attitudes but distanced themselves from a feminist label. One doesn't need to be part of a gender movement to move in the direction of changing one's own situation.

Men's Movements

Divisions of opinion and multiple perspectives on gender inequality constitute a basic similarity between women's and men's collective action. Just as there is no one perspective on how women should be or what they should do, neither is there unanimity about men's lives. Just as there are multiple feminisms, each with its own agenda, there are different viewpoints on whether, in what direction, and how men ought to change (Clatterbaugh 1997; Messner 1997).

In recent years, at one time or another, we have witnessed a variety of "men's movements": the mythopoeic men's movement, the men's rights and fathers' rights advocates, and the Christian men's movement (e.g., Promise Keepers) have received the most media attention. The closest parallel to NOW, however, is the National Organization of Men Against Sexism (NOMAS). Unlike men's rights advocates and the Promise Keepers, NOMAS is a profeminist, gay-affirmative men's organization founded in 1975 by a group of male students in a women's studies course at the University of Tennessee. Among the issues NOMAS emphasizes and acts on are child custody, fathering, ending male violence, gay rights, and reproductive rights (www.nomas.org/history).

Each of these men's movements represents just one among many. They tend to differ in what they see as men's roles in and responsibilities to their families.

Central to the **profeminist men's movement** represented by NOMAS is the issue of *fairness*. Profeminist men believe that men ought to share responsibilities within their households and that women and men ought to be equal partners. Also, profeminists argue that both men and children would benefit from closer connections between fathers and their children.

Both the Promise Keepers and the organizers of the 1995 Million Man March and rally in Washington, D.C., by African American men also stressed the idea of men's responsibilities to their families, although their versions of responsibilities included more traditional notions of men's roles as the heads of their households. They also argued that men needed to be more accountable to spouses and children. Finally, the men's rights movement has stressed supposed discrimination that men face in and out of family matters. They note, for example, that only men can be subject to compulsory military service. They also look at what they believe are inequalities in areas of divorce settlements and custody or visitation arrangements (Farrell 2001).

It is interesting to note the different positions taken on the family by the various feminist and men's movements. Although it is inaccurate and overstated

The National Organization for Women (NOW) and the Promise Keepers are two examples of organized gender movements. In the rhetoric and rallies that comprise such movements, family issues loom large.

to suggest that feminists are antifamily, the resurgent women's movement of the 1960s did grow partly out of the articulation of discontent. Similarly, early "second-wave" feminists (1960s–1970s) attempted to sever the automatic assumptions and connections people typically made among women, children, and families as a way of liberating women to pursue other aspirations. However, it is feminist organizations such as NOW that are most active in pressing for a variety of issues that would improve the situations faced by contemporary families.

Across most men's movements, there is a sense that men need to enlarge their family role, live up to or "honor" their commitments to their families, and/or share in caring for children and households. Such involvement is often seen as potentially "liberating" for men because it reconnects them to their emotional sides and broadens their lives beyond wage earning. However, there is a difference of opinion as to what men's commitments to family entail.

Looked at more closely, the women's and men's movements are really not as different as they seem. What earlier feminists railed against was not the *family* but the *gendered family*. They were less antagonistic to what women felt toward and did in the family than what men did not. Because of the differential burden carried by women in households, family life imposed constraints on women's opportunities for outside involvements in ways it did not on men's. More recently, the various men's movements have acknowledged men's lack of involvement or weaker commitments and opposed defining men solely in terms of what they do away from the family.

Contemporary gender roles are still in flux. Few men or women are entirely egalitarian or traditional. Few with egalitarian attitudes, for example, divide all labor along lines of ability, interest, or necessity rather than gender. Also, marriages that claim to be traditional rarely have wives who submit to their husbands in all things. Furthermore, some who express egalitarian attitudes, especially males, may be more traditional in their behaviors than they realize, while others share more equally without being ideologically driven. Within marriages and families, the greatest areas of gender inequality continue to be the division of housework and child care. But change continues to occur in the direction of greater gender equality, and this equality promises greater intimacy and satisfaction for both men and women in their relationships.

Summary

- Men and women are not "opposites"; they are actually more similar than different. Innate gender differences are generally minimal; differences are encouraged by socialization.

- Within any given society, there are multiple versions of masculinity and femininity; across societies, much variation exists in how gender is perceived, including the perception of how many gender categories there are.

- Gender relations are also power relations. *Patriarchies* are social structures in which men dominate. Logically, *matriarchies* would be societies in which women dominate political and economic life. Researchers have not found any true matriarchal societies.

- *Gendered roles* are those roles a person is expected to perform as a result of being male or female in a particular culture. *Gender stereotypes* are rigidly held and oversimplified beliefs that males and females possess distinct psychological and behavioral traits. *Gender identity* refers to the sense one has of being male or female.

- Some individuals are *transgendered*, having developed gender identities contrary to their biological sex characteristics. Others identify themselves as both male and female, while still others may define themselves as neither male nor female.

- According to *gender theory*, social relationships are based on the socially perceived differences between males and females that justify unequal power relationships.

- Symbolic interactionists view gender as something we actively create or "do" in everyday situations and relationships, not an internalized set of behavioral and personal attributes.

- Two important socialization theories are social learning theory and cognitive development theory.

- Parents, teachers, and *peers* (age-mates) are important agents of socialization during childhood and adolescence. Ethnicity and social class also influence gender roles. Among African Americans, strong women are important female role models.
- After many years of evidence showing how schools disadvantage female students, recent evidence indicates that males are lagging behind educationally.
- The media tend to portray traditional stereotypes of men and women as well as of ethnic groups.
- For students, colleges and universities are important sources of gender-role learning, especially for nontraditional roles. Marriage, parenthood, and the workplace also influence the development of adult gender roles.
- The gendered roles we play in adulthood are affected by situations, opportunities, and constraints that can alter the path established by socialization.
- Traditional male roles emphasize dominance and work. For women, there is greater role diversity according to ethnicity.
- Contemporary gender roles are more *egalitarian* than the traditional ones of the past. They reflect (1) the acceptance of women as workers and professionals, (2) increased questioning of motherhood as a core female identity, (3) greater equality in marital power, and (4) the expansion of male family roles.
- Changing gender-role behavior is often difficult because we evaluate ourselves in terms of fulfilling gender-role concepts. Also, gender roles have become an intrinsic part of ourselves and our roles, and the social structure reinforces traditional roles.
- There have been various social movements dedicated to challenging or changing women's or men's roles, including various feminisms and various "movements" and perspectives on men and masculinity.

Key Terms

agents of socialization 114

androgyny 110

cognition 112

cognitive development theory 113

egalitarian 105

expressive traits 104

gender 104

gender attribution 111

gender identity 108

gendered roles 109

gender-rebellion feminism 130

gender-reform feminism 130

gender-resistant feminism 130

gender stereotype 109

gender stratification 105

gender theory 111

hegemonic models of gender 105

instrumental traits 104

intensive mothering ideology 126

intersectionality 127

intersexed 108

matriarchies 105

modeling 112

patriarchies 105

peers 119

postgender relationships 126

profeminist men's movement 132

second shift 107

sex 104

sexual orientation 130

social construct 111

social feminism 131

social learning theory 112

transgendered 108

transsexuals 108

transvestites 108

RESOURCES ON THE WEB

Book Companion Website

www.cengage.com/sociology/strong

Prepare for quizzes and exams with online resources—including tutorial quizzes, a glossary, interactive flash cards, crossword puzzles, self-assessments, virtual explorations, and more.

5

Intimacy, Friendship, and Love

What Do YOU Think? Are the following statements TRUE or FALSE?
You may be surprised by the answers (see answer key on the following page).

T	F	**1** A high value on romantic love is unique to the United States.
T	F	**2** The development of mutual dependence is an important factor in love.
T	F	**3** Love and commitment are inseparable.
T	F	**4** Friendship and love share many characteristics.
T	F	**5** Men fall in love more quickly than do women.
T	F	**6** Heterosexuals, gay men, and lesbians are equally likely to fall in love.
T	F	**7** In many ways, love is like the attachment an infant experiences for a parent or primary caregiver.
T	F	**8** A high degree of jealousy is a sign of true love.
T	F	**9** Partners with different styles of loving are likely to have more satisfying relationships because their styles are complementary.
T	F	**10** Love is something experienced and expressed similarly by people regardless of their ethnic or racial backgrounds.

"Love doesn't make the world go round, love is what makes the ride worthwhile," or so said American businessman Franklin P. Jones. From a variety of indicators, it appears as though Americans share that sentiment. We are, it seems, in love with love. We can see this in the ways we live our daily lives, especially in the kinds of relationships we want, seek, and make and in the steps we take to find and keep them. It is also evident in the popular culture that we produce and consume. There, we can see our love affair with love in everything from the music we listen to the things we read and watch.

Love is *the* dominant theme of popular music, where song titles and lyrics are typically testimonies to the power, pleasure, and pain associated with falling in and out of love. In the books we read, there is a whole genre devoted to love and "romance," which, based on 2007 data, is the largest in the consumer book market. More than 8,000 titles were published and an estimated $1.375 *billion* in sales generated by romance fiction. According to the Romance Writers of America, nearly 65 million Americans read romance novels, more than 75% of whom are female (Romance Writers of America 2009).

However, more than in our music or books, our devotion to love stands out especially well in movies. Romantic movies, *love stories* as they are often appropriately called, provide us with vivid scenes and memorable lines filled with heartfelt, often poignant declarations of the depth of a character's love. Often scenes stay with us, even coming to symbolize our very idea of true love. Some of the more recent successful, acclaimed, and/or award-winning movies had love stories at their core, albeit in very different ways. In 2005's *Brokeback Mountain*, the love was between two men, ranch hand Ennis del Mar and rodeo cowboy Jack Twist, across 20 years, two marriages, children, and divorces, until Jack's tragic death. In 2008's critically acclaimed *Wall-e*, the love relationship was between two robots, but whatever other themes characterized the movie, Wall-e's feelings for Eve and his determined desire to just hold her hand are front and center. In *Twilight*, the love story is between a vampire and human. Based on the immensely successful book series by Stephanie Meyer, the film tells of the passionate yet dangerous love that drives soul mates Edward and Isabella to take great risks to be together. The Academy Award–winning Best Picture of 2008, *Slumdog Millionaire*, though many other things, is essentially a story of Jamal and Latika, two whose love triumphs over the harshest of obstacles and suffering. The film culminates with their embrace and the following exchange:

Jamal: I never forgot. Not for one day. I knew I'd find you in the end. It's our destiny.

Latika: I thought we would meet again only in death.

Jamal: This is our destiny.

Much like the culture that surrounds them, American families place high value on love. Decisions about entering or exiting a marriage, assessments of the quality and success of any particular marriage, and devotion between spouses or between parents and children all come down to love. On both an individual level and a familial level, then, it is important to consider the role love plays in our lives. This chapter is devoted to such consideration, exploring the role love plays in our personal and familial lives, what it is, and the processes through which we go about finding it. However, before we turn to love, we need to consider the broader phenomenon of intimacy, including the intimacy of friendship.

The Need for Intimacy

Humans require other humans with whom we feel close and to whom we can commit. We need to form relationships in which we can share ourselves with others, exchange affection, and feel connected. In the developmental model formulated by psychologist Erik Erikson, this was the great task facing us in young adulthood—intimacy versus isolation; either we satisfy our need for intimacy, or we remain socially and emotionally isolated (Hook et al. 2003). In psychologist Abraham Maslow's **hierarchy of needs**, after the meeting of our physiological needs and needs for safety, our social needs—for intimacy and love—are the most fundamental of human needs (Maslow 1970) But what exactly is intimacy, and why is it so important?

Answer Key to What Do YOU Think?

1 False, see p. 141; **2** True, see p. 160; **3** False, see p. 141; **4** True, see p. 139; **5** True, see p. 144; **6** True, see p. 147; **7** True, see p. 151; **8** False, see p. 161; **9** False, see p. 150; **10** False, see p. 150.

In its most general sense, intimacy refers to closeness between two people. Sometimes we associate "intimacy" or "being intimate" with sexual relations. Certainly, sexual relations are part of physical intimacy, as are kisses, caresses, and hugs. However, it is more the emotional intimacy, having someone to talk to, to share ourselves with that is such an important part of our social and psychological well-being.

Reviewing research and theory on intimacy, psychologist Misty K. Hook and colleagues (2003) suggest that intimacy consists of four key features: the presence of *love and/or affection, personal validation, trust,* and *self-disclosure.* The more one feels as though another person likes or loves him or her, the more comfortable one will be sharing one's innermost feelings and revealing one's most personal thoughts. Feeling as though one is understood and appreciated makes one feel more accepted and safer to freely open oneself to another person without the fear of being judged or betrayed. Finally, to be intimate entails self-disclosure, the sharing of both the facts of our lives and our deeper feelings (Hook et al. 2003).

Intimate relationships provide us with a variety of benefits. They buffer us against loneliness, provide us with positive feelings about ourselves and others, give us confidence that our needs will be fulfilled in the future, and enhance our self-esteem. Intimate relationships are connected to happiness, contentment, and a sense of well-being. They also offer protection from some stress-related symptoms and reduce our likelihood of illness, depression, and accidents. People who lack satisfying, positive intimate relationships are at greater risk of illness; once ill, they recover more slowly and have higher susceptibility to relapse or recurrence of their illness.

In a relationship, intimacy can be expressed in a variety of ways—talking together, listening to each other, making time for each other, being open and honest with each other, and trusting each other. The importance of intimacy in defining relationship quality cannot be stressed strongly enough. It is a more important factor in relationship satisfaction than independence (autonomy, individuality, and freedom), agreement (harmony and few quarrels), or sexuality (sexual harmony and satisfaction and physical contact) (Hassebrauck and Fehr 2002). This holds true in but also beyond the United States. In comparative research using German and Canadian samples, intimacy was the factor most highly correlated with relationship satisfaction in both countries and for both males and females (although it was somewhat more strongly correlated with women's than with men's relationship quality and may have different meanings for females and males) (Hasselbrauck and Fehr 2002; Hook et al. 2003).

The Intimacy of Friendship and Love

Although both have proved difficult to define with precision or consistency, friendship and love are the two most important sources of intimacy people have. They help preserve both our physical and mental well-being. The loss of a friend and especially a loved one can lead to illness and even suicide.

Friendship can supply the foundation for a strong love relationship. Shared interests and values, acceptance, trust, understanding, and enjoyment are at the root of friendship and form a basis for love. As much as they may provide similar benefits, love and friendship are not the same thing. Even though individuals want and value many of the same qualities in friends and lovers—such as trust, acceptance, kindness, and warmth—they exercise more selectivity in choosing romantic partners than they do friends.

Potential friends may be deemed desirable on the basis of their specific combination of unique attributes and how those attributes match our needs and wants at a given point in time. Romantic partners, on the other hand, are more carefully selected, and their desirability is more carefully evaluated on the basis of their possession of certain qualities or attributes that might indicate their commitment to the relationship, their potential reproductive success, and their eventual attachment to offspring. In terms introduced by John Scanzoni and colleagues, romantic partners are selected on the basis of their seeming ability to satisfy multiple needs that are products of the multiple "interdependencies" two people share. Interdependencies consist of shared activities, statuses, and patterned exchanges between two people. Romantic partners are expected to be able to satisfy four types of interdependencies: intrinsic (e.g., emotional support), extrinsic (e.g., money or services), sexual (sexual activity), and formal (shared legal status). Friends, however, typically provide only intrinsic resources (Scanzoni et al. 1989; Sprecher and Regan 2002).

Research by Susan Sprecher and Pamela Regan suggests that warmth, kindness, openness, expressiveness, and a sense of humor are considered the most desirable and important qualities in both friends and partners. However, romantic partners were subjected to higher standards for these attributes, suggesting that such qualities are more important in romantic partners than in friends. Romantic partners are also evaluated on the basis of attributes such as appearance or social status; friends are not (Sprecher and Regan 2002).

Why It Matters: The Importance of Love

Love is essential to our lives. Love binds us together as partners, spouses, parents and children, and friends and relatives. Individuals make major life decisions, such as marrying, on the basis of love. Love creates bonds that one hopes will enable one to endure the greatest hardships, suffer the severest cruelty, and overcome any distance. Love is both a feeling and an activity. One *feels* love for someone, and one *acts* in a loving manner. Because of its significance, individuals may torment themselves with doubts about the sincerity ("Is it really love?") or mutuality ("Do you love me as much as I love you?") of the love relationships they experience. A paradox of love is that it encompasses opposites, including both affection and anger, excitement and boredom, stability and change, and bonds and freedom. Its paradoxical quality makes some ask whether they are really in love when they are not feeling "perfectly" in love or when their relationship is not going smoothly.

We can look at love in many ways besides through the eyes of lovers, although other ways may not be as entertaining. Whereas love was once the province of lovers, madmen, poets, and philosophers, social scientists have also taken a look at love. Although there is something to be said for the mystery of love, understanding how love works in the day-to-day world may help us keep our love vital and growing.

Love and American Families

Romantic love is the basis for family formation in the United States, as it has been for most of the last two centuries (Coontz 2004). Although American marriages were never quite as formally arranged as they have been in other places in the world, throughout the eighteenth century they were guided by more practical considerations and subject to more parental, especially paternal, control. By the end of the nineteenth century, however, most active parental involvement in their children's marriage choices had dissipated (Coontz 2004; Mintz 2004). Economic developments had decreased the dependency of adult children on their parents; increasingly, economic opportunity could be found without parental assistance, which freed people from worrying about the consequences of parental disapproval of their choice of mate. With increasing economic activity among women, a spread of legal and social recognition of women's rights, and enhanced opportunity for young people to meet and mingle, American courtship was further transformed (Mintz and Kellogg 1988; Murstein 1986). Love, as experienced, perceived, and pursued by individuals, became the vehicle that drove mate selection.

In the early decades of the twentieth century, new ideals about marriage and family emerged. Although American family life had already shifted from an economic to an emotional emphasis, this was extended even further with the emergence and celebration of companionate marriage, wherein spouses were to be each other's best friends, confidants, and romantic partners (Mintz and Kellogg 1988). Love was the foundation on which marriage was to be built and the criterion by which spouses were to be chosen.

Selecting a spouse on the basis of romantic love has consequences. It may lead to a greater tendency to idealize the partner, display affection toward the partner, and attach more importance to sexual intimacy (Medora et al. 2002). Ironically, perhaps, the high emphasis placed on love as the basis for spousal choice contributes to the American patterns of divorce and remarriage. The qualities that

Popular films, such as *Slumdog Millionaire*, reflect how much American popular culture emphasizes romantic love.

© Fox Searchlight/courtesy Everett Collection

people "fall in love" with may not be easy to sustain across the lifelong duration of a marriage. Thus, marriages are more likely to be perceived as "failures" when individuals sense that those qualities are gone or diminished. They may then seek those same idealized qualities from subsequent marriage partners.

Within our marriage practices, one finds a number of distinct but related cultural beliefs about the character and place of love; however, the prevalence and extent of homogamy—the tendency for people to marry others much like themselves—casts some doubt on some of our ideas about love and marriage. Perhaps love is more controllable and rational than presumed (and therefore not blind) because people seem to fall in love with people like themselves. On the other hand, perhaps love *is* blind (i.e., uncontrollable and irrational) but choosing a partner or spouse is not. Love is not the only determinant of mate selection, and people don't necessarily marry simply because they've fallen in love. Some "loves" are recognized as unwise marriages. Finally, the social circles within which a person lives and moves limit love. Thus, our "one and only" is drawn from a smaller pool than what the romantic mystique surrounding love suggests. With these qualifications in mind, it is worth remembering that most Americans who marry say they are marrying because they are in love.

In addition, there are other beliefs making up the ideology of romanticism in U.S. culture (see the Exploring Diversity feature on cultural differences of love). Many Americans believe that love strikes powerfully on first sight, that each of us has one and only one "one and only," and that as long as we love each other, everything else will work out. As we will see, these beliefs are not as widely shared in other cultures as they are in the United States and western European societies.

Love across Cultures

Neither "falling in love" nor the experience of romantic love are unique to Americans; 90% of the 166 societies examined by William Jankowiak and Edward Fischer (1992) recognize and value love as an important element in building intimate relationships. But love appears to have a more central role in American mate selection than in other Western societies (Goode 1982; Peoples

and Bailey 2006). It fits well with and helps reinforce other features of American families and society. Love-based marriage validates the importance of individual autonomy and freedom from parental intervention and control, establishes the relative independence of the **conjugal family** from the extended family, and fits with the wider social freedoms granted to adolescents and young adults (Goode 1982). All of these make romantic love functional in industrial societies (Goode 1977). Conversely, in societies in which nuclear families are deeply embedded in extended families or in which it is important for economic or political reasons to create alliances and exchanges through marriages, romantic love is not a central factor in mate selection. In such societies, it may be entirely irrelevant (Medora et al. 2002).

Love reflects the positive factors, such as caring and attraction, that draw two people together and sustain them in a relationship. Related to love, **commitment** reflects the stable factors, including not only love but also obligations and social pressure, that help maintain a relationship, for better or for worse. Although love and commitment are related, they are not inevitably connected. It is possible to love someone without being committed, without making the sacrifices and adjustments needed to sustain the relationship. It is also possible to be committed to someone without loving that person. Even in the absence (or disappearance)

© Hulton Archive/Getty Images

While it is difficult to come up with a formal definition of love, we usually know what we mean when we tell someone we love them. Such feelings are important at the individual, relationship, and institutional level.

Exploring Diversity: Isn't It Romantic?
Cultural Constructions of Love

Although most cultures recognize and value love, the meanings and expectations attached to love vary, sometimes greatly. In individualistic cultures, such as the United States, people value *passionate love*, the kind experienced as an "intense longing for another," a "lovesickness" that often takes us on a roller coaster of "elation and despair, thrills and terror" (Kim and Hatfield 2004, 174). If reciprocal, passionate love brings us ecstasy and fulfillment; if unrequited, it can bring us emptiness and sadness. In individualistic cultures, it is expected that people marry out of such an intense love, which is to be the most important factor in finding a spouse. This is part of the greater romanticism found in such societies, where prime importance is given to the emotional element of relationships and there is a stronger belief in each of the following components associated with romanticism (Medora et al. 2002):

1. Love conquers all.
2. For each person there is "one and only one" romantic match.
3. Our beloved should and will meet our highest ideals.
4. Love can and often most powerfully does strike "at first sight."
5. We should follow our hearts and not our minds when choosing a partner.

In collectivist cultures, including many Asian societies such as Japan, China, India, and Korea, individual happiness is subordinated to group well-being. Loyalty, especially to the wider kin group and extended family, dictates decisions people make about entering marriage and who they shall marry. Higher value is placed on what Elaine Hatfield and Richard Rapson (1993) call *companionate love*, a less intense emotion in which warm affection and tenderness is felt and expressed toward those to whom our lives are deeply connected. Importance is placed on shared values, commitment, intimacy, and trust. In collectivist cultures, passionate love and marriage based on romantic love are seen negatively as potential threats to family approved and/or arranged marriages, associated with sadness and jealousy, and thought to interfere with family closeness and kin obligations (Kim and Hatfield 2004). More traditional and less developed collectivist Eastern cultures, such as China and India, are reported to attach the least importance to romantic love. The idea of baring the soul, sharing, or confiding innermost and heartfelt feelings to

a partner receives more cultural validation in the United States and other individualistic cultures than in collectivist cultures (Kito 2005).

Additional cross-cultural research compared the attitudes toward romantic love of college undergraduates from the United States, Turkey, and India. In the United States, romanticism is idealized, and topics such as love, dating, and finding a partner are openly and frequently discussed or covered by the media. India is a sexually conservative and more collectivist society in which family stability is valued above individual gratification and autonomy and marriages are frequently arranged. Turkey is a society "in transition." The ideal of romantic love was introduced as part of the processes of westernization and secularization. Comparing the attitudes of college undergraduates from the three countries, researchers found the students from the United States to be most romantic and the Indian students to be least romantic (Medora et al. 2002). The study used a 29-item, five-point scale (from 1 = strongly disagree to 5 = strongly agree), wherein individuals could score between a low of 29 and a high of 145, with higher scores indicating more romanticism. Items included statements such as the following:

"Somewhere there is an ideal mate for most people. The problem is just finding that one."
"Love at first sight is often the deepest and most enduring type of love."

The average scores were as follows:

	United States	Turkey	India
	N = 200 (86 male, 114 female)	N = 223 (114 male, 114 female)	N = 218 (98 male, 120 female)
Mean score	86.09	74.92	70.33
Standard deviation	15.6	13.6	14.4

In all three national subsamples, females scored higher than males. Overall, the gender difference was as shown here:

	Male	Female
Mean score	74.63	79.81

From Medora et al. 2002.

or love, we might remain in a relationship, such as marriage, because of perceived obligation, for the sake of the children, or because of the fear of how other aspects of our life might be negatively affected. Yet, when all is said and done, most of us long for a love that includes commitment and a commitment that encompasses love.

Gender and Intimacy: Men and Women as Friends and Lovers

As shown in the previous chapter, many areas of our lives are gendered, meaning that they are experienced differently by males and females. Intimacy, friendship, and love are among such areas. In much social scientific literature, there is a recurring theme highlighting men's supposed shortcomings as friends and partners. Unlike women, who are said to relate more easily and deeply with others and who develop a greater capacity for disclosing and sharing their inner selves, men maintain greater emotional distance, even as they experience their closest relationships.

Francesca Cancian (1985) argued that there is a gender bias in our cultural constructions of love that distorts our understanding of how both men and women love. Through the **feminization of love**, in other words, by defining or "seeing" love in largely expressive terms (telling each other how you feel), important qualities or aspects of both women's and men's intimacy get ignored or overlooked. For example, much of what women do as expressions of love (for spouses and children, especially) consists of **instrumental displays**, consisting of tasks associated with nurturing and caregiving, more than **expressive displays**, such as telling others how much we care about or love them. Although done *out of love*, such activities may not be seen as *displays of love*. Likewise, because men may believe they "show" or express love by *what they do* more than by *what they say*, conceptualizing or recognizing love largely in terms of things said renders men's sincere attempts to show intimacy invisible and leaves them looking especially inadequate as intimate partners (Hook et al. 2003).

Hook and colleagues (2003) note the following gender differences in intimacy. To women, intimacy means sharing love and affection and expressing warm feelings toward someone. To men, being intimate may mean engaging in sexual behavior and being physically close. Women display intimacy in their verbal exchanges, which can become "negotiations for closeness, during which people try to reach agreement and both give and receive support" (Hook et al. 2003, 464). Women express more empathy, being more likely than men are to come to an understanding of what others are feeling. Men are more likely to react to disclosures of negative or problematic emotions by trying to solve a supposed problem. Men also associate intimacy with "doing" things together or for another person and often find women's need or desire to "talk things through" puzzling. Although men may feel as though they show intimacy by sharing activities and interests, telling stories, and even sitting together in silence, women associate intimacy with being together and sharing themselves with another (Hook et al. 2003).

Gender and Friendship

The critique of the cultural feminization of love applies to friendship as well. We tend to conceptualize "real" or "true" friendship by such qualities as emotional support and self-disclosure—telling each other innermost feelings and sharing personal experiences (Sprecher and Hendrick 2004). Friends share their inner lives with each other, including how they feel about each other. The closer the friend, the more personal and more frequent the disclosures. This conceptualization measures friendship against a standard more consistent with female friendships and may underestimate the "real" intimacy that men's friendships contain, especially if such closeness is expressed in other, more covert ways, such as through joking behavior and shared athletic activity (Hook et al. 2003; Swain 1989; Twohey and Ewing 1995).

Indeed, there are gender differences in disclosure in same-sex friendships. If intimacy means self-disclosure, then as early as age six, female friendships are more intimate. This gender difference is accentuated in adolescence and persists into and through adulthood (Benenson and Christakos 2003). Women experience and express "closeness" with each other through conversation, disclosing more of both a positive and a negative nature (Hook et al. 2003; Sprecher and Hendrick 2004).

There are other noteworthy differences between male and female friendships. In childhood and adolescence, boys spend more of their time in groups and in group activities, especially physical activities, games, and sports; girls spend more time in dyads (groups of two) and engage in more mutual disclosure. When conflict arises between close friends, males may have an easier time reaching resolution. Within a group context, we can draw others in, drawing on third parties to act as mediators, serve as allies, or even become alternate partners. With more loyalty to the larger group, one-on-one conflict may be kept to a lower level (Benenson and Christakos 2003).

Boys spend less time sharing and less time coruminating (excessively discussing personal problems) but as a consequence may be spared some of the emotional fallout that results from dwelling on problems (Rose 2002). In addition, men display less affection, using either words or touch, than women do toward their friends (Dolgin 2001). Yet female friendships appear to be more fragile. With increasingly intense sharing comes more opportunity for misunderstanding or even for conflict. Furthermore, when females' closest friendships end, they are more likely to "find themselves alone" (Benenson and Christakos 2003).

Heterosexual men are more open and intimate in cross-sex relationships than in their friendships with other men (Dolgin 2001). Wives or romantic partners are often the closest confidants in men's lives. In those relationships, men find themselves reaping the benefits that come from greater disclosure, even if the levels at which they disclose don't match what their partners' desire. Certainly, the tendency to funnel their intimacy into one relationship, especially marriage, is consistent with the cultural expectations of marriage as best-friendship. But even outside marriage, the depth of men's disclosure to women stands in contrast to the male–male style, suggesting not so much inability as unwillingness at or discomfort with male–male intimacy.

A unique cross-sex friendship type is known as the "friends-with-benefits" relationship. Such relationships emerge out of cross-sex friendships and incorporate sexual intimacy into those relationships without the expectation or desire for greater commitment or pushing the relationship into a love relationship (Hughes, Morrison, and Asada 2005). They retain the benefits of friendship with neither the responsibilities nor the commitment of a romantic sexual relationship. Friends with benefits relationships may be most common on college campuses, where studies have reported that from 48% to 62% of college student respondents have experienced such relationships (Hughes et al. 2005).

Gender and Love

With regard to love, the genders differ in a number of ways. Men fall in love more quickly than women, describe more instrumental styles of love (i.e., love as "doing"), and are more likely to see sex as an expression of love. Because men have fewer deeply intimate, self-disclosing friendships, when they find this quality in a relationship, they are more likely to perceive that relationship as special. Having more intimates with whom they can share their feelings, women are less likely to be as quick to characterize a particular relationship as love. In addition, traditionally, women could do so less safely unless other, economic criteria were also met. Thus, men could afford to be more romantic, and women needed to be more realistic, taking into account a man's suitability as a stable provider along with his ability to satisfy needs for intimacy (Knox and Schacht 2000). In a study of 147 never-married undergraduate students designed to look for and at gender differences in timing and reason for saying "I love you," David Knox, Marty Zusman, and Vivian Daniels (2002) found the following:

- In heterosexual relationships, males say "I love you" before their partners do.
- Males say "I love you" in part to increase the likelihood that their partner will agree to have sex with them.

Other gender differences surface in the connection between love and sex. Although men are often depicted as easily separating sex and love, there is evidence that within relationships, men see sex as a means of expressing or showing love (Cancian 1987; Rubin 1983). Women's experiences of love and sexuality are different. Although sexual scripts have changed in the direction of more open and acceptable expressions of female sexuality, to feel loved requires more than sexual expression.

Gender differences may be more exaggerated in *what people say* than in *what they do*. This may be true of both friendship and love, where bigger differences show up in how the genders talk *about* relationships than in what they actually do or experience (Eastwick and Finkel 2008; Walker 1994). Despite the gendered expectations and definitions of love, within the context of heterosexual, romantic love relationships, significant gender differences in self-disclosure are absent; men and women disclose similarly (Sprecher and Hendrick 2004).

In identifying the factors that shape men's and women's intimate relationships and *could* contribute to differences between the genders, most researchers point to lessons of gender socialization. Within dominant cultural constructions of masculinity and femininity, males are inexpressive, competitive, rational, and uncomfortable with revealing their innermost feelings, especially of vulnerability or of affection toward other males. Females are allowed and encouraged to express a wider range of feelings without concern for the consequences.

Other researchers suggest that gender-specific relationship styles emerge because of differences in how males and females resolve the developmental task of early childhood identity formation (Chodorow 1978; Rubin 1985). As a result of being "mothered" and having the closest early relationship be with a female, the genders develop different ways of relating. Females develop "permeable ego boundaries" open to relationships with others, and they retain a strong connection with their mothers. Males are forced to separate from their mothers, identify with absent or less present fathers, and build boundaries around themselves in relation to their most nurturant caregivers. This haunts them throughout their later relationships because it makes them less able to "connect" intimately with others (Rubin 1985). Women experience themselves in the context of relationships, whereas men—depicted as "selves in separation"—remain oriented more toward independence and task completion (Kilmartin 1994).

Even without attention to personality formation during early development, we might also emphasize the likely role-model consequences of being "mothered" but not "fathered." Without a loving, attentive, nurturing presence from fathers or other male role models, boys come to inhibit their own emotional expressiveness, identifying such behavior as typical of mothers (and women in general) and to be avoided. Because of the relative involvement of mothers versus fathers in caring for most young children and the greater prevalence of single-mother over single-father households, boys have fewer available role models for intimacy. Furthermore, what role models they have are products of gender socialization and carry a style of relating that results from that socialization. Girls have the opportunity to observe up close a caring, loving female role model from which they learn how to relate and express love.

Finally, still others stress evolutionary explanations for gender differences. Beginning with the idea that each gender has different "reproductive strategies," differences in intimacy are linked to such sex-specific goals. For males, the objective is to reproduce as widely as possible, seeing that their genetic material is spread widely in multiple offspring. For women, the objective is to see that each child successfully survives to a healthy adulthood. Such a difference is said to explain numerous other differences, especially in areas of intimacy, love, and sexuality. For example, from an evolutionary perspective on qualities desired in romantic and sexual partners, females will desire males of high status who are ambitious and dependable. Males will desire physically attractive females who can bear them healthy children (Sprecher and Regan 2002).

Exceptions: Love in Nontraditional Relationships

The gender differences depicted earlier, although common in the literature on love and intimacy, are not inevitable. As introduced in the previous chapter, there are "postgender" marriages and intimate heterosexual relationships that depart from traditional gender patterns (Risman and Johnson-Sumerford 1997; Schwartz 1994). Pepper Schwartz's research on peer couples illustrates loving relationships that avoid the aforementioned gender patterns. Schwartz conceptualized peer marriage as a relationship built on principles of equity, equality, and deep friendship. The emphases on equity and equality resulted in shared chores, equal say in decision making, and equal involvement in child rearing. More important for now is the element of deep friendship. These couples most valued an intense companionship, "a collaboration of love and labor in order to produce profound intimacy and mutual respect" (Schwartz 1994, 2).

In peer marriages, spouses become more alike over time, and thus both husbands and wives are more likely to display and value a blend of female and male styles of intimacy. Women value and appreciate the instrumental displays of love from their partners (e.g., finding that her husband has had her car serviced) because they know what it is like to make or take time from their demanding daily routines to attend to such things. Because husbands are more involved in daily domestic and child-rearing routines, they share interests and concerns with their wives that traditional spouses do not. With enlarged identities outside the marriage and home, peer wives also need less of the conventional, conversational demonstrations of love and affection. They and their husbands have learned to "love on each other's terms" (Schwartz 1994, 58).

Couples such as these are uncommon. They represent some possibilities in love and/or marital relationships, but creating such lifestyles requires both an ideological commitment to sharing and equality and an ability to withstand scrutiny and curiosity from more typical couples. Most such lifestyles also require each spouse to have a job or career that the other recognizes as equal in importance to his or her own. Same-sex couples appear to have more equal relationships. This equality is expressed in everything from the division of domestic and wage-earning responsibilities to how often they argue, what they argue about, and how they resolve such conflicts (Parker-Pope 2008).

Showing Love: Affection and Sexuality

Within relationships based on romantic or passionate love, the emotional connection between partners is expressed in many ways, typically through displays of affection and through sexual desire and activity. The state of "being in love" is assumed by most people to include sexual desire. Two people in a relationship lacking sexual desire are assumed to not be in love (Regan 2000). Psychologist Lisa Diamond (2004) challenges this assumption, noting that sexual desire often occurs in the absence of romantic or passionate love and, more controversially, that romantic love, even in its earliest and most passionate stage, does not require sexual desire. She offers the examples of prepubertal children who have yet to undergo the hormonal changes necessary for adult levels of sexual desire who, nonetheless, experience intensely romantic infatuations as well as instances wherein people fall in love with others of the wrong gender (heterosexuals with someone of the same sex and gay men and lesbians who fall in love with someone of the other gender). Regarding the latter, Diamond notes that both cross-cultural and historical evidence documents such romantic relationships, even in the absence of sexual desire.

Although love and sex are separate phenomena, recent research shows that for both men and women, sex often includes intimacy and caring, key aspects of love, and love is most often expected to include sexual desire. Men and women who feel the greatest sexual desire for dating partners are also likely to report the strongest feelings of passionate love. Interestingly, sexual *activity* (as measured by the average weekly number of "sexual events" in which partners engaged) is not as strongly associated with amount or depth of passionate love (Regan 2000).

Besides sexual intimacy, there are many other ways we show intimacy and love. Some such displays occur openly, in public, as we say or do things that show others that we are a couple. Holding hands, being out together alone, telling others that we are a couple, and meeting our partner's parents are examples of public displays of affection and couple status (Vaquera and Kao 2005). More privately, we may exchange presents, tell each other how we feel (saying that we love each other), and just think about ourselves as a couple. Finally, the physical acts, from kissing to touching under clothes or with no clothes on, touching each other's genitals, and having sexual intercourse, are all "intimate displays" (Vaquera and Kao 2005).

Using data drawn from the 1994–1995 National Longitudinal Study of Adolescent Health with its large, nationally representative sample of high school students, Elizabeth Vaquera and Grace Kao (2005) examined how displays of affection varied between intraracial and interracial couples. Noting that interracial relationships are still a small percentage of all couple relationships in the United States and therefore may not be as openly or enthusiastically accepted as intraracial couple relationships, Vaquera and Kao looked at whether the potential "stigmatizing" of such relationships led to different ways of behaving as a couple. Other research has determined that interracial couples limit their public visibility, thus reducing their exposure to negative reactions, rejection, and pressure. Vaquera and Kao found that in terms of both public and private displays, interracial couples display lower levels of affection. However, when it comes to intimate displays of affection (e.g., kissing, touching one's genitals, and sexual intercourse), no difference was found between interracial and intraracial couples. This is consistent with the attention to stigma because only in settings that involve no others are there no differences between interracial and intraracial couples.

Vaquera and Kao (2005) also report differences in the displays of affection across racial groups (i.e., among intraracial couples of different racial backgrounds). They determined that compared with Caucasians, African American couples displayed less public affection but more intimate affection. Hispanics, Asians, and Native Americans display lower

Bob Pool/Getty Images/Photographer's Choice

Interracial relationships are increasing but are still relatively uncommon. Such couples often find themselves the recipients of negative reactions.

levels of intimate affection in public than do African American or Caucasian couples. All minority couples displayed less public affection than white couples did. In terms of "private displays," no statistically significant differences were found among racial groups (Vaquera and Kao 2005).

Gender, Sexuality, and Love

For both women and men, sexual desire—but not sexual activity—is associated with passionate love (Regan 2000). Yet gender differences have been observed in the relationship between love and sex. Men and women who are not in an established relationship have different expectations. Men are more likely than women to more easily separate sex from affection, whereas women attach greater importance to relationships as the "context" for sexual expression (Diamond 2004; Laumann et al. 1994). Lisa Diamond suggests that there are a number of possible reasons for this gender difference. First, men are more likely than women to first experience sexual arousal "in the solitary context of masturbation," whereas women are more likely to experience sexual arousal for the first time within a heterosexual relationship. Second,

as shown in the next chapter, women and men have been differently socialized about the legitimacy of sexual expression. Women have been expected and encouraged to restrict sexual desire and activity to intimate relationships in which they find themselves. Men have been raised with more "license" regarding casual sexual relationships. Finally, Diamond notes that biological factors may partly explain the gender difference. Specifically, certain neurochemicals, such as oxytocin, that mediate bonding also mediate sexual behavior. Much as oxytocin might be associated with caregiving, it is also released in greater amounts in women than in men during sexual activity. Oxytocin is also associated with orgasmic intensity (Diamond 2004).

Sexual Orientation and Love

Love is equally important for heterosexuals, gay men, lesbians, and bisexuals (Aron and Aron 1991; Keller and Rosen 1988; Kurdek 1988; Patterson 2000; Peplau and Cochran 1988). Given that men, in general, are more likely than women to separate love and sex, it is unsurprising that gay men are especially likely to make this separation. Although gay men value love, they also tend to value sex as an end in itself. Furthermore, they place less emphasis on sexual exclusiveness in their relationships (Patterson 2000). Researchers suggest, however, that heterosexual males are not very different from gay males in terms of their acceptance of casual sex. Lesbians and heterosexual couples tend to be more supportive than gay men of monogamy and sexual fidelity. This is probably because of gender more than sexual orientation; heterosexual males would be as likely as gay males to engage in casual sex if women were equally interested. Women, however, are not as interested in casual sex; as a result, heterosexual men do not have as many willing partners available as do gay men (Hendrick and Hendrick 1995).

For lesbians, gay men, and bisexuals, love has special significance in the formation and acceptance of their identities. Although significant numbers of women and men have had sexual experiences with members of the same sex or both sexes, relatively few identify themselves as lesbian or gay. An important element in solidifying such an identity is loving someone of the same sex. Love signifies a commitment to being gay or lesbian by unifying the emotional and physical dimensions of a person's sexuality (Troiden 1988). For the gay man or lesbian, it marks the beginning of sexual wholeness and acceptance. Some researchers

Love transcends sexual orientation. Heterosexual and same-gender couples value and seek loving relationships.

believe that the ability to love someone of the same sex, rather than having sex with him or her, is the critical element that distinguishes being gay or lesbian from being heterosexual (Money 1980).

Love, Marriage, and Social Class

Gender is only one variable related to how we experience love and how love is associated with marriage. In many ways, our romantic view of love-based marriage represents a middle-class version of marriage. Among upper-class families, there is a greater urgency in ensuring that their children marry the "right kind" because considerable wealth and social position may be at stake. Furthermore, upper-class families have more ability to exercise such control by the threat of withholding inheritance from the maverick child who dares act without consideration of parental preference (Goode 1982). Among the working class, marriage was often entered as a means to escape economic instability and parental authority and to be seen as an adult (Rubin 1976, 1992). This may now be less true, as working-class marriages have taken on more characteristics of the middle-class ideal (e.g., expecting more sharing and communication) (Rubin 1995). Still, the economic circumstances that define someone's life may induce different ways of linking love and marriage.

But What *Is* This "Crazy Little Thing Called Love"?

Despite centuries of discussion, debate, and complaint by philosophers and lovers, no one has succeeded in finding a single definition of love on which all can agree.

Ironically, such discussions seem to engender conflict and disagreement rather than love and harmony.

Because of the unending confusion surrounding definitions of love, some researchers wonder whether such definitions are even possible (Myers and Shurts 2002). In the everyday world, however, most people seem to have something in mind and agree on what they mean when they tell someone they love him or her. We may not so much have formal definitions of love as we do **prototypes** of love (i.e., models of what we mean by love) stored in the backs of our minds. Some researchers suggest that instead of looking for formal definitions of love, it is more important to examine people's prototypes, that is, to consider what people mean by the concept of love when they use it. When someone says "I love you," he or she is referring to a prototype of love rather than its definition. If one finds oneself thinking about one's partner all the time, feeling happy when with them and sad (or less happy) when apart, and spending all one's available time together, these thoughts, feelings, and behaviors are compared against a mental model or prototype of love (Regan 2003). If one's experiences match the different characteristics of love, individuals then define themselves as in love (Regan, Kogan and Whitlock 1998). By thinking in terms of prototypes, we can study how people actually use the word *love* in real life and how the meanings they associate with love help them define and assess the progress of their intimate relationships.

In an effort to discover people's prototypes, researcher Beverly Fehr (1988) asked respondents to rate the central features of love and commitment. There were 68 different features of love mentioned by at least two respondents. In order, the 12 *central* attributes of love they listed are as follows:

- Trust
- Caring
- Honesty
- Friendship
- Respect
- Concern for the other's well-being
- Loyalty
- Commitment
- Acceptance of the other the way he or she is
- Supportiveness
- Wanting to be with the other
- Interest in the other

Many other characteristics were also identified as features of love (euphoria, thinking about the other all the time, butterflies in the stomach, and so on). These, however, tend to be *peripheral*. As relationships progress, the central aspects of love become more characteristic of the relationship than the peripheral ones. According to Fehr (1988), the central features "act as true barometers of a move toward increased love in a relationship." Similarly, violations of central features of love are considered more serious than violations of peripheral ones. A loss of caring, trust, honesty, or respect threatens love, whereas the disappearance of butterflies in the stomach does not. Psychologist Pamela Regan, along with coauthors Elizabeth Kocan and Teresa Whitlock, undertook a prototype analysis of the more specific concept of *romantic love*. When they asked their respondents to list all the features or characteristics of "being in love," participants identified 119 different features. Of these, the central features of romantic love were trust, honesty, sexual attraction, acceptance and tolerance, happiness, spending time together, and sharing thoughts and secrets. There were few negative aspects noted, and those that were (e.g., anger, depression, and fear) were seen as peripheral characteristics (Regan et al. 1998).

Love is also expressed behaviorally in several ways, with the expression of love often overlapping thoughts of love:

- Verbally expressing affection, such as saying "I love you"
- Self-disclosing, such as revealing intimate personal facts
- Giving nonmaterial evidence, such as offering emotional and moral support in times of need and showing respect for the other's opinion
- Expressing nonverbal feelings such as happiness, contentment, and security when the other is present
- Giving material evidence, such as providing gifts or small favors or doing more than the other's share of something
- Physically expressing love, such as by hugging, kissing, and other sexual activity
- Tolerating and accepting the other's idiosyncrasies, peculiar routines, or annoying habits, such as forgetting to put the cap on the toothpaste

These behavioral expressions of love are consistent with the prototypical characteristics of love. In addition, research supports the belief that people "walk on air" when they are in love. Researchers have found that those in love view the world more positively than those who are not in love (Hendrick and Hendrick 1988).

Studying Love

A review of the research on love finds a number of definitions that are tied to a variety of research instruments that have been developed to measure love. Jane Myers and W. Matthew Shurts (2002) reviewed the instruments that researchers have developed and that other researchers and/or clinicians might use, ultimately identifying nine different instruments. Next we look briefly at some of the more influential approaches to studying love.

Styles of Love

Hendrick and Hendrick's Love Attitude Scale is a 42-item instrument based on and designed to measure sociologist John Lee's (1973, 1988) six styles of love:

- *Eros.* Romantic or passionate love. Erotic lovers delight in the tactile, the sensual, and the immediate; they are attracted to beauty (although beauty may be in the eye of the beholder). They love the lines of the body and its feel and touch. They are fascinated by every detail of their beloved. Their love burns brightly but soon flickers and dies.
- *Ludus.* Playful or game-playing love. For ludic lovers, love is a game, something to play at rather than to become deeply involved in. Love is ultimately ludicrous. Love is for fun; encounters are casual, carefree, and often careless. "Nothing serious" is the motto of ludic lovers.
- *Storge.* (pronounced STOR-gay) is the love between companions. It is, writes Lee, "love without fever, tumult, or folly, a peaceful and enchanting affection." It usually begins as friendship and then gradually deepens into love. If the love ends, it also occurs gradually, and the couple often becomes friends once again.

The first three are "primary" styles, which can be combined to generate the following secondary styles:

- *Mania.* Coming from the Greek word for "madness," mania is an obsessive love, a combination of ludus and eros, characterized by an intense love–hate relationship. For manic lovers, nights are marked by sleeplessness and days by pain and anxiety. The slightest sign of affection brings ecstasy briefly, only to have it disappear. Satisfactions last but a moment before they must be renewed. Manic love is roller-coaster love.
- *Agape.* (pronounced ah-GAH-pay) A combination of eros and storge, agape is altruistic love. Agape is

love that is chaste, patient, selfless, and undemanding; it does not expect to be reciprocated. Agape emphasizes nurturing and caring as their own rewards. It is the love of monastics, missionaries, and saints more than that of worldly couples.

- *Pragma.* A combination of storge and ludus, pragma is a practical, pragmatic style of love. Pragmatic lovers are primarily logical in their approach toward looking for someone who meets their needs. They look for a partner who has the background, education, personality, religion, and interests compatible with their own. If they meet a person who meets their criteria, erotic or manic feelings may develop.

These styles, Lee cautions, are relationship styles, not individual styles. The style of love may change as the relationship changes or when individuals enter different relationships. In addition to these pure forms, there are mixtures of the basic types: storgic–eros, ludic–eros, and storgic–ludus. According to Lee, a person must thus find a partner who shares the same style and definition of love to have a mutually satisfying love affair. The more different two people are in their styles of love, the less likely it is that they will understand each other's love. Of these six basic types, psychologists Bianca Acevedo and Arthur Aron (2009) suggest that *romantic love* is best seen as eros, without the obsessiveness of mania and as distinct from the more friendship-like attachment of storge.

Love styles are also linked to gender and ethnicity (Hendrick and Hendrick 1986). Research indicates that heterosexual and gay men have similar attitudes toward eros, mania, ludus, and storge and that gay male relationships have multiple emotional dimensions (Adler, Hendrick, and Hendrick 1989). Women endorse the storgic and pragma styles and outscore men on measures of the manic style (Hendrick and Hendrick 1995). As to cultural differences, different styles tend to characterize Asians, African Americans, Latinos, and Caucasians. Asian Americans have a more pragmatic style of love than do Latinos, African Americans, or Caucasians, and they place a high value on affection, trust, and friendship (pragma and storge). Latinos often score higher on the ludic characteristics (Regan 2003).

Hatfield and Sprecher's Passionate and Companionate Love

Hatfield and Sprecher divide love into two types: passionate and companionate. As shown earlier in the Exploring Diversity feature, **passionate love** is "an intense longing for union with another" and is familiar to us because it most fits our ideas of romantic love, or the state of being in love (Kim and Hatfield 2004). Passionate love can be seen through cognitive, emotional, and behavioral indicators. **Companionate love** refers more to the warm and tender affection we feel for close others. It includes friendship, shared interests and activities, and companionship. It is milder and less intense, may lack sexual attraction or desire, and produces less of the extreme highs and lows people experience from passionate love (Acevedo and Aron 2009; Kim and Hatfield 2004).

Sternberg's Triangular Theory of Love

Sternberg's Triangular Love Scale is based on his **triangular theory of love**. According to the theory, love is composed of three elements that can be visualized as the points of a triangle: intimacy, passion, and decision or commitment. The *intimacy* component refers to the warm, close feelings of bonding you experience when you love someone. It includes such things as giving and receiving emotional support to and from your partner, being able to communicate with your partner about intimate things, being able to understand each other, and valuing your partner's presence in your life. The *passion* component refers to the elements of romance, attraction, and sexuality in a relationship. These may be fueled by the desire to increase self-esteem, to be sexually active or fulfilled, to affiliate with others, to dominate, or to subordinate. The decision or *commitment* component consists of two parts, one short term and one long term. The short-term part refers to your decision that you love someone. You may or may not make the decision consciously, but it usually occurs before you decide to make a commitment to that person. The commitment represents the long-term aspect; it is the maintenance of love, but a decision to love someone does not necessarily entail a commitment to maintaining that love.

Each of these components can be enlarged or diminished in the course of a relationship, and their changes will affect the quality of the relationship. They can also be combined in different ways in different relationships or even at different times in the same love relationship, creating one of the following eight different outcomes:

- Liking (intimacy only)
- Romantic love (intimacy and passion)
- Infatuation (passion only)
- Fatuous love (passion and commitment)

- Empty love (decision or commitment only)
- Companionate love (intimacy and commitment)
- Consummate love (intimacy, passion, and commitment)
- Nonlove (absence of intimacy, passion, and commitment)

These types represent extremes that probably few of us experience. Not many of us, for example, experience infatuation in its purest form, in which there is no intimacy. The categories are nevertheless useful for examining love (except for empty love, which is not really love).

Romantic love combines intimacy and passion. It is similar to liking, but it is more intense and includes a physical or emotional attraction. It may begin with an immediate union of the two components—with friendship that intensifies with passion or with passion that develops intimacy. Although commitment is not an essential element of romantic love, it may develop. Sternberg also believed that throughout the duration of even successful relationships, passion lessens, while intimacy and commitment increase (Acevedo and Aron 2009).

What then results is companionate love. Companionate love often begins as romantic love, but as the passion diminishes and the intimacy increases, it is transformed. Some couples are satisfied with such love; others are not. Those who are dissatisfied in companionate love relationships may seek extrarelational affairs to maintain passion in their lives. They may also end the relationship to seek a new romantic relationship in the hope that it will remain romantic.

Love and Attachment

The **attachment theory of love** maintains that the degree and quality of attachments one experiences in early life influence one's later relationships. It examines love as a form of attachment that finds its roots in infancy (Hazan and Shaver 1987; Shaver, Hazan, and Bradshaw 1988). Phillip Shaver and his associates (1988) suggest that "all important love relationships—especially the first ones with parents and later ones with lovers and spouses—are attachments." On the basis of infant–caregiver work by John Bowlby (1969, 1973, 1980), some researchers suggest numerous similarities between attachment and romantic love (Bringle and Bagby 1992; Downey, Bonica, and Rincon 1999; Shaver et al. 1988). These include the following:

Attachment	Love
Attachment formation and quality depend on attachment object's responsiveness, interest, and reciprocation.	Feelings of love are related to lover's feelings.
When attachment object is present, infant is happier.	When lover is present, person feels happier.
Infant shares toys, discoveries, and objects with attachment object.	Lovers share experiences and goods and give gifts.
Infant coos, talks baby and "sings."	Lovers coo, talk baby talk, talk, and sing.
There are feelings of oneness with attachment object.	There are feelings of oneness with lover.

According to research by Geraldine Downey (1996), rejection by parents of their children's needs can lead to the development of **rejection sensitivity**, or the tendency to anticipate and overreact to rejection. Individuals who develop rejection sensitivity seek to avoid rejection by their partners and closely monitor, even overanalyze, the relationship dynamics for signs of potential rejection. As Pamela Regan (2003) notes, even "minimal or ambiguous" rejection cues may lead to feelings of rejection and to anger, jealousy, and despondency. Rejection-sensitive people tend to be less satisfied with their relationships and more likely to see them end.

Based on studies conducted by Mary Ainsworth and colleagues (1978, cited in Shaver et al. 1988), there are three styles of infant attachment: (1) secure, (2) anxious or ambivalent, and (3) avoidant. In *secure attachment*, the infant feels secure when the mother is out of sight. He or she is confident that the mother will offer protection and care. In *anxious or ambivalent attachment*, the infant shows separation anxiety when the mother leaves. He or she feels insecure when the mother is not present, and this insecurity results from her being inconsistently available, leaving the infant afraid to leave her side (Cassidy and Berlin 1994). In *avoidant attachment*, the infant senses the mother's detachment and rejection when he or she desires close bodily contact. The infant shows avoidance behaviors with the mother as a means of defense. In Ainsworth's study, 66% of the infants were secure, 19% were anxious or ambivalent, and 21% were avoidant.

Some researchers (Feeney and Noller 1990; Shaver et al. 1988) believe that the styles of attachment developed during infancy continue through adulthood. Others, however, question the validity of applying infant research to adults as well as the stability of attachment styles throughout life (Hendrick and Hendrick

1994). Still others found a significant association between attachment styles and relationship satisfaction (Brennan and Shaver 1995).

Secure Adults

Secure adults find it relatively easy to get close to others. They are comfortable depending on others and having others depend on them. They believe that they are worthy of love and support and expect to receive them in their relationships (Regan 2003). They generally do not worry about being abandoned or having someone get too close to them. More than avoidant and anxious or ambivalent adults, they feel that others generally like them; they believe that people are generally well intentioned and good-hearted. In contrast to others, secure adults are less likely to believe in media images of love and more likely to believe that romantic love can last. Their love experiences tend to be happy, friendly, and trusting. They are more likely to accept and support their partners. Reportedly, compared to others, secure adults find greater satisfaction and commitment in their relationships (Pistol, Clark, and Tubbs 1995).

Anxious or Ambivalent Adults

Anxious or ambivalent adults feel that others do not or will not get as close as they themselves want. They feel unworthy of love, need approval from others, and worry that their partners do not really love them or that they will leave them (Regan 2003). They also want to merge completely with the other person, which sometimes scares that person away. More than others, anxious or ambivalent adults believe that it is easy to fall in love. Their experiences in love are often obsessive and marked by a desire for union, high degrees of sexual attraction and jealousy, and emotional highs and lows.

Avoidant Adults

Avoidant adults feel discomfort in being close to others; they are distrustful and fearful of becoming dependent (Bartholomew 1990). Thus, to avoid the pain they expect to come from eventual rejection, they maintain distance and avoid intimacy (Regan 2003). More than others, they believe that romance seldom lasts but that at times it can be as intense as it was at the beginning. Their partners tend to want more closeness than they do. Avoidant lovers fear intimacy and experience emotional highs and lows and jealousy.

In adulthood, the attachment styles developed in infancy combine with sexual desire and caring behaviors to give rise to romantic love. Comparing across these three types of attachment styles indicates that women and men with secure attachment styles tend to be the preferred type of romantic partner by women and men alike. They also tend to find more satisfaction in their relationships, experience more happiness, hold more positive views of their partners, and display fewer negative emotions (Regan 2003).

Love and Commitment

We expect our romantic partner to be there for us through "thick and thin." Later, should we enter into marriage, we pledge our love, "for better for worse, for richer, for poorer, in sickness and in health, till death do us part." In other words, we expect that, along with loving us, our partners will be committed to us and to our relationship. Although we generally make commitments to a relationship because we love someone, love alone is not sufficient to make a commitment last. Our commitments seem to be affected by several factors that can strengthen or weaken the relationship. Ira Reiss (1980a) believes that there are three important factors in commitment to a relationship. These factors interact to increase or decrease the commitment.

1. *The balance of costs and benefits.* Whether we like it or not, humans have a tendency to look at romantic and marital relationships from a cost–benefit perspective. Most of the time, when we are satisfied, we are unaware that we judge our relationships in this manner. But as shown in our discussion of social exchange theory in Chapter 2, when there is stress or conflict, we might ask ourselves, "Just what am I getting out of this relationship?" Then we add up the pluses and minuses. If the result is on the plus side, we are encouraged to continue the relationship; if the result is negative, we are more likely to discontinue it.

2. *Normative inputs.* Normative inputs for relationships are the values that you and your partner hold about love, relationships, marriage, and family. These values can either sustain or detract from a commitment. How do you feel about a love commitment? A marital commitment? Do you believe that marriage is for life? Does the presence of children affect your beliefs about commitment? What are the values that your friends, family, and religion hold regarding your type of relationship?

3. *Structural constraints.* The structure of a relationship will add to or detract from commitment. Depending on the type of relationship—whether it is dating, living together, or marriage—different roles and expectations are structured. In marital relationships, there are partner roles (husband–wife) and economic roles (employed worker–homemaker). There may also be parental roles (mother–father).

Commitments are more likely to endure in marriage than in cohabiting or dating relationships, which tend to be relatively short lived. They are more likely to last in heterosexual relationships than in gay or lesbian relationships (Testa, Kinder, and Ironson 1987). The reason that commitments tend to endure in marriage may or may not have anything to do with a couple being happy. Marital commitments may last because norms and structural constraints compensate for the lack of personal satisfaction.

For most people, love seems to include commitment and commitment seems to include love. Beverly Fehr (1988) found that if a person violated a central aspect of love, such as caring, that person was also seen as violating the couple's commitment. If a person violated a central aspect of commitment, such as loyalty, it called love into question. Because of the overlap between love and commitment, we can mistakenly assume that someone who loves us is also committed to us. As one researcher points out, "Expressions of love can easily be confused with expressions of commitment. . . . Misunderstandings about a person's love versus commitment can be based on honest errors of communication, on failures of self-understanding" (Kelley 1983). Or a person can intentionally mislead the partner into believing that there is a greater commitment than there actually is. Even if a person is committed, it is not always clear what the commitment means: Is it a commitment to the person or to the relationship? Is it for a short time or a long time? Is it for better and for worse?

Finding Love and Choosing Partners

Do you know what this is?

> Real Life Juliet Seeks Romeo: I have been searching the world over, looking for my true love. I am a friendly, ambitious, compassionate, hardworking female, who enjoys music, dancing, travel, and the beach. Looking for someone who wants to share a movie, dinner, a laugh, and maybe a lifetime. I know you're out there somewhere.

Of course, most people will recognize that this is a personal ad, one of the many that can be found each day in newspapers and magazines or on multiple sites on the Internet. Along with dating services, computer matchmakers, and singles clubs, such personal ads represent some more recent ways in which Americans go about trying to find their "one and only." In recent years, even reality television programs have been added into the mix, pushing such attempts into previously uncharted water. On February 15, 2000, the Fox network aired *Who Wants to Marry a Multi-Millionaire?*, only the first of many reality shows to capitalize on our age-old fascination with how people get together.

There is considerable social science interest, too, in understanding how people find their spouses or partners. In addition, researchers have studied *who* we choose and *why* we choose those particular individuals. Although love is the major criterion used to select a spouse and most people who marry would say they are doing so out of love, many factors operate alongside and on love. In theory, most of us are free to select as partners those people with whom we fall in love, but other factors enter the process, and our choices become somewhat limited by societal norms and rules of mate selection. Once you understand some principles of mate selection in our culture, without ever having met a friend's new boyfriend or girlfriend, you can deduce many things about him or her. For example, if a female friend at college has a new boyfriend, you would be safe in guessing that he is about the same age or a little older, probably taller, and a college student. Furthermore, he is probably about as physically attractive as your friend (if not, their relationship may not last); his parents probably are of the same ethnic group and social class as hers; and he is probably about as intelligent as your friend. If a male friend has a new girlfriend, many of the same things apply, except that she is probably the same age or younger and shorter than he is. Some relationships will depart from such conventions, and many will have one or two characteristics on which the partners differ (or differ more), but you will probably be correct in most instances. These are not so much guesses as deductions based on the principle of **homogamy**, the pairing of people with similar characteristics or from similar backgrounds. This is discussed more fully in Chapter 9.

The Relationship Marketplace

The process of choosing partners is affected by bargaining and exchange. People select each other in a kind of **marketplace of relationships**. The notion of a "marketplace" effectively conveys that, as in a commercial marketplace, when people form relationships, they enter exchange relationships, much as when they exchange goods. Unlike a real marketplace, however, the "relationship marketplace" is more of a process, not a place, in which *people* are the goods exchanged. Each of us has certain resources—such as socioeconomic status, looks, and personality—that determine our marketability. As Matthijs Kalmijn (1998) puts it, "Potential spouses are evaluated on the basis of the resources they have to offer, and individuals compete with each other for the spouse they want most by offering their own resources in return." Individuals bargain with the resources they possess. They size themselves up and rank themselves as a good deal, an average package, or something "available at a discount" because of lack of interested "buyers"; they do the same with potential dates and, ultimately, mates. Such "exchanges" are more often between equally valuable goods. In other words, people tend to seek others about *as attractive* or *as intelligent* as they perceive themselves.

People tend to choose partners who are about as attractive as themselves.

Physical Attractiveness: The Halo Effect, Rating, and Dating

Pretend for a moment that you are at a party, unattached. You notice that someone is standing next to you as you reach for some chips or a drink. He or she says hello. In that moment, you have to decide whether to engage him or her in conversation. On what basis do you make that decision? Is it looks, personality, style, sensitivity, intelligence, or something else?

Most people consciously or unconsciously base this decision on appearance. If you decide to talk to the person, you probably formed a positive opinion about how he or she looked. In other words, he or she looked "cute," looked like a "fun person," gave a "good first impression," or seemed "interesting." Physical attractiveness is particularly important during the initial meeting and early stages of a relationship.

Critical Thinking

How important are looks to you? Think back. Have you ever mistakenly judged someone by his or her looks? How did you discover your error? How did you feel?

The Halo Effect

Most people would deny that they are attracted to others *just* because of their looks. However, individuals tend to infer qualities in others based on looks. This inference is called the **halo effect**—the assumption that good-looking people possess more desirable social characteristics than unattractive people. In a well-known experiment (Dion et al. 1972), students were shown pictures of attractive people and asked to describe what they thought these people were like. Attractive men and women were assumed to be more sensitive, sexually responsive, poised, and outgoing than others; they were assumed to be more exciting and to have better characters than "ordinary" people. Furthermore, attractive people are preferred as friends, candidates, and prospective employees, and they receive more leniency if they are defendants in court (Ruane and Cerulo 2004). Research indicates that, overall, the differences between perceptions of attractive and average people are minimal. It is when attractive and average people are compared to those considered to be unattractive that there are pronounced differences, with those perceived as unattractive being rated more negatively (Hatfield and Sprecher 1986).

In more casual relationships, the physical attractiveness of a romantic partner is especially important. Elaine Hatfield and Susan Sprecher (1986) suggest three reasons people come to prefer attractive people over unattractive ones. First, there is an "aesthetic appeal," a simple preference for beauty. Second, there is the "glow of beauty," in which we assume that good-looking people are more sensitive, modest, self-confident, sexual, and so on. Third, there is the "status" by association we achieve by dating attractive people.

Research has demonstrated that good-looking companions increase our status. In one study, men were asked their first impressions of a man seen alone, arm in arm with a beautiful woman, and arm in arm with an unattractive woman. The man made the best impression with the beautiful woman. He ranked higher alone than with an unattractive woman. In contrast to men, women do not necessarily rank as high when seen with a handsome man. A study in which married couples were evaluated found that it made no difference to a woman's ranking if she was unattractive but had a strikingly handsome husband. If an unattractive man had a strikingly beautiful wife, it was assumed that he had something to offer other than looks, such as fame or fortune.

Trade-Offs

In mixing with and meeting people, one doesn't necessarily gravitate to the *most attractive* person in the room but rather to those about as attractive as oneself. Sizing up someone at a party or dance, a man may say, "I'd have no chance with her; she's too good looking for me." Even if people are allowed to specify the qualities they want in a date, they are hesitant to select anyone notably different from themselves in social desirability.

We also tend to choose people who are our equals in terms of intelligence, education, and so on (Hatfield and Walster 1981). However, if two people are different in looks or intelligence, usually the individuals make a trade-off in which a lower-ranked trait is exchanged for a higher-ranked trait. A woman who values status, for example, may accept a lower level of physical attractiveness in a man if he is wealthy or powerful.

Are Looks Important to Everyone?

For the wider majority of people who are more ordinary looking, it will come as a relief to know that looks aren't everything. Looks are most important to certain types or groups of people and in certain situations or locations (e.g., in classes, at parties, and in bars, where people do not interact with one another extensively on a day-to-day basis). Looks are less important to those in ongoing relationships and to those older than young adults. Those who interact regularly—as in working together—put less importance on looks (Hatfield and Sprecher 1986). In adolescence, the need to conform and the impact of peer pressure make looks especially important, as we may feel pressured to go out with handsome men and beautiful women.

Men tend to care more about how their partners look than do women (Buunk et al. 2002; Regan 2003). This may be attributed to the disparity of economic and social power. Because men tend to have more assets (such as income and status) than women, they can afford to be less concerned with their potential partner's assets and can choose partners in terms of their attractiveness. Because women lack the earning power and assets of men, they may have to be more practical and choose a partner who can offer security and status. Unsurprisingly, then, women are more likely than men to emphasize the importance of socioeconomic factors (Regan 2003).

Most research on attractiveness has been done on first impressions or early dating. At lower levels of relationship involvement, physical attractiveness is more important. As relationship involvement increases, status and personality become more important, appearance less. For long-term relationships (e.g., marriage), women and men prefer mates about as attractive as themselves. For short-term, less involved relationships, both men and women prefer more attractive mates. Bram Buunk and colleagues (2002) interpret this pattern to reflect potential costs of having as a long-term partner someone to whom others are strongly attracted.

Researchers are finding, however, that attractiveness is not *un*important in established relationships. Most people expect looks to become less important as a relationship matures, but Philip Blumstein and Pepper Schwartz (1983) found that the happiest people in cohabiting and married relationships thought of their partners as attractive. People who found their partners attractive had the best sex lives. Physical attractiveness continues to be important throughout marriage. It is, however, joined by other qualities, and these other attributes are deemed more important.

Bargains and Exchanges

Likening relationships to markets or choosing partners to an exchange may not seem romantic, but both are deeply rooted in marriage and family customs. In some cultures, for example, arranged marriages take place only after extended bargaining between families. The woman is expected to bring a dowry in the form of property (such as pigs, goats, clothing, utensils, or land) or money, or a woman's family may demand a bride-price if the culture places a premium on women's productivity. Traces of the exchange basis of marriage still exist in our culture in the traditional marriage ceremony when the bride's parents pay the wedding costs and "give away" their daughter.

Gender Roles

Traditionally, relationship exchanges have been based on gender. Men used their status, economic power, and role as protector in a trade-off for women's physical attractiveness and nurturing, childbearing, and housekeeping abilities; women, in return, gained status and economic security in the exchange.

The terms of bargaining have changed some, however. As women enter careers and become economically independent, achieving their own occupational status and economic independence, what do they ask from men in the marriage exchange? Clearly, many women expect men to bring more expressive, affective, and companionable resources into marriage. An independent woman does not have to "settle" for a man who brings little more to the relationship than a paycheck; she wants a man who is a partner, not simply a provider.

But even today, a woman's bargaining position may not be as strong as a man's. Women earn between 75% and 80% of what men earn, are still significantly underrepresented in many professions, and have seen many of the things women traditionally used to bargain with in the marital exchange—such as children, housekeeping services, and sexuality—become devalued or increasingly available outside of relationships. A man does not have to rely on a woman to cook for him, sex is often accessible in the singles world, and someone can be paid to do the laundry and clean the apartment.

Women are further disadvantaged by the **double standard of aging**. Physical attractiveness is a key bargaining element in the marital marketplace, but the older a woman gets, the less attractive she is considered. For women, youth and beauty are linked in most cultures. Furthermore, as women get older, their field of potential eligible partners declines because men tend to choose younger women as mates.

Going Out, Hanging Out, and Hooking Up

If relationships are like products in a marketplace, how do most young people proceed to shop? Some contend that this process has become much more casual and informal than in the past. Rather than *dating*, they refer instead to terms like *getting together, hanging out,* or *hooking up.* While getting together and hanging out are intended to capture and reflect increasing informality in romantic relationships, hooking up is a particular type of relationship, prominent on college campuses, that we consider in a later section.

Research indicates that dating practices have become more diverse, but "its purposes and activities are similar to when it first emerged" (Surra et al. 2007). In fact, young people who date do so for the predictable reasons—intimacy, companionship, and support—and tend to adhere to certain gendered scripts, especially at the beginning of dating relationships. Furthermore, given increases in divorce and cohabitation, later ages at marriage, and increased prevalence of singlehood, individuals may date more people and date for more of their adult lives, making dating more important than ever (Surra et al. 2007).

Although there are some general rules of mate selection that are important in the abstract, they do not tell us how relationships begin. The actual process of beginning a relationship is discussed in the sections that follow.

Seeing

On a typical day, one may see dozens, hundreds, or even thousands of men and women. But seeing isn't enough; one must become aware of someone for a relationship to begin. It may take only a second from the moment of noticing to meeting, or it may take days, weeks, or months. Sometimes "noticing" occurs between two people simultaneously, other times it may take considerable time, and sometimes it never happens.

The setting in which you see someone can facilitate or discourage meeting each other (Murstein 1976, 1987). **Closed fields**, such as small classes or seminars, dormitories, parties, and small workplaces, are characterized by a small number of people who are likely to interact whether they are attracted or not. In such settings, you are likely to "see" and interact simultaneously. In contrast, **open fields**, such as beaches, shopping malls, bars, amusement parks, and large university campuses, are characterized by large numbers of people who do not ordinarily interact. Common venues for first encountering a dating partner are social gatherings, such as bars and parties.

Meeting Face to Face

How is a meeting initiated, and who tends to do the initiating? Among heterosexuals, does the man initiate it? On the surface, the answer appears to be yes. In reality, however, the woman often contributes or even covertly initiates a relationship by communicating nonverbally her availability and interest (Metts and Cupach 1989). A woman will glance at a man once or twice and catch his eye; she may smile or flip her hair. If the man moves into her physical space, the woman then relies on nodding, leaning close, smiling, or laughing (Moore 1985).

Regardless of who initiates contact, a variety of verbal and nonverbal signals are used to convey attraction and interest to a potential partner. Smiling, moving closer to, gazing at, laughing, and displaying "positive facial expressions" are all gestures to convey interest or "flirt" (Regan 2003). Touch is also an important element in flirting, whether the touch consists of lightly touching the arm or hand or the face or hair of the target of interest or rubbing fingers across the other's arm (Regan 2003).

If a man believes that a woman is interested, he often initiates a conversation using an opening line. The opening line tests the woman's interest and availability. You have probably used or heard an array of opening lines. According to women, the most effective are innocuous, such as "I feel a little embarrassed, but I'd like to meet you" or "Are you a student here?" The least effective are sexual come-ons, such as "You really turn me on. Do you want to have sex?" or corny lines such as "If I had a nickel for every time I met someone as beautiful as you, I'd have a nickel." Women, more than men, prefer direct but innocuous opening lines over cute, flippant ones, such as "What's a good-looking babe like you doing in a college like this?"

Much as they are more likely than women to use "a line," men are more likely to initiate a meeting directly, whereas women are more likely to wait for the other person to introduce himself or herself or to be introduced by a friend (Berger 1987). About a third or half of all relationships rely on introductions (Sprecher and McKinney 1993). An introduction has the advantage of a kind of prescreening, as the mutual acquaintance may believe that both may hit it off.

Parties are the most common settings in which young adults meet, followed by classes, work, bars, clubs, sports settings, or events centered on hobbies, such as hiking (Marwell et al. 1982; Shostak 1987; Simenauer and Carroll 1982).

Meeting in Cyberspace

The Internet continues to gain popularity as a major way for people to "meet" a potential partner. Online, people can introduce themselves in fantasy-like images. A growing number of people first "meet" in cyberspace, find common interests, and form relationships that develop and intensify rapidly, even before they ever actually meet face-to-face. What might take months to develop in traditional dating can develop online within weeks or even days. Where dating usually proceeds predictably from two people seeing, meeting, evaluating each other's attractiveness and personal appeal, sharing personal interests and information, and—if compatible and trusting—moving toward deeper and more meaningful disclosure, online dating relationships are characterized by earlier and more deeply personal sharing and self-disclosing, perhaps even in the first few e-mails. For many couples, by the time they meet face-to-face, they have already developed a strong emotional bond (Rosen et al. 2008). In fact, one might say that they fall in love "from the inside" through heart-felt disclosures online and/or on the phone.

What was once stereotyped and even stigmatized as the last resort taken by desperate people has become common. Eleven percent of all Internet-using adults in the United States state that they have gone to an Internet dating site for the purpose of meeting a potential partner (Madden and Lenhart 2006). This translates to an estimated 16 million adults. Moreover, Mary Madden and Amanda Lenhart report the following:

- Nearly a third of adults (31%), an estimated 63 million people, know someone who has used a dating Web site.
- A quarter of American adults (26%), 53 million people, claim to know someone who has gone on a date that was initiated via an Internet site
- Fifteen percent of American adults, 30 million people, claim to know someone who has had a long-term relationship or married someone whom they met on the Internet.

Single men and women also rely on personal ads, whether in print or online on such sites as Craig's List. Ads often reflect stereotypical gender roles. Men advertise for women who are attractive and deemphasize intellectual, work, and financial aspects. Women advertise for men who are employed, financially secure, intelligent, emotionally expressive, and interested in commitment. Other alternative forms of meeting others include video dating services, introduction services, and 1-900 party-line phone services.

For lesbians and gay men, the problem of meeting others is exacerbated because they cannot necessarily assume that the person in whom they are interested shares their orientation. Instead, they must rely on identifying cues, such as meeting at a gay or lesbian bar or events, wearing a gay or lesbian pride button, or being introduced by friends to others identified as being gay or lesbian (Tessina 1989). Once a like orientation is established, gay men and lesbians usually engage in nonverbal processes to express interest. Both lesbians and gay men tend to prefer innocuous opening lines. To prevent awkwardness, the opening line usually does not make an overt reference to orientation unless the other person is clearly lesbian or gay.

Dating

For many of us, asking someone out for the first time is not easy. Shyness, fear of rejection, and traditional gender roles that expect women to wait to be asked may fill us with anxiety and nervousness. (Sweaty palms and heart palpitations are not uncommon when asking someone out the first time.) Both men and women contribute, although sometimes differently, to initiating a first date. Men are more likely to ask directly for a date: "Want to go see a movie?" Women are often more indirect. They hint or "accidentally on purpose" run into the other person: "Oh, what a surprise to see you *here* studying for your marriage and family midterm!" Although women may initiate dates, they do so less often than do men (Laner and Ventrone 2000).

In addition, research indicates that both women and men believe in certain gendered roles regarding cross-gender dating, especially as pertains to first dates, and that these expectations are fairly similar to those obtained in research over the past 25 years. These include the idea that men *should* initiate first dates, pick up the woman, pay for the date, and walk or drive the woman home. Men also have a higher expectation for sexual intimacy to occur.

Men also express a desire for women to more actively participate in initiating relationships, either by asking directly for a date or by at least hinting. Interesting results from research by communications researchers Mary Morr Serewicz and Elaine Gale show that men expect more sexual intimacy on first dates when initiated by women than when initiated by men. Women, however, do not develop that same expectation. Serewicz and Gale found that men expect "more than kissing" on a female-initiated date more than on a male-initiated date. Women thought "a good night kiss" was more likely on a male-initiated date than on a female-initiated date. When it comes to physical intimacy, men interpret first dates in more sexual terms, while women interpret the first date in more social or romantic terms (Serewicz and Gale 2008).

Although both women and men have ideas about what behaviors are most likely of men and of women on first dates, their ideas don't always match. In one study of 103 males and 103 females, women are slightly more egalitarian than men; almost twice as many women as men thought that either gender could do the inviting or initiating, and 22% of women compared to only 9% of men thought that either person could pick up the bill (Laner and Ventrone 2000). Gendered dating scripts introduce potential problems for men and women. A woman who wants to see a man again faces a dilemma: how to encourage him to ask her out again without engaging in more sexual activity than she really wants. Men often felt that they didn't know what to say, or they felt anxious about the conversation dragging. Communication may be a particularly critical problem for men because traditional gender roles do not encourage the development of intimacy and communication skills among males. A second problem, shared by almost identical numbers of men and women, was where to go. A third problem, named by 20% of the men but not mentioned by women, was shyness. Although men can take the initiative to ask for a date, they also face the possibility of rejection. For shy men, the fear of rejection is especially acute. A final problem—and, again, one not shared by women—was money, cited by 17% of the men. Men apparently accept the idea that they are the ones responsible for paying for a date.

Problems in Dating

Dating is often a source of both fun and intimacy, but a number of problems may be associated with it. In dating relationships, disagreements inevitably occur. When disagreements do occur, who generally wins? Does it depend on the issue? When one person wants to go to the movies and the other wants to go to the beach, where do they end up going? If one wants to engage in sexual activities and the other doesn't, what happens?

Chapter 7 examines such questions in more detail. For now, it is sufficient to note that typical of much relationship research, the female demand–male withdraw pattern found in many marriages is common among heterosexual dating couples. Of the 108 participants in dating relationships studied by David Vogel, Stephen Wester, and Martin Heesacker (1999), 51% reported having a female demand–male withdraw communication pattern. Another 28% described their communication as male demand–female withdraw, and 21% had no pronounced pattern. The female demand–male withdraw pattern was more often the style used by couples engaged in "difficult discussions." David Vogel and colleagues suggest that either version of demand–withdraw may prove to be a problem for dating couples as far as their relationship satisfaction and cohesion are concerned (Vogel et al. 1999). They recommend reduction of the overall level of demand–withdraw behavior as an important step toward enhancing the quality of relationships.

Costs and Benefits of Romantic Relationships

As anyone who has had a romantic relationship can attest, relationships bring both positive and negative experiences. In other words, when asked, people identify both rewards (companionship, sexual gratification, feeling loved and loving another, intimacy, expertise in relationships, and enhanced self-esteem) and costs (loss of freedom to socialize or date, investment of time and effort, loss of identity, feeling worse about oneself, stress and worry about the health or durability of the relationship, and other nonsocial costs, like lower grades) of romantic relationships (Sedikedes, Oliver, and Campbell 1994, cited in Regan 2003).

Males and females differ some in what costs and rewards they identify. More males than females identify sexual gratification as a benefit of romantic relationships, and women are more likely than men to identify the benefit of enhanced self-esteem. More women than men mention loss of identity, feeling worse about themselves, or growing too dependent on their partners as relationship costs. Males, on the other hand, stress perceived loss of freedom (to socialize or date) and financial costs more than women do (Regan 2003).

Hooking Up

As sociologist Kathleen Bogle (2005) points out, as far back as the 1970s, social scientists were forecasting the end of dating as we knew it, especially on college campuses. Nonetheless, researchers continued to use the term *dating* to refer to behaviors that seemed less and less to fit what dating meant. Bogle, author of the book *Hooking Up: Sex, Dating, and Relationships on Campus* (2008), differentiates between dating and hooking up, the latter being the pattern she says is dominant among college students today.

What is "hooking up"? Bogle (2005, 2) states that it "generally involves a college man and a college woman pairing off at the end of a party or evening at a bar to engage in a physical or sexual encounter." The term is used differently by different individuals, but all uses imply some degree of physical intimacy, be that "having sex," "making out," "fooling around," or even "everything but intercourse" (Bogle 2008). Unlike the physical intimacy that *might* occur in dating, couples who hook up understand and expect that some level of physical intimacy will be shared. Furthermore, sex between dating partners typically might occur after the couple had dated (at least once), whereas in hookups, partners "become sexual first and then *maybe* someday go on a date." In addition to occurring earlier, sex

A pattern prevalent on college campuses, "hooking up" differs in a number of important ways from more typical "dating" relationships.

in hookups is unrelated to the degree of emotional closeness between the partners and implies no commitment beyond the encounter itself. As articulated by both women and men Bogle studied, the most likely outcome of a hookup is nothing. Once they part, neither pursues a relationship with the other.

Although they don't include the same gendered role expectations that dating does, hookups are differently experienced by women and men. Especially as they move beyond the freshman year, men continue to want hookups with "no strings attached" (i.e., no expectation for anything further developing), whereas many women begin to desire "some semblance of a relationship" (Bogle 2008, 96). Men have hardly any restrictions imposed on them in the hookup culture, whereas women can experience reputational damage and labeling if they hook up with too many men; hook up with two men who knew each other well, especially if there was little time between the encounters; or hang around a specific fraternity too much or if their behavior (e.g., how much they drink, how they dress, or whether they behave "wildly") is deemed excessive. Hooking up comes with a sexual double standard even as it "frees" both genders to engage in more sexual encounters (Bogle 2008).

Bogle suggests that the hookup culture makes relationship building more difficult, as there is less opportunity to get to know each other or develop feelings for each other. In other words, hooking up is a less effective method for "finding love" than conventional dating has been. The hookup model makes the transition to the "real world" beyond college more problematic. Away from college campuses, dating, not hooking up, is the norm, and dating continues to be characterized by gendered scripts regarding such tasks as initiating the date, driving, paying, and so on.

How Love Develops and Ends: Spinning Wheels and Winding Clocks

As shown earlier, one of the core beliefs that of the ideology of romantic love in the United States is the idea of love at first sight. This is often articulated by romantic partners in describing how they "just knew" that they were meant for each other on their first meeting, the first time they gazed at each other, or when they first heard the other laugh or speak. In fact, love develops through a process, beginning with first meeting but commencing through an intensification of the relationship and eventually a definition or interpretation of feelings as "love."

One of the more popular models depicting this process is Ira Reiss's **wheel theory of love** (Reiss 1960, 1980a). According to Reiss (1980a), we all have a basic need for intimacy—for someone we can be ourselves with, who will listen to and understand us. These needs are important for fulfilling our roles as a partner or parent. His "wheel theory" depicts the development of love as a spinning wheel, consisting of four spokes, each of which drives the others as the wheel spins forward. The four spokes are (1) rapport, (2) self-revelation, (3) mutual dependency, and (4) fulfillment of the need for intimacy:

- *Rapport.* When two people meet, they quickly sense if rapport—a sense of ease, the feeling that they understand each other—exists between them. People tend to feel rapport with those who share the same social and cultural background as themselves. If one or both feel as though they have much in common, they are more likely to feel as though they can understand each other; they may even experience a comfort that makes them feel like they have known each other a long time (or before).

- *Self-revelation.* The greater the rapport two people feel with each other, the more likely they are to feel relaxed and confident around each other and to develop trust about the relationship. As a result, self-revelation—the disclosure of intimate feelings—is more likely to occur. They will reveal more about themselves and more of a personal nature with greater confidence and trust. Furthermore, disclosure becomes mutual. Self-revelation may depend on more than the presence or absence of rapport. It may also depend on what is considered proper within one's ethnic group or economic class. Members of certain groups, such as Chinese spouses, have more of a tendency to be reserved about themselves. Others (e.g., middle-class Americans) feel more comfortable in revealing intimate aspects of their lives and feelings (Fitzpatrick et al. 2006).

- *Mutual dependency.* After two people feel rapport and begin revealing themselves to each other, they may become mutually dependent. Each needs the other to share pleasures, fears, and jokes as well as sexual intimacies; each becomes the other's confidant. Each person develops ways of acting and being that cannot be fulfilled alone. Going for a walk is no longer something done alone; they walk together. The two people form a couple. Here, too, social and cultural background is important. The forms of mutually dependent behavior that develop are influenced by each person's conception of the role of courtship.

- *Fulfillment of intimacy needs.* If one finds that his or her need for love and intimacy are met by his or her partner, rapport will deepen, setting the stage for more self-revelation, increased mutual dependency, and greater fulfillment of each person's intimacy needs.

Relationships, like wheels, can spin in reverse as well as forward. In other words, one can "fall out of love," and the wheel theory addresses this phenomenon. A reduction in any one of the four spokes affects the development or maintenance of the love relationship. If one feels less comfort (i.e., rapport) with one's mate, one may reveal fewer thoughts or feelings, feel less dependent on the other for a sense of happiness or contentment, and seek and fulfill one's intimacy needs elsewhere. This seems to approximate what happens through the process of divorce or the ending of intimate relationships as well (Vaughan 1986). If a couple habitually argues, the arguments will affect the partners' mutual dependency and their need for intimacy; this in turn will weaken their rapport. Thus, the model can depict falling in *or* out of love.

To capture the idea that as relationships persist, they tend to deepen—couples grow closer, and their connection "tightens"— Dolores Borland (1975) suggested thinking not so much in terms of a wheel but rather a "clockspring." (See Figure 5.1.) The "most intimate aspects of the 'real self'" are at the center of the clockspring.

Figure 5.1 The Clockspring Variation on Reiss's Wheel Theory

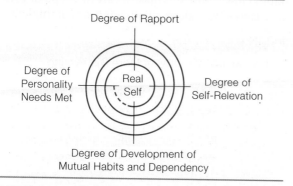

Degree of Rapport

Degree of Personality Needs Met

Real Self

Degree of Self-Relevation

Degree of Development of Mutual Habits and Dependency

SOURCE: Borland 1975, 289–292.

As rapport leads to self-revelation, revelation leads to mutual dependency, and mutual dependency leads us to seek and find satisfaction of our need for intimacy in our partner; we wind closer to a relationship with the "real inner self of the other person" (Borland 1975, 290–291). A clockspring representation can also depict the depth of a relationship (by how much of our "real self," including our vulnerabilities and sensitivities, we expose to others) as well as the difficulty and time it will take to "unwind" a relationship. Borland's clockspring is meant mostly as an aid in teaching about and better understanding the basic elements put forth by Reiss. Such elements are important if we are to fully understand how intimate relationships begin, develop, persist, and/or end.

Jealousy: The Green-Eyed Monster

In addition to bringing great joy, love relationships are often the source of painful insecurities and jealousy. What exactly is jealousy? As studied by researchers, **jealousy** is an aversive response that occurs because of a partner's real, imagined, or likely involvement with a third person (Bringle and Buunk 1985; Sharpsteen 1993). Jealousy sets the boundaries for what an individual or group feels are important relationships; others cannot trespass these limits into other emotional and/or sexual relationships without evoking jealousy.

Some people may think that jealousy proves love and, by flirting with another person, may try to test their partner's interest or affection by attempting to make him or her jealous. If one's date or partner becomes jealous, the jealousy is taken as a sign of love. But making jealousy a litmus test of love is dangerous because jealousy and love are not necessarily related. Jealousy may be a more accurate yardstick for measuring insecurity or possessiveness than love (for a discussion of changing cultural attitudes toward jealousy, see Mullen 1993).

Social psychologists suggest that there are two types of jealousy: suspicious and reactive (Bringle and Buunk 1991). **Suspicious jealousy**, which generally occurs in the early stages of a relationship, is jealousy that occurs when there is either no reason to be suspicious or only ambiguous evidence to suspect that a partner is involved with another. **Reactive jealousy** is jealousy that occurs when a partner reveals a current, past, or anticipated relationship with another person. It is a more intense jealousy. Basic trust is questioned, and the damage can be irreparable.

Jealousy is not necessarily a sign of love.

Psychologist Robert Cramer and colleagues asked a sample of undergraduate women and men to indicate which of two infidelities would distress or upset them more by circling either alternative A or alternative B:

A. Imagining your partner forming a deep emotional attachment to another person
B. Imagining your partner enjoying passionate sexual intercourse with another person

Other respondents were asked to imagine their partners committing *both* infidelities: "Imagine your partner forming a deep emotional attachment to another person *and also* enjoying passionate sexual intercourse with that person." Participants were then asked to indicate which infidelity, assuming that both had occurred, would distress or upset them more. Results for both questions show a striking gender pattern (see Table 5.1).

Like much jealousy research, Cramer and colleagues use evolutionary theory to account for these gender differences. They suggest that emotional infidelity is more distressing for women than for men because, at least in theory, it threatens a romantic partner's commitment and, therefore, continued access to material resources and economic stability needed to ensure the healthy growth and development of offspring. Men, on the other hand, are more distressed by sexual infidelity than women are because it decreases their "paternity certainty" through the loss of sexual exclusivity (Cramer et al. 2001–2002).

Both men and women react to jealousy with a host of emotions. Betrayal, anger, rejection, hurt, distrust, anxiety, worry, suspicion, and sadness are all possible. The kind of emotional reaction appears to depend on the type of infidelity that provokes it. Following emotional infidelity, such feelings as anxiety, suspicion, worry, distrust, and threat are more common. Bram Buunk and Pieternel Dijkstra call this type of jealousy *preventive* jealousy. Following sexual infidelity,

Gender Differences in Jealousy

Both men and women are susceptible to jealous fears that their partner might be attracted to someone else because of dissatisfaction with the relationship, attractiveness of a rival, or the desire for sexual variety. Women feel especially vulnerable to losing their partner to a physically attractive rival, whereas in men jealousy is evoked more by a rival's status (Buunk and Dijkstra 2004). Furthermore, men and women become jealous about different matters. Men tend to experience more jealousy when they feel their partner is sexually involved with another man. Women, by contrast, tend to experience jealousy over intimacy issues, when they feel their partner is too emotionally close to someone else (Buunk and Dijkstra 2004; Cramer et al. 2001, 2002). This gender difference has been found in research in the United States as well as in China, Germany, Japan, Korea, the Netherlands, and Sweden (Cramer et al. 2001–2002).

Table 5.1 Percentage of Women and Men Selecting Emotional or Sexual Infidelity as More Distressing

		Men	Women
Forced choice	Emotional infidelity	12.9%	54.5%
	Sexual infidelity	87.1%	45.5%
Assuming both, which is worse?	Emotional infidelity	13.3%	40.6%
	Sexual infidelity	86.7%	59.4%

jealousy was expressed more through anger, sadness, a sense of betrayal, hurt, and rejection. Buunk and Dijkstra (2004) label this *fait accompli* (after the fact) jealousy. Further differentiating the genders, following emotional infidelity, jealousy was evoked in men by a rival's dominance and was experienced mostly as a sense of threat. Following sexual infidelity, men's jealousy was evoked by his rival's physical attractiveness, not his dominance, and was experienced as betrayal or anger. For women, after emotional infidelity, a rival's physical attractiveness evoked a sense of threat, whereas after sexual infidelity, women's jealousy responses were unaffected by any particular characteristics of her rival (Buunk and Dijkstra 2004).

Managing Jealousy

Jealousy can be unreasonable or a realistic reaction to genuine threats. Unreasonable jealousy can become a problem when it interferes with an individual's well-being or that of the relationship. Dealing with irrational suspicions can often be difficult because such feelings touch deep recesses in ourselves. As noted earlier, jealousy is often related to personal feelings of insecurity and inadequacy. The source of such jealousy lies within a person, not within the relationship.

If we can work on the underlying causes of our insecurity, then we can deal effectively with our irrational jealousy. Excessively jealous people may need considerable reassurance, but at some point they must confront their own irrationality and insecurity. If they do not, they may emotionally imprison their partner. Their jealousy may destroy the very relationship they were desperately trying to preserve.

Managing jealousy requires the ability to communicate, the recognition by each partner of the feelings and motivations of the other, and a willingness to reciprocate and compromise (Ridley and Crowe 1992). If the jealousy is well founded, the partner may need to modify or end the relationship with the "third party" whose presence initiated the jealousy. Modifying the third-party relationship reduces the jealous response and, more important, symbolizes the partner's commitment to the primary relationship. If the partner is unwilling to do this—because of lack of commitment, unsatisfied personal needs, or other problems in the primary relationship—the relationship is likely to reach a crisis. In such cases, jealousy may be the agent for profound change.

It's important to understand jealousy for several reasons. First, jealousy is a painful emotion filled with anger and hurt. Its churning can turn one inside out and make one feel out of control. If one can understand jealousy, especially when it is irrational, then one can eliminate some of its pain. Second, jealousy can help cement or destroy a relationship. It helps maintain a relationship by guarding its exclusiveness, but in its irrational or extreme forms, it can destroy a relationship by its insistent demands and attempts at control. There is a need to understand when and how jealousy is functional and when and how it is not. Third, jealousy is often linked to violence. It is a factor in precipitating violence or emotional abuse in dating relationships among both high school and college students; among marital partners, it is often used by abusive partners to justify their violence (Jewkes 2002). It has been linked to abuse and violence in other countries, such as China (Wang et al. 2009). Rather than being directed at a rival, jealous aggression is often used against a partner, especially when accompanied by alcohol (Foran and O'Leary 2007).

Breaking Up

"Most passionate affairs end simply," Elaine Hatfield and G. William Walster (1981) noted. "The lovers find someone they love more." Love cools; it changes to indifference or hostility. Perhaps the relationship ends because one partner shows a side that the other partner decides is undesirable. Or couples disclose *too much*, revealing negative feelings or ideas that lead to unhappiness and the demise of the relationship (Regan 2003).

Relationships are also susceptible to outside influences. Perhaps, some new opportunity for greater fulfillment appears in someone else or in a return to a more autonomous and independent state. Even satisfying relationships may end under these circumstances (Regan 2003). Over the course of their lifetimes, most people will experience multiple relationships and endure numerous breakups (Tashiro and Frazier 2003).

Breaking up is typically painful because few relationships end by mutual consent. The extent of distress caused by breakups is revealed by research indicating that many people include breaking up among the "worst events" they can experience; they are also among the biggest risk factors for adolescent depression (Tashiro and Frazier 2003). For college students, breakups are more likely to occur during vacations or at the beginning or end of the school year. Such timing is related to changes in the person's daily living

Because of the variety of problems that plague relationships, many couples break up. In the process of breaking up, both the initiator and the rejected partner suffer.

schedule and the greater likelihood of quickly meeting another potential partner. Holidays, including—ironically—Valentine's Day, also may increase the risk of breakup (see Popular Culture feature).

Research indicates that relationships often begin to sour as one partner grows quietly dissatisfied (Duck 1994; Vaughan 1990). Steve Duck (1994) calls this the *intrapsychic* phase, and Diane Vaughan (1990) talks about "keeping secrets." One partner decides that something is wrong with the relationship, considers the possibility of ending the relationship, weighs the likely outcomes associated with being out of the relationship, and begins to build an identity as a "single." All of this may happen before the other partner learns what has happened. By the time the "initiator" informs the partner, the partner is forced to play "catch-up" in that the initiator is a few steps ahead in the exiting process. This is further discussed in Chapter 13.

Breaking up is rarely easy, regardless of whether one is on the initiating or "receiving" end. As Pamela Regan (2003) summarizes, the more satisfied one is with one's partner, the closer one feels to one's partner, and the more difficult one believes it will be to find another relationship, the harder it is to experience a breakup. Social support and self-esteem appear to be important factors in helping someone recover more quickly and completely (Regan 2003).

Also important are the **attributions** individuals make to account for the demise of a relationship. Attributions may be important factors in efforts to avoid such problems in later relationships, shielding people from experiencing the heartache that accompanies a breakup. Ty Tashiro and Patricia Frazier (2003) suggest that there are four such attributions:

- *Person.* Personal traits and characteristics are identified as causes of relationship failure ("if only I hadn't been so jealous").
- *Other.* Personal traits and characteristics of the partner are seen as the causes of relationship failure ("he or she was always so insensitive").
- *Relational.* The unique combination of person and other is perceived as the cause of the breakup ("we just wanted different things").
- *Environmental.* The social environment is identified as the cause of the breakup. It comprises many things, from familial pressure and disapproval of the relationship to work pressures to "alternative romantic partners."

According to Tashiro and Frazier (2003), "relational" attributions are usually cited by those who construct accounts to explain why their relationships failed. These are followed by "other" attributions, "person" attributions, and "environmental" attributions. Although environmental attributions are quite uncommon, environmental factors weigh heavily on relationships. Ironically, environmental factors may be the "real cause" of a breakup incorrectly attributed to something else.

Attributions are also related to how distressing a breakup is felt to be. People who apply relational attributions are happier, more confident, and more socially active. "Other" attributions are associated with greater distress, including sadness, lack of self-confidence, and greater pessimism. Research on person attributions is mixed, with some showing that it is related to less and some suggesting that it is associated with more distress.

Research demonstrates that, alongside pain and distress, breakups can induce positive changes that improve the quality of subsequent relationships you might enter. One might expect that the degree to which breakups are associated with positive rather than negative outcomes (e.g., growth rather than distress) would depend on such things as whether we initiated the breakup (noninitiators suffer greater distress), our gender (females more than males experience more positive emotions such as "growth" following breakups), and/or our personality (people high in traits such as "agreeableness" respond to stressful situations more positively; people high in "neuroticism" are more likely to suffer from distress). Of these, gender differences occur in "stress-related growth," with women reporting more growth following breakups than men. In addition, people high in

Popular Culture: Chocolate Hearts, Roses, and . . . Breaking Up? What About "Happy Valentine's Day"?

Every year on February 14, millions of Americans exchange tokens of love and affection. As the day approaches, post office branches fill with Hallmark cards. Florists take orders and send roses around the country. Chocolate hearts show up on store shelves. Jewelry is bought for and given to those we love. Millions of dollars are spent in efforts to show and tell our "one and only" how much we love them, and, collectively, the country celebrates love and romance in the name of Valentine's Day. You may be familiar with such rituals as both a giver and a recipient. You may be wondering what if anything is noteworthy about such rituals. One of the lesser-known aspects of this holiday devoted to love is the effect it can have on ongoing love relationships. This effect was provocatively captured by exploratory research undertaken by Katherine Morse and Steven Neuberg (2004) in their study following the relationship outcomes for 245 undergraduate students (99 male and 146 female; mean age of 19.5 years) from the week before to the week after Valentine's Day. The average relationship across all research participants was 18 months, suggesting that these were meaningful relationships. The results may surprise you, especially if you consider yourself a romantic at heart.

It seems that "Valentine's Day is harmful to many relationships" (Morse and Neuberg 2004, 509). Although at first this may seem hard to imagine, Morse and Neuberg remind us that, although limited, research has shown that holidays can affect behaviors. The best illustration of this is the effect major holidays (e.g., Christmas, Thanksgiving, and New Year's Eve and Day) have on suicide rates: essentially, they postpone such acts from before to after the holiday.

Morse and Neuberg (2004) predicted that during the two-week period straddling Valentine's Day (from one week before the holiday to one week after the holiday) there would be more breakups than in comparison periods from other nonholiday times of year, and, indeed, Valentine's Day posed relationship hazards. The overall odds of breaking up were 5.49 times greater during the Valentine's Day period than during the comparison months (which did not differ from one another). They further determined that the effect of the holiday on breakups was the result of a *catalyst effect*. The holiday had no effect on breakups among high-quality or improving relationships but did affect breakups among those in moderately strong and weak relationships if they were encountering relationship downswings. Already suffering from diminishing expectations and unfavorable comparisons to other relationships or potential partners, such relationships might be deemed not worth the effort and expenses associated with trying to successfully play out the Valentine's Day script, thus "making the option of relationship dissolution more attractive" (Morse and Neuberg 2004, 512).

They suggest that the catalyst effect may in part be a favor to troubled relationships in that it facilitates a breakup that was likely anyway and hence "saved these couples the psychological stress, wasted time, and wasted resources that result from perpetuating a doomed relationship" (Morse and Neuberg 2004, 524). However, they also admit that because many long-term relationships go through periods of ups and downs, it is "at least plausible that a good number of our couples might have otherwise survived the downward blip in relationship expectations and quality had it not been for the catalytic effects of Valentine's Day" (524).

SOURCE: Morse and Neuberg (2004, 509–27).

"agreeableness" reported more postbreakup growth (Tashiro and Frazier 2003).

A study of 92 undergraduates (75% female; age range 18–35 with a mean age of 20 years) found that positive change, such as personal growth, following breakups was common. On average, respondents reported positive changes that they believed would strengthen their future relationships and the chances for success in those relationships. Using the same four categories of attributions discussed earlier, the most commonly reported positive changes were "person related" (e.g., "I learned not to overreact") followed by "environmental" (e.g., improved family relationships and increased success in school), "relational" (e.g., better communication), and "other" (remember, "other" refers to characteristics of the other with whom we have a relationship). Individuals who use environmental attributions were most likely to report both distress from a breakup and having experienced growth as a result of the breakup. Potentially, those who can explain the failure of their relationship in terms of changeable environmental factors are in a better position to learn from and implement such changes in future relationships (Tashiro and Frazier 2003).

Breakups among Gay and Lesbian Couples

As we will see in Chapter 6, there are both similarities and differences between same-sex and heterosexual couples. Couple relationships, especially those that entail sharing a household, encounter many of the same day-to-day issues (e.g., housework, money management, and the effects of outsiders, such as family and friends, on relationships). Furthermore, all couples need to manage issues such as communication and conflict management. Given these similarities, how do same-sex couples and heterosexual couples compare in terms of susceptibility to breaking up?

Same-sex couples are more likely to break up than are heterosexual couples (Wagner 2006). Citing research findings by Lawrence Kurdek (1994), Cynthia Wagner reports that in comparisons of married heterosexuals, cohabiting heterosexuals, and gay and lesbian couples, married heterosexuals had the lowest rate (4%) of breaking up within 18 months of getting together ("relationship dissolution") and that lesbian couples had the highest (18%). However, Kurdek argues that the cause of differences is more likely the result of *marriage* than of *sexuality*. All cohabiting couples had similar "dissolution rates," and all were significantly higher than the rate found among married heterosexuals. Marriage is more likely to be associated with cultural acceptance and social support. Furthermore, once married, it is more difficult to simply walk away or to separate easily. Using comparative data from Sweden and Norway, Kurdek illustrates that state-sanctioned and recognized legal unions between gay men or lesbians lowered the rates at which such relationships broke up, even though they still did so at levels greater than among married heterosexuals (Wagner 2006). With only recent passage of gay marriage law and civil union legislation, we don't yet have enough data on whether breakups and dissolutions have occurred at levels similar to those among married heterosexuals.

There will still be differences between married heterosexuals and married gay or lesbian couples that are products of something other than sexual orientation. Married gay or lesbian couples will not likely benefit from the same levels of social support and acceptance as married heterosexuals. Even if their own intimate networks of family and friends are supportive (and in the case of families, that is far from automatic), the wider society doesn't offer the same climate of acceptance and support to gay and lesbian relationships. Thus, differences in rates of dissolution may follow from different levels of acceptance and support versus hostility.

What becomes of relationships once couples break up? How do former romantic partners relate to each other? Research on 298 individuals from same-sex and 272 individuals from heterosexual romantic relationships reveals some interesting similarities in "postdissolution relationships" (Lannutti and Cameron 2002). Many gay men and lesbians, as well as many heterosexuals, report remaining (or becoming) friends with former partners, especially following the "let's just be friends" type of breakup. Those friendships are different, however, from friendships in which two people have no shared romantic past. In comparing characteristics of postdissolution relationships, Pamela Lannutti and Kenzie Cameron found that heterosexuals reported moderate amounts of satisfaction and emotional closeness and low levels of interpersonal contact and sexual intimacy with former partners. Gay and lesbian respondents revealed high levels of satisfaction, moderate levels of emotional intimacy and personal contact, and low levels of sexual intimacy in their postdissolution relationships. For both same-sex and heterosexual former partners, postdissolution relationships also differ from intact or ongoing romantic relationships (Lannutti and Cameron 2002).

Unrequited Love

As most of us know from painful experience, love is not always returned. We may reassure ourselves that, as Alfred, Lord Tennyson, wrote 150 years ago, "'Tis better to have loved and lost / Than never to have loved at all." Too often, however, such words sound like a rationalization. Who among us does not sometimes think that it is better never to have loved?

Unrequited love—love that is not returned—is a common experience. It is distressing for both the would-be lover and the rejecting partner. Although would-be lovers may experience intensely negative feelings about their unlucky attempt at a relationship, such as anger and jealousy or fear of rejection, they also identify positive feelings. The rejecters, however, tend to feel more uniformly negative about the experience. They may find the other person's persistence intrusive and annoying; they wish the other would have simply gotten the hint and gone away. They are likely to feel annoyed by the unwanted attention, feel badly about having to reject the other, and experience a range of negative emotions, such as frustration and resentment (Regan 2003).

Unrequited love presents a paradox in that if the goal of loving someone is an intimate relationship, why should we continue to love a person with whom we could not have such a relationship? Arthur Aron and his colleagues (1989) addressed this question in a study of almost 500 college students. The researchers found three different attachment styles underlying the experience of unrequited love:

- *The Cyrano style.* Named after Cyrano de Bergerac, the seventeenth-century French poet and musketeer whose love for Roxanne was so great that it was irrelevant that she loved someone else, this refers to the desire to have a romantic relationship with a specific person regardless of how hopeless the love is. The benefits of loving someone are considered so great that it does not matter how likely the love is to be returned.
- *The Giselle style.* This refers to the misperception that a relationship is more likely to develop than it actually is. This might occur if we misread the other's cues, such as in mistakenly believing that friendliness is a sign of love. This style is named after Giselle, the tragic ballet heroine who was misled into believing that her love was reciprocated.
- *The Don Quixote style.* This refers to the general desire to be in love regardless of whom we love. Here, the benefits of being in love are more important than actually being in a relationship. This style is named after Cervantes's Don Quixote, whose love for the common Dulcinea was motivated by his need to dedicate knightly deeds to a lady love.

Using attachment theory, the researchers found that some people were predisposed to be Cyranos, others Giselles, and still others Don Quixotes. Anxious or ambivalent adults tended to be Cyranos, avoidant adults often were Don Quixotes, and secure adults were likely to be Giselles. Those who were anxious or ambivalent were most likely to experience unrequited love; those who were secure were least likely to experience such love. Avoidant adults experienced the greatest desire to be in love in general, yet they had the least probability of being in a specific relationship. Anxious or ambivalent adults showed the greatest desire for a specific relationship; they also had the least desire to be in love in general.

Critical Thinking

Have you experienced unrequited love? How did it differ from requited love? Do you have a "style" of unrequited love? Have you been the object of someone's unrequited love? How did you handle it?

Lasting Relationships through the Passage of Time

As discussed earlier, romantic love occupies a particularly valued place in societies such as the United States. Its emergence and presence can move two people into marriage, and its absence or disappearance can lead them to terminate their relationship. However, there is also the fairly widespread belief that romantic love

© Jean Higgins/Envision

Being physically limited does not inhibit love and sexuality any more than being able-bodied guarantees them.

cannot be sustained in relationships that last many years. Ultimately, it is believed that romantic love will be transformed or replaced by a quieter, more lasting love, and that such a transformation is probably beneficial. However, there is evidence that should cause one to question the inevitability of such a transformation (Acevedo and Aron 2009).

Changes in Intimacy, Passion, and Commitment over Time

Many of the major theories of love predict that the transformation from romantic to companionate love is inevitable. For example, returning to Robert Sternberg's triangular theory of love, love has three components: passion, intimacy, and commitment. A love that contains all three, what Sternberg called **consummate love**, or complete love, is idealized and sought after but difficult to find and even harder to sustain (Sternberg 2007). The passage of time affects our levels of intimacy, passion, and commitment though not at the same pace or in the same direction.

Intimacy over Time. When two people first meet, intimacy increases rapidly as we make critical discoveries about each other, ranging from our innermost thoughts of life and death to our preference for strawberry or chocolate ice cream. As the relationship continues, the rate of growth decreases and then levels off. After the growth levels off, the partners may no longer consciously feel as close to each other. This may be because they are beginning to drift apart, or it may be because they are becoming intimate at a different, less conscious, even deeper level. This kind of intimacy is not easily observed. It is a latent intimacy that nevertheless is forging stronger, more enduring bonds between the partners.

Passion over Time. Passion may be subject to habituation. What was once thrilling—whether love, sex, or roller coasters—becomes less so the more we get used to it. Once we become habituated, more time with a person (or more sex or more roller-coaster rides) does not increase our arousal or satisfaction.

If the person leaves, however, we experience withdrawal symptoms (fatigue, depression, and anxiety), just as if we were addicted. In becoming habituated, we have also become dependent. We fall beneath the emotional baseline we were at when we met our partner. Over time, however, we begin to return to that original level.

Commitment over Time. Unlike intimacy and passion, time does not necessarily diminish, erode, or alter commitments. Our commitment is most affected by how successful our relationship is. Even initially, commitment grows more slowly than intimacy or passion. As the relationship becomes long term, the growth of commitment levels off. Our commitment will remain high as long as we judge the relationship to be successful. If the relationship begins to deteriorate, after a time the commitment will probably decrease. Eventually, it may disappear, and an alternative relationship may be sought.

Disappearance of Romance as Crisis

The disappearance (or transformation) of passionate love is often experienced as a crisis in a relationship. But intensity of feeling does not necessarily measure depth of love. Intensity often diminishes over time. It is then that we begin to discover if the love we experience for each other is one that will endure.

Our search for enduring love is complicated by our contradictory needs. Elaine Hatfield and William Walster (1981) offer this observation:

> What we really want is the impossible—a perfect mixture of security and danger. We want someone who understands and cares for us, someone who will be around through thick and thin, until we are old. At the same time, we long for sexual excitement, novelty, and danger. The individual who offers just the right combination of both ultimately wins our love. The problem, of course, is that, in time, we get more and more security—and less and less excitement—than we bargained for.

The disappearance of passionate love, however, can also enable individuals to refocus their relationship. They are given the opportunity to move from an intense one-on-one togetherness that excludes others to a togetherness that includes family, friends, and external goals and projects. They can look outward on the world together.

The Reemergence of Romantic Love

Contrary to what pessimists believe, many people find that they can have both love and romance and that the rewards of intimacy include romance. Romantic love may be highest during the early part of romantic relationships and decline as stresses from child rearing and work intrude on the relationship. Most studies suggest that marital satisfaction proceeds along a U-shaped curve, with highest satisfaction in the early and late periods. Romantic love may be affected by the same stresses as general marital satisfaction. Romantic love begins to increase as children leave home. In later life, romantic love may play an important role in alleviating the stresses of retirement and illness.

So it is that, among those whose marriages survive, passion and romance do not necessarily decline over time. In fact, Bianca Acevedo and Arthur Aron contend that long-term romantic love is very real and that individuals who are in such relationships are likely to have higher levels of satisfaction with their relationships, better mental health, and a better sense of overall well-being. They suggest that awareness of the existence of long-term romantic relationships might give couples in committed relationships something toward which to strive. Long-term marriages or couple relationships don't inevitably lose the intensity, engagement, or sexual interest that characterize romantic love (Acevedo and Aron 2009).

Although we increasingly understand the dynamics and varied components of love, the experience of love itself remains ineffable, the subject of poetry rather than scholarship. A journal article is not a love poem,

© David Young-Wolff/PhotoEdit

As we age the dynamics that characterize our intimate relationships change even when the relationships themselves endure.

and romantics should not forget that love exists in the everyday world. Researchers have helped us increasingly understand love in the light of day—its nature, its development, and its varied aspects—so that we may better be able to enjoy it in the moonlight.

Summary

- Love is of major significance in American society for individual well-being and family formation.

- Humans have a basic need for intimacy or closeness with others. Intimacy consists of affection, personal validation, trust, and self-disclosure.

- Intimate relationships such as love and friendship offer numerous emotional, psychological, and health benefits.

- We can see the differences between friendship and love by contrasting the qualities we seek in friends versus lovers.

- In the twentieth century, love became a more central theme in our search for a mate and our expectations for marriage.

- Americans have an ideology of romanticism, in which love is seen as blind, irrational, uncontrollable, and likely to strike at first sight. In addition, it is believed that there is a "one and only" for each of us.

- In more individualistic societies like the United States, a high value is placed on passionate love. In more collectivist societies, individual happiness is subordinate to the well-being of the group (including especially the family), and companionate love is more highly valued (and romantic love is devalued and frowned on).

- Cultural expectations surrounding friendship and love in the United States define such intimacy in more feminine ways, involving heavy emphasis on self-disclosure.

- There are consistent gender differences in experiences and expectations surrounding friendship, love, and intimacy.

- Males and females do not attach the same meanings to how sexual expression fits within love relationships.

- Heterosexuals, gay men, and lesbians all value meaningful loving relationships.

- *Prototypes* of love and commitment are models of how people define these two ideas in everyday life.

- Commitment is affected by the balance of costs to benefits, normative inputs, and structural constraints.

- The *wheel theory of love* emphasizes the interdependence of four processes: (1) rapport, (2) self-revelation, (3) mutual dependency, and (4) fulfillment of intimacy needs.

- According to John Lee, there are six basic styles of love: *eros*, *ludus*, *storge*, *mania*, *agape*, and *pragma*.

- The *triangular theory of love* views love as consisting of three components: (1) intimacy, (2) passion, and (3) decision or commitment.

- The *attachment theory of love* views love as being similar in nature to the attachments we form as infants. The attachment (love) styles of both infants and adults are secure, anxious or ambivalent, and avoidant.

- Aside from the popular emphasis on love, many factors shape our choices of partners and spouses, revealing the existence of rules of mate selection.

- The *marketplace of relationships* refers to the selection activities of men and women when sizing up someone as a potential date or mate. In this marketplace, each person has resources, such as social class, status, age, and physical attractiveness.

- Initial impressions are heavily influenced by physical attractiveness. A *halo effect* surrounds attractive people, from which we infer that they have certain traits, such as warmth, kindness, sexiness, and strength.

- The setting in which you see someone can facilitate or discourage a meeting; *closed fields* allow you to see and interact simultaneously, whereas an *open field* makes meeting more difficult.

- Women often covertly initiate meetings by sending nonverbal signals of availability and interest. Men then initiate conversation with an opening line.

- *Dating scripts* prescribe certain behavior as expected of each gender. For women, problems in dating include sexual pressure, communication, and where to go on the date; for men, problems include communication, where to go, shyness, and money.

- *Hooking up* is a widespread phenomenon on college campuses. It differs from conventional dating in its focus on sexual intimacy early, before the building of a relationship, and in its lack of longer-term objectives.

- Breakups are commonplace. In accounting for breakups, we attribute the cause to one of the following: our own or our partner's personal characteristics, characteristics within the relationship, and environmental influences.

- Gay and lesbian couples are more prone to breaking up. Gay and lesbian individuals report higher levels of satisfaction with postbreakup friendships with former partners than do heterosexuals.

- *Jealousy* is an aversive response that occurs because of a partner's real, imagined, or likely involvement with a third person. Jealousy acts as a boundary marker for relationships.

- Time affects romantic relationships. The rapid growth of intimacy may level off, and individuals may become habituated to passion. Commitment tends to increase, provided that the relationship is judged to be rewarding.

- For many, romantic love diminishes over the course of a long-term relationship. It may either end or be replaced by intimate love. Many individuals experience the disappearance of romantic love as a crisis. Intimate love is based on commitment, caring, and self-disclosure.

- The transformation from romantic to more companionate love is not inevitable. There are long-term relationships that retain the intensity and sexual interest that are characteristic of passionate love.

Key Terms

attachment theory of love 151	instrumental displays 142
attributions 164	intimacy 139
closed fields 156	jealousy 161
commitment 141	marketplace of relationships 154
companionate love 150	open fields 156
companionate marriage 140	passionate love 150
conjugal family 141	peer marriage 145
consummate love 168	prototypes 148
deep friendship 145	reactive jealousy 161
double standard of aging 156	rejection sensitivity 151
expressive displays 142	self-disclosure 139
feminization of love 142	suspicious jealousy 161
halo effect 154	triangular theory of love 150
hierarchy of needs 138	unrequited love 166
homogamy 153	wheel theory of love 160

RESOURCES ON THE WEB

Book Companion Website

www.cengage.com/sociology/strong

Prepare for quizzes and exams with online resources—including tutorial quizzes, a glossary, interactive flash cards, crossword puzzles, self-assessments, virtual explorations, and more.

6

© Rubberball/GettyImage

Understanding Sex and Sexualities

What Do YOU Think? Are the following statements TRUE or FALSE?
You may be surprised by the answers (see answer key on the following page).

T	F	
T	F	**1** Unlike most human behavior, sexual behavior is instinctive.
T	F	**2** A significant number of women require manual or oral stimulation of the clitoris to experience orgasm.
T	F	**3** It is normal for children to engage in sexual experimentation with other children of both sexes.
T	F	**4** Kissing is the most common and most accepted sexual activity.
T	F	**5** A decline in the frequency of intercourse almost always indicates problems in the marital relationship.
T	F	**6** Most married women and men have had an extramarital sexual relationship.
T	F	**7** Bisexuality is more widely accepted than male homosexuality or lesbianism.
T	F	**8** Partners in cohabiting relationships are more likely to be sexually unfaithful than are married couples.
T	F	**9** Because of their knowledge, college students rarely put themselves at risk for sexually transmitted infections.
T	F	**10** Condoms are ineffective at preventing pregnancy or protecting against sexually transmitted infections.

It is now time to consider sex. For many of you, that must seem like a silly statement. Quite apart from this book, we often consider, think about, or take steps to pursue—or avoid—sexual encounters. Our popular culture is heavily sexualized. Advertising, in particular, uses sexual innuendo and image to sell us any number of products. Furthermore, being sexual is an essential part of being human. Through our sexuality, we are able to connect with others on the most intimate levels, revealing ourselves and creating deep bonds and relationships. Sexuality is a source of great pleasure and profound satisfaction. It is the typical means by which we reproduce, transforming ourselves into mothers and fathers. Paradoxically, sexuality also can be a source of guilt and confusion, a pathway to infection, and a means of exploitation and aggression. Examining the multiple aspects of sexuality helps us understand our sexuality and that of others. It provides the basis for enriching our relationships.

In this chapter, we offer an overview of sexuality and sexual issues, especially as they are interconnected with relationships, marriages, and family life. We begin by considering the sources of our sexual learning and proceed through sexual development and expression in young, middle, and later adulthood, including the gay, lesbian, or bisexual identity process. We consider the shifts in sexual scripts from traditional to modern and the social control of sexuality. When we examine sexual behavior, we cover the range of activities and relationships in which people engage. Ultimately, we look at nonconsensual sexual relations, sexual problems, and dysfunctions; birth control; sexually transmissible diseases, human immunodeficiency virus (HIV), and acquired immunodeficiency syndrome (AIDS); and sexual responsibility. We hope that this chapter will help you make sexuality a positive element in your life and relationships.

Sexual Scripts

Although we might consider sexual behavior one of the most natural parts of being human, what we do, think about, and feel regarding sexual activity is the product of our social learning. Organizing and directing our sexual impulses are culturally shared sexual scripts, which we learn and act out. A **sexual script** consists of expectations of how to behave sexually as a female or male and as a heterosexual, bisexual, lesbian, or gay male. Like a sexual road map or blueprint offering us general directions, a sexual script enables each individual to organize sexual situations and interpret emotions and sensations as sexually meaningful (Hynie et al. 1998). It may be more important than our own experiences in guiding our actions. Over time, we may modify or change our scripts, but we will not throw them away.

In looking at sexual scripts, we see how society influences our sexual attitudes and behavior. This is not unique to the United States; every society regulates and controls the who, what, when, where, and why of sexuality.

Who. We are taught to have sex with people who are unrelated, around our own age, and of the other sex (heterosexual). Less acceptable is having sex with yourself (masturbation), with members of the same sex (gay or lesbian sexuality), and with relatives (incest). In most societies, extramarital relationships are prohibited. Examining the "whos" in sexual scripts raises the question, With whom is it or would it be acceptable to engage in sexual behaviors?

What. Society classifies various sexual acts as good or bad, moral or immoral, and appropriate or inappropriate. Although these designations may seem absolute, they, like all aspects of sexual scripts, are culturally relative.

When. For teenagers and young adults, "when" might mean when parents are out of the house. If one is a parent, "when" may entail wait until one's children are asleep. Usually, such timing is related to privacy, but "when" may also be related to the age at which sexual activity is expected to start and stop, how often people are expected to engage in sexual relations, and when in a relationship sex should begin. Finally, it may pertain to times when sex is considered appropriate or inappropriate. Some societies frown on a woman engaging in sex during her menstrual flow, for a period after the birth of a child, or while nursing (Miracle, Miracle, and Baumeister 2003).

Where. Where do sexual activities occur with society's approval? In the United States, they usually occur in the bedroom, where a closed door signifies privacy. For adolescents, automobiles, fields, beaches, and motels may be identified as locations for sex; churches, classrooms, and front yards usually are not. "Where" may also extend to where it might be considered appropriate or inappropriate to discuss sex or to expose parts of your body.

Why. There are many reasons for having sex: procreation, love, passion, revenge, intimacy, exploitation, fun, pleasure, relaxation, boredom, achievement, relief from loneliness, exertion of power, and on and on. Some of these reasons are approved by society; others are not. Some we conceal; others we do not.

Gender and Sexual Scripts

Our gender socialization is also in part a socialization into sexuality as we learn what sexual behavior is appropriate, legitimate, and acceptable for each gender. They are most powerful during adolescence, when individuals are first learning to be sexual. Gradually, as one gains experience, one may modify and change these sexual scripts. As children and adolescents, we learn our sexual scripts primarily from our parents, siblings, peers, and the media. As we get older, interactions with our partners become increasingly important.

Traditional heterosexual sexual scripts are clearly "gendered," prescribing different roles and responsibilities to females and males. The traditional male script casts men as the initiators of sexual encounters. They are expected to be assertive, confident, and knowledgeable about sexual matters. They are supposed to know how to please their partners as well as how to coax their partners to share sexual intimacy. Sex is a goal to be achieved, and each sexual encounter is almost like a "conquest," enhancing the male's self-esteem and reputation.

The male script places performance expectations on men. The orgasm is the proof of the pudding; the more orgasms, the better the sex. If a woman does not have an orgasm, the male feels that he is a failure because he was not good enough to give her pleasure. Researchers who study sexual stereotypes observe that men's sexual identity may depend heavily on a capricious physiological event: getting and maintaining an erection. They note that the following traits are associated with the traditional male role: (1) sexual competence, (2) ability to give partners orgasms, (3) sexual desire, (4) prolonged erection, (5) being a good lover, (6) fertility, (7) reliable erection, and (8) heterosexuality (Riseden and Hort 1992). Common to all these beliefs, sex is seen as a performance in which men are both the directors and the principal actors.

The traditional female script prescribes females a more passive role in sexual relations. They are expected to wait for and comply with the male's initiation of sexual activity and to be pleased with how each sexual encounter progresses. Where sex is the goal to be achieved in the male sexual script, the female script focuses more on feelings than on sex, more on love than on passion. When not sanctioned by love or marriage, sexually active women risk their reputations. Many women cannot talk about sex easily because they are not expected to have strong sexual feelings. Some women may feel comfortable enough with their partners to have sex with them but not comfortable enough to communicate their needs to them. Under more traditional expectations, to keep her image of sexual innocence and remain pure, a woman does not tell a man what she wants.

Women often "learn" that the right way to experience orgasm is from penile stimulation during sexual intercourse. But there are many ways to reach orgasm: through oral sex; manual stimulation before, during, or after intercourse; masturbation; and so on. Women who rarely or never have an orgasm during heterosexual intercourse may be deprived by not sexually expressing themselves in other ways.

These gendered sexual scripts affect the ways in which many people look at a range of sexual and sexually related behaviors. For example, by portraying males as active, sexually aggressive initiators whose sexual response cannot be easily controlled or constrained and females as sexually passive, innocent, and even nonsexual, traditional scripts ignore the possibilities of sexually reluctant males or sexually coercive females. This, in turn, obscures the phenomenon of female sexual offending, especially with male victims, and leaves police, victims, and helping professionals in the dark about what may be a more common phenomenon than is understood.

Contemporary Sexual Scripts

As gender roles have changed, so have sexual scripts. To a degree, traditional sexual scripts have been replaced by more liberal and egalitarian ones. Contemporary sexual scripts include the following elements for both sexes (Gagnon and Simon 1987; Rubin 1990):

- Sexual expression is positive.
- Sexual activities are a mutual exchange of erotic pleasure.
- Sexuality is equally involving of both partners, and the partners are equally responsible.
- Legitimate sexual activities are not limited to sexual intercourse but also include masturbation and oral–genital sex.
- Sexual activities may be initiated by either partner.
- Both partners have a right to experience orgasm, whether through intercourse, oral–genital sex, or manual stimulation.
- Nonmarital sex is acceptable within a relationship context.
- Gay, lesbian, and bisexual orientations and relationships are increasingly open and accepted or tolerated, especially on college campuses and in large cities.

Contemporary scripts both give greater recognition to female sexuality and are relationship centered rather than male centered. As we said earlier, traditional scripts have been replaced to a degree. Women who have several concurrent sexual partners or casual sexual relationships, for example, are still more likely to be regarded as more promiscuous than are men in similar circumstances. The "suppression of female sexuality" is often carried out by women, whether through maternal influence or the judgments of female peer groups (Miracle et al. 2003).

For example, a 1999 study of 165 young adult, heterosexual women showed both the liberalization of attitudes and the continuation of a **sexual double standard**. On the liberalization side, 99% of the university sample "strongly agreed" or "agreed" that women can enjoy sex as much as men do. Of the women, 69% disagreed or strongly disagreed with the statement that "women are less interested in sex than men are." However, 95% of the women believed that there continues to be a double standard wherein it is less accepted for a woman to have many sexual partners than it is for a man. In addition, 93% "probably" or "definitely agreed" that women who have many partners are more harshly judged than men with many partners, 49% indicated that women were labeled and penalized, but 48% stated that men were encouraged to have and rewarded for having many partners.

From their study of 698 heterosexual couples from a northeastern college campus and surrounding community, researchers Kathryn Greene and Sandra Faulkner (2005) contend that the contemporary sexual scripts for females and males are more alike, especially within committed heterosexual relationships where there is much overlap in the behavior of women and men. However, they also suggest that the "rules and roles" regarding sexual relationships in some ways still point to a "double standard." This is most evident on issues such as how many lifetime partners are "acceptable" (as opposed to "too many"), who in the relationship suggests using condoms or other methods of contraception, who should initiate sexual activity, and how openly one should discuss sexual matters. Greene and Faulkner found that couples who expressed less belief in the sexual double standard and other traditional sexual attitudes also reported more sexual communication and sexual self-disclosure. Such communication is further associated with higher relationship satisfaction (Greene and Faulkner 2005).

Critical Thinking

Do you believe that a sexual double standard still exists? Do you believe that sexually active females and males should be held to different standards? Why or why not?

How Do We Learn about Sex?

Children and adolescents are subjected to gender-specific messages about sexuality as well as both subtle and explicit socialization into heterosexuality (and away from homosexuality and bisexuality). Sources of influence include parents, siblings, peers, media, and school.

Parental Influence

Children learn a great deal about sexuality from their parents. They learn both because their parents set out to teach them and because they are avid observers of their parents' behavior. Even in families where open and active discussion of sexual topics is avoided, lessons about sex are taking place. When silence surrounds sexuality, it suggests that one of the most important dimensions of life is off limits, bad to talk about, and dangerous to think about.

Parents convey sexual attitudes to their children in a number of ways. What parents say or do, for example, to children who touch their "private parts" or try to touch either their mother's or some other woman's breasts conveys meanings about sex to a child. Parents who overreact to children's curiosity about their bodies and others' bodies may create a sense that sex is wrong. On the other hand, parents who acknowledge sexuality rather than ignoring or condemning it help children develop positive body images, comfort with sexual matters, and higher self-esteem (Miracle et al. 2003). Research suggests that what parents teach their children about sex and sexuality is shaped by what they, themselves, were taught and reflects what they wish they had received from their parents (Byers, Sears, and Weaver 2008). Existing research also indicates that most parent–child discussion about sex is really mother–daughter discussion about sex, that sons receive much less parental insight and information than daughters do, and that what information sons do receive they tend to learn from their mothers, not their fathers (Lehr et al. 2005).

As young people enter adolescence, they are especially concerned about their own sexuality, but they may be too embarrassed or distrustful to ask their parents directly about these "secret" matters. Meanwhile, parents, too, may experience embarrassment when discussing sex with their adolescents. That embarrassment along with the fear that they might not be able to answer their adolescents' questions are among the biggest reservations mothers express about talking about sexual matters with their teens (Byers et al. 2008). Furthermore, many parents are ambivalent about their children's developing sexual nature. They are often fearful that their children (daughters especially) will become sexually active if they have too much information. They tend to indulge in wishful thinking: "I'm sure Jessica's not really interested in boys yet" or "I know Tyrese would never do anything like that." As a result, they may put off talking seriously with their children about sex, waiting for the "right time," or they may bring up the subject once, say their piece, breathe a sigh of relief, and never mention it again. Sociologist John Gagnon called this the "inoculation" theory of sex education: "Once is enough" (cited in Roberts 1983). But children may need frequent "boosters" where sexual knowledge is concerned.

Research is somewhat mixed about the consequences of parent–child communication about sexuality (Bersamin et al. 2008). There is evidence that parents' attitudes toward premarital sex and sexuality can strongly influence adolescent sexual behavior. For example, mothers' disapproval of "risky sexual behavior" delays adolescents' initial experiences of sexual intercourse and reduces the frequency with which adolescents engage in sexual intercourse. The more positive, effective, and comprehensive the communication about sexuality, the more adolescents delayed their initial sexual experiences. Some research suggests

© Ryan McVay/Getty Images/Photodisc

Parents, especially mothers, are important sources of information and advice about sexuality. Although both sons and daughters speak more to mothers than to fathers about sexual issues, most parent–child sexual communication is really between mothers and daughters.

that "early, clear communication" between parents and their teenage children leads to lower levels of teen sexual activity and, for those who become sexually active, to greater understanding and use of safe-sex practices (Lehr et al. 2005; DiIorio, Kelly, and Hockenberry-Eaton 1999). In addition, research cited by Lehr and colleagues indicates that mothers who discuss sex-related issues influence their children's later protective sexual behaviors, including condom use (Dittus, Jaccard, and Gordon 1999; Miller et al. 1998). Other researchers report (Newcomer and Udry 1985; O'Sullivan et al. 1999) little to no effect of mother–child sexual communication on subsequent teen sexual behavior, and still other research suggests that it is associated with greater involvement in sexual activity, although this may be as much a consequence as a cause—in other words, parents of teens who are sexually active or expected to soon be sexually active may then engage in more discussion of sexual matters with their adolescents (Bersamin et al. 2008; Paulson and Somers 2000).

The effectiveness of parental supervision is less equivocal than the consequences of communication. Parental monitoring—knowledge of who their children are with, where they are, and how they spend their time—is inversely associated both with adolescents becoming sexually active and with the total number of sexual partners teens have; it is positively related to adolescents' use of contraceptives. For example, high school students who received high levels of parental supervision (i.e., were unsupervised less than five hours a week) were less likely to have had sexual intercourse. They also had fewer lifetime sexual partners than teens who were unsupervised more than five hours per week (Bersamin et al. 2008).

Because parents assume that their children are (or will be) heterosexual, they may avoid—intentionally or merely without thinking it relevant—discussion about sexual orientation. In so doing, they leave their children less aware of homosexuality, which, itself, becomes invisible. In addition, even simple comments such as "When you grow up and get married someday . . ." assumes that the child is or will become heterosexual (Shibley-Hyde and Jaffe 2000).

Siblings

Parents are not the only family members that help shape one's sexual attitudes and behavior. Siblings, specifically older siblings, are influential sources of sexual socialization and by their own behavior may become role models for their younger siblings sexual attitudes and behaviors (Kowal and Blinn-Pike 2004). In fact, siblings may in some instances be more influential sources of sexual socialization than parents, though their influence can be positive or negative. Adolescents may feel less reluctance to seek information about sex or contraception from an older sister or brother than from a parent because siblings are assumed to be less judgmental, less embarrassed, and less punitive than parents. One study that followed 297 midwestern high school students over 42 months found that sibling conversation about safe(r) sexual practices was associated with less risky attitudes about sexual behavior when it is "in concert with parent-adolescent conversations about sex" (Kowal and Blinn-Pike 2004, 382).

Certain characteristics of sibling relationships appear to influence the frequency of sibling sexual communication. Where adolescents report positive relationships with older siblings, communication is more likely regardless of whether the older sibling is perceived to be sexually conservative or to take risks. Girls with older sisters described their relationships as closer, perceived their sisters to have safer attitudes about sex, and engaged in more conversations about sexual matters than any other sibling dyad (i.e., brother–brother or brother–sister) (Kowal and Blinn-Pike 2004).

Peer Influence

Adolescents garner a wealth of information, as well as much misinformation, from one another about sex. They often put pressure on one another to carry out traditional gender roles. Boys encourage other boys to become and be sexually active. Those who are pressured must camouflage their inexperience with bravado, which increases misinformation; they cannot reveal sexual ignorance. Even though many teenagers find their earliest sexual experiences less than satisfying, many still seem to feel a great deal of pressure to conform, which may mean becoming or continuing to be sexually active. For many young people, virginity may be experienced as a stigma, whereas virginity loss is seen either as a way to shed the stigma (more true for males) or merely as part of growing up (Carpenter 2002).

Evidence suggests that more teens are delaying their initial experience of sexual intercourse, thus creating a somewhat different peer culture. Findings from a Health and Human Services study indicate that

Table 6.1 Percentage of Never-Married Teens, Ages 15 to 19, Who Ever Had Intercourse by Age and Sex, United States 1995 and 2002

Female	1995	2002
15–19 years old	49.3%	45%
15–17 years old	38.0%	30.3%
18–19 years old	68.0%	68.8%
Male		
15–19 years old	55.2%	45.7%
15–17 years old	43.1%	31.3%
18–19 years old	75.4%	64.3%

SOURCE: NSFG, Series 23, No. 24, Table 1.

from 1995 to 2002, fewer teenage boys and younger teenage girls were sexually active. Among never-married females 15 to 17 years old, the percentage who ever had sexual intercourse declined from 38% in 1995 to 30% in 2002. For females 18 to 19, there was virtually no change over this same time period. As reflected in Table 6.1, among males there were even more dramatic drops. We look further at the trend in sexual involvement in a later section of this chapter.

Media Influence

Although many may believe that the media have no effect on them, the various mass media profoundly affect our sexual attitudes and may also affect our sexual choices and behaviors (Strasburger 2005). In fact, "private electronic media has become the primary sexuality educator of youth," and "young teens (ages 10–15) consider the mass media as more important sources of information about sex and intimacy than parents, peers, and sexuality education programs" (Allen et al. 2008, 518).

Certain media—most notably television and popular music—have been shown to have an effect on attitudes about sex and on sexual behavior. For example, research undertaken by social psychologist Rebecca Collins and colleagues (Collins et al. 2004) found that both teenagers' sexual beliefs and their behavior are affected by their exposure to sexual content on television. Sexual content is common in prime-time programming. After reviewing a sample of almost 1,000 programs, 7 out of 10 were found to have sexual content and convey messages about sex. Talking about sex was more common than was portraying sexual activity (Kunkel et al. 2005).

Extensive exposure to television portrayals of sexual behavior, televised talk about sex, and talk about contraception can all influence the ways teens think about sex and how they act. Collins and associates (2004) found that extensively exposed viewers of sexual content were twice as likely as those who viewed the lowest level of sexual content to initiate sexual intercourse or progress from less (e.g., kissing) to more advanced levels of sexual expression (e.g., oral sex). Other research, stretching back from the 1980s to the 2000s, finds similar associations between viewing sexual content on television and earlier involvement in sexual activity.

What we listen to has its own effects. Physician and researcher Brian Primack and colleagues (2009) examined the effect of exposure to music lyrics, especially those that are sexually degrading (sex based only on physical characteristics and in which consent is not mutual but reflects a power imbalance). In Primack et al.'s study, it wasn't just the level of exposure but also the nature of the sexual content that mattered. High levels of exposure to lyrics describing degrading sex were associated with higher levels of sexual behavior. Where overall nearly a third of the teens had had intercourse, among those with the greatest amount of exposure to sexually degrading lyrics, 45% had sexual intercourse. This effect was found for both genders. Exposure to lyrics describing nondegrading sex did not have the same effect. Only 21% of those teens least exposed to sexually degrading lyrics but exposed to nondegrading sexual lyrics had had intercourse (Primack et al. 2009).

Along with the electronic media, magazines, newspapers, and print advertising can also affect what young people learn about and learn to expect from sex. Sometimes, the messages are mixed. For example, research on the effects of magazines on young women suggests that print media can affect women's attitudes and beliefs about sex. The more "teen-focused magazines" expose young women to a contradictory message that encourages them be sexually provocative in their demeanor and dress but that discourages them from being sexually active. Encouraged to devote much time and effort toward making themselves physically appealing to boys and to presenting themselves as sexual objects, at the same time girls are discouraged from and warned about pursuing sexual relationships. Males are negatively portrayed as "either emotionally inept . . . or as sexual predators" (Kim and Ward 2004, 49), neither of which are flattering to males or encouraging for females entering the world of heterosexual relationships. More "adult-focused magazines," such as Cosmopolitan, convey a different message. Sexually

Popular magazines such as *Seventeen* or *Cosmopolitan* are part of the sexual socialization that many young women in the United States experience. Although both are widely read, they convey different messages about sexuality.

aggressive women are portrayed positively, almost in the same way as a stereotypical male is portrayed. College-age women who more frequently read magazines such as Cosmopolitan were less likely to perceive sex as risky or dangerous and "more likely to view sex as a fun, casual activity and to be supportive of women taking charge in their sexual relationships" (Kim and Ward 2004, 53).

Using a variety of designs and questions, recent research strongly suggests that teenage behavior is influenced by the amount of exposure teens have to sexual content in the media. A 2006 study looking at the effects of television along with popular music, movies, and magazines on the behavior of more than 1,000 North Carolina middle school students from more than a dozen middle schools found that white students who were in the top 20% of exposure to sexual content at ages 12 to 14 were twice as likely to have engaged in sexual intercourse by ages 14 to 16 as those in the bottom 20% of exposure. Among African Americans, media exposure had less effect. Instead, sexual experience was influenced more by their friends' sexual behavior and by their perceptions of what their parents expected of them (Brown et al. 2006).

A second recent study examined six different media, adding newspapers and Internet sites to the four media already mentioned, assessing both the amount of sexual content and the effects of that content. Communications researchers Carol Pardun, Kelly Lardin L'Engle, and Jane Brown (2005) found that popular music and movies have the largest percentage of sexual content, with television third. In terms of impact, movies and music were more strongly associated with both "light" (e.g., being alone with a romantic partner, light kissing, or "French kissing") and "heavy" (breast touching, penis or vagina touching, oral sex, or sexual intercourse) sexual behavior and with one's intentions to have sex than were the other media, including television. This study also indicated that it was really how much sexual content, not what kind of sexual content, one was exposed to that mattered in affecting one's behavior and expectations.

Sex on the Internet

From its inception in 1983 to the present, the Internet has revolutionized the way we live. By 2000, there were a reported 1 billion Web pages. A decade later, some contend that there may be hundreds of billions of Web pages and a billion or more Web sites one can access via the Internet. Beginning in the 1990s, researchers looked with increasingly critical eyes at "sex and the Net." Topics such as "addiction" to pornography, exploitation and entrapment of children, sexual harassment, and deviant pornography are among the issues and concerns that emerged.

One of the most popular innovations to emerge in our increasingly wired world is the *weblog,* or "blog." First appearing in the early 1990s, weblogs now number in the millions and have become resources for people with shared interests, people seeking

worried about negative outcomes of their first experience of intercourse. In addition, women are more worried about pregnancy, more likely to be nervous, more likely to be in pain, and less likely to experience orgasm. They are also more likely to experience postcoital guilt and express with regret the wish that they had waited.

Converging Patterns for Women and Men

As recently as the 1980s, young women were more likely to value virginity and to contemplate its loss primarily within committed romantic relationships, and men welcomed opportunities for casual sex and expressed disdain for virginity. Research in the 1990s revealed increasing similarities between women and men. More young men than before were expressing pride and happiness about being virgins. Growing numbers of young women were perceiving virginity in neither a positive nor a negative light, with a minority eagerly anticipating "getting it over with." By the 1990s, gender differences in the age at which one first engages in intercourse had all but disappeared. By 1999, age at first vaginal sex was between 16 and 17 for both females and males (Carpenter 2002).

With so much attention paid to sexuality within the media and peer culture, what motivates young people to retain virginity or maintain abstinence? There is evidence suggestive of both moral and more pragmatic reasons. For instance, some 2.5 million adolescents took public virginity pledges between 1995 and 2001. They promised to abstain from sexual intercourse until they married. These pledges, sponsored by the Southern Baptist Church, appeared to delay first intercourse (though not necessarily until marriage). More pragmatic concerns also operate. Among teenagers who were virgins, more than 90% cited concern about pregnancy, concern about HIV/AIDS or other sexually transmitted infections, and "feeling too young" as among their reasons for their decision (Kaiser Family Foundation 2003).

Gay, Lesbian, and Bisexual Identities

In contemporary America, people are generally classified as **heterosexual** (sexually attracted to members of the other gender), **homosexual** (sexually attracted to members of the same gender), or **bisexual** (attracted to both genders). Although today we may automati-

cally accept these categories, such acceptance has not always been the case, and the categories do not necessarily reflect reality. As late as the nineteenth century, there was no concept of "homosexuality." Both the label *homosexual* and the label *heterosexual* first appeared in print in the United States in a medical journal in 1892 (Katz 2004).

The familiar threefold categorization of sexual orientation used today may not accurately depict the range that exists in our sexual orientations—who we are attracted to, who we have relations with, who we fantasize about, the type of lifestyle we live, and how we identify ourselves. On any of these items we may be *exclusively* oriented toward the other sex or our sex, *mostly* drawn to the other sex or our sex, or oriented to both sexes *about equally*. In addition, the interaction of numerous factors—social, biological, and personal—leads to the unconscious formation of sexual orientation. The two most important components of sexual orientation are the gender of our sexual partner and whether we label ourselves heterosexual, gay, lesbian, or bisexual. Finally, our sexual orientation may change over time. Thus, what was true of past relationships or attractions may not fit with the present or may differ from what we envision for our future (Klein 1990; Miracle, et al. 2003).

Because *homosexual* carries negative connotations and obscures the differences between what women and men experience, we refer to gay men and **lesbians**. In addition, replacing the term *homosexual* may help us see individuals as whole people; sexuality is not the only significant aspect of the lives of gay men, lesbians, bisexuals, or heterosexuals. Love, commitment, desire, caring, work, possibly children, religious devotion, passion, politics, loss, and hope are also, if not more, important.

At different times, especially in the past, those with lesbian or gay orientations have been called sinful, sick, perverse, or deviant, reflecting traditional religious, medical, and psychoanalytic approaches. Contemporary thinking in sociology and psychology has rejected these older approaches as biased and unscientific and has focused, instead, on how women and men come to identify themselves as lesbian or gay, how they interact among themselves, and what effect society has on them (Heyl 1989).

How does one "become" gay, lesbian, bisexual, or even heterosexual for that matter? Such a question is neither easily answered nor inconsequential. If sexual orientation is biologically based, discrimination against gay men, lesbians, or bisexual women and

Public Policies, Private Lives: "Sexting" and the Law

Jessica Logan was 18, a high school senior who was planning to go to college, before she got caught up in a nightmarish situation resulting from her decision to send nude pictures of herself, by cell phone, to her boyfriend. After the two broke up, her ex-boyfriend sent some of the photos to other girls at their Ohio high school. Before long, the pictures were more widely distributed among students. Jessie Logan became an easy target for harassment and name-calling, had things thrown at her, and was pretty badly bullied. She even went on a Cincinnati television station to tell her story, saying that she wanted to warn others about the ramifications of sending such images and spare them the same sort of consequences that she was experiencing. Instead of attending college, she committed suicide, hanging herself in her bedroom closet.

This truly tragic story is extreme and uncommon. However, the situation that triggered it is not as unusual. Instances of teenagers exchanging nude or seminude pictures of themselves via cell phone have happened in many places throughout the United States. Called "sexting" to emphasize the use of text messaging for sharing sexually explicit material, the phenomenon has left parents, educators, and lawmakers in a quandary. What can or should be done about the behavior?

Efforts to answer that often lead to other questions, including the question of how common or uncommon sexting is among teens. A widely quoted statistic in the media coverage of sexting cases suggests that 20% of teens have sent or posted nude or seminude photographs or videos of themselves. This statistic is the product of a survey cosponsored by CosmoGirl and the National Campaign to Prevent Teen and Unplanned Pregnancy. Nearly 1,300 13- to 26-year-olds were asked a series of questions about their sending or receiving sexually suggestive material. Of the 1,300, half were female, and slightly more than half were teens, ages 13 to 19. Some additional findings of the survey are as follows:

- Girls were a little more likely than boys (22% vs. 18%) to have sent nude or seminude photo or video images of themselves. Among the youngest teens (13–16 years old), 11% of female respondents had sent such material.
- Just over 70% of teen girls and 67% of teen boys in the sample have posted or sent "sexually suggestive" content to a boyfriend or girlfriend.
- 21% of teen girls and 39% of teen boys said they had sent sexually suggestive material to someone they wanted to hook up with or date; 15% of teens sent such sexually suggestive messages to someone they had only met online.
- Among reasons given for why they engage in "sexting," a majority of the teenage girls cited "to be fun or flirtatious" and to send a sexy present for their boyfriends.

Disturbing as these figures are, there is also some reason to question their accuracy. The sample from which the data come was a volunteer sample of Internet users. Noting that it is likely that such an instrument would be more likely to attract heavy Internet users and that such teens may be more prone to sending electronic images and messages, sociologist David Finkelhor estimates that those who respond to online surveys such as this one may

men is especially unjustified. It becomes no different than discriminating against someone because of their age, their gender, or their race, all statuses over which we exercise no control.

Research on the self-identification process suggests that we can divide the gay, lesbian, and bisexual population into two groups of people: one group comprising men and women who say that they knew, from a much earlier age while growing up, that they were "different" from others and the second group growing up "never questioning the suitability of a heterosexual identity" until later in their lives, such as college age or middle age. Most men and many of the women in this latter group attribute this delayed identification to denial. However, many women (but not many men) reject the idea that they were driven by uncontrollable or irresistible desires, saying, instead, that they "chose" to become involved with a same-sex partner and that their choice was a political one, associated with their particular feminist politics. For others, it was a choice motivated out of the desire for more equal, more intimate relationships than they believed they could have with men (Butler 2005).

couples—especially those not in civil unions—than to lesbians or heterosexual couples (Peplau and Fingerhut 2007; Solomon et al. 2005). Gay men in civil unions are less tolerant or accepting of sex outside their relationship than are gay men not in civil unions, but both groups of gay men are less exclusive than married heterosexual men (Solomon et al. 2005). Within their relationships, sexual intimacy is less frequent among lesbian couples than among gay male or heterosexual couples, although nongenital or nonsexual affection (e.g., cuddling, kissing, and hugging) is reportedly more common. Across couple types, lesbians have less frequent sex, and gay male couples have sex most frequently. As with heterosexuals and lesbians, sexual frequency declines the longer gay couples stay together.

Monogamy and romantic love are more important to lesbians and to heterosexual women than to men in heterosexual or gay relationships (Spitalnik and McNair 2005). Lesbians and gay men who are neither married nor in civil unions also have fewer barriers than do married heterosexuals to ending their relationships once troubles surface. This makes it unlikely that lesbians and gay men will live in long-term, dissatisfying, "miserable and deteriorating" relationships, but perhaps more gay and lesbian relationships than heterosexual relationships will end that could have been saved or improved with patience and effort. In addition, gay male and lesbian couples must deal with disagreements about how much they wish to disclose their sexuality to others. Such disagreements may lead a more open partner to pressure a less open partner with the threat of disclosure or leave the more open partner feeling as though the less open partner is less committed to the relationship (Peplau et al. 2004).

Antigay Prejudice and Discrimination

Antigay prejudice is a strong dislike, fear, or hatred of lesbians and gay men because of their homosexuality. **Homophobia** is an irrational or phobic fear of gay men and lesbians, while **heterosexism** is bias and/or discrimination in favor of heterosexuals. Like other forms of bias (e.g., racism and sexism), there are both institutionalized forms of heterosexism—such as the denial of marriage rights or relationship recognition—and individualized expressions of heterosexism—as in the telling of jokes or making disparaging comments about sexual minorities. Prior research has indicated that whether in the workplace or in academic environments, such mistreatment has negative consequences for one's health, performance, and well-being. Out-

comes have included diminished self-esteem, higher rates of depression, lower levels of life satisfaction, and more substance abuse (Silverchanz et al. 2008).

Psychologist Perry Silverchanz and colleagues used Gregory Herek's definition of heterosexism—actions that deny, denigrate, or stigmatize nonheterosexual behavior, relationships, identity, or community—to assess how commonly heterosexism was experienced by a sample of more than 3,000 college students of varying sexual orientations at a northwestern U.S. university. Silverchanz et al. wanted to assess the impact of heterosexism on those who were exposed to either direct and personal (e.g., being called a "fag" or a "dyke" or being called perverted) or "ambient" heterosexist harassment (e.g., overhearing offensive jokes, crude remarks, or name-calling in one's presence). Of their sample, 41% had some experience with heterosexist harassment, including 39% of heterosexual students and 57% of sexual minorities. The impact of heterosexist harassment was worse when one experienced both personal and ambient harassment, and such effects affected both sexual minorities and heterosexuals (Silverchanz et al. 2008).

Other research confirms the continued experience of harassment and mistreatment by sexual minorities. Psychologists David Huebner and Gregory Rebchook and professor of medicine Susan Kegeles surveyed more than 1,200 young gay or bisexual men, ages 18 to 27, from three southwestern U.S. cities and found that more than a third had received antigay harassment, 11% had experienced antigay discrimination, and 5% were victims of antigay violence. Younger men suffered higher rates of such mistreatment, with half acknowledging having been harassed, 14% indicating having experienced discrimination, and 10% suffering antigay violence (Huebner et al. 2004). A 1999 study by Gregory Herek and colleagues found that among their sample of 2,259 gay men and lesbians, 28% of the men and 19% of the women had, in their lifetimes, experienced violent or other criminal victimization because of their sexual orientation (Herek, Gillis, and Cogan 1999).

In 2005, psychologist Gregory Herek undertook a study of a nationally representative sample of 662 gay men, lesbians, and bisexuals to determine their level of bias-related lifetime victimization. Among his findings are the following:

- 13% had been physically attacked, sexually assaulted, beaten or hit.
- 15% had been robbed or vandalized.

- 14% had had someone try to rob or attack them or vandalize their property but not succeed.
- 23% had been threatened with violence.
- 13% had had objects thrown at them.
- 49% had been verbally insulted or abused.

Gay men were more likely to have experienced such victimization than were lesbians or bisexuals. More than 33% of gay men compared to between 11% and 13% of lesbians or bisexual women or men had had such experiences (Herek 2009).

Typically, males harbor more prejudice and express more negative attitudes toward gays and lesbians than women do. As they move through adolescence toward young adulthood, male prejudice increases where female prejudice diminishes.

As shown in the research by Silverchanz and colleagues, antigay prejudice can have adverse effects on heterosexuals, too. It can do the following:

- Create fear and hatred—aversive emotions that cause distress and anxiety
- Alienate heterosexuals from gay family members, friends, neighbors, and coworkers
- Limit the expression of a range of behaviors and feelings, such as hugging or being emotionally intimate, with same-sex friends for fear that such intimacy may be "homosexual"
- Lead to exaggerated displays of masculinity by heterosexual men trying to prove they are not gay

Bisexuality

As we noted earlier, bisexuals are individuals attracted to members of both genders. Asked what their bisexual identities meant to them, most of Paula Rust's respondents said it meant that they had "the potential to be sexually, emotionally and/or romantically attracted to members of both sexes or genders" (Rust 2004, 216). For many, it is the capacity or potential, not necessarily the actual experience, that makes them identify themselves as bisexual. For some, bisexuality is expressed in alternating relation-

AP Images/Phil McCarten

The hate-based killing of 15-year-old Lawrence King by a fellow student illustrates the most extreme form of antigay hostility.

ships with women and men. Others have concurrent sexual relationships with women and men (e.g., "I have a girlfriend, with whom I have sex often and am very attracted to, but am still attracted to men, with whom I also have sexual relations") (Rust 2004, 218). Still others base their self-definitions more on feelings than on any actual relationships, past or present (e.g., "over 99% of my sexual interactions have been heterosexual. But I fantasize about women a great deal and enjoyed the one-on-one encounter I had") (Rust 2004, 218).

Becoming bisexual requires the rejection of two recognized categories of sexual identity: heterosexual and homosexual. In a nationwide study by Samuel and Cynthia Janus (1993), about 5% of men and 3% of women identified themselves as bisexual. Data from the 2002 National Survey of Family Growth reveal that 1.8% of 15- to 44-year-old males and 2.8% of 15- to 44-year-old females identify themselves as bisexual.

Because it is only since the 1980s that bisexuality has become more visible and bisexuals more politicized and organized, it shouldn't be surprising that research on the "bisexual experience" is less abundant than the similar literature on gays and lesbians (Herek 2002). What research we have indicates that, like gays and lesbians, bisexuals are often the targets of hostility and harassment. Herek et al. (1999) report that 15% of bisexual women and 27% of bisexual men in their sample had experienced a property or violent crime. These rates are similar to what was reported by lesbians (19%) and gay men (28%). A Kaiser Family Foundation survey of 405 lesbians, gay men, and bisexuals found that among bisexuals, because of their sexual orientation, 60% had experienced some form of discrimination, 52% had suffered verbal abuse, and 26% felt that they were not accepted by their families. These percentages were lower than the comparable percentages for lesbians and gay men, suggesting that bisexuality may be somewhat less stigmatized than homosexuality (Herek 2002).

More negative attitudes toward bisexual men and women seem to be associated with certain individual characteristics, such as frequent attendance at religious services, a conservative political ideology, and having had minimal prior contact with bisexual men or women. These same factors are associated with heterosexual attitudes toward gay men and lesbians (Herek 2002).

Because they might be perceived as rejecting both heterosexuality and homosexuality, bisexuals can also be stigmatized by gay men and lesbians who might view bisexuals as "fence-sitters" not willing to admit their homosexuality or as people simply "playing" with their orientation (Herek 2002). Thus, bisexuality may not be taken seriously by either group. Loraine Hutchins and Lani Kaahumanu (1991) believe that bisexuality arouses hostility because it challenges widely held beliefs about the permanence of sexual orientation and about the existence of discrete categories of sexuality (e.g., heterosexual, gay or lesbian, bisexual).

Becoming Bisexual

In 1994, the first model of bisexual identity formation was developed (Weinberg, Williams, and Pryor 1994). According to this model, bisexual women and men go through four stages in developing their identity:

1. *Initial confusion.* This may last years. People may be distressed by being sexually attracted to both sexes, may believe that their attraction to the same sex means an end to their heterosexuality, or may be disturbed by their inability to categorize their feelings as either heterosexual or homosexual.
2. *Finding and applying the bisexual label.* For many, discovering there is such a thing as bisexuality is a turning point. Some find that their first hetero-

sexual or same-sex experience permits them to view sex with both sexes as pleasurable; others learn of the term *bisexuality* from friends and are able to apply it to themselves.
3. *Settling into the identity.* At this stage, bisexuals begin to feel at home with and accept the bisexual label.
4. *Continued uncertainty.* Bisexuals don't have a community or social environment that reaffirms their identity. Despite being settled in, many feel persistent pressure from gay men and lesbians to relabel themselves as homosexual and to engage exclusively in same-sex activities.

Sexuality in Adulthood

Psychosexual development and change does not end in young adulthood. It continues throughout our lives. In middle age and old age, our lives, bodies, sexuality, relationships, and environment continue to change. New tasks and new satisfactions arise to replace or supplement older ones.

Developmental Tasks in Middle Adulthood

In the middle adult years, some tasks of psychosexual development begun but only partly completed or deferred in young adulthood (e.g., issues surrounding intimacy or childbearing) may continue. Because of separation or divorce, we may find ourselves facing the same intimacy and commitment tasks at age 40 that we thought we completed 15 years earlier (Cate and Lloyd 1992). But life does not stand still; it moves steadily forward, whether we're ready or not. Other developmental issues appear, including the following:

- *Redefining sex in marital or other long-term relationships.* In new relationships, sex is often passionate and intense and may be the central focus. But in long-term marital or cohabiting relationships, the passionate intensity associated with sex is often eroded by habituation, competing parental and work obligations, fatigue, and unresolved conflicts. Sex may need to be redefined as a form of intimacy and caring. Individuals may also need to decide how to deal with the possibility, reality, and meaning of extramarital or extrarelational affairs.

- *Reevaluating sexuality.* Single men and women may need to weigh the costs and benefits of sex in casual or lightly committed relationships. In long-term relationships, sexuality often becomes less central to relationship satisfaction. Nonsexual elements, such as communication, intimacy, and shared interests and activities, become increasingly important to relationships. Women who have deferred their childbearing begin to reappraise their decision: should they remain child free, "race" against their biological clocks, or adopt a child?
- *Accepting the biological aging process.* As we age, our skin wrinkles, our flesh sags, our hair grays (or falls out), our vision blurs—and we become in the eyes of society less attractive and less sexual. By our forties, our physiological responses have begun to slow noticeably. By our fifties, society begins to "neuter" us, especially if we are women who have gone through menopause. The challenges of aging are to accept its biological mandate and to reject the stereotypes associated with it.

Sexuality and Middle Age

Men and women view and experience aging differently. As men approach their fifties, they fear the loss of their sexual capacity but not their attractiveness; for women, the reverse is true. As both age, purely psychological stimuli, such as fantasies, become less effective for arousal. Physical stimulation remains effective, however.

Among American women, sexual responsiveness continues to grow from adolescence until it reaches its peak in the late thirties or early forties; it is usually maintained near the same level into the sixties and beyond. Data from both the United States and elsewhere have yielded inconsistent research findings on women's sexuality at midlife. Some studies suggest that rates of sexual intercourse, levels of sexual interest, frequency of orgasm, extent of sexual fantasizing, vaginal lubrication, and satisfaction with a partner all decline in midlife. Others show no decline in sexual interest, responsiveness, or "functioning." About the only thing that can be safely concluded is that considerable variability occurs in midlife women's sexuality.

Having emotional and psychological needs met (feeling attractive, appreciated, independent, understood, and productive) is related both to feeling attractive and to satisfaction with one's sex life. Frequency of intercourse and orgasm and finding sex pleasant, enjoyable, and satisfying are associated with higher levels of marital adjustment and contentment, although it is not clear whether marital quality causes or follows sexual satisfaction (Fraser, Maticka-Tyndale, and Smylie 2004). Data from the United States, Great Britain, and France indicate age differences that may be the result of cohort differences (based on differences in sexual socialization and changing cultural attitudes) or possible effects of aging (Table 6.2).

In addition to age differences in whether and how often women report having engaged in sexual intercourse, data from the 1994 Sex in America Survey reveal information about sexual problems or dysfunctions for women of different ages (see Table 6.3). We can see the effect of aging on women's sexuality in a number of reported dysfunctions.

As the data in Table 6.2 reveal, relative to other ages, high levels of orgasmic difficulty, lack of pleasure and interest, and trouble with vaginal lubrication are reported by 55- to 59-year-olds. Keep in mind that these data are from "sexually active women" and as such may even understate the effect of aging, as more women may become sexually inactive as they move through their forties and into their fifties. Sexually inactive women and any problems they have that might cause them to refrain from sex are not represented in these data (Fraser et al. 2004).

Around the age of 50, the average American woman begins **menopause**, which is marked by a cessation of the menstrual cycle and an end to fertility. Menopause is not a sudden event. Usually, for several years preceding menopause, the menstrual cycle becomes increasingly irregular. Menopause does not end interest in sexual activities. The decrease in estrogen, however, may cause thinning and dryness of the vaginal walls, making intercourse painful. The use of vaginal lubricants will remedy much of the problem.

There is no male equivalent to menopause. Male fertility slowly declines, but men in their eighties are

Table 6.2 Women Reporting Intercourse in Past Year

Age	Percentage in Past Year			Mean Frequency in Past Month	
	U.S.	France	Great Britain	U.S.	France
35–44	87%	96%	92%	5.8	8.1
45–54	82%	90%	78%	5.0	6.1
55–59	59%	66%	NA	3.5	4.0

NOTE: NA means data not available.

SOURCE: Fraser et al. (2004).

Table 6.3 Percentage of Sexually Active U.S. Women Reporting Various Sexual Dysfunctions

Age	Pain	Not Pleasurable	Unable to Orgasm	Lack of Interest	Trouble Lubricating
35–39	13.0	18.3	26.9	37.6	18.1
40–44	12.0	15.7	20.8	36.0	15.9
45–49	10.3	15.4	18.8	33.7	22.6
50–54	7.4	15.3	20.2	30.2	21.4
55–59	8.7	16.4	21.8	37.0	24.8

SOURCE: Laumann et al. (1994).

Sexuality among the aged tends to be sensual and affectionate. Older couples may experience an intimacy forged by years of shared joys and sorrows that is as intense as the passion of young love.

often fertile. Men's physical responsiveness is greatest in late adolescence or early adulthood; beginning in men's twenties, responsiveness begins to slow imperceptibly. Changes in male sexual responsiveness become apparent only when men are in their forties and fifties. As a man ages, achieving erection requires more stimulation and time, and the erection may not be as firm.

Psychosexual Development in Later Adulthood

As we leave middle age, new tasks confront us, especially dealing with the process of aging itself. Our health and the presence or absence of a partner are key aspects of this time in our lives.

Many of the psychosexual tasks older Americans must undertake are directly related to the aging process:

- *Changing sexuality.* As physical abilities change with age, sexual responses change as well. A 70-year-old person, although still sexual, is not sexual in the same manner as an 18-year-old. Sexuality tends to be more diffuse, less genital, and less insistent. Chronic illness and increasing frailty understandably result in diminished sexual activity and desire. These considerations contribute to the ongoing evolution of the individual's sexual philosophy.
- *Loss of partner.* One of the most critical life events is the loss of a partner. After age 60, there is a significant increase in spousal deaths. As having a partner is the single most important factor determining an older person's sexual interactions, the death of a partner signals a dramatic change in the survivor's sexual interactions.

The developmental tasks of later adulthood are accomplished within the context of continuing aging. Their resolution helps prepare us for acceptance of our own eventual mortality.

Adult Sexual Behavior

In this section we examine various sexual behaviors. For a discussion of sexual structure and the sexual response cycle, see Appendix A.

Autoeroticism

Autoeroticism consists of sexual activities such as sexual fantasies, masturbation, and erotic dreams that involve only the self. Autoeroticism is one of our earliest and most universal yet also less accepted expressions of sexual stirrings. By condemning it, our culture sets the stage for the development of deeply negative inhibitory attitudes toward sexuality.

Sexual Fantasies

Erotic fantasizing may be the most universal of all sexual behaviors, but because they may touch on feelings or desires considered personally or socially unacceptable, typically sexual fantasies are not widely discussed. Although they are normal and serve certain functions (such as escape or rehearsal for later sexual behavior), fantasies may also interfere with an individual's self-image, causing a loss of self-esteem as well as confusion.

Various studies report that between 60% and 90% of respondents fantasize during sex—the percentage depending on gender, age, and ethnicity (Knafo and Jaffe 1984; Miracle et al. 2003). A large-scale study (Michael et al. 1994) found that 54% of the men and 19% of the women thought about sex daily.

Women and men have sexual fantasies, although their fantasies often differ. Can you tell the gender of the individuals who supplied the following fantasies?

- "It's evening time, the sun is setting, I'm on a tropical island, a light breeze is blowing into my balcony doors, and the curtains [white] are fluttering lightly in the wind. The room is spacious, and there is white everywhere, even the bed. There are flowers of all kinds, and the light fragrance fills the room."
- "Have sex on the beach."

If you guessed that the first of these fantasies was from a woman and the second was from a man, you guessed correctly. These are real examples that Michael Kimmel and Rebecca Plante received from undergraduates at three New York colleges or universities. Men's and women's fantasies contained similarities (e.g., in the acts they described), but the differences were more striking: women's fantasies were longer and more vivid, using more emotional and sensual imagery, especially in describing the setting; men more often fantasized about *doing something sexual to someone*, whereas women's fantasies were often more passive and gentler, of *having something sexual done to them*; and women's fantasies tended to have more emotional and romantic content (47% of women described their fantasy partners as boyfriends or husbands; only 15% of men depicted their fantasy partners as "significant others"). Women's fantasies were also often romantic stories of love and affection, whereas men's had less romance and less emotional language or context.

Masturbation

Masturbation is the manual stimulation of one's genitals. Individuals masturbate by rubbing, caressing, or otherwise stimulating their genitals to bring themselves sexual pleasure. Masturbation is an important means of learning about our bodies. Girls, boys, women, and men may masturbate during particular periods or throughout their entire lives. An analysis of research articles on gender roles and sexual behavior found that the greatest male–female difference was in masturbation (Oliver and Hyde 1993). Males had significantly more masturbatory experience than females.

By the end of adolescence, nearly all males and a majority of females have masturbated to orgasm (Knox and Schacht 1992; LoPresto, Sherman, and Sherman 1985). Masturbation continues after adolescence. Gender differences, however, continue to be significant (Atwood and Gagnon 1987; Leitenberg, Detzer, and Srebnik 1993). Although the rate is significantly lower for those who are married, many people, especially men, continue to masturbate even after they marry. There are many reasons for continuing the activity during marriage: masturbation is a pleasurable form of sexual excitement, a spouse may be unavailable or unwilling to engage in sex, sexual intercourse may not be satisfying, the partners may fear sexual inadequacy, or one partner may want to act out fantasies. In marital conflict, masturbation may act as a distancing device, with the masturbating spouse choosing masturbation over sexual intercourse as a means of emotional protection (Betchen 1991).

Cohabitation has a different effect than marriage on frequency of masturbation. Many cohabiting men masturbate often despite the presence or availability of a sexual partner. Thus, social factors other than the presence of a partner affect masturbation. In citing reasons for why they masturbate, only a third of women and men list an unavailable partner (Laumann et al. 1994).

Interpersonal Sexuality

We often think that sex is sexual intercourse and that sexual interactions end with orgasm (usually the male's). But sex is not limited to sexual intercourse. Heterosexuals engage in a variety of sexual activities, which may include erotic touching, kissing, and oral and anal sex. Except for sexual intercourse, gay and lesbian couples engage in sexual activities similar to those experienced by heterosexuals.

Touching

Because touching, like desire, does not in itself lead to orgasm, it has largely been ignored as a sexual behavior. Sex researchers William Masters and Virginia Johnson (1970) suggest a form of touching they call **pleasuring**—nongenital touching and caressing. Neither partner tries to stimulate the other sexually; the partners simply explore each other. Such pleasuring gives each a sense of his or her own responses; it also allows each to discover what the other likes or dislikes. We can't assume that we know what any particular individual likes because there is too much variation among people. Pleasuring opens the door to communication; couples discover that the entire body rather than just the genitals is erogenous.

Kissing is probably the most acceptable premarital sexual activity.

As we enter old age, touching becomes increasingly significant as a primary form of erotic expression. Touching in all its myriad forms—ranging from holding hands to caressing, massaging to hugging, and walking with arms around each other to fondling—becomes the touchstone of eroticism for the elderly. One study found touching to be the primary form of erotic expression for married couples more than 80 years old (Bretschneider and McCoy 1988).

Kissing

Kissing as a sexual activity is probably the most common and acceptable of all premarital sexual activities, occurring in more than 90% of all cultures (Fisher 1992, cited in Miracle et al. 2003; Jurich and Polson 1985). The tender lover's kiss symbolizes love, and the erotic lover's kiss simultaneously represents passion. Both men and women in one study regarded kissing as a romantic act, a symbol of affection and attraction (Tucker, Marvin, and Vivian 1991). A cross-cultural study of jealousy found that kissing is also associated with a couple's boundary maintenance: in each culture studied, kissing a person other than the partner evoked jealousy (Buunk and Hupka 1987).

The lips and mouth are highly sensitive to touch. Kisses discover, explore, and excite the body. They also involve the senses of taste and smell, which are especially important because they activate unconscious memories and associations. Often we are aroused by familiar smells associated with particular sexual memories: a person's body smells, perhaps, or perfumes associated with erotic experiences. In some cultures—among the Borneans, for example—the word *kiss* literally translates as "smell." Among traditional Eskimos and Maoris, there is no mouth kissing, only the nuzzling that facilitates smelling.

Although kissing may appear innocent, it is in many ways the height of intimacy. The adolescent's first kiss is often regarded as a milestone, a rite of passage, the beginning of adult sexuality (Alapack 1991). Philip Blumstein and Pepper Schwartz (1983) report that many of their respondents found it unimaginable to engage in sexual intercourse without kissing. They found that those who have a minimal (or nonexistent) amount of kissing feel distant from their partners but engage in coitus nevertheless as a physical release.

The amount of kissing differs according to orientation. Lesbian couples tend to engage in more kissing than heterosexual couples, and gay male couples kiss less than heterosexual couples. As many as 95% of lesbian couples, 80% of heterosexual couples, and 71% of gay couples engage in kissing whenever they have sexual relations (Blumstein and Schwartz 1983).

Oral–Genital Sex

In recent years, oral sex has become an increasing part of our sexual scripts. It is engaged in by heterosexuals, bisexuals, gay men, and lesbians. In recent years, it has increased among teens and young adults in part as an alternative to sexual intercourse. The two types of oral–genital sex are cunnilingus and fellatio.

Cunnilingus is the erotic stimulation of a woman's vulva by her partner's mouth and tongue. **Fellatio** is the oral stimulation of a man's penis by his partner's sucking and licking. Cunnilingus and fellatio may be performed singly or simultaneously. Oral sex is an important and healthy aspect of adults' sexual selves (Wilson and Medora 1990).

According to survey data, oral–genital sex is fairly common, especially among whites. Ninety percent of 25- to 44-year-old males and 88% of 25- to 44-year-old females have engaged in heterosexual oral sex. Among 15- to 44-year-olds, the percentages are somewhat smaller and reveal notable race differences. Where 88% of 15- to 44-year-old white females and 87% of white males have had heterosexual oral sex, the percentages among African Americans and Hispanics are lower (see Figure 6.1).

Sexual Intercourse

Sexual intercourse or **coitus**—the insertion of the penis into the vagina and subsequent stimulation—is a complex interaction. As with many other types of activities, the anticipation of reward triggers a pattern of behavior. The reward may not necessarily be orgasm, however, because the meaning of sexual intercourse varies considerably at different times for different people. There are many motivations for sexual intercourse; sexual pleasure is only one. Other motivations include showing love, having children, gaining power, ending an argument, demonstrating commitment, seeking revenge, proving masculinity or femininity, or degrading someone (including oneself).

Looking at adults ages 25 to 44, data indicate that nearly all have had a sexual intercourse experience. The National Survey of Family Growth data indicate that (as of 2002), 97% of 25- to 44-year-old males and 98% of 25- to 44-year-old females report having had sexual intercourse at least once in their lives (Mosher et al. 2005). As Figure 6.1 shows, among 15- to 44-year-olds, there are fairly minimal race differences in rates of sexual intercourse. Although sexual intercourse is important for most sexually involved heterosexual couples, its significance may be different for men and women. More than any other heterosexual sexual activity, sexual intercourse involves equal participation by both partners. Ideally, both partners equally and simultaneously give and receive. Many women report that this sense of sharing during intercourse is important to them.

Men tend to be more consistently orgasmic than women in sexual intercourse. Part of the reason may be that the clitoris often does not receive sufficient

> **Matter of Fact** According to a scientific nationwide study of adults of all ages, about one-third of Americans have sexual intercourse twice a week, one-third a few times a month, and one-third a few times a year or not at all. Married couples are more likely to engage in coitus than singles; married women are more likely to be orgasmic. About 40% of married couples and 25% of singles report having coitus twice a week (Michael et al. 1994).

stimulation from penile thrusting alone to permit orgasm. Many women need manual stimulation during intercourse to be orgasmic. They may also need to be more assertive. A woman can manually stimulate herself or be stimulated by her partner before, during, or after intercourse. But to do so, she has to assert her own sexual needs and move from the idea that sex is centered around male orgasm.

Anal Eroticism

Sexual activities involving the anus are known as **anal eroticism**. The male's insertion of his erect penis into his partner's anus is known as **anal intercourse**. Both heterosexuals and gay men may participate in this activity. For heterosexual couples who engage in it, anal intercourse may be an experiment or occasional activity rather than a common mode of sexual expression. This is suggested by data from the National Survey of Men. A representative sample of more than 3,000 men, ages 20 to 39, found that 20% of men had engaged in anal intercourse, with white men (21%) and Hispanic men (24%) more likely than African American men to have done so (13%) (Billy et al. 1993).

More recent data, from the National Survey of Family Growth found that within the sample of 12,000, among 25- to 44-year-old women and men, 35% of women and 40% of men had engaged in heterosexual anal intercourse at least once. Data in Figure 6.3 illustrate that there are only modest racial differences in the percentages of 15- to 44-year-old men who have engaged in heterosexual anal sex; among females, larger race differences surface. White males and females are more likely to have had heterosexual anal sex than either African American or Hispanic males and females (Mosher and Jones 2005).

Anal intercourse is less common than oral sex but remains an important ingredient in the sexual satisfaction of many gay men (Blumstein and Schwartz 1983). From a health perspective, anal intercourse is the

Figure 6.3 Sexual Experiences of 15- to 44-Year-Old Males and Females, by Race

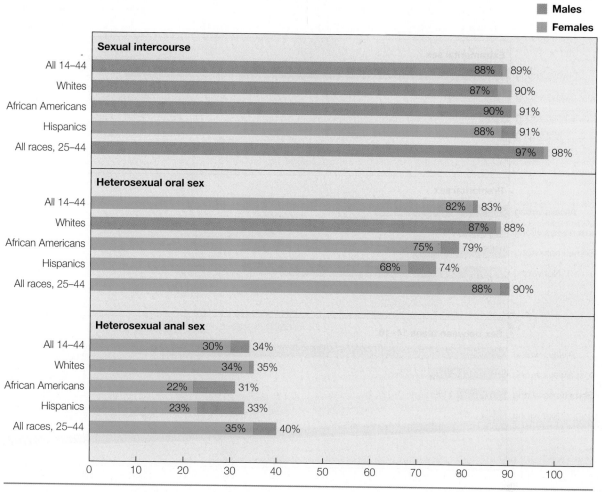

Males
Females

Sexual intercourse

All 14–44	88% 89%
Whites	87% 90%
African Americans	90% 91%
Hispanics	88% 91%
All races, 25–44	97% 98%

Heterosexual oral sex

All 14–44	82% 83%
Whites	87% 88%
African Americans	75% 79%
Hispanics	68% 74%
All races, 25–44	88% 90%

Heterosexual anal sex

All 14–44	30% 34%
Whites	34% 35%
African Americans	22% 31%
Hispanics	23% 33%
All races, 25–44	35% 40%

SOURCE: Mosher, Shandra, and Jones 2005.

riskiest form of sexual interaction and the most prevalent sexual means of transmitting the HIV among both gay men and heterosexuals. Because the delicate rectal tissues are easily torn, HIV (carried within semen) can enter the bloodstream. (HIV is discussed later in this chapter.) Using a lubricated condom significantly decreases the risk of transmitting HIV during anal sex (Bell 1999).

Sexual Expression and Relationships

Sexuality exists in various relationship contexts that may influence our feelings and activities. These include nonmarital, marital, and extramarital contexts.

Nonmarital Sexuality

Nonmarital sex encompasses sexual activities, especially sexual intercourse, that take place outside of marriage. We use the term *nonmarital sex* rather than *premarital sex* to describe sexual behavior among unmarried adults in general. Of forms of nonmarital sex, only **extramarital sex**—sexual interactions that take place outside the marital relationship between at least one married partner—continues to be consistently frowned on. General Social Survey data from 1990 to 2008 show the relative reactions to extramarital and premarital sex. More than 90% of males and females said that having sex with someone other than one's spouse while married is always or almost always wrong. However, when asked about sex before marriage, only 31% of males and 39% of

Figure 6.4 Attitudes about Nonmarital Sex, by Gender

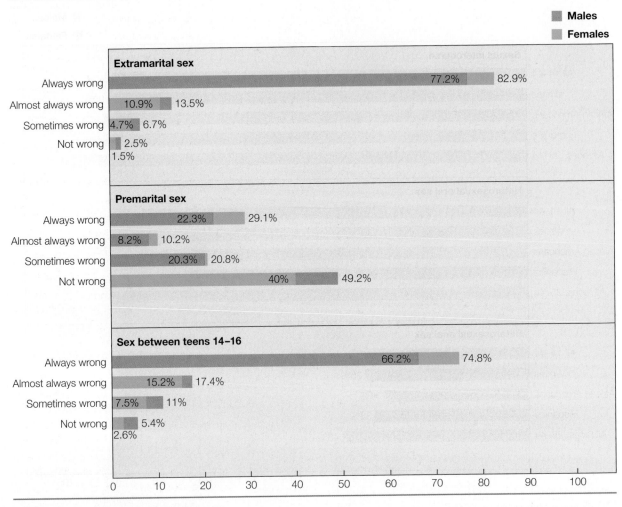

SOURCE: General Social Survey Data, 1990–2008.

females responded that premarital sex was always or almost always wrong. Only when asked specifically about teenagers 14 to 16 years old having sexual relations did respondents express strong opposition to sex before marriage, with 84% of males and 90% of females saying that such sexual activity was "always" or "almost always wrong" (see Figure 6.4).

When we use the term **premarital sex**, we are referring to never-married adults under the age of 30. There are several reasons to make premarital sex a subcategory of nonmarital sex. First, because increasing numbers of never-married adults are over 30, "premarital sex" does not adequately describe the nature of their sexual activities. Second, at least 10% of adult Americans will never marry; it is misleading to describe their sexual activities as "premarital." Third, many adults

are divorced, separated, or widowed; 30% of divorced women and men will never remarry. Fourth, between 3% and 10% of the population is lesbian or gay, and gay and lesbian sexual relationships cannot be categorized as "premarital" until gays and lesbians are given the right to marry.

Sexuality in Dating Relationships

Over the past several decades, there has been a remarkable increase in the acceptance of premarital sexual intercourse, a decline in the numbers of people who believe that premarital sex is "always wrong," and an increase in the percentages who feel it is "not wrong at all." This trend has been interpreted as a shift toward "moral neutrality" regarding intercourse before marriage (Christopher and Sprecher 2000).

Common conditions for a satisfying sexual relationship include feelings of intimacy, capability, trust, arousal, alertness, and positiveness about the environment and situation.

For adolescents and young adults, the combination of effective birth control methods, changing gender roles that permit females to be sexual, and delayed marriages have played a major part in the rise of premarital sex. For middle-aged and older adults, increasing divorce rates and longer life expectancy have created an enormous pool of once-married men and women who engage in nonmarital sex.

The increased legitimacy of sex outside of marriage has transformed both dating and marriage. Sexual intercourse has become an acceptable part of the dating process for many couples, whereas only petting was acceptable before. As a consequence, many people no longer feel that they need to marry to express their sexuality in a relationship (Scanzoni et al. 1989).

There appears to be a general expectation among students that they will engage in sexual intercourse sometime during their college careers. As we saw in the previous chapter, the emergence of "hooking up" on college campuses suggests that for many college students, sex no longer has to occur within an existing emotional or loving relationship. Females may be more likely than men to hope that such a bond develops, but the need to have an established committed relationship is no longer a prerequisite for many sexually active students.

Directing Sexual Activity. As we begin a sexual involvement, we have several tasks to accomplish:

1. *We need to practice safe sex.* Ideally, we need information about our partners' sexual history and whether he or she practices safe sex, including the use of condoms. Unlike much of our sexual communication, which is nonverbal or ambiguous, we need to use direct verbal discussion in practicing safe sex.

2. *Unless we are intending a pregnancy, we need to discuss birth control.* Condoms alone are only moderately effective as contraception, although they help prevent the spread of sexually transmitted diseases. To be more effective, they must be used with contraceptive foam or jellies or with other devices.

3. *We need to communicate about what we like and need sexually.* What kind of **foreplay** or afterplay do we like? Do we like to be orally or manually stimulated during intercourse? What does each partner need to be orgasmic? Many of our needs and desires can be communicated nonverbally by our movements or other physical cues. But if our partner does not pick up our nonverbal signals, we need to discuss them directly and clearly to avoid ambiguity.

Sexuality in Cohabiting Relationships

Cohabitation has become a widespread phenomenon in American culture. In contrast to married men and women, cohabitants have sexual intercourse more often, are more egalitarian in initiating sexual activities, and are more likely to be involved in sexual activities outside their relationship (Blumstein and Schwartz 1983; Waite and Gallagher 2001). The higher frequency of intercourse, however, may be because of the "honeymoon" effect: cohabitants may be in the early stages of their relationship, the stages when sexual frequency is highest. The differences in frequency of extrarelational sex may result from a combination of two factors: norms of sexual fidelity may be weaker in cohabiting relationships, and men and women who cohabit tend to conform less to conventional norms.

Marital Sexuality

When people marry, they discover that their sexual life is different than it was before marriage. Sex is now morally and socially sanctioned. It is in marriage that most heterosexual interactions take place, yet as a culture we seem ambivalent about marital sex. On the one hand, marriage is the only relationship in which sexuality is fully legitimized. On the other hand, marital sex is an endless source of humor and ridicule: "Marital sex? What's that?"

Sexual Interactions

A variety of large-scale studies report consistent findings in regard to how often married couples engage in sexual intercourse and in how sexual frequency changes over the course of a marriage. Married couples report engaging in sexual relations about once or twice a week, or about six to seven times a month (Christopher and Sprecher 2003).

Sexual intercourse tends to diminish in frequency the longer a couple is married. For newly married couples, the average rate of sexual intercourse is about three times a week. Data from more than 13,000 respondents in the National Survey of Families and Households reported that couples under the age of 24 had sex on average 11.7 times per month (or approximately three times per week) (Call, Sprecher, and Schwartz 1995, cited in Christopher and Sprecher 2000). As couples get older, sexual frequency drops. In early middle age, married couples make love an average of 1.5 to 2 times a week. After age 50, the rate is about once a week or less. Among couples 75 and older, the frequency is a little less than once a month (Christopher and Sprecher 2000).

This decreased frequency, however, does not necessarily mean that sex is no longer important or that the marriage is unsatisfactory. For dual-worker families and families with children, fatigue and lack of private time may be the most significant factors in the decline of frequency. Couples also report "being accustomed" to each other. In addition, activities and interests other than sex engage them. The decline in interest and frequency of sex may begin within the first two years of marriage (Christopher and Sprecher 2000).

Bringing New Meanings to Sex

Sex within marriage is significantly different from nonmarital sex in at least three ways: it is expected to be monogamous, procreation is a legitimate goal, and such sex takes place in the everyday world. These differences present each person with important tasks.

Monogamy. Before marriage or following divorce, a person may have various sexual partners, but within marriage, all sexual interactions are expected to take place between the spouses. Approximately 90% of Americans believe that extramarital sexual relations are "always" or "almost always" wrong (Christopher and Sprecher 2000; Miracle et al. 2003). This expectation of monogamy lasts a lifetime; a person marrying at 20 commits to 40 to 60 years of sex with the same person. Within a monogamous relationship, each partner must decide how to handle fantasies, desires, and opportunities for extramarital sexuality. Does one tell one's spouse about having fantasies about other people? Does one have an extramarital relationship and, if so, tell one's spouse? How should one handle sexual conflicts or difficulties with his or her partner?

Socially Sanctioned Reproduction. In most segments of society, marriage remains the more socially approved setting for having children. In marriage, partners are confronted with one of the most crucial decisions they will make: the task of deciding whether and when to have children. Having children will profoundly alter a couple's relationship. If they decide to have a child, sexual activity may change from simply an erotic activity to an intentionally reproductive act as well.

Changed Sexual Context. Because married life takes place in a day-to-day living situation, sex must also be expressed in the day-to-day world. Sexual intercourse must be arranged around working hours and at times when the children are at school or asleep. One or the other partner may be tired, frustrated, or angry.

In marriage, some emotions associated with premarital sex may disappear. For many, the passion of romantic love, especially as experienced in the earliest period of a relationship, eventually disappears as well, to be replaced with a love based on intimacy, caring, and commitment.

Celibate Marriages

The discussion of social exchange theory in Chapter 2 used the example of involuntarily celibate couples to illustrate how such couples make the decision to remain together (Donnelly and Burgess 2008). Such relationships, which may amount to around 14% to 15% of marriages, can also be instructive illustrations of the role that sex plays in marriage, and the causes and consequences of celibacy.

Using six months of desiring but not having any form of sexual contact as their measure of involuntary celibacy, sociologists Denise Donnelly and Elisabeth Burgess identified a number of factors that contribute to declines in sexual relations. In addition to the impact of the passage of time and the disappearance of novelty, they point to the following stressors that affect sexual activity:

- Late-term pregnancy and/or early postpartum adjustments

- Competing time demands, especially for dual-earning couples, opposite-shift couples, or couples who are either raising children or caring for aged or ill parents
- Chronic illness, disability, or mental illness (e.g., depression)
- Guilt or conflict resulting from one's religious beliefs

The potential consequences resulting from sexual inactivity threaten the quality or stability of the relationship. More specifically, Donnelly and Burgess identify decline in relationship quality, such as lower levels of satisfaction or happiness; sexual dissatisfaction and frustration; infidelity; decreased mental health; and relationship instability.

Relationship Infidelity and Extramarital Sexuality

As we noted, a fundamental assumption in our culture is that marriages are sexually and emotionally monogamous. This assumption is not unique to the United States. Eric Widmer, Judith Treas, and Robert Newcomb (1998) undertook comparative research using a 24-country sample of more than 33,000 respondents and found strong and widespread disapproval of extramarital sex, although people in different countries varied some in their levels of disapproval, with some being more tolerant than the majority (e.g., those in Russia, Bulgaria, and the Czech Republic). In the United States, nearly 80% of Americans believe extramarital sex is "always wrong" (Blow and Hartnett 2005).

How Much Infidelity and Extramarital Sex Is There?

Although we sometimes overstate the amount of "cheating" that goes on, it is neither an isolated phenomenon nor restricted to married couples. As we reported previously, there is more nonmonogamy among cohabiting than among married couples and among gay male couples than among lesbians or heterosexual couples. There are varying estimates of how prevalent extramarital sex is in the United States:

Critical Thinking

What would you do if you found out that your partner or spouse had been sexually unfaithful? Do you believe that relationships can withstand infidelity?

Figure 6.5 Percent of Ever-Married People Who, While Married, Had Sex with Someone Other Than Spouse, 1990–2008

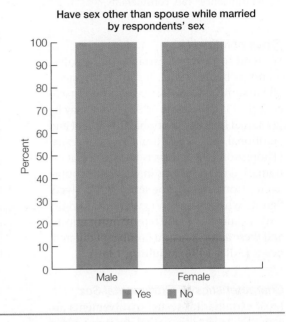

SOURCE: General Social Survey, 1990–2008.

- Looking at 1990–2008 General Social Survey data, among respondents who were or had been married (thus including divorced, separated, and widowed respondents), 22.5% of the males and 13% of the females admitted to having had sex outside of their marriages (see Figure 6.5).
- Examining only a more recent period (2003–2008), among survey respondents who were married at the time they were surveyed, 17% of the males and 10% of the females said that they had had sex with someone other than their spouse during the time they had been married.
- Findings from a study of 2,598 men and women ages 18 to 59 who had ever been married or lived with a partner suggested that 11% had been unfaithful (Olenick 2000).

Attempting to bring these disparate findings to some conclusion, Blow and Hartnett (2005, 220) suggest the following (emphasis added):

We can conclude that *over the course of married, heterosexual relationships* in the United States, [extramarital] sex occurs in less than 25% of committed relationships, and more men than women appear to be engaging in infidelity. . . . From studies of other countries, it appears that rates of infidelity are higher or lower in some places and that gender differences vary considerably.

Types of Infidelity

We tend to think of extramarital involvements as being sexual, but they may actually assume several forms (Moore-Hirschl et al. 1995; Thompson 1993). They may be (1) sexual but not emotional, (2) sexual and emotional, or (3) emotional but not sexual (Thompson 1984). Less is known about extramarital relationships in which the couple is emotionally but not sexually involved. People who engage in extramarital affairs have a number of different motivations, and these affairs satisfy a number of different needs (Adler 1996; Moultrup 1990).

Characteristics of Extramarital Sex

Most extramarital sexual involvements are sporadic. Most extramarital sex is not a love affair; it is generally more sexual than emotional. Affairs that are both emotional and sexual appear to detract more from the marital relationship than do affairs that are only sexual or only emotional (Thompson 1984). More women than men consider their affairs emotional; almost twice as many men as women consider their affairs only sexual. About equal percentages of men and women are involved in affairs that they view as both sexual and emotional. Research suggests that men are more bothered by the sexual nature of a partner's infidelity, whereas women are disturbed more by the emotional aspect (Christopher and Sprecher 2000).

An emotionally significant extramarital affair creates a complex system of relationships among the three individuals (Moultrup 1990). Long-lasting affairs can form a second but secret "marriage." In some ways, these relationships resemble polygamy, in which the outside person is a "junior" partner with limited access to the other. The involved partners, who know that their system is triadic, must try to meet each other's needs for time, affection, intimacy, and sex while taking the uninvolved partner into consideration. Such extramarital systems are stressful and demanding. Most people find great difficulty in sustaining them. If both people involved in the affair are married, the dynamics become even more complex.

Figure 6.6 Percent of Ever-Married People Who, While Married, Had Sex with Someone Other Than Spouse, 1990–2008

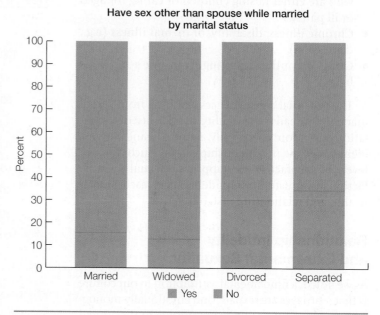

SOURCE: General Social Survey, 1990–2008.

Figure 6.6 illustrates the number of ever-married people who, between 1990 and 2008, indicated that while married they had sex with someone other than their spouses. The higher percentage among divorced and separated people probably reflects at least part of the reason for their marital breakup.

One factor consistently found to be associated with lower levels of infidelity is religiousness. Interestingly, David Atkins and Deborah Kessell analyzed General Social Survey data and determined that of the many possible measures of religiousness (e.g., feeling close to God, praying, having faith, experiencing a religious turning point, and attending religious services), only attendance at religious services was associated with lower levels of infidelity. Reporting religion to be important in one's life without attending religious services is actually positively associated with infidelity (Atkins and Kessell 2008). With attendance at religious services comes ties to a community (i.e., a congregation) from which one can draw support but also receive scrutiny and criticism should one's behavior stray from shared norms. That such attendance is typically joint attendance *with one's spouse* may make one feel closer to one's spouse, less likely to feel disengaged or to drift apart, either of which could prompt infidelity (Atkins and Kessell 2008).

Sexual Nonmonogamy and Sexual Orientation

Although the outbreak of the AIDS crisis in the 1980s and 1990s affected behavior and changed how accepting of nonmonogamy gay men were, research conducted both before and after the onset of the epidemic show that a higher proportion of gay men than lesbians or heterosexuals maintain relationships in which both partners agree to be nonexclusive (LaSala 2004). Among heterosexual couples, lesbian couples, and gay male couples, gay male couples have been and continue to be more likely to both accept and experience sexual nonmonogamy. Furthermore, in comparing monogamous and nonmonogamous (i.e., "faithful" and "unfaithful") gay male couples, research often finds no differences in relationship adjustment or satisfaction (LaSala 2004). However, more recent research examining whether and how gay male couples in civil unions and those not in civil unions differ finds that gay male couples in civil unions are less accepting of nonmonogamy than those not in civil unions. However, even among gay males in civil unions, a higher percentage expressed openness about nonmonogamy than among either lesbians or heterosexuals.

Some clinicians have gone as far as to deem those who condemn nonmonogamy as dysfunctional "heterocentrist," meaning that they are applying standards that may pertain to heterosexual relationships too broadly. Given that males think about and act differently regarding sexual relationships, we might assume that gay male couples would display these tendencies even more than heterosexual couples and lesbians (each of whom has at least one female partner). Michael LaSala (2004) reminds us that research has established that, compared to women, men are more likely to separate sex and love, to engage in sexual relationships in the absence of emotional involvement, to engage in sexual relations within even "casual relationships," and to consider having sex with strangers.

Just as a portion of gay men maintain nonexclusive relationships, many gay couples, especially those who are able to marry or enter civil unions, construct relationship boundaries that prohibit sex with others. Like heterosexual and lesbian couples, such men come to see infidelity as a breach of trust. In LaSala's (2004) sample of 121 gay male couples, 60% described their "relationship agreements" as assuming monogamy (40% were in sexually open relationships). However, among these 73 couples, 33 had breached this expectation and broken their monogamous agreement. It was this latter group that had the lowest scores on satisfaction, expression of affection toward one's partner, and relationship adjustment. Interestingly, however, when those who engaged in nonmonogamous sex in the prior 12 months were removed from the analysis, there appeared to be no real difference between those whose monogamous expectations had been upheld and those whose expectations had been violated.

Sexual Enhancement

Sexual behavior cannot be isolated from our personal feelings and relationships. Sometimes dissatisfaction arises because the relationship itself is unsatisfactory; at other times the relationship itself is good but the erotic fire needs to be lit or rekindled. Such relationships may grow through **sexual enhancement**— improving the quality of a sexual relationship.

Being aware of our sexual needs is often critical to enhancing our sexuality. Gender-role stereotypes and negative learning about sexuality often cause us to lose sight of our sexual needs. Sex therapist Bernie Zilbergeld (1993) suggests that to fully enjoy our sexuality, we need to explore our "conditions for good sex," those things that make us "more relaxed, more comfortable, more confident, more excited, more open to your experience."

Different individuals report different conditions for good sex. More common conditions include the following:

- *Feeling intimate with your partner.* Emotional distance can take the heart out of sex.
- *Feeling sexually capable.* Generally, this relates to an absence of anxieties about sexual performance.
- *Feeling trust.* Both partners may need to know that they are emotionally safe with the other and confident that they will not be judged, ridiculed, or talked about.
- *Feeling aroused.* A person does not need to be sexual unless he or she is sexually aroused or excited. Simply because your partner wants to be sexual does not mean that you have to be.
- *Feeling physically and mentally alert.* Both partners should not feel particularly tired, ill, stressed, preoccupied, or under the influence of excessive alcohol or drugs.
- *Feeling positive about the environment and situation.* A person may need privacy, to be in a place where he or she feels protected from intrusion.

Sexual Problems and Dysfunctions

Many of us who are sexually active may experience sexual difficulties or problems. Recurring problems that cause distress to the individual or his or her partner are known as **sexual dysfunctions**. Although some sexual dysfunctions are physical in origin, many are psychological. Some dysfunctions have immediate causes, others originate in conflict within the self, and still others are rooted in a particular sexual relationship.

Both men and women may suffer from hypoactive (low or inhibited) sexual desire (Hawton, Catalan, and Fagg 1991). Other dysfunctions experienced by women are orgasmic dysfunction (the inability to attain orgasm), arousal difficulties (the inability to become erotically stimulated), and **dyspareunia** (painful intercourse). The most common dysfunctions among men include **erectile dysfunction** (the inability to achieve or maintain an erection), **premature ejaculation** (the inability to delay ejaculation after penetration), and delayed orgasm (difficulty in ejaculating) (Spector and Carey 1990). Figure 6.7 shows the percentage of heterosexual adults in the general U.S. population who reported experiencing sexual problems during the previous year in response to a recent survey (Laumann et al. 1994).

Physical Causes of Sexual Problems

It is generally believed that between 10% and 20% of sexual dysfunctions are structural in nature. Physical problems may be *partial* causes in another 10% or 15% (Kaplan 1983; LoPiccolo 1991). Various illnesses may have an adverse effect on a person's sexuality (Wise, Epstein, and Ross 1992). Alcohol and some prescription drugs, such as medication for hypertension, may affect sexual responsiveness (Buffum 1992).

Among women, diabetes, hormone deficiencies, and neurological disorders, as well as alcohol and alcoholism, can cause orgasmic difficulties. Painful intercourse may be caused by an obstructed or thick hymen, clitoral adhesions, a constrictive clitoral hood, or a weak pubococcygeus muscle. Coital pain caused by inadequate lubrication and thinning vaginal walls often occurs as a result of decreased estrogen associated with menopause. Lubricants or hormone replacement therapy often resolve the difficulties.

Among males, diabetes and alcoholism are the two leading physical causes of erectile dysfunctions; atherosclerosis is another important factor (LoPiccolo 1991; Roenrich and Kinder 1991). Smoking may also contribute to sexual difficulties (Rosen et al. 1991).

Psychological or Relationship Causes

Two of the most prominent causes of sexual dysfunctions are performance anxiety and conflicts within the self. **Performance anxiety**—the fear of failure—is probably the most important immediate cause of erectile dysfunctions and, to a lesser extent, of orgasmic dysfunctions in women (Kaplan 1979). If a man does not become erect, anxiety is a fairly common response. Some men experience their first erectile problem when a partner initiates or demands sexual intercourse. Women are permitted to say no, but many men have not learned that they too may say no to sex. Women suffer similar anxieties, but they tend to center around orgasmic abilities rather than the ability to have intercourse. If a woman is unable to experience

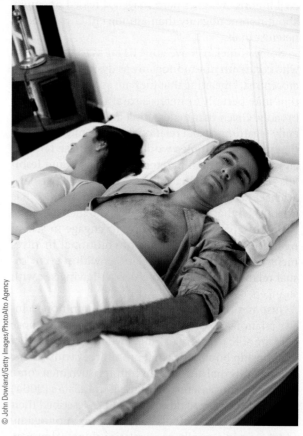

Sexual problems can become self-fulfilling, as they can cause performance anxiety or a fear of future failures to achieve erections or experience orgasms.

orgasm, a cycle of fear may arise, preventing future orgasms. A related source of anxiety is an excessive need to please one's partner.

Conflicts within the self are guilt feelings about one's sexuality or sexual relationships. Guilt and emotional conflict do not usually eliminate a person's sexual drive; rather, they inhibit the drive and alienate the person from his or her sexuality. Such inner conflicts often are deeply rooted. The relationship itself, rather than either individual, sometimes can be the source of sexual problems. Disappointment, anger, or hostility may become integral parts of a deteriorating or unhappy relationship. Such factors affect sexual interactions because sex can become a barometer for the whole relationship.

Relationship discord can affect our sexuality in several ways, such as through poor communication that inhibits our ability to express our needs and desires, power struggles in which sexuality becomes a tool in struggles for control, and sexual sabotage where partners ask for sex at the wrong time, put pressure on each other, and frustrate or criticize each other's sexual desires and fantasies. People most often do this unconsciously (Kaplan 1979).

Sex between Unequals, Sex between Equals

Sociologist Pepper Schwartz (1994) identified a number of sexual problems that plague traditional marriages because of the gender hierarchy and absence of empathy that characterize such marriages. Partners are "too distant, too different, and too inequitable" to enjoy complete sexual fulfillment. Sexual problems among traditional couples include the following:

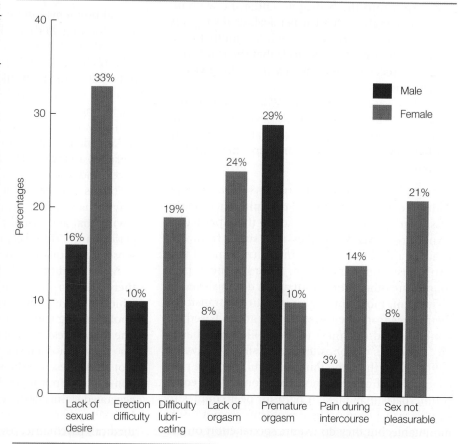

Figure 6.7 Heterosexual Sexual Dysfunctions in a Nonclinical Sample

SOURCE: Adapted from Laumann et al. 1994, 370–371.

- *Failure of timing.* This results when one person is more in charge of the couple's sexual relationship and his or her needs define when the couple has sexual relations.
- *Failure of intimacy.* If traditional couples lack the same depth of intimacy (i.e., sharing and communication) that Schwartz contends more egalitarian relationships possess, it is apparent in their sexual relationship. According to Schwartz, this can prevent them from finding complete fulfillment in their sexual relationship.
- *Failure of sexual empathy.* Some couples fail to realize that what one finds pleasing the other may not. This is particularly true in the most traditional marriages, where "men and women have little experience of each other's lives." They may show little respect for each other's sexual needs and refuse to make the effort to learn what each other wants (Schwartz 1994).

- *Failure of reciprocity.* Inequality outside the bedroom can spill into the bedroom. Often the woman, but potentially either partner, feels as if she gives more than she receives. There is less mutual massage than desired, or she feels that she is touched less or receives less oral sex than she gives or performs.
- *Failure of overromanticization.* Women in more traditional relationships may possess overly romanticized expectations of sexual relations. These are often beyond what most "ordinary" men can live up to.

Schwartz notes that peer marriages—relationships built on deep friendship and commitments to fairness, sharing, and equality—avoid these particular sexual problems. What they may suffer from, instead, is a decline in sexual intensity. Some of this results just from habituation. More specific to peer couples are other problems that can diminish sexual excitement, most notably an inability to transform themselves from their everyday identities based on sameness and openness to erotic identities based on "principles of opposites and mystery" (Schwartz 1994). Thus, the same things that differentiate peer relationships from their more common and less equal counterparts may make it hard for peer couples to sustain sexual energy. These problems are not insurmountable, but they do require special effort on the part of peer couples to create a separate and special sexual environment removed from more mundane life matters.

More recently, researchers garnered much media attention for concluding that couples who share housework have more frequent and satisfying sexual relations. This led to more than a few headlines suggesting that "men who do more (housework), get more (sex)." Neil Chethik, author of the book *VoiceMale*, based on telephone surveys with 288 married men and in-depth interviews with 70 more, devotes a chapter to the link between housework and sexual intimacy. He notes of his sample,

> Among wives who were satisfied with the division of housework, two-thirds had sex with their husbands at least once a week, and only 11 percent had sex less than once a month, according to the husbands. When the wife was unsatisfied with the housework situation, the proportion having weekly sex dropped to 50 percent, and the proportion having sex less than once a month more than doubled to 24 percent. (Chethik 2006, 124)

Resolving Sexual Problems

Sexual problems can be embarrassing and emotionally upsetting. Perhaps the first step in dealing with a sexual problem is to turn to immediate resources. Talking about the problem with one's partner, finding out what he or she thinks, discussing specific strategies that might be useful, and simply communicating feelings and thoughts can sometimes resolve the difficulty. One can also go outside the relationship, seeking friends with whom to safely share feelings and anxieties, asking whether they have had similar experiences, and learning how they handled them. Sexual problems can become self-fulfilling because couples may focus so much on the difficulties that they are having that additional pressure is placed on sexual performance. Thus, keeping perspective—and often a sense of humor—may be quite helpful.

Aside from one's circle of friends and intimates, there are ever-increasing, additional resources on which one can draw. A growing number of self-help books dealing with sexuality and relationship issues line the shelves in bookstores and libraries. There are also numerous Web sites one can access and consult. Not all of these will be sites offering help or advice. Some may be pornographic, and others may carry exaggerated claims designed to sell products, but many Web sites offer information compiled or overseen by medical, psychiatric, psychological, nursing, or educational specialists.

Cumulatively, partners, friends, Web sites, and books may provide information and grant individuals needed "permission" to engage in sexual exploration and discovery by making such inquiries normal. From these sources, we may learn that our sexual issues, problems, fantasies, and behaviors are not unique. Such methods are most effective when the dysfunctions arise from a lack of knowledge or mild sexual anxieties.

If, despite conversation with one's partner, consultation with friends, and/or reading books, magazines, or other resources one remains unable to resolve his or her sexual difficulties, seeking professional assistance is the logical next step. It is important to realize that seeking such assistance does not signal personal weakness or failure. Rather, it demonstrates an ability to reach out and a willingness to change. It is a sign of caring for one's partner, one's relationship, and oneself.

For those whose problems stem mostly from psychological or relationship causes, therapists can help deal with sexual problems on several levels. Some focus directly on the problem, such as lack of orgasm, and suggest behavioral exercises, such as pleasuring and masturbation, to develop an orgasmic response. Others focus on the couple relationship as the source of difficulty. If the relationship improves, they believe that sexual responsiveness will also improve. Still others work with the individual to help develop insight into the origins of the problem to overcome it. Therapy can also take place in a group setting. Group therapy may be particularly valuable for providing partners with an open, safe forum in which they can discuss their sexual feelings and experience and discover commonalities with others.

A relatively new development for men who suffer from sexual problems is medication. In March 1998, the Food and Drug Administration approved **Viagra**, the first oral treatment for male impotence. With as many as 50% of men, 40 and older, suffering from at least occasional and mild impotence, Viagra quickly became an economic and cultural phenomenon. In just its first year of availability, Viagra had sales of $1 billion, propelling its manufacturer, Pfizer, to the second spot among the world's largest drug companies. In 2002, Viagra had sales in excess of $1.7 billion. Optimistic forecasts predicting continued growth and sales success turned out to be exaggerated, but Viagra definitely has made its mark on the economy, society, and culture. There is still no equivalently successful prescription drug for women suffering from orgasmic difficulties or other sexual dysfunctions.

Issues Resulting from Sexual Involvement

Most of us think of sexuality in terms of love, passionate embraces, and entwined bodies. Sex involves all of these, but what we so often forget (unless we are worried) is that sex is also a means of reproduction. Whether we like to think about it or not, many of us (or our partners) are vulnerable to unintended pregnancies. Not thinking about pregnancy does not prevent it; indeed, not thinking about it may increase the likelihood of its occurring. Sex can also involve transferring sexually transmitted infections from one partner to another. Unless we practice **abstinence**, refraining from sexual intercourse, we need to think about unintended pregnancies and sexually transmitted infections and then take the necessary steps to prevent them.

Sexually Transmitted Infections, HIV, and AIDS

Americans are in the middle of the worst epidemic of sexually transmitted infections (STIs) in our history. There are an estimated 19 million new cases of STIs in the United States each year, the highest rate of infection of any industrialized nation in the world (Miracle et al. 2003). With nearly half of all infections occurring among 15- to 24-year-olds, college students are among the population at highest risk of contracting an STI. Risks to females may be more severe than risks to males; left untreated, chlamydia and gonorrhea can lead to pelvic inflammatory disease in women, a major cause of infertility.

Most people might wince if asked on a first date, "Do you have chlamydia, herpes, HIV, HPV (human papilloma virus), or any other sexually transmissible disease that I should know about?" However, given the risks of contracting an STI and the consequences associated with infection, it is a question whose answers one should know before becoming sexually involved. Just because a person is "nice" or good looking or available and willing is no guarantee that he or she does not have one of the STIs discussed later in this chapter. One can become infected through such sexual contact as sexual intercourse, oral sex, or anal sex. Unfortunately, no one can tell by a person's looks, intelligence, or demeanor whether he or she has contracted an STI. The costs of becoming sexually involved with a person without knowing about the presence of any of these diseases are potentially steep.

The most prevalent STIs in the United States are chlamydia, gonorrhea, HPV, genital herpes, syphilis, hepatitis, and HIV and AIDS. Table 6.4 briefly describes the symptoms, exposure intervals, treatments, and other information regarding the principal STIs. According to data from the Centers for Disease Control, in 2007 there were 1.1 million reported cases of chlamydia, a 7.5% increase over 2006 and three times the number of gonorrhea cases. It is likely that the real incidence of chlamydia is around 2.8 million infections, with more than half going undiagnosed. Chlamydia is more common among females than males and is highest among 15- to 19-year-olds followed by 20- to 24-year-olds.

Gonorrhea, the second most common STI, after nearly a quarter century of decline, has plateaued over the past decade. Females have a slightly higher rate of developing gonorrhea than males. Rates of gonorrhea are also highest among adolescents and young adults and among racial and ethnic minorities. African Americans have a rate 19 times higher than whites, American Indians have three times the rate of whites, and Hispanics have twice the rate of whites. In turn, whites have a gonorrhea rate nearly twice the rate among Asians.

After reaching an all-time low in 2000, the number of syphilis cases has increased from 2000 to 2007, rising 15% from 2006. Among almost all subpopulations, syphilis increased for each of the following groups: rates grew 18% among males, 10% among females, 25% among African Americans, 23% among Hispanics, 5% among whites, and 6% among Native Americans. Only among Asians did the rate remain unchanged. Two-thirds of syphilis cases are the result of male same-sex relations.

HPV is a sexually transmitted infection that is passed through genital contact. Twenty million Americans are currently affected. At some point in their lives, it is estimated that 50% of sexually active women and men acquire HPV and that at any one point 1% of sexually active adults have genital warts (a possible consequence of HPV infection). More serious consequences include certain cancers, including cervical cancer and cancers of the vagina, anus, or penis. Recently, a new vaccine has been developed to prevent the types of HPV that lead to most cervical cancer and genital warts. It is recommended for girls 11 and 12 years old and for 13- to 26-year-old females who have not been vaccinated.

HIV and AIDS

The **human immunodeficiency virus (HIV)** is the virus that causes **acquired immunodeficiency syndrome (AIDS)**. The disease is so termed because of its characteristics:

acquired—because people are not born with it
immunodeficiency—because the disease relates to the body's immune system, which is lacking in immunity
syndrome—because the symptoms occur as a group

Overall, the effects of HIV have been devastating, with the worst effects happening not in the United States but in other parts of the world. According to a joint UN program on HIV-AIDS report, more than 25 million people have died from HIV or AIDS worldwide, including 2.1 million children and adults in 2007. In sub-Saharan Africa, some 12 million children under age 18 have lost one or both parents to AIDS. It is estimated that more than a half a million people have died from HIV or AIDS in the United States, two-thirds of the deaths having occurred before age 45 (UN-AIDS 2008; www.avert.org).

Furthermore, an estimated 33 million people worldwide are living with the disease, including 2 million children and including 1.1 million or more Americans. Each year there are an estimated 2.7 million new cases of HIV AIDS worldwide. In 2007 in the United States, there were more than 40,000 newly diagnosed cases of HIV or AIDS. The Centers for Disease Control estimate that approximately 48% of the infections in the United States were transmitted through male–male sexual contact with an additional 28% resulting from high-risk heterosexual sexual contact. Among the

Gardasil, a vaccine for human papilloma virus (HPV), protects recipients against most of the HPV strains that are linked to 70% of cervical cancers and is recommended for girls 11 to 12 years of age.

latter cases, 72% of those infected were female. Overall, 5% to 7% of U.S. men are estimated to have sex with other men, yet more than two-thirds of all men living with HIV are men who have had same-sex sexual relations. In fact, men who have sex with men are 19 times more likely to be infected with HIV than is the general population ("U.S. Statistics," San Francisco AIDS Foundation, 2008).

HIV and AIDS cases have hit African Americans and Latinos especially hard, with each group infected at disproportionate rates. Despite being approximately 13% of the population of the United States, blacks make up half of all new HIV infections and 42% of new AIDS diagnoses (www.avert.org). Perhaps even more striking, among African Americans, AIDS is the fourth leading cause of death among men ages 25 to 44 and the third leading cause among women ("Black Americans and HIV/AIDS," Kaiser Family Foundation, 2008). African Americans represent 46% of the number of people in the United States infected with HIV. Another 35% are white and 17.5% are Hispanic. According to the Centers for Disease Control, African Americans with AIDS don't live as long as people of other races or ethnic groups, mostly as a result of higher rates of poverty and less access to medical care (Centers for Disease Control, "HIV/AIDS and African Americans," 2007).

Although AIDS was initially discovered in gay men and was thought of early on as a "gay disease" or the "gay plague," sexually transmitted cases among heterosexuals account for more than a quarter of new HIV infections. The Centers for Disease Control report that between 2004 and 2007, there was a 26% increase in diagnosed cases of HIV/AIDS among men who have sex with other men.

Without discounting or diminishing the devastation that the gay community suffered from AIDS and HIV or signs of resurging rates of infection, it is important to keep in mind that heterosexuals and bisexuals are also at risk and become infected. Virtually all adults in the United States are or will soon be related to, personally know, work with, or go to school with people infected with HIV or will know others whose friends, relatives, or associates test HIV positive.

As of mid-2009, there is still no surefire vaccine to prevent HIV, nor is there a cure for those who are or become infected. Significant strides have been made in fighting the disease, suppressing its symptoms, and prolonging life for those who are infected. In addition, between 1996 and 2001, AIDS death rates were reduced by 80%, and postdiagnosis survival had doubled in length. Those diagnosed after 1998 could expect to live 9 to 10 years longer than those who were diagnosed during the mid-1980s (Fallon 2005).

In addition to new treatments that can lengthen the life span of an AIDS-infected person as much as 15 years (Fallon 2005), we have considerable knowledge about the nature of the virus and how to reduce the likelihood of infection:

- *HIV attacks the body's immune system.* HIV is carried in the blood, semen, and vaginal secretions of infected people. A person may be HIV positive (infected with HIV) for years before developing AIDS symptoms.
- *HIV is transmitted only in certain clearly defined circumstances.* It is transmitted through the exchange of blood (as by shared needles or transfusions of contaminated blood), through sexual contact involving semen or vaginal secretions, and from an infected woman to her fetus through the placenta. Infected mothers may also transmit the infection during delivery or through breast milk (Miracle et al. 2003).
- *All those with HIV (whether or not they have AIDS symptoms) are HIV carriers.* They may infect others through unsafe sexual activity or by sharing needles; if they are pregnant, they may infect the fetus.
- *Heterosexuals, bisexuals, gay men, and lesbians are all susceptible to the sexual transmission of HIV.* No group owns HIV and no group is immune from the possibility of infection.
- *There is a definable progression of HIV infection and a range of illnesses associated with AIDS.* HIV attacks the immune system. AIDS symptoms occur as opportunistic diseases—diseases that the body normally resists—infect the individual. The most common opportunistic diseases are pneumocystis carinii pneumonia and Kaposi's sarcoma, a skin cancer. It is an opportunistic disease rather than HIV that kills the person with AIDS.
- *The presence of HIV can be detected through various kinds of antibody testing.* Though the most common method of testing for HIV is through blood tests, including one which tests dried blood, there are also tests to be used on oral fluid or urine (U.S. Food and Drug Administration, "Vaccines, Blood and Biologics," fda.gov).

Table 6.4 Principal Sexually Transmitted Diseases

STD and Infecting Organism	Time from Exposure to Occurrence	Symptoms	Medical Treatment	Comments
Chlamydia (*Chlamydia trachomatis*)	7–21 days	Women: 80% asymptomatic; others may have vaginal discharge or pain with urination. Men: 30%–50% asymptomatic; others may have discharge from penis, burning urination, pain and swelling in testicles, or persistent low fever.	Doxycycline, tetracycline, erythromycin	If untreated, may lead to pelvic inflammatory disease (PID) and subsequent infertility in women.
Gonorrhea (*Neisseria gonorrhoeae*)	2–21 days	Women: 50%–80% asymptomatic; others may have symptoms similar to chlamydia. Men: itching, burning, or pain with urination; discharge from penis ("drip").	Penicillin, tetracycline, or other antibiotics	If untreated, may lead to PID and subsequent infertility in women.
Human papilloma virus (HPV)	1–6 months (usually within 3 months)	Variously appearing bumps (smooth, flat, round, clustered, fingerlike white, pink, brown, and so on) on genitals, usually penis, anus, vulva. High-risk HPV can lead to cervical cancer.	Surgical removal by freezing, cutting or laser therapy. Virus remains in the body after warts are removed. Chemical treatment with podophylin (80% of warts eventually reappear).	HPV vaccine was developed and licensed for use in 2006.
Genital herpes (*Herpes simplex virus*)	3–20 days	Small, itchy bumps on genitals, becoming blisters that may rupture, forming painful sores; possibly swollen lymph nodes; flulike symptoms with first outbreak.	No cure, although acyclovir may relieve symptoms. Nonmedical treatments may help relieve symptoms.	Virus remains in the body, and outbreaks of contagious sores may recur. Many people have no symptoms after the first outbreak.
Syphilis (*Treponema pallidum*)	Stage 1: 1–12 weeks Stage 2: 6 weeks to 6 months after chancre appears	Stage 1: Red, painless sore (chancre) at bacteria's point of entry. Stage 2: Skin rash over body, including palms of hands and soles of feet.	Penicillin or other antibiotics	Easily cured, but untreated syphilis can lead to ulcers of internal organs and eyes, heart disease, neurological disorders, and insanity.
Hepatitis (hepatitis A or B virus)	1–4 months	Fatigue, diarrhea, nausea, abdominal pain, jaundice, and darkened urine due to impaired liver function.	No medical treatment available; rest and fluids are prescribed until the disease runs its course.	Hepatitis B is more commonly spread through sexual contact and can be prevented by vaccination.
Urethritis (various organisms)	1–3 weeks	Painful and/or frequent urination; discharge from penis. Women may be asymptomatic.	Penicillin, tetracycline, or erythromycin, depending on organism	Laboratory testing is important to determine appropriate treatment.

Table 6.4 Continued

STD and Infecting Organism	Time from Exposure to Occurrence	Symptoms	Medical Treatment	Comments
Vaginitis (*Gardnerella vaginalis*, *Trichomonas vaginalis*, or *Candida albicans*) (women only)	2–21 days	Intense itching of vagina and/or vulva; unusual discharge with foul or fishy odor; painful intercourse. Men who carry organisms may be asymptomatic.	Depends on organism; oral medications include metronidazole and clindamycin, and vaginal medications include clotrimazole and miconazole	Not always acquired sexually. Other causes include stress, oral contraceptives, pregnancy, tight pants or underwear, antibiotics, douching, and dietary imbalance.
HIV infection and AIDS (human immunodeficiency virus)	Several months to several years	Possible flulike symptoms but often no symptoms during early phase. Variety of later symptoms including weight loss, persistent fever, night sweats, diarrhea, swollen lymph nodes, bruiselike rash, and persistent cough.	No cure available, although many symptoms can be treated with medications and antiviral drugs may strengthen the immune system. Good health practices can delay or reduce the severity of symptoms.	Cannot be self-diagnosed; a blood test must be performed to determine the presence of the virus.
Pelvic inflammatory disease (PID) (women only)	Several weeks or months after exposure to chlamydia or gonorrhea (if untreated)	Low abdominal pain; bleeding between menstrual periods; persistent low fever.	Penicillin or other antibiotics; surgery	Caused by untreated chlamydia or gonorrhea; may lead to chronic problems such as arthritis and infertility.

SOURCE: Strong and DeVault (1997).

Anonymous testing is available at many college health centers and community health agencies. HIV antibodies develop between one and six months after infection. Antibody testing should take place one month after possible exposure to the virus and, if the results are negative, again six months later. If the antibody is present, the test will be positive. That means that the person has been infected with HIV and that an active virus is present. The presence of HIV does not mean, however, that the person necessarily will develop AIDS symptoms in the near future; symptoms generally occur 7 to 10 years after the initial infection.

Protecting Yourself and Others

As with avoiding unintended pregnancies, the safest practice to avoid STIs is *abstinence*, forgoing sexual relations. There is no chance of contracting STIs, although HIV infection can and does occur through nonsexual transmission (e.g., intravenous drug use with shared needles). If one is sexually active, however, the key to protecting oneself and others is to talk with one's partner about STIs in an open, nonjudgmental way and to use condoms. Because many people are uncomfortable asking about STIs, one can open the topic by revealing one's anxiety: "This is a little difficult for me to talk about because I like you, and I'm embarrassed, but I'd like to know whether you have herpes, or HIV, or whatever." If *you* have an STI, you can say, "Look, I like you, but we can't hook up right now because I have a herpes infection, and I don't want you to get it."

Remember, however, that not every person with an STI knows she or he is infected. Women with chlamydia and gonorrhea, for example, generally don't exhibit symptoms. Both men and women infected with HIV may not show any symptoms for years, although they are capable of spreading the infection through sexual contact. If you are or are planning to be sexually active but don't know whether your partner has an STI, use a condom. Even if you don't discuss STIs, condoms are simple and easy to use without much discussion. Both men and women can carry them. A woman can take a condom from her purse and give it to her partner. If he doesn't want to use it, she can say, "No condom, no sex." Since HIV can be transmitted through semen and vaginal secretions, when engaging in oral sex it is also recommended to use a condom (on a man) and a dental dam (on a woman).

Sexual Responsibility

Because we have so many sexual choices today, we need to be sexually responsible. Sexual responsibility includes the following:

- *Disclosure of intentions.* Each person needs to reveal to the other whether a sexual involvement indicates love, commitment, recreation, and so on.
- *Freely and mutually agreed-on sexual activities.* Each individual has the right to refuse any or all sexual activities without the need to justify his or her feelings. There can be no physical or emotional coercion.
- *Use of mutually agreed-on contraception in sexual intercourse if pregnancy is not intended.* Sexual partners are equally responsible for preventing an unintended pregnancy in a mutually agreed-on manner.
- *Use of "safer sex" practices.* Each person is responsible for practicing safer sex. Safer sex practices do not transmit semen, vaginal secretions, or blood during sexual activities and guard against STIs, especially HIV and AIDS.
- *Disclosure of infection from or exposure to STIs.* Each person must inform his or her partner about personal exposure to an STI because of the serious health consequences, such as infertility or AIDS, that may follow untreated infections. Infected individuals must refrain from behaviors—such as sexual intercourse, oral–genital sex, and anal intercourse—that may infect their partner. To help ensure that STIs are not transmitted, a condom and/or dental dam should be used.
- *Acceptance of the consequences of sexual behavior.* Each person needs to be aware of and accept the possible consequences of his or her sexual activities. These consequences can include emotional changes, pregnancy, abortion, and STIs.

Responsibility in many of these areas is facilitated when sex takes place within the context of an ongoing relationship. In that sense, sexual responsibility is a matter of values. Is responsible sex possible outside an established relationship? Are you able to act in a sexually responsible way? Sexual responsibility also leads to the question of the purpose of sex in your life. Is it for intimacy, erotic pleasure, reproduction, or other purposes?

As we consider the human life cycle from birth to death, we cannot help but be struck by how profoundly sexuality weaves its way through our lives. From the moment we are born, we are rich in sexual and erotic potential, which begins to take shape in our sexual experimentations of childhood. As children, we are still unformed, but the world around us haphazardly helps give shape to our sexuality. In adolescence, our education continues as a mixture of learning and yearning. But as we enter adulthood, with greater experience and understanding, we undertake to develop a mature sexuality: we establish our sexual orientation as heterosexual, gay, lesbian, or bisexual; we integrate love and sexuality; we forge intimate connections and make commitments; we make decisions regarding our fertility and sexual health; and we develop a coherent sexual philosophy. Then, in our middle years, we redefine sex in our intimate relationships, accept our aging, and reevaluate our sexual philosophy. Finally, as we become elderly, we reinterpret the meaning of sexuality in accordance with the erotic capabilities of our bodies. We come to terms with the possible loss of our partner and our own end. In all these stages, sexuality weaves its bright and dark threads through our lives.

Summary

- Our sexual behavior is influenced by *sexual scripts*: the acts, rules, stereotyped interaction patterns, and expectations associated with male and female sexual expression.
- Traditional sexual scripts are gendered, prescribing different sexual roles for women and men.
- Contemporary sexual scripts are more egalitarian, though evidence suggests that there is still a *sexual double standard* in which different sexual behaviors are accepted and expected of men and women.
- We learn about sexuality from multiple sources: parents, siblings, peers, the mass media, and, increasingly, the Internet.
- Even amid a longer-term trend toward more open acceptance of nonmarital sexual behavior, a recent

- trend seems to point toward a decline in sexual activity among teenagers.

- The actual percentage of the U.S. population that is lesbian, gay, or bisexual remains unknown. Estimates vary depending on whether one considers self-identity, sexual attraction, or sexual experience.

- Identifying oneself as gay or lesbian occurs in stages.

- Gay men and lesbians maintain intimate relationships that have much in common with heterosexual relationships, though they lack comparable social support and legal rights and protections.

- A continuing difference between gay male couples and both heterosexual and lesbian couples is in the acceptance of extrarelational sex. Gay couples in civil unions are more likely to expect and experience monogamy.

- Lesbian, gay, and bisexual individuals may confront homophobia and heterosexism. Many are subjected to prejudice or hostility, including verbal abuse, discrimination, or violence. Attitudes toward bisexuals may be even harsher than attitudes toward gay men and lesbians.

- Bisexuals are attracted to both sexes, though they may or may not have relationships with both women and men.

- Developmental tasks in middle adulthood include (1) redefining sex in marital or other long-term relationships, (2) reevaluating one's sexuality, and (3) accepting the biological aging process.

- The main determinants of sexual activity in old age are health and the availability of a partner.

- *Autoeroticism* consists of sexual activities that involve only the self. It includes sexual fantasies and masturbation, both of which are very common, and erotic dreams.

- The most common and acceptable of all premarital sexual activities is kissing, which occurs in more than 90% of all cultures.

- *Oral–genital sex*, which includes *cunnilingus* and *fellatio*, is practiced by heterosexuals, gay men, and lesbians. Data indicate both increasing rates of oral sex among teenagers and young adults and persisting race differences in rates of oral sex.

- *Sexual intercourse (coitus)* is the insertion of the penis into the vagina and the stimulation that follows.

- *Anal eroticism* is practiced by both heterosexuals and gay men. From a health perspective, *anal intercourse* is the most common means of sexually transmitting HIV. Among females, there are notable differences in rates of heterosexual anal sex.

- *Sexual enhancement* is based on accurate information about sexuality, developing communication skills, fostering positive attitudes, and increasing self-awareness.

- *Nonmarital sex* includes all sexual activities, especially sexual intercourse, that take place outside of marriage. *Premarital sex* has gained in acceptability, whereas *extramarital sex* has not.

- In marriage, sex takes on new and different meanings and occurs in a different day-to-day context.

- Marital intercourse tends to decline in frequency over time, but this does not necessarily signify marital deterioration.

- An estimated 14% to 15% of marriages experience prolonged periods without sex. Such celibate marriages occur for a variety of reasons and can negatively affect both the individual spouses and the marriage.

- Extrarelational sex occurs among heterosexual cohabitants, gay male couples, and lesbian couples at higher rates than among married couples.

- Sexual dysfunctions (such as orgasmic or arousal difficulties in women or erectile dysfunction or premature ejaculation in men) may be physiological or psychological in origin.

- *Sexually transmitted infections (STIs)*, especially chlamydia and gonorrhea, are epidemic. *Acquired immunodeficiency syndrome (AIDS)* is caused by the *human immunodeficiency virus (HIV)*, which attacks the body's immune system. HIV is carried in the blood, semen, and vaginal fluid of infected people.

- The rate of infection and death because of HIV is far greater in other parts of the world (e.g., Africa and Asia) than in the United States, where most HIV infections are the result of either heterosexual or male–male sexual contact and where Hispanics and African Americans have been particularly hard hit.

- If someone is sexually active, the keys to protection against STIs, including HIV and AIDS, are communication and condom use.

- Anyone sexually active should practice sexual responsibility: disclose any STD infections, engage only in mutually agreed-on activities, use mutually agreed-on methods of contraception, and engage in safer sex.

Key Terms

abstinence 211

acquired immunodeficiency syndrome (AIDS) 212

anal eroticism 200

anal intercourse 200

antigay prejudice 193

autoeroticism 197

RESOURCES ON THE WEB

Book Companion Website

www.cengage.com/sociology/strong

Prepare for quizzes and exams with online resources—including tutorial quizzes, a glossary, interactive flash cards, crossword puzzles, self-assessments, virtual explorations, and more.

7

Communication, Power, and Conflict

What Do YOU Think? Are the following statements **TRUE** or **FALSE**?
You may be surprised by the answers (see answer key on the following page).

T	F	**1** Conflict and intimacy go hand in hand in intimate relationships.
T	F	**2** Touching is one of the most significant means of communication.
T	F	**3** Avoiding conflict is the best way to sustain intimacy.
T	F	**4** Studies suggest that those couples with the highest marital satisfaction tend to disclose more than those who are unsatisfied.
T	F	**5** Negative communication patterns before marriage are a poor predictor of marital communication because people change once they are married.
T	F	**6** Partners tend to communicate sexual interest and disinterest nonverbally more than verbally.
T	F	**7** Women's power in marriage inevitably increases when she earns as much or more than her husband.
T	F	**8** The party with the least interest in continuing a relationship generally has the power in it.
T	F	**9** The meaning conveyed through nonverbal communication is precise and unambiguous.
T	F	**10** Wives tend to give more negative messages than husbands.

"Your mother called again."

That seems like such a simple, ordinary statement. It hardly seems like the kind of comment that would provoke an argument, nor does it appear particularly revealing about the tone or quality of a marriage or relationship. In fact, it sounds so routine, so "matter of fact," that we might overlook its significance and potential effect on married or coupled life.

Of course, we only have the four words; we don't know *how* they were said. What was the tone of voice, the cadence or rhythm of speech? Was it, "Your mother called again" or "Your mother called. Again." Or, combining tone and cadence, "Your mother called. Again!" We also have no information about the nonverbal signs. What was the expression on the face of the speaker—say a wife to a husband—when the statement was made? Did she smile? Roll her eyes? Frown? Shake her head? All these aspects of nonverbal communication help reveal more of the meaning and significance of such a statement. Clearly, seemingly simple comments such as this may have greater importance than whatever words they otherwise convey.

Finally, of potentially even greater significance is how the other person responds to a statement such as this one. Whether he or she responds with "an irritable groan," a laugh (as if to say, "What, again!"), a defensive explanation of the frequency of such maternal phone conversations, an expression of concern ("I hope she's all right"), or a positive discussion of his or her mother tells us a lot. A nonresponse may tell us yet more. It may suggest indifference and lack of interest in talking with the partner. Exchanges surrounding statements such as this one, "mundane and fleeting" as they may appear to be, can build and, in the process, greatly affect the quality of a relationship, the amount and nature of conflict, and the feeling of closeness and romance (Driver and Gottman 2004).

Thinking about the kinds of relationships that are the focus of this book, what is it you most want or expect from marriages, families, and other intimate relationships? Chances are, if you list the many characteristics or qualities you desire in such relationships, somewhere on that list will be "communication." We want our loved ones to share their feelings and ideas with us and to understand the ideas or feelings that we voice to them. After all, as shown in Chapter 5, that is how we expect to share intimacy. We want to be able to communicate effectively.

Chances are that "conflict" will not be included among desired relationship characteristics. After all, who wants to argue? We tend to see conflict as a negative to be avoided. Yet conflict is as much a feature of intimate relationships as are love and affection. As long as we value, care about, and live with others, we will experience occasions when we disagree, when disagreements lead to conflict, and when we find ourselves in the middle of arguments. An absence of conflict not only is unrealistic but would be unhealthy as well. How we resolve our disagreements tells us much about the health of our relationships.

Both communication and conflict are inextricably connected to intimacy. When we speak of communication, we mean more than just the ability to relay information (e.g., "Your mother called"), discuss problems, and resolve conflicts. We also mean communication for its own sake: the pleasure of being in each other's company, the excitement of conversation, the exchange of touches and smiles, and the loving silences. Through communication, we disclose who we are, and from this self-disclosure, intimacy grows.

One of the most common complaints of married partners, especially unhappy partners, is that they don't communicate. But it is impossible not to communicate—a cold look may communicate anger as effectively as a fierce outburst of words. What these unhappy partners mean by "not communicating" is that their communication is somehow driving them apart rather than bringing them together, feeding and creating conflict rather than resolving it. Communication patterns are strongly associated with marital satisfaction (Noller and Fitzpatrick 1991).

In this chapter, we explore patterns and problems in communication in marital and intimate relationships. We also examine the role of power in marital relationships, where it comes from, and how it is expressed in both the styles of communication and the outcomes of conflict. Finally, we look at the relationship between conflict and intimacy, exploring different types of conflict and approaches to conflict resolution. We look especially at three of the more common areas of relationship conflict: conflicts about sex, money, and housework.

Answer Key to What Do YOU Think?

1 True, see p. 241;   **2** True, see p. 227;   **3** False, see p. 222;   **4** True, see p. 234;   **5** False, see p. 229;   **6** False, see p. 232;   **7** False, see p. 240;   **8** True, see p. 239;   **9** False, see p. 224;   **10** True, see p. 229.

Verbal and Nonverbal Communication

When we communicate face-to-face, the messages we send and receive contain both a verbal and a nonverbal component. Verbal communication expresses the *basic content* of the message, whereas **nonverbal communication** reflects more of the *relationship* part of the message. The relationship part conveys the attitude of the speaker (friendly, neutral, or hostile) and indicates how the words are to be interpreted (as a joke, request, or command). To understand the full content of any message, we need to understand both the verbal and the nonverbal parts.

For a message to be most effective, both the verbal and the nonverbal components should be in agreement. If you are angry and say "I'm angry," and both your facial expression and your voice show anger, the message is clear and convincing. But if you say "I'm angry" in a neutral tone of voice and a smile on your face, your message is ambiguous. More commonly, if you say "I'm not angry" but clench your teeth and use a controlled voice, your message is also unclear.

Popular Culture: Staying Connected with Technology

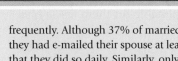

Suppose you received the following text message while in class or at work or while otherwise engaged in some activity away from your significant other:

"Luv u. Miss u. C u later."

Despite your physical separation, it is likely that you would recognize such a message as a warm and loving gesture, a means of staying connected and feeling close, despite being in different places and engaged in different activities. Such messages, made possible and more likely given the immense spread of cell phone technology, add a dimension of closeness to our romantic and familial relationships.

According to the Pew Internet and American Life Project 2008 report, "Networked Families," 89% of married (or cohabiting) households with 7- to 17-year-old children have multiple cell phones; almost half of such households have three or more phones. In households where both partners have cell phones, 70% contact each other at least once a day to say hello or chat. For perhaps the busiest couples, cell phones are also used to coordinate their "highly scheduled lives" (Kennedy, Smith, Wells, and Wellman 2008). For nearly two-thirds of couples in which each partner owns a cell phone, spouses contact each other at least once a day to coordinate their schedules and activities. Couples with children display even more frequent use of cell phones: 70% contact each other to coordinate and review their daily schedules, and 74% contact each other "to stay in touch and say hello."

In addition to cell phones (and to land-line phones), couples use other technologies—from text messages, e-mail, and instant messaging—to keep in touch. According to data gathered between December 2007 and January 2008, these other technologies are not used nearly as frequently. Although 37% of married respondents said that they had e-mailed their spouse at least once, only 8% said that they did so daily. Similarly, only 8% of married individuals said that they exchanged text messages with their spouse daily, though 21% reported having done so in the past. Respondents with children showed a greater use of text messaging to communicate with their children; 12% of fathers and 28% of mothers reported exchanging text messages with their children at least once a day (Kennedy et al. 2008). These percentages have no doubt increased as text messaging has become much more common. Although more recent couple data were unavailable, the trend among teens and young adults indicates a dramatic increase in the ownership of cell phones and in the daily use of text messaging (Lenhart 2009). In the fourth quarter of 2008, U.S. teens averaged nearly 80 text messages a day, more than double the average from the prior year (Hafner 2009).

"Texting" has special utility for those who choose to use it. For the most part, texting is possible when phone calls may not be available or desirable. With their necessary brevity, they are still a means of maintaining and displaying "connectedness" between sender and recipient (Pettigrew 2007). In a qualitative interview study, Jonathan Pettigrew quotes one husband:

When you're with a bunch of guys you don't want to be seen on the phone talking to your wife, so I might text her then. (12)

For romantic partners especially, text messages were useful as a nearly always available means of "connectedness-oriented communication," a way to create and maintain a sense of emotional and relational connection" (Pettigrew 2007, 25).

Your tone and expression make your spoken message difficult to take at face value.

In addition to both verbal and nonverbal communication, increasingly people are communicating with each other via technologies that allow some of the same qualities of verbal communication minus any of the information conveyed nonverbally or via tone of voice. Such electronically mediated communication is rapidly increasing and, along with an increase in volume of use, has become a unique tool in maintaining relationships (see the Popular Culture feature in this chapter).

The Functions of Nonverbal Communication

Whenever two or more people are together and aware of each other, it is impossible for them *not to communicate*. Even when you are not talking, you communicate by your silence (e.g., an awkward silence, a hostile silence, or a tender silence). You communicate by the way you position your body and tilt your head, your facial expressions, your physical distance from the other person or people, and so on. Take a moment, right now, and look around you. If there are other people in your presence, how and what are they communicating nonverbally?

Research supports the idea that nonverbal communication has important consequences that extend beyond the message and the moment. For example, parents can affect their children's physical and mental health by their nonverbal communication. The same is true of marital partners' effects on each other. Supportive nonverbal behavior can benefit relationship partners by actually affecting their immune system and their overall health. Negative nonverbal communication can negatively impact individual (marital) partners as well as threaten the stability of the relationship between them (Giles and LePoire 2006).

One of the problems with nonverbal communication, however, is the imprecision of its messages. Is a person frowning or squinting? Does the smile indicate friendliness or nervousness? A person may be in reflective silence, but we may interpret the silence as disapproval or distance. We may incorrectly infer meanings from expressions, eye contact, stance, and proximity that are other than what is intended. However, by acting on the meaning we read into nonverbal behavior, we give it more weight and make it of greater consequence than it initially might have been.

More than 20 years ago, an important study of nonverbal communication and marital interaction found that nonverbal communication has the following three important functions in marriage (Noller 1984): (1) conveying interpersonal attitudes, (2) expressing emotions, and (3) handling the ongoing interaction.

Conveying Interpersonal Attitudes
Nonverbal messages are used to convey attitudes. Holding hands can suggest intimacy; sitting on opposite sides of the couch can suggest distance. Not looking at each other in conversation can suggest discomfort or lack of intimacy. Rolling eyes at another's statement conveys a negative attitude or reaction to what's being said or the person saying it, even if the eye-rolling culprit claims, "What? I didn't say *anything*."

Expressing Emotions
Our emotional states are expressed through our bodies. A depressed person walks slowly, head hanging; a happy person walks with a spring. Smiles, frowns, furrowed brows, tight jaws, tapping fingers—all express emotion. Expressing emotion is important because it lets our partner know how we are feeling so that he or she can respond appropriately. It also allows our partner to share our feelings, whether that means to laugh or weep with us. It is this feature of nonverbal communication that is most lacking from phone conversations and electronic communication. Without those emotional cues that we read and come to depend on, it is sometimes a challenge to know just what the person on the other end of the phone is "really saying."

Handling the Ongoing Interaction
Nonverbal communication helps us handle the ongoing interaction by indicating interest and attention. An intent look indicates our interest in the conversation; checking one's watch or yawning can indicate boredom. Posture and eye contact are especially important. Are you leaning toward the person with interest or slumping back, thinking about something else? Do you look at the person who is talking, or are you distracted, glancing at other people as they walk by or watching the clock?

The Importance of Nonverbal Communication
According to psychologist John Gottman (1994), even seemingly simple acts, such as rolling one's eyes in response to a statement or complaint made by a spouse, can convey **contempt**, a feeling that the target of the

Even the act of gently touching hands can communicate closeness between two people.

© Christopher Thomas/Getty Images/Store

of raising your hands in front of yourself and "pushing at the air" communicates defensiveness to those you are interacting with; it is as if you were saying "back off." In fact, nonverbally, you *are* saying just that.

Shifting from negative to positive, such nonverbal behaviors as touch, proximity (i.e., physical closeness), smiling, and gazing help define the intimacy of an interaction. Such behaviors also differentiate closer from more casual relationships and, within relationships, more satisfied from less satisfied partners. In the next section, we will look specifically at some important means of communicating nonverbally.

Proximity, Eye Contact, and Touch

Three forms of nonverbal communication that are especially interesting are proximity, eye contact, and touch. Awareness of the ways in which such forms of nonverbal communication convey intimacy may enable one to observe interactions and get a fairly good sense of the closeness and warmth within that relationship.

Proximity

Nearness, in terms of physical space, time, and so on, is referred to as **proximity**. Where we sit or stand relative to another person can signify levels of intimacy or the type of relationship. In a social situation, the face-to-face distances between people when starting a conversation are clues to how the individuals wish to define the relationship. A distance of 0 to 18 inches is considered an **intimate zone** not typically found among people interacting in public settings and typically reserved for one's intimate relationships (e.g., romantic partners, close friends, parents, and young children). Within this intimate zone, multiple senses are involved, and the other person is seen "up close" wherein all details of complexion, eye color, hair roots, wrinkles around one's eyes, and so on can be observed while, at the same time, characteristics of the body and extremities may not be. Aside from visual imagery, scents and sounds (e.g., of breath and of breathing) are also accessible within this narrow spatial distance (Altman and Chemers 1984).

The intimate zone is followed by a space of 1.5 to 4 feet that Hall (1966) considered one's **personal space**. Within this area, one can access a variety of kinds of sensory information, though not with the same detail as in the intimate zone. One is close enough to touch the person, one can see details of a person's appearance, and one can still obtain some

expression is undesirable. Contempt can be displayed verbally as well through such things as insults, sarcasm, and mockery. Along with contempt, there are three other negative behaviors that indicate particularly troubled and vulnerable relationships. These others are criticism (especially when it is overly harsh), defensiveness, and stonewalling or avoiding. Together, these four behaviors make up Gottman's "four horsemen of the apocalypse," spelling potential for eventual divorce (Gottman 1994). Eventually, Gottman added a fifth—belligerence. Gottman suggested that all these are warning signs of serious risk of eventual divorce (Gottman 1994; Gottman et al. 1998). Conversely, couples who communicate with affection and interest and who maintain humor amid conflict can use such a *positive affect* to diffuse potentially threatening conflict (Gottman et al. 1998).

As you think about Gottman's danger signs, consider how easily they can be expressed and conveyed via nonverbal communication as well as by things we say to each other. For example, failing to make eye contact is a way of avoiding or stonewalling. The common gesture

(though not as much) sense of the person's scent. It is the more common distance at which people interact in public and allows them to move closer (as intimacy might dictate) or move even farther away from each other (Altman and Chemers 1984).

Although we speak here of the most personal zones as understood in the United States, all cultures have their own standards of spatial norms. In most cultures, decreasing the distance signifies an invitation to greater intimacy or a threat. Moving away denotes the desire to terminate the interaction. For example, when standing at an intermediate distance from someone at a party, one sends the message that intimacy is not encouraged. If either party wants to move closer, however, he or she risks the chance of rejection. Therefore, they must seek out and exchange cues, such as eye contact, laughter, or small talk, before moving closer to avoid facing direct rejection. If the person moves farther away during this exchange or, worse, leaves altogether ("Excuse me, I think I see a friend"), he or she is signaling disinterest. But if the person moves closer, there is the "proposal" for greater intimacy.

As relationships develop, couples do more than just narrow the spatial distance that partners maintain while interacting. They may also engage in close gazing into each other's eyes, holding hands, and walking with arms around each other. These require closer proximity and signal greater intimacy.

However, because of cultural differences, there can be misunderstandings. The neutral distance for Latinos, for example, is much closer than for Caucasians, who may misinterpret the distance as close (too close for comfort). In social settings, this can lead to problems. As Carlos Sluzki (1982) pointed out, "A person raised in a non-Latino culture will define as seductive behavior the same behavior that a person raised in a Latin culture defines as socially neutral." Because of the miscue, the Caucasian may withdraw or flirt, depending on his or her feelings. In addition, the neutral responses of people in cultures with customs of greater intermediate distances and less overt touching, such as Asian American culture, may be misinterpreted negatively by people of other cultural backgrounds.

There is more than just distance to consider. In addition to the actual space between two people, one can identify other physical signs that reveal aspects of the closeness or quality of their relationship. Two such signs are body orientation and lean. Face-to-face body orientation is associated with greater intimacy, as is a forward lean (Anderson, Guerrero, and Jones 2006).

Eye Contact and Facial Expressions

In looking for signs of the degree of closeness or intimacy shared between two people, eye contact is an important component. Much can be discovered about a relationship by watching whether, how, and how long people look at each other. Making eye contact with another person, if only for a split second longer than usual, is a signal of interest. Brief and extended glances, in fact, play a significant role in women's expression of initial interest (Moore 1985). When you can't take your eyes off another person, you probably have a strong attraction to him or her. You can often distinguish people in love by their prolonged looking into each other's eyes. In addition to eye contact, dilated pupils may be an indication of sexual interest (or poor lighting).

Research suggests that the amount of eye contact between a couple having a conversation can distinguish between those who have high levels of conflict and those who don't. Those with the greatest degree of agreement have the greatest eye contact with each other (Beier and Sternberg 1977). Those in conflict tend to avoid eye contact (unless it is a daggerlike stare). As with proximity, however, the level of eye contact may differ by culture. For example, research reveals that compared to Americans, Arabs gaze longer and more directly at their partners. Furthermore, in countries where physical touch or contact during interaction is accepted and common, individuals engage in more gazing with those with whom they are interacting than is observed among those in noncontact cultures (see Matsumoto 2006). In the United States, African Americans display less eye contact than do whites (Dovidio et al. 2006).

Eye contact may be best understood as part of a broader category of information and emotions conveyed and communicated with one's face. Via various facial expressions, the face may be the most important conveyor of the level of intimacy (or, conversely, animosity) shared between people in social interaction. "Pleasant" facial expressions, especially smiles, help convey warmth and display a sense of comfort.

Interesting research by psychologists Masaki Yuki, William Maddux, and Takahiko Masuda identified cultural differences between Japanese and Americans in the importance paid to the eyes as opposed to the mouth in conveying and interpreting emotions. They suggest that the eyes are more of a diagnostic cue for Japanese, whereas for Americans the mouth (e.g., smiles, frowns, and so on) is a more influential facial cue. When subjects were shown photographs of faces in which competing emotions were displayed (e.g., happy eyes, sad mouth), the eyes were the more influential cue for Japanese, whereas the mouth was more influential for Americans (Yuki, Maddux, and Masuda 2006).

Critical Thinking

Think about your nonverbal communication. In instances where you and another person had significant eye contact, what did the eye contact mean? As you think about touch, what are the different kinds of touch you do? What meanings do you ascribe to the touch you give and the touch you receive?

Touch

A review of the research on touch finds it to be extremely important in human development, health, and sexuality (Hatfield 1994). Touch is associated with intimacy across many different types of relationships, from close friends to romantic partners to family members. Indeed, one might state that without touch, intimacy is nearly impossible. When it is welcome, touch of the face or torso is experienced as especially intimate (Anderson et al. 2006).

Touch is the most basic of all senses; it contains receptors for pleasure and pain, hot and cold, and rough and smooth. Touch is a life-giving force for infants. If babies are not touched, they may fail to thrive and may even die. We hold hands with small children and those we love.

People vary in their responsiveness and receptiveness to touch, with some people being more "touch avoidant" (Andersen, Andersen, and Lustig 1987; Guerrero and Andersen 1991). For some, touch can be experienced as a violation. A stranger or acquaintance may touch you in a way that is too familiar. Your date or partner may touch you in a manner you don't like or want. Sexual harassment often consists of unwelcome touching.

Touching is a universal part of social interaction, but it varies in both frequency and meaning across cultures and between women and men (Dibiase and Gunnoe 2004). Often, touch has been taken to reflect social dominance. Based largely and initially on research by Nancy Henley (1977) in which men were found to touch women more than women touched men, the generalization was drawn that touch is often a privilege that higher-status, more socially dominant individuals enjoy over lower-status, more subordinate individuals. This generalization was further modified some by research that revealed that when individuals

were of close but different statuses, the lower-status person often strategically used touch as a means of "making a connection" with the higher-status person. Status differences also determined the *type of touch*; lower-status individuals were more likely to initiate handshakes, and higher-status individuals were more likely to initiate somewhat more intimate touching, such as placing a hand on another's shoulder (Dibiase and Gunnoe 2004).

What about culture? Differences surface in a number of interesting ways. For example, people in colder climates use relatively larger distance and hence relatively less physical contact when they communicate, whereas people in warmer climates prefer closer distances. Latin Americans are comfortable at a closer range (have smaller personal space zones) than North Americans. Middle Eastern, Latin American, and southern European cultures can be considered "high-contact cultures," where people interact at closer distances and touch each other more in social conversations than people from noncontact cultures, such as those of northern Europe, the United States, and Asia (Dibiase and Gunnoe 2004). In so-called high-contact cultures, the kind of touch used in greetings is more intimate, often consisting of hugging or kissing, whereas a firm but more distant handshake is an accepted greeting in noncontact cultures.

Comparing women and men in the United States, Italy, and the Czech Republic, Rosemarie Dibiase and Jaime Gunnoe found that gender differences in touch varied across the three cultures. Men tended to engage in more "hand touch" than women, and women engaged in more "non–hand touch" in all three cultures, though the extent of gender difference varied some in the three countries observed (Dibiase and Gunnoe 2004).

Touch can signify more than dominance; it often is a way to convey intimacy, immediacy, and emotional closeness. Touch may well be the most intimate form of nonverbal communication. Touching seems to go hand in hand with self-disclosure. Those who touch seem to self-disclose more; touch seems to be an important factor in prompting others to talk more about themselves (Heslin and Alper 1983; Norton 1983).

The amount of contact, from almost imperceptible touches to "hanging all over" each other, helps differentiate lovers from strangers. How and where a person is touched can suggest friendship, intimacy, love, or sexual interest.

Sexual behavior relies above almost all else on touch: the touching of self and others and the touching of hands, faces, chests, arms, necks, legs, and genitals. Sexual behavior is skin contact. In sexual interactions, touch takes precedence over sight, as we close our eyes to caress, kiss, and enjoy sexual activity. We shut our eyes to focus better on the sensations aroused by touch; we shut out visual distractions to intensify the tactile experience of sexuality.

The ability to interpret nonverbal communication correctly appears to be an important ingredient in successful relationships. The statement "What's wrong? *I can tell something is bothering you*" reveals the ability to read nonverbal clues, such as body language or facial expressions. This ability is especially important in ethnic groups and cultures that rely heavily on nonverbal expression of feelings, such as Latino and Asian American cultures. Although the value placed on nonverbal expression may vary among groups and cultures, the ability to communicate and understand nonverbally remains important in all cultures. A comparative study of Chinese and American romantic relationships, for example, found that shared nonverbal meanings were important for the success of relationships in both cultures (Gao 1991).

We convey feelings via a variety of nonverbal means— proximity, touch, and eye contact.

Gender Differences in Communication

Compared with men's nonverbal communication patterns, women smile more; express a wider range of emotions through their facial expressions; occupy, claim, and control less space; and maintain more eye contact with others with whom they are interacting (Borisoff and Merrill 1985; Lindsey 1997). In their use of language and their styles of speaking, further differences emerge (Lakoff 1975; Lindsey 1997; Tannen 1990). Women use more qualifiers ("It's *kind of* cold out today"), use more tag questions ("It's kind of cold out today, *don't you think?*"), use a wider variety of intensifiers ("It was *awfully* nice out yesterday; now it's kind of cold out, don't you think?"), and speak in more polite and less insistent tones. Male speech contains fewer words for such things as color, texture, food, relationships, and feelings, but men use more and harsher profanity ("It's so damn cold out!") (Lindsey 1997). In cross-gender interaction, men talk more and interrupt women more than women interrupt men. In same-gender conversation, men disclose less personal information and restrict themselves to safer topics, such as sports, politics, or work (Lindsey 1997).

There are two things to note about such differences between female and male styles of both verbal and nonverbal communication. The more male style fits more with positions of dominance, whereas the more female style is often found among people in subordinate positions. At the same time, women's style of communicating is characterized more by cooperation and consensus seeking; thus, it is also situationally appropriate and advantageous to relationship building and maintenance (Lindsey 1997; Tannen 1990). In light of these facts, researchers differ in their interpretations of these gender patterns: those who see women's style more as artifacts of subordination versus those who see gender patterns as reflecting difference.

Gender Differences in Partner Communication

In addition to overall gender differences in communication noted earlier, researchers have identified several gender differences in how heterosexual spouses or partners communicate (Klinetob and Smith 1996; Noller and Fitzpatrick 1991; Thompson and Walker 1989).

First, wives tend to *send clearer messages* to their husbands than their husbands send to them. Wives are often more sensitive and responsive to their husbands' messages both during conversation and during conflict. They are more likely to reply to either positive messages (e.g., compliments) or negative messages (e.g., criticisms) than are their husbands, who may not reply at all.

Second, wives tend to *give more positive or negative messages*; they tend to smile or laugh when they send messages, and they send fewer clearly neutral messages. Husbands' neutral responses make it more difficult for wives to decode what their partners are trying to say. If a wife asks her husband if they should go to dinner or see a movie and he gives a neutral response, such as "Whatever," does he really not care, or is he pretending he doesn't care to avoid possible conflict?

Third, although communication differences in arguments between husbands and wives are usually small, they nevertheless follow a typical pattern. Wives tend to *set the emotional tone* of an argument. They escalate conflict with negative verbal and nonverbal messages ("You're not even listening to me!") or deescalate arguments by setting an atmosphere of agreement ("I understand your feelings"). Husbands' inputs are less important in setting the climate for resolving or escalating conflicts. Wives tend to *use emotional appeals more than husbands*, who tend to reason, seek conciliation, and find ways to postpone or end an argument. A wife is more likely to ask, "Don't you love me?," whereas a husband is more likely to say, "Be reasonable."

Communication Patterns in Marriage

Communication occupies an important place in marriage. Research in the United States as well as in Europe (e.g., Italy), Asia (Taiwan), and Latin America (Brazil) indicates that partners' satisfaction with relationships is affected by the quality of their communication (Christensen et al. 2006). When couples have communication problems, they often fear that their marriages are seriously flawed. In addition, negative communication is associated with both less relationship satisfaction and greater instability. As shown in a subsequent section, one of the most common complaints of couples seeking therapy is about their communication problems (Burleson and Denton 1997).

There continues to be a substantial amount of research examining premarital and marital communication. Researchers are finding significant correlations between the nature of communication and satisfaction as well as differences in male versus female communication patterns in marriage.

Premarital Communication Patterns and Marital Satisfaction

"Drop dead, you creep!" is hardly the thing someone would want to say when trying to resolve a disagreement in a dating relationship. But it may be an important clue as to whether such a couple should marry. Many couples who communicate poorly before marriage are likely to continue the same way after marriage, and the result can be disastrous for future marital happiness. Researchers have found that how well a couple communicates before marriage can be an important predictor of later marital satisfaction (Cate and Lloyd 1992). If communication is poor before marriage, it is not likely to significantly improve after marriage—at least not without a good deal of effort and help. On the other hand, *self-disclosure*—the revelation of our own deeply personal information—before or soon after marriage is related to relationship satisfaction later. Talking about your deepest feelings and revealing yourself to your partner builds bonds of trust that help cement a marriage.

Whether a couple's interactions are basically negative or positive can also predict later marital satisfaction. In a notable experiment by John Markham (1979), 14 premarital couples were evaluated using "table talk," sitting around a table and simply engaging in conversation. Each couple talked about various topics. Using an electronic device, each partner electronically recorded whether the message was positive or negative. Markham found that the negativity or positivity of the couple's communication pattern barely affected their marital satisfaction during their first year. This protective quality of the first year is known as the **honeymoon effect**—which means that you can say almost anything during the first year, and it will not seriously affect marriage (Huston, McHale, and Crouter 1986). But after the first year, couples with negative premarital communication patterns were less satisfied than those with positive communication patterns. A later study (Julien, Markman, and Lindahl 1989) found that those premarital couples who responded more to each other's positive communication than to each other's negative communication were more satisfied in marriage four years later.

Cohabitation and Later Marital Communication

Researchers have revealed that cohabitation has an effect on the outcome of marriage. Specifically, couples who live together before marrying are more likely to separate and divorce than couples who don't live together before marriage. That may seem counterintuitive. Wouldn't couples who live together first find it easier to adjust to marriage? Doesn't cohabitation weed out the unsuccessful matches before marriage? In Chapter 9, we consider the range of explanations for this cohabitation effect. Here, we simply look at how communication patterns might contribute to later marital failure.

Catherine Cohan and Stacey Kleinbaum (2002) hypothesized that spouses who live together before marrying display more negative problem-solving and support behavior compared with their counterparts who marry without first living together. Why would cohabitation lead to poorer marital communication? Cohan and Kleinbaum suggest three possible reasons:

1. Couples who live together come from backgrounds that may predispose them to poorer communication abilities. Compared with couples who don't cohabit, cohabitants tend to be younger, less religious, and more likely to come from divorced homes. Cohan and Kleinbaum point out that this translates into their being less mature, less traditional, and less likely to have had good parental role models for effective communication.
2. People who cohabit may be more accepting of divorce and less committed to marriage. Thus, they may expend less effort or energy developing good marital communication skills because they are less sure that they will stay married.
3. Cohabitation is associated with factors such as alcohol use, infidelity, and lower marital satisfaction, which in turn are correlated with less effective communication.

In studying 92 couples who were in their first two years of marriage, Cohan and Kleinbaum found that premarital cohabitation was associated with poorer marital communication. Couples with one or more cohabitation experiences displayed poorer, more divisive, and more destructive communication behaviors than did couples with no prior cohabitation experience (Cohan and Kleinbaum 2002).

Marital Communication Patterns and Satisfaction

Researchers have found a number of patterns that distinguish the communication patterns in satisfied and dissatisfied marriages (Gottman 1995; Hendrick 1981; Noller and Fitzpatrick 1991; Schaap, Buunk, and Kerkstra 1988). The following characteristics tend to be found among couples in satisfying marriages:

- Willingness to accept conflict but to engage in conflict in nondestructive ways.
- Less frequent conflict and less time spent in conflict. Both satisfied and unsatisfied couples, however, experience conflicts about the same topics, especially about communication, sex, and personality characteristics.
- The ability to disclose or reveal private thoughts and feelings, especially positive ones, to a partner. Dissatisfied spouses tend to disclose mostly negative thoughts to their partners.
- Expression by both partners of equal levels of affection, such as tenderness, words of love, and touch.
- More time spent talking, discussing personal topics, and expressing feelings in positive ways.
- The ability to encode (send) verbal and nonverbal messages accurately and to decode (understand) such messages accurately. This is especially

Touching the face of one's partner is an example of especially intimate nonverbal communication.

important for husbands. Unhappy partners may actually decode the messages of strangers more accurately than those from their partners.

Additionally, the effects of communication between spouses on couples' satisfaction with their marriages can be found in many cultures, though not necessarily to the same extent that it does among marriages in the United States. Given the premium placed on intimacy and more romantic conceptualizations of love, marriages in the United States are especially susceptible to the effects of positive and negative communication on their sense of marital well-being. However, even in cultures where marriage is more "practical" and arranged, based on matching partners on a host of characteristics (e.g., Pakistan), marital satisfaction is affected by the nature of marital communication (Rehman and Holtzworth-Munroe 2006).

Demand–Withdraw Communication

One prominent type of familial communication, especially evident in conflict communication, is referred to as **demand–withdraw communication**—a pattern in which one person makes an effort to engage the other person in a discussion of some issue of importance. The one raising the issue may criticize, complain, or suggest a need for change in the other's behavior or in the relationship. The other party, in response to such overtures, withdraws by either leaving the discussion, failing to reply, or changing the subject (Klinetob and Smith 1996).

In many ways, the demand–withdraw pattern is an understandable outcome of differences in what each partner in a relationship wants. One partner, often the wife in heterosexual marriages, wants something different from the status quo. She is then left with a choice between doing nothing and confronting her husband. If she resists or avoids bringing the topic up, nothing will change, a situation she may deem unacceptable. Instead, she raises the subject, voices her complaint, and presses for change. The other partner, more often the husband in heterosexual marriages, satisfied with the way things are (or otherwise not interested in the change his partner wishes to discuss), wants no change. By agreeing to discuss the subject, there is a chance that tension will rise and conflict will result. He also runs the risk of having to ultimately agree to change in the way his wife desires. Wanting neither to argue nor to agree with his partner's wishes, he withdraws.

As the preceding illustrates, in seeking change, the person making the demand is in a potentially vulnerable, less powerful position than the person withdrawing from the interaction. The latter can choose to change or not. By withdrawing, he or she maintains the status quo. Withdrawal has other consequences. Just as it keeps the conflict from escalating, it may prevent the resolution of the conflict by curtailing needed communication and necessary relationship adjustment (Sagrestano, Heavey, and Christensen 1999).

The demand–withdraw pattern is common in the United States and cross-culturally (Christensen et al. 2006). Although most often studied within heterosexual couples, it has also been observed in parent–child conflict among parents and their adolescents (Caughlin and Malis 2004). In both versions of parent–child demand–withdraw (i.e., instances of parents demanding and adolescents withdrawing and instances of adolescents demanding and parents withdrawing), both adolescents' and parents' relationship satisfaction can be negatively affected.

Research on heterosexual couples shows a tendency for the demand–withdraw pattern to be associated with gender. As in the earlier illustration, more often, women "demand" and men "withdraw." Researchers

The demand–withdraw style of communication is a common pattern among heterosexual couples, in which typically the woman initiates conversation about an issue and the man withdraws.

have considered a variety of explanations for this more common gender pattern, including a more biologically based explanation. Men and women may have different physiological responses to conflict, and these may help produce the familiar male withdrawal that is part of the female demand–male withdraw pattern of communication. With greater tolerance for physiological arousal, women can maintain the kinds of high levels of engagement that conflict contains. John Gottman and Robert Levenson (1992) reported that compared to women, men show different physiological reactions—more rapid heartbeat, quickened respiration, and the release of higher levels of epinephrine in their endocrine systems—to disagreements. To men, this arousal is highly unpleasant; thus, they act to avoid it by withdrawing from the conflict. Withdrawal may be a means of avoiding these reactions (Gottman and Levenson 1992; Levenson, Carstensen, and Gottman 1994).

Other researchers have looked at gender socialization and at the nature of heterosexual relationships. Females are socialized to seek greater connectedness with others, whereas males to seek greater independence and autonomy. Theoretically, this could motivate women to desire greater closeness, which in turn could put them in the position of making demands of male partners. Relationship dynamics are also important. Women are more often the partners seeking to initiate some sort of change in the relationship, whether that be increased intimacy, increased male participation in housework, or increased male involvement in child care (Christensen and Heavey 1993; Christensen et al. 2006).

Researchers who have explored similarities and differences between heterosexual and same-sex couples have cast additional light on the gendered pattern of demand–withdraw communication. Psychologists Sondra Solomon, Esther Rothblum, and Kimberly Balsam compared same-sex couples and heterosexual couples on a number of relationship issues, including what they called "relationship maintenance behavior." Among married heterosexual couples, women indicated that they were more likely than their husbands to "begin to talk about what is troubling us when there is tension" and "to sense that the other is disturbed about something." Married heterosexual men reported themselves as more likely to keep their feelings to themselves. Among same-sex couples, these relationship maintenance behaviors were more equally displayed by both partners (Solomon, Rothblum, and Balsam 2005).

Although the demand–withdraw pattern is fairly common, it is neither a healthy nor an effective style of communication and conflict resolution. It is associated with less marital satisfaction and higher likelihood of relationship failure or divorce and may even be a predictor of violence within the couple relationship, especially among couples with high levels of husband demand–wife withdraw (Sagrestano et al. 1999). Psychologist John Gottman and colleagues contend that the demand–withdraw pattern is "consistently characteristic of ailing marriages" (Gottman et al. 2002, in Christensen et al. 2006).

Sexual Communication

To have a satisfying sexual relationship, a couple must be able to communicate effectively with each other about expectations, needs, attitudes, and preferences (Regan 2003). Both the frequency with which couples engage in sexual relations and the quality of their involvement depend on such communication.

As addressed in Chapter 5, among heterosexuals, in both married and cohabiting relationships, women and men often follow sexual scripts that leave the initiation of sex (i.e., the communication of desire and interest) to men, with women then in a position of accepting or refusing men's overtures. Reviewing the literature on sexual communication, Pamela Regan observes that regardless of who takes the role of initiating, the efforts are usually met with positive responses. Both attempts to initiate and positive responses are rarely communicated explicitly and verbally (Regan 2003, 84):

> A person who desires sexual activity might turn on the radio to a romantic soft rock station, pour his or her partner a glass of wine, and glance suggestively in the direction of the bedroom. The partner . . . might smile, put down his or her book, and engage in other nonverbal behaviors that continue the sexual interaction without explicitly acknowledging acceptance.

Interestingly, lack of interest or refusal of sexual initiations is more likely to be communicated directly and verbally (e.g., "Not tonight, I have a lot of work to do"). By framing refusal in terms of some kind of account, the refusing partner allows the rejected partner to save face (Regan 2003).

Effective sexual communication may be difficult, but it is important if couples hope to construct and keep mutually satisfying sexual relationships. We must trust our partner enough to express our feelings about sexual needs, desires, and dislikes, and we must be able to hear the same from our partner without feeling judgmental or defensive (Regan 2003).

Other Problems in Communication

Studies suggest that poor communication skills precede the onset of marital problems (Gottman 1994; Markman 1981; Markman et al. 1987). Even family violence has been seen by some as the consequence of deficiencies in the ability to communicate (Burleson and Denton 1997). In this section, we consider some additional sources of communication difficulties and suggest ways to develop better communication.

Topic-Related Difficulty

Some communication problems are topic dependent more than individual or relationship based. That is, some topics are more difficult for couples to talk about. As Keith Sanford (2003, 98) states, "It would seem easier to resolve a disagreement about what to do on a Friday night than a disagreement about whether one spouse is having an affair." If some topics are more difficult to discuss than others, couples are likely to display poorer communication when discussing those topics.

In an attempt to determine the difficulty of different topics, Sanford gave a sample of 12 licensed PhD psychologists a list of topics and asked them to provide their best guess as to how difficult each topic is for couples to discuss and resolve (from 1 = extremely easy to 5 = extremely difficult). The list consisted of 24 topics, generated from a sample of 37 couples who were asked to identify two unresolved issues in their relationships. The 10 topics to which the psychologists assigned the highest "difficulty scores" are listed in Table 7.1.

Other familiar relationship trouble spots and their assigned ratings include child-rearing issues (3.42), finances (3.42), lack of listening (3.08), household tasks (2.33), and not showing sufficient appreciation (2.25). Although the scores demonstrate differences in the degree of sensitivity of different marital issues, these differences do not, themselves, appear to determine how couples communicate about them (Sanford 2003).

Barriers to Effective Communication

We can learn to communicate, but it is not always easy. Traditional male gender roles, for example, work against the idea of expressing feelings. Traditional masculinity calls for men to be strong and silent, to ride off into the sunset alone. If men talk, they talk about things—cars, politics, sports, work, or money—but not about feelings. In addition, both men and women may have personal reasons for not expressing their feelings. They may have strong feelings of inadequacy: "If you really knew what I was like, you wouldn't like

Table 7.1 Ten Topics That Are Most Difficult for Couples to Discuss

Topic	Difficulty Score*
Relationship doubts (possibility of divorce)	4.58
Disrespectful behavior (lying, rudeness)	4.50
Extramarital intimacy boundary issues (use of pornography, jealousy)	4.42
Excessive or inappropriate display of anger (yelling, attacking)	4.25
Sexual interaction	4.17
Lack of communication (refusal to talk)	4.00
In-laws and extended family	3.83
Confusing, erratic, emotional behavior	3.75
Criticism	3.58
Poor communication skills (being unclear or hard to understand)	3.46

*1 = extremely easy; 5 = extremely difficult.

How partners express and handle conflict verbally, as well as nonverbally, says much about the direction in which the relationship is heading.

me." They may feel ashamed of or guilty about their feelings: "Sometimes I feel attracted to other people, and it makes me feel guilty because I should only be attracted to you." They may feel vulnerable: "If I told you my real feelings, you might hurt me." They may be frightened of their feelings: "If I expressed my anger, it would destroy you." Finally, people may not communicate because they are fearful that their feelings and desires will create conflict: "If I told you how I felt, you would get angry."

Obstacles to Self-Awareness

Before we can communicate with others, we must first know how we feel. Although feelings are valuable guides for actions, we often place obstacles in the way of expressing them. First, we suppress "unacceptable" feelings, especially feelings such as anger, hurt, and jealousy. After a while, we may not even consciously experience them. Second, we deny our feelings. If we are feeling hurt and our partner looks at our pained expression and asks us what we're feeling, we may reply, "Nothing." We may actually feel nothing because we have anesthetized our feelings. Third, we project our feelings. Instead of recognizing that we are jealous, we may accuse our partner of being jealous; instead of feeling hurt, we may say that our partner is hurt.

Becoming aware of ourselves requires us to become aware of our feelings. Perhaps the first step toward this self-awareness is realizing that feelings are simply emotional states—they are neither good nor bad in themselves. As feelings, however, they need to be felt, whether they are warm or cold, pleasurable or painful. They do not necessarily need to be acted on or expressed. It is the acting out that holds the potential for problems or hurt.

Problems in Self-Disclosure

Self-disclosure creates the environment for mutual understanding (Derlega et al. 1993). We live much of our lives playing roles—as student and worker, husband or wife, or son or daughter. We live and act these roles conventionally. They do not necessarily reflect our deepest selves. If we pretend that we are only these roles and ignore our deepest selves, we have taken the path toward loneliness and isolation. We may reach a point at which we no longer know who we are. In the process of revealing ourselves to others, we discover who we are. In the process of our sharing, others share themselves with us. Self-disclosure is reciprocal.

How Much Openness?

Can too much openness and honesty be harmful to a relationship? How much should intimates reveal to each other? Some studies suggest that less marital satisfaction results if partners have too little *or too much* disclosure; a happy medium offers security, stability, and safety. But a review of studies on the relationship between communication and marital satisfaction finds that the greater the self-disclosure, the greater the marital satisfaction, provided that the couple is highly committed to the relationship and willing to take the risks of high levels of intimacy. High self-disclosure can be a highly charged undertaking.

Research by Brant Burleson and Wayne Denton (1997) suggests that the relationship between communication skill and marital success and satisfaction is "quite complex." In a study of 60 couples, the researchers explored the importance of four communication skills in determining marital satisfaction:

- *Communication effectiveness.* Producing messages that have their intended effect
- *Perceptual accuracy.* Correctly understanding the intentions underlying a message
- *Predictive accuracy.* Accurately anticipating the effect of the message on another
- *Interpersonal cognitive complexity.* The capacity to process social information

A pivotal aspect of effective communication, self-disclosure is reciprocal.

Prior research had indicated that each of the preceding skills was important in differentiating satisfied from dissatisfied couples or nondistressed from distressed couples. On the basis of their research, Burleson and Denton suggest that *communication skill* may not adequately explain levels of distress or dissatisfaction. The intentions and feelings being communicated were more important factors separating distressed from nondistressed couples. Spouses in distressed couples had "more negative intentions" toward each other. "The negative communication behaviors frequently observed in distressed spouses may result more from ill will than poor skill" (Burleson and Denton 1997, 897). Burleson and Denton also observe that good communication skills can worsen marital relationships when spouses have "negative intentions toward one another" (900).

"Can I Trust You?"

When we talk about intimate relationships, among the words that most often pop up are *love* and *trust*. As shown in the discussion of love in Chapter 5, trust is an important part of love. But what is trust? **Trust** is the belief in the reliability and integrity of a person.

When a person says, "Trust me," he or she is asking for something that does not easily occur. For trust to develop, three conditions must exist (Book et al. 1980). First, a relationship has to exist and have the likelihood of continuing. We generally do not trust strangers or people we have just met, especially with information that makes us vulnerable, such as our sexual anxieties. We trust people with whom we have a significant relationship.

Second, we must believe that we are able to predict how the person will behave. If we are married or in a committed relationship, we trust that our partner will not do something that will hurt us, such as having an affair. If we discover that our partner is involved in an affair, we often speak of our trust being violated or destroyed. If trust is destroyed in this case, it is because the predictability of sexual exclusiveness is no longer there.

Third, the person must have other acceptable options available to him or her. If we were marooned on a desert island alone with our partner, he or she would have no choice but to be sexually monogamous. But if a third person who was sexually attractive to our partner swam ashore a year later, then our partner would have an alternative. Our partner would then have a choice of being sexually exclusive or nonexclusive; his or her behavior would then be evidence of trustworthiness—or the lack of it.

> **Matter of Fact** The happiest couples are those who balance autonomy with intimacy and negotiate personal and couple boundaries through supportive communication (Scarf 1995).

Trust is critical to communication in close relationships for two reasons (Book et al. 1980). First, the degree to which you trust a person influences the way you are likely to interpret ambiguous or unexpected messages. If your partner says that he or she wants to study alone tonight, you are likely to take the statement at face value if you have a high trust level. But if you have a low trust level, you may believe your partner is going to meet someone else while you are studying in the library. Second, the degree to which we trust someone influences the extent of our self-disclosure. Revealing our inner selves—which is vital to closeness—makes a person vulnerable and thus requires trust. A person will not self-disclose if he or she believes that the information may be misused—for example, by a partner who resorts to mocking behavior or revealing a secret.

Trust in personal relationships has both a behavioral and a motivational component (Book et al. 1980). The behavioral component refers to the probability that a person will act in a trustworthy manner. The motivational component refers to the reasons a person engages in trustworthy actions. Whereas the behavioral element is important in all types of relationships, the motivational element is important in close relationships. One has to be trustworthy for the "right" reasons. You may not care why your mechanic is trustworthy as long as he or she charges you fairly for repairing your engine, but you do care why your partner is trustworthy. For example, you want your partner to be sexually exclusive to you because he or she loves you or is attracted to you. Being faithful because of duty or because your partner can't find anyone better is the wrong motivation. Disagreements about the motivational bases for trust are often a source of conflict. "I want you because you love me, not because you need me" or "You don't really love me; you're just saying that because you want sex" are typical examples of conflict about motivation.

The Importance of Feedback

Self-disclosure is reciprocal. If we self-disclose, we expect our partner to self-disclose as well. As we self-disclose, we build trust; as we withhold self-disclosure, we erode trust. To withhold ourselves is to imply that we don't trust the other person, and if we don't, he or she will not trust us.

A critical element in communication is **feedback**, the ongoing process in which participants and their messages create a given result and are subsequently modified by the result (see Figure 7.1). If someone self-discloses to us, we need to respond to his or her self-disclosure. The purpose of feedback is to provide constructive information to increase self-awareness of the consequences of our behaviors toward each other.

If your partner discloses to you his or her doubts about your relationship, for example, you can respond in a number of ways:

- You can remain silent. Silence, however, is generally a negative response, perhaps as powerful as saying outright that you do not want your partner to self-disclose this type of information.
- You can respond angrily, which may convey the message to your partner that self-disclosing will lead to arguments rather than understanding and possible change.
- You can remain indifferent, responding neither negatively nor positively to your partner's self-disclosure.

Figure 7.1 Communication Loop

In successful communication, feedback between the sender and the receiver ensures that both understand (or are trying to understand) what is being communicated. For communication to be clear, the message and the intent behind the message must be congruent. Nonverbal and verbal components must also support the intended message. Verbal aspects of communication include not only language and word choice but also characteristics such as tone, volume, pitch, rate, and periods of silence.

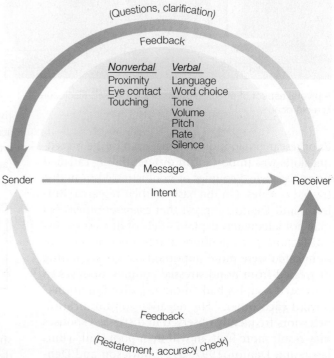

- You can acknowledge your partner's feelings as being valid (rather than right or wrong) and disclose how you feel in response to his or her statement. This acknowledgment and response is constructive feedback. It may or may not remove your partner's doubts, but it is at least constructive in that it opens the possibility for change, whereas silence, anger, and indifference do not.

Some guidelines, developed by David Johnston for the Minnesota Peer Program, may help you engage in dialogue and feedback with your partner:

1. *Focus on "I" statements.* An "I" statement is a statement about your feelings: "I feel annoyed when you leave your dirty dishes on the living room floor." "You" statements tell another person how he or she is, feels, or thinks: "You are so irresponsible. You're always leaving your dirty dishes on the living room

floor." "You" statements are often blaming or accusatory. Because "I" messages don't carry blame, the recipient is less likely to be defensive or resentful.

2. *Focus on behavior rather than the person.* If you focus on a person's behavior rather than on the person, you are more likely to secure change. A person can change behaviors but not himself or herself. If you want your partner to wash his or her dirty dishes, say, "I would like you to wash your dirty dishes; it bothers me when I see them gathering mold on the living room floor." This statement focuses on behavior that can be changed. If you say, "You are such a slob; you never clean up after yourself," then you are attacking the person. He or she is likely to respond defensively: "I am not a slob. Talk about slobs, how about when you left your clothes lying in the bathroom for a week?"

3. *Focus on observations rather than inferences or judgments.* Focus your feedback on what you actually observe rather than on what you think the behavior means. "There is a towering pile of your dishes in the living room" is an observation. "You don't really care about how I feel because you are always leaving your dirty dishes around the house" is an inference that a partner's dirty dishes indicate a lack of regard. The inference moves the discussion from the dishes to the partner's caring. The question "What kind of person would leave dirty dishes for me to clean up?" implies a judgment: only a morally depraved person would leave dirty dishes around.

4. *Focus on observations based on a continuum.* Behaviors fall on a continuum. Your partner doesn't *always* do a particular thing. When you say that he or she does something sometimes or even most of the time, you are measuring behavior. If you say that your partner always does something, you are distorting reality. For example, there were probably times (however rare) when your partner picked up the dirty dishes. "Last week I picked up your dirty dishes three times" is a measured statement. "I always pick up your dirty dishes" is an exaggeration that will probably provoke a hostile response.

5. *Focus on sharing ideas or offering alternatives rather than giving advice.* No one likes being told what to do. Unsolicited advice often produces anger or resentment because advice implies that you know more about what a person needs to do than the other person does. Advice implies a lack of freedom or respect. By sharing ideas and offering alternatives, however, you give the other person the freedom to decide on the basis of his or her own perceptions and goals. "You need to put away your dishes immediately after you are done with them" is advice. To offer alternatives instead, you might say, "Having to walk around your dirty dishes bothers me. Maybe you could put them away after you finish eating, clean them up before I get home, or eat in the kitchen. What do you think?"

6. *Focus on the amount that the recipient can process.* Don't overload your partner with your response. Your partner's disclosure may touch deep, pent-up feelings in you, but he or she may not be able to comprehend all that you say. If you respond to your partner's revelation of doubts by listing all doubts you have ever experienced about yourself, your relationship, and relationships in general, you may overwhelm your partner.

7. *Focus on responding at an appropriate time and place.* Choose a time when you are not likely to be interrupted. Turn the television off and don't answer your phone. In addition, choose a time that is relatively stress free. Talking about something of great importance just before an exam or a business meeting is likely to sabotage any attempt at communication. Finally, choose a place that will provide privacy; don't start an important conversation if you are worried about people overhearing or interrupting you. A dorm cafeteria at lunchtime, a kitchen teeming with kids, or a car full of friends are all inappropriate places.

Power, Conflict, and Intimacy

Although we may find it unusual to think about family life in these terms, day-to-day family life is highly politicized. By that we mean that the politics of family life—who has more power, who makes the decisions, and who does what—are complex and can be a source of conflict between spouses or intimate partners. Like other groups, families possess structures of power. As used here, **power** is the ability or potential ability to influence another person or group, to get people to do what you want them to do, whether they want to or not. Often, we are unaware of the power aspects of our intimate relationships. We may even deny the existence of power differences because we want to believe that intimate relationships are based on love alone. Furthermore, the exercise of power is often subtle.

When we think of power, we tend to think of coercion or force; as we show here, however, power in relationships between spouses or intimate partners takes many forms and is often experienced as neither coercion nor force. A final reason we are not always aware of power is that power is not constantly exercised. It comes into play only when an issue is important to both people and they have conflicting goals.

As a concept, power in marital and other relationships has been said to consist of power bases, processes, and outcomes. *Power bases* are the economic and personal assets (such as income, economic independence, commitment, and both physical and psychological aggression) that make up the source of one partner's control over the other. *Power processes* are the "interactional techniques" or methods that partners or spouses use to try to gain control over the relationship, the partner, or both, such as persuasion, problem solving, or demandingness. *Power outcomes* can be observed in such things as who has the final say and determines—or potentially could determine and control—the outcome of attempted decision making (Byrne, Carr, and Clark 2004; Sagrestano et al. 1999).

Power and Intimacy

The problem with power imbalances and the blatant use of power is the negative effects they have on intimacy. If partners are not equal, self-disclosure may be inhibited, especially if the powerful person believes that his or her power will be lessened by sharing feelings (Glazer-Malbin 1975). Genuine intimacy appears to require equality in power relationships. Decision making in the happiest marriages seems to be based not on coercion or tit for tat but on caring, mutuality, and respect for each other. Women or men who feel vulnerable to their mates may withhold feelings or pretend to feel what they do not. Unequal power in marriage may encourage power politics. Each partner may struggle with the other to keep or gain power.

It is not easy to change unequal power relationships after they become embedded in the overall structure of a relationship, yet they can be changed. Talking, understanding, and negotiating are the best approaches. Still, in attempting changes, a person may risk estrangement or the breakup of a relationship. He or she must weigh the possible gains against the possible losses in deciding whether change is worth the risk.

Sources of Marital Power

Traditionally, husbands have held authority over their wives. In Christianity, the subordination of wives to their husbands has its basis in the New Testament. Paul (Colossians 3:18–19) states, "Wives, submit yourselves unto your husbands, as unto the Lord." Such teachings reflected the dominant themes of ancient Greece and Rome. Western society continued to support wifely subordination to husbands. English common law stated, "The husband and wife are as one and that one is the husband." A woman assumed her husband's identity, taking his last name on marriage and living in his house.

The U.S. courts formally institutionalized these power relationships. The law, for example, supported the traditional division of labor in many states by making the husband legally responsible for supporting the family and the wife legally responsible for maintaining the house and rearing the children. She was legally required to follow her husband if he moved; if she did not, she was considered to have deserted him. But if she moved and her husband refused to move with her, she was also considered to have deserted him (Leonard and Elias 1990).

Legal and social support for the husband's control of the family declined through most of the twentieth century, especially the latter decades. Still, even into the 1970s, judicial discourse reflected an assumption of a "unitary spousal identity" wherein the husband and wife were one, represented by the husband. Ultimately, this was replaced by a more egalitarian model in which marriage was a partnership between equals, each of whom retained an independent legal existence, enjoyed the same rights, and held mutual responsibilities (Mason, Fine, and Carnochan, 2001). Especially through employment and wage earning, wives have gained more power in the family, increasing their influence in deciding such matters as family size and how money is spent.

Sociologist Jessie Bernard (1982) drew an important distinction between authority and power in marriage. Authority is based in law, but power can be derived from personality. A strong, dominant woman may, in her relationship, be as likely to exercise power over a more passive man as vice versa simply through the force of personality and temperament.

Among dating couples, power imbalances are common, whether they are measured by who makes decisions or who is perceived to be more powerful. Such imbalances tend to favor males over females.

An interesting gender pattern finds men perceiving themselves to be more powerful in decision making, whereas women are more likely to characterize decision making as equal (Sprecher and Felmlee 1997).

The relationship among gender, power, and violence is complex. Although some research suggests that men's violence is an expression of men's power over their wives (and of women's powerlessness), research also asserts that violence may be used by men who themselves feel powerless. Framed in this way, violence can be a method through which men who lack power or have a need for power attempt to control their wives. Even the threat of violence can be an assertion of power because it may intimidate women into complying with men's wishes even against their own (Kimmel 2008; Levitt, Swanger, and Butler 2008; Sagrestano et al. 1999).

If we want to see how power works in marriage, we need to look beneath gender stereotypes and avoid overgeneralizations. Women have considerable power in marriage, although they often feel that they have less than they actually do. They may fail to recognize the extent of their power; because cultural norms traditionally put power in the hands of their husbands, women may look at norms rather than at their own behavior, failing to recognize the degree to which they wield power. She may feel that her husband holds the power in the relationship because she believes that he is supposed to be dominant. Similarly, women who may believe their relationships to be egalitarian nevertheless may exercise control over domains of day-to-day life. For example, linguist Alexandra Johnston illustrated the process of **gatekeeping** in her analysis of "manager–helper interactions" between a couple she calls Kathy and Sam in their distribution of caregiving responsibility for their daughter Kira. Despite their commitment to and endorsement of shared parenting, their roles are not as equal as they believe or would like. Although they spend equal amounts of time and participate in the same caregiving tasks, Kathy is the primary decision maker for many of the issues concerning the care of two-year-old Kira (Johnston 2007). Finally, husbands may believe that they have more power in a relationship than they actually do because they see only traditional norms and expectations.

Power is not a simple phenomenon. Researchers generally agree that family power is a dynamic, multidimensional process (Szinovacz 1987). Generally, no single individual is always the most powerful person in every aspect of the family. Nor is power always based on gender, age, or relationship. Power often shifts from person to person, depending on the issue.

Explanations of Marital Power

Relative love and need theory explains power in terms of the individual's involvement and needs in the relationship. Each partner brings certain resources, feelings, and needs to a relationship. Each may be seen as exchanging love, companionship, money, help, and status with the other. What each gives and receives, however, may not be equal. One partner may be gaining more from the relationship than the other. The person gaining the most from the relationship is the most dependent.

Love itself is a major power resource in a relationship. Those who love equally are likely to share power equally (Safilios-Rothschild 1976). Such couples are likely to make decisions according to referent, expert, and legitimate power.

Principle of Least Interest

Akin to relative love and need as a way of looking at power is the **principle of least interest**. Sociologist Willard Waller (Waller and Hill 1951) coined this term to describe the situation in which the partner with the least interest in continuing a relationship enjoys the most power in it.

Quarreling couples may unconsciously use the principle of least interest to their advantage. The less involved partner may threaten to leave as leverage in an argument: "All right, if you don't do it my way, I'm going." The threat may be extremely powerful in coercing a dependent partner. It may have little effect, however, if it comes from the dependent partner because he or she has too much to lose to be persuasive. Knowing this, the less involved partner can easily call the other's bluff.

In their study of 101 heterosexual dating couples, sociologists Susan Sprecher and Diane Felmlee found that Waller's principle of least interest described the power imbalances in their sample couples. They found that the partners who perceived themselves to be more emotionally involved and invested in the relationship also perceived themselves to have less power than their partners. This pattern held true for both women and men, but men were significantly more likely to perceive themselves as the less involved partner. Women's perceptions echoed men's, as they saw their male partners as less invested in their relationships than they perceived themselves to be (Sprecher and Felmlee 1997).

Resource Theory of Power

In 1960, sociologists Robert Blood and Donald Wolfe studied the marital decision-making patterns as revealed by their sample of 900 wives. Using "final say" in decision making as an indicator of relative power, Blood and Wolfe inquired about a variety of decisions (e.g., whether the wife should be employed, what type of car to buy, and where to live) and who "ultimately" decided what couples should do. They noted that men tended to have more of such decision-making power and attributed this to their being the sole or larger source of the financial resources on which couples depended. They further observed that as wives' share of resources increased, so did their roles in decision making (Blood and Wolfe 1960).

This **resource theory of power** has been met with both criticism and some empirical support. By focusing so narrowly on resources, the theory overlooks other sources of gendered power. Specifically, it fails to explain the power that many men continue to enjoy when they are outearned by their wives or why a woman's power doesn't automatically increase when she earns more than her husband. In fact, sociologist Karen Pyke contended that when husbands perceive their wives' employment and earnings as threats to their status and identities rather than as gifts, wives derive less power from their earnings and employment (Pyke 1994). It appears that marriages in which couples are relatively equal in their earnings are the most egalitarian couples (Tichenor 2005a). As sociologist Veronica Tichenor asserted, "Money, then, is still linked to power, but only for husbands" (Tichenor 2005a, 202). Gender has influence on the balance (or imbalance) of power that is somewhat immune to the influence of spouses' earnings (Tichenor 2005b).

The resource theory has also been criticized for equating power with decision making and for ignoring that having power may sometimes mean that one is freed from having to make decisions. In still other circumstances, one may exercise power by influencing how others make decisions, by forcing them to consider the possible consequences of making a decision about which one disapproves. This may be best understood using Steven Lukes's three-dimensional view of power. To Lukes, power may be *overt*, *latent*, or *hidden*. This is nicely depicted in the following example:

> If a husband and wife struggle over domestic labor, and the husband successfully resists the greater participation the wife seeks, he has exercised overt power. If his wife then accepts the situation and avoids raising the issue again out of fear of renewed conflict, he has exercised latent power. But . . . if this issue is never raised . . . because the wife accepts it as her duty to bear the domestic labor burden even when she is employed outside the home . . . the husband has benefited from the hidden power in prevailing gendered practices and ideology. (Tichenor 2005a, 194)

Despite continuing disagreement among researchers about how to measure marital power, the most commonly used method still relies on determining who has "final say" in decision making (Amato et al. 2007).

Rethinking Family Power: Feminist Contributions

Even though women have considerable power in marriages and families, it would be a serious mistake to overlook the inequalities between husbands and wives. As feminist scholars have pointed out, major aspects of contemporary marriage point to important areas in which women are clearly subordinate to men: examples are the continued female responsibility for housework and child rearing, inequities in sexual gratification (sex is often over when the male has his orgasm), the extent of violence against women, and the sexual exploitation of children.

Feminist scholars suggested several areas that required further consideration (Szinovacz 1987). First, they believed that too much emphasis had been placed on the marital relationship as the unit of analysis. Instead, they believed that researchers should explore the influence of society on power in marriage—specifically, the relationship between social structure and women's position in marriage. Researchers could examine, for example, the relationship of women's socioeconomic disadvantages, such as lower pay and fewer economic opportunities than men, to female power in marriage.

Second, these scholars argued that many of the decisions that researchers study are trivial or insignificant in measuring "real" family power. Researchers cannot conclude that marriages are becoming more egalitarian on the basis of joint decision making about such things as where a couple goes for vacation, whether to buy a new car or appliance, or which movie to see. The critical decisions that measure power are such issues as how housework is to be divided, who stays home with the children, and whose job or career takes precedence.

Some scholars suggested that we shift the focus from marital power to family power. Researcher Marion Kranichfeld (1987) called for a rethinking of power in a family context. Even in instances where women's *marital power* may not be equal to men's, a different picture of women in families may emerge if we examine power within the entire family structure, including power in relation to children. The family power literature has traditionally focused on marriage and marital decision making. Kranichfeld, however, felt that such a focus narrows our perception of women's power. Marriage is not the same thing as family, she argues, and in the wider family context, women often exert considerable power. Their power may not be the same as male power, which tends to be primarily economic, political, or religious. But if by *power* we mean having the ability to change the behavior of others intentionally, women have a good deal of power in their families.

Research on marital violence suggests that it is the level of absolute power that has violent consequence for couples. In relationships that are *either* male dominated *or* female dominated, we find the highest levels of violence. In relationships that are "power divided," there is less violence, and in egalitarian relationships, we see the lowest levels of violence (Sagrestano et al. 1999). Aside from violence, among heterosexual couples, egalitarian couples also report the highest levels of relationship satisfaction, whereas couples in which the female has more power than the male have been found to have the lowest levels of satisfaction for both partners (Sprecher and Felmlee 1997).

One difficulty with discussions of "egalitarian relationships" is the question of whether such relationships truly are equal. Feminist research has revealed that even among self-professed equal couples, power processes seem to favor men. Carmen Knudson-Martin and Anne Rankin Mahoney's (1998) study of equal couples—in which each spouse perceives the relationship to be characterized by mutual accommodation and attention and each spouse has the same ability to receive cooperation from the other in meeting needs or wants—is a case in point. Although couples described their relationships as equal and their roles as "non–gender specific," men wielded more power

than women. Wives made more concessions to fit their daily lives around their husbands' schedules than husbands did to fit their lives around the schedules of their wives. Women were also more likely than their husbands to report worrying about upsetting or offending their spouses, to do what their spouses wanted, and to attend to their spouses' needs (Fox and Murry 2000). It appears as if characterizing an unequal marriage as equal allows a couple to ignore real if covert power differences that might otherwise threaten their relationships (Fox and Murry 2000).

Intimacy and Conflict

Conflict between people who love each other may initially seem to be a mystery. The simultaneous coexistence of conflict and love has puzzled human beings for centuries. An ancient Sanskrit poem reflected this dichotomy:

In the old days we both agreed
That I was you and you were me.
But now what has happened
That makes you, you
And me, me?

We expect love to unify us, but often it doesn't. Two people don't really become one when they love each other, although at first they may have this feeling. In reality, they retain their individual identities, needs, wants, and pasts while loving each other—and it is a paradox that the more intimate two people become, the more likely they may be to experience conflict. But it is not conflict itself that is dangerous to intimate relationships; it is the manner in which the conflict is handled. Conflict, itself, is natural.

If this is understood, the meaning of conflict changes, and it will not necessarily represent a crisis in the relationship. David and Vera Mace, prominent marriage counselors, observed that on the day of marriage, people have three kinds of raw material with which to work. First, there are things they have in common—the things they both like. Second, there are the ways in which they are different, but the differences are complementary. Third, unfortunately, there are the differences between them that are not complementary and that cause them to meet head on with a big bang. In every relationship between two people, there are a great many of those kinds of differences. So when they move closer to each other, those differences become disagreements (Mace and Mace 1979).

Experiencing Conflict

The presence of conflict within a marriage or family doesn't automatically suggest trouble or indicate that love is going or gone; it may mean quite the opposite. In the healthiest of relationships, it is common and normal for couples to have disagreements or conflicts. The important factor is not *that* they have differences or even how often or what in particular they fight about but *how* constructively or harmfully they resolve their differences. By using occasions of conflict to implement mutually acceptable behavior changes or to decide that the differences between them are acceptable, couple relationships may grow and solidify as a product of their differences. Couples who resolve conflict with mutual satisfaction and who find ways to adapt to areas of conflict tend to be more satisfied with their relationships overall and are less likely to divorce.

It seems that in couples where one or both partners keep quiet and don't vent their frustrations or express their feelings, problems may result for either the individual partners or the longer-term stability of the relationship. One study of almost 4,000 women and men found that during spousal arguments, nearly a third of men and a quarter of women said that they usually kept their feelings inside during an argument (Parker-Pope 2007). Such behavior, called **self-silencing**, had particularly harsh effects on women; women who kept their feelings to themselves during marital arguments were four times as likely to die during the 10-year span of the research compared to women who said that they always expressed their feelings. Men's health was not measurably affected by whether they did or didn't express themselves during a fight.

Conflict is an inevitable and normal part of being in a relationship. Rather than withdrawing from and avoiding conflict, we should use it as a way to build, strengthen, and deepen our relationships.

leading to the possible end of the relationship. Nonbasic conflicts are more common and less consequential; couples learn to live with them.

Basic Conflicts

Basic conflicts revolve around carrying out marital roles and the functions of marriage and the family, such as providing companionship, working, and rearing children. It is assumed, for example, that a husband and a wife will have sexual relations with each other. But if one partner converts to a religious sect that forbids sexual interaction, a basic conflict is likely to occur because the other spouse considers sexual interaction part of the marital premise. No room for compromise exists in such a matter. If one partner cannot convince the other to change his or her belief, the conflict is likely to destroy the relationship. Similarly, despite recent changes in family roles, it is still expected that the husband will work to provide for the family. If he decides to quit work and not function as a provider, he is challenging a basic assumption of marriage. His partner is likely to feel that his behavior is unfair. Conflict ensues. If he does not return to work, his wife is likely to leave him.

Nonbasic Conflicts

Nonbasic conflicts do not strike at the heart of a relationship. The husband wants to change jobs and move to a different city, but the wife may not want to move. This may be a major conflict, but it is not a basic one. The husband is not unilaterally rejecting his role as a provider. If a couple disagrees about the frequency of

> **Matter of Fact** When the communication patterns of newly married African Americans and Caucasians were examined, couples who believed in avoiding marital conflict were less happy two years later than those who confronted their problems (Crohan 1996).

Basic versus Nonbasic Conflicts

Relationships experience two types of conflict—basic and nonbasic—that have different effects on relationship quality and stability. Basic conflicts challenge the fundamental assumptions or rules of a relationship,

sex, the conflict is serious but not basic because both agree on the desirability of sex in the relationship. In both cases, resolution is possible.

Dealing with Conflict

If we handle conflicts in a healthy way, they can help strengthen our relationships. But conflicts can go on and on, consuming the heart of a relationship, turning love and affection into bitterness and hatred. In the following section, we look at ways of resolving conflict in constructive rather than destructive ways. In this manner, we can use conflict as a way of building and deepening our relationships.

Marital Conflict

Comparing sociological research on marital quality undertaken in 1980 and again in 2000 enabled Paul Amato and colleagues (2007) to identify many aspects of marriage that have changed and what factors precipitated such changes. One of the key areas they looked at was *marital quality*, a big part of which pertained to the extent of conflict between spouses. Some key findings regarding marital conflict are as follows:

- Between 1980 and 2000, marital conflict decreased.
- Gender differences surfaced in both 1980 and 2000, with wives reporting more problems, including more conflict, than husbands.
- Individuals who cohabited before marriage reported more conflict with their spouses than those who had not first cohabited, although the difference was a relatively modest one. Similarly, the overall increase in cohabitation between 1980 and 2000 was associated with an increase in the average level of marital conflict.
- In both 1980 and 2000, **marital heterogamy** (marriage of two people from different backgrounds and/or with different demographic characteristics) was associated with more marital conflict than couples who were **homogamous**. In 2000, couples who differed significantly in educational attainment, religious affiliation, or marital history (e.g., where one spouse is marrying for the first time to a partner who has been married at least once before) reported higher levels of conflict and more marital problems. Racial and age heterogamy made little to no difference in reported levels of conflict in 2000.

- Wives' employment had effects on marital conflict when (1) they were employed because they didn't like spending time at home or (2) their families needed the extra money. Wives who were satisfied with their jobs reported lower levels of conflict, whereas wives who were dissatisfied with their jobs reported higher levels of conflict.
- Couples where one or both spouses felt that their division of responsibilities and/or of child care was unfair reported higher levels of conflict (along with lower levels of happiness).
- Couples who maintained more egalitarian relationships, sharing decision making equally, expressed less marital conflict.
- People with no religious affiliation reported higher levels of marital conflict. Couples who attended religious services together reported lower levels of conflict.

On the subject of religion and marital conflict, when there are differences in a couple's religious views, conflict can ensue over marital and familial matters. Research suggests that discrepancies about the Bible are associated with more conflicts about housework and money. In the specific instance of husbands having more conservative Christian beliefs than their wives, disputes about child rearing are more likely. When wives have more conservative Christian beliefs than husbands, conflicts about in-laws or how couples should spend their time occur more often (Mahoney 2005).

Comparing Conflict in Marriage and Cohabitation

Most research has revealed no significant differences between cohabiting and married couples in their frequency of conflict over such relationship aspects as time spent together, in-laws, money, sex, decisions about childbearing, or the division of responsibilities. Nor have most studies found cohabitants to differ significantly in the likelihood of heated arguments or the level of open disagreements. Based on a recent study of more than 1,200 people who were either in dating relationships (220), cohabiting (231), or married (801), psychologists Annie Hseuh, Kristin Rahbar Morrison and Brian Doss established that compared to married individuals, those in cohabiting relationships were more likely to report problems associated with relationship conflict, including disagreement about values and goals for the future and an inability

Exploring Diversity: Gender and Marital Conflict among Korean Immigrants

"Did you bring me to this country for exploitation?"

Such is the plaintive appeal of 41-year-old Yong Ja Kim, a Korean immigrant, to her husband, Chun Ho Kim. What is it she is objecting to? In what way does she feel exploited? Sociologist Pyong Gap Min (2001) researched the consequences of immigration for marital relations among Korean immigrant couples. Existing research indicated that marital conflicts had emerged among Korean immigrants to the United States because of women's increased role in the economic support of families without concurrent changes in their husbands' gender attitudes or marital behavior. Min sought to delve more deeply into such conflicts.

Among Min's interviewees were Yong Ja Kim and Chun Ho Kim, husband and wife, who work together at their retail store six days a week from 9:30 A.M. to 6:00 P.M. On returning home, he watches Korean television programs and reads a Korean daily newspaper while she prepares dinner. Defensively, he retorts,

> It makes no sense for her to accuse me of not helping her at home at all. In addition to house maintenance, I took care of garbage disposal more often than she and helped her with grocery shopping very often. I did neither of the chores in Korea.

To his wife, however, the comparison is not between what he did in Korea and what he does in the United States but between what *he does* and what *she does*:

> I work in the store as many hours as you do, and I play an even more important role in our business than you. But you don't help me at home. It's never fair. My friends in Korea work full-time at home, but don't have to work outside. Here, I work too much both inside and outside the home.

Culturally, there are noteworthy differences between the traditional status of husbands in Korea and the situations of most immigrant Korean men in the United States. Traditionally, Korean husbands were breadwinners and patriarchal heads of their families. Wives and children were expected to obey their husbands and fathers. Women were further expected to bear children and cater to their husbands and in-laws. Although the traditional South Korean family system has been "modified," it remains a patriarchal system, justified by Confucian ideology. As they have immigrated to the United States, Korean women's involvement in paid employment has increased radically. In the process, traditional gender attitudes and male sense of self as patriarch and provider have been undermined.

Exacerbating the cultural transition are real economic adjustments. Min notes that with immigration to the United States, most Korean immigrant men encounter significant downward occupational mobility. This, in turn, results in further "status anxiety." They compensate by seeking ways to assert their authority in the household, only to find that their wives and children no longer grant them such status automatically. Min states that Mr. Kim "could not understand much and how fast his wife had changed her attitudes toward him since they had come to the United States. He did not remember her talking back to him in Korea."

Min summarizes his research findings by noting that for Korean immigrant couples, the gulf between their gender-role behavior and their traditional gender attitudes may be greater than for many other ethnic groups. If so and if such discrepancies are partly responsible for marital conflict, the situation for Korean immigrants may be harder than for other immigrant groups.

---

to resolve conflicts (Hseuh, Morrison, and Doss 2009). Cohabitants were also more likely than individuals in dating relationships to report problems with arguments and with conflict resolution.

Dealing with Anger

Differences can lead to anger, and anger transforms differences into fights, creating tension, division, distrust, and fear. Most people have learned to handle anger by either venting or suppressing it. As indicated

earlier, suppressing anger is unhealthy, especially for women. It also can be dangerous to the relationship because it is always there, simmering beneath the surface. It leads to resentment, that brooding, low-level hostility that poisons both the individual and the relationship.

Anger can be dealt with in a third way: when conflict escalates into violence. Especially in a culture that cloaks families in privacy, surrounds people with beliefs that legitimize violence, and gives them the sense that they have a right to influence what their loved ones do, escalating anger can result in assault, injury, and even death. Given the relative power of men over women and adults over children, threats against one person's supposed advantage may provoke especially harsh reactions. We look closely at the causes, context, and consequences of family violence in Chapter 12.

Finally and most constructively, anger can be recognized as a symptom of something that needs to be changed. If we see anger as a symptom, we realize that what is important is not venting or suppressing the anger but finding its source and eliminating it.

Not all conflict is overt. Some conflict can go undetected by one of the partners. As such, it will have minimal effect on him or her and is not likely to lead to anger. In addition, not all "conflicts" (i.e., of interest, goals, wishes, expectations, and so on) become *conflicts*. Spouses and partners can approach their differences in many ways short of overt conflict (Fincham and Beach 1999).

How Women and Men Handle Conflict

In keeping with observed gender differences in communication, research has identified differences in how men and women approach and manage conflict. As summarized by Rhonda Faulkner, Maureen Davey, and Adam Davey (2004), we can identify the following gender differences:

- As we saw earlier in discussing the demand–withdraw pattern, women are more likely than men are to initiate discussions of contested relationship issues. Where men have been found to be more likely to withdraw from negative marital interactions, women are more likely to pursue conversation or conflict.
- Typically, women are more aware of the emotional quality of and the events that occur in the relationship.
- In the course and processes of conflict management and resolution, men take on instrumental roles, and women take on expressive roles. Men approach conflict resolution from a task-oriented stance, as in "problem solving"; women are more emotionally expressive as they pursue intimacy.

We need to bear in mind that the research designs used to study patterns of interaction in conflict management may have exaggerated the gender connection by commonly asking couples to engage in discussion of topics of greater importance to females than to males (e.g., intimacy and child-rearing practices). When both partners in heterosexual couples were required to discuss an area in which they would like their partner to make changes, gender patterns were more varied. Significantly more woman demand–man withdraw behavior occurred when couples addressed the woman's top issue, but there was also more man demand—woman withdraw behavior during discussions of issues most important to the man. Thus, it is crucial to avoid overgeneralizing gender patterns in partners' conflict styles; importance of the issue to each party also affects conflict behavior.

Conflict Resolution and Relationship Satisfaction

How couples manage conflict is one of the most important determinants of their satisfaction and the well-being of their relationships (Greeff and deBruyne 2000). Happy couples are not conflict free; instead, they tend to act in positive ways to resolve conflicts, such as changing behaviors (putting the cap on the toothpaste rather than denying responsibility) and presenting reasonable alternatives (purchasing toothpaste in a dispenser). Unhappy or distressed couples, in contrast, use more negative strategies in attempting to resolve conflicts ("If the cap off the toothpaste bothers you, then *you* put it on").

Thus, we can talk of "constructive" and "destructive" or "helpful and unhelpful" conflict management (duPlessis and Clarke, 2007; Greef and deBruyne 2000). Constructive conflict management is characterized by flexibility, a relationship rather than individual (self-interest) focus, an intention to learn from their differences, and cooperation. Additional characteristics of "helpful" conflict management include compromise, negotiation, turn taking, calm discussions, listening carefully, and trying to understand the other's perspective.

Destructive conflict management consists of the following:

- Escalating spirals of manipulation, threat, and coercion
- Avoidance
- Retaliation
- Inflexibility
- A competitive pattern of dominance and subordination
- Demeaning or insulting verbal and nonverbal communication

Additional characteristics of "unhelpful" conflict management include confrontation, complaining, criticizing, defensiveness, and displaying contempt.

A study of happily and unhappily married couples found distinctive communication traits as these couples tried to resolve their conflicts (Ting-Toomey 1983). The communication behaviors of happily married couples displayed the following traits:

- *Summarizing.* Each person summarized what the other said: "Let me see if I can repeat the different points you were making."
- *Paraphrasing.* Each put what the other said into his or her own words: "What you are saying is that you feel bad when I don't acknowledge your feelings."
- *Validating.* Each affirmed the other's feelings: "I can understand how you feel."
- *Clarifying.* Each asked for further information to make sure that he or she understood what the other was saying: "Can you explain what you mean a little bit more to make sure that I understand you?"

In contrast, "distressed" or unhappily married couples displayed the following reciprocal patterns:

- *Confrontation.* Both partners confronted each other: "You're wrong!" "Not me, buddy. It's you who's wrong!"
- *Confrontation and defensiveness.* One partner confronted and the other defended: "You're wrong!" "I only did what I was supposed to do."
- *Complaining and defensiveness.* One partner complained, and the other was defensive: "I work so hard each day to come home to this!" "This is the best I can do with no help."

Overall, distressed couples use more negative and fewer positive statements. They become "locked in" to conflict. Thus, a major task for such couples is to find an effective or adaptive way out (Fincham and Beach 1999).

One of the strongest predictors of marital unhappiness and of the possibility of eventual divorce is whether couples engage in **hostile conflict**. Hostile conflict is a pattern of negative interaction wherein couples engage in frequent heated arguments, call each other names and insult each other, display an unwillingness to listen to each other, and lack emotional involvement with each other (Gottman 1994; Topham, Larson, and Holman 2005). Once such patterns become the normative pattern in a relationship, they are difficult to change.

Research is mixed as to the effectiveness of humor in conflict management. More satisfied couples display higher levels of nonsarcastic humor and, during discussions of problems, share laughter. The type of humor appears to make a critical difference. *Affiliative humor*, where one says funny or witty things or tells jokes in an effort to reduce tension and enhance the relationship, may in fact lead to less distress and assist discussions of and facilitate resolution of problems. *Aggressive humor*, which is used to tease, ridicule, or disparage the other, may impede such efforts and lead to greater distress (Campbell, Martin, and Ward 2008).

What Determines How Couples Handle Conflict?

Many factors might affect how couples approach and attempt to manage the inevitable conflict that relationships contain. Among these, premarital variables, including carryover effects of upbringing, may be particularly influential. Glade Topham, Jeffrey Larson, and Thomas Holman (2005) suggest that such influence may be conscious or unconscious; may affect behaviors and patterns of interaction as well as attitudes, beliefs, and self-esteem; and may remain even in the absence of contact with the family of origin.

Family-of-origin factors can be explained by social learning theory or attachment theory. Learning theory suggests that by observing parents and how they interact with each other, we develop a **marital paradigm**: a set of images about how marriage ought to be done, "for better or worse" (Marks 1986). When, as children, we fail to experience a positive model of marriage, we may develop ineffective communication or conflict resolution skills. Attachment theory suggests that our attachment style influences the way conflict is expressed in relationships (Pistole 1989). Secure parent–child relationships lead us to be more self-confident and socially confident, more likely to view others as trustworthy and dependable, and more comfortable with and within relationships. Individuals who had insecure parent–child attachments are more demanding of support and attention, more dependent on others for self-validation, and more self-deprecating and emotionally hypersensitive (Topham et al. 2005).

In contrast to anxious or ambivalent and avoidant adults, secure adults are more satisfied in their relationships and use conflict strategies that focus on maintaining the relationship. Helping the relationship stay cohesive is more important than "winning" the battle. Secure adults are more likely to compromise than are anxious or ambivalent adults, and anxious or ambivalent adults are more likely than avoidant adults to give in to their partners' wishes, whether they agree with them or not.

Although either husbands' or wives' family-of-origin experiences *could* negatively affect marital quality and conflict management, the influences are not equivalent. Wives' family-of-origin experiences—including the quality of relationships with their mothers, the quality of parental discipline they received, and the overall quality of their family environments—are more important than husbands' experiences in predicting hostile marital conflict (Topham et al. 2005).

There are two "analytically independent" dimensions of behavior in conflict situations: assertiveness and cooperativeness (Greeff and deBruyne 2000; Thomas 1976). **Assertiveness** refers to attempts to satisfy our own concerns; *cooperativeness* speaks to attempts to satisfy concerns of others. With these two dimensions in mind, we can identify five conflict management styles:

- *Competing.* Behavior is assertive and uncooperative, associated with "forcing behavior and win–lose arguing." This style can lead to increased conflict as well as to either or both spouses feeling powerless and resentful (Greeff and deBruyne 2000).
- *Collaborating.* Behavior is assertive and cooperative; couples confront disagreements and engage in problem solving to uncover solutions. Collaborative conflict management may require relationships that are relatively equal in power and high in trust. Using this style then accentuates both the trust and the commitment that couples feel.
- *Compromising.* This is an intermediate position in terms of both assertiveness and cooperativeness. Couples seek "middle-ground" solutions.
- *Avoiding.* Behavior is unassertive and uncooperative, characterized by withdrawal and by refusing to take a position in disagreements.
- *Accommodating.* This style is unassertive and cooperative. One person attempts to soothe the other person and restore harmony.

Research has yielded inconsistent ("diverse") results about the relationship outcomes of each of these styles. Some studies favor one style—collaboration—over all others as the only style displayed by satisfied couples. There is research suggesting that avoidance is dysfunctional and antisocial, yet there is research that finds avoidance associated with satisfied, nondistressed

couples. Finally, although some research suggests that when husbands and wives agree on how to manage conflict they have happier marriages, other findings indicate that discrepancies in spouses' beliefs about conflict are not predictive of how satisfied they are (Greeff and deBruyne 2000).

Conflict Resolution across Relationship Types

All couple relationships experience conflict. Using self-report and partner-report data, Lawrence Kurdek (1994) explored how conflicts were handled by 75 gay, 51 lesbian, 108 married nonparent, and 99 married parent couples. Essentially, the differences across couple type were less impressive than were the similarities. The four types of couples did not significantly differ in their level of ineffective arguing, and there were no noteworthy differences in their styles of conflict resolution as measured by the Conflict Resolution Styles Inventory (CRSI). The CRSI includes four styles of conflict resolution: (1) *positive problem solving* (including negotiation and compromise), (2) *conflict engagement* (such as personal attacks), (3) *withdrawal* (refusing to further discuss an issue), and (4) *compliance* (such as giving in). Ratings were obtained from both partners about themselves and the other partner. There was little indication that the frequency with which conflict resolution styles were used varied across couple type. As Kurdek (1994) notes, there is similarity in relationship dynamics across couple types.

Common Conflict Areas: Sex, Money, and Housework

Even if, as the Russian writer Leo Tolstoy suggested, every unhappy family is unhappy in its own way, marital conflicts still tend to center on certain recurring issues, especially communication, children and parenting, sex, money, personality differences, how to spend leisure time, in-laws, infidelity, and housekeeping. In this section, we focus on three areas: sex, money, and housework. Then we discuss general ways of resolving conflicts.

Fighting about Sex

Fighting and sex can be intertwined in several ways (Strong and DeVault 1997). A couple can have a specific disagreement about sex that leads to a fight. One person wants to have sexual intercourse, and the other does not, so they fight. A couple can have an indirect fight about sex. The woman does not have an orgasm, and after intercourse, her partner rolls over and starts

Conflict is common. Living with other people introduces numerous points of potential disagreement. Not all disagreements are equally serious or carry equal risks for the health and future of the relationship. When researchers surveyed therapists, seeking to identify the frequency, difficulty in treating, and severity of the effect of 29 problems couples might face, they found the following problems identified as the most frequent problems couples bring to therapy: unrealistic expectations, power struggles, communication problems, sexual problems, and conflict management difficulties. Problems deemed most difficult to treat included lack of loving feelings, alcoholism, extrarelational affairs, and power struggles (Miller et al. 2003; Whisman, Dixon, and Johnson 1997).

Using a clinical sample of 160 couples married between one and 20 years, Richard Miller and colleagues (2003) sought to determine whether couples at different life cycle stages experience and seek help with different kinds of problems. Couples were asked to consider as problem areas: children, communication, housecleaning, gender-role issues, financial matters, sexual issues, spiritual matters, emotional intimacy, violence, commitment, values, parents-in-law, decision making, and commitment. Couples were asked to consider where each problem ranked in frequency from "very often a problem" (5) to "never a problem" (1). Because it was a clinical sample, couples were also asked to consider from nine choices the problem that most brought them to therapy, including as possibilities communication, violence, sexual issues, financial matters, emotional intimacy, separation or divorce concerns, extramarital affairs, commitment issues, or some other problem.

Problems with communication and financial matters were the most commonly reported. Also frequently mentioned were emotional intimacy, sexual issues, and decision making. Gender-role issues, values, violence, and spiritual issues were not common problems. These tendencies can be seen in Table 7.2, reflecting the percentage of spouses who listed a problem as either "very often a problem" or "often a problem."

As far as what problem area couples were most likely to identify as their "presenting problem," by far "communication problems" were most often mentioned by both males and females, *regardless of how long they were married*. Finally, as shown in Table 7.3, there were statistically significant gender differences for six problem areas.

According to Miller and colleagues, their findings indicate that problems experienced by couples are relatively stable as opposed to varying much over the life cycle. As to gender, they remind therapists that females generally perceive more problems than males within marital relationships. Somewhat consistent with the idea of "two marriages," males and females may indeed experience relationships and problems within those relationships differently (Storaasli and Markman 1990). Women's tendencies to report problem areas as more severe or frequent suggest "a complex picture of gender-related issues." Finally, regardless of how long a couple has been married, couples' therapists must be prepared to assess and treat problems dealing with communication, financial matters, sexual issues, decision-making skills, and emotional intimacy because such problem areas are consistent features of married life over which couples encounter difficulty.

It is worth pointing out that conflict is driven not only by "what" couples fight about but also by the wider social context in which relationships exist. Taking a broader view, we need to pay attention to the effects of negative life events, essentially nonmarital stressors, that may lead to more negative communication, poorer parenting, and lower satisfaction. Likewise, the amount of social support a couple enjoys outside the marriage may influence the direction and outcomes of conflict (Fincham and Beach 1999).

---

to snore. She lies in bed feeling angry and frustrated. In the morning, she begins to fight with her partner over his not doing his share of the housework. The housework issue obscures why she is angry. Sex can also be used as a scapegoat for nonsexual problems. A husband is angry that his wife calls him a lousy provider. He takes it out on her sexually by calling her a lousy lover. They fight about their lovemaking rather than about the issue of his provider role. A couple can fight about the wrong sexual issue. A woman may berate her partner for being too quick during sex, but what she is really frustrated about is that he is not

Table 7.2 Percentage Reporting Area Is Either "Very Often" or "Often" a Problem

Problem	Males			Females		
	<3 years*	3–10 years	>10 years	<3 years	3–10 years	>10 years
Communication	56.7%	63.8%	53.2%	62.9%	67.4%	66.6%
Financial matters	37.8	54.4	56.3	26.9	55.1	67.7
Decision making	27.0	34.4	25.0	34.2	42.7	48.4
Emotional intimacy	21.6[+]	50.3	21.9	42.8	52.8	45.2
Sexual issues	21.6	34.1	28.2	37.2	38.2	29.0
Parent-in-law	27.0	24.2	19.4	28.5	31.5	22.6
Leisure activities	18.9[+]	30.1	15.7	34.3	40.4	35.5
Dealing with children	18.2	22.8	28.1	26.9	35.6	29.1
Commitment	21.6	19.8	9.4	11.4	18.2	32.2
Housecleaning	13.5	25.6	18.8	17.1	28.1	29.0
Gender-role issues	10.8	13.5	0.0	14.3	16.8	9.7
Values	13.5	15.7	9.4	5.7	17.0	10.0
Violence	8.8	1.3	3.4	9.4	3.9	3.6
Spiritual matters	0.0	5.6	3.1	5.8	6.9	9.7

*Numbers represent duration of marriage.
[+]Duration of marriage group differences for that gender significant at $p < 0.05$.

Table 7.3 Frequency of Reporting Areas

Problem	Males	Females
Dealing with children*	2.71	2.98
Emotional intimacy*	3.15	3.45
Sexual issues[+]	2.90	3.08
Parents-in-law[+]	2.62	2.84
Communication[@]	3.70	4.00
Decision making[+]	3.05	3.27

Range: (1) "never a problem" to (5) "very often a problem"
*Difference significant at $p < 0.01$
[+]Difference significant at $p < 0.05$
[@]Difference significant at $p < 0.001$

interested in oral sex with her. She, however, feels ambivalent about oral sex ("Maybe I smell bad"), so she cannot confront her partner with the real issue. Finally, a fight can be a cover-up. If a man feels sexually inadequate and does not want to have sex as often as his male partner, he may pick a fight and make his partner so angry that the last thing he would want to do is to have sex with him.

In power struggles, sexuality can be used as a weapon, but this is generally a destructive tactic (Szinovacz 1987). A classic strategy for the weaker person in a relationship is to withhold something that the more powerful one wants. In male–female struggles, this is often sex. By withholding sex, a woman gains a certain degree of power. A few men also use sex in its most violent form: they rape (including date rape and marital rape) to overpower and subordinate women. In rape, aggressive motivations displace sexual ones.

It is hard to tell during a fight if there are deeper causes than the one about which a couple is fighting. Is a couple fighting because one wants to have sex now and the other doesn't? Or are there deeper reasons involving power, control, fear, or inadequacy? If they repeatedly fight about sexual issues without getting anywhere, the ostensible cause may not be the real one. If fighting does not clear the air and make intimacy possible again, they should look for other reasons for the fights. It may be useful for them to talk with each other about why the fights do not seem to accomplish anything. In addition, it would be helpful if they step back and look at the circumstances of the fight, what patterns occur, and how each feels before, during, and after a fight.

Sexual tensions and strains arise because of these other conflicts that happen to play themselves out in the physical relationship. With a more "positive, respectful, affirming process of conflict resolution," partners may deepen the respect and admiration they feel for each other, develop a greater level of trust and of self-esteem in their relationship, and grow more confident that the relationship can withstand and grow through future conflict. These can create positive feelings and comfort with each other that facilitate sexual desire (Metz and Epstein 2002). Although the conflicts being resolved need not be sexual, positive and constructive relationship conflict resolution may provide affirmation of the love and intimacy two people share, bring emotional relief, and even serve as a sexual stimulant (Metz and Epstein 2002). Thus, the intensity of pleasure supposedly accompanying "makeup sex" is another reminder of how conflict and its resolution can affect sex regardless of whether it is about sex.

Money Conflicts

An old Yiddish proverb addresses the problem of managing money quite well: "Husband and wife are the same flesh, but they have different purses." Money is a major source of marital conflict in families in the United States and abroad. People in intimate relationships differ about spending money probably as much as or more than any other single issue.

Couples disagree or fight over money for a number of reasons. One of the most important has to do with power. Earning wages has traditionally given men power in families. A woman's work in the home has not been rewarded by wages. As a result, full-time homemakers have been placed in the position of having to depend on their husbands for money. In such an arrangement, if there are disagreements, the woman is at a disadvantage. If she is deferred to, the old cliché "I make the money but she spends it" has a bitter ring to it. As women increased their participation in the workforce, however, power relations within families have shifted some. Studies indicate that women's influence in financial and other decisions increases if they are employed outside the home.

Another major source of monetary conflict is allocation of the family's income. Not only does this involve deciding who makes the decisions, but it also includes setting priorities. Is it more important to pay a past-due bill or to buy a new television set to replace the broken one? Is a dishwasher a necessity or a luxury? Should money be put aside for long-range goals, or should immediate needs be satisfied? Setting financial priorities plays on each person's values and temperament; it is affected by basic aspects of an individual's personality.

Dating relationships are a poor indicator of how a couple will deal with money matters in marriage. Dating has clearly defined rules about money: either the man pays, both pay separately, or they take turns paying. In dating situations, each partner is financially independent of the other. Money is not pooled, as it usually is in a committed partnership or marriage. Power issues do not necessarily enter spending decisions because each person has his or her own money. Differences can be smoothed out fairly easily. Both individuals are financially independent before marriage but financially interdependent after marriage. Even cohabitation may not be an accurate guide to how a couple would deal with money in marriage, as cohabitators generally do not pool all (or even part) of their income. It is the working out of financial interdependence in marriage that is often so difficult.

Why do we find it difficult to be financially interdependent and talk about money? There may be several reasons. First, we don't want to appear to be unromantic or selfish. If a couple is about to marry, a discussion of attitudes toward money may lead to disagreements, shattering the illusion of unity or selflessness. Second, gender roles make it difficult for women to express their feelings about money because women are traditionally supposed to defer to men in financial matters. Third, because men tend to make more money than women, women feel that their right to disagree about financial matters is limited. These feelings are especially prevalent if the woman is a homemaker and does not make a financial contribution, but they devalue her child care and housework contributions.

Housework and Conflict

The division of responsibility for housework can be one of the most significant issues couples face, especially dual-earner couples (Kluwer, Heesink, and Van de Vliert 1997). It can become a source of tension and conflict within marriage (Hochschild 1989). Part of this is an understandable consequence of the inequality in each spouse's contribution; most men do not do much housework. Whether or not they are employed outside the home and whether there are children in the home or not, wives bear the bulk of housework responsibility. A husband's lack of involvement can create resentment and affect the levels of both conflict and happiness in a marriage. Longitudinal research on married couples reveals that husbands whose wives perceived that the division of housework was unfair report higher levels of marital conflict over time (Faulkner, Davey, and Davey 2005). Similarly, in her acclaimed study of the division of housework among 50 dual-earner couples, Arlie Hochschild (1989) argued that men's level of sharing "the second shift" (i.e., unpaid domestic work and child care) influenced the levels of marital happiness couples enjoyed and their relative risk of divorce. This held true whether couples were traditional or egalitarian in their views of marriage.

In a study of 54 Dutch couples, Esther Kluwer, Jose Heesink, and Evert Van de Vliert (1996) found that conflict about household work was related to wives' dissatisfaction with how much they and their spouses were contributing in terms of time and tasks and their and their husbands' relative contributions and expenditures of time. They noted that 72% of the wives preferred to do less than they actually did; that is, when they spent more time on housework than they preferred to, they were dissatisfied. They also tended to be dissatisfied if they perceived their husbands spending less time than they preferred them to spend on housework. In the study, 52% of the wives wished their husbands would do more housework than they actually did (Kluwer et al. 1996).

How much each spouse contributes to the household is only the more observable aspect of the "politics of housework." In addition, couples must reach agreements about standards, schedules, and management of housework. Conflicts about standards are struggles over whose standards will predominate: who decides whether things are "clean enough"? Similarly, disputes about schedules reflect whose time is more valuable and which partner works around the other's sense of priorities. Finally, arguments about who bears responsibility for organizing, initiating, or overseeing housework tasks are also disputes about who will have to ask the other for help, carry more responsibility in his or her head, and risk refusal from an uncooperative partner.

Thus, housework conflicts have both practical and symbolic dimensions. Practically, there are things that somehow must get done for households to run smoothly and families to function efficiently. Couples must decide who shall do them and how and when they should be done. On a more symbolic level, disputes over housework may be experienced as conflicts about the level of commitment each spouse feels toward the marriage. Because marriage symbolizes the union of two people who share their lives, work together, consult each other, and take each other's feelings and needs into consideration, resisting housework or doing it only under duress may be seen as a less-than-equal commitment. We look more in detail at

Housework is one of the more contested areas of married life.

the dynamics surrounding the division of housework in Chapter 11.

The absence of overt conflict over the allocation of tasks and time does not mean that there is no conflict. It means only that the conflict is not openly expressed. Wives in more traditional marriages are more likely than wives in egalitarian relationships to avoid conflict over housework even if they are dissatisfied with their domestic arrangements. They may withdraw from discussions of the division of labor as a way of avoiding the issue. Because egalitarian couples may engage in more open discussion and conflict over housework responsibilities, such conflict gives them more opportunity to establish a solution (Kluwer et al. 1997).

Consequences of Conflict

While conflict is a normal part of marriages and relationships, excessive conflict can have negative personal or relationship consequences. Among couples who engage in frequent conflict, spouses can suffer negative consequences to their physical and mental health and to their overall well-being (Choi and Marks 2008). Spouses in high-conflict marriages may engage in more behaviors that negatively affect their physical health, such as smoking or drinking alcohol, while at the same time lacking the benefit of social and emotional support that accompanies happier marriages. Marriages assessed as low-quality marriages expose spouses to higher risk of depression, which itself can have physical health consequences. It is even possible, health researchers suggest, that repeated exposure to marital conflict and tension can induce physiological effects that eventually affect one's physical health.

Marital conflict has effects on a host of outcomes related to individual mental and physical health, family health, and child well-being. Frank Fincham and Steven Beach's thorough review (1999) of research on marital conflict showed the following outcomes.

Mental Health

There are links between experiencing marital conflict and suffering from depression, eating disorders, being physically and/or psychologically abusive of partners, and male alcohol problems (including excessive drinking, binge drinking, and alcoholism). There is less evidence connecting marital conflict to elevated levels of anxiety.

Physical Health

Marital conflict is associated with poorer overall physical health, as well as certain specific illnesses. These include cancer, heart disease, and chronic pain. The associations are stronger for wives than for husbands and may be the result of altered physiological functioning, including endocrine, cardiac, and immunological functioning, associated with the distress introduced by marital conflict.

Familial and Child Well-Being

Marital conflict may disrupt the entire family, especially if the conflict is frequent, intense, and unresolved. Marital conflict has been shown to be connected to poorer parenting, problematic parent–child attachments, and greater frequency and intensity of parent–child or sibling–sibling conflict. Consequences for children can be particularly harmful when the conflict centers on issues about the children and child rearing. The most destructive form of marital conflict appears to be when couples engage in attacking and withdrawing (hostility and detachment). In addition, when marriage is characterized by the absence of or low levels of warmth, mutuality, and harmony between parents, along with the presence of high levels of competitiveness and conflict, children develop more externalizing and peer problems (Katz and Woodin 2002). When parental marriages lack relationship cohesiveness, are devoid of playfulness and fun, and yet have high degree of conflict, children miss out on the warmth, intimacy, and security that healthy families can provide (Katz and Woodin 2002).

Research reveals numerous problematic effects of marital conflict on children, including health problems, depression, anxiety, conduct problems, and low self-esteem. When marital conflict is frequent, intense, and child centered, it has especially negative consequences for children. Peer relations also suffer when children are exposed to early and prolonged high levels of parental conflict. This is especially severe when children have insecure parental attachment and can be observed in children as young as three years old (Lindsey, Caldera, and Tankersley 2009).

How do children react to marital conflict? Research indicates that children are distressed by both verbal and physical conflict but reassured by healthy conflict resolution. Witnessing threats, personal insults, verbal and nonverbal hostility, physical aggressiveness between parents or by parents toward objects (e.g., breaking or slamming things), and defensiveness all can give rise to "heightened negative emotionality" (Cummings, Goeke-Morey, and Papp 2003). When parental conflict leads one parent (or both) to withdraw as a means of dealing with the differences between them, children's distress is also worsened (Goeke-Morey, Cummings, and Papp 2007).

Conversely, when parents engage in calm discussion and display affection and continued support even while engaged in conflict, children react positively. Conflict resolution lessens the negative effects of parental conflict on children, especially when what children see is parents compromising with each other so as to resolve their differences (Goeke-Morey et al. 2007). Parents' displays of support, including providing validation to one another and affection during conflict, may reassure children that the marital relationship remains strong and loving even though par-

Children react to parental conflict in a variety of ways, depending on how the parents handle themselves. Although children can be hurt by outward displays of anger and especially by witnessing violence, "healthy conflict management" may be beneficial for children to witness.

ents disagree (Cummings et al. 2003). However, the absence or failure of resolution causes anger, sadness, and distress. A frequently posed question, one that we consider in Chapter 13, is whether the effects of conflict on children are worse than the effects of divorce.

Can Conflict Be Beneficial?

As we noted earlier, conflict is a normal and predictable part of living with other people, especially given the intensity of emotions that exist within marriage. Conflict, itself, is not necessarily damaging; there may be benefits of conflict in which spouses' "conflict engagement" (especially that of husbands) predicts positive change in husbands' and wives' satisfaction with marriage. It appears as though some negative behavior— such as conflict—may be both healthy and necessary for long-term marital well-being. Too little conflict (suggestive of avoidance), like too much conflict, may lead to poorer outcomes. However, the outcome of conflict varies, along with the meaning and function of conflict behavior. It can as easily reflect engagement with a problem as it can suggest withdrawal from the problem (Christensen and Pasch 1993). Furthermore, it may be part of an effort to maintain the relationship or conversely indicate that one or both partners have given up on the relationship (Holmes and Murray 1996). Thus, as Frank Fincham and Steven Beach (1999, 54) suggest, "We have to identify the circumstances in which conflict behaviors are likely to result in enhancement rather than deterioration of marital relationships."

Resolving Conflicts

There are a number of ways to end conflicts and solve problems. You can give in, but unless you believe that the conflict ended fairly, you are likely to feel resentful. You can try to impose your will through the use of power, force, or the threat of force, but using power to end conflict leaves your partner with the bitter taste of injustice. Less productive conflict resolution strategies include *coercion* (threats, blame, and sarcasm), *manipulation* (attempting to make your partner feel guilty), and *avoidance* (Regan 2003).

More positive strategies for resolving conflict include *supporting your partner* (through active listening, compromise, or agreement), *assertion* (clearly stating your position and keeping the conversation on topic), and *reason* (the use of rational argument and the consideration of alternatives) (Regan 2003). Finally, you

Figure 7.2 Family Problem-Solving Loop

Most family problem solving occurs in the ebb and flow of daily family events. Although family dynamics and transition take various forms, it is interesting to note the types that might have relevance for family issues.

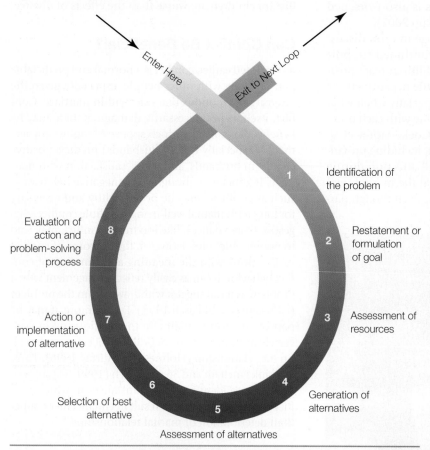

SOURCE: Kieren, Maguire, and Hurlbut 1996, 442–455. Copyright © 1996 by the National Council on Family Relations. Used by permission.

thing you don't want to do. When you agree without coercion or threats, the agreement is a gift of love, given freely without resentment. As in all exchanges of gifts, there will be reciprocation. Your partner will be more likely to give you a gift of agreement. This gift of agreement is based on referent power, discussed earlier.

Bargaining

Bargaining in relationships—the process of making compromises—is different from bargaining in the marketplace or in politics. In relationships, you want what is best for the relationship, the most equitable deal for both you and your partner, not just the best deal for yourself. During the bargaining process, you need to trust your partner to do the same. In a marriage, both partners need to win. The result of conflict in a marriage should be to solidify the relationship, not to make one partner the winner and the other the loser. Achieving your end by exercising coercive power or withholding love, affection, or sex is a destructive form of bargaining. If you get what you want, how will that affect your partner and the relationship? Will your partner feel that you are being unfair and become resentful? A solution has to be fair to both, or it won't enhance the relationship.

can end the conflict through negotiation. In negotiation, both partners sit down and work out their differences until they come to a mutually acceptable agreement (see Figure 7.2). Conflicts can be solved through negotiation in three primary ways: agreement as a gift, bargaining, and coexistence.

Agreement as a Gift

If you and your partner disagree on an issue, you can freely agree with your partner as a gift. If you want to go to the Caribbean for a vacation and your partner wants to go backpacking in Alaska, you can freely agree to go to Alaska. An agreement as a gift is different from giving in. When you give in, you do some-

Coexistence

Although unresolved conflict may, over time, wear away at marital quality, sometimes differences simply can't be resolved. In such instances, they may need to be lived with. If a relationship is sound, often differences can be absorbed without undermining the basic ties. All too often, we regard a difference as a threat rather than as the unique expression of two personalities. Rather than being driven mad by the cap left off the toothpaste, perhaps we can learn to live with it.

Forgiveness

Related to the issues of conflict and its resolution is the topic of *forgiveness*. Conceptualized as a reduction in negative feelings and an increase in positive feelings toward a "transgressor" after a transgression, an attitude of goodwill toward someone who has done us harm, and showing compassion and forgoing resentment toward someone who has caused us pain, research has determined that forgiveness has long-term physical and mental health benefits for the person forgiving. Forgiveness is associated with enhanced self-esteem, positive feelings toward the transgressor, and reduced levels of negative emotions, such as anger, grief, revenge, and depression. In a relationship context, forgiveness has been defined as "the tendency to forgive partner transgressions over time and across situations" (Fincham and Beach 2002).

Forgiveness has been found to be a crucial element of married life. It is an important aspect of efforts to restore trust and relationship harmony after a transgression. Most "forgiveness narratives" mention motivations such as a partner's well-being, restoration of the relationship, and love (Fincham and Beach 2002). Forgiveness has been shown to resolve existing difficulties and prevent future ones. It also enhances marital quality, as can be seen in the positive association between forgiveness and marital satisfaction and longevity (Kachadourian, Fincham, and Davila 2004).

Research has identified both personal and relationship qualities associated with the ability or tendency to forgive. Qualities such as agreeableness, religiosity, humility, emotional stability, and empathy are associated with forgiveness. Pride and narcissism are associated with decreased tendencies to forgive. Individuals who are more accommodating within their relationships, are more securely attached, and have more positive models of self and others are also more likely to be forgiving toward partners who have committed transgressions.

Not all relationship transgressions are equivalent. The ability to forgive relatively minor transgressions doesn't automatically guarantee forgiveness of more major transgressions. In heterosexual couples, wives who display tendencies to forgive seem able to do so in both minor and major transgressions. For husbands, on the other hand, tendencies to forgive apply more to major transgressions. It appears as though men may not consider minor transgressions important enough to warrant either receiving apologies or granting forgiveness (Kachadourian et al. 2004).

If we can't talk about what we like and what we want, there is a good chance that we won't get either. Communication is the basis for good relationships. Communication and intimacy are reciprocal: communication creates intimacy, and intimacy in turn helps create good communication.

Helping Yourself by Getting Help

Despite good intentions and communication skills, we may not be able to resolve our relationship problems on our own. Accepting the need for professional assistance may be a significant first step toward reconciliation and change. Experts advise counseling when communication is hostile, conflict goes unresolved, individuals cannot resolve their differences, and/or a partner is thinking about leaving.

Marriage and partners counseling are professional services whose purpose is to assist individuals, couples, and families gain insight into their motivations and actions within the context of a relationship while providing tools and support to make positive changes. A skilled counselor offers objective, expert, and discreet help. Much of what counselors do is crisis or intervention oriented.

It may be more valuable and perhaps more effective to take a preventive approach and explore dynamics

© Gary Conner/PHOTOTAKE/Alamy

It may be necessary to seek outside assistance to resolve conflicts effectively and preserve one's relationship.

Is marital conflict something we can be taught to avoid? Is marital communication something we can learn to do more effectively? Can we learn such things before ever spending even a day as married? Existing research on the success of premarital education offers encouraging answers to these questions. A recent study by Scott Stanley and colleagues (2007) is an especially good case in point.

After noting promising research about premarital education programs, Stanley and associates examined data from a large, multistate U.S. random sample to determine the effectiveness of premarital education in lowering marital conflict, raising marital satisfaction, and reducing the likelihood of divorce. More than 3,000 adults in Oklahoma, Kansas, Arkansas, and Texas were surveyed to see whether the use of such programs had increased, whether use varied by race and education, whether use led to desired marital outcomes, and whether the length or venue within which a program occurred made a difference in outcome.

Their findings are encouraging:

1. Participation in premarital education programs has greatly increased, from less than 10% of those married in the 1930s and 1940s to more than 40% of those married since 1990.
2. Participation in premarital education varied by race and ethnicity and with education. Latinos were most likely and African Americans least likely to participate. Those with lower levels of education as well as those with higher levels of economic distress were less likely to participate.
3. Premarital education appears to enhance marital outcomes. Participation was positively associated with marital satisfaction and commitment and negatively associated with marital conflict. Perhaps of greatest interest, participation in premarital education was associated with a 31% reduction in one's odds of divorce.
4. The more time one spends in such a program is associated with greater benefit from such training, though this benefit tops out at about 10 hours, after which additional benefits are slight.

Stanley et al.'s results support other research findings in generating optimism in the possibility of enhancing marital quality by learning to improve communication and reduce and more effectively manage conflict. At present, most people who participate in premarital programs do so through the religious organizations to which they belong as they prepare to marry. Perhaps a mechanism to make such training more widely available and incentives to make participation more likely would combine to improve the marital experience of a greater number of individuals and couples.

and behaviors before they cause more significant problems. This may occur at any point in relationships: during the engagement, before an anticipated pregnancy, or at the departure of a last child.

Each state has its own degree and qualifications for marriage counselors. The American Association for Marital and Family Therapy is one association that provides proof of education and special training in marriage and family therapy. Graduate education from an accredited program in social work, psychology, psychiatry, or human development, coupled with a license in that field, ensures that the clinician has recieved necessary education and training. It also offers the consumer recourse if questionable or unethical practices occur. However, this recourse is available only if the practitioner holds a valid license issued by the state in which he or she practices. Mental health workers belong to any one of several professions:

- *Psychiatrists* are licensed medical doctors who, in addition to completing at least six years of postbaccalaureate medical and psychological training, can prescribe medication.
- *Clinical psychologists* have usually completed a PhD, which requires at least six years of postbaccalaureate course work. A license requires additional training and the passing of state boards.
- *Marriage and family counselors* typically have a master's degree and additional training to be eligible for state board exams.
- *Social workers* have master's degrees requiring at least two years of graduate study plus additional training to be eligible for state board exams.
- *Pastoral counselors* are clergy who have special training in addition to their religious studies.

Financial considerations may be one consideration when selecting which one of the preceding to see.

Typically, the more training a professional has, the more he or she will charge for services.

A therapist can be found through a referral from a physician, school counselor, family, friend, clergy, or the state department of mental health. In any case, it is important to meet personally with the counselor to decide if he or she is right for you. Besides inquiring about his or her basic professional qualifications, it is important to feel comfortable with this person, to decide whether your value and belief systems are compatible, and to assess his or her psychological orientation. Shopping for the right counselor may be as important a decision as deciding to enter counseling in the first place.

Marriage or partnership counseling has a variety of approaches: individual counseling focuses on one partner at a time, joint marital counseling involves both people in the relationship, and family systems therapy includes as many family members as possible. Regardless of the approach, all share the premise that, to be effective, those involved should be willing to cooperate. Additional logistical questions, such as the number and frequency of sessions, depend on the type of therapy.

At any time during the therapeutic process, one has the right to stop or change therapists. Before doing so, however, one should ask oneself whether his or her discomfort is personal or has to do with the techniques or personality of the therapist. This should be discussed with the therapist before making a change. Finally, if one believes that therapy is not benefiting him or her, a change in therapists seems necessary.

If we fail to communicate, we are likely to turn our relationships into empty facades, with each person acting a role rather than revealing his or her deepest self. But communication is learned behavior. If we have learned *how not to* communicate, we can learn *how to* communicate. Communication will allow us to maintain and expand ourselves and our relationships.

Summary

- A common complaint of married couples is that they don't communicate or don't communicate well.

- Communication includes both verbal and nonverbal communication. For the meaning of communication to be clear, verbal and nonverbal messages must agree.

- The functions of nonverbal communication are to convey interpersonal attitudes, express emotions, and handle the ongoing interaction.

- Proximity, eye contact, and touch are three of the most important kinds of nonverbal communication.

- Much nonverbal communication, such as levels of touching, varies across cultures and between women and men.

- Nonverbal communication patterns can reveal whether a relationship is healthy or troubled.

- Gender differences have been identified in both verbal and nonverbal communication.

- How well a couple communicates before marriage can be an important predictor of later marital satisfaction. *Self-disclosure* before marriage is related to relationship satisfaction later.

- Some problems in marital communication first arise during cohabitation.

- Research indicates that happily married couples engage in less frequent and less destructive conflict, disclose more of their thoughts and feelings, and more accurately and effectively communicate.

- In marital communication, wives send clearer, less ambiguous messages; send more positive, more negative, and fewer neutral messages; and take more active roles in arguments than husbands do.

- *Demand–withdraw communication* is common among heterosexual couples. One partner, more often the woman, will raise an issue for discussion, and the other partner, more likely the man, will withdraw from the conversation instead of attempting to communicate.

- Demand–withdraw patterns are found commonly in the United States and many other cultures.

- Demand–withdraw patterns may be a reflection of the relative power of each partner, of gender socialization, or even of biological differences between women and men in their reactions to conflict.

- Satisfying sexual relationships requires effective sexual communication.

- Some topics are more highly charged and more sensitive to discuss.

- Barriers to communication include the traditional male gender role; personal reasons, such as feelings of inadequacy; the fear of conflict; and an absence of self-awareness.

- Self-disclosure requires trust, the belief in the reliability and integrity of a person. Trust influences the way you are likely to interpret ambiguous or unexpected messages from another person.
- *Feedback* is the ongoing process in which participants and their messages create a given result and are subsequently modified by the result.
- *Power* is the ability or potential ability to influence another person or group. There are six types of marital power: coercive, reward, expert, legitimate, referent, and informational.
- There are a variety of explanations for relationship power. One prominent idea, the *principle of least interest*, is that the person who has the least invested in the relationship is, as a result, more powerful.
- Resource-based theories of power fail to account for why women don't gain power as much as men do, even when they bring in more of the financial resources that couples require.
- Theories that focus on decision making as indicators of power may miss more covert expressions of power.
- Self-described equal (or egalitarian) couples often still reveal power differences and inequalities that more often favor men.
- Conflict is natural in intimate relationships. *Basic conflicts* challenge fundamental rules; *nonbasic conflicts* do not threaten basic assumptions and may be negotiable.
- People handle anger in relationships by suppressing or venting it. When anger arises, it is useful to think of it as a signal that change is necessary.
- Among heterosexual couples, women have greater awareness of the emotional quality of the relationship and are more likely to initiate discussion of contested issues. Men are more likely to approach conflict from a task-oriented stance or to withdraw.
- Hostile conflict, characterized by frequent heated arguments, name-calling, and/or an unwillingness to listen to each other, is a particularly strong predictor of eventual divorce.
- Major sources of conflict include sex, money, and housework.

- Conflict can have effects on the mental and physical health of spouses or partners, the health of the relationship, and the well-being of the children.
- Happily married couples use certain techniques to resolve conflict, including summarizing, paraphrasing, validating, and clarifying. Unhappy couples use confrontation, complaining, and defensiveness.
- Conflict resolution may be achieved through negotiation in three ways: agreement as a freely given gift, bargaining, and coexistence.
- *Forgiveness* is an important part of efforts to restore trust and rebuild relationship harmony. It is positively associated with both relationship satisfaction and stability (i.e., longevity).

Key Terms

assertiveness 247

basic conflicts 242

contempt 224

demand–withdraw
   communication 231

feedback 236

gatekeeping 239

homogamous 243

honeymoon effect 229

hostile conflict 246

intimate zone 225

marital heterogamy 243

marital paradigm 246

nonbasic conflicts 242

nonverbal
   communication 223

personal space 225

power 237

principle of least
   interest 239

proximity 225

relative love and need
   theory 239

resource theory of
   power 240

self-silencing 242

trust 235

RESOURCES ON THE WEB

Book Companion Website

www.cengage.com/sociology/strong

Prepare for quizzes and exams with online resources—including tutorial quizzes, a glossary, interactive flash cards, crossword puzzles, self-assessments, virtual explorations, and more.

8

Marriages in Societal and Individual Perspective

What Do You Think? Are the following statements **TRUE** or **FALSE**?
You may be surprised by the answers (see answer key on the following page).

T	F	**1** Trends in cohabitation and divorce clearly indicate a decrease in the importance of marriage in the United States.
T	F	**2** Couples who are unhappy before marriage significantly increase their happiness after marriage.
T	F	**3** Compared to adults in other Western countries, more Americans tend to agree that marriage is an outdated institution.
T	F	**4** Most interracial marriages are between African American men and Caucasian women.
T	F	**5** The advent of children generally increases a couple's marital satisfaction.
T	F	**6** Age at marriage is a strong indicator of later marital success.
T	F	**7** We are more likely to marry within our same social class than to marry above or below.
T	F	**8** Married people are less likely to socialize with friends and neighbors than are never married or previously married women and men.
T	F	**9** Older married couples report more disagreements than do younger married couples.
T	F	**10** Long-term marriages are happy marriages.

. . . to have and behold from this day on, for better or for worse, for richer, for poorer, in sickness and in health, to love and to cherish; until death do us part.

As you probably realize, those words are a traditional version of wedding vows that, in some similar form or fashion, are exchanged between two people as they enter marriage. Some may add more religious language, some may be less traditional, and some may be longer or more personal. It is likely, however, that all will convey one's intentions to: share life's ups and downs, to be as one *together*, and to so commit for as long as both people live.

When two people exchange wedding rings and vows, they make a public declaration, in a ceremony overseen by a legally recognized officiate, usually a clergyperson or justice of the peace. Once they do, each newly married couple embarks on a journey that is simultaneously intensely personal, inarguably public, and increasingly politicized. Our goals in this chapter are to examine marriage in all three ways—as a relationship between spouses, as a commitment certified by the state and celebrated by one's wider circle of friends and kin, and as a legal relationship undergoing dramatic changes and challenges.

The chapter begins by considering the ambiguous status and direction of marriage in the United States today. Although most Americans will marry at some point, fewer enter and stay in marriage today than did in the recent past. Is marriage less valued than it was in the past? As a society, are we less committed to marriage as a central life goal, and are those who marry less willing or able to work hard to make their marriages work?

We then describe how people choose their spouses and who they tend to choose and examine issues that confront couples as they enter marriage and as they share a lifetime together. Along the way, we identify some factors that predict marital success, discuss marital roles and boundaries, and look at how having and raising children affects married couples. Finally, we turn to middle-aged and later life marriages and the end of marriage with the death of one's spouse and

survey the different patterns and factors that characterize lasting marriages.

Marriage in American Society

Marriage has long been the foundation on which American families are constructed. Although we have recognized and valued our ties connecting us with our wider families, marriage has been the centerpiece of family life in the United States. As we saw in Chapter 1, in our nuclear family system, our relationships with our spouses are more important than our relationships with our extended families. In our lives as individuals, the person we marry is expected to be someone with whom we will share everything, a soul mate and partner "for as long as we both shall live."

As central as marriage has been to our family system, these are confusing times in which the current status and the direction of marriage in the United States are subjects of considerable ongoing controversy and debate. Even the "marriage experts" can't agree about whether marriage is or isn't "endangered," whether it has retained or lost its appeal and its meaning as a major life goal to which people aspire.

For example, consider the November 2004 issue of the *Journal of Marriage and the Family*, the leading scholarly journal that focuses on family life. This particular issue contained a series of articles and commentaries as part of a "Symposium on Marriage and Its Future." As article after article revealed, evidence can be marshaled on either side of what sociologist Paul Amato (2004a) called the **marriage debate**. As Amato points out, although it may seem odd, even respected family scholars cannot reach consensus; where some see marriage as "in decline," others portray it as dynamic, changing, and resilient. Why can't they agree?

To begin, let's examine the following, clearly mixed portrait of marriage in the United States today, beginning with some of the more encouraging indicators:

Behaviorally, approximately 55% of adults, 18 and older, in the United States are currently married. Another 19% are formerly married, being either widowed (6.4%), separated (2.3%), or divorced (10.4%). Although separation and divorce are not the best news about marriage, these data show that cumulatively nearly 75% of adults are or have been married. In numerical terms, an estimated 124 million Americans 18 years and older are currently

Answer Key to What Do YOU Think?

1 False, see p. 263;   **2** False, see p. 282;   **3** False, see p. 263;   **4** False, see p. 273;   **5** False, see p. 292;   **6** True, see p. 282;   **7** True, see p. 275;   **8** True, see p. 291;   **9** False, see p. 296;   **10** False, see p. 299.

married (U.S. Census Bureau, http://census.gov/pop/socdemo/hh-fam/cps2008/tabA1-all.xls). It is projected that over 80% of U.S. women and men will marry (at least once) before age 40: 81% of men and 86% of women (Jayson 2009). This is a higher percentage than what we observe in other Western societies (Cherlin 2009). Even among couples whose marriages end in divorce, most formerly married women and men remarry, suggesting that marriage is still desired and that Americans are unwilling to compromise their expectations for marital fulfillment, even if it means ending their marriages in pursuit of more fulfilling ones.

Attitudinally, marriage remains highly valued, even with increased acceptance of divorce and nonmarital lifestyles. Most young adults say they want to marry someday and recognize that marriage brings benefits to their lives (Amato 2004b). According to sociologist Andrew Cherlin, each year for more than a quarter of a century, around 80% of female high school seniors express an expectation to marry someday. Among males, the percentage expecting to marry has increased during this period from 71% to 78%. Cherlin also noted that marriage has been *and continues to be* seen as an "extremely important" part of life. Roughly 80% of young women and 70% of young men express such an attitude (Cherlin 2009).

Data drawn from the World Values Surveys of respondents from more than 60 countries found that fewer Americans (10%) agreed with a statement suggesting that marriage was an "outdated institution" as compared with adults in the other Western countries (Cherlin 2009). For example, data comparing the United States to Canada, the United Kingdom, France, Germany, Italy, and Sweden show stronger support for marriage in the United States (see Figure 8.1).

Between 1980 and 2000, the view of marriage as a lifelong relationship *received increased support* (Amato 2004b). Asked in a national survey about their agreement with a statement that, except when faced with extreme circumstances, marriage is for life, 76% agreed (only 11% disagreed, and 13% neither agreed nor disagreed) (Cherlin 2009). As a result of a host of demographic trends, married couples today are "older, more mature, and better educated at the time of marriage" even than were couples who married in the last two decades of the twentieth century (Amato et al. 2007:31). Furthermore, increases in married women's employment and the acceptance of less traditional and more equitable gender roles may have actually strengthened the couple relationship (Amato et al. 2007).

Figure 8.1 Belief That Marriage Is Outdated: International Comparison

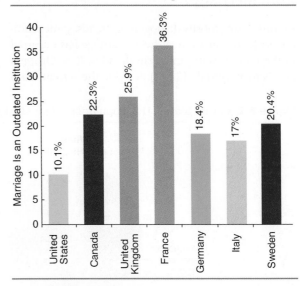

SOURCE: Stevenson and Wolfers (2007).

Curiously, alongside the many positives noted above, one can also find less encouraging indicators of the status of marriage in the United States, especially in the combination of trends regarding divorce, singlehood, cohabitation, and births outside of marriage. Consider these facts:

- With a 2007 divorce rate of 17.5 per 1,000 marriages, the United States has one of the highest divorce rates among Western societies(http://marriage.rutgers.edu/Publications/2008; www.nationmaster.com/graph/peo_div_rat-people-divorce-rate). More than 40% of new marriages are projected to end in divorce.

- Singlehood, cohabitation, and births to unmarried mothers (either single or cohabiting) all have increased over the last three decades of the twentieth century (Cherlin 2009). Pessimistically, these might suggest that marriage has become less attractive, less valued, and less essential as a prerequisite for having and raising children.

- Despite the fact that beginning in the 1980s and lasting through 2008 divorce rates decreased, they continue to be relatively high, perhaps suggesting that Americans are less committed to their marriages (www.census.gov/compendia/statab/tables/09s1292.xls).

Given the conflicting nature of the statistics described above, what are we to conclude about the status of marriage? In the next sections, we'll explore some additional factors to consider.

Has There Been a Retreat from Marriage?

In considering mostly demographic trends, some scholars contend that a retreat from marriage has occurred in the United States in recent decades. R. S. Oropesa and Nancy Landale (2004) describe the **retreat from marriage** as evident in such recent and ongoing trends as older age in first marriage for both women and men, more people never marrying, significant increases in cohabitation and nonmarital births, and continued high divorce rates. Robert Schoen and Yen-Hsin Alice Cheung (2006, 1) assert that marriage has actually "been in retreat for more than a generation," as fewer men and

Figure 8.2 Marital Status, U.S. Population 18 and Older, by Ethnicity, 2008

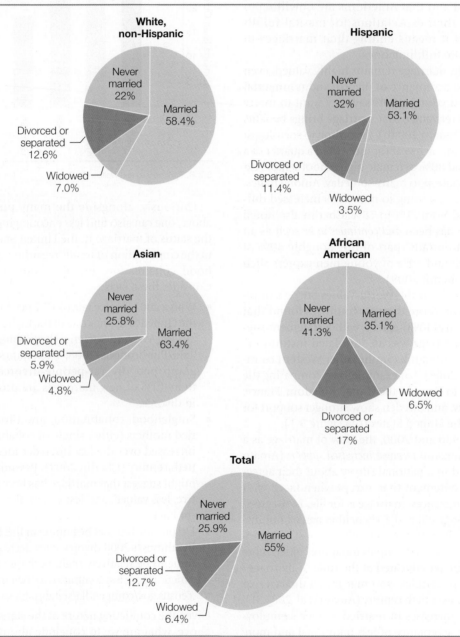

NOTE: Some percentages may not equal 100%.

SOURCE: U.S. Census Bureau, "America's Families and Living Arrangements: 2008," Table A

women "ever marry," and that the "U.S. withdrawal from marriage" persisted at least through 2003. They suggest that the retreat from marriage appears to be associated with increases in employment of women, smaller gender wage gaps in earnings, and persistent economic inequality between racial groups (Schoen and Cheung 2006).

The Economic and Demographic Aspects Discouraging Marriage

Closer inspection of marriage trends indicates that whatever retreat from marriage has occurred is not equal among all social groups. Instead, racial, economic, and educational differences can be identified. As shown earlier in Chapter 3, there are considerable differences in marital status for different racial, ethnic, and economic groups. Looking again, using 2007 census data, reveals the differences as detailed in Figure 8.2.

Where majorities of Caucasians, Asians, and Hispanics are married, only about 40% of African Americans are married. Adding the widowed and divorced to the portion married, nearly 80% of whites, 74% of Asians, and 68% of Hispanics *are or have been* married compared to 59% of African Americans (see Figure 8.2). Some other racial differences to note: Although in general young people expect to marry someday, fewer young African Americans express an expectation to ever marry, and those who do report an older desired age at marriage than whites (Crissey 2005; Manning, Longmore, and Giordano, 2007). African Americans are more likely to divorce than Caucasians. Divorced African Americans are less likely than divorced Caucasians to remarry. Blacks are also much more

Marriage patterns show significant race and ethnic differences in the likelihood of entering and remaining married.

likely to bear children outside of marriage. Although nearly 40% of all children born in the United States in 2007 were born to unmarried mothers, the racial differences were pronounced: approximately 28% of all births were to Caucasian women compared to more than 70% of births to African Americans, and more than half of the births to Hispanic women were to unmarried mothers. Seventeen percent of 2007 births to Asian American women occurred outside of marriage (Hamilton, Martin, and Ventura 2009).

What about Class?

Within the shifts in marriage rates, there are also notable socioeconomic differences, observable in differences by educational level. For example, although lifetime marriage rates among women have dropped by 5% in the United States, they have declined by 30% for women without a high school diploma (Gibson-Davis, Edin, and McLanahan 2005). Among college-educated white women, the prospect of marrying has *grown greater*, whereas among those without college degrees it has decreased (Huston and Melz 2004).

For both women and men, educational attainment is positively associated with the likelihood of marriage (Schoen and Cheng 2006). In addition, in the 1980s and 1990s, marriages among college-educated women became more stable—that is, less likely to end in divorce—than they had been in the previous decade; among women at the bottom of the educational distribution, marriage became less stable (Edin, Kefalas, and Reed 2004). In discussions of a retreat from marriage among Hispanics, R. S. Oropesa and Nancy Landale (2004) emphasize how limited economic opportunities may be major barriers or disincentives to marriage.

Does Not Marrying Suggest Rejection of Marriage?

Even if low socioeconomic status reduces the likelihood of marriage, it may not signal an attitudinal rejection of marriage. In fact, Edin and colleagues (2004, 1008) assert that "marriage has by no means lost its status as a cultural ideal among low-income and minority populations." Where less than half of college graduates disapproved of cohabitation, two-thirds of high school dropouts disapproved or strongly disapproved of living together with no intention to marry (Edin et al. 2004).

Despite what the race data on marriage appear to suggest, African Americans remain "strong believers in the value of marriage" (Huston and Melz 2004). Some researchers have even found that unmarried blacks and Hispanics express greater interest in marrying than do unmarried whites (Huston and Melz 2004). Overall, research reveals very few significant racial or class differences in attitudes regarding either the importance of marriage or people's aspirations toward marriage. In fact, even 70% of welfare recipients indicate that they hope to marry (Gibson-Davis, Edin, and McLanahan 2005).

Perhaps, then, a good portion of the "marriage retreat," at least that portion occurring among the most economically disadvantaged, is *not* a *rejection of marriage*. What, then, keeps low-income unmarried parents from marrying? Interviews with a sample of low-income unmarried couples with children led Gibson-Davis et al. (2005) to identify three barriers to marriage: financial concerns, concerns about the quality and durability of their relationships, and fear of divorce:

- *Financial concerns.* These concerns covered four aspects of financial matters: whether couples had the resources to "consistently make ends meet," whether they could exercise financial responsibility and wisely use what resources they possess, whether they could "work together toward long-term financial goals," and whether they'd saved enough or had enough money for a "respectable wedding" (Gibson-Davis et al. 2005, 1307).
- *Relationship quality.* Believing that marriage ought to be for life, couples want to make sure that their partners are suitable for marriage. One way they believe they can achieve this is by living together long enough to tell that their relationships are "up to the challenge" of marriage, that they can weather any storm, and that they have answered any doubts about whether they and their partners are ready and their relationships are strong enough for marriage.
- *Fear of and opposition to divorce.* Claiming not to believe in divorce as an option and viewing marriage as somewhat "sacred," couples wait to marry until they fully believe that their relationships will last.

Despite data indicating that they are less likely to marry or expect to marry, African Americans express strong belief in the importance of marriage.

As explained by Gibson-Davis et al. (2005, 1311), what lies

> at the heart of marital hesitancy is a deep respect for the institution of marriage. . . . The bar for marriage has grown higher for all Americans, making it increasingly difficult for those in the lower portions of the income distribution to meet the standards associated with marriage.

Religion and Marriage

Part of the supposed retreat from marriage consists of the delayed age at which women and men first enter marriage. Along with race and social class, religious affiliation is among the factors that may influence whether and when people choose to marry. Religious differences have been identified in mate choice, childbearing and child rearing; the division of housework; domestic violence; marital quality; and divorce (Xu, Hudspeth, and Bartkowski 2005). Religious traditions and denominations differ in the kinds and degree of emphasis they place on marriage.

Religion and the Importance of Marriage
Although Judeo-Christian religious groups tend to support marriage, uphold marriage and family as desirable and important lifestyles, and discourage both premarital and extramarital sex, there are differences among them, especially in the extent to which they support traditional gender roles and

In American society, the expectation is that through the process of dating, singles find their eventual life partner. Dating, or whatever else it might be called, allows us to test out our suitability for each other, develop stronger and closer relationships, fall in love, and select our life partners. Marriage without love goes against the culture of romantic love and these established patterns of mate selection. Although we might consider marriage without love an exceptional case, anthropologists tell us that in traditional cultures, most people do not consider love the basis for their entry into marriage.

Marriage customs vary dramatically across cultures, and marriage means different things in different cultures. If we consider how marriages come about—how they are "arranged"—we find that it is usually not the bride and groom who have decided to marry, as is the case in our own society today. Typically, the parents and elders have done the matchmaking, sometimes relying on intermediaries and matchmakers to locate suitable spouses for their children.

These strategies are still practiced and are not entirely restricted to other countries. New York Times journalist Stephen Henderson tells the story of Rakhi Dhanoa and Ranjeet Purewal and captures some of the motivation behind using others to arrange marriages: "Each wanted a love marriage . . . yet neither would dream of marrying someone who wasn't a Sikh." An immigration lawyer in New York whose parents emigrated from Punjab, India, 27-year-old Dhanoa decided that she wanted to marry someone of the same faith. "I began to appreciate that my religion is based on complete equality of the sexes," she said. At the same time, Purewal was beginning to think about finding a partner. His mother had approached Jasbir Hayre, a Sikh matchmaker, living nearby in New Jersey. So,

when it came time to throw a party for her own daughter, Hayre invited both Dhanoa and Purewal. Although he had firmly believed in choosing for himself based on love, like Dhanoa, Purewal came to feel as though there were important issues to take into account. "I was adamant that I'd marry whoever I wanted. . . . But seeing how different cultures treated their families, I realized the importance of making the right match." After two months of mostly covert dating, "their cover was blown, on a double date, [and] the matchmaker was quickly summoned to negotiate marital arrangements" (Henderson 2002).

The story of Dhanoa and Purewal illustrates a variation of a phenomenon common in many parts of the world. In most cultures, marriage matches do not result from individuals meeting and dating; instead, the parents of the bride and groom arrange the marriage of their children. In some cultures, mothers are the primary matchmakers, as in traditional Iroquois culture. In others, fathers have a dominant voice in arranging marriage, as in traditional Chinese society. In still other cultures, the pool of elders involved in matchmaking is more extensive, including grandparents, aunts, uncles, and even local political and religious authorities, such as tribal chiefs and clan leaders. In all these instances, though, marriage is a major event in the life of two families—both the bride's and the groom's—as well as for the clan, tribe, and community to which each family belonged. As such, important matters must be taken into account before agreeing to any particular match. Families must know how a particular marriage affects the family as a whole. The feelings and love between an individual bride and groom are subordinate to the greater interests and welfare of the family, clan, and community.

relationships and reject divorce, abortion, and homosexuality (Xu et al. 2005). Conservative Protestant denominations and Latter-Day Saints (Mormons) articulate especially strong commitments to marriage, encouraging members to marry and stay married, by portraying marriage as "part of God's plan for self-development . . . in this life, as well as . . . long term spiritual salvation" (Xu et al. 2005, 589–90). Although, traditionally strongly promarriage, the Catholic Church has a "considerably less robust" promarriage orientation, as evidenced in the

tendencies of American Catholics to move from the Church's traditional teachings about marriage and toward a viewpoint that marital matters are subjects of individual choice. Liberal and moderate Protestants do not attach the same importance to marriage as do evangelical Protestants. Among Jews, we find greater emphasis on the importance of marriage and on more traditional gender roles in marriage among Orthodox Jews and considerably less encouragement to marry and bear children as well as, among Reform Jews, less emphasis on gender differences.

Religion and the Timing of Marriage

Xu et al. (2005) found that women and men affiliated with moderate and conservative Protestant denominations and with the Mormon Church are both more likely to marry and to marry young than those unaffiliated with a religious faith. Interestingly, they may face different consequences of early marriage. Baptists, among the most conservative Protestant denominations, have the highest divorce rate in the United States; Mormons are among those with the lowest likelihood of divorce.

Catholics and liberal Protestants also differ from the unaffiliated but to a lesser extent. By emphasizing marriage as "the joining of two individuals with the goal of living a constructive, harmonious life" and "creating a good environment for rearing children," Judaism, especially Reform Judaism, may encourage people to delay marriage. Indeed, Jews are more likely than Catholics, moderate and conservative Protestants, and Mormons to delay their entry into marriage. Jewish, liberal Protestant, and unaffiliated individuals were found to marry later (Xu et al. 2005, 589–90).

Somewhere between Decline and Resiliency

How can we best understand what has happened and is happening to marriage in the United States? One way is to use sociologist Andrew Cherlin's (2004) argument that marriage has been "deinstitutionalized." The **deinstitutionalization of marriage** refers to the "weakening of the social norms that define people's behavior in a social institution such as marriage" (Cherlin 2004, 848). As a result of wider social changes, individuals can no longer rely on shared understandings of how to act in and toward marriage. In the late nineteenth and early twentieth centuries, the form of marriage known as **companionate marriage** emerged (Cherlin 2009). In companionate marriages, spouses were expected to supply each other with companionship, friendship, romantic love, and mutually gratifying sexual intimacy. Held together by love and friendship between spouses rather than social obligations, characterized by egalitarian as opposed to the earlier patriarchal ideals for marriage, and encouraging spouses to focus on self-development and expression, the companionate marriage was by the 1950s the widely shared cultural ideal.

Beginning in the 1960s and accelerating in the 1970s, the companionate marriage began to lose ground to a form of marriage Cherlin (2009) calls the **individualized marriage**. Partly the product of "cultural upheavals of the 1960s and 1970s," individualized marriages emphasize personal fulfillment and personal growth in marriage and expect that our spouses will facilitate such growth and be sources of unprecedented support (Amato 2004a). In the individualized marriage, emphasis is placed on self-development, flexible and negotiable roles, and openness and communication in problem solving. In this newer form of marriage, "spouses are free to grow and change . . . what matters most is not merely the things they jointly produce—well-adjusted children, nice homes—but also each person's own happiness" (Cherlin 2009, 90).

This is where the marriage debate centers. Some scholars see the changes and trends described here as worrisome because they undermine marriage as an institution that meets the needs of society. They believe that we have become *too* individualistic and *too* focused on personal happiness and have less commitment to making our marriages work. Such attitudes help explain the increases in cohabitation, single parenthood, and divorce, as individuals pursue what they most want regardless of their effects on others. To proponents of this viewpoint, we need to enact policies to reinstitutionalize marriage, to restrict and decrease divorce, and to strengthen values such as marital commitment, obligation, and sacrifice.

Others put more emphasis on marriage as a relationship between two individuals and stress the value of such characteristics of contemporary marriage as self-development, freedom, and equality between spouses. Rejecting the idea that we have grown too individualistic or selfish, they also challenge the idea that ongoing trends should be seen with such negativity. Even the increase in divorce may be seen as an opportunity for happiness for adults and a means of escape for children from dysfunctional or dangerous environments.

As articulated by sociologist Paul Amato, neither the **marital decline perspective** (the belief that marriage is endangered) nor the **marital resilience perspective** (the belief that marriage is changing though still highly valued) is consistently or uniformly supported by the variety of available data on marriage. Along with David Johnson, Alan Booth, and Stacy Rogers, Amato compared two national surveys of married women and men in the United States, one from 1980 the other from 2000 (Amato, et al, 2007). As expected, given some trends we have already discussed, the demographics of marriage had changed considerably. Age at first marriage had increased, as had the proportion of remarried individuals and couples marrying after first cohabiting, the proportion of wives in the labor force, and the share of household income that married women contributed. Gender relations had changed in less traditional directions. Couples also became more religious and expressed greater support for the norm that marriage was for life.

Linking these sorts of changes to shifts in marital quality, data appear to partially support both the marital decline perspective and the marital resilience perspective. In other words, some changes in late twentieth-century marriage led to declines in marital happiness and interaction and were associated with increases in likelihood of divorce. Yet other changes were associated with improved marital quality, such as an improved economic standing of married couples, the adoption of less traditional gender roles, and an increase in the belief in marriage as a lifelong relationship (Amato et al. 2007). And the overall effect? Although the average level of marital interaction declined significantly (couples became less likely to eat dinner together, go shopping together, visit friends together, and share downtime together), as Amato (2004b, 101) expresses, "In general, these changes tended to offset one another, resulting in little net change in mean levels of happiness and divorce proneness in the U.S. population."

Who Can Marry?

Not everyone can marry the partner of their choice. In Chapter 1, we looked at some restrictions imposed on one's choice of a marriage partner. As we noted then,

who we are allowed to legally marry has undergone change and challenge over the past 150 years in the United States over such issues as number of spouses in a lifetime and marriage across racial lines and, most recently, over the question of marriage between two people of the same sex.

What criteria do state marriage laws currently specify regarding eligibility to marry? Each state enacts its own laws regulating marriage, leading to some discrepancies from state to state. Although some restrictions or lack thereof are uniform across all 50 states (e.g., no state prohibits people from marrying someone of another race), others, such as those regarding same-sex marriage, marriages among cousins, and the minimum ages at which people can marry, are more variable. All states limit people to one living husband or wife at a time and will not issue marriage licenses to anyone with a living spouse. Once an individual is married, the person must be legally released from the relationship by either death, divorce, or annulment before he or she may legally remarry. Limitations that some but not all states prescribe are the requirement of blood tests, good mental capacity, and being of the opposite sex.

Marriage between Blood Relatives

Nowhere in the United States is marriage allowed between parents and children, grandparents and grandchildren, brothers and sisters, uncles and nieces, and aunts and nephews. Half siblings (e.g., children who share the same biological mother but different fathers) are similarly restricted.

Perhaps these restrictions are unsurprising. Such blood relations are commonly considered "too close," and marriage within such relationships is seen as incestuous and unacceptable. Some states disallow all "ancestor/descendant marriages," and a handful of states explicitly extend the prohibition against marriages between parents and children to marriages between parents and their adopted children. Although many state marriage statutes articulate very specific restrictions, some states, such as Ohio or Washington, more simply and generally prohibit marriage between relatives "closer than second cousins."

Along with Canada and Mexico, 20 U.S. states and the District of Columbia allow first cousins to marry. Six other states allow such marriages under certain circumstances. Some may find it surprising that so many states allow

Although most states continue to restrict marriage to heterosexual couples, as of September 2009, six states allow same-sex couples to legally marry.

© Rob Melnychuck/White/photolibrary

first cousins to marry, thinking that if they were to have children together they would face risks of passing genetic defects to their children. Furthermore, there is debate about the justification for prohibiting first-cousin marriages, common in many other parts of the world, including the Middle East, Europe, and South Asia. One genetics researcher estimates that as many as 20% of marriages worldwide are between first cousins (Willing 2002). As to the risk to offspring of such marriages, there is only a slightly elevated risk (of an additional 2% to 3%) of such children inheriting recessive genetic disorders such as cystic fibrosis or Tay-Sachs disease (Willing 2002).

Age Restrictions

State laws regulate and restrict marriage based on age requirements. Throughout the United States, 49 of 50 states require both would-be spouses to be at least 18 years old to marry without parental consent. In Nebraska, both partners must be at least 19 years old. Some states will waive the age requirement under certain circumstances (e.g., if the woman is pregnant), but in such instances the couple may need approval from a court. Many states allow couples to marry in their early to mid-teens, provided that they secure parental and/or court consent.

Number of Spouses

Recall the discussion in Chapter 1 of the controversy over polygamy in Utah. No state allows an individual to marry legally if he or she is already married. In other words, all 50 states consider monogamy the only legally accepted form of marriage. If a divorced or widowed man or woman wishes to remarry, he or she must present evidence of the legal termination of the prior marriage or of the death of his or her former spouse.

The Controversy over Same-Sex Marriage

In Chapter 1, we also briefly considered the issue of same-sex marriage. After a flurry of recent judicial or legislative decisions, six states in the United States—Massachusetts, Connecticut, Iowa, Vermont, and New Hampshire—have legalized same-sex marriage. Meanwhile, in recent years, many other states have added to their state marriage laws explicit declarations that same-sex marriage will not be recognized

Figure 8.3 Attitudes toward Same-Sex Marriage

Do you think marriages between same-sex couples should or should not be recognized by the law as valid with the same rights as traditional marriage? (Gallup poll, May 8–11, 2008)

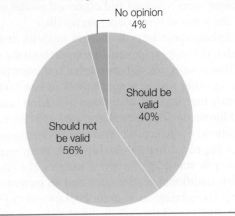

SOURCE: L. Saad, "Americans Evenly Divided on Morality of Homosexuality," Gallup Poll, June 18, 2008, gallup.com

Figure 8.4 Attitudes toward a Constitutional Amendment Barring Same-Gender Couples from Marrying

Would you favor or oppose a constitutional amendment that would define marriage as being between a man and a woman, thus barring marriages between gay or lesbian couples? (Gallup poll, May 8–11, 2008)

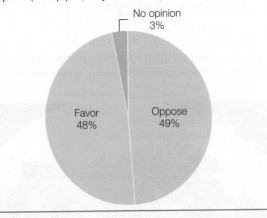

SOURCE: "Marriage," Gallup Poll, www.gallup.com/poll/117328/Marriage.aspx

within the state, even if it is legally allowed elsewhere in the United States. As of mid-2009, 41 of 50 states have laws defining marriage as between a man and a woman; 30 of those states have amended their state constitutions, defining marriage as strictly between a man and a woman.

The dynamic and divided legal status of same-sex marriage is mirrored by similarly divided and dynamic public opinion on the issue. As can be seen in recent public opinion data, Americans are divided on the issue of providing same-sex couples the same right to marry as heterosexuals (see Figures 8.3 and 8.4).

Same-Sex Marriage: A Quick Look Back

Internationally, the movement toward gay marriage began in the 1990s, when a number of countries enacted legislation extending marital rights or marriage-like protections to gay couples. Some stopped short of allowing gay or lesbian couples to legally marry, but as of mid-2009 in the Netherlands, Belgium, Canada, Spain, Sweden, Norway, and South Africa, the right to marry extends to same-sex couples. In addition, a number of countries—including Denmark, Norway, Sweden, Switzerland, Iceland, France, Germany, Finland, Luxembourg, Britain, Portugal, Slovenia, Australia, and New Zealand—granted recognition to same-sex couples who register as "domestic partners" or enter "civil unions," providing them with many if not all of the rights otherwise conferred by marriage. With the issue in such a dynamic state of change, by the time you read this, that list may well have grown.

In the United States, the issue of gay marriage has been in flux for more than a decade. In the 1990s, U.S. courts rendered decisions that appeared to pave the way toward American legalization of same-sex marriage. The two most notable cases were in Hawaii and Vermont. In 1993, the Hawaii Supreme Court ruled that denying gay men and lesbians the right to marry was unconstitutional in that it violated the equal protection clause of the state's constitution. This decision led many to anticipate the eventual legalization of same-sex marriage in the United States. It also caused opponents of gay marriage to take action. A number of state legislatures, along with the federal government, passed laws that declared marriage to be the union of one man and one woman, thus preventing the forced acceptance of gay or lesbian marriages should the Hawaiian decision stand up to an appeal.

In 1996, Congress passed the **Defense of Marriage Act**, and President Bill Clinton signed it into law. This act denied federal recognition to same-sex couples and gave states the right to legally ignore gay or lesbian marriages should they gain legal recognition in Hawaii or any other state. But the earlier Hawaiian decision did not stand. In a November 1998 ballot, 69% of Hawaiian voters chose to amend the state constitution, giving lawmakers the power to block same-sex marriage and limit legal marriage to heterosexual couples. Similar laws were passed in more than half of the 50 states by November 1998.

As 1999 drew to a close, the state of Vermont took a major step toward what some then believed, others hoped, and opponents feared would be the eventual legal recognition of gay marriage. There, three same-sex couples filed lawsuits, challenging a 1975 state ruling prohibiting same-sex couples from marrying. On December 20, 1999, the Vermont Supreme Court ruled that the state legislature had to either grant marriage rights to same-sex couples or assure them a legal equivalent to marriage, providing them the same range of state benefits enjoyed by married heterosexuals.

On April 26, 2000, Vermont Governor Howard Dean signed into law legislation recognizing same-sex "civil unions." Although they are not marriages, "civil unions" are officially entered, offer the same rights and protections as marriages, and must be officially dissolved when they fail. As of April 2008, close to 10,000 such civil unions had been recorded in Vermont, nearly 1,500 between state residents and more than another 8,000 involving residents of other states, the nation's capital, and several other countries, including Canada (Vermont Guide to Civil Unions 2005, www.sec.state.vt.us/otherprg/civilunions/civilunions.html).

In October 2001, California passed Chapter 893, a law granting gay or lesbian domestic partners many benefits (including tax benefits, stepparent adoption, sick leave, and permission to make medical decisions) otherwise restricted to married couples. Although far less sweeping in scope than Vermont's civil union legislation, Chapter 893 provided same-sex couples more benefits than found anywhere in the United States other than Vermont (Vermont Guide to Civil Unions 2005). In June 2002, Connecticut passed more limited legislation giving gay or lesbian couples certain partnership rights and responsibilities.

Of greatest significance, the Massachusetts Supreme Court ruled on November 18, 2003, that the state's ban of same-sex marriage was unconstitutional. The ruling gave the state legislature six months to remedy the situation. Although Vermont's response was to create civil unions that provided the same rights and benefits as legal marriage, the Massachusetts court's decision specified the *right to marry*. Although the Massachusetts legislature and governor remained opposed to same-sex marriage, on February 4, 2004, the state supreme court ruled four to three that a "civil union" solution was unacceptable in that it would constitute "an unconstitutional, inferior, and discriminatory status for same-sex couples." As you read these words, Massachusetts has had more than five years of fully legal gay marriage.

Where We Are and Where We're Going

In 2008 and 2009, there were a number of significant developments. In late 2008, Connecticut became the second state to legalize same-sex marriage. In April 2009, Iowa and Vermont joined Connecticut and Massachusetts and a month later were joined by Maine as the states granting full marriage rights to gay men and lesbians. Additionally, some states, such as Hawaii, New Jersey, Oregon, Washington, and California, recognize same-sex domestic partnerships or civil unions. In 2009, Washington, D.C., joined Rhode Island in recognizing same-sex marriages performed elsewhere, even though gay or lesbian couples were not able to marry in the district.

It is difficult to predict what level legalization of or opposition to gay marriage will continue or what effect it will have. Some additional states may legalize same-sex marriage. Even if no other state legalizes same-sex marriage, we may see wider recognition of civil unions or same-sex marriages performed elsewhere in the United States (as is the case in Rhode Island, New York, and Washington, D.C.). Without such reciprocal recognition (i.e., other states acknowledging and supporting same-sex marriages performed in states where they are legal), more civil suits are certain to follow. Still other state legislatures might create their own domestic partner legislation. In the opposite direction, voters in states that legalized same-sex marriage may vote to overturn such decisions, as happened in California in 2008. The only certainty is that the status of same-sex marriage will continue to change.

The Marriage Market: Who and How We Choose

People usually marry others from within their same large group—such as the nationality, ethnic group, or socioeconomic status with which they identify—because they share common assumptions, experiences, and understandings. This practice, known as **endogamy**, strengthens group structure. If people already have ties as friends, neighbors, work associates, or fellow church members, a marriage between such acquaintances solidifies group ties. But another, darker force may lie beneath endogamy: the fear and distrust of outsiders, those who are different from ourselves. Both the need for commonality and the distrust of outsiders urge people to marry individuals like themselves.

Conversely, the principle of **exogamy** requires us to marry outside certain groups—specifically, outside our own family (however defined) and outside our sex. Exogamy is enforced by taboos deeply embedded within our psychological makeup. The violation of these taboos may cause a deep sense of guilt. A marriage between a man and his mother, sister, daughter, aunt, niece, grandmother, or granddaughter is considered incestuous; women are forbidden to marry their corresponding male relatives. Beyond these blood relations, however, the definition of incestuous relations changes. One society defines marriages between cousins as incestuous, whereas another may encourage such marriages.

Homogamy

Endogamy and exogamy interact to *limit* the field of eligibles. The field is further limited by society's encouragement of **homogamy**, the tendency to choose a mate whose personal or group characteristics are similar to ours. This is also known as **positive assortative mating** (Blackwell 1998). **Heterogamy** refers to the tendency to choose a mate whose personal or group characteristics differ from our own. The strongest pressures are toward homogamy. We may make homogamous choices regarding any number of characteristics, including age and race, but also such characteristics as height (Blackwell 1998). As a result, our choices of partners tend to follow certain patterns. These homogamous considerations generally apply to heterosexuals, gay men, and lesbians alike in their choice of partners.

The most important elements of homogamy are race and ethnicity, religion, socioeconomic status, age, and personality characteristics. These elements are strongest in first marriages and weaker in second and subsequent marriages (Glick 1988). They also strongly influence our choice of sexual partners, in part because our sexual partners are often potential marriage partners (Michael et al. 1994).

Race and Ethnicity

Most marriages are between members of the same race. Of the nearly 55 million married couples in the United States in 2000, 98% of them consisted of husbands and wives of the same race. More than 7% of marriages in 2005 were between people from different racial backgrounds. Interestingly, as the overall phenomenon of interracial marriage has been increasing since the 1980s, it has especially increased among highly educated people (Harris and Ono 2005). Although racial intermarriage is sometimes taken to mean *black–white marriage*, in fact Asian Americans,

Hispanics, and Native Americans are all much more likely to marry whites than are African Americans (Batson, Qian, and Lichter 2006). Black–white marital pairings are only approximately 25% of all racial intermarriages. Nevertheless, this is the pairing most likely to be the target of hostility and prejudice (Leslie and Letiecq 2004).

Among the Hispanic population, Puerto Rican women and men are least likely to marry white spouses and are the most likely to marry African Americans. However, Puerto Ricans are more likely than African Americans to intermarry with whites. Of course, homogamy is the more prevalent pattern for all groups. As shown in Figure 8.5, data from the 2002 National Survey of Family Growth revealed homogamy estimates.

Interracial marriage varies greatly across different cities and regions in the United States. David Harris and Hiromi Ono assert that without taking into consideration "local" marriage markets, we can't completely and accurately understand racial marriage patterns. For example, the 2008 racial composition of the U.S. population is as follows: 66% non-Hispanic whites, 15% Hispanics, 14% African Americans, and 5% Asians. The racial composition of major cities exhibits substantial deviations from the national pattern and differs from one another in important ways. For example, whites are 45% of the population in Philadelphia but only 12% in Detroit. Asians are at least 25% of the population in San Jose, San Francisco, and Honolulu but no more than 2% of the population in Phoenix, San Antonio, and Detroit (Harris and Ono 2005). These differing population compositions matter because where there is greater opportunity to find spouses of the same race, rates of homogamy are higher and intermarriage is less. On the other hand, racial and ethnic heterogeneity are associated with higher levels of intermarriage (Kalmijn 1998).

Black–White Intermarriage. Approximately one-fourth of racial intermarriages are marriages between African Americans and Caucasians (Fields and Casper 2001). Racial intermarriage is more common among highly educated people than among the less educated. Thus, among the reasons for increases in both black groom–white bride and black bride–white groom marriages is the rise of a black middle class, making African American men and women more attractive spouses to middle-class whites (Gullickson 2006).

Approximately three-fourths of all marriages of African Americans and whites are between an African American man and a white woman. This same pattern persists among interracial cohabitors and holds regardless of level of education (Batson et al. 2006). Overall, interracial marriages have been found to be at greater risk of divorce than racially homogamous marriages, though the risk is not the same for all interracial pairings. Some research has found interracial marriages where one spouse is Asian to be more stable than marriages of two white spouses. In contrast, the particular combination of marriage between a black man and a white woman faces the greatest risk of instability (Bratter and King 2008; Zhang and van Hook 2009).

Qualitative research sheds some light on experiences of intermarried African American and Caucasian couples. Based on their study of 76 such intermarriages (52 black male–white female couples and 24 black female–white male couples), Leigh Leslie and Bethany Letiecq (2004) suggest that success in black–white intermarriages may depend on the degree to which the partners possess pride in their race or culture without diminishing other races. This appears to be especially true for the black spouse in such marriages and seems to influence the quality of married life well into the marriage. Those who had resolved issues of racial

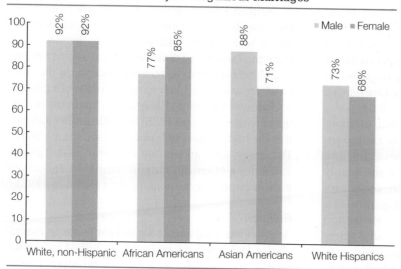

Figure 8.5 Percent of Racially Homogamous Marriages

SOURCE: L. Bratter and King (2008).

Only about one in four racial intermarriages are between an African American and Caucasian. Typically, when they occur, these are marriages of black husbands to white wives.

if not thrive, even in the absence of social support (Leslie and Letiecq 2004).

Intensive interviews with individuals involved in interracial relationships uncovered a range of harassment and hostility they endure and strategies they develop to deal with such mistreatment (Datzman and Brooks Gardner 2000). These include ignoring the harassment, limiting the settings where they would be seen as a couple to those who they knew would be supportive or staying home altogether, having others with them who are more supportive, and directly confronting any harassment. Especially when such harassment is new, the emotional impact might include shock and surprise, numbness, sadness and shame, and ultimately resentment or anger. Eventually, the anger might be replaced by pity felt toward the harasser or harassers. According to research by sociologist George Yancey, white spouses married to African Americans, especially white women married to black men, experience more racism than do white spouses of nonblack minorities (Yancey 2007).

Among European ethnic groups in this country—such as Italians, Poles, Germans, and Irish—only one in four marries within their ethnic group. The ethnic identity of these groups has decreased considerably since the beginning of the twenty-first century. Interestingly, Louisiana Cajuns have high rates of ethnic homogamy, especially for a group of their size and considering the length of time they have been in the United States. Among married Cajun women, more than 75% were married to Cajun men; among Cajun men, more than 70% were also homogamous (Bankston and Henry 1999).

Matthijs Kalmijn points out that marrying *outside* the group is not the same for all ethnic groups. For example, when Latinos marry "out," they are more likely marrying Latinos of a different cultural origin than they are white, European Americans. Asians, on the other hand, are much less likely to marry Asians of a different background and more likely when "marrying out" to marry whites (Kalmijn 1998). Kalmijn further indicates that the highest rates of homogamy are among blacks. The lowest rates are among European ethnic groups and among American Indians. Hispanics and Asians have intermediate homogamy rates (Kalmijn 1998).

Religion

Religion has long been a significant factor in marital choice and has been consistently linked to marital quality and stability. Sociologist Scott Myers (2006) summarizes past research as follows:

identity and developed a strong black identity while showing racial tolerance and appreciation of other races more positively evaluated their marriage, felt less ambivalent about it, and/or worked harder to maintain it. However, those who had more negative assessments of either black or white culture experienced lower marital quality (Leslie and Letiecq 2004).

To an extent, this challenges the idea that, for interracial couples, race becomes irrelevant or unimportant (Leslie and Letiecq 2004). How they think and feel about race is of major significance in the quality of their marital experience. Unexpectedly, social support only "modestly" predicted marital quality. This could be a by-product of the relative prevalence and acceptance of interracial marriage in the area where the research was done, the relatively comfortable economic circumstances of the couples studied, or evidence that interracial couples have learned to survive,

- Religious homogamy has a stronger effect on marital quality than does the level of religiosity of either spouse or of the couple.
- Marital conflict is higher among couples with theological differences. The greater the differences in spouses' religious beliefs, the more likely the marriage is to be deemed unhappy.
- Religious homogamy unites spouses who have similar values, thus strengthening their commitment and giving them a unified approach to marital and family matters. Myers found that religious homogamy, especially when expressed in joint church attendance, continues to be associated with higher marital quality, albeit to a lesser extent than in the past.

Most religions still oppose interreligious marriage because they believe that it weakens individual commitment to the faith. Although interreligious dating and marriage have increased, religious homogamy remains evident in American patterns of choosing spouses. Data from 1972 to 1996 indicate that religious homogamy continued to occur more often than would be predicted by chance (i.e., if religion made no difference in people's spousal choices) for Protestants, Catholics, and Jews in the United States (Bisin, Topa, and Verdier 2004).

Those who marry from different religious backgrounds do have greater risk of marital unhappiness and divorce than those from similar religious backgrounds (Myers 2006). Apparently, the *larger* the "religious distance," or disparity between two people's backgrounds, the more likely they are to characterize their marriage as "unhappy" (Ortega, Whitt, and William 1988). In a study of Jewish marriages, what matters more in predicting the amount of conflict and instability is the extent of agreement or disagreement on Jewish issues, not what self-reported labels people use to identify themselves (Chinitz and Brown 2001).

Much of the concern that is expressed about religious intermarriage has to do with the potential effects on children and the conflicts that might ensue over child-rearing issues. With data from the National Survey of Families and Households, sociologists Richard Petts and Chris Knoester found that religious heterogamy is associated with more marital conflict and with less religious participation. Children of intermarried couples are more likely to engage in substance use (e.g., underage drinking and/or marijuana use), though they are no more likely to engage in delinquency, experience academic difficulties or suffer from reduced self-esteem (Petts and Knoester 2007).

Socioeconomic Status

Most people marry others of their own socioeconomic status and of the same or similar educational background. Even if a person marries outside his or her ethnic, religious, or age-group, the selected spouse will probably be from the same socioeconomic level. Furthermore, some ethnic or racial homogamy may be increased because of tendencies toward socioeconomic homogamy (Bankston and Henry 1999). Of the various dimensions of socioeconomic status (family background, education, and occupation), the weakest appears to be between spouses' class origins (correlation of about 0.30). The correlation between husbands' and wives' occupational statuses is stronger (around 0.40). However, the strongest correlation is between spouses' educational backgrounds (approximately 0.55). This holds true in the United States as well as most other countries. In the United States, educational homogamy "strongly" increased over the latter decades of the twentieth century (Kalmijn 1998).

Not everyone marries homogamously. Men more than women marry below their socioeconomic level (**hypogamy**); women more often "marry up" (a practice known as **hypergamy**). When class *intermarriage* occurs, it is rarely a case of spouses from opposite extremes (i.e., paupers and princesses). Both the upper and the lower levels of the class spectrum appear more "closed" than the middle levels (Kalmijn 1998).

Much research reveals a tendency for individuals to select spouses within their same educational level. In fact, the likelihood that spouses share the same level of educational attainment has increased in recent decades and may now be higher than any time since 1940 (Schwartz and Mare 2005). Such marriage patterns may increase economic inequality, as more highly educated individuals tend to earn higher incomes. Marriages between two spouses with the educational resources to earn high incomes will be more likely to achieve or retain middle-class or higher status. Similarly, marriages at the lower end of the educational spectrum bring together two spouses with much more limited earnings potential, making it more difficult for them to achieve upward mobility. According to sociologists Christine Schwartz and Robert Mare, educational homogamy has increased at both ends of the education spectrum.

Occupationally, the biggest divide is between those in white-collar and those in blue-collar occupations. It appears as though the cultural status, not the economic status, of occupations is a more important factor in determining compatibility and attractiveness for marriage (Kalmijn 1998). A teacher may not find a plumber to be an attractive marriage choice, even if the teacher and the plumber have very similar incomes. More than the economic resources associated with an occupation is often considered in potential partners.

The Marriage Squeeze and Mating Gradient

An important factor affecting the marriage market is the ratio of men to women. The **marriage squeeze** refers to the gender imbalance reflected in the ratio of available unmarried women and men. Because of this imbalance, members of one gender tend to be "squeezed" out of the marriage market. The marriage squeeze is distorted, however, if we look at overall figures of men and women without distinguishing between age and ethnicity. Overall, from ages 18 to 44, the prime years for marriage, there are significantly more unmarried men than women. Combining widowed, divorced, and never-married people, in 2007 there were 113 unmarried men, aged 18 to 44, for every 100 unmarried women (U.S. Census Bureau 2008). Thus, women in this age group have greater bargaining power and are able to demand marriage and monogamy. But once ethnicity is taken into consideration, many African American women of all ages are squeezed out of the marriage market. With eligible males scarcer, African American men have greater bargaining power and are less likely to marry because of more attractive alternatives (see Figure 8.1).

"All the good ones are taken" is a common complaint of women in their mid-thirties and beyond, even if there are still more men than women in that age bracket. The reason for this is the **mating gradient**, the tendency for women to marry men of higher status. Although we tend to marry those with the same socioeconomic status and cultural background, men tend to marry women slightly below them in age, education, and so on.

Age

Americans tend to marry those of similar ages. The trend toward age homogamy commenced in the latter decades of the nineteenth century and continued through the twentieth. In 1900, husbands were an average of five years older than wives; by the end of the century. the gap was little more than two years (Rolf and Ferrie 2008).

Age is important because we view ourselves as members of a generation, and each generation's experience of life leads to different values and expectations. Furthermore, different developmental and life tasks confront us at different ages. A 20-year-old woman wants something different from marriage and from life than a 60-year-old man does. By marrying people of similar ages, we often ensure congruence for developmental tasks. As the gap between grooms' and brides' ages has narrowed in recent years, the ages at which both men and women enter marriage have climbed.

Research suggests that the importance individuals place on age varies *by age* differently for men than for women. As men age, they prefer women progressively younger than themselves. Women, on the other hand, prefer for their partners to be about the same age (ranging from slightly younger through slightly older), up to 10 years older than themselves. This does not appear to vary much, even as women age (Buunk et al. 2002). Interesting data from the United States and Australia reveal that the same age preferences that exist among heterosexuals exist among homosexual men and women—men prefer younger partners, and women prefer partners of about the same age (Over and Phillips 1997).

Despite the popular beliefs that we will be more compatible with partners similar to us in age and, conversely, that relationships with partners much older than ourselves will be plagued by problems of incompatibility, research suggests otherwise. A study by David Knox and Tim Britton of 97 female students and faculty involved with partners between 10 and 25 years older than themselves concluded that couples in "age-discrepant" relationships were, indeed, happy. In the study, 80% indicated either agreement (40%) or strong agreement (40%) with the statement, "I am happy in my current relationship." Only 4% disagreed. Furthermore, more than 60% stated that if their current relationship ends, they would enter another age-discrepant relationship. They identified the following benefits of such relationships: maturity (mentioned by 58% of the women), financial security (58%), dependability (51%), and higher status (28%). Also of note, only 25% of the women stated that their relationships had the support of their friends or parents. Fathers were most disapproving, with more than 40% identified as not being in support of the relationship (Knox and Britton 1997).

Marital and Family History

An interesting application of the concepts of homogamy and heterogamy (intermarriage) can be found with regard to marital history. Essentially, never-married people are more likely to marry other never-married people than they are to "intermarry" by marrying divorced people (Ono 2005). Hiromi Ono questions whether this is a "by-product" of other homogamous patterns (such as age, socioeconomic status, or parenthood status) or a deliberate choice that individuals make to marry someone of similar marital history. For example, a divorced person may believe that only another divorced person will similarly understand and have experience with the lingering ties to prior marriages.

Conversely, never-married individuals who marry divorced partners may find that they have to deal with lower amounts of resources because of the continued demands of former spouses and the needs of children of former marriages. This, in turn, may give rise to jealousy and impede the development of needed levels of trust (Ono 2005). Ono determined that **marital history homogamy** occurs more as a result of deliberate choices. Ono also reasonably speculated that parental status, like marital history, operates in a similar fashion. Parents make lifestyle concessions to their parenting responsibilities that nonparents don't have to make. Where children and their needs become priorities for parents, nonparents can maintain other priorities.

The structure of an individual's family of origin also turns out to be a factor in the process of mate selection. Children of divorced parents often marry other children of divorced parents. Research by Nicholas Wolfinger suggests that coming from a divorced home increases by 58% the likelihood of choosing another child of divorce as a spouse. Although homogamy often is associated with a greater chance for marital happiness and stability, family structure homogamy may be a noteworthy exception because marriages in which both spouses are children of divorce face greater odds of marital failure. Marriages in which either spouse comes from a divorced family are twice as likely to fail as those in which neither spouse is a child of divorce. When both spouses are from divorced homes, their marriages face three times the likelihood of failure as marriages between two children of intact parental marriages (Wolfinger 2003).

Residential Propinquity

An additional homogamous factor is based on the principle of **residential propinquity**—the tendency we have to select partners (for relationships and for marriages) from a geographically limited locale. Put differently, the likelihood of marriage decreases as the distance between two people's residences increases. The obvious explanation behind this is one of opportunity. In most instances, to start dating or get together with someone, you have to first meet. Our chances of meeting are greater when our daily activities (shopping, commuting, eating out, and so forth) overlap.

Although it is easy to trivialize this tendency as too obvious to be meaningful, consider the implications it has for some of the other patterns of homogamy. American communities are often segregated by class, race, or both. In some towns, they may even have religious splits (e.g., the Catholic side and the Protestant side of town or a Jewish neighborhood). Public schools, being neighborhood based, further the tendency for us to associate with others like ourselves. Thus, the types of people we are most likely to come into contact with and with whom we might develop intimate relationships or eventually marry are a lot like ourselves. Meeting at school promotes age, educational, and social class homogamy (Kalmijn and Flap 2001).

Thus, within a society somewhat residentially segregated by race or social class, residential propinquity may explain some other homogamous tendencies by how it limits our opportunity. But the story is more complicated than where we live. After all, unmarried people do not simply wander around a region looking for a spouse; they spend most of their lives in small and functional places, such as neighborhoods, schools, workplaces, bars, and clubs. Such local marriage markets are often socially segregated, which is why they are important for explaining marriage patterns. In the sociological literature, three local markets have been considered most often: the school, the neighborhood, and the workplace. Of these three, schools are considered the most efficient markets because they are homogeneous with respect to age and heterogeneous with respect to sex (Kalmijn 1998).

Critical Thinking

Keeping heterogamy and homogamy in mind, think about those who are or have been your romantic or marital partners. In what respects have your partners shared the same racial, ethnic, religious, socioeconomic, age, and personality characteristics with you? In what respects have they not? Have shared or differing characteristics affected your relationships? How?

Understanding Homogamy and Intermarriage

Factors in the choice of partner interact with one another. Ethnicity and socioeconomic status, for example, are often closely related because of discrimination. Many African Americans and Latinos are working class and are often not as well educated as Caucasians. Thus, a marriage that is endogamous in terms of ethnicity is also likely to be endogamous in terms of education and socioeconomic status.

Matthijs Kalmijn (1998) identifies three social forces that help explain marriage patterns: the preferences of individuals for resources in a partner, the influence of one's peer group, and the constraints of the marriage market. It appears that all three of these combine to produce the tendencies toward homogamy and the patterns of mate choice we observe, but it is difficult to determine the relative strength of the factors or what is "most influential" in shaping mate selection practices. What we can say with more certainty is that the presence of both opportunity constraints and outside influence (or "interference") makes it unwise to conclude that homogamy automatically reflects hostility or animosity toward others unlike oneself. It may not even illustrate an outright preference for people like oneself.

Hiromi Ono (2005) contends that—especially with regard to race, education, and social class patterns but also with reference to marital history—homogamy has the potential to widen social inequality. What about consequences of intermarriage? Kalmijn (1998) argues that intermarriage potentially has the following effects:

- Intermarriage can decrease the importance of cultural differences because the children of mixed marriages are less likely to identify themselves with a single group. Even when mixed couples socialize children into the culture of a single group, the children are less likely to identify with that group when intermarriage in society is common.
- Through intermarrying, individuals may question and lose negative attitudes they have toward other groups. Spouses and their wider networks (of kin and friends) gain the opportunity to get to know people "different" from themselves and question any biases and stereotypes they previously held.

Theories and Stages of Choosing a Spouse

At most, homogamy narrows down the pool from which we might seek a spouse. Furthermore, sharing background characteristics doesn't automatically guarantee that two people will feel close to each other, fall in love, and/or want to share their lives. Homogamy alone isn't sufficient to account for whom we choose. A range of theories has been suggested to address the question of why we select particular individuals. Do "opposites attract"? Do "birds of a feather flock together"? Do we unconsciously select people like our parents? What is more important: finding someone who seems to think as we do about things or finding someone whose behavior fits what we expect in a partner?

Each of the preceding questions illustrates the premise of an existing theory of mate selection. The commonsense notion that "opposites attract" is in keeping with **complementary needs theory**, the belief that people select as spouses those whose needs are different from their own. Thus, an assertive person who has difficulty compromising will be drawn to a less outgoing and highly adaptable person. The complementarity and subsequent interdependence are alleged to strengthen the bond between the two partners.

The notion that "birds of a feather flock together" is more in keeping with theories such as **value theory** or **role theory**, in which gratification follows from finding someone who feels and/or thinks like we do. Having someone who shares our view of what's important in life or who acts in ways that we desire in a partner validates us, and this sense of validation leads to an intensification of what we feel toward that other person.

Parental image theory suggests that we seek partners similar to our opposite-sex parent. Some versions of parental image theory draw on Freudian concepts such as the Oedipus complex, whereas others point toward the lasting impressions made by our parents (Eshelman 1997; Murstein 1986).

Bernard Murstein developed a social exchange–based sequential theory known as **stimulus–value–role theory** to depict what happens between that "magic moment" with its mysterious chemistry of attraction and the decision to maintain a long-term relationship such as marriage. Murstein's theory identifies three stages of romantic relationships: the stimulus, value, and role stages. At each stage, if the exchange seems equitable, the two will progress to the next stage and ultimately remain together (Murstein 1986):

- In the *stimulus* stage, each person is drawn or attracted to the other before actual interaction. This attraction can be physical, mental, or social. During the stimulus stage, with little other information on which to evaluate the other person, we make potentially superficial decisions. This is especially evident during first encounters.
- In the next stage, the *value* stage, partners weigh each other's basic values, seeking compatibility. As relationships continue, each person discovers the other's philosophy of life, politics, sexual values, religious beliefs, and so on. Wherever they agree, it is a plus for the relationship. However, if they disagree—for example, on religion—it is a potential minus for the relationship. Each person adds or subtracts the pluses and minuses along value lines. Depending on the outcome, the couple will either disengage or go on to the next stage.
- Eventually, in the *role* stage, each person analyzes the other's behaviors, or how the person fulfills his or her roles as lover, companion, friend, and worker and potential husband or wife and mother or father. Are the person's behaviors consistent with marital roles? Is he or she emotionally stable? This aspect is evaluated in the eighth and subsequent encounters.

Public Policies, Private Lives: What Are We Getting Into?
The Essence of Legal Marriage

On hearing the declaration "I now pronounce you husband and wife," couples may kiss, be congratulated, and celebrate. Their relationship also becomes legally binding. As a legal contract, marriage imposes certain responsibilities and obligations but also bestows considerable rights and protections on spouses. As discussed in Chapter 1, marriage confers a wide range of benefits from tax breaks to rights to care for one another if hospitalized or to inherit ("Marriage Rights and Benefits," www.nolo.com).

Marriage also imposes legal *responsibilities* and *obligations* on spouses, although these may not be spelled out. The "model marriage statute" is intended as a legislative device to provide "firmer guidance to courts and family law as a discipline about the nature and public purposes of marriage" (www.marriagedebate.com/ml_marriage/cat03-ml01.php). According to law professor Katherine Spaht, who drafted a "Model Marriage Obligations Statute," when they marry, husbands and wives owe each other mutual respect, sexual fidelity, mutual economic and emotional support, practical assistance, and mutual commitment to and responsibility for the joint care of any children they have together (www.marriagedebate.com/ml_marriage/cat03-ml01.php).

Fidelity, or sexual exclusivity, is described by Spaht as "the hallmark of marriage" and the essence that distinguishes marriage from "mere cohabitation." Cumulatively, the other designated obligations include spousal expectations about appropriate marital behavior and trust, reciprocity, and sharing. However, most states do not explicitly define marriage responsibilities and obligations in statute, relying instead on a common law understanding of marriage. Louisiana is one notable exception. According to Louisiana *Civil Code Art. 98,*

Mutual duties of married persons, "Married persons owe each other fidelity, support, and assistance" (www.marriagedebate.com/ml_marriage/cat03-ml02.php).

Before we leave the topic of legal marriage, we ought to note that there is much ongoing disagreement and debate about what marriage does or ought to mean legally, whether legal marriage should or shouldn't be made available to same-sex as well as heterosexual couples, and whether its benefits and responsibilities ought to extend to unmarried couples. One way in which the debate has been framed, albeit by those from a more conservative perspective, is as a clash between two views of marriage: a conjugal model of marriage versus a close relationship model.

The **conjugal model of legal marriage** defines marriage as "child centered" because it stresses the importance of "sustaining enduring bonds between women and men in order to give a baby its mother and father, to bond them to one another and to a baby." A conjugal marriage is "a sexual union between a man and a woman who promise each other sexual fidelity, mutual caretaking, and the joint parenting of any children they may have."

On the other hand, the **close relationship model** of legal marriage sees marriage "as one in a universe of diverse close, private relationships, with intrinsic emotional, psychological, and sexual dimensions." From the conjugal model, only heterosexual legal marriage ought to be recognized in family law. In the close relationship model, the law ought to recognize and protect all relationships in which individuals share intimacy, commitment, interdependence, mutual support, and communication, regardless of whether partners are of the same or opposite sex and regardless of whether they legally marry or not.

Although the stimulus–value–role theory has been one of the more prominent theories explaining relationship development, some scholars have criticized it, especially regarding the question of whether we actually test the degree of "fit" between us and our partners. In reality, we might underestimate the importance of certain issues or, conversely, be focused more extensively on others. For example, religious fundamentalists and atheists may sometimes believe that they are compatible. They may not discuss religion; instead, they might focus on the "incredible" physical attraction in their relationship or their mutual desire to raise a family. They may believe that religion is not that important, only to discover after they are married that it is important.

None of these theories has been proven to be *the explanation* of how two people select each other for marriage. In fact, it is unlikely that any single theory can account for the broad range of relationships that people construct and enter.

Why Marry?

If you stopped each couple just before they exchanged their vows and asked, "Why are you doing this? Why are you getting married?," you would no doubt hear many different answers. More important for the moment, however, are the many reasons people can give for why they want to marry. The greatest attraction of marriage is probably the love and intimacy that we expect to come with it. A nationally representative sample of 1,003 young adults (20–29 years old) demonstrated the extent to which our views about marriage and, perhaps, the appeal of marriage is rooted in the intimacy and love we hope to find there. More than 9 out of 10 never-married respondents endorsed the notion that "when you marry, you want your spouse to be your soul mate, first and foremost" (Whitehead and Popenoe 2001, in Cherlin 2006). In addition, more than 80% of women surveyed indicated that it was more important to "have a husband who can communicate his deepest feelings" than a husband who is financially successful (Cherlin 2004). Clearly, we are drawn to marriage in pursuit of a level of love and intimacy that we believe may not be otherwise available or possible. As sociologist Paul Amato (2004b) puts it, we tend to see marriage as "the gold standard" for relationships.

Among the many reasons for marriage, we can easily recognize the role of possible economic and social pressures (i.e., "pushes" toward marriage) as well as the strong desires to have and raise children, which, for many, seem to be best accomplished in marriage. As Amato (2004b) expresses, "Most people will continue to see marriage as the best context for bearing and raising children" and, if they desire to become parents, will marry. For many, marriage also symbolizes that two people have reached a stage in their lives as well as in their relationships and that in it they have attained "a prestigious, comfortable, stable style of life" (Cherlin 2004, 857).

If the practical importance of marriage has diminished, if marriage can no longer be counted on to cement relationships, allowing spouses to confidently invest themselves in each other without fear, invest their time and energy in raising children together, and invest financially in acquiring such goods as cars and homes, the "symbolic significance" of marriage remains considerable and attractive. It has become less a marker of conformity and more a marker of prestige (Cherlin 2004).

Benefits of Marriage

In what ways does being married benefit the women and men who marry? Marriage confers benefits in economic well-being (e.g., higher income, greater productivity, and mobility at work), physical and mental health, and personal happiness. Marriage provides clear economic benefits, and married couples are better off financially than those living in all other types of households (Hirschl, Altobelli, and Rank 2003). Marriage both reduces the risk of poverty and increases the probability of affluence. Defining *affluence* as living in a household that earns 10 times the poverty level, Thomas Hirschl, Joyce Altobelli, and Mark Rank conclude that married-couple households are more likely to attain affluence than those living outside of marriage. Women, in particular, face a much greater likelihood of attaining affluence in marriage than outside of marriage (Hirschl, Altobelli, and Rank 2003; Wu and Hart 2002/2003). Some of the health advantages come from the fact that married people tend to live healthier lifestyles than unmarried people. Researchers at the Centers for Disease Control concluded that married women and men are less likely to smoke, drink heavily, or be physically inactive and are less likely to suffer from headaches and serious psychological distress (Schoenborn 2002; Stimpson and Wilson 2009). The social and emotional support one derives from having a spouse also help improve one's health and well-being. When marriages end, women suffer increased depression, and men suffer poorer physical and mental health.

Is It Marriage?

In considering the benefits that seem to accompany marriage, researchers have been somewhat divided as to whether these benefits truly follow marriage or were instead reflections of differences in the types of people who do and don't marry. Sometimes phrased as a difference between *selection* into marriage and *protection* afforded by marriage, it raises the question of whether there is something unique and beneficial about being married (i.e., protection) or whether those who marry are somehow unique compared to those who don't marry (i.e., selection).

In research on health and well-being, selection is typically not the major factor, accounting instead for "only a small proportion of the variance in mental and physical health" (Wu and Hart 2002/2003, 421). For example, research into the effect of marriage and on depression looked to differentiate between marriage effects and differences in the types of people who do and don't marry. The researchers concluded that marrying was associated with "substantively meaningful reduction" in rates of depression and that there was no indication that marriage was selective of less depressed people (Lamb, Lee, and DeMaris 2003).

Although we have painted these as alternatives—as *either* selection *or* protection—the two are hardly mutually exclusive. It is possible that both operate simultaneously. Thus, although healthier and more stable individuals may be more attractive as marriage partners, thus bringing better mental health with them into their marriages, a good marriage also has healthful and stabilizing effects on those who marry. Such is the conclusion of research by sociologists Daniel Hawkins and Alan Booth. They contend that happier and healthier people may be more likely to marry but that marriage itself is associated with increased psychological well-being and at least half of the observed difference in health between married and unmarried individuals (Hawkins and Booth 2005).

Or Is It a *Good* Marriage?

It should be noted that we said that a *good marriage* has healthful and stabilizing effects; what happens to those in "not-so-good" marriages or marriages in which one or both spouses are unhappy, where there

are high levels of conflict and spouses' expectations remain unmet? Research indicates that such marriages do not provide the health, happiness, or other benefits that otherwise appear to separate happily married people from the unmarried. Furthermore, long-term unhappy marriages appear to have negative effects on spouses. Hawkins and Booth (2005) contend that remaining unhappily married actually lowers one's happiness, life satisfaction, and self-esteem and is associated with poorer overall health. Divorce, though associated with psychological distress, elevated levels of depression, and lower levels of self-esteem and happiness, may nonetheless improve the health and well-being of those in unsatisfying or unhappy marriages. This appears especially true for those divorced women and men who remarry, but even those who remain unmarried after divorce report greater

© Creasource/Corbis

Although marriage can bring many benefits, unhappy, high-conflict marriages may leave spouses worse off than had they not married or ended their marriage.

self-esteem, more satisfaction with their lives, and better overall health than do unhappily married people. They assert that unhappily married individuals are *moderately worse off* than those who divorce.

Predicting Marital Success

The period before marriage is especially important because couples learn about each other—and themselves. Courtship sets the stage for marriage. Many of the elements important for successful marriages, such as the ability to communicate in a positive manner and to compromise and resolve conflicts, develop during courtship. They are often apparent long before a decision to marry has been made (Cate and Lloyd 1992). Couples who are unhappy before marriage are more likely to be unhappy after marriage as well (Olson and DeFrain 1997).

Ted Huston and Heidi Melz (2004, 952) describe three "prototypical courtship experiences," each of which has different likely consequences for couples who marry. Of critical importance in differentiating these courtships are personality characteristics of partners, which affect "both the dynamics of their courtships and the success of their marriages." Some qualities, such as warmheartedness or an even temper, are important determinants of whether people create happy and stable marriages. Other qualities, such as being less stubborn, less independent minded, and more conscientious, are important factors in determining whether couples stay married. These personality characteristics are associated with the three courtships and marital outcomes that Huston and Melz identify as follows:

- *Rocky and turbulent courtships.* Such courtships are characterized by periods of upset and anger, distress and jealousy over potential rivals, and uneasiness about placing love in "undeserving hands" (950). They are more typically experienced by "difficult" personalities, people who are exceedingly independent minded, who lack conscientiousness, and who have high anxiety. If men are excessively independent, they may make poor husbands, and their marriages are likely to be "brittle." If men and women high in anxiety marry each other, their marriages tend to be unhappy but lasting marriages.
- *Sweet and undramatic courtships.* Partners are people with "good hearts" who are helpful, sensitive to the needs of others, gentle, warm, and understanding. Good-hearted couples find enjoyment and pleasure

in each other's company. Their marriages are more likely to be satisfying and enduring.
- *Passionate courtships.* These are characterized by partners "plunging into love, having sex early in the relationship, and deciding to marry one another within a few months" (950). Such couples begin marriage as "star-crossed lovers" sharing far more affection than typical of even newly married couples, "but over the first two years, much of the sizzle fizzles" (950). They are also vulnerable to divorce.

Huston and Melz (2004, 949) contend that we can tell "from the psychological make-up of partners and how their courtships unfolded, whether they would be delighted, distressed or divorced years later." How couples reach marriage, as well as what types of personal traits they bring into marriage, are important (Huston and Melz 2004).

Whether marriage is an arena for growth or disenchantment depends on the individuals and the nature of their relationship. It is a dangerous myth that marriage will change a person for the better: an insensitive single person likely becomes an insensitive husband or wife. Undesirable traits tend to become magnified in marriage because we must live with them in close, unrelenting, and everyday proximity.

Family researchers have found numerous premarital factors to be important in predicting later marital happiness and satisfaction. Although they may not necessarily apply in all cases—and when we are in love, we may ignore such factors, believing that we are the exceptions—they are worth thinking about. As enumerated by researchers Jeffrey Larson and Rachel Hickman, these include *background factors* (e.g., age at marriage, level of education, race, parental marital status, and so on), *contextual factors* (e.g., support and approval from friends and freedom from pressures to marry), *individual traits and behaviors* (e.g., level of self-esteem, interpersonal skills, physical health, or illness), and *couple characteristics* (e.g., being from similar backgrounds; possessing similar values, attitudes, beliefs, and gender-role expectations; and communication and conflict management skills) (Larson and Hickman 2004). It is to some of these that we now turn.

Background Factors

Age at marriage is an especially important, perhaps even the most important, background factor in shaping marital outcomes. People who "marry young" are at greater risk of seeing their marriages fail. Adolescent marriages (where either party is younger than 20) are especially likely to end in divorce. Such young marriages may be more divorce prone because of the immaturity and impulsivity of the partners (Clements, Stanley, and Markman 2004). In addition, early entry into marriage is associated with reduced educational attainment, which itself can curtail one's later occupational and economic success and thus contribute to additional marital stress.

As we saw earlier, the trend in the United States has been toward delaying marriage; on average, women and men are entering marriage at older ages than ever before, at age 28 for men and at age 26 for women. Nevertheless, despite this trend, more than 25% of women and 15% of men marry before they turn 23 (Uecker and Stokes 2008). Certain characteristics are associated with greater likelihood of marrying in one's early twenties, including one's gender, race, region, socioeconomic status, religion, education, and history of cohabitation. Women are twice as likely as men to marry early. Sociologists Jeremy Uecker and Charles Stokes identify a white, rural southerner from families of lower socioeconomic status (as measured by parents' incomes and educations) as most likely to marry young. Other factors associated with early marriage include having lower educational aspirations and attainment, growing up in a conservative Protestant or Mormon family, having parents who married early (22 or younger), and having cohabited (Uecker and Stokes 2008). Uecker and Stokes point out that although young, disadvantaged women and men in urban areas of the United States are the ones most likely to retreat from marriage, young, disadvantaged women and men from rural and southern areas are likely *to embrace* marriage.

Marriage age seems to have less effect as people age further into adulthood. In other words, differences between those who marry in their mid- to late twenties and those who marry in their thirties and older are slight. Length of courtship is also related to marital happiness. The longer you date and are engaged to someone, the more likely you are to discover whether you are compatible with each other. But you can also date "too long." Those who have long, slow-to-commit, up-and-down relationships are likely to be less satisfied in marriage. They are also more likely to divorce.

Level of education seems to affect both marital adjustment and divorce. Education may give us additional resources, such as income, insight, or status, that contribute to our ability to carry out our marital roles.

Similarly, level of religiousness is a factor in shaping marital outcomes; higher religiousness, especially by wives, is associated with greater probability of happy and stable marriages (Clements et al. 2004). Parental divorce may cause someone either to shy from marriage or to marry with the determination not to repeat the parents' mistakes. Parental divorce increases risks to married children; those who grew up in households where parents divorced are considerably more likely to experience a divorce themselves (Amato and De Boer 2001). This pattern has been identified in the United States and at least 15 other countries (Diekmann and Schmidheiny 2004).

Personality Factors

We bring with us into our marriages personality characteristics, attitudes and values, habits and preferences, and unique personal histories and early experiences. As you can imagine, your partner's personality will affect your life, your relationship, and your marriage considerably. Such personality characteristics are relatively stable and likely exert influence on the quality and outcomes of our marriages (Bradbury and Karney 2004).

We do know, however, that opposites do not usually attract; instead, they repel. We choose partners who share similar personality characteristics because similarity allows greater communication, empathy, and understanding. It may be that personality characteristics are most significant during courtship. It is then that those with undesirable or incompatible personalities are weeded out—or ought to be, at least in theory.

Researchers tend to focus more attention on relationship process and change than on personality. Personality seems fixed and unchanging. Nevertheless, it clearly affects marital processes. For example, a rigid personality may prevent negotiation and conflict resolution, and a dominating personality may disrupt the give-and-take necessary to making a relationship work, whereas warmth, an even temperament, and a forgiving and generous attitude toward one's spouse contribute to happy, stable marriages. In Ted Huston's longitudinal study, following couples from courtship through early marriage and at nearly 14 years after they were wed, there was notable stability to assessments of spouses' personalities made when couples were first married. These early assessments predicted the feelings that couples had and the behavior they displayed *nearly 14 years later* (Huston and Melz 2004). Thus, such attributes and characteristics matter greatly in shaping marital outcomes.

The ability to identify and communicate emotions is also associated with marital satisfaction. Such ability appears to differ by gender, with men having more difficulty identifying emotions expressed by women and with expressing and effectively communicating their own emotional state. Also gendered is the importance of effective emotional communication; women's marital satisfaction is more affected by the effective communication of emotions than is men's. This suggests that the emotional skills one brings with one into marriage are important factors in determining marital quality because they affect the level of shared intimacy (Cordova, Gee, and Warren 2005).

Relationship Factors

Besides personality characteristics, researchers have also examined aspects of premarital interaction and relationships that might predict marital success. Not all research substantiates the idea that marital success or failure is determined by how spouses communicate and solve problems (Bradbury and Karney 2004). Problem-solving skills are important but perhaps not as important as the emotional climate within which such skills are implemented. "If spouses have a reservoir of good will and they show their affection regularly, they are more likely to be able to work through their differences, to warm to each other's point of view, and to cope effectively with stress" (Huston and Melz 2004). If couples can maintain humor, express "genuine enthusiasm for what the partner is saying," and convey their continued affection for each other, couples with low levels of problem-solving ability will experience similar outcomes (in terms of shifts in marital satisfaction) as couples more skilled at problem solving (Bradbury and Karney 2004).

The same holds for conflict. As discussed in the previous chapter, the presence of conflict early in marriage does not indicate that the marriage is doomed any more than the absence of conflict guarantees positive feelings of warmth or more affection. Researchers suggest that negative interactions did not significantly affect the first year of marriage because of the *honeymoon effect*, the tendency of newlyweds to overlook problems. Failure to fulfill a partner's expectations about marital roles, such as intimacy and trust, predicted marital dissatisfaction (Kelley and Burgoon 1991).

Attempts have been made to identify and differentiate types of premarital relationships and to see whether such types differed in the outcomes of their marriages. One example of such efforts is based on the PREPARE program developed by psychologists David Olson and Peter Larson to assist couples in developing relationship skills to strengthen their relationships. Based on the degree of agreement between partners on 11 different dimensions of their relationships (personality issues, communication, conflict resolution, financial management, leisure activities, the sexual relationship, children/parenting, family/friends, realistic expectations, the role relationship, and spiritual beliefs), couples were classified as one of four types of relationships: vitalized, harmonious, traditional, and conflicted. Follow-up analysis two to three years later found that vitalized couples were the most likely to be married and satisfied and least likely to be married and dissatisfied or to have canceled their plans to marry. At the opposite end, conflicted couples were least likely to be married and satisfied and most likely either to have canceled their plans to marry or to have separated or divorced if they married (Fowers, Montel, and Olson 1996; Fowers and Olson 1992).

Two and a half million couples have taken the PREPARE/ENRICH program since 1980, and the results predict with 80% to 85% accuracy which couples will get divorced and which couples will be happily married (http://prepare-enrich.com). Psychologists Blaine Fowers, Kelly Montel, and David Olson contend that "distress and dissolution are relatively predictable on the basis of premarital relationship quality (Fowers et al. 1996).

Engagement, Cohabitation, and Weddings

For married couple families, family life begins with a serious romantic relationship that leads to engagement and/or cohabitation. Eventually, either of these is then followed by a wedding, the ceremony that signals formal entry into marriage. The first stage of the **family life cycle** may begin with engagement or cohabitation followed by a wedding, the ceremony that represents the beginning of a marriage.

Engagement

Engagement is the culmination of the premarital dating process. Today, in contrast to the past, engagement has more significance as a ritual than as a binding commitment to be married. Engagement is losing even its ritualistic meaning, however, as more couples start out in the less formal patterns of "getting together" or living together. These couples are less likely to become formally engaged. Instead, they announce that they "plan to get married." Because it lacks the formality of engagement, "planning to get married" is also less socially binding.

Engagements typically average between 12 and 16 months (Carmody 1992). They perform several functions:

- Engagement signifies a commitment to marriage and helps define the goal of the relationship as marriage.
- Engagement prepares couples for marriage by requiring them to think about the realities of everyday married life: money, friendships, religion, in-laws, and so forth. They are expected to begin making serious plans about how they will live together as a married couple.
- Engagement is the beginning of kinship. The future marriage partner begins to be treated as a member of the family. He or she begins to become integrated into the family system.
- The engagement period allows the marrying couple to plan the wedding.
- Engagement allows the prospective partners to strengthen themselves as a couple. The engaged pair begin to experience themselves as a social unit.

Men and women may need to deal with a number of social and psychological issues during engagement, including the following (Wright 1990):

- *Anxiety.* A general uneasiness that comes to the surface when you decide to marry
- *Maturation and dependency needs.* Questions about whether you are mature enough to marry and to be interdependent
- *Losses.* Regret over what you give up by marrying, such as the freedom to date and responsibility for only yourself
- *Partner choice.* Worry about whether you're marrying the right person
- *Gender-role conflict.* Disagreement over appropriate male and female roles
- *Idealization and disillusionment.* The tendency to believe that your partner is "perfect" and to become disenchanted when he or she is discovered to be "merely" human

© Israel Images/Alamy

Weddings carry multiple meanings, both about the individuals marrying and the nature of their commitment.

- *Marital expectations.* Beliefs that the marriage will be blissful and conflict free and that your partner will be entirely understanding of your needs
- *Self-knowledge.* An understanding of yourself, including your weaknesses as well as your strengths

Cohabitation

The rise of cohabitation has led to a new chapter in the story of contemporary families (Glick 1989; Surra 1991). As we will see in the next chapter, for some people cohabitation is an alternative way of *entering marriage.* More than half of first unions result from cohabitation (Seltzer 2000). For still others, cohabitation is an alternative *to marrying.*

Although cohabiting couples may be living together before marriage, their relationship is not legally recognized until the wedding, nor is the relationship afforded the same social legitimacy. For example, most relatives do not consider cohabitants as kin. As will be discussed in Chapter 9, there is evidence that marriages that follow cohabitation have a higher divorce rate than do marriages that begin without cohabitation. Cohabitation does, however, perform some of the same functions as engagement, such as preparing the couple for some realities of marriage and helping them think of themselves as a couple.

Weddings

There were approximately 2.2 million weddings in the United States in 2008. Data from the second quarter of 2009 reveal an average cost of $16,546 per wedding,

Critical Thinking

As you look at the factors predicting marital success, consider your past relationships. Retrospectively, what factors, such as background, personality characteristics, and relationship characteristics, might have predicted the quality of your relationship? Were any particular characteristics especially important for you? Why?

down 14% from the first quarter of the year (www .theweddingreport.com). Weddings are ancient rituals that symbolize a couple's commitment to each other. The word *wedding* is derived from the Anglo-Saxon *wedd*, meaning "pledge." It included a pledge to the bride's father to pay him in money, cattle, or horses for his daughter (Ackerman 1994; Chesser 1980). When the father received his pledge, he "gave the bride away." The exchanging of rings dates back to ancient Egypt and symbolizes trust, unity, and timelessness because a ring has no beginning and no end. It is a powerful symbol. To return a ring or take it off in anger is a symbolic act. Not wearing a wedding ring may be a symbolic statement about a marriage. Another custom, carrying the bride over the threshold, was practiced in ancient Greece and Rome. It symbolized the belief that a daughter would not willingly leave her father's house. The eating of cake is similarly ancient, representing the offerings made to household gods; the cake made the union sacred (Coulanges 1960). The African tradition of jumping the broomstick, carried to America by enslaved tribespeople, has been incorporated by many contemporary African Americans into their wedding ceremonies (Cole 1993).

The honeymoon tradition can be traced to a pagan custom for ensuring fertility: each night after the marriage ceremony, until the moon completed a full cycle, the couple drank honey wine. The honeymoon was literally a time of intoxication for the newly married man and woman. Flower girls originated in the Middle Ages; they carried wheat to symbolize fertility. Throughout the world, gifts are exchanged, special clothing is worn, and symbolically important objects are used or displayed in weddings (Werner et al. 1992).

Wedding ceremonies, celebrations, and rituals such as those described are rites of passage encompassing rites of separation (e.g., the giving away of the bride), aggregation, and transition. It is especially noteworthy as a rite of transition wherein it marks the passage from single to married status. The wedding may also reflect the degree to which both the bride's and the groom's "social circles" are part of the transition into marriage. As such, weddings vary. As Matthijs Kalmijn (2004) describes, they range from highly public large weddings to highly private with just a couple of witnesses.

Marriage is a major commitment, and entering marriage may provoke considerable anxiety and uncertainty. Is this person right for me? Do I really want to get and be married? What is married life going to be like? Will I be a good wife or husband? These are examples of the kinds of anxieties brides and grooms might feel as they approach marriage. Weddings may increase the commitment by creating and involving an audience that can serve as witnesses to the commitment a marrying couple is making.

Andrew Cherlin (2004, 856) suggests that where weddings had historically been celebrations of a kinship alliance between two kin groups and later a reflection of parental "approval and support" for their child's marriage, today's weddings are more a symbolic demonstration of "the partners' personal achievements and a stage in their self-development." A wedding is, in part, a statement, as is the buying of a house. It says, "Look at what I have achieved. Look at who I have become." Seen this way, we can understand why, despite the economic obstacles they face, low-income couples can honestly contend that a major barrier preventing them from marrying is insufficient money to have a "real wedding" (i.e., a church wedding and reception party). "Going down to the courthouse" is not a real or sufficient wedding (Smock 2004). A big wedding means that a couple "has achieved enough financial security to do more than live from paycheck to paycheck" (Cherlin 2004, 857). Both "the brides and grooms of middle America" and low-income, unmarried parents alike desire "big weddings," even if the nature of "big" varies between the two (Edin, Kefalas, and Reed 2004). This is all part of the deinstitutionalization of marriage raised earlier. Marriage and the wedding that signifies its beginning has become more of a symbol of individual achievement and development. If it is no longer the foundation of adult life, it still serves as a capstone (Cherlin 2004).

To other analysts, weddings are seen as mostly "occasions of consumption and celebrations of romance" (Cherlin 2004, 857). Indeed, weddings of today are big business. Not all couples, however, have formal church weddings. Data from 18 states that track such trends indicate that 40% of weddings in 2001 were civil ceremonies (typically performed by a judge, notary, or justice of the peace), up from 30% two decades earlier

(Grossman and Yoo 2003). Because of the expense, some couples opt for civil ceremonies, which have relatively minimal cost beyond the marriage license.

Whether a first, second, or subsequent marriage, a wedding symbolizes a profound life transition. Most significantly, the partners take on marital roles. For young men and women entering marriage for the first time, marriage signifies a major step into adulthood. Some apprehension felt by those planning to marry may be related to their taking on these important new roles and responsibilities. Therefore, the wedding must be considered a major rite of passage. When they leave the wedding scene, the couple leaves behind singlehood; they are now responsible to each other as fully as they are to themselves and more than they are to their parents.

However, if we focus too much on the ceremonial aspect of marriage, we overlook two important points. First, marrying is a process that begins well before and continues after the couple exchanges their vows. Second, the legal or ceremonial aspect of marrying may not be the most profound part of the transition.

The Stations of Marriage

As couples navigate their way through the transition to marriage, they experience changes in a host of areas or dimensions of their day-to-day lives. Past analyses of both divorce and remarriage have used the concept of **stations of marriage** to represent both the dynamic and the multidimensional nature of transitions out of and back into marriage (Bohannan 1970; Goetting 1982). Yet these analyses work equally well to depict the multidimensional, complex process of marrying (for further discussion of Bohannan's stations of divorce, see Chapter 13). Both Bohannan and Goetting stressed that marital transitions are thick with complexity. Applying their notions of "stations," we can say that marrying consists of the following:

- *Emotional marriage.* The experiences associated with falling in love and the intensification of an emotional connection between two people. In the love-based marriage system in the United States, it is on falling in love that many begin to contemplate marriage.
- *Psychic marriage.* This refers to the change in an individual's identity from an autonomous individual to a partner in a couple. As this occurs, priorities may shift, one's self-identity changes, and both one's perceptions of social reality and expectations for the future may be reconstructed (Berger and Kellner 1970).

- *Community marriage.* This pertains to the changes in social relationships and social networks that accompany the shift in priorities and identity. It is a two-way process of redefining and being redefined by others. Friends may perceive themselves as no longer able to make the same claims or hold the same expectations about a formerly single or unattached friend. Once married, new spouses are unquestionably looked on differently *because they are married*. They may even find their single friends becoming less interesting to or interested in them.
- *Legal marriage.* This is the legal relationship that—as we have seen—provides a couple with a host of rights and responsibilities. It also restricts the individual's right to marry again without first ending the current marriage. However, aside from these and restrictions on whom we may marry (which, granted, are not insignificant matters), there are few legal interventions into marriage as long as both parties remain content with their marriage. We may not notice any changes in our daily relationship caused exclusively by this dimension of marriage.
- *Economic marriage.* There are a variety of economic changes that people experience when they marry. If both are employed, they now have more financial resources that need to be managed and allocated in ways that differ from their single days. Whether the decision they face is which overdue bill to pay or whether to buy a Lexus or a sport-utility vehicle, they will have to change the way they previously made economic decisions and decide as part of a couple. Typically, there are stylistic differences in spending or money management that require some compromise.
- *Coparental marriage.* This refers to changes induced in marriage relationships by the arrival (birth, remarriage, or adoption) of children. Important in both Bohannan's and Goetting's analyses, coparental marriage is not part of becoming married per se, as couples can marry and never become parents. With regard to divorce, the coparental station includes attending to such issues as daily care and custody, financial support, and visitation. In the coparental remarriage, the primary issue is to establish stepparenting roles and relationships (see Chapter 14). As far as a "station of marriage," we might say that if either party has any children, both partners will need to establish routines and share responsibilities. If childless at marriage, the coparental station would refer to those issues that change married relationships once children arrive (see Chapter 10).

Although it was not included among either Bohannan's or Goetting's included *stations*, we might add a seventh station, a *domestic marriage*, encompassing all the negotiating, dividing, managing, and performing of daily household chores. Couples must establish a working division of household labor. Even if they have cohabited before marriage, there is no guarantee that their "cohabiting division of labor" will be sustained in marriage.

By conceptualizing becoming married in such terms, we can state the following important points. One may indeed feel and function as married before being legally married. That in no way guarantees success in marriage because the research on cohabitants who marry is fairly pessimistic. But it does mean that when people think about the process of marrying, if they think only in terms of before versus after the wedding ceremony (essentially the legal station), the transition may seem less sweeping than it is.

Becoming married can transform lives in all the ways depicted here. However, because one will likely encounter at least the emotional, psychic, and community (or some of it) stations of marriage by the time one enters legal marriage, one has an opportunity to begin to make the sorts of adjustments that marriage brings without yet being married. Bear in mind, too, that couples may experience these stations in different sequences. Cohabitants may experience all these stations of marriage before legally marrying. Marriages entered into because of pregnancy or as escape from a single lifestyle will encounter these dimensions in a different order than those who marry out of first dating and falling in love. What's useful, however, about the concept of stations is how it helps us appreciate how broadly and deeply marriage changes two people.

In the Beginning: Early Marriage

Ted Huston and Heidi Melz (2004) contend that early in marriage, newly married couples are affectionate, very much in love, and relatively free of excessive conflict, a state that might be called "blissful harmony." Within a year, this affectionate climate "melts" into a more genial partnership. As they point out, "One year into marriage, the average spouse says, 'I love you,' hugs and kisses their partner, makes their partner laugh, and has sexual intercourse about half as often as when they were newly wed" (Huston and Melz

2004, 951). Even though conflict is not necessarily more frequent or intense, when it occurs it is less likely to be embedded in the highly affectionate climate of new marriage. Thus, it may feel worse.

Huston and Melz (2004, 952) also found that couples establish a "distinctive emotional climate" from the outset that does not change over the initial two years of marriage; they are either happy or unhappy. Thus, it is not the case that unhappy couples begin on a blissful happy note and see things fail; instead, "most unhappy yet stable marriages fall short of the romantic ideal" from the beginning. All couples, even happy ones, have their ups and downs. Happy couples, however, typically contain two people who are both warm and even tempered.

Establishing Marital Roles

The expectations that two people have about their own and their spouse's marital roles are based on gender roles and their own experience. Traditional legal marriage contained the following four assumptions about husband or wife responsibilities: (1) the husband is the head of the household, (2) the husband is responsible for supporting the family, (3) the wife is responsible for domestic work, and (4) the wife is responsible for childrearing (Weitzman 1981).

These traditional assumptions about marital responsibilities may not have reflected what many couples actually experienced in marriage, however. Furthermore, such assumptions no longer fit contemporary marital reality. For example, the husband traditionally may have been regarded as head of the family, but today power tends to be more shared, albeit perhaps not equally. In dual-earner families, both men and women contribute to the financial support of the family, sometimes with the wife earning a larger share of the couple's income. Although responsibility for domestic work still rests largely with women, men have gradually increased their involvement in household labor, especially child care. The mother is generally still responsible for child rearing, but fathers are participating more.

Marital Tasks

Newly married couples need to begin a number of marital tasks to build and strengthen their marriages. The failure to complete these tasks successfully may contribute to what researchers identify as the **duration-of-marriage effect**—the accumulation over time of various factors such as unresolved conflicts, poor communication, grievances, role overload, heavy work schedules, and child-rearing responsibilities that might cause marital disenchantment (see the Issues and Insights feature in this section that examines marital satisfaction). These tasks are primarily adjustment tasks and include the following:

- *Establish marital and family roles.* Spouses need to discuss marital-role expectations for oneself and one's partner and make appropriate adjustments to fit each other's needs and the needs of the marriage.
- *Provide emotional support for the partner.* Couples must learn how to give and receive love and affection, support the other emotionally, and fulfill personal identity as both an individual and a partner.
- *Adjust personal habits.* Each partner faces adjusting to each other's personal ways by enjoying, accepting, tolerating, or changing personal habits, tastes, and preferences, such as differing sleep patterns, levels of personal and household cleanliness, musical tastes, and spending habits.

Spouses also need to do the following:

- Negotiate gender roles
- Make sexual adjustments
- Establish family and employment priorities and negotiate a division of labor
- Develop communication skills and learn how to effectively share intimate feelings and ideas with each other
- Manage budgetary and financial matters
- Establish kin relationships, participate in extended family, and manage boundaries between family of marriage and family of orientation
- Participate in the larger community

As you can see, newly married couples must undertake numerous tasks as their marriages take form. Marriages take different shapes according to how different tasks are shared, divided, or resolved. It is no wonder that many newlyweds find marriage harder than they expected. But if the tasks are undertaken in a spirit of love and cooperation, they offer the potential for marital growth, richness, and connection (Whitbourne and Ebmeyer 1990). If the tasks are avoided or undertaken in a selfish or rigid manner, however, the result may be conflict and marital dissatisfaction.

Identity Bargaining

People carry around idealized pictures of marriage long before they meet their marriage partners. They have to adjust these preconceptions to the reality of the partner's personality and the circumstances of the marriage. The interactional process of role adjustment is called **identity bargaining** (Blumstein 1975). Identity bargaining is a three-step process. First, a person has to identify with the role that he or she is performing. A man must feel that he is a husband, and a woman must feel that she is a wife. The wedding ceremony acts as a catalyst for role change from the single state to the married state. The process is critical to marriage.

Second, a person must be treated by the other as if he or she fulfills the role. The husband must treat his wife as a wife, and the wife must treat her husband as a husband. The problem is that partners may disagree on what constitutes the roles of husband and wife. This is especially true now as the traditional content of marital roles has changed.

Third, the two people must negotiate changes in each other's roles. A woman may not like housework (who does?), but she may be expected to do it as part of her marital role. Does she then do all the housework, or does she ask her husband to share responsibility with her? A man believes that he is supposed to be strong, but sometimes he feels weak. Does he reveal this to his wife?

Eventually, these adjustments must be made. At first, however, there may be confusion; both partners may feel inadequate because they are not fulfilling their role expectations. Although some may fear losing their identity in the give-and-take of identity bargaining, the opposite may be true: a sense of identity may grow in the process of establishing a relationship. In the process of forming a relationship, we discover ourselves. An intimate relationship requires us to define who we are.

Establishing Boundaries

When people marry, many still have strong ties to their parents. Until the wedding, their family of orientation has greater claim to their loyalties than their spouse-to-be. After marriage, the couple must negotiate a different relationship with their parents, siblings, and in-laws. Loyalties shift from their families of orientation to their newly formed family.

Of course, relationships with parents and siblings continue after marriage and continue to influence and be influenced by marriage. Difficulties between one's spouse and one's family can add considerable strain

Popular Culture: Can We Learn Lessons about Marriage from "Wife Swap" and "Trading Spouses?"

Marriage is lived behind closed doors and drawn curtains. Although we see many different married couples out and about in public settings, we really don't know how they live their daily lives. In fact, we have the opportunity to see "up close and in person" very few marriages aside from our own should we, in fact, marry. As a consequence, our understanding of just how diverse marriages really are—how many different ways people construct and experience their marital relationships—is very limited. Although researchers such as John Cuber and Peggy Harroff or Yoav Lavee and David Olson have constructed typologies (discussed later in this chapter) to illustrate different types of marriages, we may still operate with the misconception that there is one right or best way to be married or believe that most marriages are alike.

That's what makes a reality television program like ABC's *Wife Swap* (or Fox's version, *Trading Spouses: Meet Your New Mommy*) potentially so interesting. What the show does, as did its award-winning British predecessor, is take two married women with children and have them "swap lives," living in the other's house with the other's family for a two-week period of time. During the first week, the visiting wife tries to live by the other's rules and standards. During the second week, the families are supposed to adhere to what the visiting wife recommends.

Even acknowledging the distortions likely created by filming and selectively editing people's daily lives, each episode of *Wife Swap* gives us a glimpse at two very different households, often from different socioeconomic circumstances, and how difficult it is for the two women to switch places. Although the hope (and claim) is that the women and their families learn other ways in which they could structure their lives, they also learn to appreciate each other as well. For viewers, however, there is more. As American studies professor Allison McCracken observes,

> *Wife Swap* reveals the specificity of people's lives through attention to the mundane, rather than sensational, details that accompany the "wifely" role: cleaning, cooking, child care, spousal negotiations, religious practices, professional responsibilities. (FlowTV, 2005)

Writing in the Irish newspaper *The Independent* about the British version of the show, journalist John Masterson offers, "It should be compulsory viewing for anyone thinking of shacking up with anyone. Even before, 'Do you take this man/woman etc,' the cleric should ask, 'Have you both watched *Wife Swap?*' After they have both solemnly uttered, 'I did,' then the ceremony may continue" (Masterson 2008). While we wouldn't go that far, we recognize the value of such inside views of marriage and family life, especially as experienced by an outsider.

to a marriage. Indeed, the better one's relationship with one's in-laws, the more likely the marriage is to be happy and the lower the likelihood of divorce. Furthermore, in-laws are often the suppliers of needed assistance, whether that takes the form of economic assistance, practical help, emotional support, or advice. Newly married couples often have little money or credit and ask parents to loan money, cosign loans, or obtain credit. But such financial dependence can keep the new family tied to the family of orientation. The parents may try to exert undue influence on their children because their money is being spent.

The critical task is to form a family that is interdependent rather than independent or dependent. It is a delicate balancing act as parents and their adult children begin to make adjustments to the new marriage. We need to maintain bonds with our families of orientation and to participate in the extended family network, but we cannot let those bonds turn into chains. It is a

potentially delicate balancing act, made more difficult by the claim that marriage makes in people's lives.

Sociologists Natalia Sarkisian and Naomi Gerstel used two large national surveys to assess what they call the "greedy" nature of married life. Of interest, they found that married women and men are less involved with their parents and siblings than either the never married or previously married. They are less likely to visit, call or write, or provide either emotional or practical help to their parents. They are also less likely to see, call, or write their siblings, though they are slightly more likely than the formerly married to offer emotional or practical help to siblings. Interestingly, divorced women and men are less involved with their parents than are never-married people. The pattern of "diminished ties," especially to parents, persists even after marriages end. This effect appears not to result not only from such factors as time demands, needs and resources, demographic, or extended family characteristics. Even taking all these into account, marriage inhibits family ties, especially with parents (Sarkisian and Gerstel 2008).

Looking beyond the family of orientation to relationships with friends and neighbors, Gerstel and Sarkisian found that married people are less likely to socialize with friends and neighbors than are either the never married or the previously married. Once married couples have children, they show more similarity to the never married or previously married in providing friends and neighbors with practical or emotional help, though they are still less likely to hang out with either neighbors or friends (Gerstel and Sarkisian 2006).

Ideally, both families of orientation and peers (friends and neighbors) will understand, accept, and support these breaks. The new family must establish its own boundaries. The married couple should decide how much interaction with and assistance to others, especially their families of orientation, is desirable and how much influence these others may have.

Social Context and Social Stress

Even with all the attention paid to the dynamics of spousal relationships, marital success is also affected by things that happen outside of and around the married couple (Bradbury and Karney 2004). Marriages are affected by the wider context in which we live, including "the situations, incidents, and chronic and acute circumstances that spouses and couples encounter," as well as the developmental transitions they undertake (Bradbury and Karney 2004). Changes in employment, the transition to parenthood, health concerns, friends, finances, in-laws, and work experiences can all affect the quality of marriage relationships. As Thomas Bradbury and Benjamin Karney (2004, 872) express, "Theoretically identical marriages are unlikely to achieve identical outcomes if they are forced to contend with rather different circumstances."

Similarly, they contend that marriages that are "rather different" in their internal dynamics may reach similar outcomes in quality, depending on whether the wider context is especially healthy or especially "toxic" (Bradbury and Karney 2004). From their research on married couples, Bradbury and Karney offer the following points to consider:

- Marital quality was lower among couples experiencing higher average levels of stress.
- During times of elevated stress, more relationship problems were perceived, and partner's negative behaviors were more often viewed as selfish, intentional, and blameworthy.

Incorporating research findings from other studies, they also offer the following especially supportive evidence of the importance of social context on marital interaction and quality:

- Observational research found that because of greater job stress, blue-collar husbands were more likely than white-collar husbands to respond with negative affect to negative affect from their wives in problem-solving discussions.
- Among married male air-traffic controllers, on high stress days in which they received support from their wives, they expressed less anger and more emotional withdrawal.
- Among a sample of more than 200 African American couples, those living in more distressed neighborhoods (as measured by a composite that included such things as income and the proportion of the neighborhood on public assistance, living in poverty, being unemployed, and living in single-parent households) experienced less warmth and more overt hostility.

Cumulatively, findings such as these remind us that improving the quality of marriage may take more than educating couples about marriage or even providing therapeutic intervention. It may be necessary, too, to attend to and "fix" contextual circumstances, even if it means "bypassing couples and lobbying for change in environments and conditions that impinge on marriages and families" (Bradbury and Karney 2004, 876). This doesn't lessen the importance or potential benefit of either premarital education for marriage or therapeutic assistance; it simply puts social contexts and circumstances onto an equally significant level of importance.

Marital Commitments

What keeps us in a relationship? Is it something internal to an individual (a reflection of attitudes, values, and beliefs), or is it something external (the outcome of constraints)? Just what does the commitment to marriage entail?

Trying to sort out the meaning and experience of **marital commitment**, Johnson, Caughlin, and Huston (1999) identify three major types of commitment, each of which operates within marriage:

- *Personal commitment.* In essence, this is the degree to which one wishes to stay married to his or her spouse. As such, it is affected by how strongly one is attracted to one's spouse, how attractive one's relationship is, and how central the relationship is to one's concept of self.

- *Moral commitment.* This is the feeling of being "morally obligated" to stay in a relationship, resulting from one's sense of personal obligation ("I promised to stay forever and I will"), the values one has about the lifelong nature of marriage (a "relationship-type obligation"), and a desire to maintain consistency in how one acts in important life matters ("I am not a quitter, I have never been a quitter, I won't quit now").

- *Structural commitment.* This is our awareness and assessment of alternatives, our sense of the reactions of others and the pressures they may put on us, the difficulty we perceive in ending and exiting from a relationship, and the feeling that we have made "irretrievable investments" in a relationship and that leaving the relationship would mean we had wasted our time and lost opportunities all for nothing.

Personal commitment is more a product of love, satisfaction with the relationship, and the existence of a strong couple identity. Moral commitment is the product of our attitudes about marriage and divorce, our sense of a personal "contract" with our spouse, and the desire for personal consistency. Finally, structural commitment is a product of attractive alternatives, social pressures, fear of termination procedures, and the feeling of sacrifices we have made and can't recover. Johnson et al. (1999) contend that in our efforts to understand why marriages do or don't last, we tend to look mostly at personal commitment. We need to move beyond that narrower focus and look at how all three types are experienced and how each influences the outcome and experience of marriage.

How Parenthood Affects Marriage

Marriages change over time. Circumstances change, individuals age, couples may become parents, and spousal roles and responsibilities are subject to renegotiation. Marital satisfaction may ebb and flow, as marital conflicts may increase or decrease depending on the situations and circumstances couples face. Much research attention has been devoted to examining the effects of children on marriage. The presence of children in the household appears to lower marital satisfaction and increase marital conflict (Hatch and Bulcroft 2004). In fact, there has been considerable research that reveals a "deteriorating of marital functioning" when couples become parents. Although it is sometimes unclear how long this lasts or whether it is much different than what occurs among couples without children, after becoming parents, spouses appear to devote less time and energy to their marital relationships.

As a consequence of the demands imposed by new parenthood, the arrival of children is also frequently followed by an immediate decline in both independent and shared leisure activities. Although some of the decrease is gradually reversed, neither spouse fully returns to the level of either the independent or the shared leisure that they maintained before becoming parents. Given that shared leisure time affects the way spouses feel about their relationship and each other, a reduction in those activities may contribute to a lower level of marital satisfaction or happiness after spouses become parents (Claxton and Perry-Jenkins 2008). In fact, with all that accompanies the transition to parenthood (see Chapter 10), it is unsurprising that more frequent conflict and tension ensue, that couples often change the ways in which they handle or resolve conflict, and that marital satisfaction drops.

Interestingly, marital satisfaction declines in the early years of marriage even among nonparents. Known as the duration of marriage effect, such a pattern suggests that as couples move beyond the honeymoon phase, as they grow more familiar, they assess each other and their relationship more realistically. Any unrealistic premarital or even newly married expectations about married life or about their partners can translate into disappointment as couples experience the reality of marriage (VanLaningham, Johnson, and Amato 2001).

Marriages are also affected by changes experienced by each spouse. Husbands and wives do not carry

The arrival and presence of children profoundly affect marital relationships.

© David/Hanover/Getty Images/Stone

identical parenting or wage-earning responsibilities and thus may have very different experiences if and as children are born and grow and as jobs increase in demandingness or peak. In addition, changes in employment, income, or residence can potentially affect how spouses evaluate their marriages (VanLaningham et al. 2001).

Middle-Aged Marriages

Middle-aged marriages, in which couples are in their forties and fifties, are frequently families with adolescents and/or young adults leaving home. Adolescence brings with it a number of cognitive, emotional, physical, and social changes, and adolescents may trigger considerable family reorganization on the part of parents: they stay up late, play loud music, infringe on their parents' privacy, and leave a trail of empty pizza cartons, popcorn, and dirty socks in their wake. Conflicts over tidiness, study habits, communication, and lack of responsibility may emerge. Adolescents want rights and privileges but have difficulty accepting responsibility. Such conflict can and often does spill over into the husband–wife relationship. Furthermore, the research literature indicates changes in parent–child relationships, including less time spent together, less mutual acceptance, increasing emotional distance, and increasing conflict—particularly between adolescent daughters and their mothers (Whiteman, McHale, and Crouter 2007)—can affect how parents relate as spouses.

Married couples with adolescent children may disagree about such issues as their teens' dating, increasing autonomy, and their assumption of more adult roles and responsibilities. Such disagreements may be at least part of the cause of declines in spouses' reports of marital love and satisfaction during this stage of parenting. Shawn Whiteman, Susan McHale, and Ann Crouter found that among the 188 families that they studied over a seven-year period, both fathers and mothers seemed especially affected by the onset of adolescence of their firstborn children, with both expressing declines in positive assessments and increases in negativity (Whiteman et al. 2007).

Whiteman et al. (2007) remind us of the importance of recognizing that other things may occur during the period of parenting adolescents that can also cause marital changes. For parents at midlife, their own aging process may cause them distress. As their hair grays or their vision weakens, they may feel distress that can spill over into their feelings about their marriages.

Families as Launching Centers

Some couples may be happy or even grateful to see their children leave home, some experience difficulties with this exodus, and some continue to accommodate their adult children under the parental roof. As children are "launched" from the family (or "ejected," as some parents wryly put it), the parental role becomes increasingly less important in daily life. The period following the child's exit is commonly known as the **empty nest**.

Much research found what appeared to be an increase in marital satisfaction after children leave. Such studies were mostly cross-sectional in design, studying marriages of different durations at the same single point in time. From those studies, it appeared that satisfaction trended upward among couples whose children grew and left home. Longitudinal studies, following a sample of couples over a span of time, find no such bump in satisfaction. Instead, marital satisfaction appears to be relatively stable over time or declines (VanLaningham et al. 2001).

With children gone, couples must re-create their family, minus their children. Some couples may divorce at this point if the children were the only reason the pair remained together. The outcome is more positive when parents have other, more meaningful roles, such as school, work, or other activities, to turn to (Lamanna and Riedmann 1997).

The Not-So-Empty Nest: Adult Children and Parents Together

Just how empty homes are after children reach age 18 is open to question. More than 3.6 million parents lived in the same house with an adult son or daughter in 2007. This represents a 67% increase from 2000 and doesn't reflect changes since 2007 caused by the worsening economy. Census data revealed that in 2008, 20 million 18- to 34-year-olds, 30% to 34% of individuals in that age-group, were living at home with their parents (Shellenbarger 2008; Trejos 2009). Just a decade earlier, only 15% of men and 8% of women in that same age range were living at home with their parents. The average stay lasts between 12 and 18 months. Many come home right after completing their educations. Roughly half of 2008 college graduates expected to move back home after graduating (Dunleavy 2008). Others are doing an extra rotation through their family home after a temporary or lengthy absence. This latter group is sometimes referred to as the **boomerang generation**, though, in

CATHY *Cathy Guisewite*

YOU'RE MOVING HOME WITH YOUR MOTHER ?? ARE YOU CRAZY, ALEX ???

DO YOU WANT YOUR MOTHER DECIDING WHAT YOU EAT FOR DINNER ?? DO YOU WANT HER PICKING OUT YOUR CLOTHES ??

DO YOU WANT SOMEONE HOVERING OVER YOUR EVERY MOVE AND TREATING YOU LIKE A BABY ??!!

WELL, SURE!

MEN ARE TIED TO THE APRON STRINGS. WOMEN ARE STRANGLED BY THEM.

fact, either young adults or their parents may be the ones to move in with the other, leading some to now refer to "boomerang parents" (Gross 2008).

Researchers note that there are important financial and emotional reasons for this trend. The high unemployment rates and a tightened job market related to the U.S. recession are among the factors causing adult children to return home. High divorce rates, as well as personal problems, push adult children back to the parental home for social support and child care as well as for cooking and laundry services.

Young adults at home are such a common phenomenon that one of the leading family life cycle scholars suggested an additional family stage: *adult children at home* (Aldous 1990). Sue Shellenbarger, in the *Wall Street Journal*, referred to this stage as an "open nest" (Shellenbarger 2008). This new stage generally is not one that parents have anticipated. Almost half reported serious conflict with their children. For parents, the most frequently mentioned problems were the hours of their children's coming and going and their failure to share in cleaning and maintaining the house. Most wanted their children to be "up, gone, and on their own."

Reevaluation

Middle-aged people find that they must reevaluate relations with their children who have become independent adults and must incorporate new family members as in-laws. Some must also begin considering how to assist their own parents who are becoming more dependent as they age. Any and all of these issues may take away from a marriage.

Couples in middle age tend to reexamine their aims and goals. The man may decide to stay at home or not work as hard as before. The woman may commit herself more fully to her job or career, or she may remain at home, enjoying her new child-free leisure. Because the woman has probably returned to the workplace, wages and salary earned during this period may represent the highest amount the couple will earn.

As people enter their fifties, they probably have advanced as far as they will ever advance in their work. They have accepted their own limits, but they also have an increased sense of their own mortality. They not only feel their bodies aging but also begin to see people their own age dying. Some continue to live as if they were ageless—exercising, working hard, and keeping up or even increasing the pace of their activities. Others become more reflective, retreating from the world. Some may turn outward, renewing their contacts with friends, relatives, and especially their children and grandchildren.

Aging and Later-Life Marriages

There are some 37 million Americans over age 65, representing an estimated 13% of the U.S. population. Nearly 2% of the population is 85 or older. Fifty-eight percent of those over 65 are females. More than half of those over age 65 are married, and nearly a third (31%) are widowed. Most men over 65 are married, including 78% of 65- to 74-year-olds, 74% of 75- to 84-year-olds, and even 60% of men over 85. Given the gender differences in life expectancy and the tendency for women to marry men who are older than themselves, the pattern among women differs from that seen among men (see Figure 8.6). Among women

65 to 74, a majority are married. However, as reflected in Figure 8.6, most women 75 and older are widowed (U.S. Census Bureau 2007). Americans age 65 can expect to live another 19 years on average, 17 for males and 20 for females (Centers for Disease Control, Life Expectancy at Birth, 65 and 85 Years of Age, U.S. Selected Years: 1900–2005).

Beliefs that the elderly are neglected and isolated tend to reflect myth more than reality. According to 2005–2007 census data, among those over 65, an estimated 10% were below poverty. This is a smaller percentage than the overall percentage of the population living in poverty. Rather than being isolated, those over 65 are also more likely to be living with either a spouse or other relatives than alone, though this is more evident among men than among women (see Figure 8.7). Finally, as individuals age, most appear to remain fairly well connected to their families. An estimated 42% of adults report that they see or speak with a parent, usually their mother, every day; another 44% say that they see or speak with a parent at least once a week. Two-thirds of adults with a living parent say that they live within an hour's drive of a parent, while 72% of parents with an independent child say that they live within an hour of their child (Pew Research Center: A Social Trends Report, 2006).

AP Images/The Index-Journal, Shavonne Potts

Marriage relationships continue to face new challenges and circumstances as couples age.

Figure 8.6 Marital Status of U.S. Population Age 65 and Over, by Age Group and Sex, 2007

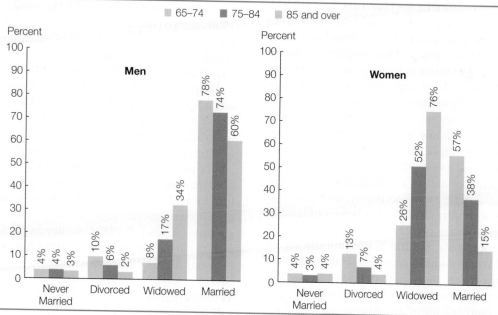

■ 65–74   ■ 75–84   ■ 85 and over

Men

Percent

	Never Married	Divorced	Widowed	Married
65–74	4%	10%	8%	78%
75–84	4%	6%	17%	74%
85 and over	3%	2%	34%	60%

Women

Percent

	Never Married	Divorced	Widowed	Married
65–74	4%	13%	26%	57%
75–84	3%	7%	52%	38%
85 and over	4%	4%	76%	15%

NOTE: Married includes married, spouse present; married, spouse absent; and separated. Reference population: These data refer to the civilian noninstitutionalized population.
SOURCE: "Older Americans 2008: Key Indicators of Well-Being," www.agingstats.gov/agingstatsdotnet/Main_Site/Data/2008.

Figure 8.7 Living Arrangements of U.S. Adults, 65 and Older, by Sex and Race, 2007

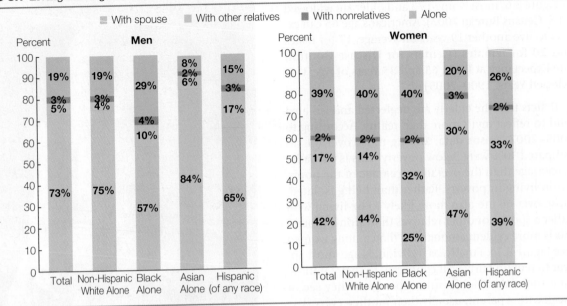

NOTE: Living with other relatives indicates no spouse present. Living with nonrelatives indicates no spouse or other relatives present. The term "non-Hispanic white alone" is used to refer to people who reported being white and no other race and who are not Hispanic. The term "black alone" is used to refer to people who reported being black or African American and no other race, and the term "Asian alone" is used to refer to people who reported only Asian as their race. The use of single-race populations in this report does not imply that this is the preferred method of presenting or analyzing data. The U.S. Census Bureau uses a variety of approaches.

Reference population: These data refer to the civilian noninstitutionalized population.

SOURCE: "Older Americans 2008: Key Indicators of Well-Being," www.agingstats.gov/agingstatsdotnet/Main_Site/Data/2008.

Marriages among Older Couples

Compared with middle-aged couples, older couples engage in less conflict. Laurie Hatch and Kris Bulcroft studied the frequency of marital conflict among women and men ages 20 to 79 and found that older husbands and wives reported less frequent conflict in areas including amount of time spent together, sex, money, and household tasks. They note that there are a number of possible explanations for the difference in frequency of conflict, including effects of aging, stage in the family life course, and birth cohort. They ultimately emphasize cohort effects (i.e., one's age or generation and its effect on one's expectations and experience of marital conflict) along with family life course stage. More specifically, they note that less frequent conflict was found in households without children, where neither spouse was employed or the couple had a traditional division of marital roles, and where couples spent more time together. As to cohort effects, they found that regardless of the duration of their marriages, older respondents, ages 60 to 79, reported fewer disagreements with their spouses than did younger respondents (Hatch and Bulcroft 2004).

Older couples also have greater potential for engaging in pleasurable activities together and separately, such as dancing, travel, or reading (Levenson, Carstensen, and Gottman 1993). Research in the 1990s showed that older people without children experienced about the same level of psychological well-being, instrumental support, and care as those who have children (Allen, Bleiszner, and Roberto 2000).

During this period, the three most important factors affecting late middle-aged and older couples are health, retirement, and widowhood (Brubaker 1991). In addition, these women and men often take on roles as caretakers of their own aging parents or adjust to adult children who have returned home. Later-middle-aged men and women tend to enjoy good health, are firmly established in their work, and have their highest discretionary spending power because their children are gone (Voydanoff 1987). As they age, however, they tend to cut back on their work commitments for both personal and health reasons.

Widowhood

Marriages are finite; they do not last forever. Eventually, every marriage is broken by divorce or death.

Despite high divorce rates, most marriages end with death, not divorce. "Till death do us part" is a fact for most married people.

In 2007, 8% of men and 26% of women between ages 65 and 74 were widowed. Among those 75 to 84 years old, 17% of men and 52% of women were widowed. Finally, a third of men and three-fourths of women 85 and older were widowed. Because women live about seven years longer on average than men, as shown above, most widowed people are women (U.S. Census Bureau 2007).

Demographic facts of life expectancy yield many more widows than widowers, thus creating for men "many more opportunities to date and remarry should they choose to" (Carr 2004, 1052). Indeed, greater proportions of older men than older women are married. Widowhood is often associated with a significant decline in income, plunging the grieving spouse into financial crisis and hardship in the year or so following death. This is especially true for poorer families. Feelings of well-being among both elderly men and elderly women are related to their financial situations. If the surviving spouse is financially secure, he or she does not have the added distress of a dramatic loss of income or wealth.

Recovering from the loss of a spouse is often difficult and prolonged. A woman may experience considerable disorientation and confusion from the loss of her role as a wife and companion. Having spent much of her life as part of a couple—having mutual friends, common interests, and shared goals—a widow suddenly finds herself alone. Whatever the nature of her marriage, she experiences grief, anger, distress, and loneliness. Physical health appears to be tied closely to the emotional stress of widowhood. Widowed men and women experience more health problems over the 14 months following their spouses' deaths than do those with spouses. Over time, however, widows appear to regain much of their physical and emotional health (Brubaker 1991).

One common response of widowed women and men is to glorify or "sanctify" their marriages and their deceased spouses. This is especially true shortly after a spouse's death (Carr 2004). One way in which women and men differ in their reactions is that women who had close marriages may feel less open to seeking and forming a new relationship with another man, retaining the feeling that they are "still married" to their late husbands. Men who were in close marriages may be especially motivated to establish another marriage. Having experienced and grown dependent on the emotional support and intimacy of their marriages, they may have few other alternative sources for support to whom they can turn. Thus, they have greater incentive to form a new emotionally supportive marriage or partner relationship (Carr 2004). In addition to the loss of their confidant and chief source of emotional support, they may have limited experience managing households, cooking, and cleaning and as a result suffer from poor nutrition and distress over the conditions in which they live (Carr 2004). Widows certainly suffer, too, although they may be beneficiaries of more practical help from their children and draw on emotional support from a wider and deeper network of friends.

Eventually widowed women and men must in some way adjust to the loss. Some remarry; 2% of older widows and 20% of older widowers remarry. Each year, 3 of every 1,000 widows and 17 of every 1,000 widowers marry (Carr 2004). Men may seek remarriage because they were emotionally, socially, and domestically more dependent on their wives. Widows typically draw on wider support networks for emotional and social support (Carr 2004).

© Rolf Bruderer/CORBIS

The loss of one's spouse confronts women and men with a variety of deep and painful losses. Although both women and men lose their chief source of emotional support, women typically have wider and deeper friendship networks to turn to for support.

Many others adjust by learning to enjoy their new freedom. Others believe that they are too old to date or remarry; still others cannot imagine living with someone other than their former spouse. A large number of elderly men and women live together without remarrying.

Enduring Marriages

Examining marriages over time is an important way of exploring the different tasks we must undertake at different times in our relationships. Which marriages last? What researchers find is what many of us already know: little correlation exists between happy marriages and stable ones. Many unhappily married couples stay together, and some happily married couples undergo a crisis and break up. In general, however, the quality of the marital relationship appears to show continuity over the years. Much of the discrepancy between happiness and stability results because happiness or satisfaction is an evaluative judgment of a marriage relative to what we expected from marriage and what better alternatives are available. Stability results more from assessments of the costs and rewards of staying in or leaving a marriage. Unhappy marriages may be enduring ones because there are no better alternatives, because the costs of leaving exceed the costs of staying married, or both.

Long-term marriages are not immune to conflict. As many as one-fourth of middle-aged couples and between 12% and 20% of older couples acknowledge engaging in conflict over such issues as children, money, communication, recreation, sex, and in-laws (Levenson et al. 1993). Surviving together does not require couples to eliminate or avoid conflict.

A study by Robert and Jeanette Lauer (1986) used a more modest definition of *long term* to look at marriages that last. Their study of 351 couples married at least 15 years (most were married a good deal longer) found the following to be the "most important ingredients" identified by men and women to explain their marital success:

1. Having a spouse who is a best friend and whom you like as a person
2. Believing in marriage as a long-term commitment and sacred institution
3. Consensus on such fundamentals as aims and goals and philosophy of life
4. Shared humor

When assessing marriages, keep in mind that there is considerable diversity in married life. Thus,

attempts have been made to document some types of marriages that couples construct (Cuber and Harroff 1965; Schwartz 1994; Wallerstein and Blakeslee 1995). One popular typology details five types of marriage, each of which could either last "till death do us part" or end in divorce. Thus, these are not degrees of marital success but rather different kinds of marriage relationships (Cuber and Harroff 1965):

- **Conflict-habituated marriages** are relationships in which tension, arguing, and conflict "permeate the relationship" (Cuber and Harroff 1965). It may well be that conflict is what holds these couples together. It is at least understood to be a basic characteristic of this type of marriage.
- **Passive-congenial marriages** are relationships that begin without the emotional "spark" or intensity contained in our romantic idealizations of marriage. They may be marriages of convenience that satisfy practical needs in both spouses' lives. Couples in which both spouses have strong career commitments and value independence may construct a passive-congenial marriage to enjoy the benefits of married life and especially parenthood.
- **Devitalized marriages** begin with high levels of emotional intensity that dwindles over time. From the outside looking in, they may closely resemble passive-congenial relationships. What sets them apart is that they have a history of having been in a more intimate, sexually gratifying, emotional relationship that has become an emotional void. Obligation and resignation may hold such couples together, along with the lifestyle they have built and the history they have shared.
- **Vital marriages** appeal more to our romantic notions of marriage because they begin and continue with high levels of emotional intensity. Such couples spend much of their time together and are "intensely bound together in important life matters" (Cuber and Harroff 1965). The relationship is the most valued aspect of their lives, and they allocate their time and attention on the basis of such a priority. Conflict is not absent, but it is managed in such a manner as to make quick resolution likely.
- **Total marriages** are relationships in which characteristics of vital relationships are present but to a wider and deeper degree, with the "points of vital meshing" extended across more aspects of daily coupled life. Spouses appear to share everything.

Differentiating between these five types, John Cuber and Peggy Harroff noted that the first three types were more common than the last two. As many as 80% of the relationships among their sample were of one of the first three types. Both vital and total marriages (what they called *intrinsic marriages*) were relatively rare. Again, we must remember that the researchers were not sorting relationships into "successful" versus "unsuccessful" or "good" versus "bad." Marriages of all five types were enduring marriages, and any of the five types could end in divorce, although the reasons for divorce would differ.

A seven-type marriage typology was constructed by Yoav Lavee and David Olson (1993) from an analysis of the marriages of more than 8,000 couples voluntarily in marriage enrichment programs or marital therapy. Although such a sample may be more difficult to generalize, Lavee and Olson suggested that we could differentiate couples on the basis of their satisfaction or dissatisfaction with nine areas of married life: personality issues, conflict resolution, communication, sexual intimacy, religious beliefs, financial management, leisure, parenting, and relationships with friends and family. Of their types, *vitalized couples* (9% of sample) reported themselves satisfied with all nine areas. At the opposite end, *devitalized couples* reported problems in all nine areas. Keeping in mind that the sample was drawn from either clinical or enrichment intervention, the devitalized were by far the most common type, representing 40% of their sample.

The remainder of the sample was relatively evenly divided across the other types: *balanced, harmonious, traditional, conflicted,* and *financially focused.* All types except the vitalized reported problems, although the areas and extent of problems differed across these types. For example, the financially focused (11%) had problems in all areas but financial matters. Traditional couples (10%) reported problems in their handling of conflict, communication, sexual intimacy, and parenting. The conflicted (14%) reported themselves generally satisfied with only their parenting, leisure activities, and religious beliefs. Even those couples designated as harmonious (8%) tended to have difficulties in areas such as religious beliefs, parenting, and relations with family and friends. Balanced couples (8%) were generally satisfied with all areas except financial matters.

For different reasons, we need to be cautious about generalizing too far from either Cuber and Harroff or Lavee and Olson. Nonetheless, in both typologies, 75% or more of the sample couples were in marriages that many would define as unattractive, seeming to be held together by something other than a deep emotional connection. In addition, both typologies underscore that marriage has to be free of conflict to last. Most obviously, both typologies illustrate that not all marriages are alike. This is a simple and obvious but important point.

Throughout marriage, from the earliest most hopeful and optimistic beginning until death or divorce do us part, we are presented with opportunities for growth and change as we enter our roles as husbands or wives, become parents or stepparents, and still later become grandparents. As we have seen, marriages and families never remain the same. They change as we change, as we learn to give and take, as children enter and exit our lives, and as we create new goals and visions for ourselves and our relationships. In our intimate relationships, we are offered the opportunity to discover ourselves.

As marriage continues to undergo changes, we are left to wonder about what the future holds. We will not likely see a return to traditional marriages any more than we should expect a disappearance of marriage. As Paul Amato (2004b) assesses, alternatives to marriage will be accepted and widespread. People will continue to have sex prior to marriage, live together without being married, have children outside of marriage, avoid marriage altogether, and divorce if their marriages are flawed. However, sociologist Andrew Cherlin (2009) offers a more cautionary note to at least consider. Cherlin, himself, is optimistic about marriage. Like Amato and colleagues, he believes that marriage will continue at least to have symbolic value in the United States, serving as the highest achievement—a capstone—for adult family life if no longer the foundation of such a life and acting as a marker to demonstrate that one has acquired the resources needed for marriage—a good education, a decent job, and a partner who is willing to pledge to stay with you indefinitely. However, using sociologist William Ogburn's concept of **cultural lag**, he warns of another, more pessimistic interpretation. Cultural lag is the outcome of rapid social change, when part of the culture (e.g., technology) changes more rapidly than another part (e.g., behavior). In this vein, he acknowledges that it is at least possible that people are entering marriage because that is exactly what their parents and grandparents before them did. Perhaps they haven't yet realized that they can have most if not all of the benefits of marriage without marrying, that they can cohabit with many of the legal rights and benefits formerly reserved for marriage, that they can enjoy sexual relations without marrying by making use of the availability of more effective birth control, or that they can have and raise children successfully and acceptably outside of marriage. Perhaps at some point they will realize that there is no longer a good reason to marry and most will stop doing so. Only time will tell.

Summary

- Marriage is the foundation and centerpiece of the American family system and the subject of much public discussion and concern.

- There is an ongoing *marriage debate* over the status and future of marriage. The two extreme positions in this debate are the *marital decline* and *marital resilience positions*.

- Behavioral indicators of a *retreat from marriage* include increasing percentages of adults remaining unmarried, living together, having children outside of marriage, and divorcing. However, more than 80% of Americans are expected to someday marry.

- The retreat from marriage varies considerably by race, education, and socioeconomic status.

- Even those most likely not to marry continue to articulate support for and a desire to marry but face barriers to marriage, such as financial concerns, concerns about relationship quality, and fear of divorce.

- There are religious differences in the importance placed on marriage and the push toward early marriage. Conservative Protestants and Latter-Day Saints are most likely to marry young and to articulate strong endorsement of and commitment to marriage.

- The *deinstitutionalization of marriage* refers to weakening of the social norms that define people's behavior in a social institution such as marriage. In *individualized marriage*, new emphases on personal self-fulfillment and freedom of choice become more important than marital commitment and obligation.

- Sociological research supports aspects of both the decline and the resilience positions.

- Legal limits imposed on choice of marriage partner include gender, age, family relationship, and number of spouses.

- As the current edition was being completed, five states had legalized same-sex marriage.

- Public opinion remains divided, with increasing percentages of people opposing restricting marriage to only heterosexual couples.

- Restricting people to spouses from within their same group (e.g., tribe, race or religion) is known as endogamy. Exogamy compels people to select partners from outside their same group.

- Within the population of eligible potential partners, we tend to marry others like ourselves on criteria such as age, race, religion, social class, and family history, a practice known as homogamy.

- Because of uneven sex ratios and the tendencies for women to marry men who are older and of higher socioeconomic backgrounds, some are squeezed out of marriage.

- We tend to select partners who live within a limited and shared geographical area.

- Our choice of partners results from factors such as choice, opportunity, and social pressure.

- There are multiple theories to account for why we choose the person we do.

- Reasons to marry include both attractions of marriage and rejection of singlehood. Marriage also retains symbolic value. Marital intimacy is the biggest attraction of marriage.

- Marriage provides various benefits to married people, including economic benefits, health benefits, and psychological benefits. Research supports both a selection effect (healthier and better-adjusted people are more likely to marry) and a protection effect (marriage provides a range of protective resources enabling people to prosper). The benefits depend on the quality of the marriage.

- The relationships that precede marriage often predict marital success because marital patterns emerge during these times. Premarital factors correlated with marital success include (1) background factors, (2) personality factors, and (3) relationship factors.

- Engagement prepares the couple for marriage by involving them in discussions about the realities of everyday life, involving family members with the couple, and strengthening the couple as a social unit.

- A wedding is a ritual that symbolizes a couple's commitment to each other.

- The process of marrying and becoming spouses consists of different dimensions of experience that can be classified as the stations of marriage: emotional, psychic, community, economic, legal, and parental. The domestic responsibilities that marriage introduces is yet another part of becoming married.

- Marital tasks include establishing marital and family roles, providing emotional support for the partner, adjusting personal habits, negotiating gender roles, making sexual adjustments, establishing family and employment priorities, developing communication skills, managing budgetary and financial matters, establishing kin relationships, and participating in the larger community.

- Marital success is affected by the wider social context and the extent and kind of social stresses couples face.

- Marital commitments consist of personal commitments, moral commitments, and structural commitments.

- A critical task in early marriage is to establish boundaries separating the newly formed family from the couple's families of orientation. Married women and men experience diminished ties to their families of origin.

- The arrival and raising of children leads to a decline in shared leisure activities by spouses, reduced marital satisfaction, and increased marital conflict. Marriages without children also experience a reduction in satisfaction.

- In middle age, many married couples must deal with issues of independence in regard to their adolescent children. Most women do not suffer from the *empty-nest* syndrome. For many families, there is no empty nest because of the increasing presence of adult children in the home. As children leave home, parents reevaluate their relationship with each other and their life goals.

- Despite familiar stereotypes, Americans over 65 years old are neither impoverished nor isolated from family.

- In later-life marriages, usually no children are present. The most important factors affecting marriage during this life cycle stage are health, retirement, and widowhood.

- Older husbands and wives engage in less frequent conflict.

- Most marriages end with a spouse's death, not divorce.

- Most widowed people are women. Widowers have greater opportunity to date, repartner, or remarry than do widows.

- Recovering from the loss of one's spouse is difficult and prolonged.

- Some factors associated with long-term marriages are liking your spouse as a person, thinking of your spouse as your best friend, believing in marriage as a commitment, spousal agreement on life's goals, and a sense of humor.

- Marriages differ from one another. One popular typology contrasts five types of marriage: *conflict-habituated*, *devitalized*, *passive-congenial*, *vital*, and *total*. These reflect different conceptualizations and experiences of marriage, not different degrees of marital success.

- The future of marriage is unclear. Despite changes in the nature or necessity of marriage, it has retained considerable symbolic value. Most people still value marriage, and most unmarried people expect to marry someday.

Key Terms

boomerang generation 293
close relationship model of legal marriage 279
companionate marriage 268
complementary needs theory 278
conflict-habituated marriages 298
conjugal model of legal marriage 279
cultural lag 299
Defense of Marriage Act 271
deinstitutionalization of marriage 268
devitalized marriages 298
duration-of-marriage effect 288
empty nest 293
endogamy 272
exogamy 272
family life cycle 284
heterogamy 272
homogamy 272
hypergamy 275
hypogamy 275
identity bargaining 289

individualized marriage 268
marital commitment 291
marital decline perspective 268
marital history homogamy 277
marital resilience perspective 268
marriage debate 262
marriage squeeze 276
mating gradient 276
parental image theory 278
passive-congenial marriages 298
positive assortative mating 272
residential propinquity 277
retreat from marriage 264
role theory 278
stations of marriage 287
stimulus–value–role theory 278
total marriages 298
value theory 278
vital marriages 298

RESOURCES ON THE WEB

Book Companion Website

www.cengage.com/sociology/strong

Prepare for quizzes and exams with online resources—including tutorial quizzes, a glossary, interactive flash cards, crossword puzzles, self-assessments, virtual explorations, and more.

9

Unmarried Lives: Singlehood and Cohabitation

What Do YOU Think? Are the following statements TRUE or FALSE?
You may be surprised by the answers (see answer key on the following page).

T	**F**	**1** There are more unmarried women than there are unmarried men.
T	**F**	**2** A majority of African American adults are unmarried.
T	**F**	**3** Both the pressures to marry and the attractions of marriage have greatly increased over the past few decades.
T	**F**	**4** Unmarried women and men are subjected to numerous unflattering stereotypes and often encounter discriminatory treatment.
T	**F**	**5** Cohabitation has become part of the courtship process among many young adults.
T	**F**	**6** Most cohabiting couples choose to live together because they don't perceive marriage as a desirable goal.
T	**F**	**7** Cohabitation is more common among college educated adults.
T	**F**	**8** Cohabitation before a remarriage improves one's likelihood of having a successful remarriage.
T	**F**	**9** Only same gender cohabiting couples can enter into a legal domestic partnership.
T	**F**	**10** Most same gender cohabiting couples are dual earners.

The next time you are in class, take a look around the room. It is likely that most of the people who share that classroom space with you expect to someday marry. After all, this is true of the vast majority of young women and men. It is also probable that some of your classmates don't intend to marry, and that some who do intend to wed, won't. Despite the title of this book and the emphasis we place on marriage, not everyone is or desires to be part of a couple, and not every couple either intends to or is able to marry. Some people prefer the freedom and flexibility they associate with being unattached and unmarried. Others might prefer to be part of a couple but are unable to find partners with whom they wish to share their lives. In either of those instances, some of your classmates will find themselves making a life of and on their own. Still others find themselves in a relationship, perhaps even living with a partner but remaining unmarried, either by choice or because they don't feel ready to marry. And, of course, as we saw in the previous chapter, not everyone *can* marry despite whatever intent or desire they might have to do so. These are the lifestyles that are the focus of this chapter: singlehood and heterosexual and same-sex cohabitation.

Singlehood

Even a casual inspection of demographics in this country illustrates the increasing phenomena of singlehood and cohabitation. The trends, which have taken root and grown substantially since 1960, include an eclectic combination of divorced, widowed, and never-married individuals. Each year, more adult Americans are among the unmarried (see Table 9.1).

According to U.S. Census estimates, in 2008 there were approximately 101 million Americans age 18 or older who were either never married, divorced, widowed, or separated. Of this group, 58% had never married, 23% were divorced, 14% were widowed, and 5% were separated. If one adds the 3.3 million women and

---

Answer Key to What Do YOU Think?

1 True, see p. 304; **2** True, see p. 305; **3** False, see p. 307; **4** True, see p. 309; **5** True, see p. 311; **6** False, see p. 313; **7** False, see p. 311; **8** False, see p. 314; **9** False, see p. 323; **10** True, see p. 326.

---

men who were married but whose spouse was absent, 48.7 million males and 55.7 million females were living without a spouse. They represented almost 47% of all U.S. residents age 18 and over (Current Population Survey 2008). Of this population, 56% had never married, 22% were divorced, 5% were separated, and 14% were widowed. Three percent were married, but their spouses were absent from the home.

Even if one considers only the unmarried, 45% of Americans age 18 and older were either never married, widowed, separated, or divorced. Looking at households instead of individuals, there were 58.4 million households headed by unmarried men or women in 2008, representing slightly over half of all U.S. households.

The Unmarried Population

There are more single women than men; a ratio of 86 men to every 100 women in the U.S. population, 18 and older, who are not currently married. Women make up nearly 54% of all unmarried Americans. Sixteen percent of the unmarried population, representing approximately 15 million people, are age 65 years old or older (U.S. Census Bureau 2008).

Table 9.1 Percentage of Population 15 and Older Who Are Unmarried

Year	Men	Women
1890	48%	45%
1900	47%	45%
1910	46%	43%
1920	42%	43%
1930	42%	41%
1940	40%	40%
1950	32%	34%
1960	30%	34%
1970	34%	39%
1980	37%	41%
1990	39%	43%
2000	42%	45%
2008	46%	50%

SOURCE: Marital status data for 1890–1970 from U.S. Census Bureau (1989). Data for 1980–2000 from U.S. Census Bureau (2001). Data for 2008 from U.S. Census Bureau (2008, table A-1).

Popular Culture: Celebrating and Studying Singlehood

Throughout much of the United States, by the third week of September we are surrounded by signs of summer's end. Leaves start to change color, vacations become memories, and students from preschool to professional school have returned to classes. Along with the official arrival of autumn that same week comes a weeklong celebration, now known as Unmarried and Single Americans Week. Unmarried and Single Americans Week acknowledges and recognizes single women and men, celebrates the lifestyles of the unmarried, and acknowledges the contributions unmarried people make to American society.

First started in Ohio by the Buckeye Singles Council in 1982, the week was originally called National Singles Week. Taken over by the American Association for Single People in 2001, the weeklong "celebration" was renamed in recognition that many unmarried people are in relationships or are widowed and don't identify with the "single" label (www.census.gov/Press-Release/www/releases/archives/facts_for_features_special_editions/005384.html). In 2002, the association changed its name to Unmarried America. Although the designated week has been around now for more than a quarter of a century and is recognized by mayors, city councils, and governors in some 33 states, as of 2005 it had yet to be "legitimized" and incorporated into mainstream American culture, as indicated by both the absence of greeting cards for the occasion and the number of people (including millions of unmarried people) unaware that the weeklong recognition exists (Coleman 2005).

Another sign of increasing attention being paid to the growth in numbers and importance of single women and men can be seen in the efforts to broaden the academic curriculum so as to keep pace with the changing demographics. The best example of this effort can be seen in the development of "singles studies" or, in the absence to date of such a discipline, in the call for scholars to consider such a venture seriously. Psychologist Bella DePaulo, law professor Rachel Moran, and sociologist and gender studies scholar E. Kay Trimberger have put together a singles studies Web site at the University of California at Berkeley's Institute for the Study of Social Change. Created to "advance the work of scholars who are integrating the study of singles into their research and teaching," it contains an exhaustive bibliography of research on all varieties of singles—formerly married, never married, gay, lesbian, bisexual, heterosexual, single parents, and "singles of different races, ethnicities, and social classes" (http://issc.berkeley.edu/singlesstudies/index.html).

DePaulo, Moran, and Trimberger advocate applying a "singles perspective" to a host of topics of study (e.g., friendship, stereotyping and discrimination, and laws and policies that privilege married couples) and in disciplines such as law, sociology, psychology, women's and gender studies, and ethnic studies. Such an effort would, they claim, "broaden and deepen scholarship while enriching the intellectual life of the classroom."

A historical comparison in Table 9.1 reveals a decline in the percentage of the population that was unmarried from the late nineteenth century until the 1960s. The trend then reversed, and across the past four decades of the twentieth century and the first decade of the twenty-first, the percentage of women and men age 15 and older who were unmarried grew steadily (we use age 15 and older because of data available for purposes of historical comparison).

The percentage of unmarried Americans varies widely by race and ethnicity. Based on 2008 data, 65% of African Americans, 37% of Asians, 47% of Hispanics, and 42% of non-Hispanic whites were unmarried. The percentage of people who had never married also shows considerable racial or ethnic variation: 22% of non-Hispanic whites, 41% of African Americans, 32% of Hispanics, and 26% of Asian and Pacific Islanders, age 18 and older, had *never married*. Furthermore, an additional 11% of non-Hispanic whites, 12% of African Americans, 8% of Hispanics, and 4% of Asian Americans were divorced (U.S. Census Bureau 2008, table A-1). Thus, the population of singles is both very large and quite diverse.

The varieties of lifestyles that single or unmarried people maintain are really too numerous to fit under one "umbrella," too complex to be understood within any one category, and too diverse to be accounted for by any one explanation. The lifestyles include the never married and the divorced; young, middle-age, and old, single parents; gay men,

The proportion of the population that is unmarried varies considerably by race.

lesbians, and bisexuals; and widows and widowers. In addition, as noted above, more than 5 million married women and men were separated and living apart from their spouses in 2008. Some singles live with other people, whereas others live alone. In 2008, 27.5% of the almost 117 million households were men and women living alone (U.S. Census Bureau, 2008).

These different populations experience diverse living situations that affect how singleness is experienced. However, when we think about those generally regarded as "single," we often think mostly of young or middle-age women and men who are heterosexual, not living with someone, and working rather than attending school or college. Although there are numerous single lesbians and gay men, they have not traditionally been included *as singles* in research on unmarried men and women, in part because of their long-standing inability to marry and because of the association of marriage with heterosexuals.

Never Married Singles in America: An Increasing Minority

The growth in the percentage of *never-married adults*, from 20.3% in 1980 to 30% in 2008, has occurred across all population groups. In part, this increase (like the creation of National Unmarried and Single Americans Week) reflects a change in the way in which society views this way of life. Many singles appear to be postponing marriage to an age that makes better economic and social sense (U.S. Census Bureau 2000). The growing divorce rate, which peaked in the early 1980s and has since decreased, also contributed to the numbers of singles. In 2008, 9.3% of men and 11.7% of women age 18 and over were divorced (U.S. Census Bureau 2008). The proportion of widowed men and women has declined somewhat but remains similar to past numbers. Among older people, singlehood most often occurs because of the death of a spouse rather than by choice. Nevertheless, as society continues to value individualism and choice, the numbers of singles will most likely continue to grow.

The increases in the numbers of single adults are the result of several factors:

- *Delayed marriage.* With a median age at first marriage of 27.4 years for men and 25.6 years for women in 2007 (U.S. Census Bureau 2009), Americans are waiting longer than ever to first enter marriage. Thus, more young adults find themselves among the never married, even if many of them are really "yet to be married." The longer one postpones marriage, the greater one's likelihood of never marrying. Thus, delaying the age at which one marries increases the population of singles in two different but related ways. As shown in Table 9.2, the percentage of never-married men and women of typical "marrying ages" dramatically increased between 1970 and 2000. Based on 2007 estimates, 62% of men and 52% of women *between the ages of 20 and 34* had never married.
- *Increasingly expanded educational, lifestyle, and employment options open to women.* Such changes have reduced women's economic need to be married and expanded their lifestyle options outside of marriage.
- *Increased rates of divorce.* Even with the decline and later stabilization of the divorce rate, the fear of divorce remains high. Thus, more people

The delay in age at marriage is one of the biggest factors in the increase in both singlehood and cohabitation.

may avoid marrying out of the fear that their marriages will fail. Joining them are the still high numbers of women and men whose marriages do end in divorce. Combine these with a somewhat decreased likelihood of remarriage, especially among African Americans, and one has additional ingredients for increasing the unmarried population.

- *More liberal social and sexual standards.* As nonmarital sex became and remains more acceptable, one major motive for marrying has been reduced if not removed.
- *Uneven ratio of unmarried men to unmarried women.* As we have touched on previously, the opportunity to marry is not the same for all groups. Some women (e.g., African American women) are *squeezed out* by the imbalanced sex ratio.

Table 9.2 Percentage of Never-Married Women and Men by Age, 1970–2008

Age	Male 1970	Male 2008	Female 1970	Female 2008
20–24	35.8	87.3	54.7	78.9
25–29	10.5	58.8	19.1	45.5
30–34	6.2	34.5	9.4	26.1
35–39	5.4	23.2	7.2	16.0
40–44	4.9	19.7	6.3	13.5

SOURCE: Fields and Casper (2001) and U.S. Census Bureau (2008, table A-1).

Relationships among the Unmarried

When intentionally single heterosexual people form relationships within the singles world, both the man and the woman tend to remain highly independent. Unmarried women and men are typically employed and, thus, tend to be economically independent of each other. They may also be more emotionally independent because their energy may already be heavily invested in their work or careers. Their relationships consequently tend to more greatly emphasize autonomy and egalitarian roles. Employed single women tend to be more involved in their work, either from choice or from necessity, than are married women, but the result is the same: they are accustomed to living on their own without being supported by a man. Early analysis of the various factors that draw people to singlehood or marriage identified the various "pushes" and "pulls" of each lifestyle. These are illustrated in Table 9.3.

Readers will notice how many of these have changed since first enumerated by sociologist Peter Stein in the mid-1970s. Many of the pressures toward and attractions associated with marriage have diminished, if they haven't completely disappeared. For example, cultural expectations to marry and marry young have lessened, sex outside of marriage has grown in acceptability and availability, increasing numbers of women and/or couples have embarked on parenthood outside of marriage, and one is likely to feel less economic pressure to marry because of an increasing ability to find economic security outside of marriage. Singlehood has become less stigmatized, while media images (from news stories through entertainment media) have cast marriage in a less attractive and more precarious light. Finally, as more people forgo or delay marriage, one can do so with less guilt, greater peer support, and less parental pressure. Such changes can help us understand much of what is responsible for the growth in the single population reflected in the data in Table 9.2.

These changes can be said to have reduced the need to marry. Men can obtain many of the "services" long provided by wives—such as cooking, cleaning, intimacy, and sex—outside of marriage without being tied down by family demands and obligations. Thus, men may not have a strong incentive to commit, marry,

Table 9.3 Pushes and Pulls toward Marriage and Singlehood

Pushes/pulls toward Marriage	Cultural norms
	Love and emotional security
	Loneliness
	Physical attraction and sex
	Parental pressure
	Desire for children
	Economic pressure
	Desire for extended family
	Social stigma of singlehood
	Economic security
	Fear of independence
	Peer example
	Media images
	Social status as "grown up"
	Guilt over singlehood
	Parental approval
Pushes/pulls toward Singlehood	Perceived problems of marriage and fear of divorce
	Freedom to grow
	Stagnant relationship with spouse
	Self-sufficiency
	Feelings of isolation with spouse
	Expanded friendships
	Poor communication with spouse
	Mobility
	Unrealistic expectations of marriage
	Career opportunities
	Sexual problems
	Sexual exploration
	Media images

SOURCE: Adapted from Stein (1975).

With greater economic opportunity and success, women's economic need to marry is less today than in the past.

to make commitments. Commitment requires sacrifice and obligation, which may conflict with ideas of "being oneself." A person under obligation can't necessarily do what he or she "wants" to do; instead, a person may have to do what "ought" to be done (Bellah et al. 1985).

Types of Never-Married Singles

In examining the experiences of unmarried women and men, much depends on whether people are single by choice or circumstance and whether they consider being single a temporary or a permanent situation (Shostak 1987). If the person is voluntarily single, his or her sense of well-being is likely to be better than that of a person who is involuntarily single. With this in mind, one can identify the following types of singles:

- *Voluntarily and temporarily unmarried.* These are usually younger men and women actively pursuing education, career goals, or "having a good time." Voluntarily single, they consider their singleness temporary. Although they are not actively seeking marital partners, they remain open to the idea of marriage. Sociologist Arthur Shostak referred to such women and men as *ambivalents.* Some ambivalents are cohabitors.
- *Involuntarily and temporarily unmarried.* Women and men in this category are actively and consciously seeking marital partners. Their desire to marry someday led Shostak to call them *wishfuls.*

or stay married. Nevertheless, "flying solo at midlife" appears to be more problematic for men than for women (Marks 1996).

Single women appear to have better psychological well-being than do single men. For those socialized during an era of traditional gender roles and family values with marriage as the norm, there seemed to be a degree of mental health risk associated with singlehood, especially for men. With increased educational attainment and expanded occupational opportunities, women have greater opportunity to succeed professionally and live independently.

The emphasis on independence and autonomy blends with an increasing emphasis on self-fulfillment, which, some critics argue, makes it difficult for some

- *Resolveds.* Individuals who regard themselves as permanently single might be called *resolveds*. This would include priests and nuns, but of greater relevance, it would also account for many single parents who prefer rearing their children alone. Most, however, are "hard-core" singles who simply prefer to be single or cohabitors who live with a partner but do not intend to marry.

- *Involuntarily unmarried.* Individuals in this category might be considered *regretfuls* in that they would prefer to marry but are otherwise resigned to their "fate." These include well-educated, high-earning women over age 40 who may find a shortage of similar men as a result of the marriage gradient, divorced or widowed women and men who can't find a partner with whom they could remarry, and any other women and men whose circumstances, health, or finances don't allow them to enter or reenter marriage.

Individuals may shift from one type to another at different times. All but the resolveds share an important characteristic: they want to move from a single status to a romantic couple status. The dramatic increase in the single population may be most accounted for by increases in the numbers of resolved singles, those we might also simply call single by choice.

Singlism and Matrimania

There are many myths and misconceptions about those who are unmarried (DePaulo 2006). As articulated and systematically debunked by psychologist Bella DePaulo, these include the following:

1. All singles, especially single women, want to be coupled.
2. Singles are miserably lonely, bitter, and envious of their coupled friends.
3. Singles are self-centered and immature.
4. Single women will someday regret not having married and not having families. They are alternately and unflatteringly portrayed as either sexually deprived or sexually promiscuous.
5. Single men are frequently portrayed as threatening, irresponsible, sexually obsessed, or gay.
6. Children of single parents are destined to suffer emotionally, socially, academically, financially, and behaviorally.
7. Singles lack a partner and, therefore, lack a purpose.
8. Singles will age and die alone.
9. Married people with families deserve special benefits, perks, higher pay, and other resources that singles don't need.

DePaulo considers such myths and stereotypes evidence of what she calls **singlism**. Singlism, like racism, sexism, and heterosexism, embodies stereotyping, discrimination, and negative, dismissive treatment, though without the segregation, hatred, and violence that other, more "brutally stigmatized" groups face (DePaulo 2006). According to DePaulo, singles (i.e., unmarried women and men) are stereotyped on the basis of the myths enumerated above.

Beyond stereotyping, though, singlism entails discriminatory treatment of unmarried people. They can be asked and expected to stay late at work, travel over the holidays, and pay more (per person) for vacation packages, club memberships, and even meals in restaurants and often must do without discounted health benefits, greater Social Security options, lower tax bills, and higher salaries. According to DePaulo, it is almost as though single people were worth less than married women and men. She notes that surviving spouses can collect the Social Security benefits of their deceased spouses, including small amounts to assist with funeral expenses. When unmarried women and men die, "no other adult can receive your benefits. Your money goes back into the system" (DePaulo 2006, 6).

Alongside the stereotyping of and discrimination against singles comes the glorifying of couples, especially married couples. DePaulo calls this

"Matrimania" casts marriage as the only way to achieve intimacy and personal fulfillment. As a consequence singles are assumed—incorrectly—to be miserable and want what only marriage is said to bring.

matrimania, "the over-the-top hyping of marriage and coupling" (www.psychologytoday.com/blog/living-single/200804). She notes that the matrimania mythology glorifies marriage as "the gold standard," supposedly offering the power to transform an immature single into a mature spouse—to create commitment, sacrifice, selflessness, intimacy, and loyalty where before (i.e., when single) none existed. Marriage is seen as the source of happiness—"not just garden variety happiness, but deep and meaningful well-being. A sense of fulfillment that a single person cannot even fathom" (DePaulo 2006, 13).

Between the glorification and celebration of marriage and the stereotyping and discrimination faced by singles, unmarried women and men suffer needlessly. Their status as unmarried, however, does not by itself cause them suffering and misery.

Cohabitation

Few changes in patterns of marriage and family relationships have been as dramatic as changes in cohabitation. What in the 1960s was rare and relegated to hushed whispers and secrets from families is now a common experience (King and Scott 2005) (see Table 9.4).

Table 9.4 Numbers of Individuals Cohabiting by Age

Age	Number of People Cohabiting
<30	3.6 million
30–39	2.6 million
40–49	1.7 million
>50	1.2 million

SOURCE: King and Scott (2005) and 2000 Census Public Use Microdata Samples.

The Rise of Cohabitation

Over the past 40 years, cohabitation has increased more than 10-fold. It has increased across all socioeconomic, age, and racial groups. Using a combination of census reports it is estimated that there are now more than 7.5 million cohabiting couples in the United States, including 6.8 million heterosexual couples (as of 2007) and almost 800,000 gay and lesbian couples (as of 2005) (American Community Survey 2005; U.S. Census Bureau 2009). Forty years ago, there were only approximately 400,000 cohabiting heterosexual couples. Thus, we can see how steep an increase has occurred, especially since 1970 (see Figure 9.1).

Figure 9.1 Cohabitation: 1960 to 2008

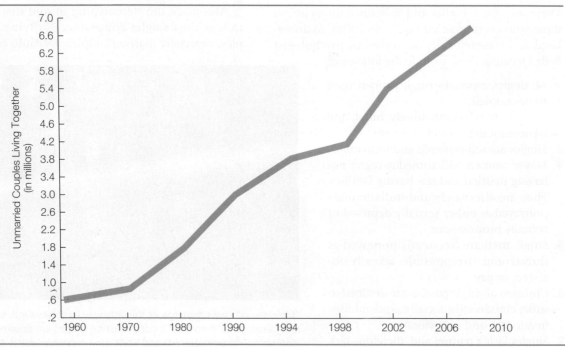

SOURCE: U.S. Bureau of the Census (2001).

Looking at the percentage of people who cohabit at some point in their lives reveals what an increasingly common experience cohabitation has become. As reported by sociologists Pamela Smock, Lynne Casper, and Jessica Wyse (2008), more than 60% of women between the ages of 25 and 39 have cohabited, and 65% of women and men who first entered marriage since 1995 had cohabited beforehand. Just in the past two decades, the number of cohabiting unmarried partners increased by 88% (U.S. Census Bureau 2007). Cohabitation appears no longer to be as much of a moral issue as it once was; instead, it has become a fairly common and relatively accepted family lifestyle. It also appears to be a lifestyle that is still increasing and is here to stay (Manning and Smock 2009).

> **Matter of Fact** Although cohabitation has increased for all educational groups and for Caucasians, Latinos, and African Americans, it is more common among those with lower levels of education and income (Smock et al. 2008).

In accounting for the increase in cohabitation, we can return to some of the same factors that help explain the increase in singlehood (Smock et al. 2008):

- *The general climate regarding sexuality is more liberal than it was a generation ago.* Sexuality is more widely considered to be an important part of a person's life, whether or not he or she is married. Unmarried couples can more freely engage in an ongoing sexual relationship with less stigma.
- *The meanings of marriage and divorce have changed.* Because of the dramatic increase in divorce for most of the last quarter of the twentieth century, marriage is no longer thought of as a necessarily permanent commitment. Permanence is frequently replaced by serial monogamy—a succession of marriages—and as a consequence, the difference between marriage and living together has lost some of its sharpness.
- *Men and women are delaying marriage longer.* For many couples, especially lower-income couples, marriage is seen as a desirable outcome but one that requires them to first accumulate sufficient resources and display their suitability to marry. While putting off marriage until they are ready, many choose to cohabit.
- *Women are less economically dependent on marriage.* With greater labor force participation, more

opportunity to enter fields previously closed to them, and a slowly diminishing though stubbornly persistent wage gap, marriage is less central and essential to women's economic well-being.

- Cohabitation has become normalized as a stage or phase in one's life course. Research on teenagers and young adults shows that approval of cohabitation is common and that many teenagers expect to someday cohabit with someone, perhaps along the way toward marriage.

Examining these items reveals both economic and cultural factors at work. Many of these same factors help account for other familial changes, including the previously discussed increase in the population of unmarried women and men. According to Smock et al. (2008), the combination of cultural and economic factors changed the societal context within which people make lifestyle choices. They contend that these same two influences operate on a more individual or micro level in terms of the resources one possesses and the attitudes and values one embraces. Looking first at socioeconomic influences, they note the following:

- Cohabitation is more likely among those with less education.

© Blend Images/Alamy

Hispanic and African American couples are more likely to cohabit than are whites or Asian Americans.

- Men's economic situations, including their employment, earnings, and work experience, all influence their likelihood of marrying versus cohabiting. Men with less than full-time, year-round employment are more likely to cohabit than to marry.
- The childhood experience of parental divorce or family instability increase the likelihood that one will enter a nontraditional lifestyle such as cohabitation. Although such experiences occur at all class levels, they are more common among those who have fewer economic resources.
- There are racial differences that often overlap with class differences regarding the likelihood of growing up in cohabiting parent households, with African American and Hispanic children overrepresented in such households. Additionally, black and Hispanic women are more likely to become pregnant while cohabiting and to continue to cohabit rather than marry after their children are born.

Critical Thinking

Do you think the increase in cohabitation is an indication of a lesser commitment to marriage? Would you think people who are living together would be more likely or less likely to succeed if they were to marry later? Why?

Exploring the role of attitudes and values, Smock et al. (2008) note that cohabitation is more common among those who have more liberal attitudes—those who are more likely to support nontraditional families and egalitarian gender roles and who are less religious.

Types of Cohabitation

Although the picture sketched so far has addressed the overall phenomenon of and increase in cohabitation, there is considerable diversity among cohabitors. There is no single reason to cohabit, no one type of person who cohabits, and no one type of cohabiting relationship. One popular typology differentiates among four different types of cohabiting relationships: *substitutes or alternatives for marriage, precursors to marriage, trial marriages,* and *coresidential dating* (Casper and Bianchi 2002; Casper and Sayer 2002; Phillips and Sweeney 2005). These can be distinguished by whether partners anticipate a married future, their perceptions of the stability of the relationship, and their general attitudes toward cohabiting relationships:

- *Trial marriage.* In **trial marriages**, the motive for living together *outside marriage* is to assess whether partners have sufficient compatibility to successfully *enter marriage.* They are undecided as to their likelihood of marriage, are uncertain about their suitability for marriage, and, by cohabiting, hope to determine whether they should proceed to marry.
- *Precursor to marriage.* When the relationship is a **precursor to marriage**, partners share an expectation that eventually they will marry.
- *Substitute for marriage.* When the relationship is a *substitute for marriage,* partners are not engaged and have no intention or expectation to marry but they do anticipate staying together.
- *Coresidential dating.* In *coresidential dating,* the relationship is more like a serious boyfriend–girlfriend relationship, lacking any intention or expectation to marry; cohabitation offers such couples greater convenience than does living apart (Bianchi and Casper 2000).

In neither substitutes for marriage nor coresidential dating is there any expectation of eventually marrying, even though many such cohabitors do ultimately marry. As to the expected duration of the relationship, a coresidential dating situation is understood to last for only a relatively short time. In either a substitute for marriage or a precursor to marriage, couples expect to be together a long time. In the case of trial marriages, couples don't know whether and how long they will stay together (Heuveline and Timberlake 2004).

A second typology separates cohabitation into the following *five types* (Heuveline and Timberlake 2004):

- *Prelude to marriage.* Cohabitation is used as a "testing ground" for the relationship. Cohabitants in this type of situation would likely marry or break up before having children. The duration of this type is expected to be relatively short, and couples should decide to transition into marriage or to part ways relatively soon after beginning to cohabit.
- *Stage in the marriage process.* Unlike the prior type, couples may reverse the order of marriage and childbearing. They cohabit for somewhat longer periods, typically in response to opportunities that they can pursue "by briefly postponing marriage" (Heuveline and Timberlake 2004, 1216). It is understood by both partners that they intend to eventually marry.

- *Alternative to singlehood*. Considering themselves too young to marry and with no immediate intention to marry, such couples prefer living together to living separately. Having a commitment level more like a dating couple than that of a married couple, such relationships will be prone to separation and breaking up.
- *Alternative to marriage*. Couples choose living together over marriage but choose, as married couples would, to form their families. Greater acceptance of out-of-wedlock births and child rearing will increase the numbers of couples experiencing this type of cohabitation. Such couples would not likely transition into marriage but would likely build lasting relationships.
- *Indistinguishable from marriage*. Such couples are similar to the previous type but are more *indifferent* rather than *opposed* to marriage. As cohabitation becomes increasingly accepted and parenting receives support regardless of parents' marital status, couples lack incentive to formalize their relationships through marriage.

Finally, consider a third typology of cohabitation, highlighting the factors that couples consider in deciding to live together, the "tempo of relationship advancement" into cohabitation, and the language used, or story told, by couples in accounting for their cohabiting (Sassler 2004). Sharon Sassler (2004, 498) identifies six broad categories of reasons that couples decide to cohabit—finances, convenience, housing situation, desire, response to family or parents, and as a trial—out of which she constructs a three-category typology:

- *Accelerated cohabitants* decided to move in together quickly, typically before they had dated six months. Emphasizing the strength and intensity of their attraction and their connection and the fact that they were spending a lot of time together and identifying finances and convenience as major reasons for their decision, they contended that moving in together felt like "a natural process."
- *Tentative cohabitants* admitted to some uncertainty about moving in together. Together for 7 to 12 months on average before living together, they typically saw each other less often than the "accelerateds" did before moving in together (e.g., three or maybe four nights a week) or had experienced disruptions in their relationships with one of the partners being gone for a period, slowing their pro-

gression into cohabitation. They often mentioned "unexpected changes in their residential situation" as a reason for their decision. Absent such a situation, they might not have moved in together when they did.
- *Purposeful delayers* were the most deliberate in the decision-making process. Their relationships progressed more gradually, taking more than a year before they decided to live together and allowing them an opportunity to discuss future plans and goals. They most often mentioned housing arrangements and finances as the reasons they moved in together.

As the three typologies reveal, not all cohabitors desire, intend, or expect to marry. Although cultural attitudes and values—as well as ideas about singlehood, dating, marriage, and cohabitation—somewhat determine whether someone expects to marry, socioeconomic criteria are also of importance. Wendy Manning and Pamela Smock (2002) suggest that the percentage of cohabiting women who expect to marry their partners remained fairly stable from the late 1980s through the mid-1990s, with 74% of cohabitants expressing an expectation to marry. Those women with a higher probability of expecting marriage are those who live with partners of high socioeconomic status. In addition, men's age and religiosity (strength of religious involvement and identification) make a difference in women's expectations of marriage. In terms of race, black women have a lower probability of expecting marriage than either white or Hispanic women. Furthermore, despite relatively worse economic circumstances, Latinos have higher marriage rates than whites. Among cohabiting women, Latinos and whites have similar expectations regarding the likelihood that they will marry their partners, and, more generally, research suggests that women are more likely than are men to see cohabitation as a step

Matter of Fact Only about half of the births to cohabiting couples are the result of planned pregnancies. Among married couples, roughly 80% of births were planned. Although cohabiting couples who become pregnant are more likely to marry than are "singles" who become pregnant, cohabiting couples are less likely today than in the past to marry because of pregnancy (Lichter 2009).

toward marriage (Hartwell-Walker 2008; Manning and Smock 2002).

What Cohabitation Means to Cohabitors

The meaning of cohabitation varies for different groups. For African Americans, cohabitation is more likely to be a substitute for marriage than a trial marriage, and blacks are more likely than whites to conceive, give birth, and raise children in a cohabiting household. Indirectly, this implies that cohabitation is a more committed relationship for blacks than for whites, a more acceptable family status, and a more acceptable family form within which to rear children—even though blacks are no more likely than whites to say that they approve of cohabitation (Phillips and Sweeney 2005).

The same appears to be true among Hispanics. The idea of "consensual unions" outside of marriage goes back a long way in Latin America, especially among the economically disadvantaged. More than among whites, for Hispanics cohabitation is more likely to become an alternative to marriage. Again, we draw this conclusion from rates of nonmarital pregnancy and childbearing. Julie Phillips and Meghan Sweeney (2005, 299) report that for Hispanics, cohabitation "may be a particularly important context for *planned* childbearing" (emphasis added).

One of the most notable social effects of cohabitation is that it delays the age of marriage for those who live together. *Ideally*, then, cohabitation could encourage more stable marriages because the older people are at the time of marriage, the less likely they are to divorce. However, as we shall soon see, cohabitation does *not* ensure more stable marriages.

Although there may be a number of advantages to cohabitation, there are also disadvantages. Young cohabitors—say, of college age—may find that parents may refuse to provide support for school as long as their child is living with someone, or they may not welcome their child's partner into their home. As cohabitation continues to grow in frequency and acceptability, this outcome will likely lessen.

In the United States, cohabiting couples still lack most of the rights that married couples enjoy, a topic we will return to shortly. According to Judith Seltzer (2000), children of cohabiting couples may also be disadvantaged unless they have legally identified fathers. For example, in the absence of legally established paternity, children are not guaranteed financial support from nonresidential fathers. This situation differs greatly from what exists in many other countries. In Sweden, for instance, the law treats unmarried cohabitants and married couples the same in such areas as taxes and housing. In many Latin American countries, cohabitation has a long and socially accepted history as a substitute for formal marriage (Seltzer 2000).

There are additional disadvantages. Cohabiting couples may also find that they cannot easily buy houses together because banks may not count their income as joint; they also usually don't qualify for health insurance benefits. If one partner has children, the other partner is usually not as involved with the children as he or she would be if they were married. Cohabiting couples may find themselves pressured to marry if they have a child, though again this may be a diminishing likelihood. Finally, cohabiting relationships generally don't last more than two years; couples either break up or marry.

Cohabitation and Remarriage

Roughly half of those who remarry after a divorce cohabit before formally remarrying, and postdivorce cohabitation is now more common than premarital cohabitation (Xu, Hudspeth, and Bartkowski 2006). In fact, one reason that remarriage rates have dipped is because postmarital cohabitation has increased (Seltzer 2000). Living together takes on a different quality among those who have been previously married. Some postdivorce cohabitors may choose to live together because of their experiences in their earlier failed marriage. Sociologist Xiaohe Xu and colleagues (2006) suspect that this may be especially true if those postdivorce cohabitors didn't cohabit prior to their first marriage. They may see cohabitation as a way to test the relationship that they missed—and suffered the consequences—in their earlier marriage. Additionally, postdivorce cohabitation has likely delayed the process of remarrying and become a period more like extended courtship or dating.

Marital quality and happiness appear to be lower among postdivorce (premarriage) cohabitors. Happiness in remarriage is nearly 30% less for couples who cohabited before remarrying than for those who did not live together before remarriage. Similar outcomes resulted regarding the effects of postdivorce cohabitation on the stability of remarriages (Xu et al. 2006).

Cohabitation and Marriage Compared

Perhaps the major difference between cohabitation and marriage results from the fact that marriage is institutionalized and cohabitation isn't. Even with all the changes to marriage discussed in the previous chapter, there are still more "shared understandings" of what marriage means—its rules and roles and rights and responsibilities—than there are about cohabitation.

Different Commitments

Marriages begin with spouses pledging a lifelong commitment to each other ("till death do we part"). A lesser level of commitment characterizes cohabiting couples as a whole when compared to married couples. Generally, when a couple lives together, their primary commitment is to each other, but it is a more transitory commitment. As long as they feel they love each other, they will stay together. In marriage, the couple makes a commitment not only to each other but also to their marriage. Frequently, cohabitants are less committed to the certainty of a future together (Forste and Tanfer 1996; Waite and Gallagher 2001).

However, there is a wider range in how much commitment one finds among cohabiting couples compared to married couples, with some cohabiting couples being as committed as most married couples. Michael Pollard and Kathleen Mullan Harris used survey data from 18- to 26-year-olds who were part of the National Institute of Child Health and Human Development's National Longitudinal Study of Adolescent Health (Add Health) to explore the dynamics of cohabiting relationships. They looked specifically at differences among heterosexual cohabiting couples and between cohabitors and married women and men on such relationship characteristics as commitment and closeness. In general, cohabitors were somewhat less likely than married respondents to claim that they were "completely committed" to their relationships, but there was interesting variation between types of cohabiting relationships. People who were in a *precursor to marriage relationship* were as likely to report themselves as "completely committed" to their partners as were married respondents and were more likely to characterize their relationships at the maximum level of closeness. On the other hand, those in more casual *coresidential dating* relationships were less than half as likely as either married or other cohabitors to express the same level of commitment (see Figure 9.2).

Living together tends to be a more temporary arrangement than marriage, and in recent years heterosexual cohabiting relationships have become even more short term and less linked to marriage (Seltzer 2000; Smock et al. 2008). A man and woman who are living together may not work as hard to save their relationship. Less certain of a lifetime together, they are more likely to continue to live somewhat more autonomous lives. In marriage, spouses will typically do more to save their marriage, giving up dreams, work, ambitions, and extramarital relationships for marital success.

Figure 9.2 Type of Relationship

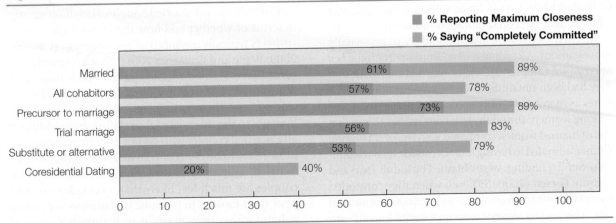

Legend:
■ % Reporting Maximum Closeness
■ % Saying "Completely Committed"

Type	% Reporting Maximum Closeness	% Saying "Completely Committed"
Married	61%	89%
All cohabitors	57%	78%
Precursor to marriage	73%	89%
Trial marriage	56%	83%
Substitute or alternative	53%	79%
Coresidential Dating	20%	40%

SOURCE: Pollard and Harris (2007).

Unmarried couples are also less likely than married couples to be encouraged by others to make sacrifices to save their relationships. Parents may even urge their children who are "living together" to split up rather than give up plans for school or a career. If a cohabiting couple encounters sexual difficulties, it is more likely that they will split up. It may be easier to abandon a problematic relationship than to change and fix it. Again, among cohabitants who intend to marry, relationships—including the aspect of commitment—are not significantly different from marriages. Cohabitating partners who are highly educated and employed and enjoy higher levels of income are more likely to express an intention to marry, are more likely to marry, and are less likely to break up (Brown and Booth 1996; Smock et al. 2008).

Sex

There are differences in the sexual relationships and attitudes of cohabiting and married couples. Linda Waite and Maggie Gallagher (2001) suggest that married couples experience more fulfilling sexual relationships because of their long-term commitment to each other and their emphasis on *exclusivity*. Because they expect to remain together, married couples have more incentive to work on their sexual relationships and discover what most pleases their partners.

Heterosexual cohabitants, however, have more frequent sexual relations. Whereas 43% of married men reported that they had sexual relations at least twice a week, 55% of cohabiting men said that they had sex two or three times a week or more. Among married women, 39% said that they had sex at least twice a week, compared with 60% of never-married cohabiting women. Sex may also be more important in cohabiting relationships than in marriages. Waite and Gallagher (2001) go as far as calling it the "defining characteristic" of cohabitants' relationships.

Married couples are also more likely to be sexually monogamous. According to data from the National Sex Survey (see Chapter 6), 4% of married men said that they had been unfaithful over the past 12 months; four times as many, 16%, of cohabitants reported infidelity. Among women, the equivalent comparison shows that 1% of married women compared with 8% of cohabiting women admitted to having had sex outside their relationship. Similar findings were obtained by Judith Treas and Deirdre Giesen (2000, 59) even when they controlled for how permissive individuals were toward extramarital sex: "This finding suggests that cohabitants' lower investments in their unions, not their unconventional values,

accounted for their greater risk of infidelity." Importantly, however, they find that in the vast majority of cohabiting relationships, sexual exclusivity *is* expected.

Finances

Overall, cohabiting women and men have more precarious economic situations than married couples. The latter have higher personal earnings and higher household incomes and are much less likely to live in poverty. There is also evidence that cohabitation carries an "economic premium" over remaining single and living apart that is comparable to what accompanies marriage; entering cohabiting relationships alleviates some financial distress, especially for Hispanic and African American women and their children (Avellar and Smock 2005). Unfortunately, as with the end of a marriage, when cohabiting relationships end, there is considerable economic suffering, especially for women. Sarah Avellar and Pamela Smock (2005) contend that where cohabiting men suffer modest effects when their relationships end, cohabiting women suffer "dramatic declines" in their standards of living. Men suffer declines of roughly 10% in their household income. For women, there is a more notable loss of household income (33%) and a striking spike in the level of poverty (reaching nearly 30% in poverty) following breakups.

Perceptions of current and future financial stability are more instrumental in decisions that couples make to marry than in decisions to live together. In ongoing relationships, partners' relative financial contributions differently affect married and cohabiting couples. For cohabitors, relative equality in partners' earnings seems to stabilize relationships. Conversely, among married couples in which each spouse contributes equally to the family income, the risk of divorce is heightened (Raley and Sweeney 2007).

Cohabiting and married couples also differ greatly in terms of whether and how they pool their money, which is typically a symbol of commitment (Cannon 2009; Waite and Gallagher 2001). People generally assume that a married couple will pool their money, as it suggests a basic trust or commitment to the relationship and a willingness to sacrifice individual economic interests to the interests of the relationship. Additionally, married couples who do not pool their money are more likely to be dissatisfied with their marriages than married couples who do (Dew 2008). Cohabiting couples are much less likely than married couples to pool their money. In fact, one of the reasons couples cohabit rather than marry is to maintain a sense of financial independence (Waite and Gallagher 2001).

As this chapter has addressed, an increasing number of couples in long-term, loving and committed relationships forgo marriage in favor of living together. In recent years, there also has been an increase in the number of couples in long-term, loving and committed relationships who forgo living together and instead maintain their relationships across different households. Characterized by labels such as "home alone together," "dual-dwelling duos," and, most commonly, "living apart together" (or LAT), such arrangements have become more common in the United States and abroad. They are on the rise in Canada, Britain, Holland, Sweden, Norway, and France. In fact, in Canada, it is estimated that 1 in every 12 adults age 20 and older is in a LAT relationship (Milan and Peters 2003).

Like marriage and cohabitation, main features of LAT relationships include a long-term commitment, public identification as a couple, open acknowledgment of love for one's partner, and involvement in regular sexual relations with one's partner (Connidis 2006). Unlike marriage and cohabitation, these features are shared and expressed while living apart from each other.

LAT relationships can be the choice of couples of a variety of ages and statuses. For younger couples, LAT relationships might be chosen to allow one or both partners to continue to live at home, supported in part or full by parents, while maintaining a committed, monogamous, couple relationship. For those who are older, such as in midlife, a variety of practical reasons might motivate them to maintain separate residences while, at the same time, considering their lives to be joined together. For example, for those who have children, living apart together can allow them to focus on the things that need to be done for and with their children without the added complication of a live-in partner. In instances where one or both partners own their residence, this arrangement allows them to maintain the value of two residences. For those who have been married and divorced, the LAT arrangement allows them to avoid subsequent marital failure.

There are aspects of LAT relationships that make them especially attractive to older women and men. For older women, a LAT relationship allows them to maintain the freedom and independence that accompanies having one's own space. It also frees them from the potential obligation to provide long-term care for a partner whose health diminishes. Also potentially important to older couples, a LAT relationship can more clearly protect their financial resources for future inheritance to be left to one's children and grandchildren. In marriage or cohabitation, such money might be contested by a surviving spouse or cohabiting partner.

Of course, across all age-groups, by living together apart, partners can maintain their independence and autonomy, avoid the need for frequent compromise, and live life "on one's own terms." Advocates contend that living apart together can be the best of both worlds, as partners can both "be themselves" and "be a couple." As sociologist Ingrid Arnett Connidis (2006) suggested, LATs can allow one to simultaneously satisfy one's need for intimacy and autonomy and companionship and independence.

More negative assessments cast the choice to enter and maintain a LAT relationship as self-centered and indicative of an inability or unwillingness to compromise (Hart 2006). Like so many other issues that surface throughout this text, the ultimate meaning and significance of living apart together can be interpreted in more than one way.

---

or cohabitors who were engaged first. As expressed by family researchers Scott Stanley and Galena Rhoades (2009, F3), "The greatest risk is for those who do not have mutual clarity about the future together because they are increasing the likelihood of marriage [by cohabiting] before clarifying ... important matters of fit, intention, and commitment." Research by Wendy Manning and Pamela Smock found that for over 50% of the cohabiting couples they studied, cohabitation "just sort of happened. One thing led to another and bingo, the couple was living together" (Stanley and Rhoades 2009, F3).

According to Stanley and Rhoades (2009), many couples similarly "slide" into cohabitation rather than actively deciding to live together. Once a couple is living together, they may find themselves on a path to marriage simply because it is harder to end their relationship than if they were just dating. Such things as "the idea of moving out, splitting things and friends up, and finding another place to live" may lead many cohabitors to remain together, and remaining together increases the likelihood that they will eventually marry. Once married, such "sliders" are more likely to experience marital difficulties that increase their likelihood of divorce (Stanley and Rhoades 2009).

Related to the matter of *type of cohabitation* is the question of one's past *history of cohabiting relationships*. Individuals who cohabit with only their future spouse are unlikely to experience the negative outcomes otherwise associated with cohabitation. On the other hand, **serial cohabitation**, where individuals have cohabited with more than one partner, has more problematic outcomes. Research by Daniel Lichter and Zhenchao Quan revealed that female cohabitors who had prior cohabiting relationships were much less likely to marry their partners than were women who had cohabited only with their current partners. Those serial cohabitors who did enter marriage experienced more than double the risk of divorce compared to women who cohabited only with their future husbands (Lichter and Quan 2008).

There is yet another question worth asking. For those former cohabitors who marry and then divorce (or face other negative marital outcomes), what is it about cohabitation that leads to these difficulties? Is it that the *types of people* who choose to live together before marrying are also more likely to divorce should they face marital difficulties, or is there something about the *experience of living together* that causes problems later? Susan Brown and Alan Booth (1996) suggest that the characteristics of people who cohabit are more influential than the cohabiting experience itself. People who live together before marriage tend to be more liberal, more sexually experienced, and more independent than people who do not live together before marriage. They also tend to have slightly lower incomes and are slightly less religious than noncohabitants (Smock 2000).

At the same time, there is some evidence that cohabitation itself may affect individual partners and their relationships. Compared with married couples, cohabiting partners tend to have more similar incomes and divide household tasks more equally. Egalitarian arrangements may be harder to sustain once married, especially once couples have children, and strain or conflict may occur. Sociologist Judith Seltzer (2000) suggests that marriage presses couples toward a gendered division of labor. As cohabitation continues to increase among people from varying backgrounds, we will be better positioned to see whether the *experiences of cohabitation* or *characteristics of cohabitants* have greater effect on later marriage. As cohabitation grows in number and acceptability, its effects on marriage also may be changing. The negative effect that cohabitation has on marriage may have diminished and now be, as some report, "trivial" at best (Smock et al. 2008). Perhaps all we can suggest at this point is that at least some poorly chosen relationships break up at the cohabitation stage. Thus, although cohabitation may not afford couples protection from later marital failure, it does show some high-risk couples that they were not meant for each other. This can at least spare them the later experience of a divorce (Seltzer 2000).

Common Law Marriages and Domestic Partnerships

Before the nineteenth century, U.S. couples who lived together without marrying would, after a short period of living together, enter what is known as **common law marriage**. A couple who "lived as husband and wife and presented themselves as married" was considered married. Originating in English common law, as practiced in the United States, common law marriage was seen as a practical way to enable couples who wanted to be married but were too geographically removed from both an individual with the authority to marry them and a place where they could obtain a marriage license to marry (Willetts 2003). Common law marriage became less necessary in the nineteenth century as the availability of officials who could perform marriage ceremonies grew. Although most states no longer allow or recognize common law marriage, in the United States the following 11 states and the District of Columbia still do:

Alabama	Oklahoma
Colorado	Pennsylvania
District of Columbia	Rhode Island
Iowa	South Carolina
Kansas	Texas
New Hampshire (for inheritance purposes only)	Utah

If you happen to be reading this in one of these states and you meet the requirements described earlier, congratulations, you have just been pronounced married! Although we are being facetious, common law marriage does unite into legal marriage two people who never sought and never obtained a marriage license. Once in such a marriage (Solot and Miller 2005),

> if you choose to end your relationship, you must get a divorce, even though you never had a wedding. Legally, common law married couples must play by

all the same rules as "regular" married couples. If you live in one of the common law states [and live together] and don't want your relationship to become a common law marriage, you must be clear that it is your intention not to marry.

In states that recognize common-law marriages, the amount of time a couple must live together before being considered married varies. What is essential is that they have presented themselves as if married by acting like they are married, telling people they are married, and doing the things married people do (including referring to each other as "husband" and "wife"). In states that don't recognize common law marriages, no matter how long you live together or how married you act, you are not married. Of note, *no state* in the United States recognizes same-sex common law marriage.

Domestic partners—cohabiting heterosexual, lesbian, and gay couples in committed relationships— are gaining some legal rights. Domestic partnership laws that grant some of the protection of marriage to cohabiting partners are increasing the legitimacy of cohabitation. In some ways, domestic partnerships are alternative forms of cohabitation with certain formal rights and protections. Civil unions are more like alternative versions of marriage (Willetts 2003).

Domestic partnerships can be regulated by the state or be recognized by private employers. Domestic partnership benefits can include a variety of the following (http://family.findlaw.com/living-together):

- Medical, dental, and vision insurance
- Sick leave
- Bereavement leave
- Death benefits
- Accident and/or life insurance
- Parental leave
- Tuition reduction or remission at universities where a partner is employed
- Housing rights
- Access to recreational facilities

In 1982, the *Village Voice* newspaper (based in New York City) was the first private business to offer domestic partnership benefits to employees and their partners. Two years later, in 1984, Berkeley, California, was the first U.S. city to enact a domestic partnership ordinance and extend it to both heterosexual and same-sex couples (Willetts 2003). In 1997, San Francisco extended health insurance and other benefits to their employees' domestic (which includes same-sex) partners, requiring all companies contracting with the

© AP Images/Sandra Chereb

A steadily increasing number of employers, including a majority of Fortune 500 companies, now offer benefits to domestic partners of their employees.

city or county to offer the same benefits to same-sex couples that they provided married couples. Individual employers, such as the Gap, Levi Strauss, and Walt Disney Company, soon followed suit, introducing domestic partner policies that have now become fairly common in the private sector as well as in many local and state governments, colleges, and universities. As of 2008, 57% of Fortune 500 (and 83% of Fortune 100) companies offered employees domestic partnership benefits, up from 40% of Fortune 500 companies and 60% of Fortune 100 companies in 2003 (www.hrc .org/issues/workplace/benefits).

A number of states—including California, Oregon, Washington, Hawaii, Maine, and Wisconsin—have domestic partnership laws in place or pending. In Connecticut, Vermont, New Jersey, and New Hampshire, civil unions are available to same-sex couples, and this status grants them some of the same benefits as married couples. Additionally, some municipal governments provide domestic partner benefits even when their state governments do not. In New York City, for example, domestic partnership benefits extend to heterosexual or same-sex couples, whereas the New York statewide laws are more narrowly framed and available only to gay or lesbian couples. In Cleveland, Ohio, same-sex couples can register as domestic

"I get kind of upset when people say that a domestic partnership is an alternative to marriage.... That's not what it is.... It's a different approach to looking at partnerships in sort of a legal sense."

The preceding comment is from 26-year-old Marie as she talked to sociologist Marion Willetts (2003) about her four-year-long licensed relationship. Willetts interviewed 22 other licensed heterosexual domestic cohabitants in the first study that attempted to uncover and document the motives for embarking on a domestic partnership instead of marriage. Although the rights and benefits bestowed by domestic partnership recognition could be obtained by marrying, some couples opt instead to enter licensed domestic partnerships. Typically, they must sign an affidavit declaring that they are not married to someone else and that they are not biologically or legally related to each other. They further pledge to be mutually responsible for each other's well-being and to report to authorities any change in their relationship—either marriage or dissolution. Willetts notes that motives behind heterosexual couples' choice to *cohabit* rather than marry may include, for some couples, economic benefits. This is especially true if partners have similar incomes and can take larger standard deductions on their income taxes when each files as single than they would as a married couple. Older women and men may avoid remarriage and cohabit instead, thus retaining their individual Social Security benefits. Other cohabiting heterosexual couples are motivated by more personal and philosophical benefits, such as rejecting the assumptions that are part of legal marriage, not wanting the state to intervene in their relationships, and wanting to avoid past marital failures. But what about motives to license partnerships?

Economic benefits, including health insurance coverage, access to university-owned family housing or in-state tuition benefits, or access to family membership rates in outside organizations, made up the motive most cited by Willetts's interviewees with domestic partners. For formerly married cohabitants, licensed partnerships allowed them to avoid reentering marriage yet obtain the protection and recognition of documentation. For others, such as 31-year-old Leslie, obtaining a domestic partnership license with her partner, Alan, was a means to obtain recognition in the eyes of others that they had made a deep and meaningful commitment to each other, even in the absence of a wedding: "I guess [we wanted] to sort of be counted. There's [sic] relationships that mean a lot that aren't recognized by law and to sort of be counted in that count in the city."

When Willetts posed the question of why that mattered, Leslie continued,

It's difficult to be in a relationship where people are like, "Oh, aren't you married?" or "Are you not married?" ... It's like an issue all the time. "Why aren't you married, you've been together for 10 years?" ... so we were like, "We'll get a domestic partnership [to have some sort of documentation in response to these questions]." But it wasn't really something that meant a great deal to us.... It wasn't a big deal.

Although Leslie and Alan desired recognition, they wished to avoid too much interference, such as what accompanies a marriage license: "We didn't want to have any law interfere in our relationship, or we didn't feel we needed to have a legal stamp on our relationship."

Other respondents stressed wanting to avoid the trappings of the patriarchal institution that they perceived marriage to be or wanting to demonstrate support for friends whose same-sex relationships were denied the right to marry. Licensed partnerships did not, however, give heterosexuals the same recognition and support with their families or friends that they would have had if they had married. Marie comments on what her four-year-long licensed partnership has lacked:

With a marriage license, there's that sort of social and economic and political legitimacy involved in it.... With our domestic partnership, nobody gave us any sort of crockery, nobody bought us a house, nobody sends us anniversary cards, and nobody sort of celebrated, or has celebrated that, you know, that special day [when she and her partner obtained their certificate].

Marie did not feel that legally defining licensed partners as though they were married was desirable: "Once the court says, 'Well, we're going to define this as marriage' ... once you start having courts that intervene in using words that this is like a marriage, it takes away from, once again, the legitimacy of these other sorts of different types of families that can come about."

Willetts suggests that the wider implementation of civil union laws like those in Vermont may cause states and municipalities that already have domestic partnership ordinances to deem them no longer necessary and abandon them. Once same-sex couples can enter civil unions, and given that heterosexuals can legally marry, why continue to offer this other legal category?

partners despite the fact that Ohio has some of the most restrictive language in its state constitution regarding marriage rights and sexuality and offers no recognition to same-sex domestic partners. Furthermore, because heterosexual couples *can* marry whereas same-sex couples cannot in most states, some domestic partnership protections, such as those provided by the state of New Jersey, are restricted primarily to same-sex couples (and to opposite-sex couples in which one partner is at least 62 years old). Note that even in the absence of laws recognizing domestic partnerships, many employers offer benefits to domestic partners of their employees. As Marion Willetts (2003) details, thousands of private companies, along with hundreds of colleges and universities, provide employees' domestic partners with health benefits.

Domestic partners, whether heterosexual, gay, or lesbian, may still lack some of those legal rights and benefits that come automatically with marriage. Recalling only some of the rights and benefits noted in Chapter 1, these include the right to do the following:

- File joint tax returns
- Automatically make medical decisions if your partner is injured or incapacitated
- Automatically inherit your partner's property if he or she dies without a will
- Collect unemployment benefits if you quit your job to move with a partner who has obtained a new job
- Live in neighborhoods zoned "family only"
- Obtain residency status for a noncitizen partner to avoid deportation

Keep in mind that heterosexual domestic partnerships and same-sex domestic partnerships and civil unions frequently result from different motivations. Among heterosexuals, domestic partnership is a *deliberately chosen alternative to marriage*. This is illustrated in the following Real Families feature. For at least some gay and lesbian couples, domestic partnerships or civil unions are the closest approximation to legal marriage available to them. Some same-sex couples would marry if marriage were an available and/or accessible option.

Gay and Lesbian Cohabitation

The 2000 U.S. Census reported nearly 600,000 gay or lesbian couples living together. More recent census

estimates, based on the 2005–2007 American Community Survey, put the figure at nearly 770,000 same-sex-partner households, with 53% of the households shared by gay male couples and 47% by lesbian couples (U.S. Census Bureau 2007). Who are these couples, and how do they compare to heterosexual couples?

Utilizing the 2000 census and 2005 American Community Survey, researchers Adam Romero, Amanda Baumle, M. V. Lee Badgett, and Gary Gates, 2007, assembled the following "snapshot" of how same-sex cohabitors compare to married heterosexual couples:

- On average, partners in same-sex couple households are younger than married women and men (40 vs. 48 years old).
- Nearly one in four same-sex couples (24%) are nonwhite, making them slightly more racially diverse than married couples (22% nonwhite).
- A greater percentage of partners in same-sex households have earned college degrees (40% vs. 27%).

© Jack Slomovits/Getty Images/Photodisc

Same-sex cohabitors are more likely than married heterosexual couples to be dual earners.

Figure 9.3 Median and Mean Incomes for Same-Sex and Married Heterosexual Couples

SOURCE: Romero et al. (2007).

Figure 9.4 Average Individual Earnings by Gender, Marital Status, and Sexuality

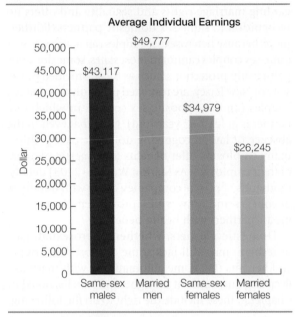

SOURCE: Romero et al. (2007).

- On average, incomes of households headed by same-sex couples are higher than those of married couples (see Figure 9.3). However, annual incomes of men in same-sex couples are lower than the incomes of married men. Gay male partners earn on average $43,117, more than $6,000 less than the average incomes of married men ($49,777). They also have median incomes (the midpoint in income distributions) 15% lower than the median for married men ($32,500 vs. $38,000).
- Partners in lesbian households tend to earn substantially more than married women; women in same-sex couples average $34,979 in annual earnings, with a median of $28,600, compared to $26,245 earned by married women, with a median of $21,000 (see Figure 9.4).
- Same-sex couples are more likely to be dual earners than married couples. Nearly a third (31%) of married couples have only one wage-earning spouse. Less than a quarter (23%) of same-sex couples depend on the earnings of only one partner.
- The 20% of same-sex couples who are raising children, like married couples with children, have an average of two children. However, they are raising those children with substantially fewer financial resources than married parents have. Their median incomes are 23% less than the median incomes of married parents, and they are much less likely to own their own home than are married parents.

The relationships of gay men and lesbians have been stereotyped as less committed than heterosexual couples for a few different reasons, including the following:

1. Until recently, lesbians and gay men could not legally marry anywhere in the United States and even currently can marry in only 5 of the 50 states (as of mid-2009). As we saw in Chapter 6, evidence suggests that the inability to marry has contributed to the higher rate of breaking up that some have attributed to the lack of strength of the love and commitment that same-sex couples feel for each other.
2. Same-sex couples, especially gay male couples, tend not to have the same expectation for or place the same emphasis on sexual exclusivity as heterosexual couples do. Higher rates of extrarelational sex have then been interpreted as reflections of the degree of commitment shared between partners. However, as we saw earlier, lesbian couples tend to value and expect fidelity as strongly as heterosexual women. In addition, gay couples in civil unions are less accepting of partners having sexual relationships with others. Thus, at best, the stereotype pertains mostly to some portion of gay male couples and even among them fails to recognize that sexual exclusivity has had a different meaning among the gay population than it has in the wider society.

3. Heterosexuals misperceive love between gay couples and between lesbian couples as being somehow less "real" than love between heterosexual couples. In fact, however, the experience of love is more similar than different regardless of sexual orientation.

As we have already seen, numerous relationship similarities exist between same-sex and heterosexual couples. Regardless of their sexual orientation, most people want a close, loving relationship with another person. For same-sex and heterosexual couples alike, intimate relationships provide love, romance, satisfaction, and security. Of course, there are also some obvious and important differences. Heterosexual couples tend to adopt a gender-divided model, whereas for same-sex couples these traditional gender divisions make no sense. Tasks are often divided pragmatically, according to considerations such as who likes cooking more (or dislikes it less) and work schedules. Most same-sex couples are dual-earner couples; furthermore, because gay and lesbian couples are the same gender, the economic discrepancies based on greater male earning power are absent. Although gay couples emphasize flexibility and egalitarianism, if there are differences in power, they are attributed to personality or to dependency on the relationship. Household tasks are either shared or done by the partner who has the time, interest, and/or expertise (Peplau, Veniegas, and Campbell 1996).

For heterosexual couples, marriage and cohabitation are choices that partners make. Only in the past decade have same-sex couples had the opportunity to choose whether to marry, and even those who wished to marry had very limited options as to where they could marry and whether their marriages would be recognized in their own states of residence. They also often cope with less familial support of and potentially more opposition to their relationships. Partly as a response to the often less than supportive reaction of others, Peplau and colleague (1996) describe gay and lesbian partners as maintaining more of a "friendship model" of relationships: "In best friendships, partners are often of relatively similar age and share common interests, skills and resources ... best friendships are usually similar in status and power."

Same-Sex Couples: Choosing and Redesigning Families

In addition to the couple relationship itself, gay men and lesbians actively construct their wider families in ways that broaden our conception of family life. Lesbian, gay, bisexual, and transgendered women and men are "acting in creative ways to plan families, keep them together, and redefine what it means to 'do family'" (Allen 2007, 175). As masterfully introduced by anthropologist Kath Weston in her book *Families We Choose*, gay men and lesbians construct **families of choice**, whose boundaries are fluid and cross household lines. They are "people who are there for you" and who can be counted on to provide support and assistance, whether of an emotional or a material nature. In addition to one's partner, such chosen families include varying combinations: children (biological or adopted), a lover's children and biological or adopted kin, friends, and ex-lovers. They can coexist simultaneously with or function in the absence of one's biological or adopted family.

Psychologist Valory Mitchell (2008b, 309) suggests that LGBT (i.e., lesbian, gay, bisexual, and transgender) families "have both the freedom and the need to define their commitment boundaries and expectations of one another. Speaking of lesbians, though true of same-sex couples more broadly, Mitchell asserts the following:

- Lesbians *make families*, independent of legal and blood ties, that are enduring and satisfying.
- Lesbians blend members of their families of origin (families into which they were born), families of procreation (families they make via marriage, cohabitation, and childbearing), and families of choice *to create* a "fabric of family life" that extends over one's life span and across generations.
- Many lesbians are parents, raising children who do at least as well as children of heterosexual parents.

As emphasized above, lesbians and gay men construct and count on their families of choice, which may become the only family or the more important ones they have.

In their efforts to have and raise children, further evidence of the chosen and constructed nature of family for many gay men and lesbians can be seen. A decade ago, when the last available decennial census data were collected, 34% of lesbian couples and 22% of gay male couples were raising children. Although we look somewhat more closely at gay men's and lesbians' experiences as parents in Chapter 10, it is instructive to at least consider the range of households and relationships that same-sex couples with children construct in order to become and be parents. As summarized by journalist John Bowe (2006, 69).

Lesbians and gay men who have children often raise them in arrangements in which each of up to four partners act as parents.

For many gay parents, the family structure is more or less based on a heterosexual model: two parents, one household.... Then there are families ... that from the outset seek to create a sort of extended nuclear family, with two mothers and a father who serves, in the words of one gay dad, as "more than an uncle and less than a father." How does it work when [a child] has two mommies, half a daddy, two daddies, or one and a half daddies?

Sociologist Judith Stacey has suggested that the gay men and lesbians who construct such innovative family forms are motivated by a strong desire to have children, "drawing from all kinds of traditional forms [of family], but at the same time ... inventing new ones" (Bowe 2006, 69). The potential complexity of such new forms can be considerable, as illustrated in the following Real Families feature.

Critical Thinking

Do you think that "families of choice" should be recognized somehow in official family policies in the United States?

Along with their novelty and complexity comes the issue of legal recognition and protection. What rights do the nonbiological parents have in relationships such as those described? Should relationships end, individuals may find themselves deeply attached to yet denied custody of or even access to children

whom they were once raising, along with their partners, as their own. As sociologist Katherine Allen (2007, 181) describes, "There are no community or institutional supports (e.g., legal adoption) available to sustain non-biological family ties after a mutual partnership agreement is severed." These "chosen families" lack legal safeguards, and "in the absence of legal protections, lesbian co-mothers, if they are not the biological mother, are likely to lose access to the child they helped to parent" (Allen 2007, 175). Given the informal and unprotected nature of the relationships, Allen warns that their ending creates a situation that at least some members may experience as ambiguous loss (see Chapter 2). If a same-sex couple who has been coparenting breaks up, the nonbiological parent may find his or her tie to the child whom he or she has been raising severed, with little he or she can do to alter that outcome. As a consequence, the boundaries that define members of the family may get redrawn.

When Friends Are Like Family

Sociologist Margaret K. Nelson authored a sensitive and thoughtful account of the illness and later death of a close friend and colleague (whom she calls "Anna Meyers"). Noteworthy in Nelson's account is how a group of friends came together to become Anna's support system. Without a lover, partner, or spouse; having had no children; and with a biological family of orientation that "was so dysfunctional ... that their input was not only inappropriate but irrelevant," Anna's care was left in the hands of Nelson, a second colleague, and later a "small group of others whom [we] consulted before making decisions because they also cared about and took good care of Anna" (Nelson 2006, 78).

Nelson (2006, 78) uses Anna's illness and death to ask two fundamental questions: Were [we] supposed to act as if we were family?" and "Just what does acting like family mean?" Ultimately, Nelson's account is testament to the important, family-like role often played by friends. She cites the work of anthropologist Carol Stack, whose urban ethnography *All Our Kin* revealed the extent of interdependence among

Becoming Parents and Experiencing Parenthood

What Do YOU Think? Are the following statements TRUE or FALSE?
You may be surprised by the answers (see answer key on the following page).

T	F	
T	F	**1** The number of births in the United States in 2007 was the highest total ever recorded.
T	F	**2** It is estimated that a third of women who marry will forego having children.
T	F	**3** More than one-third of pregnancies in the United States were unplanned.
T	F	**4** Adopted children tend to be poorer than children who live with their biological parents.
T	F	**5** Men and women both can suffer from "postpartum blues."
T	F	**6** There is often a decline in marital happiness following the transition to new parenthood.
T	F	**7** Egalitarian marriages usually remain so after the birth of the first child.
T	F	**8** Behavior of fathers has changed more than cultural beliefs about fatherhood.
T	F	**9** Compared to nonparents, parents are less likely to experience depression.
T	F	**10** Children of gay or lesbian parents are likely to be gay themselves.

"It's unbelievable.". . . There's really no way a nonparent can think like a parent. It's really knocked me for a loop. And in my wildest dreams, I never thought of it. . . . Something just creeps into your life and all of a sudden it dominates your life. It changes your relationship with everybody and everything, you question every value and every belief you ever had. And you say to yourself, "This is a miracle." It's like you take your life, open up a drawer, put it all in a drawer, and close the drawer.

These comments convey a 33-year-old man's thoughtful reactions to becoming a first-time father. As he reflects on it, becoming a parent is life defining and life altering. He is not alone. Along with new and demanding responsibilities, becoming a parent introduces profound changes in how we see ourselves, how we live, what we think about, and how we feel. Simultaneously, parents experience changes in their social relationships and how they are viewed by others. These changes are neither minor nor temporary. Becoming a parent is as profound a life change as any other we make.

Not everyone decides to become a parent. With widespread availability of effective contraception and access to legal abortion, women and men can decide whether and when to have children. The first part of this chapter focuses on the choices people make whether to have children and the range of factors that figure in to the decision-making process. We examine the characteristics of those who decide to or are forced to forgo parenthood. But those who embark on parenthood face other choices. *How* should they become parents? For some, bearing a child is difficult or impossible, leading them to attempt to adopt or take advantage of the ever-expanding options presented by advances in reproductive technology. And *when* should they become parents? Is there an optimal time or age for entering motherhood or fatherhood? Throughout this chapter, we explore these choices.

Our focus then shifts to how women and their partners experience pregnancy, the transition to parenthood, and the changes parenthood introduces into our lives. Special attention is paid to the responsibilities and challenges confronting mothers and fathers

in the United States today and the diversity of parent–child relationships.

Fertility Patterns and Parenthood Options in the United States

There were more than 4.3 million births in the United States in 2007, an increase of 1% over 2006 and *the highest total recorded to date*. Table 10.1 shows the birthrate and general fertility rate for 2007, both of which reflect a 1% increase over 2006. The 2007 fertility rate of 69.5 births per 1,000 women ages 15 to 44 was the highest level since 1990. Of course, a long-term view would show that in the midst of the baby boom, the U.S. fertility rate was considerably higher. In 1960, for example, the fertility rate was 118 (National Vital Statistics Reports 2009).

Fertility and birthrates vary considerably according to social and demographic characteristics, such as race, ethnicity, education, income, and marital status. Figure 10.1 shows variation by ethnicity. Although it is not reflected in the figure, within the Hispanic population, rates vary from a high among Mexican Americans to a low among Cuban Americans. Cultural, social, and economic factors play a significant part in influencing the number of children a family has. Because of a combination of higher fertility rates and continuing immigration patterns, Hispanics have become our nation's largest minority group, thereby surpassing African Americans.

Fertility rates also vary by education and income. In 2006, women with the highest level of education (e.g., graduate or professional school) had the highest fertility rate (66.9 births per 1,000 women

Table 10.1 Births, Birth Rates and Fertility Rates, United States, 2006 and 2007

	Total Births, 2006	Total Births, 2007
	4.27 million	4.32 million
Birthrate (births per 1,000 population)	14.3	14.2
Fertility rate (births per 1,000 women, 15–44)	69.5	68.5

SOURCE: Hamilton, Martin, and Ventura (2009, table 1).

Figure 10.1 Fertility Rates, Race, and Ethnicity, 2007

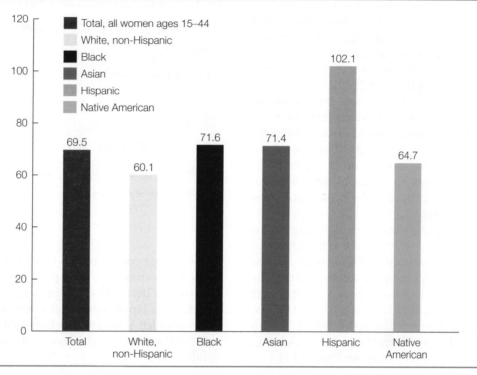

SOURCE: Hamilton, Martin, and Ventura, 2009 (table 1).

ages 15–44). College graduates had the next highest rate (59.1), followed by women with high school educations (57.5). Women with less than a high school education (51.2) and women with some college or associates degrees (49.9) had the lowest levels (U.S. Census Bureau, *Current Population Reports*, 2008).

In addition to education level, income affects rates of childbearing (see Table 10.2). Birthrates tend to decline, and the percentage of women who are childless increases as income level goes up.

Approximately 20% of American women between the ages of 40 and 44 have not had children. Among women of that same age who had ever married, 14.6% were reported to be childless (U.S. Census Bureau, *Current Population Survey, June 1976–2006*, table 1). As will be shown below, in addition to income, being childless (or child free) also varies by characteristics such as race and education.

Unmarried Parenthood

In 2007, a record number of unmarried women gave birth. There were 1.7 million nonmarital births, a 26% increase since 2002. Taking a longer view, the num-

Table 10.2 Women Who Had a Child in the Last Year per 1,000 Women, by Family Income: 2006 (numbers in thousands)

Family Income	Percent Childless	Births per 1,000 Women
Under $10,000	39.4	87
$10,000–$19,999	41.4	77
$20,000–$24,999	41.7	77
$25,000–$29,999	44.7	62
$30,000–$34,999	42.8	63
$35,000–$49,999	44.5	64
$50,000–$74,999	45.0	66
$75,000 and over	47.9	62
Total	45.1	64

SOURCE: Dye (2008).

ber of births to unmarried women in 2007 was two and a half times higher than the number in 1980 and 19 times higher than the 1940 level. The birthrate among unmarried women was 52.9 in 2007, a 5% increase over 2006 and more than a 20% increase since 2002. Births to unmarried women represented almost 40% of all births. Although a large majority of births to teenagers are outside of marriage, such births represent less than a quarter (23%) of all births to unmarried mothers (Ventura 2009).

Clearly, there has been a dramatic increase in unmarried childbearing. As reflected in Table 10.3, the percentage of births to unmarried women differs for different racial or ethnic groups. Although there have been increases among women of all ethnic backgrounds, among African Americans, Native Americans, and Hispanics, a majority of births in 2007 took place outside of a marriage.

In acknowledging these trends, some cross-national comparisons may be instructive. In every one of 13 other countries included in a Centers for Disease Control report, there has been an increase—and typically a steep increase—in the percentage of births that were to unmarried women from 1980 to 2007. The U.S. percentage (40%) was exceeded by Iceland (66%), Sweden (55%), Norway (54%), France (50%), Denmark (46%) and the United Kingdom (44%) (see Figure 10.2).

Table 10.3 Percentage of Births to Unmarried Mothers by Ethnic Origin, 2006 and 2007

Ethnic Origin of Mother	2006	2007
All ethnic groups	38.5	39.7
Non-Hispanic whites	26.6	27.8
Non-Hispanic blacks	70.7	71.6
American Indian	64.6	65.2
Asian or Pacific Islander	16.5	16.9
Hispanic	49.9	51.3

SOURCE: Hamilton et al. (2009, table 1).

Even those countries with lower levels of unmarried childbearing have seen proportional increases that rival or dwarf what has occurred in the United States. Only Japan remains at a very low level, but it has doubled since 1980. A later section addresses some of the consequences associated with nonmarital childbearing.

Figure 10.2 Percentage of Births to Unmarried Women, Selected Countries, 1980 and 2007

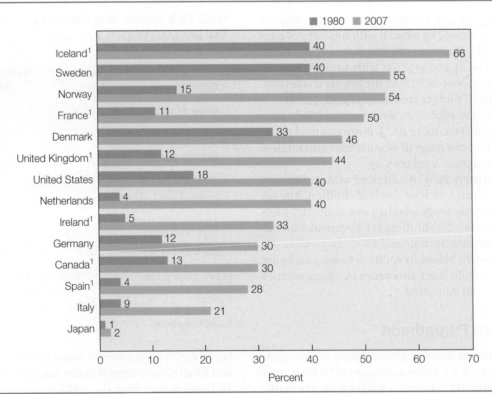

Forgoing Parenthood: "What If We Can't?" "Maybe We Shouldn't"

Increasingly, there are couples who don't become parents. What should we call such couples? In the past, the most common way to describe them was as *childless*, meaning simply that they had no children. However, the term *childless* carries the connotation that such women or couples were "less something" that they wanted or were *supposed to have*. This description no doubt describes the experiences of those women 15 to 44 years old who have an "impaired ability" to have children or those couples who seek help for infertility. Both are *involuntarily childless*.

Worldwide, it is estimated that over 70 million of the more than 800 million women in married or consensual relationships face an impaired ability to conceive, going more than 12 months without becoming pregnant despite not using contraception. In developed countries, the estimated prevalence of infertility ranges from 3.5% to 16.7%; in less developed countries, the prevalence ranged from 6.9% to 9.3% of women. According to a 2007 study, approximately 40 million are currently seeking treatment (Boivin et al. 2007).

In the United States, the American Society for Reproductive Medicine estimates that 10% to 12% of the U.S. population of reproductive age experiences infertility. This amounts to more than 6 million women and men (American Society of Reproductive Medicine; www.protectyourfertility.org/infertility_stats.html; Andrews 2009).

The 2002 National Survey of Family Growth estimated that 12 percent of U.S. women ages 15 to 44 had **impaired fecundity**, a broader term referring to difficulty conceiving or carrying a pregnancy to term. It includes those who had some physical impairment that prevented them from having a baby or made it physically difficult or dangerous to have a child and those who were continuously married or cohabiting and had gone three years without using contraceptives and had not gotten pregnant. This amounts to more than 7 million women and represents an increase from earlier estimates from 1988 and 1995. Looking specifically at married women, an estimated 2.1 million (7.4%) were infertile, and 4.3 million (15%) experienced impaired fecundity (Chandra et al. (2006); National Center for Health Statistics 2005). Of the approximately 62 million U.S. women of reproductive age, approximately 2%, had had an infertility-related medical appointment within the previous year, and an additional 10% had received infertility services at some time in their lives. (Infertility services include medical tests to diagnose infertility, medical advice and treatments to help a woman become pregnant, and services other than routine prenatal care to prevent miscarriage.)

In more recent years, the term *childless* has been joined by *child free* when referring to those without children. In the United States, we have experienced a cultural and a demographic shift toward more voluntarily childless women and **child-free marriages**—couples who *expect and intend* to remain nonparents. The term *child free* suggests that those who do not choose to have children need no longer be seen objects of sympathy, as though they are lacking something essential for personal and relationship fulfillment. Instead, they consider themselves "free" of the responsibilities of parenting. Their childlessness is by choice.

Using data from the most recent National Survey of Family Growth, it is estimated that 8.7% of women ages 15 to 44 expect to have no children in their lifetimes. Among women who were married at the time of the survey, 5.5% expected to have no children, while 11.4% of unmarried women indicated the same expectation.

These estimates do not necessarily differentiate women who actually remain without children from those who intend not to have children, nor does "expecting not to have children" necessarily indicate that such an expectation is by choice. Sociologist Kristin Park (2002) suggests that as many as 25% of the childless population is truly child free in that they are intentionally without children. Looking at data from the 2002 National Survey of Family Growth, researchers Joyce Abma and Gladys Marinez (2006) estimate that among childless women ages 35 to 44, 42% were voluntarily childless, down from 50% of childless 35- to 44-year-old women in 1988.

Who are the women who remain child free? Research indicates that compared to mothers, the child free are women with the highest levels of education, those employed in high-status occupations such as managerial and professional occupations, and those with high levels of household income in dual-earner or dual-career marriages. They are also less religious, have less traditional family attitudes, are more likely to be firstborn or only children, and have less traditional ideas about gender. A higher percentage of white women than African American or Hispanic expect to not have children (Abma and Martinez 2006; Chandra et al. 2006; Park 2002).

Furthermore, Abma and Martinez (2006) drew comparisons between voluntarily and involuntarily childless women ages 35 to 44. Compared to those who are involuntarily or temporarily childless, voluntarily childless women are characterized as follows:

- Less ethnically diverse—they are disproportionately white, whereas involuntarily and temporarily childless women are more representative of the overall population
- Less likely to be or ever have been married
- Less likely to believe that staying home and caring for children makes women happier
- More likely to report no religious affiliation
- More likely to be employed, work full time, and be employed in professional and managerial occupations, and have higher individual and family incomes

What social scientist Rosemary Gillespie (2003, 133) feels has changed over the past quarter century is the emergence of "a more radical rejection or push away from motherhood." She asserts that an increasing number of women are resisting and rejecting the cultural expectations that automatically associate women with motherhood. Instead, she suggests, "modernity has given rise to wider possibilities for women" (134). One can see evidence of Gillespie's claim in the numbers of books, organizations and Web sites—such as www.nokidding.net, www.childfree.net, or www.happilychildfree.com—advocating or defending a child-free lifestyle.

Couples usually have *some* idea that they will or will not have children before they marry. If the intent isn't clear from the start or if one partner's mind changes, the couple may have serious problems ahead. Many studies of child-free marriages indicate a higher degree of marital adjustment or satisfaction than is found among couples with children. Given the time and energy required by child rearing, these findings are not particularly surprising. It has also been observed that divorce is more probable in child-free marriages, perhaps because child-free couples do not stay together "for the sake of the children," as do some other unhappily married couples.

Today, although greater in number than in the past, child-free women and couples may still find themselves stereotyped, perceived as career oriented, materialistic, individualistic, or selfish, with child-free women more negatively perceived than child-free men. Of these stereotypes, only career oriented seems to accurately apply to the child free, especially the women (Park 2002). From in-depth interviews with 24 voluntarily child-free women and men, Park (2002) identified a variety of strategies they used to reduce the stigma attached to not wanting children. These strategies included the following:

- *Passing.* This involves pretending to intend someday to become parents.
- *Identity substitution.* This includes feigning an involuntary childless status as well as letting other statuses (e.g., as a voluntary single or an atheist) dominate a social identity.
- *Condemning the condemners.* This may mean suggesting that some people have children for the wrong or for selfish reasons or that they do so without thinking fully about the responsibility.
- *Asserting their "right" to self-fulfillment.* Park contends that this is a modern type of justification.
- *Claiming a biological "deficiency."* The individual lacks the desire or lacks the nurturing "instinct."
- *Redefining the situation.* This turns potential accusations around by showing how the lifestyle allows nurturing qualities to be used in other ways or allows the individual to be productive. Some also claim that their careers just don't allow for the inclusion of children.

An inspection of the preceding strategies shows that some are more defensive than others, suggesting that would-be parents feel that they must justify their child-free status. Others are more proactive, redefining childlessness as something socially valuable (Park 2002).

Waiting a While: Parenthood Deferred

Although most women who become mothers still begin their families while in their twenties, we can expect the trend toward later parenthood to continue to grow, especially in middle- and upper-income groups. A number of factors contribute to this. More career and lifestyle options are available to single women today than in the past. Marriage and reproduction are no longer economic or social necessities. People may take longer to search out the "right" mate (even if it takes more than one marriage to do it), and they may wait for the "right" time to have children. Increasingly effective birth control (including safe, legal abortion) has also been a significant factor in the planned deferral of parenthood.

Besides giving parents a chance to complete education, build careers, and firmly establish their own relationship, delaying parenthood can be advantageous for other reasons. Older parents may be more emotionally mature and thus more capable of dealing with parenting stresses (although age isn't necessarily indicative of emotional maturity). Speaking economically, raising children is expensive. Delaying parenthood until one's economic position is more secure makes good sense for many people given the economic effect of parenthood.

How Expensive Are Children?

Cost estimates of raising a "typical" child, born in 2007, to age 18, range between $196,000 (for those in the lower third of the income range) and $393,230 (upper-third income bracket). These estimates do not include costs associated with prenatal medical care or childbirth, nor do they include costs of college or other postsecondary education. Furthermore, some contend, such as author Pamela Paul (2008), that the USDA estimates are unrealistically low. The government estimate of childcare costs of $2,850 for parents in the wealthiest group are significantly less than the average obtained from the National Association of Child Care Resource and Referral Agencies (NACCRA), a network of more than 805 child care resource and referral centers. Paul notes that according to the NACCRA, the average cost of child care for an infant ranges from $3,803 to over $13,000 regardless of how much parents earn.

Paul (2008) also notes how much variation there is throughout the United States in many of the costs associated with raising children, such as the aforementioned cost of child care. In cities such as New York and San Francisco, she suggests that one may have difficulty finding caregivers who will accept less than $500 a week for full-time (40 hours of) child care. That amounts to $26,000 just on child care.

Other cost estimates to consider include the cost of higher education. The College Board reports that the average 2008–2009 cost of tuition and fees for a year of college at a private college or university was $25,143. At a public school, the costs are considerably less (tuition costs average $6,585), but multiplying by four or more years, with likely increases over that time, would still add a significant amount to the USDA estimated costs (www.collegeboard.com/student/pay). Add to these figures the cost of room and board, books, and other expenses, and the costs grow significantly. Then there is the matter of wages lost or reduced because of cutting back one's time at work. For example, a parent who earned $40,000 a year but reduced to half-time employment until her or his child turned 18 is incurring a cost of at least $360,000 (assuming an unlikely static income) on top of whatever direct expenses one must assume.

When including costs of higher education, costs for prenatal care and childbirth, as well as estimated wages lost while raising children, the total cost of raising one child from birth to age 22 reaches a much greater, much more daunting total. If one adds those into the total costs, including estimates for inflation, the amounts can easily exceed a million dollars (more than $2,000,000 for the upper-third income bracket). The costliness of raising children is certainly one factor in a number of related trends, including the delaying of parenthood, the decision to have fewer children, and the decision not to have children.

© Matt Carr/Getty Images/Riser

Many couples today (especially those in middle- and upper-income brackets) defer having children until they have established their own relationships and built their careers. These parents are usually quite satisfied with their choice.

Choosing When: Is There an Ideal Age at Which to Have a Child?

Although we have briefly addressed the question of delayed or deferred parenthood, we should point out that delaying parenthood "too long," like having children "too young," carries risks and brings costs. For example, research on the health effects for mothers caused by their age at first birth reveals that both "unusually young" and "unusually old" mothers face health risks.

Teen Mothers

Mothers who bear their first child in their teens face nearly twice the risk of anemia as women who have their first child between ages 30 and 35 (Mirowsky 2002). Young parents are at greater risk for other difficulties. Teen mothers (and fathers) experience worsened educational outcomes, though some of that may be due to "preexisting socioeconomic and other factors" that make both teen parenthood and dropping out more likely (Mollborn 2007). Once they do become parents, given the increased economic needs introduced by the birth of a baby, along with their limited resources, teen parents are likely to have less material resources, which in turn also affects their educational attainments (Mollborn 2007).

Furthermore, children born to teen mothers have been found to be more likely to experience a variety of negative consequences, including delinquency, depression, anxiety, poor performance in school, dropping out of school, higher incarceration rates, and becoming teen parents themselves. They also face greater risks of poverty and unemployment as adults. Although some problems, such as higher levels of drug use, gang membership, and unemployment, were greater for boys than for girls born to women who had their first child while still teenagers, both boys and girls experienced a greater likelihood of becoming teen parents themselves (Campa and Eckenrode 2006; Levine, Emery, and Pollack 2007; Pogarsky, Thornberry, and Lizotte 2006).

What about teen parenting might contribute to these negative outcomes? Teen motherhood may lead many young mothers to end their schooling "early," assume parenting responsibilities before having completed their own social development, delay their entry into the labor market, and

be less likely to marry the fathers of their children. These, in turn, reduce the household resources, lessen the father's involvement, and create a poorer context within which to raise one's children. It is possible that such social and economic conditions are among the consequences associated with teen motherhood, but it is also possible that such conditions lead to teen motherhood. In fact, it is quite likely that both are true.

Older Mothers

But at the opposite end, there are significant risks for pregnancy- and labor-related distress among older first-time mothers, too. For example, pregnancy-related hypertension rates are highest for mothers under 20 and over 40 years of age. Fetal mortality is higher for teens and for women over age 35 (NCHS Data Brief #16, 2009). Women over 35 face increased risk of pregnancy complications, miscarriages, birth defects, and cesarean-section deliveries. Late first births and the care associated with infants can take their physical toll on women. The kind of physical energy required to care for children tends to decline with age (Mirowsky 2002).

In recent decades, both in the United States and elsewhere, women have increased their age at first motherhood. In the United States, the mean age at first childbirth for women in 2006 actually declined for the first time in the almost 40 years of tracking the statistic, from 25.2 in 2005 to 25.0. There is ethnic and racial variation in the age at first birth, with Native Americans having the youngest average age (21.9) and

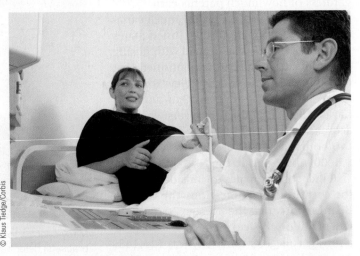

Having a child in one's teens or after age 35 can be risky for both mothers and children.

Table 10.4 Women's Mean Age at First Childbirth, United States, 2006

Racial or Ethnic Group	Mean Age
White, non-Hispanic	26.0
African American	22.7
Hispanic	23.1
Asian American	28.5
Native American	21.9
All groups	25.0

SOURCE: *National Vital Statistics Reports* (2009, tables 14 and 15).

Asian Americans the oldest (28.5). The age pattern by race is illustrated in Table 10.4.

In Italy, the Netherlands, Denmark, and Spain, the typical age at first motherhood is near or beyond 30 years (Mirowsky 2002). Zheng Wu and Lindy MacNeill (2002) report that, in Canada, too, delaying childbearing has become increasingly popular. Probable factors to explain these widespread trends are the increasing age at which women and men marry, women's increased labor force participation and educational attainment, and the increasingly effective measures of reproductive control (Wu and MacNeill 2002).

Pregnancy in the United States

Women and men who become parents enter a new phase of their lives. For those who bear their own children, this phase begins with pregnancy. From the moment it is discovered, a pregnancy affects people's feelings about themselves, their relationship with their partner, and the interrelationships of other family members.

Critical Thinking

If you don't have any children now, do you want to have them in the future? When? How many? What factors do you need to take into consideration when contemplating a family for yourself? Does your partner (if you have one) agree with you about having children?

According to a 2008 by the Centers for Disease Control report on pregnancy in the United States, in 2004, there were more than 6.4 million pregnancies, down 6% from the 1990 peak of 6.8 million but up from 6.3 million in 1999. Interesting age patterns in pregnancies and their outcomes can be identified. Between 1990 and 2004, the teen pregnancy rate dropped steadily from 116.8 to a rate of 72.2 pregnancies per 1,000 women 15 to 19 years old. The 2004 rate was the lowest recorded teen pregnancy rate since 1976. It has since increased. Between 2005 and 2007, the teen birthrate rose 5%. More than 750,000 teens get pregnant each year, with about 445,000 giving birth. The Centers for Disease Control estimate that one-third of American girls will get pregnant before age 20.

Pregnancy rates remain highest for women in their twenties. About one out of six women in their twenties was pregnant in 2004, with 25- to 29-year-old women having the highest rates, followed by 20- to 24-year-olds. Trend data for women in their twenties show that pregnancy rates decreased among 20- to 24-year-olds (from 198.5 in 1990 to 163.7 in 2004), but among 25- to 29-year-olds, pregnancy rates have been increasing since 1997 after a steady decrease between 1990 to 1997. Among women ages 30 to 44, pregnancy rates have increased steadily from 1990 to 2004.

Of the nearly 6.4 million pregnancies in the United States in 2004, 4.1 million (64%) resulted in births, 1.2 million (19%) in induced abortions, and 1 million (17%) in fetal loss (National Center for Health Statistics 2008). As shown in Figure 10.3, pregnancy outcomes are affected by both marital status and race. In 2004, 75% of pregnancies among married women resulted in a live birth, while only 51% of pregnancies to unmarried women resulted in births. Unmarried women were much more likely to end their pregnancies via abortion than were married women (National Center for Health Statistics 2008).

There are race differences in the number of expected lifetime pregnancies and in pregnancy outcomes. Both African American and Hispanic women are expected to have 4.2 pregnancies in their life times compared to 2.7 pregnancies each for white women. African American women were less likely than either white or Hispanic women to have their pregnancies result in childbirths. They were nearly twice as likely as Hispanic women and more than three times more likely than white women to end their pregnancies by abortions (see Figure 10.3).

Figure 10.3 Pregnancy Outcomes by Marital Status and Race, 2004

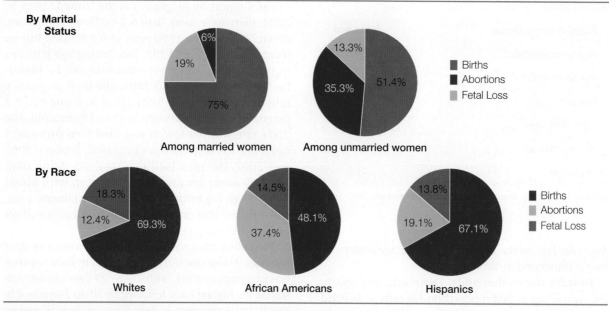

By Marital Status

Among married women
- 75%
- 19%
- 6%

Among unmarried women
- 51.4%
- 35.3%
- 13.3%

Legend: ■ Births ■ Abortions ■ Fetal Loss

By Race

Whites
- 69.3%
- 12.4%
- 18.3%

African Americans
- 48.1%
- 37.4%
- 14.5%

Hispanics
- 67.1%
- 19.1%
- 13.8%

Legend: ■ Births ■ Abortions ■ Fetal Loss

SOURCE: Ventura, S. J., Abma, J. C., Mosher, W. D., Henshaw, S. K. Estimated pregnancy rates by outcome for the United States, 1990–2004. *National vital statistics reports*, vol 56, no 15. Hyattsville, MD: National Center for Health Statistics, 2008.

Being Pregnant

A woman's feelings during pregnancy will vary dramatically according to who she is, how she feels about pregnancy and motherhood, whether the pregnancy was planned, whether she has a secure home situation, and many other factors. Her feelings may be ambivalent; they will probably change over the course of the pregnancy.

Planned versus Unplanned: Was It a Choice?

It is estimated that more than a third (35.1%) of all births to women ages 15 to 44 were either unwanted (14.1%) or unexpected (20.8%) at the time of conception. The experience of unwanted

Matter of Fact It is fairly simple to figure out the date on which a baby's going to be born. Add 7 days to the first day of the last menstrual period. Then subtract 3 months and add 1 year. For example, if a woman's last menstrual period began on July 17, 2006, add 7 days (July 24). Next subtract 3 months (April 24). Then add 1 year. This gives the expected date of birth as April 24, 2007. Few births actually occur on the date predicted, but 60% of babies are born within 5 days of the predicted time.

or unexpected pregnancy is more common among poorer women, minority women, women with lower levels of education, and women who were younger at the time they gave birth (Chandra et al. 2006). If we assume that a much greater percentage of pregnancies that end in abortions were also unplanned, the true percentage of all unplanned pregnancies (unwanted or mistimed) is likely much higher than the percentage resulting in births (Bouchard 2005). Comparing planned and unplanned pregnancies reveals the following:

- Unplanned pregnancies present greater risks for problems associated with lack of readiness or preparedness for parenting on the part of pregnant women or expectant couples.
- Babies born from unintended pregnancies are more likely to suffer both physical and social disadvantages.
- Mothers who give birth to "unintended" babies are more likely to report experiencing psychological problems, such as postpartum depression.

A woman's first pregnancy is especially important because it has traditionally symbolized the transition to maturity. Even as social norms change and it becomes more common and acceptable for women to defer childbirth until they have established a career or to choose not to have children, the significance of first pregnancy should not be underestimated. It is a major developmental milestone in the lives of mothers—and fathers (Marsiglio 1991; Notman and Lester 1988; Snarey et al. 1987).

How Pregnancy Affects Couples' Relationships

A couple's relationship is likely to undergo changes during pregnancy. It can be a stressful time, especially if the pregnancy was unanticipated. Communication is particularly important at this time because each partner may have preconceived ideas about what the other is feeling. Both partners may have fears about the baby's well-being, the approaching birth, their ability to parent, and the ways in which the baby will affect their relationship. All these concerns are normal (Kitzinger 1989). If the pregnant woman's partner is not supportive or if she does not have a partner, it is important that she find other sources of support—family, friends, or women's groups—and that she not be reluctant to ask for help.

The first trimester (three months) of pregnancy may be difficult physically and emotionally for the expectant mother. She may experience nausea, fatigue, and painful swelling of the breasts. She may also fear that she will miscarry. Her sexuality may undergo changes, resulting in unfamiliar needs (for more, less, or differently expressed sexual love), which may in turn cause anxiety. (Sexuality during pregnancy is discussed later in this chapter.) Education about the birth process, information about her body's functioning, and support from partner, friends, relatives, and health care professionals are the best antidotes to her fear.

During the second trimester, most nausea and fatigue disappear, and the pregnant woman can feel the fetus move within her. Worries about miscarriage will probably begin to diminish because the riskiest part of fetal development has passed. The pregnant woman may look and feel radiantly happy. Some women, however, may be concerned about their increasing size; they may fear that they are becoming unattractive. A partner's attention and reassurance can help ease this fear.

The third trimester may be the time of the greatest difficulties in daily living. The uterus, originally about

Both expectant parents may feel that the fetus is already a member of the family. They begin the attachment process well before birth.

the size of the woman's fist, has now enlarged to fill the pelvic cavity and is pushing up into the abdominal cavity, exerting increasing pressure on the other internal organs. Water retention (edema) is a fairly common problem during late pregnancy; it may cause swelling in the face, hands, ankles, and feet. It can often be controlled by reducing salt and refined carbohydrates (such as bleached flour and sugar) in the diet. If dietary changes do not help this condition, however, the woman should consult her physician.

Another problem is that the woman's physical abilities are limited by her size. She may be required by her employer to stop working at some point during her pregnancy. A family dependent on her income may suffer hardship. And the woman and her partner may become increasingly concerned about the upcoming birth.

Some women experience periods of depression in the month preceding their delivery; they may feel physically awkward and sexually unattractive. Many, however, feel an exhilarating sense of excitement and anticipation marked by energetic bursts of industriousness. They feel that the fetus is a member of the family. Both parents may begin talking to the fetus and "playing" with it by patting and rubbing the expectant mother's belly.

Sexuality during Pregnancy

It is not unusual for a woman's sexual feelings and actions to change during pregnancy, although there is great variation among women in these expressions of sexuality. Some women feel beautiful, energetic, sensual, and interested in sex; others feel awkward and decidedly "unsexy." A woman's feelings may also fluctuate during this time. Generally, by the third trimester of pregnancy, approximately 75% of first-time mothers report loss of sexual desire, and between 83% and 100% report reduced frequency of sexual intercourse (De Judicibus and McCabe 2002).

Men may feel confusion or conflicts about sexual activity during pregnancy. They, like many women, may have been conditioned to find the pregnant body unerotic. Or they may feel deep sexual attraction to their pregnant partner yet fear that their feelings are "strange" or unusual. They may also worry about hurting their partner or the baby.

A couple, especially during their first pregnancy, may be uncertain as to how to express their sexual feelings. The following guidelines may be helpful (Strong and DeVault 1997):

- Even during a normal pregnancy, sexual intercourse may be uncomfortable. The couple may want to try positions such as side by side or rear entry to avoid pressure on the woman's abdomen and to facilitate more shallow penetration.
- Even if intercourse is not comfortable for the woman, orgasm may still be intensely pleasurable. She may wish to consider masturbation (alone or with her partner) or cunnilingus.
- Both partners should remember that there are no rules about sexuality during pregnancy. This is a time for relaxing, enjoying the woman's changing body, talking a lot, touching each other, and experimenting with new ways—both sexual and nonsexual—of expressing affection.

Men and Pregnancy

Obviously, pregnancy is something men do not experience directly. It is the woman's body that carries the fetus and undergoes profound change along the way. For men, pregnancy is only accessible vicariously. Still, how men navigate the pregnancy process has consequences for their later conceptualization of and involvement in fathering (Marsiglio 1998).

The roles men play in supporting their partners, participating in the preparation for parenthood, and at the birth also are significant. Not all men act in similar ways. Some may be relatively detached, others fully involved, and still others practical in their participation in the pregnancy (Marsiglio 1998). The way men act during pregnancy (reading material, attending prenatal classes, involving themselves in the birth process, and so on) may affect how they later relate with their newborns. When men are involved prenatally—supporting his partner (e.g., by helping with chores, taking her to the doctor, and buying needed items) and experiencing the unborn child (e.g., listening to the child's heartbeat or examining ultrasound images)—they are more likely to be involved post-birth with their partners and their infants (Cabrera, Fagan, and Farrie 2008).

Men's anxieties during pregnancy cover a number of areas, including the health of both fetus and partner, whether they will be a good father, how fatherhood will affect their lives, and how well they will manage their economic responsibilities, especially given new expenses and reduced spousal income. Although a man's traditional role as father centered on providing, the concern over competence as a provider is not the major source of men's pregnancy anxieties. Men whose employment is unstable or whose incomes are insufficient will experience more provider anxiety than will men who simply take for granted that they can meet their financial responsibilities (Cohen 1993).

Experiencing Childbirth

Women and couples planning the birth of a child have decisions to make in a variety of areas—birthplace, birth attendants, medications, preparedness classes, circumcision, and breastfeeding, to name but a few. In the past few decades, there was much criticism directed at what was seen as excessive and intrusive institutionalized control of women's birth experiences.

The Critique against the Medicalization of Childbirth

The concept of the **medicalization of childbirth** depicts women receiving impersonal, almost assembly line–quality care during labor and delivery and lacking much input or control over their childbirth experiences. During one of the most profound experiences of her life, a woman may find herself surrounded by strangers to whom birth is merely business as usual. The concept of control is central to the critique of medicalization: these critics contend that women have less say and control over the process than they should and less than the medical professionals whose expertise they seek.

The critique also takes into account the environment and medical procedures used. Lighting, noise, routine use of monitoring devices, administering of enemas, rates of episiotomies (a surgical procedure to enlarge the vaginal opening by cutting through the perineum toward the anus), rates of cesarean-section deliveries, use of forceps or vacuum suction to assist in pulling the fetus from the womb—all of these reflect what critics have suggested is society's increasing dependence on technology and medical control. Critics of medicalization contend that episiotomies, the use of forceps, and/or vacuum extraction are often employed more for the convenience and control of the obstetrician than because of medical necessity. In general, critics recognize and value the potential lifesaving use of such interventions *when needed* but question procedures that seem to place medical convenience above women's interests or needs.

The Feminist Approach

The question of what women most want or need is central to what became a feminist critique of contemporary childbirth. Along with consumer advocates and government policymakers, feminists and activists in the "women's health movement" raised concerns and objections about the medicalization of childbirth. For feminists, a woman had a right "to be informed, fully conscious, and to experience childbirth as a 'natural' process" (Treichler 1990). Feminists raised questions about how much medical intervention and control are necessary to reduce risks associated with the "normal, natural physiological process" of childbirth. They asserted that most pregnant women are essentially healthy and require minimal medical management during the birth process.

Many of the above criticisms have been heard and addressed by hospitals and medical practitioners. For example, most hospitals have responded to the need for family-centered childbirth. Fathers and other relatives or close friends typically participate today. "Birthing rooms," with softer lighting and more comfortable birthing chairs, are increasingly common. Today, most hospitals encourage rooming-in (the baby stays with the mother rather than in the nursery) or a modified form of rooming-in. Moreover, there are more choices for women as to how they wish to give birth. Nevertheless, as the following sections show, the birth process continues to be characterized by considerable use of medical procedures.

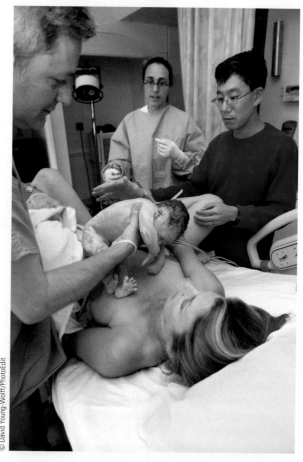

© David Young-Wolff/PhotoEdit

Family-centered childbirth allows fathers to participate alongside mothers in the birth process.

What Mothers Say

With the preceding feminist critique and the critique of medicalization in mind, we might predict women to express high levels of discontent with their experiences and treatment during pregnancy, while in labor and giving birth, and after they give birth. We see something much more mixed from data collected in the "Listening to Mothers" surveys, conducted by Harris Interactive and the Maternity Center Association. The reports are based on telephone interviews and online surveys with a combined sample of 1,583 women who gave birth during the two-year period between May 2000 and June 2002 and a second sample of 1,573 women who gave birth in U.S. hospitals in 2005. Some key findings are as follows:

1. According to the 2002 survey, most mothers felt "quite positive" about their birthing experiences:
   - 95% felt that they generally understood what was happening to them.

- 93% felt comfortable asking questions.
- 91% felt that they had received the necessary amount of attention.
- 89% felt as involved as they desired in decision making about their deliveries.

2. Nearly all women (97%) reported giving birth in a hospital, and most (80% in 2002, 79% in 2005) were attended to by obstetricians. In both samples, around 10% of the births were attended by midwives, with the remainder divided among family physicians, other physicians, nurses, or physician's assistants.

3. Qualitative assessments of the overall care and treatment women received from their physicians were mostly positive. In the 2002 study, approximately 9 out of 10 women reported that their doctor or midwife had been polite, supportive, and understanding. The biggest complaint was that physicians or midwives seemed "rushed."

Real Families: Men and Childbirth

I remember thinking, "Nobody else cares." My wife was knocked out, everybody else in the room was taking care of my wife and the baby, and the baby was wet, cold, red, really an ugly looking thing. Truthfully, I was instantly bonded; it was like a marriage. She [my daughter] opened her eyes a little bit and I immediately began to relate to what she saw. . . . I was thrilled. I followed her after she opened her eyes and tried to imagine what she saw and what she might be seeing. . . . It was real exalting.

Mark, 37-year-old father of one

My wife had the shakes and couldn't hold the baby. I held her [my daughter] and sang her a lullaby. She was looking at my face, wasn't focusing, but I could see something going on. She could obviously hear too. I got to hold her for like fifteen minutes. It was all so exciting and incredible . . . and strange.

Bill, 36-year-old father of one

The preceding comments are the recollections of two fathers to having witnessed the births of their daughters. Told to sociologist Theodore Cohen, they reveal how deeply some men are moved by their involvement at birth. In the United States and many other countries, it is now common for fathers to attend, witness, and often even actively assisting in the birth of their children. It may be so common that we forget how relatively recent it is for men to enjoy such access. In 1960, only about 15% of fathers attended the birth of their child in the delivery room. Although estimates vary, by the first years of the twenty-first century, between 75% and 80% of fathers were present at childbirth (Washington Post 2006).

We see the same trend elsewhere. In the United Kingdom, fathers are now in attendance at 80% of births (Johnson 2002). Similar trends have been observed in other European countries and in Canada. Attendance at birth offers fathers an opportunity to feel part of the birth process and to offer support to their partners. Among men in the United Kingdom, the most frequently cited motivations for attendance at birth are out of support for their partners, out of curiosity, or because of pressure. In the United States and Canada, there is a fourth reason: men often play the role of "coach," assisting their partners to implement what they have been taught in prenatal classes (Johnson 2002). Where once hospital practice and cultural expectations kept men out of the delivery room, now they are expected to be present. To illustrate this coercive element, in Martin Johnson's exploratory study of 53 British fathers, 57% of the men said they felt pressured to be there through labor and delivery. For example, "You don't get a choice, not really. It is assumed that you want to be there; I mean I did, but that is not the point. It's like not having a choice."

Finally, in Johnson's (2002) study, men's reactions to what they saw and experienced were both positive and negative.

On the negative side, 56% of the men identified as their most overwhelming memory the pain they witnessed their partners suffering. One man, Ben, claimed that he felt as though he ought to be experiencing pain himself: "In a strange way, when she dug her nails into my hands, I wanted to embrace the pain; it was like my share."

On the positive side, and unsurprisingly, men were deeply moved by the birth and awed by their partners' strength and resilience.

4. In both surveys, around 90% of women reported receiving "supportive care" or attention during labor and delivery from their spouses or partners. Regarding the quality of supportive care they received from medical personnel, 90% of the mothers surveyed described the quality of supportive care they received from all sources while in labor as either good or excellent.

5. In support of the issues most related to the critique of "medicalization," women reported experience with the following medical interventions during labor and delivery:

   - In both surveys, more than 90% reported being monitored by an electronic fetal monitor.
   - In the 2002 survey, 63% were given epidural analgesias to relieve pain, 30% were given a narcotic pain reliever, and 5% were given general anesthesia. Of the 2005 sample, 86% used some form of pain medication, with 76% receiving an epidural.
   - Of the 2002 sample, one in four women had cesarean sections. In the 2005 sample, the percentage of cesarean-section deliveries increased to nearly a third (32%).

Authors of the report on the 2005 survey assert that large segments of women experience potentially invasive interventions when giving birth, even when they were healthy, their pregnancies were normal, and their labor was uncomplicated. Such interventions may not be medically necessary. Many women did not have the knowledge they needed or were unable to make the choices they desired (e.g., only 18% of women who received episiotomies said that they had a choice in the decision).

Infant Mortality

The rate of **infant mortality** in the United States remains far higher than the rates in most of the developed world. The Centers for Disease Control's National Center for Health Statistics reported preliminary data indicating 6.7 deaths for every 1,000 live births in 2006, a slight (2%) decline from 2005. Overall, the rate has fluctuated between 6.8 and 6.9 since 2000 (MacDorman and Mathews, 2008). Among developed nations, the United States does not fare well in low infant mortality. Twenty-two other countries had rates below 5, and in certain Scandinavian and East Asian countries (e.g., Sweden, Norway, Finland,

Japan, Hong Kong, and Singapore), the rate was below 3.5 per 1,000.

In 2004, the most recent year for which cross-national data are available, the United States infant mortality rate tied for twenty-ninth in the world, along with Poland and Slovakia, meaning that 28 countries had *lower* infant mortality rates than the United States. A similar cross-national comparison in 1960 found the United States ranked twelfth (MacDorman and Mathews 2008). In the United States, the nation's capital has a higher infant mortality rate than any of the 50 states, at 12.5 per 1,000 live births (U.S. Census Bureau 2001, table 104).

Data for 2005 reveal noteworthy racial and ethnic variations, from rates below 5.0 among Asians, Cuban Americans, and Central or South Americans to a high of 13.6 among African Americans (see Figure 10.4). Neither the U.S. targeted goal rate for 2010 of 4.5 nor the determination to eliminate racial and ethnic disparities has been successfully reached.

Of the thousands of American babies less than one year old who die each year, many are victims of poverty, which often hits racial minorities harder. The United States is far behind many other countries in providing health care for children and pregnant women. In France, Sweden, and Japan, for example, all pregnant women are entitled to free prenatal care. Free health care and immunizations are also provided for infants and young children. Working Swedish mothers are guaranteed one year of paid maternal leave, and French families in need are paid regular government allowances (Scott 1990). One in six children born in the United States is born to mothers who received no prenatal care through the first trimester of pregnancy. Almost one in eight children has no health insurance (Ruane and Cerulo 2004).

Although many infants die of poverty-related conditions, others die from congenital problems (conditions appearing at birth) or from infectious diseases, accidents, or other causes. Sometimes, the causes of death are not apparent. Data from the Centers for Disease Control and the National Center for Health Statistics for 2005 attribute 2,230 infant deaths to **sudden infant death syndrome (SIDS)**, a perplexing phenomenon wherein an apparently healthy infant dies suddenly while sleeping (www.cdc.gov/nchs/fastats/pdf/mortality/nvsr52_03t32.pdf). SIDS is among the top three causes of infant deaths.

Figure 10.4 Racial and Ethnic Comparison of Infant Mortality Rate, United States, 2000–2005

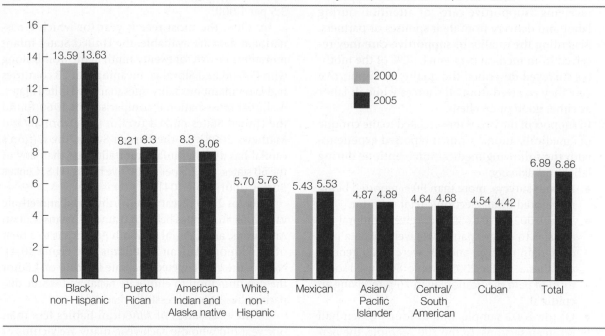

The Mayo Clinic (www.mayoclinic.com) reports that SIDS results from some combination of the following:

- The infant having some biological vulnerability, such as a heart or brain defect
- Some environmental stressor, such as stomach sleeping
- The infant being at a critical developmental period
- Some possible influence of maternal health (e.g., low weight gain or inadequate prenatal care) or behaviors (e.g., smoking or drug use) during pregnancy

Boys are more vulnerable to SIDS than girls. SIDS also is more common among infants who were born premature or of low birth weight, African American and Native American infants, and among infants one to six months of age. Other environmental risks include soft bedding (e.g., pillows, mattresses, or crib padding), sleeping in an adult bed, and sleeping in bed with other children who could roll over in their sleep and suffocate an infant (www.sidscenter.org).

Coping with Loss

The depth of shock and grief felt by many who lose a child before or during birth is sometimes difficult to understand for those who have not had a similar experience (Layne 1997). What they may not realize is that most women form a deep attachment to their children even before birth. The loss of the child must be acknowledged and felt before psychological healing can take place. Instead, however, women typically find that friends, relatives, and coworkers want to pretend that "nothing has happened" (Layne 1997).

Equally problematic are the common reactions from medical personnel and midwives. Medical personnel, especially physicians, may perceive pregnancy loss as "medically unimportant" and as evidence of normal and natural processes at work. Not surprisingly, this reaction may exacerbate the pain of couples who lose a child before or during birth.

Women (and sometimes their partners) who lose a pregnancy or a young infant generally experience similar stages in their grieving process. Their feelings are influenced by many factors: supportiveness of the partner and other family members, reactions of social networks, life circumstances at the time of the loss, circumstances of the loss itself, whether other losses have been experienced, the prognosis for future childbearing, and the woman's unique personality. Physical exhaustion and, in the case of miscarriage, hormone imbalance often compound the emotional stress of the grieving mother.

The initial stage of grief is often one of shocked disbelief and numbness. This stage gives way to sadness, spells of crying, preoccupation with the loss, and perhaps loss of interest in the rest of the world. It is not unusual for parents to feel guilty, as if they had somehow caused the loss, although this is rarely the case. Anger (toward the physician, perhaps, or God) is also a common emotion.

Experiencing the pain of loss is part of the healing process (Vredevelt 1994). This process takes time—months, a year, perhaps more for some. Support groups and counseling are often helpful, especially if healing does not seem to be progressing or depression and physical symptoms do not appear to be diminishing. Planning the next pregnancy may be curative, too, although we must keep in mind that the body and spirit need some time to heal.

Giving Birth

The 2005 "Listening to Mothers" survey asked women about their feelings during childbirth. Of positive feelings, over 40% of women said that they felt "alert" (45%), "capable" (43%), and "confident" (42%). As depicted in Figure 10.5, the most common negative feelings were feeling "overwhelmed" (44%), "frightened" (37%), or "weak" (30%). Interestingly, 14% of women reported feeling both "confident" and "overwhelmed." Overall, the method of birth, vaginal versus cesarean, affected what women reported feeling. Women who delivered their babies vaginally were more likely to feel confident and capable and less likely to feel frightened than were women who had cesarean sections (Declercq et al. 2006).

Sociologist Karin Martin (2003) conducted intensive interviews with a small sample of first-time mothers. The 26 mostly white heterosexual women, ranging in age from 20 years to over 40 were interviewed within three months of having given birth. Instead of exploring the macro-level and institutional dimensions of childbirth, Martin wanted to know how women experienced childbirth and how their experiences were shaped by gender identity. Deep within us, even in "seemingly natural experiences like birth," are our culturally constructed gender identities (Martin 2003, 57). Martin found that even during childbirth, women are "doing gender," acting compliant, nice, and kind. Martin's informants recalled trying not to "bother" strangers in adjoining rooms, remembered trying hard to remain attentive during conversations, and described doing things that indicated they were putting the needs of others ahead of their own. Even though they had to impose on others (doctors, nurses, midwives, husbands, and so on) for things (back rubs, quiet, patience, information, and so on), they recalled

Figure 10.5 Feelings While Giving Birth

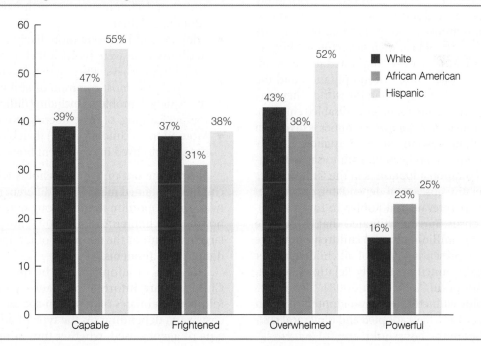

SOURCE: Declercq et al. (2006).

feeling badly about doing so. They found it hard not to feel "rude" or "selfish" for making the demands and imposing on others.

Martin (2003) suggests that the feminist critique of the medicalization of childbirth may be correct in highlighting how institutional control over birth shapes the experience. But it is only part of the story. She contends that women's birth experiences are also regulated and controlled "from within" by internalized gender identities. Even when "given permission" to depart from gender expectations, to act in gender-deviant ways, they found themselves at odds with such behavior. It was not "how they are" or "who they are" (Martin 2003).

Choosing How: Adoptive Families

Parenthood is not only entered biologically. Although adoption is being examined here as the traditionally acceptable alternative to pregnancy for infertile couples, it may also include the adoption of stepchildren in a remarriage, the adoption of a child by a relative, the adoption of adolescents, the adoption of two or more siblings, and the adoption of foster children who have been removed from their parental homes (Grotevant and Kohler, in Lamb 1999). Many people—married and single, with or without biological children—choose to adopt not because they are unable to conceive or bear their own children but because they are ideologically committed to adoption. Some have concerns about overpopulation and the number of homeless children in the world. They may wish to provide families for older or disabled children. Thus, the population of adoptive families is diverse in terms of both motivation and circumstances.

Until recently, it has been difficult to say with certainty how common adoption is in the United States given the relative absence of dependable or comprehensive data (Grotevant and Kohler, in Lamb 1999). The U.S. Census Bureau estimated that there were more than 1.6 million adopted children under age 18 in 2004, representing 2.5% of all children under 18 (National Council on Family Relations Report, Focus on Adoption, September 2008). Nearly 2% of households with children have adopted children only. Another 1.8% have adopted and biological children together in the household, and 0.1% have adopted children with stepchildren together or adopted children, biological children, and stepchildren together in the same household.

Characteristics of Adoptive Families

An earlier census report attempted for the first time to count and construct a profile of adoptive families (Kreider 2003). It found the following:

- Adopted children are more likely to be female than male, which demographer Rose Kreider suggests results from both desirability and availability. Specifically, "women in general express a preference for adopting girls, and single women more frequently have adopted girls than boys" (Kreider 2003, 8). In addition, with regard to international adoptions, more female children are available for adoption from those countries that are "leading sources for adopted children" (8).
- Economically, families with adopted children are somewhat better off than those without. Smaller percentages of households with adopted children than biological children are poor (11.8% vs. 16%). "Adoptive households" had higher median incomes than households with only biological children ($56,138 compared to $48,200), and 33% of adopted children as opposed to 27% of biological children lived in households with incomes of $75,000 or more. Table 10.5 shows some additional characteristics of households with adopted children.
- Adopted children were more likely to have some disability than were biological children (15% vs. 7% of boys, 9% vs. 4% of girls). "Mental disabilities" were the most common disability, consisting of a variety of problems including difficulty concentrating, learning, or remembering (Kreider 2003).
- More adopted children (78%) than biological children (74%) lived in two-parent households.

Nearly one out of five households with adopted children contained members of different races. This is twice the proportion found among households with no adopted children. Kreider (2003) notes that this is largely a result of the adoption of foreign-born children by U.S. residents.

The costs of adoption can be quite steep. The Child Welfare Information Gateway provides the following estimates of adoption-specific costs. Such costs vary, depending on the type of adoption, the type of agency used, whether they adopt domestically or internationally, and so on. The costs range

may force changes in other nonfamily roles and relationships, such as at work or in friendships. Although some of these changes may be temporary (e.g., leaving work only for the length of a parental leave), they nonetheless compound other things to which new parents are adjusting.

- *New parenting roles and relationships.* Couples must arrive at an agreeable division of child care. One parent may feel put on or taken advantage of in the way that the couple allocates their individual time and energy to child care tasks.

The Cowans suggest that the difficulties associated with the parental transition are more difficult for contemporary parents because of some major features of the social climate in which they parent. First, contemporary parenthood is more *discretionary* or optional, making decisions about whether and when to have children subject to more discussion, negotiation, and potential dispute. Second, many new parents, especially middle-class parents, are relatively *isolated* geographically from their wider kin groups and other long-term social supports. Third, changes in women's roles have introduced more *role conflict* for new mothers and have increased women's need and legitimate demand for more sharing by their partners. Fourth, the social policies that address the needs of parents are weak to nonexistent. Fifth, there are few enviable or attractive role models for effective parenting. Sixth, today's families are supposed to fulfill all our emotional needs. Parenting is stressful and requires mutual effort and sacrifice. But effort and sacrifice don't fit compatibly with individual emotional fulfillment. Thus, the difficulties may become sources of resentment and estrangement (Cowan and Cowan 1992, 2000).

Stresses of New Parenthood

Many of the stresses felt by new parents closely reflect gender roles. Overall, mothers seem to experience greater stress than fathers. Although a couple may have an egalitarian marriage before the birth of the first child, the marriage usually becomes more traditional once a child is born. If the mother, in addition to the father, continues to be employed outside the home or if the woman is single, she will have a dual role as both homemaker and provider. She will also probably have the responsibility for finding adequate childcare, and it will most likely be she who stays home to take care of a sick child. Multiple role demands are the greatest source of stress for mothers.

There are various other sources of parental stress. Fathers often describe severe stress associated with their work. Both mothers and fathers must be concerned about having enough money. Other sources of stress involve infant health and care, infant crying, interactions with the spouse (including sexual relations), interactions with other family members and friends, and general anxiety and depression. Changes in marital quality, increased marital conflict, and decreased interaction between spouses are also common as couples transition to parenthood (Crohan 1996).

Although the first year of child rearing is bound to be stressful, the partners experience less stress if they (1) have already developed a strong relationship, (2) are open in their communication, (3) have agreed on family planning, and (4) originally had a strong desire for the child. Despite planning, the reality for most is that this is a stressful time. Accepting this fact while developing time management skills, patience with oneself, and a sense of humor can be most beneficial.

Parental Roles

By this point one might be wondering, "But what is it like to be a parent and to raise children?" Having a child is unlike any other experience we undertake. The changes in our lives are wide ranging and irreversible, the *potential* rewards are great, and the sacrifices are many. Most people continue to decide to embark on this journey and take on the challenging tasks, to which this chapter now turns.

Over the past four decades or so, major changes in society have profoundly influenced parental roles. Parents today cannot necessarily look to their own parents as models. Most mothers and fathers of today's children have some things in common with mothers and fathers throughout history, such as the desire for their children's well-being. But in many areas, they have had to chart a new course. Here we briefly review motherhood and fatherhood, highlighting some major changes of the past quarter century that have transformed the meaning and experience of each.

Critical Thinking

If you have children, did you plan to have them? What considerations led you to have them? What adjustments have you had to make? How did your relationship with your partner change?

Motherhood

To many, a chapter about parenting might be assumed to be about mothers and children since "parenting" and "nurturing" are treated as though they are synonymous with "mothering." Furthermore, many women see motherhood as their "destiny." Given the choice of becoming mothers or not (with "not" made possible and more controllable through birth control and abortion), most women would choose to become mothers at some point in their lives, and they would make this choice for positive reasons. Some women make no conscious choice; they become mothers without weighing their decision or considering its effect on their own lives and the lives of their children and partners. The potential negative consequences of a nonreflective decision—bitterness, frustration, anger, or depression—may be great. Yet it is possible that a woman's nonreflective decision will turn out to be "right" and that she will experience unique personal fulfillment as a result.

Although researchers are unable to find any purely instinctual motives for having children among humans, they recognize many social motives impelling women to become mothers. When a woman becomes a mother, she may feel that her identity as an adult is confirmed. This may be especially true for those women with fewer marketable skills, more limited education, fewer financial resources, and more restricted opportunities. For such women, having a baby is a desirable and achievable goal. In the words of sociologists Kathryn Edin and Maria Kefalas (2005, 46), "Unlike their wealthier sisters, who have the chance to go to college and embark on careers . . . poor young women grab eagerly at the surest source of accomplishment within their reach: becoming a mother."

Having a child of her own proves her womanliness because, from her earliest years, she has been trained to assume the role of mother. The stories a girl has heard, the games she has played, the textbooks she has read, the religion she has been taught, the television she has watched—all have socialized her for the mother role. The idea of a maternal instinct reflects a belief that mothering comes naturally to women. For women who struggle with the new roles and responsibilities that motherhood brings, such an idea can be frustrating and can produce guilt. Add to this the assumption that mothers instinctively or intuitively "know" how to nurture children and the inherent ability of women to breastfeed, and we can quickly see the enormous pressures that can face new mothers more than new fathers.

Compounding the situation are those ambiguous cultural expectations alluded to earlier. "Too much" mothering? "Not enough" mothering? What do children really need, and what should mothers give and do? Women receive unclear, often contradictory, messages. Popular culture is the source of many mixed messages that women receive about motherhood. To uncover some wider cultural messages about motherhood, sociologists Deirdre Johnston and Debra Swanson examined the portrayals of mothers in five magazines targeted to mothers: *Good Housekeeping*, *Family Circle*, *Parents*, *Working Mother*, and *Family Fun*. These portrayals put both at-home mothers and employed mothers on the defensive, in difficult, no-win situations. Johnston and Swanson found that popular culture depicts at-home mothers as somewhat incompetent and yet underrepresents employed mothers, rendering them less visible models of motherhood. This occurs despite the fact that more than 60% of mothers are in the paid labor force (Johnston and Swanson 2003). Clearly, neither employed nor at-home mothers are well served by their portrayal in popular culture.

Furthermore, the standards against which mothers are judged (and come to judge themselves) are often unrealistic and idealized, putting women in a situation of comparing themselves to a model to which it is difficult, if not impossible, to fully "measure up." Sociologist Sharon Hays refers to our cultural expectations of mothers as the **ideology of intensive mothering**. This ideology portrays mothers as the essential caregivers, who should be child centered, guided by experts, and emotionally absorbed in the labor-intensive and financially demanding task of child rearing. As a result, mothers "see the child as innocent, pure, and beyond market pricing. They put the child's needs first, and they invest much of their time, labor, emotion, intellect, and money in their children" (Hays 1996, 130). In today's cultural climate, this view of motherhood contrasts with the business market ideology of efficiency, rationality, time saving, and profit.

The intensive-mothering ideology confronts mothers and women who contemplate motherhood with cultural contradictions. Living up to its standards is difficult even for stay-at-home mothers. For women employed outside the home, the ideology can provoke self-doubt, guilt, and a sense of being negatively judged by others. As Hays (1996) notes, there is almost no woman who can resolve this cultural no-win situation. Women who forgo childbearing may be perceived as "cold" and "unfulfilled." An employed woman with children may be

© Juergen Hasenkopf/imagebroker.net/photolibrary

The ideology of intensive mothering casts mothers as essential and defines motherhood as labor intensive and self-satisficing.

told that she is selfishly neglecting her children. If she scales back her workload but stays in a job, she may be "mommy tracked," put in a less demanding but also less important and less upwardly mobile position. Finally, at-home mothers, in meeting the intensive-mothering mandates, will be seen by some as "useless" or "unproductive" (Hays 1996). No matter what she chooses or does, a woman will find that there are some who question her decision and object to her lifestyle.

Motherhood affects women's employment experiences, as shown in Chapter 12. One notable way that women are affected is in their earnings. Estimates differ, but it is clear that women with children earn less than their counterparts without children (Budig and England 2001). In addition, regardless of their employment status, the responsibilities of parenthood continue to fall more heavily on women than on men, even as children age and move into their teens (Kurz 2002). Such responsibilities give employed mothers more to do and less time to do it. At the same time that women's parental status affects their employment experiences, a mother's employment status is also likely to affect the amount of time she spends with children. Women who are employed full time spend much less time with their children than do mothers who are not employed (Kendig and Bianchj 2008).

The time mothers spend with their children is affected by more than whether they are employed. A mother's marital status is associated with the amount of time spent with children; single mothers spend less time with their children than married mothers spend, though the gap—estimated by sociologists Sarah Kendig and Suzanne Bianchi (2008) at about three to five hours per week—is not huge (single mothers spend between 80% and 90% the amount of time that married mothers spend). It isn't necessarily marital status per se that is operating but rather characteristics or experiences that accompany marital status. For example, because single mothers, especially divorced single mothers, often have the need to spend more hours at work, their time with children is affected.

Still other factors that have been found to *increase* mothers' time with children are level of education, economic status, and characteristics of children (such as the number of children a mother has, their ages, and their genders). College-educated mothers spend more overall time, including both routine time and more interactive activities, with their children. Mothers with preschool-age children in the home spend more time in child care activities. Most (but not all) research also suggests a positive relationship between the number of children a mother has and her time spent in child care (the more children a mother has, the more time she spends on child care) (Kendig and Bianchi 2008).

Fatherhood

When we speak of "mothering a child," everyone knows what we mean: a process that involves nurturing and caring for the physical and emotional well-being of a child, almost daily, for at least 18 consecutive years. Fathering is more ambiguous. Nurturing behavior by a father toward his child has not typically been referred to as *fathering*. As used today, the term *parenting* is intended to describe the child-tending behaviors of *both* mothers and fathers (Atkinson and Blackwelder 1993). In fact, the popular meaning of *fathering* is sometimes reduced to the act of impregnating the child's mother. Of course, cultural expectations of fathers are much broader and deeper than that, and fathers today participate more than they did in the past in a broader range of child care activities. However, the cultural expectations of fathers are not the same as the expectations of mothers, nor do fathers participate to the same extent or kind that mothers participate.

Beginning in the 1980s and 1990s, the degree of male involvement or absence in the lives of their children became—and remains today—a popular topic of scholarly interest and societal concern (Eggebeen and Knoester 2001). In all the commentary and analysis, however, we are still left with something short of a consensus about the state of fatherhood in America. Just what are fathers supposed to do, and how well or poorly are they doing it?

Fathers are increasingly involved in parenting roles—not just playing with their children but also changing their diapers, bathing, dressing, feeding, and comforting them.

In analyzing today's fathers and today's families, we find diverse opinions and a range of experiences of fathering. There is evidence indicating that fathers have become more emotionally connected to and involved in the lives of their children (Eggebeen and Knoester 2001). Some commentators point proudly to our embracement of this "new father" model against which many men now measure themselves (Lamb 1993; Smith 2009). Feminist ideology is credited with being influential in shifting the emphasis to a more expressive model of fathering, but many men pursue more involved versions of fatherhood as part of their own quest for deeper relationships with their children (Daly 1993; Griswold 1993). When pressed, most men today compare themselves favorably with their own fathers in both the quality and the quantity of involvement they have with their children.

The "nurturant father," as Michael Lamb (1997) referred to him, can participate in virtually all parenting practices (except, of course, gestation and lactation) and experience similar emotional states to those experienced by mothers. It is clear that fathers can feel a connection to their infants that men were often thought to lack (Doyle 1994). Furthermore, *father involvement* has been reconceptualized by researchers in recognition that there are many ways in which fathers are influential participants in their children's development. Fathering activities such as communicating, teaching, caregiving, protecting, and sharing affection are ways in which fathers might be "involved" with their children, and all are viewed as beneficial to the development and well-being of both children and adults (Hawkins and Dollahite 1997; Palkovitz 1997).

Although this new standard of fatherhood has been widely hailed, it is unclear how much it reflects actual behavior (Gerson 1993; LaRossa 1988). As described by Ralph LaRossa, the **culture of fatherhood** has clearly changed in the directions described here; it is less clear how much the **conduct of fatherhood** has kept pace. In addition, when we look at how fathers compare to mothers, on average, fathers are neither as involved with nor as close to their children, including their teenaged children, as mothers are (Kurz 2002). This an important reminder that reality may be different from rhetoric when it comes to what people actually do or believe they should do in their families.

Of course, in trying to determine what and how much fathers and mothers actually do, it is important to consider the source of those reports. Researchers have found discrepancies between mothers' and fathers' accounts of how involved fathers are and in what aspects of child care and child rearing they are involved. Compared to what fathers report, mothers' reports of father involvement tend to be lower. Although there may be as many reasons for fathers to exaggerate their estimates of how involved they are as there are reasons mothers might underestimate father involvement, the mother–father discrepancy points to the complexity of measuring father involvement and to the need to consider which parent is providing the estimates (Mikelson 2008).

As the preceding examples show, it is difficult and potentially risky to generalize too widely about today's fathers. Although more of today's fathers aim to be more broadly involved with their children than what they perceive fathers to have been in the past and although most may recognize father involvement as beneficial, many fathers may feel confused and uncertain about what is expected of them. Because models of highly involved fathers are still fairly new and men's wage-earning responsibilities are still strongly felt, fathers themselves may not know exactly what they're supposed to do.

Sociologists David Eggebeen and Chris Knoester (2001) looked at how fatherhood affected men, comparing fathers and nonfathers and examining different "versions" or "settings" of fatherhood: men living with their own (biological or adopted) dependent children, men living apart from their dependent children, men whose children are independent adults, and men who are stepfathers. Eggebeen and Knoester found that on psychological and health dimensions, fathers and nonfathers did not differ significantly. In three other areas—social, intergenerational/familial, and occupational—there were "clear and compelling differences between fathers and non-fathers," as well as interesting differences across fatherhood settings (Eggebeen and Knoester 2001, 390). Men living with dependent children were significantly less likely to participate in social activities with friends or leisure pursuits. Men without children, men who lived away from their children, and men who lived with stepchildren attended church much less often than men who lived with their own biological or adopted children.

Fathers who lived with their own biological or adopted children were more likely to have regular contact with aging parents and adult siblings than were men without children or men with stepchildren. Even fathers who lived apart from their children had more frequent contact with parents and siblings, suggesting that "fatherhood tightens intergenerational family ties" (Eggebeen and Knoester 2001, 389).

Even with all the evidence pointing to the difficulty of measuring fathers' behavior, the differing views of what fathers should do, and the complexity of capturing the "two sides" of contemporary fatherhood, it is important to remember that today's fathers and mothers are still held to different parenting standards or expectations for involvement with their children. The ideology of intensive mothering has not led to an ideology of intensive fathering or even a somewhat gender-neutral ideology of intensive parenting (Hook

and Chalasani 2008). For example, a sample of college students was told of a hypothetical employed parent who showed a lack of involvement in caring for his or her child. They rated fathers and mothers lower for behavior described as "home but uninvolved" compared to "uninvolved because of business trips." However, mothers were rated even more negatively than fathers for the lack of involvement at home (Riggs 2005). The cultural stereotype is that mothers are *supposed to be involved*.

What Parenthood Does to Parents

One might wonder about the consequences of parenthood in the lives of fathers and mothers. Much of the research literature focuses attention rather narrowly on negative outcomes associated with becoming and being parents, while wider cultural attitudes and beliefs might be said to overstate the positives and downplay the costs (Simon 2008). Therefore, a brief summary of both positive and negative outcomes, of the "costs and rewards of children," seems necessary.

Among the positives, children can produce and deepen one's feelings of joy, strengthen one's social ties (especially to family but potentially to friends, neighbors, and community institutions), increase one's self-esteem, and afford one the opportunity to fulfill an expected role and transition to adulthood. Parenthood brings with it a shift in one's priorities and in the importance one attaches to other social roles. It also presents the opportunity for a feeling of **generativity**, of being committed to guiding or nurturing others (Nomaguchi and Milkie 2003). Parents feel a greater sense of purpose, meaning, and life satisfaction than nonparents (Simon 2008).

There is another side to the story. Among the negatives, research on women and men in the United States indicates an association between being a parent and such negatives as depression, emotional distress, marital strains, higher levels of stress, more work and housework, and depleting demands on one's time and energy (Nomaguchi and Milkie 2003; Simon 2008).

The Effects of Parenthood on Marriage and Mental Health

Early research depicted the transition to parenthood as a crisis leading to a decline in marital quality and satisfaction. We now know, however, that the impact of parenthood is variable. Although marital satisfaction declines for many new parents, it also declines for couples without children during the early years of marriage. Thus, what may have appeared to be an effect of parenthood may just reflect the ebbs and flows of marital satisfaction (Helms-Erikson 2001). That doesn't mean that parenthood has no effect on marriage; indeed, it does, but its effects depend at least somewhat on when couples become parents and on how couples negotiate the new responsibilities. As Heather Helms-Erikson (2001, 1100) puts it, parenthood leaves "some couples faring better following the birth of their first child, others worse, and still others seemingly unchanged."

New parents show more traditional divisions of duties and lower levels of companionship compared to couples without children, but marital discontent is by no means inevitable. Even these outcomes—traditionalization and declining marital quality—depend on the circumstances under which they become parents. Couples who become parents "early" (i.e., in their late teens or early twenties) are more likely to divide their household tasks on "traditional gender lines, with wives being responsible for the bulk of housework and childcare" (Helms-Erikson 2001, 1101) and men becoming more involved only when pushed. Couples who become parents in their late twenties and thirties tend to display more "collaborative" divisions of roles, and fathers' involvement tends both to be more self-determined and to reflect more liberal gender ideals.

Ranae Evenson and Robin Simon used data from the 2005 National Survey of Families and Households to examine effects of parenthood. They demonstrate that the picture is complicated and cannot be summarized by a generalization about happiness levels or mental health outcomes. There are both positive and negative outcomes from parenthood; there is gratification as well as an added sense of purpose and meaning to life from being parents. But there are stresses and demands, especially when parents have young children, that may overshadow the benefits and undermine parents' mental health. Furthermore, the wider social and cultural context has reduced the significance, social value, and esteem attached to the parental role and left parents without the institutional supports that could make parenting less stressful (Evenson and Simon 2005).

Looking specifically at depression, Evenson and Simon (2005) compared childless adults with parents in different circumstances. After controlling for the effects of other demographic and social characteristics, compared to nonparents, parents reported *significantly higher* levels of depression. Sociologist Robin Simon

notes that "no group of parents—married, single, step, or even empty nest—reported significantly greater emotional well-being than people who never had children" (Ali 2008, 62). Contrary to their expectations, gender did not affect the relationship between parental status and depression. Among parents, those with minor and dependent children at home report *fewer, not more*, symptoms of depression than those with older children.

Outcomes of parenthood differ depending on the circumstances under which one parent (married, never married, divorced; employed or unemployed; level of economic well-being) and gender. A later section looks at the effect of parents' marital status on parent–child relationships. The next chapter examines how employment affects parents. What about gender? The most obvious difference between the impact of parenthood on women and men is in the child-rearing responsibility that women acquire when they become mothers. Most active, hands-on parenting is done by mothers. Women also assume more responsibility for thinking about, worrying, and seeking information about children and their needs as well as planning and managing the division of responsibility for child care.

Child-to-Parent Influence

Although the focus of this chapter is more about on how parents' shape their children, children also are socializers in their own right. When an infant cries to be picked up and held, to have a diaper changed, or to be burped or when he or she smiles when being played with, fed, or cuddled, the parents are being socialized. The child is creating strong bonds with the parents. Although the infant's actions are not at first consciously directed toward reinforcing parental behavior, they nevertheless have that effect. In this sense, even very young children can be viewed as participants in creating their own environment and in contributing to their further development (see Peterson and Rollins 1987).

Critical Thinking

How should child-rearing tasks be delegated between spouses (or partners)? Are there any particular tasks that you believe either men or women should not do? How are tasks delegated in your household? What was the role of your father in the care and nurturing of you and your siblings?

Among the many familiar stereotypes that persist in the United States is that of the "dysfunctional and deviant young African American male" (Smith et al. 2005). The image of young African American fathers "as sexual predators likely to abandon their children and the child's mother" has "seeped into the nation's conscience … influencing public policy on public assistance and related issues" (Smith et al. 2005, 977). Yet there are men like 18-year-old Terrell Pough, named by *People* magazine as an "outstanding father" in a feature story in August 2005. Pough was described as a "rare breed of teenaged dads who are trying to raise children." A devoted father to his daughter, Diamond, who was not yet two years old, Pough juggled finishing high school, working, and caring for Diamond, of whom he had custody, when featured by the magazine. As he told the magazine, "She's what I work for, what I live for, why I wake up.… She's everything." Pough asserted his determination that "if something ever happens to me…no one can ever tell her that her dad didn't take care of her." Tragically, something did happen to Pough. He was shot to death while returning home from work November 17, 2005.

According to research by Carolyn Smith, Marvin Krohn, Rebekah Chu, and Oscar Best, although Pough may have been exceptional in his dedication and sacrifice, his commitment to his daughter may be more representative of young, single African American fathers than the negative stereotypes. Using data from the Rochester Youth Development Study, a longitudinal study that followed 1,000 seventh- and eighth-grade adolescents over a number of years, Smith and colleagues focused on the experiences of 193 young fathers, 67.4% of whom were African Americans, 20.7% Hispanics, and 11.9% whites. Interested in the extent of a father's contact and involvement and the matter of financial support of his child or children, Smith and colleagues offered the following findings.

Approximately 33% of the African American fathers reported that they lived with their child. Although the ethnic differences were not statistically significant, this percentage was higher than that of Hispanics (25.9%) but less than that of Caucasians (45.5%). Even among the nonresident fathers, 61.8% of the African American men reported "at least weekly" contact, an amount not widely different from that of Caucasians (67.7%) or Hispanics (54.3%). Only 11.4% of African American fathers reported "no contact," slightly more than the percentage among Caucasians (9.3%) but less than among Hispanics (15.5%).

Looking at the extent to which nonresident fathers provide financial support for their children revealed the following patterns. As summarized in Table 10.6, although again not statistically significant (largely because of sample sizes), the data suggest that the levels of support provided by nonresident African American fathers was about the same as that of Hispanic fathers. Combining this finding with the data on contact reveals two important points: African American fathers are more similar to than different from Hispanic fathers and, in terms of contact, not that different from Caucasians. In both the amount of contact with and financial support for their children, these nonresident fathers *do not fit the racial stereotype.*

Based on research findings such as these, we need to reconsider the stereotype of uninvolved and irresponsible young black fathers. Even when the majority of fathers were not living with their oldest child, many had regular contact, and two-thirds provided some to all of the financial support as arranged. No doubt, there are still men who make and maintain no commitment to their children. However, they can be found among all races and are not the norm among men of any particular race.

Table 10.6 Financial Support for Children Provided by Nonresident Fathers, by Race

	African American	Hispanic	Caucasian
No support arranged or 0% paid	33.2%	36.4%	17.7%
1% to 99% of arranged support	12.6%	9.5%	—
100% of arranged support	54.2%	54.1%	82.3%

SOURCE: Smith et al. (2005, 975–1101).

Strategies and Styles of Child Rearing

Twentieth-century parenting was shaped by child-rearing advice from such notable authorities as Benjamin Spock, T. Berry Brazelton, and Penelope Leach. These three authors sold well over 40 million copies of their books advising parents, especially mothers, as to the best ways to raise their children. Building on psychological theories of development, they stressed the importance of parents understanding their child's cognitive and emotional development.

So what do these experts advocate as effective parenting? Sharon Hays (1996) suggests that they all advocate the ideology of intensive mothering, discussed earlier in this chapter. Aside from the belief in the special nurturing capacities of mothers, this ideology contains the following assumptions about what children need from parents:

- Raising children is and should be an emotionally absorbing experience characterized by affectionate nurture. Emotional attachment is essential for healthy development; parental unconditional love and loving nurture are seen as critical to the child, no less essential, Spock asserts, than "vitamins and calories" (Spock and Rothenberg 1985, quoted in Hays 1996).
- It is the mother's job to respond to the needs and wants of her child. Parents should follow the cues given by their child, and this requires knowledge of

One of the most important things parents can do is show their children that they are loved and that their company is desired.

children's needs and developmental phases as well as great parental sensitivity.

- Parents must develop sensitivity to the particular needs of their child. This includes, for example, recognizing the different meanings of the child's crying and understanding the unique and individual developmental pattern of the child.
- Physical punishment is frowned on. Instead, setting limits, providing a good example of what parents expect from their child, and giving the child lots of love are preferred ways to convince the child to internalize and act on parents' standards. Punishment consists of "carefully managed temporary withdrawal of loving attention," a labor-intensive, emotionally absorbing method of discipline. Once a child can question, parents are urged to reason with the child, negotiate, and discuss motives and alternative ways of acting. This strategy obviously involves more time and effort than spanking.

Contemporary Child-Rearing Strategies

One of the most challenging aspects of child rearing is knowing how to change, stop, encourage, or otherwise influence children's behavior. We can request, reason, command, cajole, compromise, yell, or threaten with physical punishment or the suspension of privileges; alternatively, we can just get down on our knees and beg. Some of these approaches may be appropriate at certain times; others clearly are never appropriate. The techniques of child rearing currently taught or endorsed by educators, psychologists, and others involved with child development differ somewhat in their emphasis but share most of the tenets that follow:

- *Respect.* Mutual respect between children and parents must be fostered for growth and change to occur. One important way to teach respect is through modeling—treating the child and others respectfully.
- *Consistency and clarity.* Consistency is crucial in child rearing. Without it, children become hopelessly confused, and parents become hopelessly frustrated. Patience and teamwork (maintaining a united front when there are two parents) on the parents' part help ensure consistency. Parents should beware of making promises or threats they won't be able to keep, and a child

needs to know the rules and the consequences for breaking them.

- *Logical consequences.* One of the most effective ways to learn is by experiencing the logical consequences of our actions. Some of these consequences occur naturally—if you forget your umbrella on a rainy day, you are likely to get wet. Sometimes parents need to devise consequences appropriate to their child's misbehavior. Rudolph Dreikurs and Vicki Soltz (1964) distinguish between logical consequences and punishment. The "three Rs" of logical consequences dictate that the solution must be *related* to the problem behavior, *respectful* (no humiliation), and *reasonable* (designed to teach, not to induce suffering).

- *Open communication.* The lines of communication between parents and children must be kept open. Numerous techniques exist for fostering communication. Among these are active listening and the use of "I" messages. In *active listening*, the parent verbally reflects the child's communications to confirm that they have a mutual understanding. *"I" messages* (e.g., "I am disappointed that you aren't being nicer to your sister") are important because they impart facts without placing blame and are less likely to promote rebellion in children than are "you" messages (e.g., "You are such a bully to your sister"). In addition, regular weekly *family meetings* provide an opportunity to be together and air gripes, solve problems, and plan activities.

- *No physical punishment.* Many physicians, psychologists, and sociologists have become harsh and vocal critics of physical punishment. Both the American Psychological Association and the American Medical Association oppose physical punishment of children. Many sociologists, most notably scholars who study family violence, such as Murray Straus, oppose corporal punishment; they note that it is related to later aggressive behavior from children, including later perpetration of spousal violence (Straus and Yodanis 1996). However, such punishment is used widely. Straus and Yodanis estimate that more than 90% of parents of toddlers use corporal punishment in the United States. Psychologist Elizabeth Gershoff (2008) contends that 80% of U.S. five-year-olds have ever been spanked; by high school the percentage reaches 85%. Critics contend that spanking may "work" in the short run by stopping undesirable behavior; its long-range results—anger, resentment, fear, hatred, aggressiveness, and family violence—may be extremely problematic (McLoyd and Smith 2002; Straus and Yodanis 1996).

- *Behavior modification.* Effective types of discipline use some form of behavior modification. Rewards (hugs, stickers, or special activities) are given for good behavior, and privileges are taken away when misbehavior is involved. Good behavior can be kept track of on a simple chart listing one or several of the desired behaviors. Time-outs—sending the child to his or her room or to a "boring" place for a short time or until the misbehavior stops—are useful for particularly disruptive behaviors. They also give the parent an opportunity to cool off (Dodson 1987; see also Canter and Canter 1985).

Styles of Child Rearing

A parent's approach to training, teaching, nurturing, and helping a child will vary according to cultural influences, the parent's personality, the parent's basic attitude toward children and child rearing, and the role model that the parent presents to the child.

One popular formulation contrasts four basic styles of child rearing: authoritarian, permissive or indulgent, authoritative, and uninvolved (Baumrind 1971, 1983, 1991). *Style of parenting* refers to variations between parents in their efforts to socialize and control their child (Baumrind 1991). All four styles are found among parents. However, research tends to identify one of the following—authoritative parenting—as more effective than the others (Davis 1999).

Parents who practice **authoritarian child rearing** typically require absolute obedience. The parents' ability to maintain control is of primary importance. "Because I said so" is a typical response to a child's questioning of parental authority, and physical force may be used to ensure obedience. Working-class families tend to be more authoritarian than middle-class families. Diana Baumrind (1983) found that children of authoritarian parents tend to be less cheerful than other children and correspondingly more moody, passively hostile, and vulnerable to stress.

Permissive child rearing or **indulgent child rearing** is a more popular style in middle-class families than in working-class families. The child's freedom of expression and autonomy are valued. Permissive parents rely on reasoning and explanations. Yet permissive parents may find themselves resorting to manipulation and justification. The child is free from external restraints but not from internal ones. The child is supposedly free because he or she conforms "willingly," but such freedom is not authentic. Although children of permissive parents are generally

Popular Culture: Calling Nanny 911

There are a number of reality television programs that depict families in crisis. Sometimes it is the marital relationship that is problematic; other times it is the relationship between parents and children that is the focal point. One such program, *Nanny 911*, takes us inside families where overwhelmed parents have reached a crisis point in raising and disciplining their children. They are desperate and looking for answers.

According to the program's Web site, the show follows a team of nannies (Nanny Deb, Nanny Stella, and Nanny Yvonne under the supervision of Head Nanny Lillian) as they attempt to apply the "dos and don'ts of child rearing" to solve problems ranging from bad manners or sibling rivalry to severe temper tantrums.

Just what are the "dos and don'ts" espoused by the nannies? The following lists "The 11 Commandments of *Nanny 911*":

1. Be consistent.
2. Actions have consequences.
3. Say what you mean and mean it.
4. Parents work together as a team.
5. Don't make promises you can't keep.
6. Listen to your children.
7. Establish a routine.
8. Respect is a two-way street.
9. Positive reinforcement works much better than negative reinforcement.
10. Manners are universal.
11. Define your roles as parents.

They also advise parents to avoid corporal punishment.

Promising results, Carroll and Reid (2005, xx) state, "You, too can learn how to pinpoint what's gone wrong in your household, see what's not working, and devise a plan. Do this and your family can make amazing progress in a relatively short time. Our quick fixes will give permanent results—if you follow the plan."

It is interesting to see the overlap between the philosophy espoused by *Nanny 911* and some of the other expert child-rearing advice discussed in this chapter. It is equally, if not more, interesting to consider what the popularity of both *Nanny 911* and *Supernanny* reveal about the state of parenting and parental anxiety today.

---

cheerful, they exhibit low levels of self-reliance and self-control (Baumrind 1983).

Parents who favor **authoritative child rearing** rely on positive reinforcement and infrequent use of punishment. They direct the child in a manner that shows awareness of his or her feelings and capabilities. Parents encourage the development of the child's autonomy within reasonable limits and foster an atmosphere of give-and-take in parent–child communication. Parental support is a crucial ingredient in child socialization. It is positively related to cognitive development, self-control, self-esteem, moral behavior, conformity to adult standards, and academic achievement (Gecas and Seff 1991). Control is exercised in conjunction with support by authoritative parents.

Finally, **uninvolved parenting** refers to parents who are neither responsive to their children's needs nor demanding of them in their behavioral expectations. Children and adolescents of uninvolved parents suffer consequences in each of the following areas or domains: social competence, academic performance, psychosocial development, and problem behavior (Davis 1999).

Much research points to the authoritative style as especially effective. Children raised by authoritative parents tend to approach novel or stressful situations with curiosity and show high levels of self-reliance, self-control, cheerfulness, and friendliness (Baumrind 1983). Even bigger differences, however, are found between children of more involved parents as opposed to unengaged parents (Davis 1999).

What Do Children Need?

Parents often want to know what they can do to raise healthy children. Are there specific parental behaviors or amounts of behaviors (say 12 hugs, three smiles,

Critical Thinking

In your family, what child-rearing attitudes (authoritarian, permissive, or authoritative) predominated? Do you think these attitudes influenced your development? If so, how? Which might (or do) you find useful in raising your child?

a kiss, and a half hour of conversation per day) that all children need to grow up healthy? Of course not. Apart from saying that basic physical needs must be met (adequate food, shelter, clothing, and so on), along with some basic psychological ones, experts cannot give parents such detailed instructions.

Noted physician Melvin Konner (1991) listed the following needs for optimal child development—which, he wrote, "parents, teachers, doctors, and child development experts with many different perspectives can fairly well agree on":

- Adequate prenatal nutrition and care
- Appropriate stimulation and care of newborns
- Formation of at least one close attachment during the first five years
- Support for the family "under pressure from an un-caring world," including child care when a parent or parents must work
- Protection from illness
- Freedom from physical and sexual abuse
- Supportive friends, both adults and children
- Respect for the child's individuality and the pre-sentation of appropriate challenges leading to competence
- Safe, nurturing, and challenging schooling
- Adolescence "free of pressure to grow up too fast, yet respectful of natural biological transformations"
- Protection from premature parenthood

In today's society, especially in the absence of adequate health care and schools in so many lower-income communities, it is difficult to see how even these minimal needs can be met. Even when the necessary social supports are present, parents may find them-selves confused, discouraged, or guilty because they do not live up to their own expectations of perfection.

Yet children have more resiliency and resourceful-ness than we may ordinarily think. They can adapt to and overcome many difficult situations. A mother can lose her temper and scream at her child, and the child will most likely survive, especially if the mother later apologizes and shares her feelings with the child. A father can turn his child away with a grunt because he is too tired to listen, and the child will not necessarily grow up neurotic, especially if the father spends some "special time" with the child later.

Self-Esteem

High self-esteem—what Erik Erikson called "an opti-mal sense of identity"—is essential for growth in rela-tionships, creativity, and productivity in the world at large. Low self-esteem is a disability that afflicts chil-dren (and the adults they grow up to be) with feelings of powerlessness, poor ability to cope, low tolerance for differences and difficulties, inability to accept re-sponsibility, and impaired emotional responsiveness. Self-esteem has been shown to be more significant than intelligence in predicting scholastic performance. A study of 3,000 children found that adolescent girls had lower self-images, lower expectations from life, and less self-confidence than boys (Brown and Gil-ligan 1992). At age nine, most of the girls felt posi-tive and confident, but by the time they entered high school, only 29% said that they felt "happy" the way they were. The boys also lost some sense of self-worth but not nearly as much as the girls.

Ethnicity was an important factor in this study. African American girls reported a much higher rate of self-confidence in high school than did Cauca-sian or Latina girls. Two reasons were suggested for this discrepancy. First, African American girls often have strong female role models at home and in their communities; African American women are more likely than others to have a full-time job and run a household. Second, many African American parents specifically teach their children that "there is noth-ing wrong with them, only with the way the world treats them" (Daley 1991). According to researcher Carole Gilligan, their study "makes it impossible to say that what happens to girls is simply a matter of hormones. . . . [It] raises all kinds of issues about cultural contributions, and it raises questions about the role of the schools, both in the drop of self-esteem and in the potential for intervention" (quoted in Daley 1991, B1).

Parents can foster high self-esteem in their children by (1) having high self-esteem themselves, (2) accept-ing their children as they are, (3) enforcing clearly defined limits, (4) respecting individuality within the limits that have been set, and (5) responding to their child with sincere thoughts and feelings. It is also im-portant to single out the child's behavior—not the whole child—for criticism (Kutner 1988). Children (and adults) can benefit from specific information about how well they've performed a task. "You did a lousy job" not only makes us feel bad but also gives us no useful information about what would constitute a good job.

Misusing the concept of self-esteem with superfi-cial praise is probably the most common way parents mishandle the issue. Children notice when praise is insincere. If, for instance, Martha refuses to comb her hair yet we continually tell her how good it looks, Martha quickly realizes that we either have low expectations or do not have a clue about hair care. Instead, parents can accomplish more by giving children timely, honest, specific feedback. For example, "I like the way you discussed Benjamin Franklin's inventions in your essay" is more effective than "You're wonderful!" Each time parents treat their child like an intelligent, capable person, they increase the child's self-esteem.

What Do Parents Need?

Although some needs of parents are met by their children, parents have other needs. Important needs of parents during the child-rearing years are personal developmental needs (such as social contacts, privacy, and outside interests) and the need to maintain marital satisfaction. Yet so much is expected of parents that they often neglect these needs. Parents may feel varying degrees of guilt if their child is not happy or has some "defect," an unpleasant personality, or even a runny nose.

However, many forces affect a child's development and behavior. Accepting our limitations as parents (and as humans) and accepting our lives as they are (even if they haven't turned out exactly as planned) can help us cope with some of the many stresses of child rearing in an already stressful world. Contemporary parents need to guard against the "burnout syndrome" of emotional and physical overload. Parents' careers and children's school activities, organized sports, Scouts, and music, art, or dance lessons compete for the parents' energy and rob them of the unstructured (and energizing) time that should be spent with others, with their children, or simply alone.

Diversity in Parent–Child Relationships

The diversity of family forms in our country creates a variety of parenting experiences, needs, and possibilities as well as a range of parent–child relationships. The problems and strengths of single-parent and stepfamilies are discussed in more detail in Chapter 14 but are touched on here, along with the influences of ethnicity, sexuality (i.e., lesbian and gay parenthood), and aging.

Effects of Parents' Marital Status

Parental marital status affects children's upbringing and well-being. Consistently, researchers have found that—whether because of economic advantage, social resources, amount and kind of parental attention and commitment, or some other factors—children who live with both of their biological parents benefit in a variety of ways when compared to peers in single-parent households, remarried parent or stepparent households, and cohabiting-parent households.

Sociologist Yongmin Sun notes that children in stepfamilies and single-parent families are more likely than children living with their married biological parents to have behavior and drug problems, show lower rates of graduation from high school, report lower levels of self-esteem, and perform worse on standardized tests (Sun 2003). On a few measures, such as levels of delinquency and academic achievement, teens in married stepfamilies are somewhat advantaged compared to teens from cohabiting stepfamilies (i.e., unmarried couples with one partner functioning as a stepparent) (Manning and Lamb 2003).

On matters such as economic well-being, children of cohabiting parents benefit when compared to children of single parents but are less well off when compared to children of married parents. However, these outcomes vary by race. White children benefit more than either African American or Hispanic children from their biological parents being married (Manning and Brown 2006). Interestingly, data from the 2003–2004 American Time Use Survey show that mothers' time with children does not differ significantly between married and cohabiting mothers in terms of either amount or type of time (e.g., routine child care, interactive child care, total child care) spent with children. Single mothers spend less total time with children than either married or cohabiting mothers (Kendig and Bianchi 2008), perhaps because they are solely responsible for all household tasks, stretching them thin.

In accounting for differences that surface between married and cohabiting stepfamilies and among families with two biological parents, single-parent families, and stepfamilies, economic factors (e.g., family income and parents' level of education) are especially important. Compared to married mothers, cohabiting mothers tend to have lower levels of education and lower earnings and are more likely to be unemployed (Reed 2006). Economic disadvantages faced by single mothers as well as by stepfamilies may explain why children in such households do less well (Sun 2003).

Not only can parental cohabitation affect children but, according to interesting qualitative interview research with a sample of "parent cohabitors," children can alter the meanings that cohabiting couples attach to their relationship as well (Reed 2006). In interviewing 44 cohabiting couples with children, sociologist Joanna Reed found that pregnancy or the birth of a child is the factor that leads some couples to what she calls "shotgun cohabitations." Furthermore, the presence of children "prompts couples to construct their relationship in a new way and orient them around the child" (Reed 2006, 1128). Pregnancy and pending parenthood make the relationship more serious, increase couples' commitment to their relationships, and signal to them that they ought to seriously entertain the idea of marriage. At minimum, Reed's research reveals the different meanings that cohabitors bring to their relationships and how those meanings are differently constructed for those who are or are about to become parents.

Ethnicity and Parenting

A child's ethnic background can affect how he or she is socialized. According to some researchers, minority families socialize their children to more highly value obligation, cooperation, and interdependence (Demo and Cox 2000). It has been suggested that Mexican American parents tend to value cooperation and family unity more than individualism and competition. Asian Americans and Latinos traditionally stress the authority of the father in the family. In both groups, parents command considerable respect from their children, even when the children become adults. Older siblings, especially brothers, have authority over younger siblings and are expected to set a good example (Becerra 1988; Tran 1988; Wong 1988). Many Asian Americans tend to discourage aggression in children and expect them to sacrifice their personal desires or interests out of loyalty to their elders and to family authority more generally (Demo and Cox 2000). In disciplining their children, Asian parents tend to rely on compliance based on the desire for love and respect.

African Americans, too, may have group-specific emphases in the ways they socialize their children. As reported in Chapters 3 and 4, African American parents tend to socialize their children into less rigid, more flexible gender roles. They reinforce certain traits, such as assertiveness and independence, in both their sons and their daughters. They also seek to promote such values as pride, closeness to other African Americans, and racial awareness (Demo and Cox 2000).

Groups with minority status in the United States may be different from one another in some key ways, but they also have much in common. Such groups often emphasize education as the means for the children to achieve success. Studies show that immigrant children tend to excel as students until they become acculturated and discover that it's not "cool." Minority groups are often dual-worker families, meaning that the children may have considerable exposure to television while the parents are away from home. This may be viewed as a mixed blessing: on the one hand, television may help children who need to acquire English-language skills; on the other, it can promote fear, violence, and negative stereotypes of women and minority-status groups. Some American children are raised with a strong positive sense of ethnic identification; however, that can also result in a sense of separateness that is imposed by the greater society.

Discrimination and prejudice shape the lives of many American children. Parents of ethnic minority children may try to prepare their children for the harsh realities of life beyond the family and immediate community (Peterson 1985). According to Mary Kay DeGenova (1997), to reduce an environment of racism, it is important for us to identify the similarities among various cultures. These include people's hopes, aspirations, desire to survive, search for love, and need for family—to name just a few. Although superficially we may be dissimilar, the essence of being human is very much the same for all of us.

Gay and Lesbian Parents and Their Children

Researchers believe that the number of gay families is in the millions. They estimate that there are more than 7 million gay, lesbian, bisexual, or transgender parents just with school-age children in the United States (Kosciw and Diaz 2008). Other research has estimated that 14 million children have at least one gay parent (Stacey and Biblarz 2001). The high ends of these estimates include parents with adult children no longer in the home and use generous definitions of sexual orientation (including anyone with homoerotic desires) (Stacey and Biblarz 2001).

According to psychologist Charlotte Patterson, a leading authority on gay and lesbian parenting, the current research on the subject has some limitations. It has focused mostly on lesbian mothers and on young children (preadolescents). In addition, it has been rare to have longitudinal studies in which researchers follow a sample of gay and lesbian parents and/or their children over time (Patterson 2005). These limitations aside, existing research fails to support the notion that children of lesbian mothers or gay fathers are negatively affected (Patterson and Friel 2000, Patterson 2005; Stacey and Biblarz 2001).

In fact, most gay or lesbian parents have been in heterosexual marriages (Patterson and Chan 1998). Concerns about gay and lesbian parents tend to center on questions about parenting abilities, fear of sexual abuse, and worry that the children will become gay or lesbian themselves. Research has failed to support such concerns or identify any significant negative outcomes for children. In fact, much research has failed to identify any meaningful differences between children of gay and heterosexual parents. Sociologists Judith Stacey and Timothy Biblarz's (2001) and psychologist Charlotte Patterson's (2000, 2005) reviews of existing research on the effect of parental sexual orientation on children find that most research supports either a "no-effects" or a "beneficial-effects" interpretation.

In summarizing the research on children of gay and lesbian parents as they compare with children of heterosexual parents, Patterson notes that there are no significant differences in their gender identities, gender-role behaviors, self-concepts, moral judgment, intelligence, success with peer relations, behavioral problems, or successful relations with adults of both genders (Patterson 2000, 2005). Stacey and Biblarz (2001) suggest that there may be some defensiveness on the part of researchers, especially from those sympathetic to gay and lesbian parents. Aware of the social stigma and lack of support gay and lesbian families face, there may be a tendency to minimize differences. In so doing, some differences that might be strengths of gay and lesbian families may go underemphasized.

Fears about Gay and Lesbian Parenting

Heterosexual fears about the parenting abilities of lesbians and gay men are exaggerated and unnecessary. There are minimal differences between lesbians and heterosexual women in their "approaches to child rearing" or their mental health (Patterson 2005). No studies identify ways in which lesbian mothers or gay fathers are "unfit parents" or less fit than heterosexual parents.

Fears about gay parents' rejecting children of the other sex also are unfounded. Such fears reflect the popular misconception that being gay or lesbian is a rejection of members of the other sex. Many gay and lesbian parents go out of their way to make sure that their children have role models of both sexes (Kantrowitz 1996). Gay and lesbian parents also tend to say that they hope their children will develop heterosexual identities to be spared the pain of growing up gay in a homophobic society. Research finds children of gay males and lesbians to be well adjusted and no more likely to be gay as adults (Flaks et al. 1995; Goleman 1992; Kantrowitz 1996).

Ultimately, it is the quality of parenting and the harmony within the family—not the sexuality of the parents—that matters most to children. Like children of heterosexual parents, children whose gay or lesbian

Families headed by lesbians or gay men generally experience the same joys and pains as those headed by heterosexuals, but they are also likely to face insensitivity or discrimination from society.

Real Families: Having a Gay Parent

Consider the following account by Abigail Garner, author of *Families Like Mine: Children of Gay Parents Tell It Like It Is* (2005) and creator of the Web site FamiliesLikeMine.com:

When I was 5, my father came out as gay to his family and friends and moved in with another man. By the time I entered elementary school, I was learning about the cruelty of homophobia. "Faggot" was the favorite put-down among the boys in my class. I didn't know what it meant until my parents explained that it was a mean way of saying someone was gay. Since my classmates seemed to be so hostile about gay people, I decided I should keep quiet about my family.

I remember when I was about 8, I was walking down the street between my father and his partner and holding both of their hands. It felt dangerous, because by standing as a link between them I was "outing" them. What would happen if others realized my dad was gay? Would he lose his job? Get beaten up? Be declared an unfit parent?

Fortunately, my mother (who is heterosexual) made no attempt to limit my father's custody rights. If she had, she probably would have gained full custody. Our courts have a history of favoring straight parents over gay ones in custody battles.

My parents did their best to make me feel good about where I came from. They told me that even though they were divorced and my dad was gay, we were no less valid than any other family.

College marked a significant change in my life. The 1,500 miles between home and school gave me the distance I needed to figure out who I was, separate from my parents. I thought I had outgrown the label of "daughter from a gay family." Soon after I graduated, however, I connected with a group of teens with gay and lesbian parents while volunteering for a youth organization. When I realized how similar their stories were to mine, I was inspired to start talking openly about my own experiences.

. . . it wasn't having a gay father that made growing up a challenge, it was navigating a society that did not accept him and, by extension, me.

Summarizing the research on parenting by and children of gays and lesbians in a report for the American Psychological Association, Charlotte Patterson (2005, 15) makes the following strong assertion:

There is no evidence to suggest that lesbian women or gay men are unfit to be parents. . . . Not a single study has found children of lesbian or gay parents to be disadvantaged in any significant respect relative to children of heterosexual parents. Indeed, the evidence to date suggests that home environments provided by lesbian and gay parents are as likely as those provided by heterosexual parents to support and enable children's psychosocial growth.

parents are in "warm and caring relationships," experiencing less stress and conflict, and receiving more support from partners (as well as from other family members) tend to fare better.

What about Nonparental Households?

Yet another way to see the effects of parents on children is to examine the experiences of children in households with *no* biological parents. As indicated in Table 10.7, in 2008, 2.8 million children—nearly 3.8% of the more than 74.1 million American children under 18 years of age—lived in households with neither biological parent (U.S. Census Bureau 2008, table C3). Three-quarters of children in **nonparental households** live with relatives, most with a grandparent.

Sociologist Yongmin Sun (2003) reports that older children (ages 15–17) are twice as likely as children under age five to live in one of these nonparental households. In addition to age, ethnicity makes a difference: in 2008, 2.4% of Asian, 2.6% of Caucasian, 3.9% of Hispanic, and 8.1% of African American children live in a household without either biological parent (U.S. Census Bureau 2008).

Generally, research has documented that children in nonparental households suffer when compared to children who live with at least one parent. Comparisons of children in foster care, albeit only one type of nonparental care, show negative effects in areas ranging from children's mental health, academic achievement, drug use, and behavioral problems (Sun 2003). Likewise, children in nonparental "kinship care" have

Table 10.7 Percent of U.S. Children Living with Both Parents and with Neither Parent, by Race, 2008 (numbers in thousands)

Group	Total Children	% with Both	% with Neither
All	74,104	70%	3.8%
White, non-Hispanic	42,051	78%	2.6%
African American	11,342	38%	8.1%
Asian American	2,980	85%	2.4%
Hispanic	15,644	70%	3.9%

been found to have poorer health, mental health, and school achievement than children in "parent present" families, whether single- or two-parent families (Sun 2003). Sun suggests that it is likely that the absence of mothers has the greatest impact. In accounting for the observed effects in nonparental households, Sun argues that the differences result mainly from the resource differences between these family structures and those with at least one parent. Key resources include income and parents' education, parents' expectations for their children's education, frequency of conversations between parents and children about school, involvement of parents with the schools and with other parents, and children's experiences of various cultural activities. No differences of note existed between kinship care and nonrelative care, and no differences were observed between girls and boys in how they fare in nonparental environments (Sun 2003).

Parenting and Caregiving in Later Life

As parents and their adult children age, the relationship between them undergoes a number of changes. The needs of adult children are different than the needs of younger, dependent sons and daughters. At the same time, the aging of parents often introduces a need for someone to provide them with or oversee their care.

Parenting Adult Children

Many years ago, a Miami Beach couple reported their son missing (Treas and Bengtson 1987). Joseph Horowitz still doesn't understand why his mother became so upset. He wasn't "missing" from their home in Miami Beach: he had just decided to go north for the winter. Etta Horowitz, however, called authorities. Social worker Mike Weston finally located Joseph in Monticello, New York, where he was visiting friends. Etta, 102, and her husband, Solomon, 96, had feared that harm had befallen their son Joseph, 75. As the Horowitz story reminds us, parenting does not end when children grow up.

By some measures, children are "growing up" later than at any time in the past. They lack the means to be financially independent and delaying entry into marriage, parenthood, and independent living, away from their families. In one study that compares 1960 census data to 2000 census data, researchers noted that there has been a significant decrease in the percentage of young adults who, by age 20 or 30, have completed all of the following five traditionally defined major adult transitions: leaving the parental home, completing their schooling, achieving financial independence (being in the labor force and/or—for women—being married and a mother), marrying, and becoming a parent. In 1960, more than three-fourths of women and two-thirds of men had reached all five of these markers by age 30, yet in 2000, less than half of women and less than a third of men had achieved all five of these (Furstenberg et al. 2004). As a consequence, parents are being asked to provide support or assistance sometimes well into their children's young adulthood.

Most parents with adult children still feel themselves to be parents even when their "children" are middle aged. However, their parental role is considerably less important in their daily lives. They generally have some kind of regular contact with their adult children, usually by letters, phone calls, or e-mails; parents and adult children also visit each other fairly frequently and often celebrate holidays and birthdays together. Financially, they may make loans, give gifts, or pay bills for their children. Further assistance may come in the form of shopping, house care, and transportation and help in times of illness.

Parents tend to assist those whom they perceive to be in need, especially children who are single or divorced. Parents perceive their single children as being "needy" when they have not yet established themselves in occupational and family roles. These children may need financial assistance and may lack intimate ties; parents may provide both until the children are more firmly established. Parents often assist divorced children, especially if grandchildren are involved, by providing financial and emotional support. They may also provide child care and housekeeping services.

Parents tend to be deeply affected by the circumstances in which their adult children find themselves. Adult children who seem well adjusted and who have fulfilled the expected life stages (becoming independent, starting a family, and so on) provide their aging parents with a vicarious gratification. On the other hand, adult children who have stress-related or chronic problems (e.g., with alcohol) cause higher levels of parental depression (Allen, Blieszner, and Roberto 2000). Unsurprisingly, in a study of African American mothers of incarcerated adult sons, mothers experienced increased levels of anxious and depressed moods, much of which was the result of resulting financial difficulties and taking on the caregiving responsibility for their grandchildren (Green et al. 2006).

Family researchers Emily Greenfield and Nadine Marks examined national survey data (National Survey of Midlife in the United States) to examine the effects of adult children's problems on their parents' psychological well-being. Consistent with other research (Pillemer and Suitor 1991), parents who reported that their adult children had a greater number and more types of problems experienced poorer well-being, suffering negative outcomes in areas such as self-acceptance (e.g., liking oneself and being satisfied with one's life), positive affect (e.g., feeling cheerful, happy, calm, and peaceful), and parent–child relationship quality. They also reported more family relationship strain (e.g., feeling as if family members other than one's spouse demand too much, let them down, or get on their nerves) and more negative affect (e.g., feeling sad, hopeless, worthless, or nervous). Both mothers and fathers were affected. Parents' marital status made a difference. Single parents suffered greater negative impact on their positive affect, while married parents experienced more negative consequences in their parent–child relationships (Greenfield and Marks 2006).

Some elderly parents never cease being parents because they provide home care for children who are severely limited either physically or mentally. Many elderly parents, like middle-aged parents, are taking on parental roles again as children return home for financial or emotional reasons. Although we don't know how elderly parents "parent," presumably they are less involved in traditional parenting roles.

Critical Thinking

Think about your grandparents. How many are alive? What kind of relationship do you (or did you) have with them? What role do they (or did they) play in your life and your family's life?

Grandparenting

The image of the lonely, frail grandmother in a rocking chair needs to be discarded. Grandparents are often not old, nor are they lonely, and they are certainly not absent in contemporary American family life. Grandparents are often a very active and meaningful presence in the family lives of young children as well as young adults (Gregory Kennedy 1990).

Grandparenting is expanding tremendously these days, creating new roles that relatively few Americans played a few generations back. Grandparents play important emotional roles in American families; the majority appears to establish strong bonds with their grandchildren (Kennedy 1990; Strom et al. 1992–1993). They help achieve family cohesiveness by

Grandparents are imprtant to their grandchildren as caregivers, playmates, and mentors.

conveying family history, stories, and customs. Grandparents influence grandchildren directly when they act as caretakers, playmates, and mentors. They influence indirectly when they provide psychological and material support to parents, who may consequently have more resources for parenting (Brooks 1996).

Grandparents seem to take on greater importance in single-parent and stepparent families and among certain ethnic groups. Children of single mothers are three times as likely as children of married parents to live in the same household as a grandparent. Approximately 13% of children of single mothers also live with a grandparent (Dunifon and Kowaleski-Jones 2007). There is little difference in the percentage of African American (13%) and white children (14%) who live with a single mother and a grandparent, though the outcomes for children of such household structure can differ by race.

Researchers Rachel Dunifon and Lori Kowaleski-Jones note that there are many ways that living with a grandparent could be beneficial to children. Grandparents might reduce the stress felt by single parents, provide additional income, help monitor the children, and engage in activities with grandchildren that benefit them. Dunifon and Kowaleski-Jones found that for white children, living in a single-mother household with a grandparent present was associated with greater cognitive stimulation when compared with children in single-mother households without a grandparent. For African American children, such household structure is associated with less cognitive stimulation. These different outcomes may reflect different skills that grandparents bring with them to the household as well as different circumstances that lead to the sharing of households (Dunifon and Kowaleski-Jones 2007).

According to the U.S. Census Bureau, in 2006 more than 6 million grandparents lived in the same home as one of their grandchildren, with 2.5 million having primary caregiving responsibility for their grandchildren age 18 or younger. Of these "grandparent caregivers," 64% were grandmothers.

Grandparent–grandchild relationships vary. Sociologists Andrew Cherlin and Frank Furstenberg (1986) identified three distinct styles of grandparenting. *Companionate* grandparents' relationships with their grandchildren are marked by affection, companionship, and play. Companionate grandparents do not perceive themselves as rule makers or enforcers and rarely assume parentlike authority. Because these grandparents tend to live relatively close to their grandchildren, they can have regular interaction with them. *Remote* grandparents are not intimately involved in their grandchildren's lives. Their remoteness, however, is geographic rather than emotional. Geographic distance prevents the regular visits or interaction with their grandchildren that would bind the generations together more closely. *Involved* grandparents are actively involved in what have come to be regarded as parenting activities: making and enforcing rules and disciplining children. Involved grandparents (most often grandmothers) tend to emerge in times of crisis—for example, when the mother is an unmarried adolescent or enters the workforce following divorce. Some involved grandparents may become overinvolved, however, and cause confusion as the family tries to determine who is the real head of the family.

Single parenting and remarriage have made grandparenthood more painful and problematic for many grandparents. Stepfamilies have created stepgrandparents who are often confused about their grandparenting role. Are they really grandparents?

The grandparents whose sons or daughters do not have custody often express concern about their future grandparenting roles (Goetting 1990). Although research indicates that children in stepfamilies tend to do better if they continue to have contact with both sets of grandparents, it is not uncommon for the parents of the noncustodial parent to lose contact with their grandchildren (Bray and Berger 1990).

A variety of circumstances may lead to situations in which the grandparent role and the relationships with grandchildren are strained if not disrupted. Divorce and single parenthood may be the most prominent of such circumstances, but death of a spouse, distance, or estrangement between parents and children can all impede grandparent–grandchild relationships (Keith and Wacker 2002). Over the past 40 years, grandparent visitation statutes have been enacted in all 50 states, and grandparents' visitation rights have been increased. Generally, courts have not wanted to expand grandparents' rights at the expense of parents' rights, especially parents' rights to control the custody of their children (Keith and Wacker 2002).

Children Caring for Parents

Parent–child relationships do not flow just in one direction. A common experience faced by many American families is the need to provide care for aging or ill parents. The previously noted idea of the **sandwich generation** captures the experience of many adults, sandwiched between the simultaneous demands of raising their own children and caring for their parents. Certain circumstances create **parentified children**—children forced to become caregivers for their parents well before adulthood (Winton 2003). In situations of *parentification*, children may be pressed into taking care of parents who have become chronically ill, chemically dependent, mentally ill, incapacitated after a divorce or widowhood, or socially isolated or incapacitated (Winton 2003).

Much of the psychological and sociological literature depicts parentification as problematic. Taking on caregiving responsibilities for a parent or parents while still a child or adolescent can disrupt normal developmental processes. Sociologists tend to focus on the nonnormative nature of children being responsible for their parents. However, definitions of normative and nonnormative vary by culture. Among many populations other than white, middle-class European Americans, parentification is expected and obligatory. Additionally, under certain circumstances, parentification may be beneficial for the development of certain personality traits, the maintenance of certain family relationships, and the acquisition of particular skills. Chester Winton (2003) suggests that parentification may be a normative part of childhood in many contemporary American families. *Destructive parentification* occurs when the circumstances become extreme and long term and the responsibilities that children carry are age inappropriate (Jurkovic 1997, cited in Winton 2003).

Winton (2003) suggests the following as possible consequences of parentification:

- *Delayed entry into marriage.* Children who have had to care for parents (or siblings) over a number of years may decide to delay taking on the caretaking that comes with marriage and choose, instead, to take time for themselves and concentrate on their own needs.
- *Acquisition of certain personality characteristics.* Over time, parentification might lead to the development of such traits or tendencies as feeling excessive responsibility for the well-being of others, making it difficult to set limits, say "no" to others, or focus on oneself.
- *Relationship and intimacy problems.* Parentified children may seek as adult partners people who they can be caretakers for—in other words, "dependent, needy people" who have emotional or physical disabilities or who have been emotionally "wounded" by past experiences.

- *Career choices.* The "caretaker syndrome" associated with parentification may lead people to jobs where they can physically or emotionally take care of people, such as jobs in social work, medicine, nursing, teaching, or preschool child care.

Caring for Aging Parents

There are strong norms of **filial responsibility**—responsibility to see that one's aging parents have support, assistance, company, and are well cared for—in the United States (Gans and Silverstein 2006). Such norms appear to be more strongly felt by younger adults (who are typically further removed from having to provide such support), women, those with no college education, and those who have had experience providing support or assistance to their parents. Acceptance of these norms neither reflects nor guarantees that one will provide such support, though it is logical to assume that support is more likely among those who accept such norms (Gans and Silverstein 2006).

One factor that affects actual caregiving behavior is the adult child's marital status. Although most Americans want to and do maintain relationships with their aging parents, those relationships are often constrained by the "more pressing demands of marriage" (Sarkisian and Gerstel 2008, 373). Married people are less likely than never-married or divorced adults to share a household with their parents; to have frequent contact by letter, phone, or in person; to give and receive emotional support or practical help; or to give financial help to or receive financial help from their parents. Other factors affecting the level of involvement between adults and their parents include parental marital status (unmarried parents receiving more assistance and contact from adult sons and daughters), degree and nature of parents' needs, and gender.

Most actual elder care is provided by women, generally daughters or daughters-in-law (Mancini and Blieszner 1991). Women do more "kinkeeping" than men, staying in touch with and providing care and assistance to parents and in-laws (Sarkisian and Gerstel 2008). Elder caregiving seems to affect men and women differently. Women report greater distress, greater decline in happiness, more hostility, less autonomy, and more depression from caregiving than do men (Marks, Lambert, and Choi 2002). This may partly be because men approach their daily caregiving activities in a more detached, instrumental way. Another factor may be that women often are not only mothers but also workers; an infirm parent can sometimes be an overwhelming responsibility to an already burdened woman (Rubin 1994). Interestingly, when caring for a parent out of the household, many women feel a *caregiver gain*, a greater sense of purpose in life than might be felt by noncaregiving women (Marks et al. 2002). Fortunately, most adult children participate in parental caregiving in some fashion when needed, whether it involves doing routine caregiving, providing backup, or giving limited or occasional care (Mancini and Blieszner 1991).

A study of 539 older participants found that although there are psychological benefits associated with intergenerational support, excessive support received from adult children may be harmful, eroding competence and imposing excessive demands (Silverstein, Chen, and Heller 1996). In balancing personal needs with those of families, it is important to define the level of care that is both appropriate and necessary.

Caregivers too may experience both positive and harmful outcomes. Caregiving enables caregivers to express love, fulfill obligations, and maintain or even deepen relationships. However, it can also be the source of profound stress and conflicting feelings. Primary caregivers can experience the resumption of earlier unresolved antagonisms and conflicts, conflicting loyalties between spousal or child-rearing responsibilities and caring for the elderly relative, and resentment either toward the older relative for disrupting family routines and patterns or toward other family members for their lack of involvement.

Having and raising children are among the most fulfilling and satisfying activities in which most people eventually take part. They are also among the most frustrating and stressful. Some of the stress and frustration is inevitable given the breadth, depth, and length of the commitment one makes when raising children. However, in keeping with one of the wider themes that underlie much of this textbook, a good portion of the stresses of parenting result from the wider social context in which families live and the need to balance or juggle parenting with other roles that one plays, paid and unpaid work roles. It is to those issues that Chapter 11 turns.

Summary

- There were more than 4.3 million births in the United States in 2007, including a record number of births to unmarried women. Both the birthrate and the fertility rate increased slightly in 2007. These rates vary by race, ethnicity, education, income, and marital status.

- Approximately 20% of 40- to 44-year-old women have not had children.

- Approximately 70 million women worldwide have an impaired inability to conceive. Roughly 40 million are currently seeking treatment.

- Even with greater acceptance of voluntary childlessness, women and men who choose to forgo parenthood experience social pressure to justify or change their statuses.

- Having children both at younger and at older ages exposes women and their children to greater risks to their health and well-being.

- Teen pregnancy rates declined through the 1990s and in the first years of the twenty-first century before beginning to rise again in 2005 and 2006.

- There were more than 6 million pregnancies in 2006, the outcomes of which varied according to mothers' marital status and race.

- More than a third of pregnancies are unplanned, being either mistimed or unwanted. The proportion of pregnancies that are unplanned is greater among younger women, lower-status women, and minorities.

- Babies born from unplanned pregnancies face greater health and social risks, and their mothers have greater risks of postpartum depression.

- Feelings about sexuality are likely to change during pregnancy for both women and men.

- Men's involvement in the pregnancy and birth process may affect their later parenting.

- Critics, including especially feminist critics, have alleged that the *medicalization of childbirth*—making this natural process into a medical "problem"—has caused an overdependence on medical professionals, an excessive use of technology, and an alienation of women from their bodies and feelings.

- Research with new mothers documented mixed reactions to the treatment they received, the care they were given, and the experiences of childbirth.

- Miscarriages, which typically occur during the first trimester of pregnancy, are the most common form of pregnancy loss.

- *Infant mortality* rates in the United States are higher than in other industrialized nations and vary by race and ethnicity.

- Women report experiencing both positive and negative feelings during birth.

- According to the U.S. Census Bureau, there were more than 1.6 million adopted children under age 18 in the United States in 2004, 2.5% of all children living with a parent.

- Costs of adoption vary depending on the type of adoption, type of agency, and whether the adoption is domestic or international.

- The trend has been toward open adoption, where there is contact between the adoptive parents and the birth mother or parents during pregnancy and/or continuing after birth.

- The transition to parenthood is unlike other role transitions. It is irreversible and sudden, and it comes with little preparation.

- Reduced sexual desire and depression during the *postpartum period* are among the potential problematic reactions to childbirth. Teenage mothers are much more likely than adult mothers to suffer from post-partum depression.

- Both motherhood and fatherhood have changed, leaving new parents to chart a new course.

- Although there is no concrete evidence of a biological maternal drive, it is clear that socialization for motherhood, as well as an *ideology of intensive mothering*, affect women's expectations and intentions to have children.

- Employed mothers earn less than employed women without children.

- The traditional expectations of fathers are being supplemented and perhaps supplanted by expressive ones. This may be truer of our beliefs about fathers (the *culture of fatherhood*) than of fathers' real behavior (the *conduct of fathers*).

- Parenthood affects many areas of men's and women's lives, including their social activities, intergenerational family ties, occupational behavior, and free time.

- Most hands-on child care is done by mothers. Fathers are less engaged with and accessible to their children than are mothers. When directly engaged with children, fathers more often play or assist in personal care activities.

- Children have a number of basic physical and psychological needs, including the need for adequate prenatal care, formation of close attachments, protection from illness and abuse, and respect, education, and support from family, friends, and community.

- Parents differ in terms of their styles of parenting. Four styles are *authoritarian, permissive* (or *indulgent*), *authoritative,* and *uninvolved.* Of these, most research portrays the authoritative as most effective.

- Contemporary strategies for child rearing include the elements of mutual respect, consistency and clarity, logical consequences, open communication, and behavior modification in place of physical punishment.

- Parenthood has effects on marital relations and on the mental health, especially depression rates, of parents.

- Economic, cultural, and political institutions have neglected to adopt policies that would allow parents and children deeper and more frequent contact with each other.

- Parents' marital status, ethnicity, and sexuality all influence parenting and child socialization.

- Children who live in households without any parents (either foster care or "kinship care" from other relatives) suffer academically, psychologically, and behaviorally.

- Parents of ethnic minority status may try to give their children special skills for dealing with prejudice and discrimination.

- Most gay and lesbian parents are or have been married. Studies indicate that children of both lesbians and gay men are as well adjusted as children of heterosexual parents.

- Increasingly, older parents provide financial and emotional support to their adult children; they often take active roles in child care and housekeeping for their daughters who are single parents.

- Grandparents often provide extensive child care for grandchildren. Grandparenting can be divided into three styles: companionate, remote, and involved.

- When parents are chronically ill, chemically dependent, mentally ill, or incapacitated after a divorce or widowhood, children often become caregivers to their parents.

- Family caregiving activities often begin when an aged parent becomes infirm or dependent. Caregiving responsibilities may create conflicts involving previous unresolved problems, the caregiver's inability to accept the parent's dependence, conflicting loyalties, resentment, anger, and money or inheritance conflicts.

Key Terms

RESOURCES ON THE WEB

Book Companion Website

www.cengage.com/sociology/strong

Prepare for quizzes and exams with online resources—including tutorial quizzes, a glossary, interactive flash cards, crossword puzzles, self-assessments, virtual explorations, and more.

11

Marriage, Work, and Economics

What Do YOU Think? Are the following statements TRUE or FALSE?
You may be surprised by the answers (see answer key on the following page).

T	F	**1** In contrast to single-worker couples, dual-career couples tend to divide household work almost evenly.
T	F	**2** More than 1 million American men are full-time homemakers with no outside employment.
T	F	**3** More women report a desire to stay home than to work outside the home.
T	F	**4** Couples who work different shifts have more satisfying and stable marriages.
T	F	**5** Women in the United States currently make 90 cents for every dollar that men earn.
T	F	**6** When companies offer policies to reduce work-family conflict, most women and men make full use of them.
T	F	**7** The U.S. recession has taken a greater toll on male employment than on female employment.
T	F	**8** Most families are dual-earner families.
T	F	**9** Women tend to interrupt their work careers for family reasons far more often than do men.
T	F	**10** Compared to employed women and men in Europe and Japan, Americans work fewer hours per week.

Imagine yourself at a party put on by your school's alumni association. As you float around the room, trying to meet and mingle with some people who graduated in recent years, you overhear the following exchanges among some of the other guests. Each snippet of conversation illustrates some unspoken assumptions people have about work and family. Can you recognize the assumptions and identify what is wrong in each exchange?

- *Exchange No. 1.* A trio of women is in a corner. "What do you do?" one of the women inquires politely while being introduced by a second woman to the third. "I don't work; I'm a stay-at-home mom," the third responds. "Oh, that's . . . nice," the first woman replies, seeming to lose interest and turning toward the woman handling the introduction.
- *Exchange No. 2.* A bearded man is talking to a couple. "So, what do you two do?" the man asks. "I'm a doctor," the woman responds as she picks up her son, who is impatiently tugging on her. "And I'm an accountant, but like a lot of other people, I was laid off recently. So, in the meantime I'm home with them," her husband says, nodding toward their fidgeting son while feeding their second child with a bottle.

Although they are subtle, we can observe the following assumptions being made and attitudes being displayed. In the first exchange, both women ignore that the woman who identified herself as a stay-at-home mother does, indeed, work. They also appear to devalue such unpaid work in comparison with paid work. In the second exchange, the woman identifies herself as a physician without acknowledging that she is also a parent. Her husband defines himself in terms of a job he no longer has while downplaying his role as a full-time caregiver for their children. In fact, as husband and wife, father and mother, both the physician and the accountant are unpaid family workers making important—but generally unrecognized—contributions to the family's economy.

Answer Key to What Do YOU Think?

1 False, see p. 393;   **2** True, see p. 401;   **3** False, see p. 390;   **4** False, see p. 399;   **5** False, see p. 403;   **6** False, see p. 407;   **7** True, see p. 411;   **8** True, see p. 392; **9** True, see p. 391;   **10** False, see p. 382.

Because it is unpaid and perhaps because it is done mostly by women, family work is often ignored and looked on as less important than paid work, regardless of how difficult, time consuming, creative, rewarding, and essential it is for our lives and future as humans. This is not surprising because in the United States employment takes precedence over family.

Understanding the role of work in families may require reconsidering what one thinks families do. People ordinarily think of families in terms of relationships and feelings—the family as an emotional unit. But families are also economic units that happen to be bound by emotional ties (Ross, Mirowsky, and Goldsteen 1991). Both paid work and unpaid family work, as well as the economy itself, profoundly affect the way we live in and as families. Our most intimate relationships vary according to how we participate in, divide, and share paid work and family work.

One's paid work helps shape the quality of family life: it affects time, roles, incomes, spending, leisure,

© Myrleen Ferguson Cate/PhotoEdit

Often, when we think about "work," we fail to include the domestic work and child care that families need.

and even individual identities. Whatever time individuals have for one another, for fun, for their children, and even for sex is time not taken up by paid work. Work regulates the family, and for most families, as in the past, a woman's work molds itself to her family, whereas a man's family molds itself to his work (Ross et al. 1991). Work roles and family roles must both be fulfilled despite the difficulties and complexity involved in meeting the demands they collectively impose. These facts are the focus of this chapter.

Workplace and Family Linkages

Outside of sleeping, the activity to which most employed men and women devote the most time is their jobs. There are numerous ways in which paid employment, or lack thereof, can and does affect family life.

It's about Time

Of the nearly 138 million Americans employed in 2008, 76% were employed full time. These full-time workers worked an average of 42.5 hours per week, with men averaging approximately three hours more per week than women, 43.7 to 40.9, among men and women 16 years of age and older (Bureau of Labor Statistics, ftp://ftp.bls.gov/pub/special.requests/lf/aat22.txt). Hours spent at work also vary by race and by marital status, as illustrated in Figure 11.1.

If one looks specifically at married parents, based on the American Time Use Survey, during the period 2003–2006, 88% of married fathers and 43% of married mothers were employed full time (35 or more hours per week). Among those parents who were employed full time, 65% of those with a preschool-age child and 71% of those whose youngest child was between ages 6 and 17 worked on an average day. As revealed in Figure 11.2, the presence and number of children in the household has different effects on women's and men's employment. Where 63% of married women with no children were employed full time, with the addition of each child (from one child to four or more), the percentage of women employed full time decreases. For men, there are only slight differences in the percentage who are employed full time, and those with children were more, not less, likely to be employed full time.

Data reveal that, in contrast to declines throughout Europe, at least until recently Americans were

Figure 11.1 Average Hours of Work per Week for Those Age 16 and Over Employed Full Time, 2008

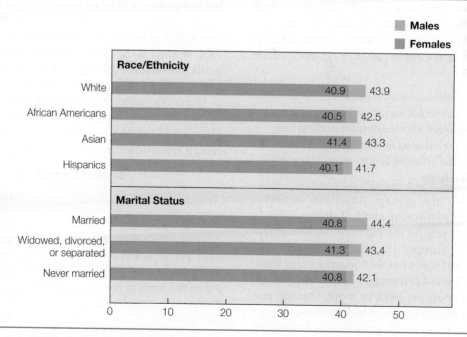

Figure 11.2 Percentage of Married Women and Men Employed Full Time by Presence and Number of Children, 2003–2006

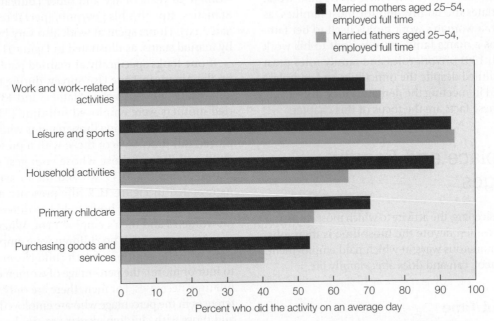

Legend:
- ■ Married mothers aged 25–54, employed full time
- ■ Married fathers aged 25–54, employed full time

Categories (vertical axis):
- Work and work-related activities
- Leisure and sports
- Household activities
- Primary childcare
- Purchasing goods and services

Horizontal axis: Percent who did the activity on an average day (0 to 100)

NOTE: Data are averages of all days of the week. All activity categories include associated travel. Data refer to parents with biological, step-, or adopted children aged 17 or younger living in the household.

SOURCE: Bureau of Labor Statistics, 2008. Married Parents' Use of Time, 2003–2006.

increasing their hours spent at work (Jacobs and Gerson 2004). Although European and American workers face similar "time dilemmas," the societal responses to these pressures have been vastly different. Jerry Jacobs and Kathleen Gerson (2004, 124) asserted the following:

> Several European countries, especially those in Northern Europe, have made sustained, highly publicized, and well-organized efforts to reduce working time as a strategy for reducing unemployment, increasing family time, and reducing gender inequalities in the market and at home.

Conversely, "the average American worker—including both part-timers and full-timers—puts in more hours per year on the job than the typical full-time worker in Europe" (Jacobs and Gerson 2004, 127). The United States has the longest average workweek and the highest percentages of men and women who work 50 hours per week or more. This was true of married women and men as well as those who are unmarried parents as well as people without children (Jacobs and Gerson 2004).

Data from July 2009 show that more than 15% of the nonagricultural labor force averaged 49 hours or more at work per week. More than 8 million workers, 6% of all men and women employed in nonagricultural jobs, worked 60 or more hours per week. The more hours individuals work, the less time they have for their families, friends, communities, and themselves.

Overwork carries consequences. Comparing those employed women and men who experience high levels of overwork to those who experience low levels reveals the following:

Levels of Overwork

	High	Low
Feel very angry toward employers	39%	1%
Often/very often resent coworkers	34%	12%
Make a lot of mistakes at work	20%	0%
Feel highly stressed	36%	6%
Report their health as good	52%	65%
Feel like they take good care of themselves	41%	68%

SOURCE: Galinsky (2005).

Time spent at one's job directly affects time spent with one's family.

It bears mentioning that high levels of overwork are not experienced by all workers. Some categories of workers—for example, professionals, executives, managers, and small business owners—are more likely to experience the greatest time demands. Others are underemployed and would prefer to work more, while still others are unemployed and would prefer any work at all (Jacobs and Gerson 2004; Perry-Jenkins, Repetti, and Crouter 2000). This **bifurcation of working time**, wherein some work longer and longer days and weeks while others work fewer hours than they need or want, is revealed by findings from the National Study of the Changing Workforce. Well before the recent recession in the U.S. economy, 60% of both men and women said that they would prefer to work less; however, about one in five men (19.3%) and 18.5% of women said that they would prefer to work more hours than they were working at the time (Jacobs and Gerson 2004).

Time Strains

Whether a person loves, loathes, or merely learns to live with it, one's job structures the time that can be spent with one's family (Hochschild 1997). The time demands faced at work can create a feeling of **time strain**, in which individuals feel that they do not have or spend enough time in certain roles and relationships. Kei Nomaguchi, Melissa Milkie, and Suzanne Bianchi (2005) found interesting gendered patterns in their investigation into the psychological effects of time strains:

- More fathers than mothers reported feeling that they did not have enough time with their children or their spouses. More mothers than fathers felt they had too little time for themselves.
- Life satisfaction was significantly reduced *for mothers but not for fathers* when they felt that they had or spent "too little time with children."
- Feelings of insufficient time with a spouse were associated with significantly higher levels of distress *for women but not for men*.
- Feelings of insufficient time for oneself were associated with reduced levels of family and life satisfaction and with increased feelings of distress *for men but not for women*.
- Fathers articulated feeling strained for time both with their spouses and their children, but these feelings did not affect them as much psychologically as they did women.

Work and Family Spillover

In addition to the time we have available to our families, paid work affects home life in other ways. We can call this **work spillover**—the effect that work has on individuals and families, absorbing their time and energy and impinging on their psychological states. It links our home lives to our workplace (Small and Riley 1990). Work is as much a part of our marriages and home lives as love is. What happens at work—frustration or worry, a rude customer, an unreasonable boss, or inattentive students—has the potential to affect our moods, perhaps making us irritable or depressed. Often, we take such moods home with us, affecting the emotional quality of our relationships.

Not surprisingly, research demonstrates that work-induced energy depletion, fatigue, or, in more extreme cases, exhaustion can affect the quality of our family relationships. Fatigue and exhaustion can make us angry, anxious, less cheerful, and more likely to complain and can cause us to experience more difficulty interacting and communicating in positive ways. Yet according to one study, although both stress and exhaustion from work affect marital relationships, "stress is far more toxic" (Roberts and Levenson 2001, 1065). These researchers suggest that although common, job stress can seriously and negatively affect marital happiness, creating dynamics that unchecked may even contribute to divorce.

Scholars have increasingly looked at how and how often negative spillover affects us. Although negative work spillover occurs neither every day nor to everyone, it is accurate to consider it fairly common (Roehling, Jarvis, and Swope 2005). This is revealed in Figure 11.3, based on data from the 1997 National Study of the Changing Workforce.

In general, women report more negative spillover than do men, though college-educated men are the most likely to indicate that work interferes with their family lives (Ammons and Kelly 2008; Sloan Work and Family Research Network, www.bc.edu/wfnetwork). Such work–family tensions are greater for employed parents than they are for employed women and men without children. Furthermore, the effects seem to be greater on mothers than on fathers, just as the differences between parents and nonparents are greater among women than among men. Studying young adults ages 21 to 31 who were part of the Youth Development Study in St. Paul, Minnesota, sociologists Samantha Ammons and Erin Kelly report that although the percentages themselves are small, twice as many young women as young men reported that work affected their childbearing decisions (7% of women vs. 3.6% of men). Among those with children, mothers were 1.5 times more likely than are fathers to say that

work interfered with their relationships with their children (23% vs. 14%) (Ammons and Kelly 2008).

Sociologists Jerry Jacobs and Kathleen Gerson note that children's ages make little difference in parents' experiences of work–family stress (Jacobs and Gerson 2004). Workplace stress often causes us to focus on our problems at work rather than on our families, even when we are home with our families. It can lead to fatigue, stomach ailments, and poorer health as well as depression, anxiety, increased drug use, and problem drinking (Crouter and Manke 1994; Roehling et al. 2005).

Family-to-Work Spillover

As many employed parents can attest, the relationship between paid work and family life cuts both ways. The emotional climate in our homes can affect our morale and performance in our jobs. Positively, family can help alleviate some workplace stress. More research has focused on how the demands of our home lives may impinge on our concentration, energy, or availability at work (Jacobs and Gerson 2004).

Yet Jennifer Keene-Reid and John Reynolds (2005) argue that workers feel more success in juggling work and family when they have some control in scheduling their time at work. Furthermore, because family demands and needs can and do arise unexpectedly, the ability of employed parents to adjust

Figure 11.3 Work-to-Family Spillover

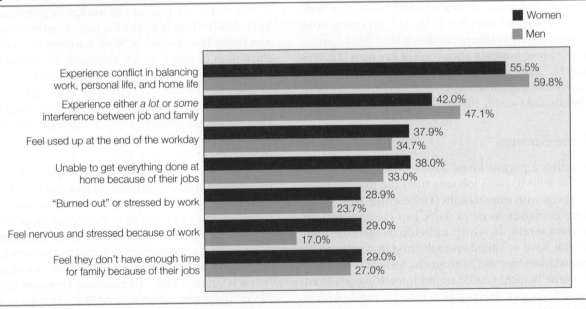

SOURCE: Jacobs and Gerson, 2004; 85.

their schedules accordingly is a useful and important family-friendly benefit.

The research by Samantha Ammons and Erin Kelly (2008) showed family-to-work spillover to be a more common problem than work-to-family spillover. Furthermore, young adults with less than college educations were more likely to report experiencing family-to-work spillover than were those with BA degrees or higher (50% vs. 34%). Women reported higher levels of family-to-work spillover, with large differences between women with (41%) and without (62%) college degrees. Because women more often than men face or anticipate the intrusion of their family responsibilities into their work lives, they are forced to make more work-related adjustments because of family needs (Keene-Reid and Reynolds 2005). Additionally, higher levels of family-to-work spillover have been found among parents compared to nonparents (Roehling et al. 2005).

Meeting family demands such as assuming more household and child care responsibility often comes with hidden or unanticipated work-related financial costs. Regardless of gender, those who carry responsibility for traditionally female housework chores are likely to suffer reduced wages. This is probably the result of having less effort and energy available to spend on paid work activities (Noonan 2001).

Role Conflict, Role Strain, and Role Overload

Two-parent families in which both partners are employed face more severe work-related problems than do nonparents. Being an employed parent usually means performing at least three demanding roles simultaneously: worker, parent, and spouse or partner. In juggling these roles, we might experience role conflict, role strain, role overload, or a combination of these.

When the multiple social statuses or positions that we occupy (e.g., spouse, parent, and worker) present us with competing, contradictory, or simultaneous role expectations, we experience **role conflict**. When the role demands attached to any particular status (e.g., mother, husband, or employee) are contradictory or incompatible, we experience **role strain**. Finally, when the various roles we play require us to do more than we can comfortably or adequately handle or when we feel that we have so much to do that we will never "catch up" or have enough time for ourselves, we experience **role overload** (Crouter et al. 2001; MacDermid 2006).

In the specific case of family and paid work roles, when we feel torn between spending time with our spouses or children and finishing work-related tasks, we are experiencing role conflict. We cannot be in two places at once.

Men who see themselves as traditional providers may experience role strain when pressed into higher levels of housework or child care. Employed wives, exhausted by their combination of paid work, housework, and child care, may also experience role strain and not enjoy sexual intimacy with their spouses.

Professor of family studies and human development Ann Crouter and colleagues (2001) found that both fathers' role overload and the amount of hours they worked affected the quality of their relationships with their adolescent children. When fathers worked long hours but did not experience overload, their relationships with their adolescents did not seem to suffer. It appears that for fathers and children, the combination of hours and overload have the greatest effect (Crouter et al. 2001).

Parental "availability" to children is affected by the levels of stress that parents experience. Particularly stressful days at work may be followed by parents being withdrawn at home. This may sometimes prove beneficial because, by withdrawing, less negative emotion is brought into the relationships (Perry-Jenkins et al. 2000).

However, the more important sources of role conflict and overload are not within the person but rather *within the person's role responsibilities*. Men experience role conflict when trying to balance their family and work roles. Because men are expected to give priority to their jobs over their families, it is not easy for men to be as involved in their families as they may like. A study examining role conflict among men (O'Neil and Greenberger 1994; see also Greenberger 1994; Marks 1994) found that men with the least role conflict fell into two groups. One group consisted of men who were highly committed to both work and family roles. They were determined to succeed at both. The other group consisted of men who put their family commitments above their job commitments. They were willing to work at less demanding or more flexible jobs, spend less time at work, and put their family needs first. In both instances, however, the men received strong encouragement and support from their spouses.

The various issues surrounding spillover, role conflict, role overload, and role strain vary depending on the household structure and division of labor. Single-parent households with full-time working parents are easily susceptible to role overload and role conflict. Two-parent, dual-earner households also face versions of work-to-family spillover different from those of households with one provider and a partner at home full time.

Comparing levels of expressed work–family interference from two large survey sources, the Quality of Employment Study in 1977 and the National Study of the Changing Workforce in 1997, sociologist Sarah Winslow (2005) offered the following conclusions about work–family conflict. Compared with respondents in 1977, respondents in 1997 reported greater difficulty balancing work and family. This was greatest among parents regardless of whether they were in dual-earner or single-earner households. In addition, women and men reported similar levels of work–family interference.

Crossover

Yet another aspect of the work–family relationship is conveyed by the concept of **crossover**. Crossover refers specifically to the effects of one's job-related feelings on one's spouse or intimate partner. It is therefore different from though related to the previously discussed idea of *spillover*. Spillover occurs when work comes home with the worker, affecting the quality and quantity of his or her interactions with family members. It is a within-person/between domains (e.g., home and workplace) phenomenon. *Crossover* occurs when a worker's work-related stress (negative) or engagement (positive) affects one's partner in similar ways. According to organizational psychologist Arnold Bakker and colleagues (2008, 2009) in their *spillover-crossover model*, workplace experiences first spill over into the household and then cross over to one's partner. *Direct crossover* occurs through the empathy of one spouse for what the other is experiencing. When one spouse feels a lot of pressure or overload at work, the other spouse may begin to feel depressed or overloaded as well. *Indirect crossover* results more from the conflict between spouses that might result from negative spillover. Job demands can create work–family conflict, which then leads to marital conflict from which one's spouse's well-being may then suffer.

Anxiety, stress, depression, and burnout can cross over, as can engagement, vigor, enthusiasm, life satisfaction, and well-being (Bakker et al. 2009; Demerouti et al. 2005). Examining a Netherlands sample of dual-earner couples with children, psychologists Evangelia Demerouti, Arnold Bakker, and Wilmar Schaufeli (2005) found that wives' job induced exhaustion crossed over to their husbands, leading to husbands' exhaustion, while husbands' job-derived

Crossover refers to when one's emotional state comes home with one from work and affects one's spouse's or partner's mood.

life satisfaction crossed over to wives' life satisfaction. Mina Westman and colleagues found that the unemployment of a spouse can also lead to crossover, as the anxiety created by unemployment felt by the out-of-work spouse crosses over to the other, still employed spouse (Westman, Etzion, and Horovitz 2004). Less clear is how much "crosses over" from parents to children or whether parent–child relationships are affected in similar ways as marriages.

The Familial Division of Labor

Families divide their labor in a number of ways. Some follow more traditional male–female patterns, the majority share wage earning, and a small number

Exploring Diversity: Industrialization "Creates" the Traditional Family

In the nineteenth century, industrialization transformed the face of America. It also transformed American families from self-sufficient farm families to wage-earning urban families. As factories began producing farm machinery such as harvesters, combines, and tractors, significantly fewer farm workers were needed. Workers migrated to the cities, where they found employment in the ever-expanding factories and businesses.

Because goods were now bought rather than made in the home, the family began to shift from being primarily a production unit to being a consumer and service-oriented unit. With this shift, a radically new division of labor arose in the family. Men began working outside the home in factories or offices for wages to purchase the family's necessities and other goods. Men became identified as the family's sole providers or "breadwinners." Their work began to be identified as "real" work and was given higher status than women's work because it was paid in wages.

Industrialization also created the housewife, the woman who remained at home attending to household duties and caring for children. With industrialization, because much of what the family needed had to be purchased with the husband's earnings, the wife's contribution in terms of unpaid work and services went unrecognized, much as it continues today (Ferree 1991).

In earlier times, the necessities of family-centered work gave marriage and family a strong center based on economic need. The emotional qualities of a marriage mattered little as long as the marriage produced an effective working partnership. Without its productive center, however, the family focused on the relationships between husband and wife and between parent and child. Affection, love, and emotion became the defining qualities of a good marriage (Coontz 2005).

---

reverse roles. Even within a single family, there will likely be a number of divisions of labor over time as family members move through various life stages. Figure 11.3 shows the percentage of women and men involved in various activities in an average day among married parents with children under age 17 who were employed full time. As can be seen by the graph, a greater percentage of women than men reported involvement in child care, housework, and making purchases, while a greater percentage of men than women reported involvement in work activities and in leisure activities.

What is not reflected in the data is the variation in couples' division of labor, yet how families allocate tasks and divide paid and unpaid work has a tremendous effect on how a family functions.

The Traditional Pattern

In what we often consider the "traditional" division of labor in the family, work roles are complementary: the husband is expected to work outside the home for wages, and the wife is expected to remain at home caring for children and maintaining the household. In this pattern, a man's family role is satisfied by his performance of the provider role, whereas even when

employed outside the home, a woman's employment role is secondary to her family role (Blair 1993). As the preceding Exploring Diversity feature reveals, this pattern was a product of industrialization.

In 2008, in 19.5% of U.S. married-couple families, the husband was the sole wage earner. Among married couple families with children, the percentage of such an arrangement was greater, as can be seen in the following data:

Married Couple Families with Males as Sole Wage Earners
Families with children under 18: 29.5%
Families with children 6 to 17 years old: 23.3%
Families with children under 6: 37.4%

Overall, then, the traditional division of labor is evident in less than 20% of married-couple families and less than 40% of such families with preschool-age children. However, this does not mean that less than 40% of families with children desire the traditional pattern, nor does it mean that 37.4% prefer this pattern. Sociologist Arlie Hochschild's classic study *The Second Shift* illustrated that, for some couples, the actual division of roles and responsibilities may not reflect their preferred role arrangement or their gender ideologies. One cannot assume that what couples do is an expression of their **gender ideologies**, their beliefs about what women and men *ought* to do (Hochschild 1989). Some male-breadwinner/female-homemaker heterosexual couples might possess more egalitarian attitudes than their household arrangement suggests. Many two-earner couples might desire an arrangement more in keeping with the traditional pattern but be unable to afford it financially.

Whether a reflection of choice or circumstance, the difference in primary roles between men and women in traditional households profoundly affects the most basic family tasks, such as who cleans the toilet, mops the floors, does the laundry or grocery shopping, and mows the lawn. Whether or not they are employed outside the home, women remain primarily responsible for day-to-day household tasks.

The division of family roles along stereotypical gender lines varies by race and class. It is more characteristic of Caucasian families than of African American families. African American women, for example, are less likely than Caucasian women to be exclusively responsible for household tasks. Latino and Asian families are more likely to be closer to the traditional than are African Americans or Caucasians (Rubin 1994).

Class differences are somewhat ambiguous. Among middle-class couples, greater ideological weight is given to sharing and fairness. Working-class couples, although less ideologically traditional than in the past, are still not as openly enthusiastic about more egalitarian divisions of labor. However, in terms of *who does what*, working-class families are more likely than middle-class families to piece together work-shift arrangements that allow parents to take turns caring for the children and working outside the home. Such "opposite-shift" arrangements may force couples to depart from tradition, even if they neither believe they should nor boast that they do (Rubin 1994).

Men's Traditional Family Work

The husband's traditional role as provider has been perhaps his most fundamental role in marriage. As sociologists Barbara Arrighi and David Maume (2000, 470) put it, "It is the activity in which they spend most of their time and depend on most for their identity." In the traditional equation, as long as the male was a good provider, he was considered a good husband and a good father (Bernard 1981). Conversely, if a man failed as a provider, in his eyes as well as the eyes of others (including his family), he was not living up to expectations held for husbands and fathers.

In the traditional model, men are expected to contribute to family work by providing household maintenance. Such maintenance consists primarily of repairs, light construction, mowing the lawn, and other activities consistent with instrumental male norms. Although men often and increasingly contribute to housework and child care, their contribution may not be notable in terms of the total amount of work to be done. Men tend to see their role in housekeeping or child care as "helping" their partner, not as assuming equal responsibility for such work. Husbands become more equal partners in family work when they, their wives, or both have egalitarian views of family work or when such a role is pressed on them by either circumstantial necessity or ultimatum (Greenstein 1996; Hochschild 1989). Men who believe that they should act as traditional providers may resist performing more housework or do so only when necessary (though perhaps reluctantly), whether their wives are employed outside the home or not. If both spouses share a traditional gender ideology (traditional beliefs about what each should contribute to paid and family work), men's low level of household participation is not problematic.

Looking at marriages in which spouses are "mutually dependent"—with wives earning between 40% and 59% of the family income—such couples increased nearly 300% between 1970 and 2001. As many as 30% of dual-earner couples and 20% of all married couples fit such a pattern. In 26% of dual-earner couples, wives outearn husbands (an increase of more than 40% between 1987 and 2006). In 12% of dual-earner couples, wives earn *at least* 60% of the total income (Winslow-Bowe 2006). Interestingly, neither pattern has a uniform effect on married life. Only when men have traditional gender attitudes despite finding themselves in nontraditional life situations and marriages do such income differentials negatively affect men (Brennan, Barnett, and Gareis 2001).

Women's Traditional Family Work

Although the majority of women now earn salaries as paid employees, contributing more than 40% of family income in dual-earner households, neither traditional women nor their partners regard employment as a woman's fundamental role (Coontz 1997). For those with traditional gender ideologies, women are not duty bound to provide, though their incomes are likely to be essential in maintaining a livable lifestyle. They are, however, expected to perform household tasks (Thompson and Walker 1991).

No matter what kind of work the woman does outside the home, there is seldom equality when it comes to housework. Although this may cause tension and distress among more egalitarian-minded couples, traditional couples do not expect housework to be divided equally. Women's family work is considerably more diverse than that of men, permeating every aspect of the family. It ranges from housekeeping to child care, maintaining kin relationships to organizing recreation, socializing children to caring for aged parents, and cooking to managing the family finances. Ironically, family work is often invisible, even to the women who do most of it (Brayfield 1992).

Sociologist Ann Oakley (1985) described four primary aspects of the **homemaker role**:

- Exclusive allocation to women rather than to adults of both sexes
- Association with economic dependence
- Status as nonwork, which is distinct from "real," economically productive paid employment
- Primacy to women—that is, having priority over other women's roles

Many stay-at-home wives are dissatisfied with housework, perceiving it to be routine, monotonous, unpleasant, unappreciated, and unstimulating. However, such dissatisfaction doesn't necessarily mean that they are dissatisfied with all aspects of the stay-at-home role, which offers them some degree of autonomy. Young women, for example, may find increasing pleasure as they experience a sense of mastery over cooking, entertaining, or rearing happy children. If homemakers have formed a network among other women—such as friends, neighbors, or relatives—they may share many of their responsibilities. They discuss ideas and feelings and give one another support. They may share tasks as well as problems.

Full-time male houseworkers may call themselves *househusbands*, but they are less likely to do so than full-time female homemakers are to call themselves *housewives*. Instead, they identify themselves as retired, unemployed, laid off, or disabled (Bird and Ross 1993). Increasingly, if they have children at home, they may call themselves "stay-at-home dads" (Smith 2009).

Women in the Labor Force

Women have always worked outside the home. Like many of today's families, early American families were **coprovider families**—families that were economic partnerships dependent on the efforts of both the husband and the wife. Although women may have lacked the economic rights that men enjoyed, they worked with or alongside men in the tasks necessary for family survival (Coontz 1997). Beginning in the early nineteenth century, "work" and "family work" were separated. Men were assigned the responsibility for the wage-earning labor that increasingly occurred away from the home in factories and other centralized workplaces.

Women stayed within the home, tending to household tasks and child rearing. But this gendered division of labor was never total. Single women have traditionally been members of the paid labor force. There have also been large numbers of employed mothers, especially among lower-income and working-class families, African Americans, and many other ethnic minorities. Economic and other societal circumstances (e.g., World Wars I and II) also influence the numbers of employed mothers.

As of July 2009, more than 72 million women age 16 and over were employed in the civilian labor force. Women made up 47% of the labor force, and 61% of adult women were employed. In comparison, 72% of males age 16 and over were employed. Data for 2008 reveal that among those age 16 and older, African American women were more likely to be employed (61.3%) than were Caucasian (59.2%), Asian (59.4%), or Hispanic women (56.2%) (Bureau of Labor Statistics, www.bls.gov/cps/tables.htm#charemp, tables 5 and 6).

The most dramatic changes in women's labor force participation have occurred since 1960, resulting in the emergence of a family model in which both husbands and wives are employed outside the home. Between 1960 and 2008, the percentage of married women in the labor force rose from 32% to about 58%. Among this 58% of employed women, more than 71% were married women with children under age 18 in the labor force in 2008. There is some variation in maternal employment based on children's ages and mothers' marital status. The 2008 rates of maternal labor force participation (i.e., meaning that one was employed or actively seeking employment) were the following:

- 77% among mothers of 6- to 17-year-olds
- 64% for mothers whose youngest child was age six or younger
- 56% among mothers of infants under a year of age

- 69.5% for married mothers
- 76% among unmarried mothers (including never-married, divorced, separated, or widowed mothers) whose youngest child was under 17 years of age

In 2007, in 26% of dual-earner families, wives' incomes exceed husbands' incomes, up from 16% in 1981 (U.S. Census Bureau, www.census.gov/hhes/www/income/histinc/f22.html, table F-22). Among married couples with children, in 62% both parents were employed. In just under 7% of all married couple families, women were sole wage earners compared to nearly 20% in which men were the only ones who were employed. In 2008, there were more than 8 million families with children under 18 headed by single women who were in the labor force; of these, almost 60% had preschool-aged children.

Why Has Women's Employment Increased?

Looking at the wider societal and cultural context, a number of social and economic trends contributed to the steady increase in women's employment between 1960 and 1990 (Cotter, England, and Hermsen 2007). What accounts for this increase? Sociologists David Cotter, Paula England, and Joan Hermsen point to each of the following:

- Increases in the numbers of single mothers, resulting from increasing divorce rates and births to unmarried women
- Increases in women's educational attainment
- Proemployment messages and the equal opportunity emphasis of the women's movement
- Better job opportunities for women
- Decline in men's wages and the ability to support families on one income

As a result of these factors, more women needed employment, and more opportunities awaited women in the paid labor force. After the 1970s, social norms changed, making it more than acceptable for mothers to hold a job. In fact, one might contend that by the social norms of today, women are expected to be employed outside the home and to return to employment should they leave to have and raise children.

As far as individual women's motives for seeking employment, there are at least a couple worth noting. Among the factors that lead women to seek employment, the most prominent factor is the same as it is for men—economic need. For unmarried women and single mothers, their income from employment may be the only income on which they live. The incomes of married women may be primary or secondary to their husbands' incomes, but increasingly couples depend on and require income from both partners in order to achieve and maintain a decent if not desirable standard of living.

Economic pressures traditionally have been powerful influences on African American women. Among many married women and mothers, entry into the labor force or increased working hours are attempts to compensate for their husbands' loss in earning power because of inflation or unemployment. In addition, the social status of the husband's employment often influences the level of employment chosen by the wife (Smits, Ultee, and Lammers 1996).

Yet as is also true of men, women have other reasons for working outside the home. Among the psychological reasons for employment are an increase in a woman's self-esteem and sense of control. Additionally, employed women may find social support, recognition and appreciation at work that they don't get or feel at home (Hochschild 1997). As a consequence, married women with children have more positive and less negative emotional experiences at work. In contrast, men report feeling more positive and less negative at home, thus creating a situation wherein spouses live in "different realities" and have different emotional reactions to being at work and being at home (Willhelm and Perrez 2004).

Related to the question of why women work outside the home is what women would prefer to do—that is, be employed or stay at home. After polling 1,000 U.S. adults, the Gallup Organization found that 58% of U.S. men and women responded that they would prefer to work outside the home, while 37% said that they would prefer to stay home (3% said that they would prefer to do both). Among women, the percentage preferring employment was higher—at 50%—than the percentage who would prefer to stay at home (45%) but not by that much. Predictably, among men this gap was much greater: 68% said they would prefer paid employment compared to 29% who said they would prefer staying home. What is noteworthy in these comparisons is that the percentage of women preferring outside employment over staying home increased between 2002 and 2007. Both

Women seek the same gratifications from paid work that men seek. These include but go beyond wages.

Figure 11.4 What Working Situation Would Be Ideal for You?

Considering everything, what would be the ideal situation for you—working full-time, working part-time, or not working at all outside the home?

| | Ideal Situation Would Be | | | | |
	Not working	Part-time work	Full-time work	Don't know	N
	%	%	%	%	
Have childern under 18					
Fathers	16	12	72	*=100	343
Mothers	29	50	20	1=100	414
Mothers with childern under 18					
Employed full-time	21	49	29	1=100	184
Employed part-time	15	80	5	*=100	75
Not employed	48	33	16	3=100	153

SOURCE: Pew Research Center.

employment status and being a parent affected peoples' preferences. Among employed individuals, 69% of men and 55% of women would rather work outside the home. Among women with children ages 5 to 18, there was a greater preference for staying home than for working outside the home (48% to 45%) (Saad, Gallup.com, 2007).

In 2007, Pew Research also surveyed a sample of U.S. adults about their preferences regarding employment, including the choices of full-time employment, part-time employment, and staying home. As shown in Figure 11.4, among adults with children, the results indicate that half the mothers sampled considered part-time employment the ideal. In addition, more mothers indicated a greater preference for not working than for full-time work. Among employed mothers, both those employed part time and those employed full time responded that part-time employment would be preferred over either full-time employment or staying home.

Women's Employment Patterns

The employment of women has generally followed a pattern that reflects their family and child care responsibilities. Because of the family demands they face, women must consider the number of hours they can work, what time of day to work, and whether adequate child care is available. Traditionally, women's employment rates dropped during their prime childbearing years from 20 to 34 years. But this is no longer true; a significant number of women with children are in the labor force regardless of age of child, marital status, and racial or ethnic affiliation.

Women no longer automatically leave the job market when they become mothers. Among first-time mothers, more than half return to their jobs within six months of giving birth, and two-thirds have returned by the time their child celebrates his or her first birthday. Looking only at women who worked during their pregnancies, only 20% stayed at home for the entire first year of motherhood. For those who returned to work for the same employer as before childbirth, 89% worked at least as many hours—if not more—than they had before they became mothers (Johnson and Downs 2005).

Marriage, Work, and Economics **391**

Because of family responsibilities, many employed women work part time or work shifts other than the nine-to-five workday. Furthermore, when family demands increase, wives, not husbands, are more likely to cut back their job commitments, and, because of family commitments, women tend to interrupt their job and career lives far more often than do men.

Dual-Earner and Dual-Career Families

Since the 1970s, inflation, a dramatic decline in real wages, the flight of manufacturing, and the rise of a low-paying service economy have altered the economic landscape. These economic changes have reverberated through families, altering the division of household roles and responsibilities. Today, 62% of married couple families with children under 18 years of age are two-earner families. This includes 66% of two-parent families with children ages 6 to 18 and 55% of families with children under age six (U.S. Census Bureau, "America's Families and Living Arrangements: 2008," table FG1).

In 2007, the median income among married couples who depended solely on the wages of a full-time, year-round male breadwinner was $62,558. Families in which both husbands and wives were employed full time and year-round had median incomes of $99,140. Families in which wives worked full-time and year-round and husbands didn't had median incomes of $57,255 (U.S. Census Bureau 2008). Looking at married-couple families with children under age 18, the median income of those who had only a male wage earner was $60,274. The median income of dual-earner couples with children was $95,987. Finally, those married couples with children in which only wives were employed had a median income of $51,670 (U.S. Census Bureau, www.census.gov/hhes/www/income/histinc/f22.html).

Economic changes have led to a significant increase in dual-earner marriages. Most employed women are still segregated in lower-paying, lower-status jobs—administrative assistants, clerks, child care workers, factory workers, waitresses, and so on. Rising prices and declining wages pushed most of them into the job market.

Dual-career families are a subcategory of dual-earner families. They differ from other dual-earner families in that both husband and wife have high-achievement orientations, a greater emphasis on gender equality, and a stronger desire to exercise their capabilities. Unfortunately, these couples may find it difficult to achieve both their professional and their family goals. Often they have to compromise one goal to achieve the other because the work world generally is still not structured to meet the family needs of its employees.

Typical Dual Earners

Although we traditionally separate housework, such as mopping and cleaning, from child care, in reality the two are inseparable (Thompson 1991). Although fathers have increased their participation in housework and child care, they have made smaller increases in the frequency with which they swing a mop or scrub a toilet. If we continue to separate the two domains, men will take the more pleasant child care tasks of playing with the baby or taking the children to the playground, and women will take the more unpleasant duties of washing floors, cleaning ovens, and ironing. Furthermore, someone must do behind-the-scenes dirty work for the more pleasant tasks to be performed. Alan Hawkins and Tomi-Ann Roberts (1992) note the following:

> Bathing a young child and feeding him/her a bottle before bedtime is preceded by scrubbing the bathroom and sterilizing the bottle. If fathers want to romp with their children on the living room carpet, it is important that they be willing to vacuum regularly. . . . Along with dressing their babies in the morning and putting them to bed at night comes willingness to launder jumper suits and crib sheets.

If we are to develop a more equitable division of domestic labor, we need to see housework and child care as different aspects of the same thing: domestic labor that keeps the family running.

Housework

On an average day, a greater percentage of women than men spend at least some time performing household activities such as housework, cooking, lawn care, and money or household management. According to data from the American Time Use Survey, 83% of women compared to 64% of men reported spending some time in such activities. Looking at *how much time* women and men spend in household activities reveals that in 2008, women spent an average of just over two and a half hours to men's two hours a day. Half of women did such tasks as cleaning or laundry, compared to 20% of men. Differences in meal preparation or after-meal cleanup were also notable; 65% of women performed such tasks daily compared to 38% of men (Bureau of Labor Statistics 2009).

The American Time Use Survey can shed light, too, on the time *married* women and men spend on housework. Married men reported spending an average of 1.5 hours a day on household activities compared to 2.5 hours a day reported by married women. This doesn't include time spent "purchasing goods and services" or "caring for household members," activities in which married women also spend more time than married men spend (Bureau of Labor Statistics 2009).

Research on the division of responsibility for housework among married couples tells a somewhat mixed story. Housework still falls more heavily on women's shoulders regardless of whether they are employed outside the home. Evidence continues to show that although men do "pitch in" much more often than in the past, they are not yet near sharing the burden of housework equally.

However, as sociologists Oriel Sullivan and Scott Coltrane (2008) report, in the United States as well as the industrialized world, there are indications of movement in the direction of more sharing of housework, the result of both men doing more and women doing less. About U.S. heterosexual marriages, Sullivan and Coltrane assert the following:

- Men's relative share of housework has increased over the past three decades (from 15% to more than 30%), as has their actual time spent doing housework.
- In the almost 40 years between 1965 and 2003, men's time in child care tripled, leaving fathers in two-parent households more engaged in child care than at any time for which we have data.
- Comparative data from 20 industrialized countries reveal that men's share of housework and child care increased from less than one-fifth to more than one-third between 1965 and 2003.

As indicated in Chapter 9, there are differences between cohabiting couples and married couples. One such difference is that cohabiting couples have more equitable divisions of household labor than do

Data suggest that men's share of housework has increased both in the United States and in other industrialized societies.

married couples (Baxter 2005). Cohabiting women also do significantly less housework than married women do (Shelton and John 1993). It seems that marriage, rather than living with a man, transforms a woman into a homemaker (Baxter 2005). Marriage seems to change the house from a space to keep clean to a home to care for.

Much as cohabiting couples have more equitable divisions of housework than married couples, so too do same-sex couples display more sharing of tasks than do heterosexual couples. As addressed in Chapter 9, absent a gender difference, such couples must negotiate and construct their division of labor. Such active negotiation pushes them in the direction of more sharing (see Chapter 9).

Various factors seem to affect men's participation in housework. Men appear to share more when their female partners are employed more hours, earn more money, and have more years of education (Sullivan and Coltrane 2008). Earnings appear to be an especially important factor in the allocation of housework. Sociologist Sanjiv Gupta reports that how much women earn and not how their earnings compare to those of their partners is what affects how much housework they do. Using data from the National Survey of Families and Households, Gupta (2007) reports that every additional $7,500 in married women's incomes is associated with one fewer hours of household work.

Men tend to contribute more when their hours and their wives' hours at work do not overlap (see the discussion of shift work later in this chapter). As their income rises, wives report more participation by their husbands in household tasks; increased income and job status motivate women to try to ensure that their husbands share tasks. However, research by Julie Brines reviewed by Coltrane (2000) suggests that men who are economically dependent on their wives do less housework. Likewise, Barbara Arrighi and David Maume (2000) found that men whose wives earn the same or greater amounts of income may attempt to restore their masculinity by avoiding housework. Thus, the benefit of greater sharing of housework associated with women's earnings has its limits (Bittman et al. 2003; Tichenor 2005).

Other factors that appear to influence men's involvement in housework include the following:

- *Men's socialization experience and modeling of parents.* Although it does not seem to influence women's participation in those same tasks, early parental division of labor acts as a strong predictor of men's involvement in the "female tasks" of housework (Cunningham 2001).
- *Men's status in the workplace.* Men who have their "masculinity challenged" at work may reduce their involvement in housework as a way of avoiding feminine behavior (Arrighi and Maume 2000).
- *Men's age and generation.* Older men do less housework than younger men do. Arrighi and Maume (2000) speculate that this may be a reflection of generational change, with younger men having been socialized toward more participation than older men were.

Whether a couple has children is a factor affecting men's and women's participation in household labor. Even though the presence of young children increases women's and men's housework, it also skews the division of housework in even more traditional directions. Men tend to work more hours in their paid jobs, and women tend to work fewer hours at paid work and more in the home. Women then end up with a larger share of housework than before the arrival of children.

As noted earlier, one factor that may not be as strong a determinant as one might predict is the husband's gender ideology—what he believes he ought to do as a husband and how paid and unpaid work should be divided. As Arlie Hochschild's (1989) research showed, even traditional men can become more egalitarian if wives successfully use direct and indirect gender strategies. In some instances, repeated requests might be enough. In other cases, ultimatums may be necessary. Aside from these direct strategies, more indirect strategies—helplessness, withholding sexual intimacy, and so on—may work with husbands who otherwise would not do more.

Furthermore, necessity may create more male involvement. Wives with particularly demanding jobs or who work unusual hours (described later) may force their husbands to share household work more, simply because they are not available (Gerson 1993; Rubin 1994). Women appeared to be more satisfied if their husbands shared traditional women's chores (such as laundry) rather than limiting their participation to traditional male tasks (such as mowing the lawn). African Americans are less likely to divide household tasks along gender lines than Caucasians.

We might assume that the stresses and inequalities of juggling paid work and domestic work among employed mothers undermine women's well-being, but research on consequences related to marital, mental health, and physical health tells a much different story. Analyzing more than a quarter century of General Social Survey data, sociologist Jason Schnittker finds that "women who are employed, regardless of the number of hours they work or how they combine work with family obligations, report better health than do those who are unemployed" (American Sociological Association 2004).

Women in dual-earner families appear to be mentally healthier than full-time stay-at-home wives are (Crosby 1991). In juggling multiple roles, they suffer less depression, experience more variety, interact with a wider social circle, and have less dependency on their marital or familial roles to provide all their needed gratification. These psychological benefits accrue despite the unequal division of labor.

Emotion Work

Although we might not typically think about them as "work" or include them in a discussion of "family work," there are other tasks that need to be performed to generate and maintain successful and satisfying relationships and families. Such tasks are often referred to as **emotion work** and include the following (Stevens, Kiger, and Riley 2001):

- Confiding innermost feelings
- Trying to bring our partner out of a bad mood
- Praising our partner
- Suggesting solutions to relationship problems
- Raising relationship problems for consideration and discussion

- Taking initiative to begin the process of "talking things over"
- Monitoring the relationship and sensing when our partner is disturbed about something

Although these might not cleanly fit many people's notions of "tasks," they may be experienced as work by those who feel unevenly burdened by them. According to research by Stevens et al. (2001), women do more of the emotion work in their relationships and report being less than satisfied with how these "responsibilities" are divided. This has important consequences because both men's and women's satisfaction with the division of emotion work in their relationships was significantly and positively associated with their marital satisfaction (Stevens et al. 2001).

Caring for Children

As noted in the previous chapter, men increasingly believe that they should be more involved as fathers than men have been in the past, yet this shift in attitudes has not been matched by equivalent changes in men's caregiving behavior. Child care responsibility varies according to the marital status of parents and their employment roles and schedules. In a two-parent family, care for children is more the responsibility of mothers than fathers (Yeung et al. 2001). When we examine data on actual involvement in tasks associated with child care or time spent with children, mothers are more involved in such tasks than fathers. In making such comparisons, it is helpful to differentiate between **engagement** with children—time spent in *direct interaction* with a child across any number of different activities—and **accessibility**—or *availability to a child*, when the parent is at the same location but not in direct interaction (Yeung et al. 2001). Even though fathers' proportional involvement with children has increased, it is estimated that fathers' engagement with children is less than 45% that of mothers' and that their accessibility to children is less than 66% that of mothers' (Yeung et al. 2001).

Generalizing from other research on parental involvement in two-parent families, we find the following:

- Mothers spend from three to five hours of active involvement for every hour fathers spend, depending on whether the women are employed.
- Mothers' involvement is oriented toward practical daily activities, such as feeding, bathing, and dressing. Fathers' time is generally spent in play.

- Mothers are almost entirely responsible for child care: planning, organizing, scheduling, supervising, and delegating.
- Women are the primary caretakers; men are the secondary.

Circumstances affect how much time fathers spend with children. One study, based on analyses of data from the Panel Study of Income Dynamics (Yeung et al. 2001), noted that in two-parent homes a child's direct engagement with biological fathers ranged from a daily average of 1 hour and 13 minutes on weekdays to 2 hours and 29 minutes on weekends. The total time (engagement plus accessibility) these fathers are involved with their children 12 years and younger is roughly 2.5 hours a day on weekdays and 6.5 hours a day on weekends.

Active Child Care

Active, or "hands-on," child care is more "in the hands" of mothers than fathers. Mothers take care of and think about their children more than fathers do (Walzer 1998). In most two-parent households, mothers' child care responsibility and involvement greatly exceed fathers' involvement (Aldous and Mulligan 2002; Bird 1997).

What do mothers and fathers do in the time they spend with children? Research from the 1960s through the 1980s suggested that fathers spend more time in interactive activities, such as play or helping with homework, whereas mothers spend time doing custodial child care, such as feeding and cleaning (Yeung et al. 2001).

In addition, fathers are more involved with sons than with daughters, with younger children more than with older children, and with firstborn more than with later-born children (Pleck 1997, cited in Doherty, Kouneski, and Erickson 1998). Fathers are engaged with or accessible to their infants and toddlers an average of a little over three hours per day during the week. By ages 9 to 12, the combined (engagement and accessibility) weekday time between fathers and children declined to 2 hours and 15 minutes (Yeung et al. 2001). Fathers spend 18 minutes more per day in play and companionship activities with sons than with daughters during the week.

Research suggests that fathers who work more hours and who have prestigious, time-demanding occupations tend to be less engaged in child rearing (NICHD Early Child Care Research Network 2000). On weekends, fathers become somewhat more equal caregivers, and their involvement is greater when mothers contribute a "substantial" portion of the family income (Yeung et al. 2001). Although fathers "help" mothers with the caregiving work and supervision involved in raising teenaged children, most fathers do less of such work than most mothers (Kurz 2002).

Mental Child Care

Responsibility for child care doesn't consist only of what we *do* with and for our children. In her book *Thinking about the Baby: Gender and Transitions into Parenthood*, sociologist Susan Walzer (1998) examines the division of responsibility for infants in 25 two-parent households. Her focus was less on "who did what" with their children than on "who thought what and how often" about their children. Walzer identifies this "invisible" parenting as **mental labor**—the process of worrying about the baby, seeking and processing information about infants and their needs, and managing the division of infant care in the household (i.e., seeking the "assistance" of their spouse). It might be thought of as similar to the "emotion work" that surrounds marriage.

Sociologist Demie Kurz reports similar kinds of mental labor among mothers of teenaged children. Fearful for their adolescents' safety and especially fearful about the sexual vulnerability of their teenaged daughters, mothers worry (Kurz 2002). Thus, mothers continue to worry as children grow. One key to understanding this mental labor at both the earliest and the late adolescent or young adult stages is that mothers feel responsible for and judged by what happens to their children in ways that most fathers do not.

How the Division of Household Labor Affects Couples

Aside from the obvious matters of unequal responsibilities and differences in the numbers of tasks and roles that spouses and partners play, how couples divide household labor affects their relationship. The balance and distribution of power, the amount of satisfaction partners feel with their relationship, their levels of sexual intimacy, and the likelihood of their remaining together are all influenced by how they divide and allocate household labor.

Marital Power

An important consequence of women's employment is a shift in the decision-making patterns in a marriage. As noted earlier in Chapter 7, decision-making power

in a family is not based solely on economic resources, but economics influence how couples make decisions and divide responsibilities. A number of studies suggest that employed wives exert greater power in the home than that exerted by nonemployed wives (Blair 1991; Schwartz 1994). Marital decision-making power is greater among women employed full time than among those employed part time. Conversely, full-time stay-at-home wives may find themselves taken for granted and, because of their economic dependency on their husbands, relatively powerless (Schwartz 1994).

Some researchers are puzzled about why many employed wives, if they do have more power, do not demand greater participation in household work on the part of their husbands. Joseph Pleck (1985) suggested several reasons for women's apparent reluctance to insist on their husbands' equal participation in housework. These include (1) cultural norms that housework is the woman's responsibility, (2) fears that demands for increased participation will lead to conflict, and (3) the belief that husbands are not competent.

Satisfaction, Sex, and Stability

How do patterns of employment and the division of family work affect the quality of marital relationships? Does the division of labor affect a couple's level of marital satisfaction? Oriel Sullivan and Scott Coltrane (2008) point out that when men share more of the housework, women's marital satisfaction increases and the frequency of marital conflict decreases. In 2008, a number of media sources publicized a report by psychologist Joshua Coleman of the Council on Contemporary Families suggesting a connection between the amount of housework men do and the amount of sex in marriage. Coleman asserts that women are more prone to depression, more likely to contemplate divorce, and less interested in sexual intimacy with their husbands when they do a disproportionate amount of the housework. He contends that wives are "more sexually interested" in husbands who do more housework (Council on Contemporary Families, www.contemporaryfamilies.org).

How does a woman's employment affect marital satisfaction? There does not seem to be any straightforward answer when comparing dual-earner and single-earner families (Piotrkowski et al. 1987). This may be partly because there are trade-offs: a woman's income allows a family a higher standard of living, which compensates for the lack of status a man may

Discrimination against Women

A woman's earnings significantly affect family well-being regardless of whether the woman is the primary or secondary contributor to a dual-earner family or the sole provider in a single-parent family. Furthermore, as we have seen, women's family responsibilities significantly affect their earnings. Given the importance, however, of women's wage contributions to their families, we need to consider at least briefly economic discrimination against women and sexual harassment. By affecting women's employment status and experiences in their jobs, these become important family issues as well as economic ones.

Economic Discrimination

The effects of economic discrimination can be devastating for women. Data from the second quarter of 2009 showed that women who worked full time had median weekly earnings of $652, 80% of men's median of $815 per week. This overall gap varies by race and ethnicity, as shown in Table 11.2.

The differences in women's and men's incomes translate into more women than men experiencing poverty and in need of state or federal assistance. Wage differentials are especially important to single women.

Although employment and pay discrimination are prohibited by Title VII of the 1964 Civil Rights Act, the law did not end the pay discrepancy between men and women. Much of the earnings gap is the result of occupational differences, gender segregation, and women's tendency to interrupt their employment for family reasons and to take jobs that do not interfere extensively with their family lives. Earnings are about 30% to 50% higher in traditionally male occupations, such as truck driver or corporate executive, than in predominantly female or sexually integrated occupations, such as administrative assistant or schoolteacher. The more an occupation is dominated by women, the less it pays.

Sexual Harassment

Sexual harassment is a mixture of sex and power, with power often functioning as the dominant element. **Sexual harassment** can be defined as two distinct types of harassment: (1) the abuse of power for sexual ends and (2) the creation of a hostile environment. In **abuse of power**, sexual harassment consists of unwelcome sexual advances, requests for sexual favors, or other verbal or physical conduct of a sexual nature as a condition of instruction or employment. Only a person with power over another can commit the first kind of harassment. In **hostile environment**, someone acts in sexual ways to interfere with a person's performance by creating a hostile or offensive learning or work environment. Sexual harassment is illegal.

In fiscal year 2008, 13,867 reports of sexual harassment were filed with the Equal Employment Opportunity Commission (EEOC). This was the largest number of complaints since 2002 and followed six years of steadily declining numbers of complaints. Of these charges, 84% were brought by women. These numbers need to be approached with some caution. In a poll undertaken by Louis Harris and Associates, 31% of women replied that they had experienced sexual harassment at work, yet 62% of those who had been harassed took no action regarding their experience. Thus, there are likely more—perhaps considerably more—women harassed than the EEOC data would indicate.

Some estimate that as many as half of employed women are harassed during their working years. Experiencing sexual harassment can lead to a variety of serious consequences. Some people quit their jobs, and others may be dismissed as part of their harassment. Victims also often report feeling depressed, anxious, ashamed, humiliated, and angry (Paludi 1990).

The Need for Adequate Child Care

As we saw in the previous chapter, even though mothers continue to enter the workforce in ever-increasing numbers, high-quality, affordable child care remains an important but uncertain support. For many women, especially for those with younger children and for single mothers, the availability of child care is critical to their employment. With nearly 60% of married mothers of children under the age of six and 64% of single

Table 11.2 Median Weekly Earnings, Full-Time Employees by Gender and Race

Group	Male	Female	Female as % of Male
Total	$815	$652	80%
White	$842	$666	79.1%
African American	$620	$567	91.5%
Asian American	$969	$781	80.6%
Hispanic	$575	$511	88.9%

mothers employed, it is unsurprising that the demand for affordable and trusted child care is high. In 2008, over half of all U.S. mothers worked full time, and an additional 16% worked part time. Less than 30% were neither employed nor looking for work ("Women in the Recession," 2009).

Approximately two-thirds of children under age five are in some form of nonparental child care regularly (Morrissey 2008). Among preschoolers with employed mothers, the percentage regularly in some form of child care jumps to 89%. For most employed mothers with children 5 to 14 years old, school attendance is their primary day care solution. Women with preschool children, however, do not have that option. Instead, they tend to use in-home care by their child's other parent, grandparents or other relatives, day care centers, family day care (care provided in someone else's home), and preschools or nursery schools as their most important resources. Perhaps as many as two-fifths of preschool-age children experience on a regular basis a combination of multiple arrangements. Psychologist Taryn Morrissey points out that the decision to use multiple arrangements may be made by choice—so as to expose children to a variety of situations and people and enrich them in the process—or by necessity because of practical constraints, such as work schedules and limited availability of options. Morrissey also notes that at least some research casts the use of multiple arrangements in a somewhat negative light in terms of its impact on children's social adjustment and the stress it induces for parents who have to coordinate schedules and arrange for transportation from one care arrangement to another. This, in turn, can affect parents' performance at work (Morrissey 2008).

Frustration is one of the most common experiences in finding or maintaining day care. Changing family situations, such as unemployed fathers' finding work or grandparents' becoming ill or overburdened, may lead to these relatives being unable to care for the children. Family day care homes and child care centers may close because of low wages or lack of funding. Roughly 1 in 10 mothers with children under age 15 report experiencing some problem with child care, such as having to cut back or restrict their hours at work, having to change child care arrangements, or being placed on a waiting list to obtain child care. Studies show that younger parents, poor parents, and never-married parents experience more of these sorts of constraints, and nonrelative child care has been found to produce more child care problems (Laughlin, www.census.gov/population/www/socdemo/child/child-care-constraints.pdf).

Furthermore, child care is expensive. Data on the costs of child care reveal that whether in a center-based or an in-home (family) child care setting, child care is costly. Costs range throughout the United States, depending on where one lives and the type of care one uses. Full-time infant care in an in-home family setting ranges from an annual average of $3,582 in South Carolina to a high cost of $10,324 in Massachusetts. Center-based infant care ranges from a low average yearly cost of $4,560 in Mississippi to a high of $15,895 in Massachusetts. In 37 states, the average yearly cost of full-time, center-based care for an infant is more than $7,000. Costs are less for older preschoolers but again cover a wide range. The average cost of full-time in-home family child care for a four-year-old ranges from $3,380 per year in Mississippi to $9,805 in Massachusetts. Average costs of center-based care ranges between the same two states, from $4,056 in Mississippi to $11,678 in Massachusetts (www.naccrra.org/randd/docs/2008_Price_of_Child_Care.pdf).

Of course, many employed parents have more than one child who requires care. The estimated costs of care for two preschool-age children range between $6,760 and $19,610 for in-home family child care and between $8,112 and $23,356 for center-based care. In the South, the Northeast, and the West, the costs associated with caring for two children are exceeded only by the median cost of monthly mortgage payments, while in the Midwest the costs for child care for an infant and a four-year-old would make child care the most costly monthly expense for families. In all regions, the average cost of child care for an infant is greater than what families spend each month on food (National Association of Child Care Resource and Referral Agencies 2009).

Despite the clear need for more affordable, dependable, and accessible quality child care, the United States compares poorly to many European countries. Take, for example, France, where child care is publicly funded as part of early education (Clawson and Gerstel 2002). Nearly all three- to five-year-olds are enrolled in full-day programs taught by well-paid teachers.

Most experts agree that the ideal environment for raising a child is in the home with the parents and family. Intimate daily parental care of infants for the first several months to a year is particularly important. Because this ideal is often not possible, the role of day care needs to be considered. Day care homes and centers, nursery schools, and preschools can relieve parents of some of their child-rearing tasks and furnish them with some valuable time of their own. Among children in nonrelative care, about 7% are looked after in their own homes. Family day care enrolls about 27% and day care centers about 66% (National Household Education Survey 2001, cited in Wrigley and Dreby 2005).

What is the effect of early outside child care on children? The results of research are mixed. In evaluating such data, it is important to keep in mind the family's education, the personalities involved, and the family interests—key factors that play a part in which parents choose to return to work and which must return to work once a child is born (Crouter and McHale 1993). Furthermore, a child's personality, the child's age when the custodial parent reentered the workforce, the involvement of the other parent in the home, the quantity of time spent working or with the child, the nature of the work, and the quality of care all contribute to how child care affects the child.

Evidence is mixed as to the effects of maternal employment on mother–child attachment. Although earlier research suggested that infants of employed mothers developed an insecure attachment to their mothers, later research indicates that instead of whether the mother is employed—and therefore away more hours from her infant—it is her sensitivity to her infant's needs when together that affects mother–infant attachment (Olendick and Schroeder 2003). Other subsequent consequences of child care—such as behavior problems, lowered cognitive performance, distractibility, and inability to focus attention—have been noted. These negative effects are not necessarily the consequence of being cared for by outside caregivers. Rather, they may be the result of *poor-quality* child care. It has been noted that high-quality care, given by sensitive, responsive, and stimulating caregivers in a safe environment with a low teacher-to-student ratio, can actually facilitate the development of positive social qualities, consideration, and independence (Field 1991). In school-age and adolescent children, maternal employment is associated with self-confidence and independence, especially for girls whose mothers become role models of competence (Hoffman 1979).

National concern periodically is focused on day care by revelations of sexual abuse of children by their caregivers or by stories of children being injured or killed while in nonparental care. However, children have a far greater likelihood of being sexually abused by a father, stepfather, or other relative than by a day care worker. Furthermore, parents with children in child care should take some degree of comfort from the evidence demonstrating that children in outside, especially in organized, child care facilities are safe. Overall, all types of child care are safer than care within children's own families (Finkelhor and Ormrod 2001; Wrigley and Dreby 2005).

Older Children, School-Age Child Care, and Self-Care

Although there are particularly acute needs when children are young, employed parents' child care needs are not restricted to families of preschoolers; parents of children in middle school also have child care needs. A number of terms are used to refer to caregiving for these older children, including *after-school*, *around-school*, *out-of-school*, and *school-age care* (Polatnik 2002).

Many children express strong opposition to after-school programs, seeing them as geared toward "little kids," but they find activities such as sports or other recreational or artistic programs more appealing (Polatnik 2002). Unfortunately, even when the programs and activities are free or when the costs are affordable, they are neither consistent nor continuous enough to cover the whole time children are out of school before parents return from work. Many parents of these children feel pressed to allow them to stay home alone. Research indicates that approximately a third of 11- to 12-year-olds are in **self-care**—that is, care for themselves without supervision by an adult or older adolescent (Casper and Smith 2002; Hochschild 1997; Polatnik 2002).

U.S. Census Bureau data on child care arrangements reveal that self-care is very rare for young children (five to eight years old). Only around 2% of five- to eight-year-old children are in self-care, regardless of race, mother's education, mother's employment status, or household income. When one looks at older children, one sees that 10% of 9- to 11-year-olds and 33% of 12- to 14-year-olds were in self-care during the spring of 2005. Census data also indicate that whites, those

The shortage of affordable, high-quality child care is an obstacle for many employed parents.

- It keeps many women in jobs for which they are overqualified and prevents them from seeking or taking job promotions or training necessary for advancement.
- It sometimes conflicts with women's ability to perform their work.
- It restricts women from participating in education programs.

For women, lack or inadequacy of child care is a major barrier to equal employment opportunity. Many women who want and need to work are unable to find adequate child care or to afford it. Child care issues may also play a significant role in women's choices concerning work schedules, especially among women who work part time.

with some college, and those with the highest incomes are most likely to have had their children in self-care during the spring of 2005 (U.S. Census Bureau, *Who's Minding the Kids? Child Care Arrangements, Spring 2005*, table 4). The Census Bureau also looked at child care arrangements during the summer of 2006. Without the availability of school as a means of child care, the percentages in self-care increased among all groups, for children five to seven and 8 to 11 years of age. Among 12- to 14-year-olds, the percentages in self-care were lower in the summer than was the case in the spring 2005 data (U.S. Census Bureau, *Who's Minding the Kids? Child Care Arrangements, Summer 2006*, table 4).

As with a number of critical services in our society, those who most need supplementary child care are those who can least afford it. The United States is one of the few industrialized nations without a comprehensive national day care policy. In fact, beginning in 1981, the federal government dramatically cut federal contributions to day care; many state governments followed suit.

Effect on Employment and Educational Opportunities

The lack of child care or inadequate child care has the following consequences:

- It prevents many mothers from taking paid jobs.
- It keeps many women in part-time jobs, most often with low pay, few or no benefits, and little career mobility.

Inflexible Work Environments, Stressful Households, and the Time Bind

In dual-worker families, the effects of the work environment stem from not only one workplace but two. Although many companies and unions are developing programs that are more responsive to family situations, the workplace in general has failed to recognize that the family has been radically altered during the past 50 years. Most businesses are run as if every worker were male with a full-time wife at home to attend to his needs and those of his children or wealthy enough to have paid domestic help to fill the needs of their families. But the reality is that women make up a significant part of the workforce, and they do not have wives at home. Nor do the vast majority of Americans have nannies for their children or domestic help to assist them in their households. Allowances are not made in the American workplace for flexibility in work schedules, day care, emergency time off to look after sick children, and so on. Many parents would reduce their work schedules to minimize work–family conflict. Unfortunately, many do not have that option.

More than 20 years ago, Carol Mertensmeyer and Marilyn Coleman (1987) contended that our society provides little evidence that it esteems parenting. It appears that little has changed. This seems to be especially true in the workplace, where corporate needs are placed high above family needs. Mertensmeyer and Coleman suggested that family policymakers encourage employers to be more responsive in providing parents with alternatives that alleviate forced choices that are incongruent with parents' values. For example, corporate-sponsored child care may offset the conflict a mother feels because she is not at home with her child. Flextime and paid maternal and paternal leaves are additional benefits that employers could provide employees. These benefits would help parents fulfill self- and family expectations and would give parents evidence that our nation views parenting as a valuable role.

Unfortunately, the presence of policies cannot guarantee that employees will use them. In her book *The Time Bind: When Work Becomes Home and Home Becomes Work*, sociologist Arlie Hochschild (1997) described the official policies and corporate culture at a large corporation that she called Amerco (so as to protect its anonymity). At Amerco, workers had access to a number of family-friendly time-enhancing policies, including job sharing, part-time work, parental leave, flextime, and "flexplace," now more widely referred to as "telecommuting" (where workers could work from home). Despite the apparent availability of such opportunities, Hochschild notes that Amerco employees rarely made use of them.

Hochschild noted that Amerco employees are typical of employees at other large corporations. Fortune 500 manufacturing companies tended to offer family-friendly policies, yet few employees used them. This lack of use was especially puzzling given that Amerco employees acknowledged not having enough family time.

In accounting for the lack of utilization of workplace policies, Hochschild considered a variety of explanations. Can employees afford to work fewer hours? Do they fear being laid off? Do employees even know about policies? Do they have insensitive and insincere supervisors?

These explanations had partial validity. Some hourly employees feared potential layoffs or reduced wages. There were some supervisors who seemed reluctant to embrace and resentful at having to accommodate family-friendly policies. But Hochschild contended that the biggest reason employees made little use of potential family time-enhancing initiatives is because they would rather be at work.

Hochschild maintained that in recent decades, with the dramatic changes in the division of labor and the growth of dual-earner families, home life has become more stressful and tightly scheduled. There is too much to do, too little time to do it, and not enough appreciation or recognition for what is done. On the other end of the work–family divide, many workplaces in the United States have implemented "humanistic management" policies designed to enhance worker morale and productivity and to reduce turnover. Thus, at work, people can find social support, appreciation, and a sense of control and competence, which makes them feel better about themselves. In other words, for some, work has become homelike, and home often feels like a job (Hochschild 1997).

Because Hochschild (1997) studied only one company, we can't generalize just from her research. In fact, other researchers have failed to support Hochschild's conclusions, at least to the same extent. For example, a study by Susan Brown and Alan Booth (2002b) that used the National Survey of Families and Households and is based on more than 1,500 dual-earner couples with children indicates that Hochschild's findings may not be generalizable.

Job status seems to be an important determinant of whether individuals see their jobs as more satisfying than their home lives. Brown and Booth claim that this is true only among workers in positions of lower occupational status. In addition, respondents who have high satisfaction with work and low satisfaction at home do not necessarily work significantly more hours at work. Only those who are satisfied with work, unsatisfied with home, and have adolescent children work more hours (Brown and Booth 2002b).

Another study by K. Jill Kiecolt (2003), based on General Social Survey data from 1973 to 1994, challenged several of Hochschild's (1997) conclusions. She argued, for example, that a "cultural reversal" in favor of work over home had not taken place and that employed parents with children under age six actually are more likely to find home rather than work to be a haven.

Even if Hochschild's findings are somewhat limited, her study is important for showing that the presence of policies does not guarantee their use (see also Blair-Loy and Wharton 2002). People must take advantage of policies. This suggests that people's values must be directed more toward home and family. Furthermore, cultural support for using family-friendly policies must be more widespread and reflected in *company* "cultures." If the message one learns at work is that one's commitment to one's job is being measured by how much time they put in, then reducing work time for family needs makes one appear undercommitted.

Change must also occur at home. Dual-earner family life must be made less stressful. One way in which this can occur is by men doing more of the second-shift work discussed earlier, thereby reducing the overload and time drain that their wives more consistently feel.

Employees who feel supported by their employer with respect to their family responsibilities are less likely to experience work–family conflict. A model corporation would provide *and support* the use of family-oriented policies that would benefit both its employees and itself, such as flexible work schedules, job-sharing alternatives, extended maternity and/or paternity leaves and benefits, and child care programs or subsidies. Such policies could increase employee satisfaction, morale, and commitment. Later in this chapter, we discuss policies such as these, alongside personal strategies that individuals and couples might chose to use.

Living without Work: Unemployment and Families

Since 2007, the U.S. economy has suffered, ultimately entering a recession in 2008. One consequence of these difficult circumstances is a dramatic increase in unemployment. In June 2007, the U.S. unemployment rate stood at 4.5%. Seven million Americans were counted as unemployed. By June 2009, the unemployment rate had more than doubled, reaching 9.5%. More than 14.5 million people were unemployed, and 5 million of them were considered "long-term unemployed," having been jobless for 27 or more weeks.

The real employment situation was even worse than the official numbers. To be counted among the "unemployed," one must have looked for a job within the four weeks prior to being surveyed. As a consequence of this measurement criterion, 2.3 million Americans who were not in the labor force, who wanted and were available for work, and who had looked for a job within the prior 12 months but not within the most recent four-week period were not included in the count of the unemployed. Instead, they were classified as "marginally attached" to the labor force.

Unemployment rates vary across lines of age, gender, race and ethnicity. Among adult women and men, the 2009 rate was 9.4. Among teenagers (ages 16–19), the rate was 23.8%. Although the 2007 unemployment rates of women (4.0%) and men (4.1%) were nearly the same, the recession has hit men's employment harder than it has women's. Almost 10% of men were officially unemployed in June 2009, compared to 7.5% of women. The biggest variations are ethnic and racial. Although 8.6% of whites were unemployed in June 2009, 12.3% of Hispanics and 14.5% of African Americans were unemployed (see Figure 11.5).

Unemployment is a major source of stress for individuals, with its consequences spilling over into their families (Cottle 2001). The number of families with an unemployed member rose from 4.9 million (6.3%) in 2007 to 6.1 million (7.8%) in 2008, with African Americans (12.8%) and Hispanics (11.0%) being more likely than whites (7.1%) or Asians (6.3%) to have an unemployed family member. Even employed workers suffer anxiety about possible job loss caused by economic restructuring and downsizing (Larson, Wilson, and Beley 1994). Job insecurity leads to uncertainty that affects the well-being of both worker and spouse. They feel anxious,

Figure 11.5 U.S. Unemployment, June 2007 and July 2009

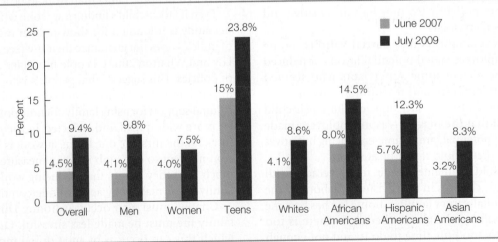

SOURCE: Bureau of Labor Statistics (2009).

depressed, and unappreciated. For some, the uncertainty before losing a job causes as much or even more emotional and physical upset than the actual job loss.

Economic Distress

Those aspects of economic life that are potential sources of stress for individuals and families make up **economic distress** (Voydanoff 1991). Major economic sources of stress include unemployment, poverty, and economic strain (such as financial concerns and worry, adjustments to changes in income, and feelings of economic insecurity).

In times of hardship, economic strain increases; the rates of infant mortality, alcoholism, family abuse, homicide, suicide, and admissions to psychiatric institutions and prisons also sharply increase. Patricia Voydanoff (1991), one of the leading researchers in family–economy interactions, notes the following:

> A minimum level of income and employment stability is necessary for family stability and cohesion. Without it, many are unable to form families through marriage and others find themselves subject to separation and divorce. In addition, those experiencing unemployment or income loss make other adjustments in family composition such as postponing childbearing, moving in with relatives, and having relatives or boarders join the household.

The emotional and financial cost of unemployment to workers and their families is high. A common public policy assumption, however, is that unemployment is primarily an economic problem. Joblessness also seriously affects health and the family's well-being.

The families of the unemployed experience considerably more stress than that experienced by those of the employed. The economic strain due to unemployment is related to lower levels of marital satisfaction as a result of financial conflict, the husband's psychological instability, and marital tensions. Mood and behavior changes cause stress and strain in family relations. As families adapt to unemployment, family roles and routines change. The family spends more time together, but wives often complain of their husbands' "getting in the way" and not contributing to household tasks. Wives may assume a greater role in family finances by seeking employment if they are not already employed. After the first few months of their husbands' unemployment, wives of the unemployed begin to feel emotional strain, depression, anxiety, and sensitiveness in marital interactions. Children of the unemployed are likely to avoid social interactions and tend to be distrustful; they report more problems at home than do children in families with employed fathers. Families seem to achieve stable but sometimes dysfunctional patterns around new roles and responsibilities after six or seven months. If unemployment persists beyond a year, dysfunctional families become highly vulnerable to marital separation and divorce; family violence may begin or increase at this time (Teachman, Call, and Carver 1994).

The types of families hardest hit by unemployment are single-parent families headed by women, African American and Latino families, and young families. Wage earners in African American, Latino, and female-headed, single-parent families tend to remain unemployed longer than other types of families. Because of discrimination and the resultant poverty, they may not have important education and employment skills. Young families with preschool children often lack the seniority, experience, and skills to regain employment quickly. Therefore, the largest toll in an economic downturn is paid by families in the early years of childbearing and child rearing.

Emotional Distress

Aside from the obvious economic effect of unemployment, job loss can have profound effects on how family members see each other and themselves (see the Real Families feature in this chapter). This in turn can alter the emotional climate of the family as much as lost wages alter the material conditions. Men are particularly affected by unemployment because wage earning is still a major way men satisfy their family responsibilities. Thus, when men fail as workers, they may feel they failed as husbands, fathers, and men (Cottle 2001). As Lillian Rubin (1994, 103) poignantly conveyed in her book *Families on the Fault Line*, when men lose their jobs, "it's like you lose a part of yourself." Psychologist Thomas Cottle (2001, 21) suggests that a man's unemployment can strip "away his pride, his self-definition, his sense of personal power, and his belief in his ability to control outcomes, not the least of which is his own personal destiny." Unemployed men may display a variety of psychological and relationship consequences, including emotional withdrawal, spousal abuse, marital distress, increased alcohol intake, and diminished self-identity (Cottle 2001; Rubin 1994). Long-term unemployment can bring with it feelings of despair, desperation, confusion, impotency, and powerlessness. Katherine Newman

(1998) suggested that when families suffer downward mobility as a result of male unemployment, relations between spouses or between fathers and children are likely to be strained. Although children and spouses may be initially supportive, their support may wear thin or run out if joblessness persists and other resources are unavailable, thus preventing families from maintaining their previous economic lifestyle.

Of course, women, too, suffer nonmaterial losses when they lose their jobs, but those losses are different in degree and kind from those that men are likely to suffer. Men have more of their self-identities—and especially their gendered identities—tied up in working; success at work comes to define successful masculinity and symbolizes having satisfied a central element of one's roles as husbands and fathers (Arrighi and Maume 2000). Women have other acceptable ways of maintaining or achieving adult status

Unemployed men may suffer losses in self-esteem, and feel as though they are failing in their roles as spouses and parents.

© Thinkstock/Getty Images

(e.g., as mothers). Thus, although both women and men will suffer from lost work relationships, lost gratification, and even lost structure and purpose to their day, women have not put as many of their "identity eggs" into the "work basket" as have most men.

Coping with Unemployment

Economic distress does not necessarily lead to family disruption. In the face of unemployment, some families experience increased closeness (Gnezda 1984). Families with serious problems, however, may disintegrate. Individuals and families use a number of coping resources and behaviors to deal with economic distress. Coping resources include an individual's psychological disposition, such as optimism; a strong sense of self-esteem; and a feeling of mastery. Family coping resources include a family system that encourages adaptation and cohesion in the face of problems and flexible family roles that encourage problem solving. In addition, social networks of friends and family may provide important support, such as financial assistance, understanding, and willingness to listen.

Several important coping behaviors assist families in economic distress caused by unemployment. These include the following:

- *Defining the meaning of the problem.* Unemployment means not only joblessness but also diminished self-esteem if the person feels that the job loss was his or her fault. If a worker is unemployed because of layoffs or plant closings, the individual and family need to define the unemployment in terms of market failure, not personal failure. During the period of time between December 2007 and June 2009, some 4.1 million people filed unemployment claims as a result of the nearly 40,000 mass layoffs that occurred (Bureau of Labor Statistics, 2009).
- *Problem solving.* An unemployed person needs to attack the problem by beginning the search for another job, dealing with the consequences of unemployment (e.g., by seeking unemployment benefits and cutting expenses), or improving the situation (e.g., by changing occupations or seeking job training or more schooling). Spouses and adolescents can assist by increasing their paid work efforts.
- *Managing emotions.* Individuals and families need to understand that stress may create roller-coaster emotions, anger, self-pity, and depression. Family members need to talk with one another about their feelings; they need to support and encourage

Real Families: A Whole Street Full of Out-of-Work Dads

It has been called the "man-cession" to highlight the fact that the U.S. recession has taken a greater toll on men's employment than on women's employment. Economist Heather Boushey points out that men have experienced three out of every four job losses and that, as of July 2009, 15.6% of employed married women had husbands who were not employed. Among employed wives with children under age six, 5.9% had unemployed husbands. Given that, on average, women in two-earner households bring in 36% of the household income, losing the man's income can drive families into dire economic straits. Worsening the problem is that in two-earner households, health insurance is usually obtained through the man's job (Boushey 2009). Thus, when men lose their jobs, the economic impact can be severe, but there are also other problems that arise.

New York Times journalist Jennifer Steinhauer describes 41-year-old Phil Winkler's transition from "primary bread-winner" to family "bus driver, disciplinarian, schedule organizer and head chef" (Steinhauer 2009). A former production worker for Nestlé, he had never before experienced job loss. Now Winkler has been out of work for a year. He has sold his favorite car, canceled the family cable television package, and is now scavenging junkyards for auto parts he can resell on e-Bay. As a consequence of his job loss, Winkler has experienced wounds to his

self-esteem, felt strains in his marriage, and has had to impose sacrifices on his family. No summer swimming lessons for his daughters, no "shopping at random," no more eating out on Friday nights—these are just some of the adjustments his family has had to make.

They are not alone. In fact, as Steinhauer reports, in their neighborhood of Beth Court, a cul-de-sac in Moreno Valley, California, hardships such as these are part of life, as "job losses swept the neighborhood like an unstoppable illness." As he saw neighbor after neighbor lose their jobs, Mr. Winkler said that where once he'd seen men going off to work, now dads were picking kids up at school, dropping them at practice, overseeing homework, and depending on their wives' jobs and incomes as well as unemployment benefits.

There are some silver linings in this cloudy economic situation. Men who are unemployed are at home and can (theoretically) take on enlarged roles in the household and with their children. Boushey (2009) notes that as difficult as families may find it, today's couples are less likely locked into traditional notions of men as the family breadwinners. Thus, a man's unemployment won't necessarily trigger "life altering role reversal" (Goodman 2009). Some men seize the opportunity to become primary parent. Some, however, continue to desperately search for work, seeing it as more than just their economic contribution to their families.

one another. They also need to seek individual or family counseling services to cope with problems before they get out of hand.

Reducing Work–Family Conflict

Before ending this chapter, it is worthwhile to consider ways in which many of the problems resulting from the conflict between work and family might be

Critical Thinking

Have you or your family experienced unemployment or job insecurity? How did it affect you? Your family? What coping mechanisms did you use?

reduced. Although individuals and/or couples can craft personal strategies for managing their busy work and family lives—such as the opposite-shift arrangement discussed earlier—more external support, from employers and the government, is needed. Individuals alone cannot design their jobs in such a way as to best fit with their family needs.

Family policy is a set of objectives concerning family well-being and the specific government measures designed to achieve those objectives. Given the host of issues raised in this chapter, we might argue that if families were truly the national priority we claim them to be, we would entertain and enact policies to reduce the complexities associated with combining work and family life.

Of the following policies, some emphasize increasing people's time and availability to meet their family needs by making their work time or place flexible; others are designed to provide assistance in securing safe and affordable alternative, non-parental care arrangements. Still others are aimed at guaranteeing a certain economic standard and equal opportunity so that employed women and men and their families can meet their economic needs. Consider the following examples of different policies:

- *Policies designed to make work more flexible.* There are a number of different ways to increase flexibility in the scheduling, amount, or location of work. Most familiar to us is **flextime**, or flexible work schedules, where workers can choose and change their start and quit times to meet their personal and familial needs. A 2002 study by the National Study of the Changing Workforce found that workers with higher levels of flexibility were more likely to report *no* interference between job and family life than workers with low levels of flexibility (32% compared to 19%), while 50% of those with highly flexible work arrangements reported higher levels of life satisfaction than those with low levels of workplace flexibility (31%).

  At present, although nearly 80% of workers would like and would use more flexible options as long as they experienced no negative consequences in potential advancement, less than 30% of workers work flexible schedules. Forty percent believe that their chances of advancing would suffer if they used a flexible option.

  According to data reported by the Families and Work Institute, 2008 National Study of Employers (http://familiesandwork.org/site/research/reports/2008), 40% of small employers (50–99 employees) and 37% of large employers (more than 1,000 employees) allow employees to have flexible schedules, with 11% of small and 7% of large employers allowing employees to change their start and quit times on a daily basis.

  In addition to flexible daily work schedules, however, each of the following is also a means of increasing the flexibility of work, thus enabling workers to more easily fit work and home life together.

  Job sharing: 9% of small and 5% of large employers allow two employees to share a full-time job, usually splitting the benefits between the two.

Part-time work: 12% of both small and large employers allow employees to move from full time to part time and back while retaining their same positions.

Compressed workweeks: 10% of small and 5% of large employers allow employees to work shorter weeks by working longer days.

Telecommuting: Data on telecommuting are problematic because while many employees might work from home at least occasionally, fewer do so on a regularly scheduled and/or formally arranged basis.

- *Policies to provide, support, or assist working parents in locating alternative nonparental child care arrangements for children of employees.* Although fewer than 10% of employers with 50 or more employees provide on-site child care (Bond et al. 2005), there are other policies to assist employed parents in their search for and access to affordable and dependable child care. These include employer-provided child care resource and referral services as well as assistance or reimbursement programs for some of the expenses incurred for child care.
- *Family leave policies.* These are used for pregnancy, for childbirth, and/or to care for sick children or time off to meet other familial emergencies and paid personal days for parenting and other family responsibilities.
- *Policies to ensure that employees receive adequate wages to provide for their families and to protect them from discrimination (by gender, ethnicity, sexual orientation, family status, or disability) in hiring, advancement or pay.* Such initiatives include increasing the minimum wage so that workers can support their families, policies to ensure fair employment, and measures to ensure pay equity between men and women for the same or comparable jobs and affirmative action programs for women and ethnic groups.

Policies such as these must then be supplemented by sincere cultural support for families and children. People must believe that if they commit themselves to their families and take advantage of such measures, they will not suffer unfair economic consequences. This may be harder to convey and carry out than are most specific workplace policies.

Public Policies, Private Lives: The Family and Medical Leave Act

In the United States, the passage of the **Family and Medical Leave Act of 1993** (FMLA) provided employees with unpaid, job-protected leave of up to 12 weeks to care for an ill family member or to take time off after childbirth or adopting a child. The job protection was an important provision, as it meant that employees could not be let go just because they took time off at the birth or adoption of a child or for a family medical emergency. However, because the leave is unpaid, many workers cannot afford to lose the income they would have to sacrifice for three months and therefore are unable to use it. In addition, such leaves are guaranteed only if one works in a workplace with 50 or more employees. As a consequence, as many as half of U.S. workers are left unprotected by the policy. In addition, in order to be eligible for the FMLA guaranteed leave, one must have been employed for 12 months by the same company, during which time they must have worked 1,250 hours (Brown 2009).

The FMLA specifies only the minimum required for parental (and family) leaves. Roughly 16% of companies with at least 100 employees go beyond what is required by providing fully paid maternity leave. This is down from 1998, when 27% of companies provided fully paid leaves (Brown 2009). In 2004, California became the first state in the United States to enact its own paid family leave policy, providing up to six weeks of leave per year, during which

employees are eligible to earn 55% of their regular wages, up to $728 per week. California's leave policy does not impose a minimum workplace size criterion for eligibility as does the FMLA, nor are employees prevented from taking the state leave simultaneously with an FMLA leave, thus leading to 12 weeks of leave with 55% of one's salary for half the period that provides employees paid maternity leave (Gonzalez 2006).

On a global scale, the family leave policies in the United States compare unfavorably to the leave policies in most other developed nations. For example, some 168 out of 173 countries studied offer women guaranteed leave with pay associated with childbirth. Nearly 60% of these countries offer paid leaves of 14 or more weeks. For example, in Japan, women receive about three months of leave with 60% of their pay. In Germany and France, new mothers receive three years of paid leave. In Italy, this same three-year paid leave is followed by up to five years of unpaid leave (Henneck 2003).

In fact, only five countries offer new mothers no paid leave: the African countries of Swaziland, Liberia, and Lesotho; Papua New Guinea; and the United States. Sixty-five countries either offer men paid paternity leave or include men in their parental leave policies. For example, in Germany, the three years of paid leave can be alternated between mothers and fathers (Henneck 2003).

Our marriages and families are not simply emotional relationships—they are also work relationships in which we divide or share many household and child-rearing tasks, ranging from changing diapers, washing dishes, cooking, and fixing running toilets and leaky faucets to planning a budget and paying the monthly bills. These household tasks are critical to maintaining the well-being of our families. They are also unpaid and insufficiently honored. In addition to household work and child rearing, there is our employment, the work we do for pay. Our jobs usually take us out of our homes from 20 to 80 hours a week. They not only are a source of income, but they also help our self-esteem and provide status. They may be a source of work and family conflict as well.

Now a decade into the twenty-first century, we still need to rethink the relationship between our work and our families. Too often, household work, child rearing, and employment are sources of conflict within our relationships. We need to reevaluate how we divide household and child-rearing tasks so that our relationships reflect greater mutuality. For many, poverty and chronic unemployment lead to distressed and unhappy families. Because most individuals do not have direct control over the number of hours they spend at work or the timing of those hours, to reduce the extent of work–family conflict they need outside help from either the government or the private sector. Designing and supporting policies that help reduce the conflict between paid employment and families will, in return, build stronger families.

Summary

- Families are economic units bound together by emotional ties. Families are involved in two types of work: *paid work* at the workplace and unpaid *family work* in the household.

- At least until the recent recession, Americans were working more and facing family *time strains*. Fathers and mothers experience these strains differently.

- Feelings of being overworked lead to a variety of negative personal and occupational consequences.

- The *bifurcation of work time* refers to the fact that while many feel overworked, others are underemployed or lacking work.

- *Work-to-family spillover* is the effect that employment has on the time, energy, and psychological well-being of workers and their families at home. *Family-to-work spillover* is when the demands from home life reduce the time and energy available to succeed at work.

- *Role strain* refers to difficulties that individuals have in carrying out the multiple responsibilities attached to a particular role, *role overload* occurs when the total prescribed activities for one or more roles are greater than an individual can handle, and *role conflict* occurs when roles conflict with one another. *Crossover* refers to when one's feelings about work affect one's spouse's or partner's feelings.

- Evidence indicates that balancing work and family has become more difficult, especially for employed parents.

- The traditional division of familial labor is complementary: husbands work outside the home for wages, and wives work inside the home without wages. A minority of U.S. families display a traditional allocation of paid work and family roles.

- The "male as good provider" role made men's paid work their means of fulfilling their family responsibilities.

- There are four characteristics that define the *homemaker role*: (1) its exclusive allocation to women, (2) its association with economic dependence, (3) its status as nonwork, and (4) its priority over other roles for women.

- The level of women's participation in the paid labor force increased as a result of social and economic changes. Individual women enter the workforce for economic, social, and psychological reasons.

- Employed women tend to have better physical and emotional health than do nonemployed women.

- Recent public opinion data indicate that more women prefer employment than staying at home and that part-time employment is especially attractive to women with children.

- More than half of all married women are in dual-earner marriages. Husbands whose wives are employed perform a greater share of the housework than do husbands of at-home wives.

- Women perform a majority of the daily housework and carry more responsibility for managing the division of housework. Women's household tasks tend to include the daily chores (such as cooking, shopping, cleaning, and so on) and child care. Men's household tasks tend to be more occasional and often outdoors.

- Men's involvement in routine housework is less affected by their gender-role attitudes than by the more immediate circumstances in which they find themselves. Their involvement is also influenced by their upbringing, their experiences and status at work, and their age.

- A variety of indicators point to increases in men's share of housework and decreases in the amount of time women spend on housework.

- The division of paid and unpaid labor and the allocation of housework affect marital power, marital satisfaction, sexual intimacy, and marital stability (i.e., the risk of divorce).

- Fathers' involvement in caregiving is influenced by the age and gender of children, age and gender attitudes of fathers, and fathers' occupations and earnings.

- Mothers caregiving responsibility includes doing more of the *mental labor* of child care, including worrying about, gathering information, and managing fathers' involvement as well as monitoring where their children go, who they are with, and what they do.

- Supplementary child care outside the home is a necessity for many families. Most children who receive outside care are in child care centers. Overall, child care is safe, and center-based care is safer than "family day care" or paid care by others in the child's home.

- The effect of child care on children depends on the quality of care. The development and maintenance of quality day care programs should be a national priority.

- Increasing attention has been paid to school-age child care. Many communities provide after-school care through the schools. A common alternative to such care is *self-care*.

- Two contemporary arrangements are (1) *shift couples*, with spouses who work opposite shifts and alternate domestic and caregiver responsibilities, and (2) households in which men stay home with children while women support the family financially.

- Nonstandard shift work has increased because of changes in the economy, demographic changes, and technological changes. It affects family experiences in both negative and positive ways.

- There are approximately 1 million fathers of children under 15 years who stay home full time. In such households, we can identify marital, parental, economic, and social consequences that follow from this arrangement.

- Among the problems women encounter in the labor force are economic discrimination and *sexual harassment*. Families suffer from lack of adequate child care and an inflexible work environment.

- Unemployment can cause both economic and emotional distress. Unemployment most often affects female-headed single-parent families, African American and Latino families, and young families.

- Family policy is a set of objectives concerning family well-being and the specific government measures designed to achieve those objectives.

Key Terms

abuse of power 403

accessibility 395

bifurcation of working time 383

coprovider families 389

crossover 386

dual-career families 392

economic distress 409

emotion work 394

engagement 395

Family and Medical Leave Act of 1993 413

family policy 411

flextime 412

gender ideology 388

homemaker role 389

hostile environment 403

mental labor 396

role conflict 385

role overload 385

role reversal 401

role strain 385

self-care 405

sexual harassment 403

shift couples 399

time strain 383

work spillover 383

RESOURCES ON THE WEB

Book Companion Website

www.cengage.com/sociology/strong

Prepare for quizzes and exams with online resources—including tutorial quizzes, a glossary, interactive flash cards, crossword puzzles, self-assessments, virtual explorations, and more.

12

Intimate Violence and Sexual Abuse

What Do YOU Think? Are the following statements TRUE or FALSE? You may be surprised by the answers (see answer key on the following page).

T	F	**1** Intimate relationships of any kind increase the likelihood of violence.
T	F	**2** Rape by an acquaintance, date, or partner is less likely than rape by a stranger.
T	F	**3** The value placed on family privacy contributes to a reluctance to report suspected family violence.
T	F	**4** Studies of family violence have helped strengthen policies for dealing with domestic offenders.
T	F	**5** Physically abused children are often perceived by their parents as "different" from other children.
T	F	**6** Sibling violence is the most widespread form of family violence.
T	F	**7** More than 2 million elderly Americans are emotionally or physically abused by a family member.
T	F	**8** Boys are more often victims of child sexual abuse.
T	F	**9** Sexual and other forms of physical abuse occur more frequently in households with stepparents than in households with only biological parents.
T	F	**10** Brother–sister incest is generally harmless.

Like most Americans, you might feel safest once you've locked your door and are home for the night. It is then that you might believe you have protected yourself from violence by locking out any would-be intruders. Unfortunately, for many people the sad reality is that they also *lock in* violence once they close and lock their doors to the outside world. It may seem a cruel irony, but the relationships people most value are also the relationships that become the most violent. The people one loves and lives with are often the people one is most likely to be hurt or assaulted by. It is an unhappy fact that intimacy or relatedness increases the likelihood of experiencing abuse, violence, sexual abuse, or even homicide.

Consider these examples. Twenty-seven year-old Rachel Entwhistle and her nine-month-old daughter Lillian; 34-year-old Lisa Underwood and her seven-year-old son Jayden; 28-year-old Jasmine Fiore; and 31-year-old Matthew Winkler were all victims of homicides committed by their spouse, partner, or parent. In January 2008, 38-year-old Lam Luong threw his four children (ages three, two, and one and four months) to their deaths off an 80-foot-high Alabama bridge after an argument with his wife. The mother, brother, and nephew of actress and singer Jennifer Hudson were killed by her brother-in-law in October 2008.

Such crimes are not unique to the United States, as intimate partners are estimated to commit between 40% and 70% of homicides against women worldwide (National Coalition Against Domestic Violence, 2007a). Such cases as the ones above are neither typical of intimate violence and abuse nor representative of most homicides in the United States, but they are chilling reminders of the worst of violence among family members.

Now, consider too the following items:

- One out of every four women will likely experience domestic violence in her lifetime.
- In 2005, more than 1,500 people—1,181 women and 329 men—were murdered by an intimate partner.

Answer Key to What Do YOU Think?

1 True, see p. 418; **2** False, see p. 431; **3** True, see p. 423; **4** True, see p. 443; **5** True, see p. 437; **6** True, see p. 438; **7** True, see p. 440; **8** False, see p. 445; **9** True, see p. 445; **10** False, see p. 446.

- Twenty percent of teenagers in a serious dating relationship report having been pushed, hit, or slapped by a partner. Approximately a fourth (26%) of teenage girls report having been subjected to repeated verbal abuse from their dating partners, and one in four teenage girls have been pressured by their partners to engage in oral sex or sexual intercourse (Teenage Research Unlimited 2005).
- In 2007, more than 1,700 children died from abuse or neglect (Centers for Disease Control 2009).
- Each year more than 2 million older Americans, typically women, are subjected to some form of abuse or neglect; 90% of the perpetrators are family members (National Coalition Against Domestic Violence 2008b).
- Almost 1 million parents are physically assaulted by their adolescents or younger children every year.
- The prevalence of same sex intimate partner violence is about the same (estimated as between 25% and 35%) as the prevalence of intimate violence experienced by heterosexual women (McLennen 2005).

In addition, as many as 90% of American parents spank their children. Although spankings are clearly different from beatings, assaults, physical and sexual abuse, or homicide, they still are acts of violence and therefore merit attention and consideration in this chapter.

Think for a moment about who our society "permits" us to shove, hit, kick, or spank. If we assault a stranger, push a coworker or employer, or spank or slap a fellow student or professor, we would run great risk of being arrested. It is with our intimates that we are "allowed" to do such things.

Those closest to us are the ones we are most likely to slap, punch, kick, bite, burn, stab, or shoot. And our intimates are the most likely to do these things to us (Gelles and Cornell 1990; Gelles and Straus 1988). Furthermore, living together provides people more opportunity to disagree, get angry at one another, and hurt one another. In effect, families and households can be very dangerous places.

To understand intimate violence and abuse requires consideration of a range of behaviors and examination of the various factors—social, psychological, and cultural—that shed light on why it is that people often hurt the ones they most love. In this chapter, we look at violence between husbands and wives (including marital rape), between gay and lesbian partners, between dating partners (including acquaintance rape), and between siblings as well as violence committed against children by parents and against parents by grown

Relationships can become violent or abusive regardless of the gender of partners.

or **intimate partner abuse** to address the full scope of violence among intimate couples. Other forms of family violence, such as those between siblings or between parents and children, still most often fall under the broader umbrella term *family violence*. They are addressed later in this chapter.

Researchers differentiate between violence and abuse. For the purpose of this book, we use the definition of **violence** offered by sociologists Richard Gelles and Claire Pedrick Cornell (1990): "an act carried out with the intention or perceived intention of causing physical pain or injury to another person." Abuse includes acts such as neglect and emotional abuse, including verbal abuse, that are not violent. Thus, abuse is broader than family violence.

Violence may best be understood along a continuum, with "normal" and "routine" violence at one end and lethal violence at the other extreme (Gelles and Straus 1988). Thus, family violence ranges from spanking to homicide. If one wishes to understand acts that fall anywhere between those extremes, it is necessary to examine the full continuum: at "families who shoot and stab each other as well as those who spank and shove, . . . [as] one cannot be understood without considering the other" (Straus, Gelles, and Steinmetz 1980). In this chapter, we focus most of our attention on physical violence and sexual abuse that occurs between intimate partners and between family members.

Types of Intimate Violence

Even limiting the discussion to violence in intimate couple relationships leaves a range of behaviors that require some kind of differentiation. Michael Johnson and Kathleen Ferraro (2000) offer the following widely used typology of partner violence:

- **Common couple violence** (sometimes called **situational couple violence**) is violence that erupts during an argument when one partner strikes the other in the heat of the moment. Such violence is not part of a wider relationship pattern; it is as likely to come from a woman as a man or to be mutual. It rarely escalates, and it is less likely to lead to serious injury or fatality.
- **Intimate terrorism** occurs in relationships where one partner tries to dominate and control the other. Violent episodes that escalate and emotional abuse are two common traits. Victims are left "demoralized and trapped" as their sense of self and their place in the world is greatly diminished by their

children. We look, too, at the various models researchers use in studying intimate violence, and we discuss the dynamics of battering relationships. We also discuss prevention and treatment strategies. In the last section of the chapter, we discuss child sexual abuse and focus on the types, perpetrators, and victims.

Intimate Violence and Family Violence: Definitions and Prevalence

In exploring the violent and abusive underside of families and intimate relationships, researchers have used different and changing terminology, trying to keep pace with increasing knowledge about the phenomenon (McHugh, Livingston, and Ford 2005). Many now use the terms **intimate partner violence**

partner's dominance. The violence in intimate terrorism is likely to recur, escalate, and lead to injury. It is also less likely to be mutual.

- **Violent resistance** encompasses what is often meant by "self-defensive" violence. It tends to be more commonly perpetrated by women than men and can signal that the victim is moving toward leaving the abusive partner.
- **Mutual violent control** refers to relationships in which both partners are violently trying to control each other and the relationship.

Distinctions such as these are important if one is to make sense of the data on who commits violence against a partner or spouse. Of the four types, common couple violence seems to be slightly more typical of men than of women, though it is the form that most appears to reflect a pattern of **gender symmetry**, a term that refers to the similarity in survey research estimates of male-on-female and female-on-male intimate partner violence. Intimate terrorism is usually perpetrated by men, and violent resistance is typically committed by women (Johnson and Ferraro 2000).

This typology is useful because it differentiates motives and outcomes of violence. Not all intimate violence is an attempt to control a partner, and injuries and fatalities do not occur equally in all types. Other outcomes—economic, psychological, and health related—also differ by the type of violence. For example, posttraumatic stress disorder is more likely suffered by a victim of intimate terrorism than by a victim of situational couple violence. However, the attempt to control a partner need not be attempted or achieved violently. Sociologist Kristin Anderson points out that some people who suffer the extreme subordination and control associated with intimate terrorism but without the violence suffer similar emotional, psychological, and even physical consequences. Thus, not all acts of intimate partner abuse that are control driven may be captured within the category of intimate terrorism (Anderson 2008).

Prevalence of Intimate Violence

It is impossible to know exactly how much violence there is in families and relationships in the United States. Part of the difficulty results from methodological limitations in the various data we gather. Depending on *how* we gather the information, estimates of *how much* there is and of *where it happens* will vary. You

© Stockbyte/Getty Images

Tension and conflict are normal features in intimate relationships but can escalate into situational or common couple violence.

might think that there are "official statistics" we could use, such as arrest records or emergency room visits. Yet so much family violence is unreported that the official data are incomplete (U.S. Bureau of Justice Statistics 1998). Perhaps as much as three-quarters of all physical assaults, four-fifths of all rapes, and half of all stalking incidents suffered by females at the hands of their intimate partners are never reported to police (National Institute of Justice and the Centers for Disease Control, 2000). Aside from underreporting by victims, some people are better positioned to hide their abusive behavior from authorities. Upper- and middle-class abusers also may be given more credibility by police as well as medical and social service professionals. People who can afford to use nonhospital medical resources (such as family doctors to treat injuries) may avoid suspicion since the incident won't show up in hospital records.

Data from domestic violence shelters are even more severely limited since most victims don't seek out a shelter. In addition, most women who use shelters are from lower economic backgrounds and have been victims of the severest forms of mistreatment (Cunradi, Caetano, and Schafer 2002). Thus, the information about shelter populations does not reflect the extent or distribution of the wider problem.

Most research on intimate and family violence is based on survey data. Many discussions of intimate violence rely on surveys of large random samples

Issues and Insights: Does Divorce Make You Safer?

There are two schools of thought on the relationship between the rate of intimate partner violence and divorce rates. Criminologists Lisa Stolzenberg and Stewart D'Alessio (2007) refer to them as the **safety-valve thesis** and the **retaliation thesis**. According to the safety-valve thesis, because divorce reduces the amount of physical contact between "warring spouses," it leads to a reduction in the opportunity for violence to occur. Terminating the relationship is thought to lower the probability of spousal violence. If, as this thesis suggests, women use divorce as a way to remove themselves from violent or abusive marriages, high divorce rates should be accompanied by lower rates of domestic violence.

The retaliation thesis is based on the idea that although a divorce dissolves a marriage, it doesn't necessarily lead to a reduction in violence between ex-spouses. It suggests that men may become *more violent* after divorce, either to intimidate their ex-spouses into reconciliation or to regain and exercise control over their former wives. A number of empirical studies indicate that violence follows women after they divorce, intensifying and escalating along the way.

Stolzenberg and D'Alessio analyze data from the 2002 National Incident-Based Reporting System and from the census on 244 cities in the United States with populations of 25,000 people or more. Based on nearly 36,000 incidents between spouses and more than 4,600 incidents between ex-spouses, they calculated both a spouse victimization rate and an ex-spouse victimization rate and assessed whether and how they were associated with the divorce rate. Finding a significant association between the divorce rate and the ex-spouse victimization rate, they conclude that as the divorce rate in a city rises, so too does the spouse victimization rate. They also found that cities with longer mandatory waiting periods between separation and legal divorce tended to have higher spouse victimization rates. From this they suggest that states consider shortening their mandatory separation periods because this will reduce the time during which couples seeking divorces have to interact with each other.

drawn from the wider U.S. population. Such studies include the National Family Violence Resurvey, the National Survey of Families and Households, the National Violence Against Women Survey, and the National Longitudinal Couples Survey. In addition, broader studies of crime and victimization, such as the National Crime Victimization Survey, the FBI's Supplemental Homicide Reports, and the Study of Injured Victims of Violence, have been used to better estimate the prevalence of intimate violence and to understand the influence of social and economic factors (Field and Caetano 2005).

Of course, reports and estimates based on survey data are themselves prone to problems. In asking people to admit to family violence, researchers may receive underreports. Even in anonymous surveys, individuals may downplay their involvement in socially undesirable behavior. Nevertheless, the estimates from such large-scale national surveys give us our best ideas of the frequency and spread of family violence. It is on such data that most estimates in this chapter are based. Keep in mind, however, that it is probable that these estimates are underrepresentative of the reality of violence and victimization.

What the Data Reveal

Based on survey data from large nationally representative samples of heterosexual couples in the United States, the rate of intimate partner violence ranges from 17% to 39% of couples each year (Caetano, Vaeth, and Ramisetty-Mikler 2008). The National Violence Against Women Survey found that 22% of women report physical assault from an intimate partner (Cherlin et al. 2004). Roughly one out of five couples in the general population report having experienced intimate partner violence according to 25 years of survey data summarized by Craig Field and Raul Caetano (2005).

Using multiple sources of data, the Bureau of Justice Statistics (2009) produced a report on violence between intimates. Key findings are as follows:

- Between 2001 and 2005, 22% of nonfatal violent victimizations of females over age 12 were committed by intimate partners. Among males 12 and older, intimate partners were responsible for 4% of nonfatal violent victimizations.
- Between 2001 and 2005, 30% of the homicides committed against females over age 12 and 5% of

those committed against males were committed by intimate partners.

- Between 2001 and 2005, on average 30% of female murder victims and 5% of male murder victims were killed by an intimate.

Why Families Are Violent: Models of Family Violence

All families have their ups and downs, and all family members at times experience anger toward one another. But why does violence erupt more often and with more severe consequences in some families than in others? To better understand violence within the family, we must look at its place in the larger sociocultural environment. The principal models used in understanding family violence are discussed in the following sections.

Individualistic Explanations

An individualistic approach emphasizes how the abuser's violence is related to a personality disorder, mental or emotional illness, or alcohol or drug misuse (O'Leary 1993). The idea that people are violent because they are crazy or drunk is widely held (Gelles and Cornell 1990), although research indicates that fewer than 10% of family violence cases are attributable to psychiatric causes and that only about 25% of cases of wife abuse are associated with alcohol.

Sociologists Richard Gelles and Claire Pedrick Cornell (1990) suggested that individualistic explanations may be especially appealing to abusers. If they can attribute the violence and abuse they inflict as due to an aberration or illness, then abusers can believe that their acts are not deliberately hurtful or abusive. But besides looking at the abuser, we must step back and look at the big picture—at the family and society that influence the abuser.

Ecological Model

The ecological model uses a systems perspective to explore child abuse. Psychologist James Garbarino (1982) suggested that cultural approval of physical punishment of children combines with lack of community support for the family to increase the risk of violence within families. Under this model, a child who doesn't "match" well with the parents (such as a child with emotional or developmental disabilities) and a family that is under stress (such as from unemployment or poor health) and that has little community support (such as child care or medical care) can be at increased risk for child abuse.

Feminist Model

The feminist model stresses the role of gender inequalities or cultural concepts of masculinity as causes of violence. Using a historical perspective, this approach holds that most social systems have traditionally placed women in a subordinate position to men, thus supporting male dominance even when that includes violence (Toews, Catlett, and McKenry, 2005; Yllo 1993).

There is no doubt that violence against women and children—and indeed violence in general—has had an integral place in most societies throughout history. Feminist theory must be credited for advancing our understanding of domestic violence by insisting that the patriarchal roots of domestic relations be taken into account. However, the patriarchy model alone does not adequately explain the variations in degrees of violence among families in the same society (Yllo 1993). Women are sometimes violent toward their husbands and partners. More mothers are implicated in child abuse than fathers. Although the latter fact has much to do with women's responsibility for and time with children, it does illustrate that the capacity for violence is not limited to men or completely explained by concepts such as masculinity or patriarchy. Finally and most telling, rates of violence between lesbian partners are such that the National Coalition of Anti-Violence Programs estimates that approximately half the lesbian population in the United States has or will sometime experience domestic violence (National Coalition of Anti-Violence Programs, 2006). Furthermore, like heterosexual partner violence, when same-gender partner violence occurs, it is more likely to be a recurrent feature of the relationship than a one-time event. The feminist approach may have its greatest use in efforts to understand and explain male-on-female intimate partner violence among heterosexual couples.

Social Stress and Social Learning Models

The two social models discussed here can be related to the ecological and feminist models in that they view violence as originating in the social structure.

First, the social *stress* model views family violence as arising from two main factors: (1) structural stress such as low income or illness and (2) cultural norms such as the "spare the rod and spoil the child" ethic (Gelles and Cornell 1990). Groups with few resources, such as the poor, are seen to be at greater risk for family violence.

Second, the social *learning* model holds that people learn to be violent from society and their families (Wareham, Boots, and Chavez 2009). The core premise is that children, especially boys, learn to become violent when they are a victim of or witness to violence and abuse (Bevan and Higgins 2002). This is even more likely if the child experiences positive reinforcement for displaying violence. Although it is true that many perpetrators of family violence were abused as children, it is also true that many victims of childhood violence do not become violent parents. These theories do not account for this discrepancy.

Resource Model

William Goode's (1971) resource theory can be applied to family violence. This model assumes that social systems are based on force or the threat of force. A person acquires power by mustering personal, social, and economic resources. Thus, according to Goode, the person with the most resources is the least likely to resort to overt force. Although family violence occurs among all income levels, Gelles and Cornell (1990) describe the typical situation: "A husband who wants to be the dominant person in the family but has little education, has a job low in prestige and income, and lacks interpersonal skills may choose to use violence to maintain the dominant position."

Exchange–Social Control Model

Richard Gelles (Gelles 1993, Gelles and Cornell 1990) posits the two-part exchange–social control theory of family violence. The first part, exchange theory, holds that in our interactions, we constantly weigh the perceived rewards against the costs. When Gelles says that "people hit and abuse family members because they can," he is applying exchange theory.

The expectation is that "people will only use violence toward family members when the costs of being violent do not outweigh the rewards." The possible rewards of violence might be getting their own way, exerting superiority, working off anger or stress, or exacting revenge. Costs could include being hit back, being arrested, being jailed, losing social status, or dissolving the family. Three characteristics of families

that may reduce those costs of violence and thus reduce social control are the following:

- *Inequality.* Men are stronger than women and often have more economic power and social status. Adults are more powerful than children.
- *Private nature of the family.* People are reluctant to look outside the family for help, and outsiders (e.g., the police or neighbors) may hesitate to intervene in private matters. The likelihood of family violence decreases as the number of nearby friends and relatives increases (Gelles and Cornell 1990).
- *"Real man" image.* In some American subcultures, aggressive male behavior brings approval.

A violent man may gain status among his peers for asserting his "authority." The exchange–social control model is useful for looking at treatment and prevention strategies for family violence, discussed later in this chapter.

The Importance of Gender, Power, Stress, and Intimacy

Each of the previously mentioned perspectives has valuable insight to offer concerning some aspects of a complex problem with no easy or single solution. Looking across the theories, several factors surface repeatedly.

Gender

Although there is female-on-male violence and female-on-female violence (discussed later), violence by males tends to be more extreme, often has different causes (power and control vs. self-defense), and typically results in different consequences (in terms of both physical injuries and domination). Thus, gender matters a lot with family violence.

Power

Central to many theories of intimate violence is the idea of power. Power is a central motive in much intimate violence, especially the long-term and extreme forms of spousal violence that Michael Johnson and Janel Leone (2005) called intimate terrorism. In addition, powerlessness can be linked to violence when those who feel dominated and unable to legitimately assert their rights may turn to violence as a last resort.

Stress

As individuals are subjected to a variety of stresses (such as unemployment, underemployment, illness, pregnancy, work-related relocations, and difficult or disabled children), tensions among family members may rise. Stress-based explanations help account for the greater prevalence of violence among lower-income families and households facing unemployment, but stress alone cannot account for the breadth and depth of family violence (McCaghy, Capron, and Jamieson 2000; Straus et al. 1980). Stress may raise the likelihood of violence, but it is not the cause. Somewhere, the individual must have learned that acting violently toward loved ones is appropriate and acceptable (Gelles and Straus 1988).

Intimacy

The heightened emotions and long-term commitments that characterize family relationships are qualities we value about those relationships. Those same qualities lead to a greater likelihood that we will have disagreements and that those disagreements will be more emotional. Furthermore, cultural beliefs promote the idea that we have the right to influence our loved one's behavior. Some abusive men explain that they assault their spouses "because they love them." This indicates that the cultural expectations surrounding love and intimacy—one's sense of what he or she can and should do within their closest relationships—contribute to the worst aspects of those same relationships.

In addition, as discussed in Chapter 3, we grant and expect privacy and even secrecy in intimate and family relationships. Even when family conflict is in a public setting, others are reluctant to intervene in such "domestic disputes." In some ways, our society thus legitimizes violence and force within families and then turns the other way when they occur.

Women and Men as Victims and Perpetrators

It is not uncommon to hear the phrase "battered women" to refer to women who have been subjected to abuse from their spouses or partners. **Battering**, as used

Critical Thinking

Have you ever been in a public setting where you witnessed abusive behavior between romantic partners? How about from a parent to a child? How did you feel in the situation? What, if anything, did you say or do? Why do you think you reacted the way you did?

in the literature on family violence, includes slapping, punching, knocking down, choking, kicking, hitting with objects, threatening with weapons, stabbing, and shooting. In fact, the term *battering* does not specify the gender of the batterer, yet most people most likely assume that the batterer is male and that the victim is female. Interestingly, survey research has found that the number of women who report expressing violence toward their male partners is the *same as or greater than* the number of men who report expressing violence toward their female partners. This is true of research on spousal, cohabiting, and dating relationships. This was described earlier as *gender symmetry*.

Ignored or rejected by many researchers through the 1970s and 1980s or interpreted as signs of "self-defensive" or reactive violence by female victims, we now know that women use violence with male partners about as often as men do with female partners (Frieze 2005). One analysis of more than 80 studies of physical aggression between intimate partners found similar proportions of male and female violence (Archer 2000, cited in Graham-Kevan and Archer 2005).

However, even when the rates of violence are similar for males and females, the motives and outcomes of male-on-female and female-on-male violence may not be. Most violence perpetrated by women on men (as well as most male-on-female partner violence) is of the more situational, routine, and relatively minor variety. It is not the sort of violence that typically leads to hospitals or shelters. The less common and more extreme violence that escalates and causes serious injury or even death is usually committed by men against women (Johnson 1995).

There is also reason to suspect that women and men use violence for different reasons. As Maureen McHugh, Nicole Livingston, and Amy Ford (2005) assert, men's violence tends to be instrumental: they use violence to get what they want and to assert control and gain power over their partners. Women's motives include self-defense, retaliation, expression of anger, attention seeking, stress or frustration, jealousy, depression, and loss of self-control.

Historically and culturally, women have unfortunately been considered "appropriate" victims of domestic violence (Gelles and Cornell 1990). Many mistakenly accept the misogynistic idea that women sometimes need to be "put in their place" by men, thus providing a disturbing cultural basis for the physical and sexual abuse of women. There is no comparable cultural justification for the physical or sexual abuse of men.

As far as outcomes are concerned, more female victims than male victims are injured from partner violence, and their injuries tend to be more severe than those received by male victims. Even the same acts are not really the same: a slap that breaks the victim's jaw is not the same as a slap that reddens the victim's face. In other words, men's slaps (or punches, shoves, kicks, and so on) are not identical to those of women (McHugh et al. 2005).

Female Victims and Male Perpetrators

No one knows with certainty exactly how many women are victims of partner violence each year, but, as shown earlier, the data we have paint a less-than-optimistic picture. Consider, too, these facts from the Bureau of Justice Statistics (2009):

- Of all violent crime experienced by women between 2001 and 2005, 22% was from an intimate (spouse, ex-spouse, or boyfriend). During that same period, intimates accounted for 4% of nonfatal violence against men.
- Females between the ages of 20 and 24 were at the greatest risk of nonfatal intimate partner violence. Males between the ages of 12 and 15 or those over age 65 experienced the lowest rates of nonfatal partner violence.
- Between 2001 and 2005, half the female victims of nonfatal intimate partner violence suffered an injury, with 5% suffering a serious injury (e.g., stab or gunshot wound, broken bones, or internal injury).
- Unmarried women have higher rates of intimate partner violence victimization than do married women. Separated women experienced the highest rates compared to females of all other marital statuses.

In addition to these facts, there are data that illustrate the vulnerability and victimization of pregnant women due to abuse from their spouses or intimate partners (National Coalition Against Domestic Violence 2008). For example, it is estimated that almost 10% of females are subjected to intimate partner violence during pregnancy. Most women (70%) who are abused before pregnancy continue to experience abuse during pregnancy, and violence tends to intensify *after* the offender learns of the woman's pregnancy. Women abused during pregnancy have been found to have a greater likelihood of miscarriages, low-birth-weight babies, high blood pressure, vaginal bleeding, and kidney or urinary tract infections, and they are much more likely to delay prenatal care. Finally, homicide is one of the leading causes of death among pregnant women (National Coalition Against Domestic Violence 2008).

Which Women Are Victimized?

Women of all races, ages, and socioeconomic statuses experience intimate violence, although they are not victimized equally. Younger women (ages 20–24), Native American and African American women, and lower-income women are more frequent victims of partner violence. Black women suffered higher rates of nonlethal violence (5 per 1,000 persons 12 or older) than did Hispanic (4.3), white (4.0), or Asian (1.4) women. The rate of violence experienced by Native American women (11.1) was more than twice the rate experienced by African American women.

© Bob Daemmrich/The Image Works

Battered women's shelters provide safe havens for women in abusive relationships. Shelters provide counseling and emotional support, as well as temporary lodging, meals, and other necessities for women and their children.

Socioeconomic differences can also be noted. As income increases, the rate of female victimization decreases (Bureau of Justice Statistics 2009). Although no social class is immune to partner violence, it is more likely to occur in lower-income, low-status families (Gelles and Cornell 1990).

Although early studies of battering relationships seemed to indicate a cluster of personality characteristics constituting a typical battered woman, more recent studies have not borne out this viewpoint. Factors such as low self-esteem or childhood experiences of violence do not appear to be necessarily associated with a woman being in an assaultive relationship (Hotaling and Sugarman 1990). Two characteristics, however, do appear to be highly correlated with wife assault. First, a number of studies have found that wife abuse is more common and more severe in families of lower socioeconomic status. However, keep in mind that higher-income adults have greater privacy and thus greater ability to conceal domestic violence (Fineman and Mykitiuk 1994). Second, marital conflict—and the inability to resolve conflict—is a factor in many battering relationships.

Characteristics of Male Perpetrators

A heterosexual man who systematically inflicts violence on his wife or girlfriend is likely to have some or all of the following traits (Edelson et al. 1985; Gelles and Cornell 1990; Goldstein and Rosenbaum 1985; Margolin, Sibner, and Gleberman 1988; Vaselle-Augenstein and Ehrlich 1992; Walker 1979, 1984):

- He believes in the "traditional" home, family, and gender-role stereotypes and in the moral rightness of his violence (although he may acknowledge "accidentally" going too far).
- He has low self-esteem and may use violence as a means of demonstrating power or adequacy.
- He may be sadistic, pathologically jealous, or passive-aggressive and may use sex as an act of aggression.

The Centers for Disease Control identify a number of individual factors that contribute to perpetrating intimate violence, including low self-esteem of the abuser, depression, antisocial or borderline personality traits, emotional dependency and insecurity, and heavy drug and alcohol use. In addition to such characteristics, they add being unemployed, being socially isolated (i.e., having few friends), and having a history of experiencing poor parenting, physical discipline, or physical or psychological abuse as a child (Centers for Disease Control, National Center for Injury Prevention

and Control, 2008). Psychologist Maureen McHugh and colleagues (2005) note that in addition to perpetrating violence, violent men are likely to be the target of violence, either in the present or in their past. In other words, they are either victims of mutual violence or have histories of being abused themselves.

We often read or hear that a major factor in predicting a man's violence is if he experienced family violence as a child. According to research, a childhood troubled by parental violence accounts for only 1% of adult dating violence and approximately the same proportion of violence in marriage or marriagelike relationships (see review by Johnson and Ferraro 2000). Although it is true that sons of the *most* violent parents have a 1,000% greater rate of wife beating than sons of nonviolent parents, the majority of these sons, including sons of the most violent parents, are not violent (Johnson and Ferraro 2000).

Female Perpetrators and Male Victims

The incidence and experiences of males victimized by violence committed by their wives are poorly understood. Although it is undoubtedly true that some heterosexual men are injured in attacks by wives or girlfriends, most injured victims of severe intimate partner violence are women. The males most likely to be injured or killed as a result of intimate partner violence are gay men, who are assaulted or killed by their same-gender partner (Greenwood, 2002). Thus, we may not consider violence by women as significant as that committed by heterosexual men (Straus 1993). Often, even if a woman attempts to inflict damage on a man in self-defense or retaliation, her chances of prevailing in hand-to-hand combat with a man are slim. A woman may be severely injured simply trying to defend herself. Remember, though, that when we combine common couple violence and violent resistance, about the same rate of female-on-male acts of violence occur.

Suzanne Steinmetz (1987) suggested that some scholars downplay the extent and importance of women's violence against male partners. As such, there is a "conspiracy of silence [that] fails to recognize that family violence is never inconsequential." Sociologist Murray Straus (1993) offered four reasons to take the study of female violence seriously:

- Assaulting a spouse—either a wife or a husband—is an "intrinsic moral wrong."
- Not doing so unintentionally validates cultural norms that condone a certain amount of violence between spouses.

- There is always the danger of escalation. A violent act—whether committed by a man or a woman—may lead to increased violence.
- Spousal assault is a model of violent behavior for children. Children are affected as strongly by viewing the violent behavior of their mothers as by viewing that of their fathers.

Familial and Social Risk Factors

In addition to the kinds of characteristics noted above, the relationship dynamics, the wider social environment, and the society at large also affect rates of family violence. As far as relationship factors are concerned, the presence of persistent tension, conflict, and fighting between spouses creates a context that heightens the likelihood of intimate violence. If the family is experiencing economic stress, the risk of violence also increases.

If one is living within an environment where there are weak sanctions against violence, where neighbors are unwilling to intervene even when they witness acts of violence, and where the relationships and interactions within the surrounding community are weak and infrequent, the risk of intimate partner violence grows. Traditional gender norms along with beliefs about family privacy, the legitimacy of violence as a means to solve problems, and the right of partners to influence or control each other are additional cultural or societal risk factors (Centers for Disease Control, National Center for Injury Prevention and Control 2008).

Socioeconomic Class and Race

We often hear about how "democratic" intimate violence is, occurring among all groups, regardless of economic status, race, or sexual orientation. Indeed, there is truth to that statement: intimate partner violence *can* be found among all ethnic and economic groups; however, the amount of violence varies greatly. More than three decades of research demonstrates an association between socioeconomic status and partner violence.

Socioeconomic Class

Using data from the 1995 National Alcohol Survey, Carol Cunradi, Raul Caetano, and John Schafer (2002) found that household income had the greatest influence on intimate partner violence across racial and ethnic lines. Both females and males of higher-income households experience less partner violence than their counterparts from lower-earning households. Noteworthy, too, are the following facts: within each income level, females were at higher risk than males of victimization, and low-income men were victimized at lower rates than high-income women (see Figure 12.1).

Although there are consistent and strong associations between low economic status and violence, research also reveals partner violence and abuse among high-status couples as well (Weitzman 2000). Their

Figure 12.1 Average Annual Nonfatal Intimate Partner Victimization Rate per 1,000 Persons Age 12 or Older by Income and Gender, 2001–2005

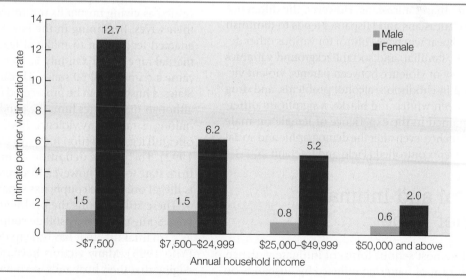

SOURCE: Bureau of Justice Statistics, www.ojp.gov/bjs/intimate/table/incgen.htm

Abuse and violence do exist among all social classes, but they tend to be more common among those with lower incomes and greater economic stress.

economic position may even create unique problems for women who are victimized. They may find other people less sympathetic to their circumstances, more skeptical of their allegations, and less willing to help because they perceive the women as having considerable means with which to help themselves.

Race

According to data from the National Family Violence Survey and the National Longitudinal Couples Survey, African Americans have higher rates of violence than either Caucasians or Hispanics, and Hispanics have a higher rate than Caucasians. However, the difference between Caucasians and Hispanics tends to diminish if not disappear when we control for various other demographic, familial, and social background variables (e.g., history of violence between parents, violent victimization in childhood, alcohol problems, and drug use). Between whites and blacks, a significant difference remained in the experience of female-on-male partner violence, even after the demographic and social variables were controlled (Field and Caetano 2005).

Marital and Intimate Partner Rape

One of the most serious forms of intimate violence, **rape** is a form of battering inflicted by husbands on wives and male-on-female or male intimate partners,

often as parts of a pattern of intimate terrorism. Most legal definitions of rape include "unwanted sexual penetration, perpetrated by force, threat of harm, or when the victim [is] intoxicated" (Koss and Cook 1993). Rape may be perpetrated by males or females and against males or females; it may involve vaginal, oral, or anal penetration, and it may involve the insertion of objects other than the penis.

Based on data gathered in the National Violence Against Women Survey, 7.7% of the 8,000 women age 18 and older who were sampled had been raped by an intimate partner. This translates to almost 8 million women having been raped by a boyfriend, ex-boyfriend, husband, or ex-husband at some point in their lives. Based on the same survey, more than 200,000 women endure more than 320,000 rapes by an intimate partner each year. This leads to an estimated rate of 3.2 intimate partner rapes per 1,000 women in the United States. The National Violence Against Women Survey also revealed that approximately 23% of gay men reported being raped, physically assaulted, and/or stalked by a same gender partner. However, the data did not allow for a reliable estimate of the rate of same-gender intimate partner rapes among men. A four-city probability sample of more than 2,800 men who have sex with men produced a rate of sexual abuse (being forced to have sex) of 5%, though with 18% indicating having suffered "multiple forms" of abuse, it is possible that the prevalence is greater (Greenwood et al. 2002).

Historically, heterosexual legal marriage was regarded as giving husbands unlimited sexual access to their wives. Beginning in the late 1970s, most states enacted legislation to make at least some forms of marital rape illegal. On July 5, 1993, marital rape became a crime in all 50 states. Throughout the United States, a husband can be prosecuted for raping his wife, although many states limit the conditions, such as requiring extraordinary violence. Less than half the states offer full legal protection for wives (Muehlenhard et al. 1992). The precise definition of marital rape differs from state to state, however. In several states, wife rape is illegal only if the couple has separated.

There still remains the problem of enforcing the laws. Some people may still discount rape in marriage as a "marital tiff" that has little to do with "real" rape (Yllo 1995). Many victims have difficulty acknowledging that their husbands' sexual violence is indeed rape. Caucasian females are more likely than African

American females to identify sexual coercion in marriage as rape (Cahoon et al. 1995), and all too often judges seem more sympathetic with the perpetrator than with the victim, especially if he is intelligent, successful, and well educated.

There is also the "notion that the male breadwinner should be the beneficiary of some special immunity because of his family's dependence on him" (Russell 1990). Because of deeply entrenched attitudes and beliefs about what constitutes rape and about marital and sexual relationships, it is estimated that two-thirds of sexual assault victims do not report the crime (U.S. Department of Justice 1997).

Violence in Gay and Lesbian Relationships

Until fairly recently, little was known about violence in lesbian and gay relationships. One reason is that such relationships have not been given the same social status as those of heterosexuals. In addition, long-term same-sex relationships are less common than long-term heterosexual relationships. Finally, many gays and lesbians are likely to be reluctant to identify their sexuality for fear of resulting stigma or mistreatment. However, understanding violence in same-sex relationships is important for at least two reasons: people are being victimized, and their victimization is mostly invisible and unaddressed. Relationships between gay men or lesbians obviously lack the gender differences that otherwise reflect male dominance and female subordination. Clearly, neither male dominance nor male socialization toward dominance, aggressiveness, or violence can account for physical abuse in lesbian relationships.

Recent research indicates that the rate of abuse in gay and lesbian relationships is comparable to that in heterosexual relationships. A recent estimate placed the range between 25% and 50% for lesbian couples (McClennen, Summers, and Daly 2002, in Frieze 2005). A study by Kimberly Balsam and Dawn Syzmanski (2005) found that of the 272 lesbian and bisexual women in their sample, 40% reported being violent, and 44% reported being victims of violence within relationships with female partners (in Frieze 2005). In their four-city study mentioned earlier, of 2,881 gay or bisexual men, psychologist Gregory Greenwood and colleagues found nearly 40% of their sample had experienced some form of abuse. Most common was psychological abuse (34%), but 22% experienced physical abuse, 5% indicated that they had been sexually abused by their partner, and 18% suffered multiple forms of abuse (Greenwood et al. 2002). Greenwood and colleagues report that such rates exceed those experienced by heterosexual men and are comparable or greater than those experienced by heterosexual women. Furthermore, violence in same-sex relationships is rarely a one-time event; once violence occurs, it is likely to recur. It also appears to be as serious as violence in heterosexual relationships, including physical, psychological, and/or financial abuse. Michael Johnson and Kathleen Ferraro (2000) note that intimate terrorism can be found among lesbian couples. One additional form of abuse, unique to same-sex couples, is the threat of "outing" (revealing another's gay orientation without consent). Threatening to out a partner to coworkers, employers, or family may be used as a form of psychological abuse in same-sex relationships.

For battered partners in same-sex relationships, there is often nowhere to go for support. Services for gay men and lesbians are often nonexistent or uninformed about the multifaceted issues that face such victims. Renzetti (1995) points out several policy issues that must be addressed among service providers and domestic violence agencies:

- Consider how homophobia inhibits gay and lesbian victims of abuse from self-identifying as such
- Recognize that battered gay men and lesbians of color experience a triple jeopardy: as victims of domestic violence, as homosexuals, and as racial or ethnic minorities
- Address the issue of gay men and lesbians as both batterers and victims who may seek services at the same time from the same agency

Dating Violence and Date Rape

In the past few decades, researchers have grown increasingly aware that violence and sexual assault can take place in all forms of intimate relationships. Violence between intimates is not restricted to spouses or family members. Even casual or dating relationships can be marred by violence or rape. In fact, although all 50 states have laws that prohibit the kinds of behaviors that make up dating violence (e.g., stalking, sexual

According to a national online survey of 615 13- to 18-year-olds and 414 parents with teens in that age-group. teenage dating violence has also gone high tech. Cell phones and social networking sites have become tools for abusive partners. The data are disturbing in the portrait they paint of such victimization:

- 25% of teens who responded have been called names or harassed by their partners in text messages or cell phone calls.
- 22% of respondents have been pressed via cell phone or Internet contact to have sex.
- 24% of surveyed teens who were in dating relationships communicated via cell phone with their dating partner every hour between midnight and 5:00 A.M.
- 30% of those teens who responded said that they receive frequent (between 10 and 30) text messages per hour by partners asking where they are, who they are with, and what they are doing.

- 68% of teens acknowledged that boyfriends or girl-friends sharing embarrassing or private pictures or videos of them with others through cell phones and computers is a serious problem. A slightly higher percentage report that spreading rumors about a boyfriend or girlfriend via social networking sites (e.g., MySpace or Facebook) or cell phones was a serious problem.

In addition to the data from teens, information from parents reveals their lack of awareness of the magnitude of the problem. For example, 82% of parents of teens who had been e-mailed or texted 30 times per hour were unaware that their sons or daughters were experiencing such harassment. Additionally, 67% of parents were unaware that their teen was being pressured via e-mail, text message, or cell phone to have sex, while 71% did not know that their teen was afraid of what their partners might do if they failed to respond to a text, an instant message, or a cell phone call.

SOURCE: Picard (2007).

---

assault, and assault), evidence has long indicated that the level of dating violence exceeds the level of marital violence (Lloyd 1995).

Tweens, Teens, and Young Adults: Dating Violence and Abuse

The incidence of physical violence and emotional or verbal abuse in dating relationships, including those of teenagers, is alarming. It is also hard to pin down with precision. The Centers for Disease Control (2009) compiled the following statistics:

- One in 11 teens report being victims of physical dating violence.
- One in four adolescents report experiencing emotional, verbal, physical, or sexual violence each year. One in five high school females has been sexually or physically abused by a dating partner.

Meanwhile, surveys offer varying estimates of both the prevalence and the outcomes associated with dating violence. Data gathered by Teenage Research Unlimited reported a rate of physical violence experienced by teenage girls in dating relationships of 13%. Research that has looked into the percentages

of teens who *know someone* who has been victimized by dating violence reveals rates between 24% and 40% among 14- to 17-year-olds. More than a quarter (26%) of teenage girls who were in dating relationships reported having endured repeated verbal abuse (Teenage Research Unlimited 2005).

The National College Women Sexual Victimization Study surveyed more than 4,000 women during the 1996–1997 academic year. Asked about victimization just in the seven months since school began in the fall, 1.7% of the women had been raped. Another 1.1% had experienced an attempted rape. Nine out of 10 of these women knew their offenders. When asked about ever having experienced forced sex while with a dating partner, 13% of college women studied by criminologists Ida Johnson and Robert Sigler said that they had (Johnson and Sigler 2000).

Dating violence and abuse may be found at very young ages, in fact as soon as young people begin relationships. In one survey, among "tweens" (11–14-year-olds), 62% knew friends who had suffered verbal abuse by a girlfriend or boyfriend. As they move into high school and then for those who continue on to college, levels of dating violence victimization increase. One study of relationships among college students found that of the sample of 572, 21% had engaged in "physically aggressive" behavior, acts that included throwing something; pushing, grabbing, or hitting; slapping; kicking, biting, or punching; beating up; choking; and threatening to or using a gun or a knife on a partner.

In two studies of undergraduate couples (18–25 years old) in ongoing relationships, Jennifer Katz and colleagues found that a third to nearly half of the students were in relationships in which their partners had acted violently toward them. In both studies, rates at which men and women were victimized were similar, although men experienced higher levels of moderate violence (Katz, Kuffel, and Coblentz 2002). Other research suggests that about a third of college students report experiencing dating violence in a previous relationship, while 21% indicated that they had been victimized by a current partner (Sellers and Bromley, 1996).

Only about half of those victimized by dating violence report their victimization to anyone else; the vast majority (88%) of those who report tell a friend. Only 20% tell criminal justice officials (Sellers and Bromley, 1996).

Some of the issues involved in dating violence appear to be different than those generally involved in spousal violence. Whereas marital violence may erupt over domestic issues such as housekeeping and child rearing (Hotaling and Sugarman 1990), dating violence is far more likely to be precipitated by jealousy or rejection (Lloyd and Emery 1990; Makepeace 1989).

Although females and males may sustain dating violence at comparable levels, they do not appear to suffer comparable consequences. As in the case of marital violence, women react with more distress than men do to relationship violence, even within mutually violent relationships (Katz et al. 2002). They also sustain more physical injuries from dating violence. More surprising is the finding that "partner violence generally is unrelated to decreased relationship satisfaction" (Katz et al. 2002, 250). Many teen victims remain in relationships even after experiencing violence from a partner, even in the absence of a legal tie or shared residence.

Those who experience dating violence are at greater risk for a variety of health consequences, including increased risk of injury, attempted suicide, binge drinking, and physical fights. They also have higher rates of alcohol, tobacco, and illegal drug use. Teen victims are also more likely than nonvictims to engage in unhealthy and unsafe sexual practices, putting themselves at greater risk of sexually transmitted infections, HIV, and unintended pregnancy (Centers for Disease Control 2009).

Many women do leave a dating relationship after one violent incident; others stay through repeated episodes. Women who have "romantic" attitudes about jealousy and possessiveness and who have witnessed physical violence between their own parents may be more likely to stay in such relationships (Follingstad et al. 1992). Women with "modern" gender-role attitudes are more likely to leave than those with traditional attitudes (Flynn 1990). Women who leave violent partners cite the following factors in making the decision to break up: a series of broken promises that the man will end the violence, an improved self-image ("I deserve better"), escalation of the violence, and physical and emotional help from family and friends (Lloyd and Emery 1990). Apparently, counselors, physicians, and law enforcement agencies are not widely used by victims of dating violence.

Date Rape and Coercive Sex

Sexual intercourse with a dating partner that occurs against his or her will with force or the threat of force—often referred to as **date rape**—is the most common form of rape. Date rape is also known as **acquaintance rape**. One study found that women were more likely than men to define date rape as a crime. Disturbingly, date rape was considered less serious when the woman was African American (Foley et al. 1995).

Date rapes are usually not planned. Two researchers (Bechhofer and Parrot 1991) describe a typical date rape. He plans the evening with the intent of sex, but if the date does not progress as planned and his date does not comply, he becomes angry and takes what he feels is his right—sex. Afterward, the victim feels raped but the assailant believes that he has done nothing wrong. He may even ask the victim out on another date.

Alcohol or drugs are often involved. When both people are drinking, they are viewed as more sexual. Men who believe in rape myths are more likely to see drinking as a sign that females are sexually available (Abbey and Harnish 1995). In many instances, either the female rape victim or her assailant had been drinking or taking drugs before the rape occurred (Caponera 1998). There are also high levels of alcohol and drug use among middle school and high school students who have unwanted sex (Rapkin and Rapkin 1991).

In recent years, certain "date-rape drugs," most often either gamma-hydroxybutyrate (GHB) or Rohypnol (flunitrazepam, popularly known as "roofies"), have surfaced as major public safety concerns. Both drugs have sedative effects, especially when combined with alcohol. They may reduce inhibitions, and they affect memory. Both are used by some men to sedate and later victimize women, many of whom wake up unaware of where they are, how they got there, or what they have done. Samantha Reid, a 15-year-old, died as a result of drinking a soft drink that had been laced with GHB. Knowing only that the drink tasted funny, she died just hours later. Her friend, Melanie Sindone, recovered after entering a coma that lasted less than a day. In Reid's death, three men were convicted of involuntary manslaughter, punishable by 15 years in prison (Bradsher 2000). In 2000, then President Bill Clinton signed into law the Hillory J. Farias and Samantha Reid Date-Rape Drug Prohibition Act of 2000, named for Reid and another teenage victim who died after unknowingly drinking a beverage mixed with GHB. It is a federal crime, punishable by up to 20 years in prison, to manufacture, distribute, or possess GHB (http://abcnews.go.com).

"No" Really Means "No"

There is considerable confusion and argument about sexual consent. Much sexual communication is done nonverbally and ambiguously. That we don't necessarily give verbal consent for sex indicates the importance of the nonverbal clues we do give off. However, as we saw in Chapter 6, nonverbal communication is imprecise. It can be misinterpreted easily if it is not reinforced verbally. For example, some men may even mistake a woman's friendliness for sexual interest (Johnson, Stockdale, and Saal 1991; Stockdale 1993). Others may misinterpret a woman's cuddling, kissing, and fondling as wishing to engage in sexual intercourse (Gillen and Muncher 1995; Muehlenhard 1988; Muehlenhard and Linton 1987). Our sexual scripts often assume "yes" unless a "no" is directly stated (Muehlenhard et al. 1992). This makes individuals "fair game" unless a person explicitly says "no."

The assumption of consent puts women at a disadvantage. First, because men traditionally initiate sex, men may feel it is legitimate to initiate sex without women explicitly consenting. Second, women's resistance can be considered insincere because consent is always assumed. Such thinking reinforces a common sexual script in which men initiate and women refuse so as not to appear promiscuous. In this script, the man continues believing that her refusal is token. One study found that almost 40% of the women had offered a "token no" at least once for such reasons as not wanting to appear "loose," uncertainty about how the partner feels, inappropriate surroundings, and game playing (Muehlenhard and Hollabaugh 1989; Muelhenhard and McCoy 1991).

There are a variety of suggestions made to women about how to protect themselves from date rape. These suggestions include examining how one communicates one's desires (e.g., one should be clear and unambiguous about one's intent and consent) and recommends that one avoid using drugs and alcohol (and, if one does drink, be careful about who one accepts a drink from and where one puts it down). Beyond these strategies and suggestions, however, is an important and unpleasant reality. As with avoidance of stranger rapes, a woman can do everything right and still be victimized.

When and Why Some Women Stay in Violent Relationships

Violence in relationships generally develops a continuing pattern of abuse over time. We know from systems theory that all relationships have some degree of mutual dependence, and battering relationships are certainly no different. Despite the mistreatment they receive, some women stay in or return to violent situations for many reasons. However, we need to be careful not to overstate the tendency for abuse victims to stay with their abusers. Michael Johnson and Kathleen Ferraro (2000) point out, the focus is often misplaced on answering "why women stay," even when a study finds that two-thirds of women have left the violent relationship. They suggest that it would be more appropriate to ask questions such as "How and why women leave." For those women who do stay in violent or abusive situations, their reasons include the following:

- *Economic dependence.* Even if a woman is financially secure, she may not perceive herself as being able to cope with economic matters. For low-income or poor families, the threat of losing the man's support—if he is incarcerated, for example—may be a real barrier against change.

- *Religious pressure.* She may feel that the teachings of her religion require her to keep the family together at all costs, to submit to her husband's will, and to try harder.
- *Children's need for a father.* She may believe that even a father who beats the mother is better than no father. If the abusing husband also assaults the children, the woman may be motivated to seek help (but this is not always the case).
- *Fear of being alone.* She may have no meaningful relationships outside her marriage. Her husband may have systematically cut off her ties to other family members, friends, and potential support sources. She has no one to go to for any real perspective on her situation (for the relationship between social isolation and abuse, see Nielsen, Endo, and Ellington 1992).
- *Belief in the American dream.* The woman may have accepted without question the myth of the perfect woman and happy household. Even though her existence belies this, she continues to believe that it is how it should (and can) be.
- *Guilt, pity, and shame.* She feels that it is her own fault if her marriage isn't working. She worries about who else will take care of her husband. If she leaves, she believes, everyone will know that she is a failure, or her husband might kill himself.
- *Duty and responsibility.* She feels that she must keep her marriage vows "till death us do part."
- *Fear for her life.* She believes that she may be killed if she tries to escape.
- *Love.* She loves him; he loves her. Love may make one want to believe that one's partner will change, that he really is a good person, and so on.
- *Cultural reasons.* Certain minority women may face greater obstacles to leaving a relationship. She may not speak English, may not know where to go for help, and may fear she will not be understood. She often fears that her husband will lose his job, retaliate against her, or take the children back to the country of origin (Donnelly 1993). Recent immigrants from Latin America, Asia, and South Asia may be especially fearful that their revelations will reflect badly on the family and community.
- *Nowhere else to go.* She may have no alternative place to live. Shelter space is limited and temporary. Relatives and friends may be unable or unwilling to house a woman who has left, especially if she brings children with her.

- *Learned helplessness.* Lenore Walker (1979, 1993) theorizes that a woman stays in a battering relationship as a result of **learned helplessness**. After being repeatedly battered, she develops a low self-concept and comes to feel that she cannot control the battering or the events that surround it. Through a process of behavioral reinforcement, she "learns" to become helpless and feels that she has no control over the circumstances of her life.

Michael Johnson and Kathleen Ferraro's (2000) distinction between common couple violence and intimate terrorism is important to add here. Women subjected to situational violence are less likely to leave than victims of intimate terrorism. Victims of intimate terrorism leave their partners more often, most commonly seeking friends and relatives for help, and look for destinations that are safe and secret (Johnson and Leone 2005).

The Costs of Intimate Violence

The cumulative financial costs associated with intimate violence are considerable. The Centers for Disease Control reports an estimated cost in excess of $8 billion per year (Max et al. 2004). This includes both costs for direct medical and mental health services for victims of partner violence, rape, assault and stalking and the millions of dollars worth of broken or stolen property and the wages lost to victims due to time out of work. The "bottom line" is indeed steep.

Then there are the nonfinancial costs. These include the actual health and mental health effects with which victims of violence must cope. DeMaris (2001) reports that thousands of women and men are treated in emergency rooms each year for injuries suffered in partner violence. Victims of intimate partner violence also suffer twice the rate of depression and four times the rate of posttraumatic stress disorder as nonvictims (Zink and Putnam 2005). According to the Centers for Disease Control and Prevention (2003), victims of severe intimate violence lose a cumulative total of nearly 8 million days of paid work—the equivalent of more than 32,000 full-time jobs—and almost 5.6 million days of household productivity each year.

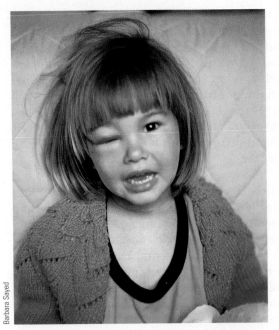

Barbara Sayed

Children are the least protected members of our society. Much physical abuse is camouflaged as discipline or as the parent "losing" his or her temper.

Children as Victims: Child Abuse and Neglect

Although it is an all-too-familiar concept today, child abuse was not recognized as a serious problem in the United States until the early 1960s. At that time, C. H. Kempe and his colleagues (1962) coined the medical term **battered child syndrome** to describe the patterns of injuries commonly observed in physically abused children. Looking at the variety of forms of potential abuse and adding the issue of neglect, the U.S. Centers for Disease Control provided the following information regarding various aspects of **child maltreatment** in the United States in 2007. *Child maltreatment* includes the following:

- *Neglect.* Failing to meet a child's basic needs for such things as food, housing, clothing, education, and access to medical care.
- *Physical abuse.* Actions such as hitting, kicking, shaking, or burning a child, resulting in the child sustaining an injury or dying.
- *Sexual abuse.* Such actions as fondling, raping, or exposing a child to other sexual activities.

- *Emotional abuse.* Subjecting a child to such behaviors as name-calling, threatening, withholding of affection, and shaming, all of which can harm the child's emotional well-being and sense of self-worth (see Table 12.1).

Prevalence of Child Maltreatment

Based on data collected in the National Child Abuse and Neglect Data System, a federal government effort to collect annual data on child abuse and neglect, the following picture of child maltreatment in the United States in 2007 emerges:

- In 2007, more than 3 million referrals to Children's Protective Services agencies were made, involving 5.8 million children. Roughly three-fourths of the investigations determined that child maltreatment had not, in fact, occurred; in 25.2% of the investigations, it was determined that at least one child had been the victim of some form of abuse or neglect.
- Most (58%) reports to Children's Services of abuse or neglect were made by professionals who had contact with the victim through their jobs as social

Table 12.1 Forms of Emotional Child Abuse

Based on the work of psychologists Stuart Hart and Marla Brassard, all of the following six categories of psychological maltreatment (or emotional abuse) convey to the child victim that she or he is unloved, unwanted, endangered, worthless, or flawed:

- *Spurning.* Ridiculing or belittling a child
- *Terrorizing.* Threatening a child or placing the child in a dangerous situation
- *Isolating.* Denying a child the opportunity to interact with others, confining a child, or imposing unreasonable limitations on a child's freedom of movement
- *Exploiting or corrupting.* Permitting the child to drink alcohol while underage or to use tobacco or illegal drugs; encouraging a child to engage in prostitution or other criminal activities; or exposing a child to criminal activities
- *Denying emotional responsiveness.* Refusing or failing to express affection or ignoring a child's attempts to interact
- *Neglect of a child's mental or medical health or educational needs*

SOURCE: www.childwelfare.gov

services staff, teachers, or police officers. The remainder were made by friends, relatives, neighbors, or coaches.

- Nearly 800,000 children were determined by Children's Protective Services to have experienced abuse or neglect.
- Rates of victimization appear to vary by race and gender, with African American, Native American, and multiracial children at greater risk. African American children had the highest victimization rate, at 16.7 per 1,000 children, followed by Native Americans (14.2 per 1,000), biracial or multiracial children (14 per 1,000), Hispanic children (10.3 per 1,000), whites (9.1 per 1,000) and Asian American children (2.4 per 1,000).
- Girls suffer slightly higher risk for all forms of maltreatment than do boys, making up 52% of victims of maltreatment.
- Despite the racial differences in rates of victimization, whites made up almost half (46%) of victims; an additional 22% were African American, and 21% were Hispanic.
- Children under one year of age had the highest rate of victimization (22.2 per 1,000 among boys and 21.5 per 1,000 among girls). Nearly a third of all victims were under four years of age (see Figure 12.2).

- The most common form of child maltreatment was neglect, which accounted for 60% of the cases of substantiated maltreatment. Of the other forms of maltreatment, in 11% of the cases the child suffered physical abuse, 8% experienced sexual abuse, and 4% were victims of emotional abuse. Thirteen percent of victims experienced multiple forms of maltreatment, and an additional 4% experienced some "other" form of maltreatment (e.g., threats to harm child, abandonment, or congenital drug addiction). One percent experienced medical neglect.
- Nearly 80% of the 859,000 perpetrators of child maltreatment were parents, and 7% were other relatives. Women represented 56.5% of perpetrators. Perpetrators were relatively young; 75% were under 40 years of age.
- Of parent perpetrators, almost 90% were biological parents. Mothers acted as the sole perpetrator in 39% of the cases. In 17% of the cases, mothers and fathers were both perpetrators, and in 6%, mothers and "others" together were perpetrators. Fathers were the sole perpetrators in 18% of the cases and acted with someone other than the child's mother in an additional 1% of the cases of maltreatment.

Considering the state of American children in 2007, the Children's Defense Fund (2008) reported the following:

Figure 12.2 Rates of Victimization among Children by Age

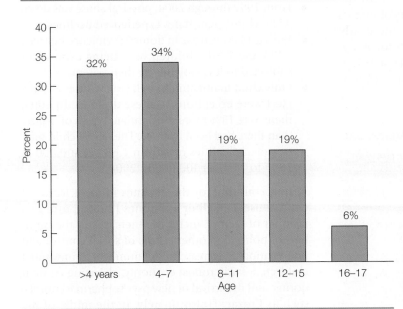

SOURCE: www.acf.hhs.gov/programs/cb/pubs/cm07/cm07.pdf

- Every 35 seconds, a child suffers from abuse or neglect.
- 2,479 children each day are abused or neglected.
- Each day, four children are killed by abuse or neglect.
- In 2007, more than 1,700 children died from abuse and neglect. Of those who died, three-fourths (76%) were younger than four years old. Another 13% were four to seven years old, 5% were 8 to 11 years old, 5% were 12 to 15 years old, and 2% were 16 to 17 years old.

Like intimate partner violence, the real prevalence of child abuse may be beyond our ability to measure. A recent article in *The Lancet*, a British medical journal, estimates that as few as 10% of incidents of child abuse are reported to and confirmed by social service agencies (Sharples 2008). As is also true of partner relationships, children are subjected to nonphysical forms of mistreatment by parents. In examining the national prevalence of **psychological aggression** by parents, Murray Straus and Carolyn Field (2003) found that verbal attacks on children are so common as to be "just about universal." Based on nearly 1,000 interviews with a nationally representative sample of households with at least one child under 18 years old living at home, Straus and Field explored the prevalence of psychological aggression. They define psychological aggression as consisting of the following kinds of behaviors, with the latter three constituting "more severe" psychological aggression:

- Shouting, yelling, or screaming at one's child
- Threatening to spank or hit one's child but not actually doing it
- Swearing or cursing at one's child
- Threatening to send one's child away or kick him or her out of the house
- Calling one's child dumb or lazy, or making some other disparaging comment

Of the sample parents, 89% reported having committed at least one of the five kinds of psychological aggression, and 33% reported at least one instance of the more severe forms. The prevalence of the various forms of psychological aggression is illustrated in Table 12.2.

Use of psychological aggression varies with the age of the child. A total of 43% of parents of infants reported using psychological aggression, and nearly 90% of parents of two-year-olds use some form of psychological aggression. The percentage peaks at 98% at age seven, and as late as age 17, the rate still remains a high 90%.

Table 12.2 Prevalence of Psychological Aggression

Prevalence	Measure (% in last year)
Overall	88.6
Severe	33.4
Shouting, yelling, screaming	74.7
Threatening to spank or hit	53.6
Swearing or cursing	24.3
Name-calling	17.5
Threatening to kick out of house	6.0

SOURCE: Straus and Field (2003).

Conversely, research on corporal (physical) punishment shows it declining with the age of the child; only 12% of parents of 17-year-olds report still using corporal punishment (Straus and Field 2003). However, more than 90% of toddlers in the United States are reportedly spanked (Straus and Field 2003). Most child-rearing experts currently advise that parents use alternative disciplinary measures.

Parents' ages matter, too. Younger parents (ages 18–29) reported the most frequent use of psychological aggression (22 times in the past 12 months) compared to parents 30 to 39 (19 times in the past 12 months) and parents over 40 (15 times in the past 12 months). Aside from age differences, there was "a lack of demographic differences in use of psychological aggression; this means that nearly all parents, regardless of sociodemographic characteristics, used at least some psychological aggression as a disciplinary tactic" (Straus and Filed 2003, 805).

In the midst of all these distressing facts, is there any good news to report? According to sociologists David Finkelhor and psychologist Lisa Jones, there is. Between 1993 and 2004, a variety of forms of maltreatment and abuse of children declined. Finkelhor and Jones (2006) report the following:

- In the early 1990s, sexual abuse began to decline after a 15-year-long period of increases. From 1990 through 2004, substantiated sexual abuse was down 49%.
- From 1992 through 2004, physical abuse was down 43%. Thirty-eight states experienced declines.
- With a 43% decrease in domestic violence between 1993 and 2001, children were being exposed or subjected to less violence by their parents.
- Only child maltreatment–related fatalities and neglect were exceptions to these declines. In 2003, there were 13% more substantiated cases of neglect than there had been in 1990. The rate of child maltreatment fatalities rose from 1.68 per 100,000 in 1995 to 2.03 per 100,000 in 2004.

In accounting for the declines in most forms of child abuse, Finkelhor and Jones (2006) point to a number of factors, including increases in the numbers of police and other agents of social control and intervention, a decrease in the numbers of unwanted children, a fairly robust economy, changing cultural norms, and the arrival of new psychopharmaceuticals, such as Prozac. Unfortunately, in the midst of the U.S. economic downturn, a number of state and local agencies around the United States (e.g., Arizona,

New York, Georgia, Washington, California, Ohio, and Massachusetts) have reported increases in reports to Children's Services, calls to hotlines, hospital emergency room visits, and arrests. This suggests that economic matters may have been large factors in the decreases that Finkelhor and Jones describe as well as in more recent increases.

Families at Risk

Research has established that the following sets of factors put families at risk for child abuse and neglect: (1) parental characteristics, (2) child characteristics, (3) family factors, and (4) the family ecosystem—that is, the family system's interaction with the larger environment. The characteristics described in the next sections are likely to be present in abusive families (Goldman et al. 2003; Straus et al. 1980).

Parental Characteristics

Some or all of the following characteristics are likely to be present in parents who abuse their children:

- The abusive parent was physically punished by his or her parents, and his or her father physically abused his or her mother. However, it is important to emphasize that a history of childhood maltreatment does not guarantee that one will become a maltreating parent. Likewise, many parents who abuse or neglect their children were not themselves abused.
- The parents believe in corporal discipline of children.

Children need to have someone, such as a teacher who they trust, in whom they can confide about their suffering.

- Parents have unrealistic expectations of their children and less understanding of age-appropriate behaviors. Either of these can lead to parental frustration or disappointment, both of which may lead them to take frustrations out on their child.
- Certain personality traits or characteristics are often identified among abusers. These include low self-esteem, poor impulse control, anxiety, or depression.
- The marital relationship itself may not be valued by the parents. There may be spousal violence.
- The parents appear unconcerned about the seriousness of a child's injury, responding, "Oh well, accidents happen."
- Parental substance abuse.

Child Characteristics

Who are the battered children? Are they any different from other children? Surprisingly, the answer is often yes; they are different in some way or at least are perceived to be so by their parents. Children who are abused are often labeled by their parents as "unsatisfactory," a term that may describe any of the following:

- A "normal" child who is the product of a difficult or unplanned pregnancy, is of the "wrong" sex, or is born outside of marriage.
- An "abnormal" child, one who was premature or of low birth weight, possibly with congenital defects or illness. Children with disabilities suffer rates of maltreatment that may be almost twice as high as children without disabilities.
  - A "difficult" child, one who shows such traits as fussiness or hyperactivity. Researchers note that all too often, a child's perceived difficulties are a result (rather than a cause) of abuse and neglect.

In addition to these, the child's age has been found to be associated with certain increased risks. Young children, especially infants to toddlers, are at greatest risk for neglect. They are also more vulnerable to "shaken-baby syndrome." Teens are at greater risk of sexual abuse.

Family Characteristics

In addition to the aforementioned characteristics of adult offenders and child victims, the following family characteristics are associated with greater risks of child abuse and maltreatment.

- Children in single-parent homes may have higher risk of victimization, although it may be less a matter of the number of parents than of other characteristics that often accompany single parenthood (e.g., lower income and increased stress). For example, the elevated rate of abusive violence found in homes headed by single mothers is likely a function of the poverty that characterizes such families.
- Marital conflict., especially if it becomes frequent, intense, and violent, is often accompanied by child maltreatment.
- Families experiencing high levels of stress (e.g., from unemployment, serious illness, or death of a family member) may be susceptible to higher risks of child abuse or neglect.
- The kind and quality of parent–child interaction may be a factor leading to harsh discipline and more use of corporal punishment or verbal aggression. Parents who abuse or neglect their children are typically less affectionate, playful, or supportive and focus more on children's negative behaviors than on their positive behaviors.

Family Ecosystem

As discussed earlier in this chapter, the community and the family's relation to it may be relevant to the existence of domestic violence. The following characteristics may be found in families that experience child abuse:

- The family experiences poverty and/or unemployment.
- The family is socially isolated, with few or no close contacts with relatives, friends, or groups.
- The family has a low level of income, creating economic stress.
- The family lives in an unsafe neighborhood that is characterized by higher-than-average levels of violence.
- The home is crowded, hazardous, dirty, or unhealthy.

Notice the clustering of such socioeconomic characteristics as unemployment, low income, neighborhood, and housing. This combination tells an important story. Like spousal or partner violence, the mistreatment of children can be found across the socioeconomic spectrum. But, like spousal violence, it happens more often at the lower levels. As noted earlier, the culprit in these associations is most likely stress.

Children who experience abuse or neglect often endure long-lasting adverse effects, including physical, emotional, cognitive, and social effects. They are more likely than children who are not maltreated to experience poor physical health, including such outcomes as hypertension and chronic fatigue; poor emotional and mental health, including depression anxiety or suicidal thoughts; social relationship difficulties; and cognitive deficits. They are also more likely to engage in more high-risk behaviors, such as younger age at onset of sexual activity; to have more sexual partners; to become pregnant as a teen; and to engage in substance use and to display such behavioral problems as aggression, delinquency, or violent behavior as adults (Wang and Holton 2007).

Hidden Victims of Family Violence: Siblings, Parents, and the Elderly

Most studies of family violence have focused on violence between spouses and on parental violence toward children. There is, however, considerable violence between siblings, between teenage children and their parents, and between adult children and their aging parents. These are the "hidden victims" of family violence (Gelles and Cornell 1990).

Sibling Violence

More than a quarter of a century of research illustrates that violence between siblings is by far the most common form of family violence (Hoffman, Kiecolt, and Edwards 2005; Straus et al. 1980). Estimates in different studies have ranged from 40% to 80% of children under age 18 experiencing sibling violence or abuse. Although violence declines as children age, no less than two-thirds of teenagers annually commit an act of violence—pushing, slapping, throwing, or hitting with an object or something more severe—against a

Critical Thinking

If you became (or are) a parent, would you consider it violent to spank your child with an open hand on the buttocks if the child were disobedient? To slap your child across the face? Is it acceptable to spank your small child to teach him or her not to run into a busy street? To spank because you are angry?

Public Policies, Private Lives: "Nixzmary's Law"

Nixzmary Brown was seven years old when she died at the hands of her abusive stepfather in 2006. Details of her sad life and tragic death reveal abuses that included many nights tied to a chair, being forced to use a litter box rather than the toilet, and frequent beatings from her stepfather. On January 11, 2006, she was beaten to death by her stepfather for "stealing a cup of yogurt" from the refrigerator. According to prosecutor Ama Dimwoh, "After being beaten, battered, broken, and thrown naked onto a cold wooden floor, the last words of seven-year-old Nixzmary Brown—moaning in pain, gasping for air— were, 'Mommy, Mommy, Mommy.'" Yet her mother, Nixzaliz Santiago, did not protect or help her daughter and, in fact, may have helped initiate some of the prolonged abuse the child endured. She was convicted of manslaughter and sentenced to more than 40 years in prison. The stepfather, Ceasar Rodriguez, was sentenced to 29 years in prison. As details of this case unfolded, it was learned that the system designed to protect children had failed Nixzmary. A social worker at her school had made numerous pleas with the Administration for Children's Services (ACS) to save Nixzmary and had even tried to visit Nixzmary at home. The school principal had expressed concern for Nixzmary's safety. Despite such expressions of concern to ACS, caseworkers for the city agency failed to act.

On September 12, 2009, the New York state legislature sent a bill, Nixzmary's Law, on to New York governor David Paterson. If signed into law, it would make those who torture a child and intentionally kill a child in an "especially cruel and wanton manner" eligible for life sentences without parole.

The case and the legal legacy it leaves is reminiscent of the 1995 murder of six-year-old Elisa Izquierdo by her mother, Awilda Lopez. In addition to sharing similar kinds of horrific abuse, the two girls were cases that somehow slipped through the cracks in the agency intended to protect children and prevent such tragedies. After the outrage surrounding Elisa's death, New York passed Elisa's Law, which—among other things—was designed to increase disclosure of child abuse allegations made by requiring "authorities to disclose this information in several situations, among them: when a person is charged with child abuse and when an abused child dies. It thus seeks to make investigators more accountable for their mistakes" (Hernandez 1996, B6). The law also mandated that records of investigated but unsubstantiated allegations of abuse be kept, thus enabling authorities to more effectively detail the history of alleged abuse victimization.

Unfortunately, neither Elisa's Law nor Nixzmary's Law— nor for that matter any statutory change, can perfectly protect children and prevent abuse. However, they can facilitate a more effective reaction from those whose job it is to investigate complaints and/or punish perpetrators.

SOURCES: Blain (2009), Haberman (2009), Hernandez (1996), and Shifrel (2008).

---

sibling. A recent study of 651 college undergraduates found that nearly 70% acknowledged having acted violently toward their closest-age sibling while seniors in high school. The violence most commonly consisted of hitting with a hand or object, pushing or shoving, and throwing things but often included slapping, punching, and pulling hair (Hoffman et al. 2005). Most of this type of sibling interaction is simply taken for granted by our culture—"You know how kids are!" However, in 2002, siblings were responsible for 6% of all intrafamilial murders in the United States (Kiselica and Morrill-Richards 2007). Thus, sibling violence, like other forms of familial violence, can reach extreme levels.

Counseling psychologists Mark Kiselica and Mandy Morrill-Richards review the literature on sibling abuse from a wide range of sources. In their review, they note the following:

- A study of 150 adult survivors of sibling abuse found that 78% had experienced emotional abuse.
- A national study of family violence reported that 80% of 3- to 17-year-olds had experienced sibling violence and that more than 50% had experienced severe violence—such acts as stabbing, striking with an object, punching, or kicking.
- The rates of physical violence tended to be higher for males and also to decline with age.
- As to sexual abuse, research suggests that as many as 2.3% of women have been sexually victimized by a sibling.

Despite the fact that sibling violence and abuse remain the most common form of intrafamilial violence, they receive surprisingly less attention and seem to generate considerably less concern than other forms of family abuse.

Parents as Victims

Many people might find it difficult to imagine children attacking their parents because it so profoundly violates our image of parent–child relations. Parents possess the authority and power in the family hierarchy. Furthermore, there is greater social disapproval of a child striking a parent than of a parent striking a child; it is the parent—not the child—who has the "right" to hit.

Nevertheless, a variety of studies have examined the phenomenon and produced estimated levels of violence that range between 7% and 96%. You might wonder how such wildly different results could be obtained. Estimates vary depending on the ages of the children studied, the length of time one inquires about (e.g., the previous year or ever), the type of violence one includes, and the nature of the sample itself (i.e., studies of delinquent youth produce higher estimates) (Ulman and Straus 2003). Sociologists Arina Ulman and Murray Straus found that those studies that focus on the prior 12-month period produce rates ranging from 7% to 18% for 3- to 17-year-olds. As Ulman and Straus report, studies indicate that more boys than girls hit their parents and that mothers were more likely to be hit than were fathers. Analyzing data from the 1975 National Family Violence Survey, Ulman and Straus indicate the following:

- 20% of the mothers and 14% of the fathers report having been hit in the 12 months before being surveyed.
- Younger children are more likely to hit a parent than are older children. A third of three- to five-year-olds but 10% of 14- to 17-year-olds were violent toward a parent.
- Both girls and boys are more likely to be violent toward their mothers than toward their fathers.

Ulman and Straus contend that even though the survey data are old, the patterns they reveal most likely still pertain.

Elder Abuse

Of all the forms of hidden family violence, only the abuse of elderly parents by their grown children (or, in some cases, by their grandchildren) has received considerable public attention. Elder mistreatment may be an act of commission (abuse) or omission (neglect) (Wolf 1995). It is estimated that more than 2 million older people (over 50) are physically abused annually (American Psychological Association 2005). Perhaps as many as 10% of the elderly population have been abused (National Center on Elder Abuse 2005). In approximately 90% of elder abuse cases, the victim is abused by a family member—a spouse or partner, adult children and grandchildren, or other family members (National Coalition Against Domestic Violence 2008).

Like the abuse of children and intimate partners, elder abuse can be physical, psychological, sexual, verbal, or financial. Sexual abuse of elders includes unwanted sexual contact as well as forced viewing of pornography or forced listening to sexual accounts. It most commonly involves a male caregiver as perpetrator and a female victim, over 70 years of age, who is either functioning at a low level or totally dependent on her caregiver(s).

Financial abuse encompasses a range of acts. Examples include such acts as coercing or deceiving an elder into signing wills, contracts, or other similar documents; taking advantage of an elder suffering from dementia by taking control over his or her money or financial decisions; and forcing a victim to part with resources or property.

Physical and verbal forms of elder abuse are much like physical and verbal abuse of other family members. Physical abuse includes any acts of physical contact intended to cause pain or injury to the victim. Signs of physical abuse include bruises, abrasions, burns, fractures, and dislocations, along with any injuries that are either unexplained or whose explanations don't fit the injury. Verbal abuse would encompass such acts as name calling, embarrassing or scaring a victim, along with other acts that damage the emotional well-being or self-worth of the recipient (National Coalition Against Domestic Violence, 2008b; Centers for Disease Control and Prevention, 2008).

Most abuse of the elderly goes unnoticed, unrecognized, and unreported. Even though mandatory reporting of suspected cases of elder abuse is the law in at least 44 of the 50 states (Wei and Herbers 2004), sociologists Karl Pillemer and David Finkelhor estimate that only 1 of every 14 cases is reported to authorities (Pillemer and Finkelhor 1998). Elderly people are often confined to bed or a wheelchair, and many do not report their mistreatment out of fear of institutionalization or other reprisal.

The abuse of the elderly by children, grandchildren, and others can be physical, psychological, sexual, or financial.

Among the factors that can protect elderly men and women from potential victimization is having a number of strong relationships with people of different social statuses, thus diminishing the likelihood of any single caregiver or contact being able to mistreat someone without such mistreatment being discovered. The more contacts one has, the less dependent one is, and the greater chance for someone to prevent the abuse or protect the potential victim (National Coalition Against Domestic Violence 2008).

The most likely victims—in most cases, women—of elder abuse are suffering from physical or mental impairments, especially those with Alzheimer's disease. Their advanced age renders them dependent on their caregivers for many if not all of their daily needs. It may be their dependency that increases their likelihood of being abused. Other research indicates that many abusers are financially dependent on their elderly parents; they may resort to violence out of feelings of powerlessness.

The Economic Costs of Family Violence

The economic impact of family violence is staggering. Whether one assesses the effects in terms of dollars and cents, impaired work performance, lost time and wages in one's job, or some other measure, the costs are great. The Centers for Disease Control estimate an annual cost from lost productivity due to domestic violence victimization of $728 million. Add to this amount the health care costs for direct medical and mental health care for victims, estimated to be as great as $4.1 billion, and the amount climbs to nearly $5 billion.

Other economic "costs" can be seen in the following information on job performance and workplace experiences:

- The Bureau of Labor Statistics Survey of Workplace Violence Prevention indicates that victims of domestic violence collectively lose almost 8 million days of paid work, an amount that translates to more than 32,000 full-time jobs (U.S. Bureau of Labor Statistics 2006).
- Of the 1 million women who are victims of stalking, about 250,000 report that stalking led them to miss work; they missed an average of 11 days of work per year. Seven percent never returned to their jobs.
- Between 35% and 56% of employed abused women were harassed by an intimate partner while at work.
- 37% of women personally affected by domestic violence report that the abuse has had an impact on their work performance in the form of tardiness, missed work, a lost job, or missed career promotions (EDK Associates 1997).

Attitudes and opinions of corporate leaders are especially instructive. More than 90% of Fortune 1000 corporate leaders believe that both the private and the working lives of employees are affected by domestic violence: 56% are aware of employees within their companies who are affected by intimate violence against women; 32% of corporate leaders contend that violence against women has damaged their company's "bottom line," while 66% believe that their company's

An estimated 1,400 children a year are murdered by their parents or guardians. Some cases remain relatively unknown to the wider public, reported in small articles in mostly local newspapers if reported at all. Others become major news stories, the focus of not only local but also wider regional or even national attention. Both kinds of cases can be seen in the following list of cases that occurred over the past 20 years. The list includes Eli Creekmore, age three, beaten to death by his father in 1986; Elizabeth "Lisa" Steinberg, age six, beaten to death by her adopted father in 1987; Joseph Wallace, age three, hanged by his mother in 1993; Elisa Izquierdo, age six, beaten to death by her mother in 1995; Nadine Lockwood, age four, intentionally starved to death by her mother in 1996; and James Pack, age three, beaten to death by his father in 2003. In just a three-month period, between late 2005 and early 2006, Sierra Roberts, age seven; Dahquay Gillians, age 16 months; and Joziah Bunch, age one, died at the hands of their parents.

This is but a partial list of child abuse homicides, selected because in each instance some agency or individuals in a position to intervene didn't—despite what in retrospect looked like clear and unambiguous evidence of severe abuse. Many of these cases were met by public outcry and led to changes in the policies used by the relevant protective agencies. Typically, the most extreme outrage is expressed at the parent perpetrators. Often there is also intense anger and blame directed at the agency or caseworkers who failed to rescue the child from his or her abusive, lethal surroundings.

performance would be better if they could address violence against women; and 48% believe that worker productivity in their organization has been negatively affected by intimate violence against women, while 42% believe that such violence has contributed to high levels of employee turnover (Roper Starch Worldwide 2002).

Researchers Ching-Tung Wang and John Holton estimated that the dollar costs associated with child abuse and neglect in the United States in 2007 reached nearly $104 billion. As high as this amount sounds, Wang and Holton (2007) argue that it is actually a conservative estimate because it includes only costs associated with victims. Costs for intervention into families or treatment costs for offenders are not included. Of these costs, $33 billion were direct costs for such things as hospitalization, child welfare services, and law enforcement, while another $70 billion were indirect costs for such things as special education, mental health and health care, and lost productivity.

Responding to Intimate and Family Violence

Based on the foregoing evidence, you may by now have concluded that the American family is well on its way to extinction as family members bash, thrash, cut, shoot, and otherwise wipe themselves out of existence. Statistically, the safest family homes are those with one or no children in which the husband and wife experience little life stress and in which decisions are made democratically. By this definition, most of us probably do not live in homes that are particularly safe. What can we do to protect ourselves (and our posterity) from ourselves?

Professionals who deal with domestic violence have long debated the most appropriate strategy: control and deterrence versus compassion (Mederer and Gelles 1989). Both approaches have their place. Controlling measures such as arrest, prosecution, and imprisonment, as well as compassionate measures such as shelters, education, counseling, and support groups, have been shown to be successful to varying degrees. Used together, these interventions may be quite effective. Helen Mederer and Richard Gelles (1989) suggest that controlling measures may be effective in motivating perpetrators to take part in treatment programs.

Intervention and Prevention

The goals of intervention in domestic violence include protecting victims, rehabilitating offenders through therapeutic intervention, and assisting and strengthening their families. In dealing with intimate and family violence, especially with child abuse, professionals and government agencies may be called on to provide medical care, counseling, and services such as day care, child care education, telephone crisis lines, and temporary foster care. Many of these services are costly, and many of those who require them cannot afford to pay. Our system does not currently provide the human and financial resources necessary to deal with these problems.

Prevention strategies usually take one of two paths: (1) eliminating social stress or (2) strengthening families (Swift 1986). Family violence experts make the following general recommendations:

- Reduce societal sources of stress, such as poverty, racism and inequality, unemployment, and inadequate health care
- Reduce sexism and provide employment and educational opportunities for women and men
- Furnish adequate day care
- Promote sex education and family planning to prevent unplanned and unwanted pregnancies
- End social isolation and explore means of establishing supportive networks that include relatives, friends, and community
- Break the family cycle of violence, eliminate corporal punishment and promote education about disciplinary alternatives, and support parent education classes to deal with inevitable parent–child conflict
- Address the cultural norms that legitimize and glorify violence

Intimate Partner Violence and the Law

Early family violence studies and feminist pressure spurred a movement toward the implementation of stricter policies for dealing with domestic offenders. Once long ignored, in the past two decades intimate violence has become a top concern for legislators and law enforcement agencies throughout the country (Wilson 1997). Today, many of the largest U.S. police forces have implemented **mandatory arrest** policies in which discretion is removed from police officers responding to a call about intimate violence. Under such policies, "if an officer finds probable cause that a crime occurred, he or she must arrest" (Goodman and Epstein 2005, 480). In addition, the adoption of **no-drop prosecution** policies compels prosecutors to proceed in the prosecution of an intimate violence case as long as evidence exists, regardless of a victim's expressed wishes (Goodman and Epstein 2005).

For police to play any effective role in combating intimate partner violence, they must first *know of the violence*. According to a "Fact Sheet on Intimate Partner Violence" put out by the National Center for Injury Prevention and Control, only about 20% of rapes or sexual assaults by a partner, 25% of physical assaults, and 50% of the incidents of stalking directed toward women are reported (www.cdc.gov/ ncipc/factsheets/ ipvfacts.htm). The rate at which men report their victimization is even less.

Even when incidents are reported, we have reason to question whether police are sufficiently and consistently committed to becoming involved in domestic disputes. This has long been a complaint of women who are victimized and who may find police reluctant to intervene, even under mandatory arrest policies. Male victims of female perpetrators find that police are often dismissive of their concerns (Migliaccio 2002).

Aside from the sincerity of the commitment of criminal justice personnel, the innovations in policy have potentially mixed consequences. Lisa Goodman and Deborah Epstein (2005) note some unintended and potentially unavoidable outcomes of no-drop policies, including a loss of a spouse's earnings due to his or her arrest and subsequent prosecution or suffering future victimization from a spouse who has been provoked into retaliatory abuse.

Working with Offenders: Abuser Programs

Treatment services for abusers provide one important component of a coordinated response to domestic violence. Such services might include psychotherapy, group discussion, stress management, or communication skills classes, all of which may be available through mental health agencies, women's crisis programs, or various self-help groups. Such interventions can also be made mandatory as part of the sanctions imposed on convicted offenders.

The most common intervention combines a criminal justice response (i.e., jail) with a mandated group intervention program. The extent to which attending batterers' groups changes the violent behavior of abusing men is difficult to measure, but based on evaluations of many different programs, effects associated with different intervention strategies are "typically small," with reoffending rates ranging from 21% to 35%. One review of the effects of mandatory offender programs concluded that, based on victim reports, the average effect was zero (Day et al. 2009). A more optimistic assessment from a National Institute of Justice Report was that intervention strategies—especially coordinated, multiagency interventions—yield "modest but statistically significant reductions in recidivism (repeat offending)" among those who participate (Healey and Smith 1998).

Strategies of intervention are shaped by how one attempts to explain or account for the problem in the first place. Rehabilitative interventions tend to be based more on individualistic and psychological approaches than on more structural explanations of violence. Perhaps the most widely used program is based on the **Duluth model**, the curriculum of which emphasizes helping batterers develop critical thinking skills around such themes as nonviolence, respect, partnership, and negotiation. The program proceeds in stages from intake and assessment to orientation and group sessions to program completion. To successfully complete the intervention, many programs require batterers to meet specific criteria—such as writing a "responsibility letter" acknowledging one's behavior that is read before the whole group.

What has become apparent is the ineffectiveness of the "one-size-fits-all" approach and the need to adopt a more sophisticated understanding of an individual's violent behaviors (Day et al. 2009). In addition, coordinated community response that includes proactive police and criminal justice strategies, advocacy and services for victims and their children, and responses by other community institutions that promote safety for victims and sanctions for those who batter are necessary interventions (Tolman 1995).

Yet, as Michael Johnson and Janel Leone (2005) warn, failure to differentiate types of violence can be problematic. For example, such common interventions as couples counseling or mediation may work effectively with more situational couple violence. However, given the more extreme nature of intimate terrorism, a woman who has been the victim of such abuse may be at considerable risk of injury or worse if she continues to live with her abuser while they partake in such common interventions.

Confronting Child and Elder Abuse

The first step in treating child abuse is locating the children who are threatened. Mandatory reporting of suspected child abuse is now required of professionals such as teachers, doctors, and counselors in all 50 states. Reported incidents of child abuse have increased greatly since mandatory reporting went into effect, but the actual number of incidents appears to have decreased. This is good news as far as it goes. Still, levels of violence against children remain unacceptably high, and not nearly enough resources are available to assist children. Child welfare workers are notoriously overburdened with cases, and adequate foster placement is often difficult to find (Gelles and Cornell 1990).

Society must address this tragedy of continued child abuse from a variety of levels:

- Parents must learn how to deal more positively and effectively with their children.
- Children need to be taught skills to recognize and report abuse as soon as it occurs.
- Professionals working with children and families should be required to receive adequate training in child abuse and neglect and to be sensitive to cultural norms.
- Agencies should coordinate their efforts for preventing and investigating abuse.
- Public awareness of child abuse needs to be created by methods such as posters and public service announcements.

While researchers continue to sort out the whys and wherefores of elder abuse, battered older people have a number of pressing needs. Karl Pillemer and Jill Suitor (1988) recommended the following services for elders and their caregiving families:

- Housing services, including temporary respite care to give caregivers a break and permanent housing (such as rest homes, group housing, and nursing homes)
- Health services, including home health care, adult day care centers, and occupational, physical, and speech therapy
- Housekeeping services, including shopping and meal preparation
- Support services, such as visitor programs and recreation
- Guardianship and financial management

Child Sexual Abuse

It is estimated that between 100,000 and 500,000 children are sexually abused annually. By the time they reach age 18, between 15% and 25% of children will likely have been sexually abused; among females, the likelihood is greater—30% to 40%—whereas for males the percentage is between 10% and 15% (Bahroo 2003).

Whether it is committed by relatives or nonrelatives, **child sexual abuse** is defined as any sexual interaction (including fondling, erotic kissing, or oral sex as well as genital penetration) between an adult or older adolescent and a prepubertal child. It does not matter whether the child is perceived by the adult as freely engaging in the sexual activity. Because of the child's age, he or she cannot legally give consent; the activity can be considered only as self-serving to the adult.

For a variety of reasons, as the American Psychological Association (APA) reports, definitive statistics "are difficult to collect because of problems of underreporting and the lack of one definition of what constitutes such abuse." In lieu of specific statistics, the APA states that child sexual abuse is "not uncommon and is a serious problem in the United States" (www .apa.org/releases/sexabuse).

Child sexual abuse is generally categorized in terms of the involvement or noninvolvement of kin. **Extrafamilial sexual abuse** is conducted by nonrelated individuals. **Intrafamilial sexual abuse** is conducted by related individuals, including step-relatives. The child's victimization may include force or the threat of force, pressure, or the taking advantage of trust or innocence. The most serious forms of sexual abuse include actual or attempted penile–vaginal penetration, fellatio, cunnilingus, and anal sex with or without the use of force. Other serious forms range from forced digital penetration of the vagina to fondling of the breasts (unclothed) or simulated intercourse without force. Sexual abuse can also include acts ranging from kissing to intentional sexual touching of the clothed genitals, breasts, or other body parts with or without the use of force (Russell 1984).

Children at Risk

Girls are the more common victims of child sexual abuse, making up between 70% and 89% of abuse survivors (Snyder 2000). More of the abuse that boys experience is extrafamilial. Lower-income children are at greater risk (www.unh.edu/ccrc/factsheet/pdf/ CSA-FS20.pdf). Males are the more common perpetrators, perhaps as many as 90% or more, though research estimates on the percentage of female perpetrators range between 5% and 20% of child sexual abuse committed by females.

At higher risk appear to be children who have poor relationships with their parents (especially mothers) or whose parents are absent or unavailable and have high levels of marital conflict. A child in such a family may be less well supervised and, as a result, more vulnerable to manipulation and exploitation by an adult. Finally, children with stepfathers are at greater risk for sexual abuse. The higher risk may result from the weaker incest taboo in stepfamily relationships and because stepfathers have not built inhibitions resulting from parent–child bonding beginning from infancy. As a result, stepfathers may be more likely to view their stepdaughters sexually.

Forms of Intrafamilial Child Sexual Abuse

The incest taboo, though in varying forms, is nearly universal in human societies. It prohibits sexual activities between closely related individuals. **Incest** is generally defined as sexual intercourse between people too closely related to marry legally (usually interpreted to mean father–daughter, mother–son, or brother–sister). Sexual abuse in families can involve blood relatives (most commonly uncles and grandfathers) and step-relatives (most often stepfathers and stepbrothers).

There is general agreement that the most traumatic form of sexual victimization is father–daughter abuse, including that committed by stepfathers. Some factors contributing to the severity of reactions to father–daughter sexual relations include fathers being more likely to engage in penile–vaginal penetration than other relatives, fathers sexually abusing their daughters more frequently, and fathers being more likely to use force or violence. The presence of a stepfather increases the risk of sexual abuse for girls, making them twice as likely to be abused as girls who live with their fathers (Putnam, 2003). In fact, comparative data from a number of countries indicates that sexual and other forms of abuse occur substantially more frequently in households with stepparents (stepmothers or stepfathers) than in households with only biological parents, a pattern psychologists Martin Daly and Margo Wilson call "the Cinderella effect" (Daly and Wilson, 1998, 2002, 2009). Sexual abuse by a stepfather represents a violation of the basic parent–child relationship.

Sibling Sexual Abuse

One resource (www.psychpage.com/family/library/ sib_abuse.htm) defines all brother–sister (or cousin) sexual interaction as *abuse*, largely in terms of the use of force or coercion and when penetration occurs or injury results. Vernon Wiehe (1997) suggests that official estimates of sibling sexual abuse are severely underestimated because most goes undetected by parents and unreported to authorities. The abuse is usually committed by an older brother (or sister) on a younger sibling.

The criteria of force and/or coercion may be the aspect most highly associated with negative outcomes, regardless of the specific sexual behavior (e.g., kissing, fondling, simulated intercourse, or exhibition). Studies assert that the circumstances, characteristics, and potential outcomes of brother–sister incest are as serious as, if not more than, those of father–daughter incest (Cyr et al. 2002; Rudd and Herzberger 1998). Furthermore, according to the U.S. Department of Health and Human Services (2002), sibling sexual abuse is more common than abuse committed by an adult relative.

Effects of Child Sexual Abuse

There is extensive research indicating that potential "profound, long-term consequences for an adult's sexual behavior and intimate relationships" can result from child sexual abuse (Cherlin et al. 2004, 770). Among the numerous well-documented consequences of child sexual abuse are both initial and long-term consequences. Many abused children experience symptoms of posttraumatic stress disorder (McLeer et al. 1992). Other initial effects include emotional disturbances, such as fear, anxiety, guilt and shame; social disturbances, such as running away or truancy; physical consequences, such as changes in eating or sleeping; and sexual disturbances, such as open masturbation and sexual preoccupation.

Long-Term Effects of Sexual Abuse

Although the initial effects of child sexual abuse can subside to some extent, the abuse may leave lasting scars on adult survivors who often have significantly higher incidences of psychological, physical, and sexual problems than the general population. Cherlin et al. (2004) list such outcomes as feelings of betrayal, lack of trust, feelings of powerlessness, low self-image, depression, and a lack of clear boundaries between self and others. Abuse as a child may predispose some women to early onset of sexual involvement, more involvement in sexually risky behavior, multiple partners, and sexually abusive dating relationships (Cate

and Lloyd 1992; Cherlin et al. 2004). Cherlin and colleagues also identify the following:

- Greater anxiety and less pleasure from sex
- Behaviors such as using drugs and/or alcohol with sex that increase risk of sexually transmitted disease or HIV infection
- Engaging in more frequent sexual encounters

As Cherlin et al. (2004, 771) point out, childhood sexual abuse victimization may also affect the ability to maintain long-term intimate relationships in adulthood. "Overall, the relationship difficulties associated with childhood sexual abuse would seem to be more consistent with frequent, short-term unions than with long-term unions."

Traumatic sexualization refers to a process in which a sexually abused child's sexuality develops inappropriately and the child becomes interpersonally dysfunctional. Sexually traumatized children learn inappropriate sexual behaviors (such as manipulating an adult's genitals for affection), are confused about their sexuality, and inappropriately associate certain emotions—such as loving and caring—with sexual activities.

As adults, sexual issues may become especially important. Survivors may suffer flashbacks, sexual dysfunctions, and negative feelings about their bodies. They may also be confused about sexual norms and standards. A fairly common confusion is the belief that sex may be traded for affection. Some women label themselves as "promiscuous," but this label may be a result more of their negative self-image than of their actual behavior.

Children feel betrayed when they discover that someone on whom they have been dependent has manipulated, used, or harmed them. Children may also feel betrayed by other family members, especially mothers, for not protecting them from abuse. As adults, survivors may experience depression, a sense of distrust, hostility and anger, or social isolation and avoidance of intimate relationships. Anger may express a need for revenge or retaliation.

Children experience a basic kind of powerlessness when their bodies and personal spaces are invaded against their will. In adulthood, such powerlessness may be experienced as fear or anxiety; a person feels unable to control events. It may also be related to increased vulnerability or revictimization through rape or marital violence; survivors may feel unable to prevent subsequent victimization. Other survivors, however, may attempt to cope with their earlier powerlessness by an excessive need to control or dominate others.

Ideas about being a bad person as well as feelings of guilt and shame about sexual abuse are transmitted to abused children and then internalized by them. Stigmatization is communicated in numerous ways. The abuser conveys it by blaming the child or, through secrecy, communicating a sense of shame. As adults, survivors may feel extreme guilt or shame about having been sexually abused. They may also feel different from others because they mistakenly believe that they alone have been abused.

Obviously, the violence and abuse discussed in this chapter are complex phenomena. They are products of individual characteristics of perpetrators and victims, relationship dynamics, and certain social and cultural factors. Not every home becomes a center of violence and abuse, and most families are not embattled. We need to realize that those families and relationships that are violent or abusive are products of a blend of qualities and are affected on multiple levels. This understanding is important if we hope to reduce the prevalence of violence and abuse and if we care to help those who are most at risk or already victimized.

Summary

- Abuse and violence are separate although certainly related and overlapping phenomena. *Violence* is defined as an act carried out with the intention or perceived intention of causing physical pain or injury to another person. Not all abuse is violent, and some intimate violence is considered appropriate and not abusive.

- Violence between intimate partners ranges from routine to extreme, from *common couple violence*, which is typically less severe, to *intimate terrorism*, which is a more severe, most often male-on-female form of violence and abuse in which power and domination are key motives.

- *Violent resistance*, often considered under the idea of "self-defense," is more often used by women.

- Different perspectives are used to study sources of family violence: (1) individualistic explanations, (2) the feminist model, (3) the social situational model, (4) the social learning model, (5) the resource model, and (6) the exchange–social control model.

- Researchers have stressed the roles played by gender, power and control, stress, and intimacy in explaining intimate violence.

- It is difficult to know exactly how much violence there is in intimate relationships. Either the use of official records and/or survey data give us underestimates of how much intimate violence there is in the United States.

- Intimate violence against women is one of the most common and most underreported crimes in the United States and throughout the world. Two characteristics that correlate highly with intimate violence against women in the United States are low socioeconomic status and a high degree of marital conflict.

- *Gender symmetry* refers to the survey data findings of similarity in both expressing and experiencing violence between the genders. However, the context and consequences of partner violence are not the same for men and women.

- Age, race, and social class all factor into domestic violence. Younger women, Native and African American women, and lower-income women experience more intimate violence than do other women.

- Research on male victims shows both similarities and differences with what research has revealed about male perpetrators and female victims.

- *Marital rape* is a form of battering. Many people, including victims themselves, have difficulty acknowledging that forced sex in marriage is rape, just as it is outside of marriage.

- Violence among same-sex couples is similar to the levels of violence among heterosexuals. Because such relationships lack the social supports that heterosexual couples can draw on, the experience of victimization may be worse.

- Dating violence, including verbal abuse, physical violence, and coercive sex, is often precipitated by jealousy or rejection.

- Dangerous date-rape drugs such as Rohypnol (flunitrazepam) and gamma hydroxybutyrate are sometimes used by offenders to sedate and sexually victimize unsuspecting women, prompting the passage of laws prohibiting date-rape drugs.

- Reasons some women may stay in or return to abusive relationships include economic dependency, religious pressure or beliefs, the perceived need for a father for the children, a sense of duty, fear, love, and reasons pertaining to their particular culture.

- Most child abuse cases are unreported. Parental violence is one of the five leading causes of childhood death.

- Families at risk for child abuse often have specific parental, child, and family ecosystem characteristics.

- Nearly 90% of parents acknowledged using some form of psychological aggression with at least one child during the prior 12-month period, with younger parents using such aggression more often.

- Intimate violence generates high costs in terms of time lost at work, mental health, and medical expenses for injuries or trauma sustained.

- The hidden victims of family violence include siblings (who have the highest rate of violent interaction), parents assaulted by their adolescent or youthful children, and elderly parents assaulted by their middle-aged children.

- Recommendations for reducing family violence include reducing sources of societal stress, such as poverty and racism; establishing supportive networks; breaking the family cycle of violence; and eliminating the legitimization and glorification of violence.

- Domestic violence intervention can be based on either control or compassion. Arrest, prosecution, and imprisonment are examples of control; shelters and support groups (including abuser programs) are examples of compassionate intervention.

- Recent legal innovations such as *mandatory arrest* and *no-drop prosecution* have had mixed results. In some ways, they raise the costs for victims of reporting the violence.

- Mandatory reporting of suspected child abuse may be helping to decrease the number of abused children in the United States. Early intervention and education also may help reduce abuse.

- *Incest* is defined as sexual intercourse between people too closely related to marry. Sexual victimization of children may include incest, but it can also involve other family members and other sexual activities.

- Children most at risk for sexual abuse include females, preadolescents, children with absent or unavailable parents, children with poor parental relationships, children with parents in conflict, and children living with a stepfather.

- *Child sexual abuse* has both initial and long-term effects, including traumatic sexualization, betrayal, powerlessness, and stigmatization.

Key Terms

acquaintance rape 431

battered child syndrome 434

battering 424

child maltreatment 434

child sexual abuse 445

common couple violence 419

date rape 431

Duluth model 444

extrafamilial sexual abuse 445

gender symmetry 420

incest 445

intimate partner abuse 419

intimate partner violence 419

intimate terrorism 419

intrafamilial sexual abuse 445

learned helplessness 433

mandatory arrest 443

mutual violent control 420

no-drop prosecution 443

psychological aggression 436

rape 428

retaliation thesis 421

safety-valve thesis 421

situational couple violence 419

traumatic sexualization 446

violence 419

violent resistance 420

RESOURCES ON THE WEB

Book Companion Website

www.cengage.com/sociology/strong

Prepare for quizzes and exams with online resources—including tutorial quizzes, a glossary, interactive flash cards, crossword puzzles, self-assessments, virtual explorations, and more.

13

Coming Apart: Separation and Divorce

What Do YOU Think? Are the following statements TRUE or FALSE?
You may be surprised by the answers (see answer key on the following page).

T	F	**1** The divorce rate is at its lowest point since 1970.
T	F	**2** Both annulment and divorce are ways of terminating a marriage in which both spouses feel that something has gone wrong in their marriage.
T	F	**3** Couples with children are less likely to divorce than couples without children.
T	F	**4** The critical emotional event in a marital breakdown is separation rather than divorce.
T	F	**5** Age at marriage is the best predictor of the likelihood of divorce.
T	F	**6** Divorce is an important element of the contemporary American marriage system because it reinforces the significance of emotional fulfillment in marriage.
T	F	**7** The higher an individual's income the greater the likelihood of divorce.
T	F	**8** Many of the problems experienced by children of divorced parents are present before the marital disruption.
T	F	**9** Those whose parents are divorced have a significantly greater likelihood of divorcing themselves.
T	F	**10** Marital conflict in an intact two-parent family is generally more harmful to children than living in a tranquil single-parent family or stepfamily.

As one woman told sociologist Joseph Hopper (2001), there was nothing she and her husband could do:

> It's something that had to happen, and it wasn't something that either one of us really controlled. It was just an awful situation that we had to get out of, and I recognized it and he didn't.

A second person offered the following:

> I had wanted that forever—the white picket fence and the whole dream. But it didn't come true. But I was at least smart enough to realize it wasn't happening and no matter what I did it wasn't going to.

In reading those words, some people will feel empathy for the speakers, sense their despair and understand their decisions. Others are likely to question such assumptions as "it had to happen" or "no matter what I did" it wasn't going to work. To some, marriage is supposed to bring us our deepest sense of fulfillment and intimacy, and if those feelings are absent, we ought to move on and try to find them elsewhere. To others, marriage is a commitment that is for life, even if it means enduring periods of disappointment or, worse yet, never feeling all of what one hoped to feel.

As we posed in Chapter 8, this raises such questions as the following: Are Americans promarriage? Are we too soft on divorce? Do we believe in the importance of marriage and the commitment we make when we exchange wedding vows? Or when we say "I do," are we really adding, perhaps not under our breath but in our heads, "at least for now"?

This chapter addresses many aspects of divorce, including the relative risk factors associated with divorce; the process of separating; the experiences and outcomes of divorce for men, women, and children; and such policy matters as joint custody, no-fault divorce, mediation, and covenant marriage. This exploration will help you better understand what parents, children, and families experience and how they cope with what increasingly has become part of our marriage system—divorce.

Answer Key to What Do YOU Think?

1 True, see p. 456;   **2** False, see p. 453;   **3** True, see p. 461;   **4** True, see p. 463;   **5** True, see p. 459;   **6** True, see p. 452;   **7** False, see p. 458;   **8** True, see p. 473;   **9** True, see p. 461;   **10** True, see p. 470.

The Meaning of Divorce

Some scholars suggest that divorce represents an idealization of marriage, not a devaluation or rejection. They reason that individuals would not divorce if they did not have such high expectations of marriage—so much hope about finding the fulfillment of their various needs in their spouses. According to Frank Furstenberg and Graham Spanier (1987), divorce may well be a critical part of our contemporary marriage system, which emphasizes emotional fulfillment and personal satisfaction.

A high divorce rate also tells us that many people may no longer believe in the permanence of marriage. Norval Glenn (1991) suggested that there has been a "decline in the ideal of marital permanence and . . . in the expectation that marriages will last until one of the spouses dies." Instead, marriages disintegrate when love goes or a potentially better partner comes along. Divorce is a persistent fact of American marital and family life and one of the most important forces affecting and changing American lives today (Furstenberg and Cherlin 1991).

The Legal Meaning of Divorce

Aside from the more debatable question about the meaning of divorce within the culture and family system in the United States, there is a more straightforward question about the legal meaning of divorce. Just what does divorce represent? Obviously, divorce is a means of terminating one's marriage. Where it once meant that the actions of one of the spouses had permanently damaged the marriage to such a degree that the marriage could not be fixed or endured, now it more often means that the marriage suffers from irreconcilable differences. This reflects the distinction between **fault-based divorce** and **no-fault divorce**. In a fault-based divorce, a person alleges that his or her spouse is responsible for the failed marriage through such actions as adultery, cruel and inhuman treatment, mental cruelty, habitual drunkenness, and desertion. In a no-fault divorce, the couple can divorce without either having to accuse the other or prove the other responsible for the failure of their marriage. They can simply claim that irreconcilable differences make it impossible for them to continue as married. Some states permit fault-based divorce, but every state allows for no-fault divorce. A later section considers the issue of no-fault divorce in more detail.

At present, however, it is important to note that divorce is not the only way a marriage can be ended.

Annulment and desertion are two other means of ending or exiting an existing marriage. The difference between annulments and divorces are many, but perhaps the most important is what each says about the terminated marriage. In a divorce, spouses agree that the relationship has failed, or one spouse proves that the other committed an act that caused the marriage to fail. Typically, an annulment is granted when it is determined that the marriage never met the legal requirements of marriage. For example, in instances where one married when underage, entered marriage already married, or married incestuously, an annulment is likely. In addition, if it can be demonstrated that at the time of marriage one lacked the mental capacity to understand the commitment they were making—or perhaps was under the influence of drugs or alcohol—the marriage is likely to be annulled. The literal legal meaning of an annulment is that the marriage has been nullified because it never really existed as a legitimate marriage. In other words, it invalidates a marriage because the marriage is considered to have never been a valid one.

In addition to divorce and annulment, some marriages end when one spouse simply leaves. Such acts of desertion leave the other spouse in some legal limbo because technically he or she remains married to the absent spouse. In the latter decades of the nineteenth century as well as during the Depression, desertion occurred more often. Throughout the latter decades of the twentieth century and the first decade of the twenty-first, divorce is the more common termination of marriage (Faust and McKibben 1999). In most states today, desertion is grounds for divorce, but until the marriage is officially terminated, the deserted spouse lives with considerable uncertainty and ambiguity. As difficult as divorce is to experience, at least it conveys a clearer message about one's marital status and one's eligibility for reentering marriage.

The Realities of Divorce

Yet another way of considering the "meaning" of divorce is to recognize the complex and multidimensional nature of the divorce experience in the lives of those who divorce. This is what the late anthropologist Paul Bohannan had in mind by his concept the **stations of divorce** (noted in Chapter 8). Bohannan (1970) developed an influential descriptive model of the divorce process that focused on the multiple experiences people have as their marriages end. He emphasized the emotional, legal, economic, coparental, community, and psychic stations of divorce. These stations neither have a particular order nor begin and end simultaneously. Some occur before one is legally divorced, and others can stretch on for years after one's divorce is final. The level of intensity of these different divorces varies at different times and for different couples. In no specific order, the stations are the following:

- The *emotional divorce*, when one spouse (or both) begins to disengage from the marriage, to feel that "something isn't quite right," begins well before the legal divorce. But even as divorce papers are filed, the partners may find themselves feeling ambivalent. Because the emotional divorce is not complete, they may try to reconcile.

- The *legal divorce* is the court-ordered termination of a marriage. By the time a couple is legally divorced, much has happened. The legal decree permits divorced spouses to remarry and conduct themselves in a way that is legally independent of each other. It also sets the terms for the division of property and child custody, issues that may lead to bitterly contested divorce battles. Many other unresolved issues surrounding the divorce, such as feelings of hurt and betrayal, may be acted out during the legal divorce. No-fault divorce was in part intended to minimize these issues.

- The *economic divorce* makes the economic aspect of marriage painfully apparent. Property acquired during a marriage is considered joint property and must somehow be divided between the divorcing spouses. The settlement is based on the assumption that each spouse contributes to the estate. This contribution may be nonmonetary, as in the case of traditional homemakers whose support and practical assistance enabled their husbands to work outside the home. As part of the economic divorce, child support and, less often, alimony may be ordered by the court. As the partners go their own ways, husbands and wives often experience different consequences in their standards of living as they set up separate households and no longer pool their resources. Women usually experience a decline in their standards of living, and men sometimes see theirs increase.

- The *coparental divorce* is experienced because even though marriages end, when there are children in the household, parenthood does not. Ex-spouses do not divorce their children. (Even those parents who never see their children remain fathers and mothers.) This is among the most complicated aspects of divorce because it also gives rise to single-parent families and, in most cases, stepfamilies, considered in more detail in Chapter 14. As parents divorce, issues of child custody, visitation, and support must be dealt with. With divorce, new ways of relating to the children and former spouses must be developed, ideally keeping the children's best interest foremost in mind.

- The *community divorce* means that when people divorce, their social world changes. In-laws become ex-laws; often they lose (or stop) contact. (This is particularly troublesome when in-laws are also grandparents.) Friends may choose sides or drop out; they may not be as supportive as desired. New friends may replace old ones as divorced men and women begin dating again. They may enter the singles subculture, where activities center on dating. Single parents may feel isolated from such activities because child rearing often leaves them little or no time or leisure and diminished income leaves them little or no money.

- The *psychic divorce* is accomplished when one once again feels like a separate individual and no longer feels like part of a couple. The former spouse becomes irrelevant to one's sense of self and emotional well-being. As part of the psychic divorce, each partner develops a sense of independence, completeness, and stability. Navigating through the psychic station may be more difficult and take a good deal longer than it does to experience the other stations of divorce.

Divorce in the United States

Before 1974, the view of marriage as lasting "till death do us part" reflected reality. However, a surge in divorce rates that began in the mid-1960s did not level off until the 1980s. In 1974, a watershed in American history was reached when more marriages ended by divorce than by death. Today, approximately 35% to 45% of all new marriages are likely to end in divorce (Stevenson and Wolfers 2007).

Measuring Divorce: How Do We Know How Much Divorce There Is?

How common is divorce, and how likely are we to experience it should we marry? According to the U.S. Census Bureau, there were nearly 23.3 million divorced people age 18 and older in the United States in 2008, representing more than 10% of the population. This included 13.6 million women and 9.7 million men. Race and gender patterns are illustrated in Figure 13.1.

There are a variety of ways to measure and represent the prevalence of divorce in the United States. The most common measures are discussed next.

Ratio Measure of Divorces to Marriages

It is likely that many of you have heard the gloomy—and misleading—statement that *one out of two marriages ends in divorce*. What exactly does that statistic

Figure 13.1 Percentage of Population All Age 18 and Over Currently Divorced

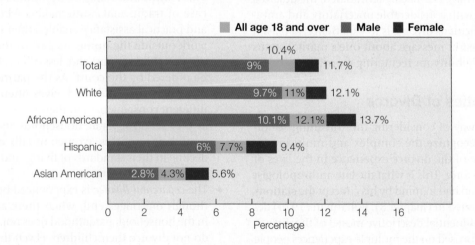

SOURCE: U.S. Census Bureau (2008).

mean? On what is it based? The **ratio measure of divorce** is calculated by taking the number of divorces and the number of marriages in a given year and producing a ratio to represent how often divorce occurs relative to marriage. Each year, there are about twice as many marriages that take place as there are divorces granted—recently, roughly 2.2 million marriages and approximately 1.1 million divorces. But recognize the difference between that statistic and a statement indicating that one of every two marriages *will end* in divorce. What the ratio measure truly reflects is the relative popularity or commonality of marriage and divorce in a given year.

Crude Divorce Rate

The **crude divorce rate** represents the number of divorces in a given year for every 1,000 people in the population. For 2008, there were 3.5 divorces for every 1,000 Americans. There were also 7.1 marriages per 1,000 people in the population, reinforcing the ratio of one divorce for every two marriages (Tejada-Vera and Sutton 2009).

Crude divorce or marriage rates have certain problems. Obviously, when calculating the crude divorce rate, counting every 1,000 people in the population means including many unmarried people, children, the elderly, the already divorced, and so on. These people cannot become divorced. It is therefore a statistic that is highly susceptible to the age distribution, proportions of married and single people in the population, and changes in such population characteristics.

Refined Divorce Rate

Considered the most useful measure of divorce, the **refined divorce rate** measures the number of divorces that occur in a given year for every 1,000 marriages (as measured by married women age 15 and older). In 2005, the refined rate was 16.7 divorces per 1,000 marriages. This is the lowest refined rate in nearly 40 years (Stevenson and Wolfers 2007).

Note that the range of available statistics produces different impressions about the reality of divorce in the United States. The ratio measure gives the most alarming impression, the one most closely approximating "one out of two marriages," or 50% of marriages, ending in divorce. When we use the refined rate of 1.7% of marriages ending in divorce annually, the picture seems much less bleak. The reality represented by each statistic is the same, but the meanings we attach to each statistic and therefore the understanding that each creates vary significantly.

Predicting Divorce

Another divorce statistic worth mentioning is the **predictive divorce rate**. This calculation (too complicated for our purposes) allows researchers to estimate how many new marriages will likely end in divorce. The prevailing estimate is that somewhere between 35% and 45% of marriages entered into in a year are likely to become divorces, though some continue to estimate as many as 50%.

Estimating future trends is a tricky business. Because this estimate is based on the experiences of prior birth cohorts (people born between specific years), we cannot be confident that current and future cohorts will make the same choices or face the same circumstances as their predecessors.

But even these predictions need to be more carefully assessed. As we show in subsequent sections describing factors associated with divorce, not everyone faces the same risk of divorce. As explained by Barbara Dafoe Whitehead and David Popenoe (2004), the risk of divorce is greatly affected—either increased or diminished—by the background characteristics that spouses bring with them into marriage. They go on to report the decreases in vulnerability to divorce during the first 10 years of marriage that are shown in Table 13.1.

Ultimately, Dafoe Whitehead and Popenoe (2005, 19) offer the following, more reassuring assessment of the likelihood of experiencing divorce: "So if you are a reasonably well-educated person with a decent income, come from an intact family and are religious,

Table 13.1 Vulnerability to Divorce in First 10 Years of Marriage

Divorce	Percentage Reduction in Risk of Divorce
Annual income over $50,000 (vs. under $25,000)	−30
Having a baby seven months or more after marriage (vs. before marriage)	−24
Marrying after 25 years of age (vs. under 18)	−24
Own family of origin intact (vs. divorced parents)	−14
Religious affiliation (vs. none)	−14
Some college (vs. high school dropout)	−13

and marry after age twenty-five without having a baby first, your chances of divorce are low indeed."

Divorce Trends in the United States

If we look at long-term divorce trends, the unmistakable conclusion is that from the start to the end of the twentieth century, the United States saw dramatic increases in marital breakups. If we look, instead, over the past 30 years, a different picture emerges. In more recent decades, the divorce rate has dropped, returning to levels not seen since the early 1970s (see Table 13.2). As we show shortly, this did not occur equally for all groups.

Both marriage and divorce rates have declined. The marriage rate is at its lowest point since the 1930s, and the 2.16 million marriages in 2008 reflect a recent decline from the 2.38 million marriages performed in 1997 (Munson and Sutton 2003). As to divorce, we can see that after three-quarters of a century of increases (minus, of course, the "time-out" of the 1950s), in recent decades the rate of divorce has declined.

Factors Affecting Divorce

Sometimes it is easy to point to the cause of a particular divorce. Perhaps one spouse was unfaithful or abusive, and the marriage was brought to a quick end.

In other instances, even the divorcing parties can't identify the exact cause or causes that led to divorce. Furthermore, every instance of divorce brings a unique combination of such causes. Researchers have looked at factors affecting both wider societal divorce rates and individual divorce decisions.

Some analyses address the complex sets of changes that make divorce rates hard to predict. For example, Heaton (2002) notes that there have been increases in the prevalence of premarital sex, premarital births, cohabitation, and both racial and religious intermarriage. All these tend to be associated with higher likelihood of marital instability, especially divorce. Yet there also have been increases in age at marriage and in educational attainment that tend to be associated with higher rates of stable marriage. In this section, we look at both the larger societal or demographic factors and the individual and couple characteristics that may be related to the likelihood of divorce.

Societal Factors

As seen earlier, even the reduced divorce rates especially evident in the 1990s and beyond were considerably higher than the rates early in the twentieth century. The 2008 divorce rate was *five times* the rate at the beginning of the twentieth century. It was 60% higher than the rates in 1960. In addition, divorce rates in the United States are higher than rates elsewhere in the industrialized world.

Sociologist Matthijs Kalmijn (2007) contrasted the average annual crude divorce rates from 1990 to 1999 for countries throughout Europe. As can be seen from Table 13.3, they range from extremely low rates throughout southern Europe (e.g., Italy, Spain, Greece, and Portugal) and southeastern Europe (e.g., Albania, Croatia, Bulgaria, and Bosnia-Herzegovina) to higher rates in northern Europe (e.g., Finland, Denmark, and Sweden) and central eastern Europe (e.g., the Russian Federation, the Czech Republic, Belarussia, and Ukraine).

Keeping in mind the U.S. rates during this same period (4.7 in 1990 and 4.0 in 2000) is a reminder that, with few exceptions, U.S. rates are higher than elsewhere. Not shown in the table but worth noting is the fact that rates throughout Africa,

Table 13.2 Divorce and Marriage through the Twentieth Century and Beyond

Year	Marriages	Rate per 1,000	Divorces	Rate per 1,000	Rate per 1,000 Marriages
1900	709,000	9.3	55,751	0.7	3
1920	1,274,476	12.0	170,506	1.6	8
1940	1,595,879	12.1	264,000	2.0	9
1960	1,523,000	8.5	393,000	2.2	9.2
1970	2,158,802	10.6	708,000	3.5	14.9
1980	2,406,708	10.6	1,189,000	5.2	22.6
1985	2,413,000	10.2	1,178,000	5.0	21.7
1990	2,448,000	9.8	1,182,000	4.7	20.9
1995	2,336,000	8.9	1,169,000	4.4	19.8
2000	2,315,000	8.2	NA*	4.0	NA
2005	2,249,000	7.6	NA	3.6	16.7
2008	2,162,000	7.1	NA	3.5	NA

*NA means data not available.

Table 13.3 International Variation in Crude Divorce Rate (divorces per 1,000 people, average for the period 1990–1999)

Southern Europe 0.84
Italy 0.51
Spain 0.76
Greece 0.78
Portugal 1.30

Western Europe 2.24
France 1.96
Germany 2.05
Switzerland 2.28
Belgium 2.47
United Kingdom 2.89

Northern Europe 2.36
Iceland 1.90
Norway 2.33
Sweden 2.42
Denmark 2.51
Finland 2.64

Southeastern Europe 0.74
Bosnia-Herzegovina 0.39
Albania 0.65
Croatia 0.91
Bulgaria 1.15

Central Eastern Europe 2.93
Poland 0.97
Hungary 2.33
Czech Republic 3.03
Ukraine 3.86
Russian Federation 4.04
Belarussia 4.19

SOURCE: Kalmijn (2007, 243–63).

Asia, South and Central America, and Canada were also lower than U.S. rates (United Nations Statistics Division, Demographic Yearbook, Table 25, 2006).

Changed Nature of the Family

In the United States, the shift from an agricultural society to an industrial one undermined many of the family's traditional functions. Schools, the media, and peers are now important sources of child socialization and child care. Hospitals and nursing homes manage birth and care for the sick and aged. Because the family pays cash for goods and services rather than producing or providing them itself, its members are no longer interdependent.

As a result of losing many of its social and economic underpinnings, the family is less of a necessity. It is now simply one of many choices we have: we may choose singlehood, cohabitation, marriage, or divorce—and if we choose to divorce, we enter the cycle of choices again: singlehood, cohabitation, or marriage and possibly divorce for a second time. A second divorce leads to our entering the cycle for a third time and so on.

Social Integration

Social integration—the degree of interaction between individuals and the larger community—is a potentially important factor related to the incidence of divorce. The social integration approach regards such factors as urban residence, church membership, and population change as especially important in explaining divorce rates (Breault and Kposowa 1987; Glenn and Shelton 1985; Glenn and Supancic 1984).

Among African Americans, the lowest divorce rate is found among those born and raised in the South; African Americans born and raised in the North and West have higher divorce rates. Similarly, those who live in urban areas, where the divorce rate is higher than in rural areas, are less likely to be subject to the community's social or moral pressures. They are more independent and have greater freedom of personal choice.

Individualistic Cultural Values

American culture has traditionally been individualistic. We highly value individual rights, we cherish images of an individual battling nature, and we believe in individual responsibility. It should not be surprising that many view the individual as having priority over the family when the two conflict. As marriage and the family lost many of their earlier social and economic functions, their meaning shifted. Marriage and family are viewed as paths to *individual* happiness and fulfillment. We marry for love and then expect marriage and our partners to bring us happiness. When individual needs conflict with family demands, however, we no longer automatically submerge our needs to those of the family. We often struggle to balance individual and family needs. But if we are unable to do so, divorce has emerged as an alternative to an unhappy or unfulfilling marriage and as an escape from a mean-spirited or violent marriage.

Demographic Factors

A number of demographic factors appear to have a correlation with divorce, including employment status, income, education level, ethnicity, and religion.

Exploring Diversity: Divorced and Seeking Remarriage in India

Yuvraj Raina is a businessman in New Delhi, India, the founder of the Aastha Center for Remarriage, a matrimonial agency for women and men seeking to *reenter* marriage after a divorce or the death of a spouse. On his list of 5,000 prospective brides and grooms, one finds Inderbir Singh, Savi Nagpal, and Anhuba Suri.

Inderbir Singh is a 35-year-old man, divorced a little more than a year. His comments, reported by journalists Saher Mahmood and Somini Sengupta, reveal the state of being divorced in India. He notes that his friends recommended to him that he find a new wife, but they have not helped him by initiating dates with available women:

Suddenly, I'm an outcast from my society. . . . People are okay with divorce. Nobody forces you to stay in a marriage.

. . . But the attitude towards a divorce is still the same: "They're outcasts."

Savi Nagpal is a 39-year-old mother with an eight-year-old daughter. She wants to marry again for "purely practical consideration(s)." She mentions the need that her daughter has for a father and that she has for someone to help with raising her child.

Meanwhile, Anhuba Suri, a 29-year-old divorced woman, was pressed by her parents to find someone new, even before her divorce was finalized.

That each has come to the Aastha Center is itself a statement of the changing Indian attitudes toward divorce and remarriage. As Yuvraj Raina, himself remarried after divorce, sees it, "In general, it's no longer taboo."

Employment Status

Among Caucasians, a higher divorce rate is more characteristic of low-status occupations, such as factory worker, than of high-status occupations, such as executive (Greenstein 1985; Martin and Bumpass 1989). Unemployment, which contributes to marital stress, is also related to increased divorce rates. Studies conflict as to whether employed wives are more likely than nonemployed wives to divorce; overall, however, the findings seem to suggest that female employment contributes to the likelihood of divorce since the wife is less dependent on her husband's earnings (White 1991). Wives' employment may also lead to conflict about the traditional division of household labor, child care stress, and other work spillover problems that, in turn, create marital distress.

Employment also creates more opportunities for spouses to meet someone else and to embark on an extramarital sexual relationship. The presence and numbers of attractive alternative partners positively influences the risk of divorce. Scott South, Katherine Trent, and Yang Shen (2001) call this the *macrostructural opportunity perspective*, calling attention to the importance of opportunities for spouses to form potentially destabilizing opposite-sex relationships that are embedded within macrosocial structures, such as the workplace.

Also related to employment effects are the hours worked. Harriet Presser (2000) estimated that among men married less than five years and with young children, working night shifts increased their likelihood of divorce or separation by six times compared to men with similar families who worked days. Women with similar families who work nights face three times the likelihood of separation and divorce compared to those who work days. In the absence of children, the same effects are not found.

Income

The higher the family income, the lower the divorce rate for both Caucasians and African Americans (Clarke-Stewart and Brentano 2006). It is interesting, however, that at least some research indicates that the higher a woman's individual income, the greater her chances of divorce. Reasons for this could include the fact that with greater incomes, women are not economically dependent on their husbands or that conflict over inequitable work and family roles increases marital tension. However, not all research indicates a positive association between women's income and their risk of divorce (Sayer 2006).

Each spouse's income alone does not explain divorce, nor does the relative income earned by each spouse. Stacy Rogers (2004) found that the highest risk of divorce occurred in marriages in which wives contributed between 50% and 60% of the family's resources if spouses were at low or moderate levels of happiness. However, "happier spouses have little incentive to divorce, irrespective of spouses' relative economic contributions" (Rogers 2004, 71). Thus, neither higher-earning wives nor lower-earning husbands are automatically prone to divorce.

Educational Level

The decline in divorce that occurred in the 1980s and 1990s happened mostly for college-educated women and men (Martin 2004). The positive effect of education appears to be greatest in early marriage. During the first three years of marriage, the predicted risk of divorce among married women with less than a high school diploma is more than twice that for high school–educated women and nearly four times the risk faced by women who have been to college (South 2001).

Of course, educational attainment is usually linked with other factors that affect marital success. For example, men and women pursuing higher education tend to delay marriage and children until they're older. In addition, increased education may lead to acquiring values more conducive to marital success (Heaton 2002). One study concluded that college graduates had the most restrictive attitudes toward divorce, believing that "it should be more difficult to obtain a divorce than it is now" (Martin and Parashar 2006). Women who haven't completed high school have the least restrictive attitudes.

Ethnicity

About a third of first marriages end in separation or divorce within the first 10 years of marriage for white (32%) and Hispanic (34%) women; for non-Hispanic black women, the figure reaches nearly half of first marriages (47%) (Phillips and Sweeney 2005). Bulanda and Brown (2004) estimate that blacks face a risk of divorce nearly 1.5 times that of whites.

In Julie Phillips and Megan Sweeney's (2005) carefully controlled, multivariate analysis of the risk of divorce among a sample of more than 4,500 white, black, and Mexican American women, black women have a 54% greater risk of experiencing a marital separation or divorce than do white women. Foreign-born Mexican women have a 76% reduced risk compared to white women. American-born Mexican American women had risks of divorce that fell between those of Caucasian and African American women. These differences persist even when comparing women with similar experience in premarital cohabitation, with similar family backgrounds, and of similar education, employment, and age at marriage.

Religion

According to sociologists Vaughn Call and Timothy Heaton (1997, 391), "No single dimension of religion adequately describes the effect of *religious* experience on marital stability." Both *religiosity* (strength of religious commitment and participation) and religious affiliation have been linked to risk of divorce. Frequency of attendance at religious services (not necessarily the depth of beliefs) tends to be negatively associated with the divorce rate. That is, the greater the involvement in religious activities, the less the likelihood of divorce. But interestingly, a difference between spouses in frequency of attendance is a risk factor, too. Marriages in which wives attend services weekly and husbands don't attend have a greater risk of divorce than those marriages in which neither spouse attends religious services. The lowest risk is found among couples in which both spouses attend services regularly (Call and Heaton 1997).

Since all major religions discourage divorce, highly religious men and women are less likely to accept divorce because it violates their values. It may also be that a shared religion and participation in organized religious life affirms the couple relationship (Call and Heaton 1997; Guttman 1993; Wineberg 1994). Religiosity even seems to influence the likelihood of divorce when marital problems arise, suggesting that religion plays a role in the decision of whether to seek a divorce (Lowenstein 2005).

Life Course Factors

Different aspects of the life course may affect the probability of divorce. These include age at time of marriage, premarital pregnancy and childbirth, cohabitation, remarriage, and intergenerational transmission.

Age at Time of Marriage

The age at which people marry is "the most consistent predictor of marital stability identified in social science research" (Heaton 2002). Young, especially adolescent, marriages are more likely to end in divorce than are marriages that take place when people are in their twenties or older. Close to 50% of those who marry before age 18 and 40% of those who marry before turning 20 years old divorce. Younger partners are less likely to be emotionally mature, younger marriages may be more likely to involve premarital pregnancy, and marrying "young" may be associated with curtailment of education, which has

Marrying young, especially in one's teens, significantly increases one's risk of divorce.

economic consequences that can undermine marital stability. Only 25% of those who marry when older than 25 end up divorced. The effect of age at marriage is not the same for all ethnic groups, however. Marrying in their teens has a "destabilizing effect" on Caucasian and African American marriages but not on Mexican American marriages (Phillips and Sweeney 2005).

Premarital Pregnancy and Childbirth

Premarital pregnancy or birth significantly increases the likelihood of divorce, the risks being 1.2 to 1.3 times greater than for women without such experiences (Kposowa 1998). Risks are especially high if the pregnant woman is an adolescent, drops out of high school, and faces economic problems following marriage. If a woman gives birth before marriage, the likelihood for divorce in a subsequent marriage increases, especially in the early years. This negative, "destabilizing" effect of a premarital conception on marriage is stronger for African Americans than for Caucasians (Phillips and Sweeney 2005); that is, African Americans who have a baby before marriage are more likely to divorce than are Caucasian couples who have a baby before marrying.

Cohabitation

As shown in Chapter 9, premarital cohabitation has been associated with an elevated risk of a later divorce, though this may vary depending on the nature of one's cohabiting relationship. What is unclear is whether the greater risk results from the experience of cohabitation—say, by altering people's attitudes toward marriage and divorce—or is more of a reflection of the less traditional attitudes toward marriage and family, *including attitudes toward divorce*, that cohabitors bring with them into cohabitation.

Remarriage

It might seem reasonable to expect that having been married and divorced at least once would make people better at making a subsequent marriage succeed. One might think that formerly married people would know what to avoid in reentering marriage and that they would reenter with greater commitment and determination for success this time around. Although this may sound logical, the reality is that the divorce rate among those who remarry is *higher* than it is for those who enter first marriages.

It is not entirely clear why there is a higher divorce rate in remarriages. Some researchers suggest that the cause may lie in a "kinds-of-people" explanation. The probability factors associated with the kinds of people who divorced in first marriages—everything from low levels of education to unwillingness to settle for unsatisfactory marriages—are present in subsequent marriages, increasing their likelihood of divorce. Similarly, people bring their same personality problems to any new relationship. Others argue that the unique dynamics of subsequent marriages, especially the presence of stepchildren, increase the chances of divorce. In fact, subsequent marriages that involve stepchildren have twice the likelihood of divorce as first marriages (Schoen 2002/2003).

Intergenerational Transmission

Those whose parents divorce are subject to **intergenerational transmission**—the increased likelihood that divorce will later occur to them (Amato 1996; Cavanagh and Sullivan 2009; Li and Wu 2008). If both the husband's and the wife's parents have been divorced, the odds of divorce increase dramatically. Sociologist Nicholas Wolfinger (2005) states that when both spouses are children of divorces, their risk of divorce is three times greater than if neither are children of divorced parents. Wolfinger also indicates that men and women contribute equally to intergenerational transmission. How can we explain this intergenerational cycle?

Paul Amato (1996) noted that children of divorced parents are more likely to marry younger, cohabit, and experience higher levels of economic hardship. They become more pessimistic about lifelong marriage and develop more liberal attitudes toward divorce. In addition, females whose parents divorce develop less traditional attitudes about women's family roles, value self-sufficiency, and possess stronger attachments to paid employment. Each of these could raise susceptibility to divorce. Interestingly, parental "marital discord" in the absence of divorce has been found to have little consequence for their children's risk of divorce. Furthermore, high-discord marriages that ended in divorce only minimally raised their children's risk of divorce. However, where low-discord marriages ended in divorce, the children were especially vulnerable to divorce themselves (Amato and De Boer 2001).

Keep in mind that, as with intergenerational cycles of family violence, this relationship is neither automatic nor inevitable. It is, however, an important factor that can undermine marital success. One way in which parental divorce may be assumed to affect children's risk of divorce is in shaping their attitudes toward divorce. Children of divorced parents, especially daughters of divorced parents, are more likely to possess prodivorce attitudes (Kapinus 2004). Research that examined the effect of parents' attitudes on more than 400 children of divorce (Kapinus 2004) concludes the following:

- There appears to be a "critical period," namely, the late teens, when parents' attitudes toward divorce have special influence on their children. This is the period of time when one is likely to be dating and forming expectations about relationships.
- Parental divorce affects sons' and daughters' attitudes toward divorce differently. Daughters of divorce are more likely to express "prodivorce attitudes" than are sons of divorce.
- Diminished relationships with fathers after divorce and continued postdivorce conflict between parents may lead sons toward negative attitudes toward divorce. Yet postdivorce conflict between parents does not have the same effect on daughters.

Family Processes

Among other factors that one might expect to influence one's likelihood of divorce are family processes that affect couples' levels of happiness and their experience of problems. The actual day-to-day marital processes of communication—handling conflict, showing affection, and other marital interactions—may be among the most important factors affecting marital outcomes (Gottman 1994).

Marital Happiness

Although it seems reasonable that there would be a strong link between marital happiness (or, rather, the lack of happiness) and divorce, this is true only during the earliest years of marriage. Low levels of liking and trusting a partner are associated with long-term outcomes such as reduced satisfaction and elevated risk of divorce. The strength of the relationship between low marital happiness and divorce decreases in later stages of marriage, however (White and Booth 1991).

Eventually, alternatives to marriage and barriers to divorce appear to influence divorce decisions more strongly than does marital happiness. With nothing better to leave for or if there are too many obstacles to overcome in leaving, a couple might stay married even if unhappy. Although the opposite is also true—even if happy, one partner might leave for a more attractive alternative—it is probably less common. The presence of alternatives to a spouse has an effect on marital stability that can be observed among both high- and low-risk couples (i.e., among those with other predisposing factors and those without).

The importance of the availability of attractive alternatives to a spouse has sometimes been overlooked as a factor accounting for divorce. Scott South, Katherine Trent, and Yang Shen (2001, 753) note that "satisfied and dissatisfied spouses alike remain, consciously or not, in the marriage market." As explained earlier, the workplace is a central component of such a market.

Children

Although 60% of divorces involve children, couples with children divorce less often than couples without children. The birth of the first child reduces the chance of divorce to almost nil in the year following the birth (White 1991). Furthermore, couples with two children divorce less often than couples with one child or no children (Diekmann and Schmidheiny 2004). This does not mean that having children will spare parents from a divorce or that troubled spouses should

© Fancy/Photolibrary

The more people one is exposed to who could be attractive alternatives to one's spouse, the greater the risk of divorce.

become parents so that their troubles will disappear. It may well be that troubled spouses hold off having children or, if they have a child, resist having more because of their troubles. Thus, the quality of the marriage may lead to childbearing more than vice versa.

There are some situations in which the presence of children may be related to higher divorce rates. Children conceived during adolescence and physically or mentally challenged children are associated with divorce, as are children from prior marriages or relationships. At the same time, however, women without children have considerably higher divorce rates than women with children.

Marital Problems

If you ask divorced people to give the reasons for their divorce, they are not likely to say, "I blame the changing nature of the family" or "It was demographics." They are more likely to respond, "She was on my case all the time" or "He just didn't understand me"; if they are charitable, they might say, "We just weren't right for each other." Personal characteristics leading to conflicts are important factors in the dissolution of relationships.

Studies of divorced men and women cite such problems as alcoholism, drug abuse, marital infidelity, sexual incompatibility, and conflicts about gender roles as relationship factors leading to their divorces. They also often cite external events—problems with in-laws or the effect of jobs (Amato and Previti 2003).

Paul Amato and Denise Previti (2003) found the most common reasons given by their sample were infidelity, incompatibility, alcohol or drug use, growing apart, personality problems, lack of communication, and abuse (physical or mental).

Gender differences in reasons for divorce indicate that, in general, women cite emotional or relationship reasons, incompatibility, infidelity, unhappiness, and insufficient love as well as aspects of their former husband's personality or behaviors (such as abusiveness, neglect of children or home, and substance use). They are less likely to blame themselves. Men more often cite external factors or claim ignorance—they say they do not know what happened (Amato and Previti 2003).

People of high socioeconomic status are more likely to stress communication problems, incompatibility of or changes in values or interests, and their former spouse's self-centeredness. People of low socioeconomic status more often mention such things as financial problems, physical abuse, going out frequently with "the boys" or "with the girls," employment problems, neglect of home responsibilities, and drinking.

We know from studying enduring marriages that marriages often continue in the face of such problems. Recent research (Amato and Rogers 1997) on the connections between marital problems and divorce reveals that reports of marital problems in 1980 were associated with later divorce between 1980 and 1992. Based on interviews with almost 2,000 people, Paul Amato and Stacy Rogers (1997) found the following:

- Although men's and women's reports differed in the particular problems they emphasized, both predicted divorce equally well.
- Certain problems such as jealousy, moodiness, anger, poor communication, and drinking increased the odds of later divorce; sexual infidelity was an especially strong predictor of divorce.
- People who later divorce report a higher number of problems as early as 9 to 12 years before their divorce. Thus, their assessments of problems are not after-the-fact justifications concocted to account for or justify their divorce.

- Marital problems are **proximal causes** of later divorce. They are those things experienced within the daily life of married couples—such as high levels of conflict—that directly raise the probability of divorce. There are also background characteristics, such as age at marriage, prior cohabitation, education, income, church attendance, and parental divorce, that operate as more **distal causes**. These are brought by each spouse to the relationship and raise the likelihood that marital problems will later arise.

No-Fault Divorce

Since 1970, beginning with California's Family Law Act, all 50 states have adopted no-fault divorce—the legal dissolution of a marriage in which guilt or fault by one or both spouses does not have to be established. It is unclear exactly how or how much no-fault divorce has affected divorce rates. Some contend that liberalization of divorce law led to increases in the divorce rate in both the United States and in other countries (e.g., Scotland, England, and Wales) (Lowenstein 2005). It is debatable that this has, by itself, affected the divorce rate. Unambiguously, however, liberalization of divorce law has altered the process of divorce by decreasing the time involved in the legal process, and it has altered the grounds for determining postdivorce financial responsibility.

Uncoupling: The Process of Separation

Divorce is not caused by a single event. You don't wake up one morning and say, "I'm getting a divorce," and then leave. It's a far more complicated process (Kitson and Morgan 1991). It may start with little things that at first you hardly notice—a rude remark, thoughtlessness, an unreasonable act, or a "closedness." Whatever the particulars may be, they begin to add up. Other times, however, the sources of unhappiness are more blatant—yelling, threatening, or battering. For whatever reasons, the marriage eventually becomes unsatisfactory; one or both partners become unhappy.

It bears noting that we know less about the process of marital breakdown and divorce than we ought to, especially given its prevalence in the United States. We understand more about falling in love and courtship than we do about falling out of love and divorce (Furstenberg and Cherlin 1991).

Perhaps the crucial event in a marital breakdown is the act of separation. Although separation generally precedes divorce, not all separations lead to divorce. Furthermore, those that do may first involve attempts at reconciliation in that about one-third of the divorced women become divorced after attempting at least one marital reconciliation (Wineberg 1999). A statistic now more than a decade old indicates that perhaps 1 in 10 marriages experiences a separation and reconciliation (Wineberg and McCarthy 1993). Those who reconcile may have separated to dramatize their complaints, create emotional distance, or dissipate their anger (Kitson 1985).

Initiators and Partners

People do not suddenly separate or divorce. Instead, they gradually move apart through a set of fairly predictable stages. Sociologist Diane Vaughan (1986) calls this process *uncoupling*. The process appears to be the same for married or unmarried heterosexual couples and for gay or lesbian relationships. The length of time together does not seem to affect the process.

"Uncoupling begins," Vaughan observes, "as a quiet, unilateral process." Usually one person, the **initiator**, is unhappy or dissatisfied but keeps such feelings to himself or herself. Because the dissatisfied partner is unable to find satisfaction within the relationship, he or she begins turning elsewhere. This is not a malicious or intentional turning away; it is done to find self-validation without leaving the relationship. In doing so, however, the dissatisfied partner "creates a small territory independent of the coupled identity" (Vaughan 1986).

Eventually, the initiator decides that he or she can no longer go on. He or she may go through a process of mourning the demise of what is still an intact marriage (Emery 1994, cited in Amato 2000). After the relationship ends, initiators have better adjustment to divorce and carry less postdivorce attachment to their former spouses (Wang and Amato 2000).

Uncoupling does not end when the end of a relationship is announced or even when the couple physically separates. Acknowledging that the relationship cannot be saved represents the beginning of the last stage of uncoupling. Diane Vaughan (1986) describes the process:

> Partners begin to put the relationship behind them. They acknowledge that the relationship is unsaveable. Through the process of mourning they, too,

In the early phases of the process of separation, estrangement can grow before both parties are fully aware of what has happened.

makes each almost a physical part of the other. This dynamic is true even if two people are locked in conflict; they, too, are attached to each other (Masheter 1991).

Almost everyone suffers **separation distress** when a marriage breaks up. The distress is real, results from the absence of one's spouse, but, fortunately, does not last forever (although it may seem so). The distress is situational and is modified by numerous external factors. About the only men and women who do not experience distress are those whose marriages were riddled by high levels of conflict. In these cases, one or both partners may view the separation with relief (Raschke 1987).

During separation distress, almost all attention is centered on the missing partner and is accompanied by apprehensiveness, anxiety, fear, and often panic. "What am I going to do?" "What is he or she doing?" "I need him . . . I need her . . . I hate him . . . I love him . . . I hate her . . . I love her."

Social support is positively correlated with lower distress and positive adjustment. Additionally, as with other stressors in a person's life, it is often the individual's perception of the event, not the stress itself, that influences how a person adjusts to change. If those experiencing separation and divorce can begin to view and accept their changing circumstances as presenting new challenges and opportunities, there is a greater likelihood that the physiological and psychological symptoms of stress that follow divorce can be reduced.

eventually arrive at an account that explains this unexpected denouement. "Getting over" a relationship does not mean relinquishing that part of our lives that we shared with another but rather coming to some conclusion that allows us to accept and understand its altered significance. Once we develop such an account, we can incorporate it into our lives and go on.

The New Self: Separation Distress and Postdivorce Identity

Examining the experiences of those who divorce may be as good a way as any to see how much our married self becomes part of our deepest self. When people separate or divorce, many feel as if they have "lost an arm or a leg." This analogy, as well as the traditional marriage rite in which a man and a woman are pronounced "one," reveals an important truth of marriage: the constant association of both partners

Establishing a Postdivorce Identity

A person goes through two distinct phases in establishing a new identity following marital separation: *transition* and *recovery* (Weiss 1975). The transition period begins with the separation and is characterized by separation distress and then loneliness. In this period's later stages, most people begin functioning in an orderly way again, although they still may experience bouts of upset and turmoil. The transition period generally ends within the first year. During this time, individuals have already begun making decisions that provide the framework for new selves. They have

Critical Thinking

From your experience, how well does "uncoupling" describe the process of separating from someone you care about? Are there missing elements that should be emphasized? What about separation distress? In your experience, what was it like? What things were you able to do to alleviate it? What advice would you give others about the distress of separating?

Public Policies, Private Lives: How Can You Get a Divorce
If They Don't Recognize Your Marriage?

With same gender marriage now legal in five states, one would anticipate an increase in the numbers of same-gender couples who decide to enter marriage. However, as more same-gender couples enter marriage, increasing numbers of breakups are sure to follow. This illustrates one of the sometimes overlooked arguments for same-gender marriage—the right to divorce (Allen 2007b).

Being that issues such as child custody and visitation as well as spousal support are tied to how states handle divorce and dissolution, if one can't divorce, one lacks such protections. The situation is complicated not only by the lack of marital rights faced by same gender couples in 44 states of the United States but also by the fact that even if one legally marries in one of the small number states that allow same-gender marriages, one still may find oneself without the right to divorce. When two women or two men marry, their ability to divorce hinges on whether they continue to reside in a state that grants gays and lesbians the right to marry. If, for example, a same-gender couple marry and reside in the state in which they were legally married, they can seek a divorce in that state. If they live in one of the other states that allows same-gender marriage (e.g., Massachusetts, New Hampshire, Vermont, Connecticut, or Iowa) or in either New York State or the District of Columbia (both of which acknowledge same-gender marriages performed elsewhere), their situation may not be any more complicated than divorce is for heterosexuals. They can simply use the same policies as are used in heterosexual divorces or dissolutions. Although divorce and dissolution may be experienced as anything

but uncomplicated, where they are available, couples at least have a means to legally end their relationships. But what if the relationships aren't legally recognized and legally protected in the first place?

Consider the predicament faced by Cassandra Ormiston and Margaret Chambers, a lesbian couple who married in Massachusetts in 2004 but filed for divorce two years later while living in Rhode Island. Rhode Island is not among the states that recognize same-gender marriages. Thus, Ormiston and Chambers could not legally end a relationship that the state refuses to recognize as a legally binding marriage. In a sense, they could neither enter nor exit marriage if residents of Rhode Island. Meanwhile, Massachusetts allows married gays and lesbians to divorce, but only if state residents for at least a year. Massachusetts let Ormiston and Chambers marry, but it won't let them divorce. Instead, Ormiston and Chambers first must establish Massachusetts residency by living there for a year.

As Cassandra Ormiston stated, "I find that an unbelievably unfair burden. I own a home here [Rhode Island], my friends are here, my life is here." Still, to bring her marriage to a legal end, she must go live in Massachusetts for a year. Ninety other same-gender Rhode Island couples would be in the same situation as Ormiston and Chambers should they want out of their marriages. Married same-gender couples who now live in any of the more than 40 states that have placed explicit bans on same-gender marriage would also face such hardships.

SOURCES: Allen (2007b) and Ray (2008).

people happier or whether it is happier people who get married and *stay* married. Citing research that compared the emotional health of a sample of people over time—some who married and stayed married, some who never married or remained divorced, and others who married and divorced—they report the following. When people married, their mental health substantially improved. When people separated and divorced, they suffered declines in their emotional and mental well-being.

Waite and Gallagher (2000) also note that compared to married people, divorced (and widowed) women and men were three times as likely to commit suicide.

Among the divorced, as among the general population, more men than women commit suicide. However, among women, those who are divorced are "the most likely to commit suicide, followed by widowed, never-married and married, in that order." As parents, divorced individuals have more difficulty raising children. They display more role strain, whether they are custodial or noncustodial parents, and they display less authoritative parenting styles (Amato 2000).

Despite the stark picture that surfaces, for some people divorce is associated with positive consequences. These include higher levels of personal growth, greater autonomy, and—for some women—improvements in self-confidence, career opportunities, social lives, and happiness as well as a stronger sense of control (Amato 2000). In addition, it would be remiss not to stress the fact that evidence indicates that remaining unhappily married is worse than divorcing. People who find themselves in "long-term low-quality marriages" are less happy than those who divorce and remarry. They also have lower overall life satisfaction, lower overall health, and lower self-esteem than those who divorce and remain single (Hawkins and Booth 2005).

Children and Divorce

Divorce not only ends marriages and breaks up families; it also can create new family forms from the old ones. It can lead one or both former spouses into remarriages (which are different from first marriages). If there are children involved, it gives birth to single-parent families, stepfamilies, and the **binuclear family**, the family created when parents divorce, making their children now members of two separate households. Today, about one out of every five American families is a single-parent family; more than half of all children will become stepchildren (U.S. Census Bureau 1996). Within the singles subculture is an immense pool of divorced men and women (most of whom are on their way to remarriage). Or consider the numbers of marriages that are truly remarriages for one or both spouses. Nearly 1 in 10 marriages in the United States consists of two people who have both been married before to other spouses. As noted earlier, such marriages carry increased risk of divorce.

The greatest concern that social scientists express about divorce is its effect on children (Aldous 1987; Wallerstein 1997; Wallerstein and Blakeslee 1989). But even in studies of the children of divorce, the research may be distorted by traditional assumptions about marriage in which divorce is clearly frowned on (Amato 1991). For example, problems that children experience may be attributed to divorce rather than to other causes, such as personality traits. Although some effects are caused by the disruption of the family itself, others may be linked to the new social environment—most notably poverty and parental stress—into which children are thrust by their parents' divorce (McLanahan and Booth 1991; Raschke 1987). Some therapists suggest that we begin looking at those factors that help parents and children successfully adjust to divorce rather than focusing on risks, dysfunctions, and disasters (Abelsohn 1992).

Slightly more than half of all divorces involve children. Popular images of divorce depict "broken homes," but it is important to remember that an intact nuclear family, merely because it is intact, does not guarantee children an advantage over children in a single-parent family or a stepfamily. An intact family wracked with frequent and intense conflict between spouses, spousal violence, sexual or physical abuse of children, alcoholism, neglect, or psychopathology creates a destructive environment likely to inhibit children's healthy development.

Although living in a two-parent family with severe marital conflict is often more harmful to children than living in a tranquil single-parent family or stepfamily, what about what researchers consider low-discord marriages? In such relationships, parents avoid overt conflict, cooperate with each other, behave in a civil—even respectful—manner toward each other, but feel as though their marriages are lacking in some way. Researchers contend that children in such households may be better off if their parents stay together, even if only until they are grown and out of the household and even if parents are less than fulfilled by their marriage. By keeping conflict limited and away from the children, the risk of children having developmental and emotional problems is low. Children suffer more interpersonal and psychological problems when their parents end such "good-enough" marriages (Amato 2003). For them, divorce may represent "an unexpected, unwelcome, and uncontrollable event." They face the loss of one parent, the emotional distress of the remaining parent, and perhaps a decline in standard of living (Booth and Amato 2001).

Unmistakably, children living in happy, two-parent families appear to be the best adjusted, and those from conflict-ridden two-parent families appear to be the worst adjusted. Children from stable, well-functioning single-parent families are in the middle (Coontz, 1997).

How Children Are Told

Telling children that their parents are separating is one of the most difficult and unhappy events in life. Whether or not the parents are relieved about the separation, they often feel extremely guilty about their children. Many children may not even be aware of the extent of parental discord, especially in low-discord marriages (Amato 2006; Furstenberg and Cherlin 1991). Even those that are may be upset by the separation, but their distress may not be immediately apparent.

Qualitative research by Heather Westberg, Thorana Nelson, and Kathleen Piercy (2002) indicates that children's reaction is influenced by how the news is disclosed and is shaped by the perception that life will be relatively better or relatively worse afterward. For those to whom the news was disclosed long before the divorce occurred, by the time it "finally" happened, it was experienced as relief. For those who had lived with frequent and intense conflict between parents, divorce may also be welcome. For those who fear that they will have to move or that they won't see their noncustodial parent, the news will cause distress.

As psychologist Judith Wallerstein suggested in her book *Second Chances* (Wallerstein and Blakeslee 1989), divorce is differently experienced within the family. For at least one of the divorcing spouses, divorce is welcomed as an escape from an unpleasant or unfulfilling relationship. Both spouses may come to appreciate the "second chance" they receive with divorce: the opportunity to make a better choice and build themselves a better relationship. Children may not see the breakup of their parents' marriage as an "opportunity." However, under certain circumstances—most notably when parental conflict has been long-term, overt and unresolved—children are at risk of developing emotional and developmental problems so long as their parents stay together (Booth and Amato 2001). For such children, divorce may, indeed, come as a relief.

Lisa Strohschein (2005) found that children's antisocial behaviors such as bullying and lying were reduced after the divorce of parents who had been experiencing high levels of dysfunction. The stress relief that comes with divorce may, however, become apparent only after enough time passes (Strohschein 2005).

The Three Stages of Divorce for Children

Part of the difficulty in determining the effect of divorce on children is a failure to recognize that, just as it is for adults, divorce is a process as opposed to a single event. Divorce comprises a series of events and changes in life circumstances. Many studies focus on only one part of the process and identify that part with divorce itself. Yet at different points in the process, children are confronted with different tasks and adopt different coping strategies. Furthermore, the diversity of children's responses to divorce is the result, in part, of differences in temperament, gender, age, and past experiences.

A study by psychologist Judith Wallerstein (1997) found that children from divorced families suffered both emotionally and developmentally. Young children fared worse than older children. Depending on the point in the process, boys tend to do less well than girls. In the "crisis period" of the two years following separation, boys' suffering is especially evident. This may be because they must internalize different gendered styles of reacting to distress. It is also the case, however, that after separation, most boys live with their mothers and not their fathers. This, too, can exacerbate their suffering, as they suffer diminished contact with their fathers (Furstenberg and Cherlin 1991).

According to Wallerstein (1983), children experience divorce as a three-stage process. Studying 60 California families during a five-year period, she argued that divorce consisted of the initial, transition, and restabilization stages:

- *Initial stage.* The initial stage, following the decision to separate, was extremely stressful; conflict escalated, and unhappiness was endemic. The children's aggressive responses were magnified by the parents' inability to cope because of the crisis in their own lives.
- *Transition stage.* The transition stage began about a year after the separation, when the extreme emotional responses

© Pixland/Photolibrary

Notifying children of a decision to divorce is difficult for both the parents and children.

of the children had diminished or disappeared. The period was characterized by restructuring of the family and by economic and social changes: living with only one parent and visiting the other, moving, making new friends and losing old ones, financial stress, and so on. The transition period lasted between two and three years for half the families in the study.

- *Restabilization stage.* Families had reached the restabilization stage by the end of five years. Economic and social changes had been incorporated into daily living. The postdivorce family, usually a single-parent family or stepfamily, had been formed.

Children's Responses to Divorce

Decisive in children's responses to divorce are their age and developmental stage (Guttman 1993). A child's age affects how one responds to one parent leaving home, changes (usually downward) in socioeconomic status, moving from one home to another, transferring schools, making new friends, and so on.

Developmental Tasks of Divorce

Judith Wallerstein (1983) suggested that children must undertake six developmental tasks when their parents divorce. The first two tasks need to be resolved during the first year. The other tasks may be worked on later; often they may need to be reworked because the issues may recur. How children resolve these tasks differs by age and social development. The tasks are as follows:

- *Acknowledging parental separation.* Children often feel overwhelmed by feelings of rejection, sadness, anger, and abandonment. They may try to cope with them by denying that their parents are "really" separating. They need to accept their parents' separating and to face their fears.
- *Disengaging from parental conflicts.* Children need to psychologically distance themselves from their parents' conflicts and problems. They require such distance so that they can continue to function in their everyday activities without being overwhelmed by their parents' crisis.
- *Resolving loss.* Children lose not only their familiar parental relationship but also their everyday routines and structures. They need to accept these losses and focus on building new relationships, friends, and routines.

- *Resolving anger and self-blame.* Children, especially young ones, often blame themselves for the divorce. They are angry with their parents for disturbing their world. Many often "wish" their parents would divorce, and when their parents do, they feel responsible and guilty for "causing" it.
- *Accepting the finality of divorce.* Children need to realize that their parents will probably not get back together. Younger children hold "fairy-tale" wishes that their parents will reunite and "live happily ever after." The older the child is, the easier it is for him or her to accept the divorce.
- *Achieving realistic expectations for later relationship success.* Children need to understand that their parents' divorce does not condemn them to unsuccessful relationships as adults. They are not damaged by witnessing their parents' marriage; they can have fulfilling relationships themselves.

Younger Children. Younger children react to the initial news of a parental breakup in many different ways. Feelings range from guilt to anger and from sorrow to relief, often vacillating among all of these. Preadolescent children, who seem to experience a deep sadness and anxiety about the future, are usually the most upset. Some may regress to immature behavior, wetting their beds or becoming excessively possessive. Most children, regardless of their age, are

Children react differently to divorce depending on their age. Most feel sad, but the eventual outcome for children depends on many factors, including having competent and caring custodial parent, siblings, and friends and their own resiliency. The postdivorce relationship between parents and the custodial parent's economic situation are also important factors.

angry because of the separation. Very young children tend to have more temper tantrums. Slightly older children become aggressive in their play, games, and fantasies—for example, pretending to hit one of their parents.

A 1996 study using longitudinal data collected over a 12-year period examines parent–child relationships before and after divorce. Researchers found that marital discord may exacerbate children's behavior problems, making them more difficult to manage (Amato and Booth 1996). Because discord between parents often preoccupies and distracts them from the tasks of parenting, they appear unavailable and unable to deal with their children's needs. This study reinforced a growing body of evidence showing that many problems assumed to be caused by divorce are present before marital disruption.

School-age children may blame one parent and direct their anger toward him or her, believing the other one innocent. But even in these cases, the reactions are varied. If the father moves out of the house, the children may blame the mother for making him go, or they may be angry at the father for abandoning them regardless of reality. Younger schoolchildren who blame the mother often mix anger with placating behavior, fearing that she will leave them. Preschool children often blame themselves, feeling that they drove their parents apart by being naughty or messy. They beg their parents to stay, promising to be better. It is heartbreaking to hear a child say, "Mommy, tell Daddy I'll be good. Tell him to come back. I'll be good. He won't be mad at me anymore." A study of 121 white children between the ages of 6 and 12 found that about 33% initially blamed themselves for their parents' divorce. After a year, the figure dropped to 20% (Healy, Stewart, and Copeland 1993). The largest factor in self-blaming was being caught in the middle of parental conflict. Children who blamed themselves displayed more psychological symptoms and behavior problems than those who did not blame themselves.

When parents separate, children want to know with whom they are going to live. If they feel strong bonds with the parent who leaves, they want to know when they can see him or her. If they have brothers or sisters, they want to know if they will remain with their siblings. They want to know what will happen to them if the parent they are living with dies. Will they go to their grandparents, their other parent, an aunt or uncle, or a foster home? These are practical questions, and children have a right to answers. They need to know what lies ahead for them amid the turmoil of a family split-up so that they can prepare for the changes.

Some parents report that their children seemed to do better psychologically than they themselves did after a split-up. Children often have more strength and inner resources than parents realize. The outcome of separation for children, Robert Weiss (1975) observed, depends on several factors related to the children's age. Young children need a competent and loving parent to take care of them; they tend to do poorly when a parenting adult becomes enmeshed in constant turmoil, depression, and worry. With older, preadolescent children, the presence of brothers and sisters helps because the children have others to play with and rely on in addition to the single parent. If they have good friends or do well in school, this contributes to their self-esteem. Regardless of the child's age, it is important that the absent parent continue to play a role in the child's life. The children need to know that they have not been abandoned and that the absent parent still cares (Wallerstein and Kelly 1980b). They need continuity and security, even if the old parental relationship has radically changed.

Adolescents. Many adolescents find parental separation traumatic. Studies indicate that much of what appear to be negative results of divorce (personal changes, parental loss, economic hardships, and psychological adjustments) are often more likely the result of parental conflict that precedes and surrounds the divorce (Amato and Booth 1996; Amato and Keith 1991; Morrison and Cherlin 1995). A study by Youngmin Sun (2001) found that such problems as poor psychological well-being, academic difficulties, and behavioral problems are present among adolescents from divorced families *at least a year before* the divorce.

Adolescents may try to protect themselves from the conflict preceding separation by distancing themselves. Although they usually experience immense turmoil within, they may outwardly appear cool and detached. Unlike younger children, they rarely blame themselves for the conflict. Rather, they are likely to be angry with both parents, blaming them for upsetting their lives. Adolescents may be particularly bothered by their parents' beginning to date again. Some are shocked to realize that their parents are sexual beings, especially when they see a separated parent kiss someone or bring someone home for the night. The situation may add greater confusion to the adolescents' emerging sexual life. Some may take the attitude that if their mother or father sleeps with a date, why can't they? Others may condemn their parents for acting "immorally."

Kathleen Boyce Rodgers and Hillary Rose (2002) assert that the negative effects of divorce on adolescents can be tempered. They suggest that strong peer support, a strong attachment to school, and high levels of support and monitoring by parents can lessen the negative consequences adolescents otherwise encounter.

Helping Children Adjust

Psychologist Helen Raschke's (1987) review of the literature on children's adjustment after divorce found that the following factors were important:

- Before separation, open discussion with the children about the forthcoming separation and divorce and the problems associated with them
- The child's continued involvement with the noncustodial parent, including frequent visits and unrestricted access
- Lack of hostility between the divorced parents
- Good emotional and psychological adjustment to the divorce on the part of the custodial parent
- Good parenting skills and the maintenance of an orderly and stable living situation for the children

Continued involvement with the children by both parents is important for the children's adjustment. Research indicates that children are more likely to suffer long-term psychological damage—well into adulthood—if the parents do not consider their emotional needs during the divorce process (Wallerstein 1997).

Betwixt and Between: Children Caught in the Middle

Recently divorced women and men often suffer from a lack of self-esteem and a sense of failure. One means of dealing with the feelings caused by divorce is to blame the other person. To prevent further hurt or to get revenge, divorced parents may try to control each other through their children. One consequence of this for children is the sense of being caught in the middle, forced to choose sides, and pulled in different directions by their parents. Some have even suggested that feeling caught between parents may be one of the factors that differentiate children's reactions to divorce, explaining why some do better and some do worse. Such feelings may also lead to adolescent depression

and deviant behavior. Evidence indicates that older adolescents are more likely than younger adolescents and children to feel caught. In addition, such feelings may extend well into adulthood, although reduced contact with both parents may lessen the intensity of such feelings.

When caught in the middle, children may opt for one of three strategies: try to maintain positive relationships with both parents, form an alliance with one parent over and against the other, or reject both parents. Trying to remain close to two embattled parents may exact costs that outweigh the benefits of such relationships. Choosing sides comes at the expense of a relationship with one parent and can trigger guilt toward the abandoned parent and resentment toward the custodial parent. Rejecting both parents means losing closeness to both—a steep price to pay. Paul Amato and Tamara Afifi (2006) also found that parents put more pressure on daughters than on sons to take sides in their disputes and that feeling caught in the middle is of more negative consequence for mothers and daughters than for mothers and sons.

Perspectives on the Long-Term Effects of Divorce on Children

There are multiple perspectives on how and why divorce affects children (Amato 1993). Specified outcomes range from negative through neutral to positive (Coontz 1997; Whitehead 1996). There is enough divergent information that we could selectively cite research to make either a more pessimistic or a more optimistic generalization. We review some of these mixed findings here.

A variety of studies reviewed by Barbara Dafoe Whitehead in her strongly antidivorce book *The Divorce Culture* (1997) suggest multiple ways in which children suffer after their parents divorce. First, across racial lines, children of divorce suffer substantial reduction in family income as a direct result of divorce. Second, most children experience a weakening of ties with their fathers, suffering damage when and after fathers leave. She suggests that separation and later divorce induce a "downward spiral" in father–child relationships wherein distance between them grows and children eventually lose their fathers' "love, support, and substantial involvement." Third, children suffer a loss of "residential stability," often having to move from the family home because of drops in their economic standing.

Whitehead (1997) goes on to detail other measurable ways in which children suffer: reduced school performance, increased likelihood of dropping out, worsened and increased behavioral problems, and a greater likelihood of becoming teen parents. Many of these same outcomes were identified as among the "risks and problems associated with stepfamily life" (Whitehead 1996).

In her more optimistic book, *The Way We Really Are: Coming to Terms with America's Changing Families*, Stephanie Coontz (1997) tempers some of this distressing news. While acknowledging the "agonizing process" that accompanies divorce and the ways in which children, especially, can be hurt by divorce, Coontz qualifies the more pessimistic interpretations. In a subtle but important comparison, she notes that research shows "*not* that children in divorced families have more problems but that *more* children of divorced parents have problems" (Coontz 1997, 99, emphasis in original). In other words, all children of divorce do not suffer the negative consequences identified by researchers and reported by people such as Whitehead. Coontz reminds us that although more children in divorced homes drop out or become pregnant than do children whose parents stay married, "divorce does not account for the majority of such social problems as high school dropout rates and unwed teen motherhood" (Coontz 1997, 100). Finally, Coontz goes even further in an optimistic direction, noting that there are some measures on which large proportions of children of divorced homes score higher than do children from homes with two parents. She reports that children of single parents (usually single mothers) spend more time talking with their custodial parent, receive more praise for their academic successes, and face fewer pressures toward conventional gender roles. Thus, she argues, in some ways single-parent households may be beneficial environments within which to be raised (Coontz 1997).

Just How Bad Are the Long-Term Consequences of Divorce?

The message about the long-term consequences varies according to the research examined. Influential longitudinal research conducted by Judith Wallerstein highlights fairly extensive, long-term trauma and distress that stays with and affects children of divorce well into adulthood. Beginning with *Surviving the Breakup: How Children and Parents Cope with Divorce* (Wallerstein and Kelly 1980b) through *Second*

Chances: Men, Women, and Children a Decade after Divorce (Wallerstein and Blakeslee 1989) and culminating with *The Unexpected Legacy of Divorce: A 25-Year Landmark Study* (Wallerstein, Lewis, and Blakeslee 2000), Wallerstein has followed a sample of (originally) 60 families, with 131 children among them, as they divorced and went through the subsequent adjustment processes at 18 months, 5 years, 10 years, 15 years, and ultimately 25 years. Seventy-five percent of the original families, and 71% of the 131 children were studied for all three books.

Wallerstein found that at the five-year mark, more than a third of the children were struggling in school and experiencing depression, had difficulty with friendships, and continued to long for a parental reconciliation. At the 10-year follow-up, she indicated that almost half the children carried lingering problems and that they had become worried, sometimes angry, underachieving young adults. Three-fifths of the children of divorce retained a persistent sense of rejection by one or both parents and suffered especially poor relationships with their fathers. Finally, at the 25-year point, Wallerstein asserted that the effects of divorce on children reached their peak in adulthood, where the ability to form and maintain committed intimate relationships was negatively affected (see Amato 2003).

A more moderate view of the long-term effects of divorce emerges from other studies (Amato 2003; Hetherington and Kelly 2002). E. Mavis Hetherington undertook the Virginia Longitudinal Study of Divorce and Remarriage, which initially consisted of following a sample of 144 families with a four-year-old "target child." Half the sample families were divorced, and half were married. Initially, they were to be followed and restudied at two years to compare how those who divorced fared in comparison to those who did not. Eventually, the sample was expanded, and subsequent research was conducted at 2, 6, 11, and 20 years after divorce. As the "target children" (i.e., the initial four-year-olds) married, had a child, or cohabited for more than six months, they were further studied (Hetherington and Kelly 2002). Meanwhile, families were added to the sample at each wave to reach a final sample of 450 evenly split between nondivorced, divorced, and remarried families.

Throughout the research, a variety of qualitative and quantitative data were collected on personalities of parents and children, adjustment, and relationships within and outside the family (Hetherington 2003). The impression that Hetherington's research leaves is more encouraging than the one received from Wallerstein's studies. For example, most adults and children adapt to the divorce within two to three years. Although at the one-year mark 70% of the divorced parents were wrestling with animosity, loneliness, persistent attachment, and doubts about the divorce, by six years most were moving toward building new lives. More than 75% of the sample said that the divorce had been a good thing, more than 50% of the women and 70% of the men had remarried, and most had embarked on the postdivorce paths they would continue to take (Hetherington 2003).

In considering the effects of divorce on children, Hetherington reports that 20% of her sample of youths from divorced and remarried families was troubled and displayed a range of problems, including depression and irresponsible, antisocial behavior. They had the highest dropout rate, had the highest divorce rate (as they themselves married), and were the most likely to be struggling economically. But perhaps more important, "80 percent of children from divorced homes eventually are able to adapt to their new life and become reasonably well adjusted" (Hetherington and Kelly 2002, 228). Given that 10% of youths from nondivorced homes also were struggling, the difference for children from divorced as opposed to nondivorced homes was fairly small (10%).

As Hetherington points out, the optimal outcome for adults and their children is to be in a happily married household. Nevertheless, her research indicates that we may overstate the risks and fail to recognize the resilience of men, women, and children of divorce.

Paul Amato (2003) suggests that much of the divorce research supports Wallerstein's claims that divorce is "disruptive and disturbing" in the lives of children, but he fails to find the same strength and pervasiveness of the supposed effects. Using still other longitudinal data gathered as part of the Marital Instability over the Life Course Study, Amato reports that 90% of children with divorced parents achieve the same level of adult well-being as children of "continuously married parents" (Amato 2003). Amato further suggests that children who experience multiple family transitions (parental divorce, remarriages, subsequent divorces, and so on) are the ones who most suffer. He found that children who experienced only a single parental divorce (without any additional parental transitions) were no different in their psychological well-being than children of continuously married parents.

Child Custody

Of all the issues surrounding separation and divorce, custody issues are particularly poignant because they represent continued versus strained or even severed ties between one parent and his or her children. Historically, until the mid-nineteenth century, child custody tended to go to fathers. The pendulum then swung in the direction of granting mothers custody of children, unless they were considered unfit parents. Where mothers were typically awarded custody, fathers were granted visitation. Beginning then in the 1970s, custody decisions were made on a more gender-neutral basis, taking into account the "best interests of the child" (Elrod and Dale 2008).

Despite the more recent trend of gender-neutral custody considerations and assignments, custody of the children is awarded to the mother in about 90% of the cases. Three reasons can be given for this: (1) women usually prefer custody, and men do not; (2) giving custody to the mother is traditional; and (3) the law reflects a bias that assumes women are naturally better able to care for children.

Sexual orientation has also been a traditional basis for awarding custody (Baggett 1992; Beck and Heinzerling 1993). In the past, a parent's homosexuality has been sufficient grounds for denying custody, but increasingly courts are determining custody on the basis of parenting ability rather than sexual orientation. Interviews with children whose parents are gay or lesbian testify to the children's acceptance of their parents' orientation without negative consequences (Bozett 1987).

Types of Custody

The major types of custody are sole, joint, and split. In **sole custody**, the child lives with one parent who has sole responsibility for physically raising the child and making all decisions regarding his or her upbringing. Until the 1970s, sole custody was the norm (Elrod and Dale 2008). In **split custody**, the children in the family are divided between the divorcing parents, with each parent receiving physical custody of at least one of the children. Split custody can have especially harsh effects on relationships between siblings. For the most part, courts are reluctant to use split custody without compelling reasons to do otherwise (Oliphant and ver Steegh 2007).

To keep both parents involved with children after a divorce, custody decisions moved toward shared or joint custody. If and when parents can cooperate and effectively coparent, joint custody is the preferred form of custody arrangements in the United States today. There are two forms of **joint custody**: legal and physical. In **joint legal custody**, the children live primarily with one parent, but both parents share jointly in decisions about their children's education, religious training, and general upbringing. In **joint physical custody**, the children actually live with both parents, dividing time between the two households.

Even though joint custody is not a guarantee that the child's time will be evenly divided between parents, it does give children the chance for a more normal and realistic relationship with each parent (Arditti and Keith 1993). Joint physical custody, however, requires considerable energy from the parents in working out both the logistics of the arrangement and their feelings about each other. Joint physical custody is particularly problematic in instances where conflict is high or domestic violence has occurred (Elrod and Dale 2008).

Any custody arrangement has both benefits and drawbacks, and joint custody is no exception. Despite the appeal of keeping both parents engaged and involved with their children that is at the heart of joint custody, it appears as though its effectiveness depends on the kind and quality of relationships between ex-spouses. Parental conflict appears to cause greater suffering among children in joint custody arrangements. Furthermore, imposed joint custody, over the strong objections of one of the parents may be more harmful to the children and their relationships with parents than would sole custody (Elrod and Dale 2008). Joint custody may force two parents to interact (*cooperate* is too benign a word) when they would rather never see each other again, and the resulting conflict and ill will may be detrimental to the children. Parental hostility may make joint custody the worst form of custody (Opie 1993).

Parental satisfaction with court-imposed custody arrangements depends on many factors (Arditti 1992; Arditti and Allen 1993). These include how hostile the divorce was, whether the parent without physical custody perceives visitation as lengthy and frequent enough, and how close that parent feels to his or her children. In addition, the amount of support payments affects satisfaction. If parents feel that they are paying too much or were "cheated" in the property settlement, they are also likely to feel that the custody arrangements are unfair. Unfortunately, custodial satisfaction is not necessarily related to the best interests of the child.

Critical Thinking

What form of custody do you believe is the most advantageous to a child? What factors would you consider important in deciding which is the best type of custody for a particular child? If two parents constantly battled over their children, what are some of the consequences you might expect for the children? How do children cope in such circumstances?

Noncustodial Parents

Much public attention has been directed at "deadbeat" parents who are depicted as absent, noncaring, and failing to honor and maintain their child support obligations. In fact, noncustodial parents demonstrate varying degrees of involvement (Bray and Depner 1993; Depner and Bray 1993). Noncustodial parent involvement exists on a continuum in terms of caregiving, decision making, and parent–child interaction, from highly involved to completely removed. Involvement also changes depending on whether the custodial family is a single-parent family or a stepfamily (Bray and Berger 1990).

Typically, one hears more about "deadbeat dads" than either deadbeat parents or "deadbeat moms." The label hardly captures the reality for most noncustodial fathers. Many noncustodial fathers suffer grievously from the disruption or disappearance of their father roles following divorce. They feel depressed, anxious, and guilt ridden; they feel a lack of self-esteem (Arditti 1990). The change in status from full-time father to noncustodial parent leaves fathers bewildered about how they are to act; there are no norms for an involved noncustodial parent.

Some men act irresponsibly after a divorce, failing to pay child support and possibly becoming infrequent parts of their children's lives. This lack of norms makes it especially difficult if the relationship between the former spouses is bitter. Without adequate norms, fathers may become "Disneyland Dads," who interact with their children only during weekends, when they provide treats such as movies and pizza, or they may become "Disappearing Dads," absenting themselves from all contact with their children. For many concerned noncustodial fathers, the question is simple but painful: "How can I be a father if I'm not a father anymore?"

Noncustodial parents often weigh the costs of continued involvement with their children, such as emotional pain and role confusion, against the benefits, such as emotional bonding (Braver et al. 1993a, 1993b). Those who perceive that they have a say in the decisions affecting their children and who are stably employed and thus have the means to contribute tend to stay involved and to pay support (Braver et al. 1993a, 1993b). However, a lack of sufficient income is one significant source of nonsupport, prompting social work professor Ronald Mincy and economist Elaine Sorensen to refer to them as "turnips" in recognition of the cliché that "you can't get blood from a turnip" (Mincy and Sorensen 1998).

Children may eventually have little contact with their nonresidential parents. This reduced contact seems to weaken the bonds of affection. Divorced fathers are less likely to consider their children sources of support in times of need (Amato 1994; Cooney 1994). Although perhaps better than Frank Furstenberg and Christine Nord's (1985) claim of more than two decades ago that "marital dissolution involves either a complete cessation of contact between the nonresidential parent and child or a relationship that is tantamount to a ritual form of parenthood," noncustodial parents certainly see their relationships suffer considerably.

Factors that affect visitation by noncustodial parents include geographic proximity, feelings about one's former spouse, and the remarriage of either parent. Noncustodial mothers in many ways behave similarly to noncustodial fathers, indulging their children during visitations and reduced involvement in children's daily lives. Research indicates that compared to noncustodial fathers, noncustodial mothers speak more often with their children by phone, write them more letters, and have more frequent extended visits. However, they experience the same difficulty maintaining consistent visitation and, like noncustodial fathers, blame the other parent (Pollack and Mason 2004).

A recommendation worth considering is one made by social work professors Daniel Pollack and Susan Mason. They advocate **mandatory visitation**, suggesting that rather than *award* or *grant* visitation rights to noncustodial parents, we should treat visitation as an obligation of parents and an expectation to be enforced by authorities. If in fact children gain socially, emotionally, and psychologically from continued contact with and involvement by both parents, then there could be much gained by mandating contact and visitation, as long as there is sufficient cooperation between ex-spouses and an absence of high levels of ongoing conflict (Pollack and Mason 2004).

© Pictor Images/ImageStates

It is usually important for a child's postdivorce adjustment that he or she have continuing contact with the noncustodial parent. Noncustodial parents are involved with their children in varying degrees.

Divorce Mediation

The courts are supposed to act in the best interests of the child, but they often victimize children by their emphasis on legal criteria rather than on the children's psychological well-being and emotional development (Schwartz 1994). There is increasing support for the idea that children are better served by those with psychological training than by those with legal backgrounds (Miller 1993). Growing concern about the effect of litigation on children's well-being has led to the development of divorce mediation as an alternative to legal proceedings (Walker 1993).

Critical Thinking

If you were divorcing, what would be the pros and cons of entering divorce mediation? What would you personally do? Why?

Divorce mediation is the process in which a mediator attempts to assist divorcing couples in resolving personal, legal, and parenting issues in a cooperative manner. More than two-thirds of U.S. states offer or require mediation through the courts over such legal issues as custody and visitation. Mediators act as facilitators to help couples arrive at mutually agreed-on solutions. Although some mediators are attorneys by profession, in the divorce process they neither act as lawyers for nor give advice to either party. Mediators can be either private or court ordered. Mediators generally come from marriage counseling, family therapy, and social work backgrounds, although increasing numbers are coming from other backgrounds and are seeking training in divorce mediation (DeWitt 1994).

Mediation has many goals. A primary goal is to encourage divorcing parents to see shared parenting as a viable alternative and to reduce anxiety about shared parenting (Kruk 1993). Their role is not to save the marriage but to see that couples exit the marriage with less conflict, feeling that their interests were represented. Data on satisfaction indicate that those who use mediators as part of their divorce process have greater levels of satisfaction than those who divorce through adversarial means. They also spend less to end their marriages because divorce mediation is less financially costly than divorce that relies on litigation (through lawyers) alone (www.divorceinfo.com/doesmediationwork.htm).

When mediation is court mandated, topics are generally limited to custody and visitation issues. Divorcing parents often find mediation helpful for these issues. In contrast to court settings, mediation provides an informal setting to work out volatile issues. Both men and women report that mediation is more successful at validating their perceptions and feelings than is litigation. Furthermore, women, poor families, and those from ethnic groups are less likely to experience bias in mediation than in a courtroom setting (Rosenberg 1992).

Some courts order parents to participate in seminars covering the children's experience of divorce as well as problem solving and building coparent relationships (Petersen and Steinman 1994). Parents report that these seminars help them become more aware of their children's reactions and give them more options for resolving child-related disputes.

Divorcing parents also report that mediation helps decrease behavioral problems in their children (Slater, Shaw, and Duquesnel 1992). If parents can work through their differences apart from their children, the children are less likely to react to the anger and fear they might otherwise observe.

What to Do about Divorce

As the previous pages have illustrated, divorcing is a painful process for those involved, and it leaves families and individuals changed forever. Most people will agree that we would be better off reducing the rate of divorce, but how can that goal be achieved? First, we must decide on the most important cause of the high divorce rates in the United States.

If we believe that divorce rates rose partly because we made it easier and more acceptable to divorce, should we restigmatize divorce? Make exiting a marriage more difficult? If divorce rates rose with the increasing economic independence of women, how can we reduce divorce? Do we need to encourage employed women to stay home? How, then, do their families survive without their incomes (see Chapter 11)? If part of the explanation for rising divorce rates is in the increasing importance given to self-fulfillment and the decline of both familistic self-sacrifice and religious constraints, how can we reduce divorce? Can we change people's values? Finally, if increases in divorce result from the weakening of all but the emotional function of marriage and the reduction, especially of the family's economic role, can *anything* be done about divorce?

Part of the dilemma has to do with how we perceive divorce. Is divorce the *problem*, or is it a *solution* to other problems? Do we want to impose restrictions on divorce that require people to remain in unfulfilling, possibly dangerous relationships? The societal reactions to reducing divorce have been largely of two kinds: cultural and legal. From a cultural perspective, some commentators bemoan the popular cultural attitudes about marriage (Popenoe 1993; Whitehead 1993, 1997). They suggest that we "dismantle the divorce culture" that we have constructed by more consistently championing and effectively demonstrating the benefits of stable, lifelong marriage. Instead of celebrating "family diversity" and glorifying single-parent households, they believe that we should consistently reiterate the idea that marriage is a lifelong commitment involving considerable sacrifice. If that means that we must "restigmatize divorce," then that is what we should do (Whitehead 1997).

The other emphasis has been a legal one. Believing that marriage was weakened and divorce increased by

Issues and Insights: Covenant Marriage as a Response to Divorce

In 1996, as a way of trying to strengthen marriage and reduce divorce rates, Louisiana became the first state in the United States to establish a two-tiered system of marriage.

Marrying couples could choose either a "standard marriage" or a covenant marriage (Hewlett and West 1998; see also Chapter 9). Following Louisiana's lead, other states have enacted their own covenant marriage legislation. Regardless of the state in question, covenant marriage usually consists of something close to the following, which is drawn from the Louisiana law:

> We do solemnly declare that marriage is a covenant between a man and a woman who agree to live together as husband and wife for so long as they both may live. We have chosen each other carefully and disclosed to one another everything which could adversely affect the decision to enter into this marriage.
>
> We have received premarital counseling on the nature, purposes, and responsibilities of marriage. We have read the Covenant Marriage Act, and we understand that a Covenant Marriage is for life. If we experience marital difficulties, we commit ourselves to take all reasonable efforts to preserve our marriage, including marital counseling.
>
> With full knowledge of what this commitment means, we do hereby declare that our marriage will be bound by Louisiana law on Covenant Marriages and we promise to love, honor, and care for one another as husband and wife for the rest of our lives.

This is supplemented by an affidavit by the parties that they have discussed with a religious representative or counselor their intent to enter a covenant marriage. Included is their agreement to seek marital counseling in times of marital difficulties and their agreement to the grounds for terminating the marriage.

We cannot say whether covenant marriage will "work" to reduce the prevalence of divorce. It may have no effect because the people who elect to enter such a marriage may already perceive marriage as a relationship to keep "till death do us part."

Research indicates that covenant marriage likely has and will have modest effects on divorce rates. Few couples appear to choose a covenant over a standard marriage. Those who do opt for covenant marriages are more conservative and more religious and tend to see the covenant marriage as a public expression of their conviction of the sacred commitment that one makes in marrying. Thus, these same people would be less likely to divorce even in the absence of covenant marriage statutes. For those women and men who have concerns about inequalities in traditional marriage or who worry about women's rights in families, covenant marriage will be unappealing.

To do more than "preach to the choir"—appealing to those who already share the covenant marriage philosophy—will be more difficult for proponents of such reform. However, sociologists Steven Nock, Laura Sanchez, and James Wright (2008) argue that noncovenant marriages could benefit from the same kind of mandatory premarital counseling or education that covenant marriage requires. Furthermore, they suggested that if nothing else, there is a symbolic value to covenant marriage, independent of religious symbolism, in that it reflects a society-wide concern about the meaning and importance of marriage.

---

no-fault divorce legislation, some have argued that we make divorce *harder to obtain*. Some states have contemplated repealing no-fault divorce legislation or raising marriage ages. Some states have enacted a two-tiered system of marriage in which couples are allowed and encouraged to consider *covenant marriage*—marriage under laws that require couples to undergo premarital counseling, swear to the lifelong commitment of marriage, and promise to divorce only under extraordinary circumstances and only after seeking marriage counseling (see the Issues and Insights on covenant marriages in this chapter; see also Chapter 9). Too new to yet evaluate, the covenant marriage system has appealed to both those who wish to reduce divorce and those who wish to establish a more traditional, even religious, understanding of marriage commitments.

The difficulty behind both cultural and legal efforts is that in attempting to make divorce harder or less attractive, they do little to make staying married easier. This, too, could be done. It might entail enacting some work–family policy initiatives to ease the stress and strain facing two-earner households. On the subject of financial resources, because we know that divorce hits hardest at lower- and working-class levels, bolstering the economic stability and security of low-income families might also lead to less divorce.

If we can't reduce or eliminate divorce, we should at least do what we can to protect those who go through divorce, especially children (Coontz 1997; Furstenberg and Cherlin 1991). We should devote resources that will help custodial parents raise their children more effectively. This means, among other things, ensuring their access to quality child care when they are at work, guaranteeing their receipt of financial obligations (such as child support and alimony) from their former spouses, and helping them avoid the devastating plunge into poverty. In addition, ex-spouses must be instructed in how to display more amicable relationships with each other and should be expected to do so. Because at least some effects of divorce are tied to the level of postdivorce conflict and adjustment, taking steps to reduce conflict and ensure more effective adjustment will benefit children and their parents. Early and aggressive intervention into the postdivorce family (such as teaching anger management or instructing fathers about the vital roles they can still play) constitutes such intervention (Coontz 1997; Furstenberg and Cherlin 1991).

There is no denying that separation or divorce is typically filled with pain for all involved—husband, wife, and children. Furthermore, as we have seen, both the process and its outcomes are often different for husbands and wives and for parents and children. It is hoped that this chapter has increased your understanding of how much divorce there is, the multiple factors that have led to shifts in the divorce rate and that expose individuals to greater or lesser risk of divorce, and the different perspectives on what we can and should do about divorce. Keep in mind that as one family ends, new family forms emerge. These include new relationships and possibilities, new circumstances and responsibilities, and new families with unique relationships: the single parent or the stepfamily. These are the families that we explore in the next chapter.

Summary

- Divorce is an integral part of the contemporary American marriage system, which values individualism and emotional gratification. The divorce rate increased significantly in the 1970s but leveled off in the early 1990s.

- Divorce can be viewed as a process involving six *stations*, or processes: emotional, legal, economic, coparental, community, and psychic. As people divorce, they undergo these stations simultaneously, but the intensity level of these stages varies at different times.

- Among the statistics researchers use to measure divorce are the *ratio* of marriages to divorces, the *crude rate* of divorces per 1,000 people in a population, the *refined rate* of divorces per 1,000 marriages, and the *predictive rate* of the future likelihood of divorce within a cohort.

- The likelihood of divorce is lower for those who earn more than $50,000, marry after age 25, come from an "intact" parental marriage, have some religious affiliation, and have attended college.

- The trend in divorce has been downward since the 1980s.

- Compared to other countries, the U.S. divorce rate is among the highest.

- A variety of societal, demographic, and life course factors can affect the likelihood of divorce.

- *No-fault divorce* revolutionized divorce by eliminating fault finding and the adversarial process and by treating husbands and wives as equals. An unintended consequence of no-fault divorce is the growing poverty of divorced women with children.

- Uncoupling, the process by which couples drift apart in predictable stages, is differently experienced by the initiator and his or her partner.

- In establishing a new identity, newly separated people go through transition and recovery. They may experience *separation distress*, often followed by loneliness. The more personal, social, and financial resources a person has at the time of separation, the easier the separation generally will be.

- Women generally experience downward mobility after divorce. The economic effect on men is more mixed and depends on what proportion of the marital income they were responsible for before the divorce.

- *Child support* often goes unpaid despite a number of legal initiatives to increase compliance by parents who owe support. A major determinant of compliance is what percentage of the parent's income is expected in support.

- Psychological distress, reduced self-esteem, less happiness, more isolation, and less satisfying sex lives are among noneconomic consequences of divorce.

For some, the consequences of divorce are more positive than negative and include higher levels of personal growth, more autonomy, and—for women—improvements in their social lives, career opportunities, and self-confidence.

- Remaining in an unhappy marriage reduces life satisfaction, mental and physical health, and self-esteem.

- Children are typically told about the divorce by mothers. Children's overall reactions are usually negative, though for some, news of the divorce may be experienced as relief.

- Consequences for children depend on the nature of their parents' marriage.

- Children in the divorce process go through stages: turmoil is greatest in the initial stage. By the restabilization stage, changes have been integrated into the children's lives.

- A significant factor affecting the responses of children to divorce is their age. Younger children tend to act out and blame themselves, whereas adolescents tend to remain aloof and angry at both parents for disrupting their lives.

- Many problems assumed to be caused by divorce are present before marital disruption.

- Factors affecting a child's adjustment to divorce include (1) open discussion before divorce, (2) continued involvement with the noncustodial parent, (3) lack of hostility between divorced parents, (4) good psychological adjustment to divorce by the custodial parent, and (5) a stable living situation and good parenting skills. Continued involvement with the children by both parents is important for the children's adjustment.

- Although divorce has been said to put children in the middle of parental conflict, this seems to occur more in intact, high-conflict parental marriages.

- Longitudinal studies following children of divorce over decades have come to different conclusions about how bad the long-term consequences of divorce are and how long they last.

- Custody is generally based on one of two standards: the best interests of the child or the least detrimental of the available alternatives. The major types of custody are *sole*, *joint*, and *split*. Physical custody is generally awarded to the mother. Joint custody has become more popular because men are becoming increasingly involved in parenting.

- Noncustodial parent involvement exists on a continuum from absent to intimately and regularly involved.

- *Divorce mediation* is a process in which a mediator attempts to assist divorcing couples in resolving personal, legal, and parenting issues in a cooperative manner.

- Recent legislative initiatives such as *covenant marriage* are attempts to reduce the divorce rate by strengthening the marriage commitment.

Key Terms

alimony 467	mandatory visitation 478
binuclear family 470	no-fault divorce 452
child support 467	predictive divorce rate 455
crude divorce rate 455	
distal causes 463	proximal causes 463
divorce mediation 479	ratio measure of divorce 455
fault-based divorce 452	
initiator 463	refined divorce rate 455
intergenerational transmission 460	separation distress 464
	social integration 457
joint custody 477	sole custody 476
joint legal custody 477	split custody 476
joint physical custody 477	stations of divorce 453

RESOURCES ON THE WEB

Book Companion Website

www.cengage.com/sociology/strong

Prepare for quizzes and exams with online resources—including tutorial quizzes, a glossary, interactive flash cards, crossword puzzles, self-assessments, virtual explorations, and more.

14

New Beginnings: Single-Parent Families, Remarriages, and Blended Families

What Do YOU Think? Are the following statements TRUE or FALSE?
You may be surprised by the answers (see answer key on the following page).

T	F	
T	F	**1** Researchers are increasingly viewing stepfamilies as normal families.
T	F	**2** Divorce does not end families.
T	F	**3** Single-parent families today are as likely to be headed by fathers as by mothers.
T	F	**4** Second marriages are significantly happier than first marriages.
T	F	**5** Women are more likely than men are to remarry.
T	F	**6** Children tend to have greater power in single-parent families than in traditional nuclear families.
T	F	**7** Becoming a stepfamily is a process.
T	F	**8** Stepmothers generally experience less stress in stepfamilies than stepfathers because stepmothers are able to fulfill themselves by nurturing their stepchildren.
T	F	**9** Researchers are increasingly finding that remarried families and intact nuclear families are similar to each other in many important ways.
T	F	**10** People who remarry and those who marry for the first time tend to have similar expectations.

When Paige was six and Daniel eight, their parents separated and divorced. The children continued to live with their mother, Sophia, in a single-parent household while spending weekends and holidays with their father, David. After a year, David began living with Jane, a single mother who had a five-year-old daughter, Lisa. Three years after the divorce, Sophia married John, who had joint physical custody of his two daughters, Sally and Mary, aged seven and nine. Some eight years after their parents divorced, Paige and Daniel's family included two biological parents, two stepparents, three stepsisters, one stepbrother, and two half brothers. In addition, they had assorted grandparents, step-grandparents, biological and step aunts, uncles, and cousins.

Today's families mark a definitive shift from the traditional family system, based on lifetime marriage and the intact nuclear family, to a pluralistic family system, including families created by divorce, remarriage, and births to single women. This new pluralistic family system consists of three major types of families: (1) intact nuclear families, (2) single-parent families (either never married or formerly married), and (3) stepfamilies. As we discussed in Chapter 9, cohabiting couples who may or may not have children are also part of the wider family system.

Single-parent families are families consisting of one parent and one or more children; the parent can be divorced, widowed, or never married. **Stepfamilies** are families in which one or both partners have children from a previous marriage or relationship. Stepfamilies are sometimes referred to as **blended families**.

About a third of Americans are expected to marry, divorce, and remarry at some point in their lives (Sweeney 2002). In more than 40% of current marriages, one or both spouses are remarrying (Goldscheider and Sassler 2006). If one considers all possible types of stepfamilies, it is expected that a majority of Americans are or will be connected with a stepfamily (Stewart 2007).

To better understand the world that such children as Paige and David live in, one needs to examine some major patterns in our evolving pluralistic family system. This chapter examines single-parent families, binuclear families, remarriage, and stepfamilies. Given the wide diversity that characterizes families today, researchers can no longer consider such families to be unusual family forms (Coleman and Ganong 1991; Pasley and Ihinger-Tallman 1987). Shifting attention from family structure or form to function, the important question becomes whether a specific family—regardless of whether it is a traditional family, a single-parent family, or a stepfamily—succeeds in performing its functions. In a practical sense, as long as a family is fulfilling its functions and meeting its needs, it is one kind of normal family. This chapter considers these versions of normal families.

Single-Parent Families

In the United States, as throughout the world, single-parent families have increased and continue to grow in number. Although no other family type has increased in number as rapidly, single-parent families are not always accurately or adequately understood. They may still be treated negatively in the popular imagination, considered to be "broken homes" as opposed to "intact families." Single-parent families created by unmarried motherhood, especially those that are headed by teenagers or young adults, are stereotyped as well, with the image being one of young women who casually bear children "out of wedlock." These images are clearly inadequate, based on ideas and stereotypes that misdirect us from a more accurate understanding. The "broken-home" image is based on the ideal of the "happy" intact family; the assumed irresponsibility of single mothers is based on misunderstandings and moralism, occasionally mixed with racism, condemning women for bearing children outside of marriage; and the "promiscuous teenage

Critical Thinking

When you think or hear about single-parent families and stepfamilies, do you think of them as "normal" families? As "abnormal" families? If you were reared in a single-parent family or a stepfamily, did your friends, relatives, schools, and religious groups treat your family as normal? Why?

Answer Key to What Do YOU Think?

1 True, see p. 486;   **2** True, see p. 493;   **3** False, see p. 488;   **4** False, see p. 497;   **5** False, see p. 495;   **6** True, see p. 491;   **7** True, see p. 500;   **8** False, see p. 502;   **9** False, see p. 499;   **10** False, see p. 497.

mother" stereotype ignores the reality that most births to single mothers are to women older than 20. Finally, although more than 85% of single, custodial parents are female, these images overlook the situations and experiences of single fathers.

Between 1970 and 2008, the percentage of children living in single-parent families doubled, increasing from 13% to 26%. Almost 20 million children lived with an unmarried parent in 2008.

In previous generations, the life pattern most women experienced was (1) marriage, (2) motherhood, and (3) widowhood. Single-parent families existed in the past, but they were typically the result of widowhood rather than either divorce or births to unmarried women. Significant numbers were headed by men. But a new marriage and family pattern later emerged. Its greatest effect has been on women and their children. Divorce and births to unmarried mothers are the key factors creating today's single-parent family.

The life pattern many married women today experience is (1) marriage, (2) motherhood, (3) divorce, (4) single parenting, (5) remarriage, and (6) widowhood. For those who are not married at the time of their child's birth, the pattern may be (1) dating or cohabitation, (2) motherhood, (3) single parenting with the later possibility but no certainty of (4) marriage, and (5) widowhood. Finally, some who marry, divorce, and remarry may experience subsequent divorces and or remarriages; they embody the characteristics that make up **serial monogamy**—having a number of monogamous marriages across one's life.

Characteristics of Single-Parent Families

Single-parent families share a number of characteristics, including the following: they are created by either divorce, births to unmarried mothers, or the death of a spouse. They are usually female headed. They are characterized by a diversity of living arrangements. Some single-parent families are created intentionally through planned pregnancy, artificial insemination, and adoption. Others are headed by lesbians and gay men (Miller 1992). Many single-parent households contain two cohabiting adults and are therefore not *single-adult* households (Fields 2003). The prevalence and experiences of single parents vary across racial or ethnic lines. Finally, although there is considerable

economic diversity among single-parent families, they are often economically disadvantaged, perhaps even impoverished.

Creation by Divorce or Births to Unmarried Women

Single-parent families today are usually created by marital separation, divorce, or births to unmarried women rather than by widowhood. Throughout the world, including the United States, single-parent families created through births to unmarried women have increased at a higher rate than single-parent families created through divorce (Burns and Scott 1994). In 2007, nearly 40% of U.S. births were to unmarried women (Ventura 2009). The number of children living with an unmarried couple more than tripled between 1980 and 2000. Today, 19.5 million children under age 18 live in 10.5 million households with either the mother only or the father only (U.S. Census Bureau 2008a).

In comparison to single parenting by widows, single parenting by divorced or never-married mothers receives considerably less social support. Widowed mothers often receive social support from their husband's relatives. A divorced mother may receive some assistance from her own kin but typically receives considerably less (or none) from her former partner's relatives. Our culture is still ambivalent about divorce and tends to consider divorce-induced, single-parent families as less certain environments for protecting children's well-being. It is even less supportive of families formed by never-married mothers.

© Design Pics Inc/Alamy

Although there is a tendency to assume that single parent means single-mother household, there are many single fathers raising children.

Headed by Mothers (and Sometimes Fathers)

More than 85% of single-parent families are headed by women (U.S. Census Bureau 2008a). This has important economic ramifications because of gender discrimination in wages and job opportunities, as discussed in Chapter 12. Still, at least 2.2 million unmarried men are raising one or more children under 18 years of age (U.S. Census Bureau 2008a). Like women, men take a range of paths to single parenthood. They either divorce or separate from their children's mothers, or they raise children from relationships in which they were never married. They are less likely to have never been married than are single mothers. They are also less likely to be poor (Fields 2006).

Significance of Ethnicity

Ethnicity remains an important demographic factor in single-parent families. In 2007, among Caucasian children, nearly 20% lived in single-parent families; among African American children, 54% lived in such families; among Hispanics, 26% lived in single-parent families; and among Asian and Pacific Islander children, 13% lived in such households (U.S. Census Bureau 2008a). White single mothers were more likely to be divorced than their African American or Latino counterparts, who were more likely to be unmarried at the time of the birth or widowed.

Poverty

Married women tend to experience a drop—frequently steep—in their income when they separate or divorce (as discussed in Chapter 13). Among unmarried single mothers, poverty and motherhood often go hand in hand. In 2008, 37% of single-mother-headed families with children were living below the poverty level (Shierholz 2009). Because they are women, because they are often young, and because they are often from ethnic minorities, single mothers have few financial resources. They are under constant economic stress in trying to make ends meet (McLanahan and Booth 1991). They work for low wages, endure welfare, or both. They are unable to plan because of their constant financial uncertainty. They move more often than two-parent families as economic and living situations change, uprooting themselves and their children. They accept material support from

kin but often at the price of receiving unsolicited "free advice," especially from their mothers.

Both mother-only and father-only families are more likely to be poor than are two-parent families; in 2008, 5.5% of married-couple-headed families lived in poverty compared to 13.8% of those in families headed by unmarried men and 28.7% of those in female-headed families. Looking specifically at families with children under age 18 shows higher rates of poverty exists for all three types of families (see Figure 14.1) (U.S. Census Bureau 2009).

Clearly, the association between single parenthood and poverty is greater for mothers than for fathers (Zhan and Pandey 2004). Among fathers, single fathers are less well off compared to married fathers. They tend to be younger, less educated, less likely to have jobs, and more likely to receive public assistance and to live in poverty.

Even aside from poverty, the overall economic well-being of single parents is not as good as it is for married parents. Half of families headed by single parents earned less than $30,000 in 2008. Among married couple households with children under age 18, 11% had incomes under $30,000 in 2008 (U.S. Census Bureau 2009). Among single mothers, divorced single mothers are economically better off than never-married single mothers (Fields 2003).

Diversity of Living Arrangements

There are many different kinds of single-parent households. While 26% of all children under age 18 live in families headed by a single parent, 22.8% live in mother-headed households, and only 3.5% live in households headed by an unmarried father (U.S. Census 2008b). Single-parent families need greater flexibility in managing child care and housing with

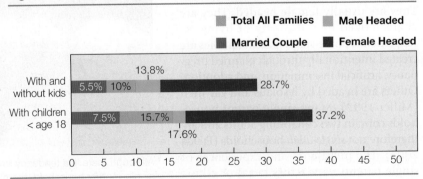

Figure 14.1 Families in Poverty, 2008, by Type of Family

SOURCE: U.S. Census Bureau, www.census.gov/hhes/www/poverty/histpov/famindex

The economic well-being of single parents, especially single mothers, is much less than that of two-parent families.

Even in the absence of parents' *live-in* partners, parents' romantic partners may play important roles in their children's lives. For example, many children of single mothers and nonresidential biological fathers have a **social father**—a male relative, family associate, or mother's partner—"who demonstrates parental behaviors and is like a father to the child" (Jayakody and Kalil 2002).

Along these same lines, single parents, especially mothers, often rely on a combination of state or federal assistance and **private safety nets**: support from their social networks on which they can fall back in times of economic need (Hamer and Marchioro 2002; Harknett 2006). Social support, whether from family or friends, can lead to enhanced well-being and self-esteem among economically disadvantaged single mothers. These, in return, may lead to more effective parenting, even under difficult and highly stressful conditions. Without such support, mothers raising children on their own in economically distressed, potentially dangerous, urban neighborhoods are more likely to experience psychological distress, which then negatively affects their parenting behavior (Kotchik, Dorsey, and Heller 2005).

Transitional Form

Single parenting is usually a transitional state. A single mother has strong motivation to marry or remarry because of cultural expectations, economic stress, role overload, and a need for emotional security and intimacy. The increasing presence of social fathers, including mothers' live-in romantic partners, may be part of the reason low-income families increasingly cohabit rather than marry. The presence of such men can reduce the various pushes toward marriage or remarriage (Jayakody and Kalil 2002).

Intentional Single-Parent Families

For many single women in their thirties and forties, single parenting has become a more accepted, intentional, and less transitional lifestyle (Seltzer 2000). Some older women choose unmarried single parenting because they have not found a suitable partner and are concerned about declining fertility. They may plan their pregnancies or choose donor insemination or adoption. If their pregnancies are unplanned, they decide to bear and rear the child. Others choose single parenting because they do not want their lives and careers encumbered by the compromises necessary in marriage. Still others choose it because they don't want a husband but do want a child.

limited resources. In doing so, they rely on a greater variety of household arrangements than is suggested by the umbrella heading "single-parent household." For example, many young African American mothers live with their own mothers in a three-generation setting.

Of perhaps more interest is that many "single-parent households" actually contain the parent and his or, more often, her unmarried partner. In 2002, for example, 11% (1.8 million) of the 16.5 million children living with single mothers also lived with their mothers' unmarried partners. A third (1.1 million) of the 3.3 million children living with an unmarried father also lived with their fathers' unmarried partners (Fields 2003). A Census Bureau report indicated that in 2004, two of five children in households headed by unmarried mothers and three of five children in father headed households had another adult in the house. Thus, though such households were headed by unmarried parents, those parents were not necessarily alone in the tasks of child rearing (Fields 2004).

Lesbian and Gay Single Parents

Although estimates from the research literature were provided in Chapter 10, in fact it is nearly impossible to confidently assert how many children in the United States are being raised by gay or lesbian couples. Most of these parents were married either before they were aware of their sexual orientation or in order to "fit in" with social or familial expectations. Such men and women become single parents as a result of divorce. Others were always aware of being lesbian or gay; they chose adoption or donor insemination to have children. However, it must be noted that the lack of marriage rights for same-gender couples in 45 of the 50 states in the United States imposes single parenthood—at least in a legal sense—on those same-gender couples who are raising children but live where they cannot marry. Couple the lack of marriage rights with the restrictions that many states impose on gay adoption, and one immediately sees that gay or lesbian single parents—especially those who are nonbiological parents—are in a vulnerable position as far as their parental rights are concerned.

Limited marriage and adoption rights make same-gender families with children into legal single-parent families.

Although in many states in the United States gay men and lesbians have adopted children (the American Academy of Pediatrics lists 22 and the District of Columbia), second-parent adoptions are not necessarily available to the non–birth parent or partner of the primary adoptive parent. Furthermore, many states—including some who previously allowed same-gender couples to adopt—have since restricted adoption rights to heterosexuals or are attempting to impose such restrictions via proposed legislation (Pawelski et al. 2006).

Children in Single-Parent Families

Children born outside of marriage tend to suffer economic disadvantages that may then lead to other educational, social, and behavioral outcomes. Their disadvantages tend to be worse than those experienced by children of divorced parents or by children in two-parent, married households (Seltzer 2000). They are more likely to engage in high-risk, "health-compromising" behaviors, such as cigarette smoking, drug and alcohol use, and unprotected sex; are less likely to graduate from high school and college; are more likely to have a child outside of marriage and/or during their teens; are more likely to be "idle" (out of school and out of work), have lower earnings, and suffer lower levels of psychological well-being; and are more vulnerable to divorce and marital instability as adults (King, Harris, and Heard 2004).

The bulk of research on the effects of divorced, single-parent households have on children points to some negative outcomes in areas such as behavioral problems, academic performance, social and psychological adjustment, and health. The gaps between children in such households and those whose parents remain continuously married are relatively small but consistent. As Paul Amato (2000) reported, especially when exposed to associated negative life events such as having to move or change schools, the effects of living in a divorced, single-parent home can create particular adjustment difficulties. The consequences appear to be linked to the lack of economic resources but also to the reduced money, attention, guidance, and social connections—what researchers call **social capital**—that fathers provide.

Parental Stability and Loneliness

After a divorce, single parents are usually glad to have the children with them. Everything else seems to have fallen apart, but as long as divorced parents have their children, they retain their parental function. Their children's need for them reassures them of their own importance. A custodial parent's success as a parent becomes even more important to counteract the feelings of low self-esteem that result from divorce.

Feeling depressed, the mother knows that she must bounce back for the children. Yet after a short period, she comes to realize that her children do not fill the void left by her missing spouse. The children are a chore, as well as a pleasure, and she may resent being constantly tied down by their needs. Thus, minor incidents with the children—a child's refusal to eat or a temper tantrum—may be blown out of proportion. A major disappointment for many new single parents is the discovery that they are still lonely. It seems almost paradoxical. How can a person be lonely amid all the noise and bustle that accompany children? However, children do not ordinarily function as attachment figures; they can never be potential partners. Any attempt to make them so is harmful to both parent and child. Yet children remain the central figures in the lives of single parents. This situation leads to a second paradox: although children do not completely fulfill a person, they rank higher in most single mothers' priorities than anything else.

Changed Family Structure

A single-parent family is not the same as a two-parent family with one parent temporarily absent. The permanent absence of one parent dramatically changes the way in which the parenting adult relates to the children. Generally, the mother becomes closer and more responsive to her children. Her authority role changes, too. A greater distinction between parents and children exists in two-parent homes. Rules are developed by both mothers and fathers. Parents generally have an implicit understanding to back each other up in child-rearing matters and to enforce mutually agreed-on rules. In the single-parent family, no other partner is available to help maintain such agreements; as a result, the children may find themselves in a more egalitarian situation with more power to negotiate rules. They can be more stubborn, cry more often and louder, whine, pout, and throw temper tantrums. Any parent who has tried to convince children to do something they do not want to do knows how soon an adult can be worn down.

Additional "handicaps" faced by single-parent families include the following:

- With only one adult in the household, if that adult is distressed, overwhelmed, or angry, the tone of the whole house is affected (Coontz 1997).
- Facing more intense time pressures, single parents are less able to participate in their children's schooling and spend less time monitoring their children's homework (Coontz 1997).

- Parental depression, especially among custodial mothers, can affect their abilities to parent effectively and thus exposes their children to more "adjustment problems" (Amato 2000).
- Single mothers with higher levels of life stresses and less time for themselves are more likely to be anxious and to transmit their anxiety to their children. Repeated experiences of transmitted anxiety from mother to child can lead to chronic distress in children (Larson and Gillman 1999).

On the "plus side," children in single-parent homes may also learn more responsibility, spend more time talking with their custodial parent, and face less pressure to conform to more traditional gender roles (Coontz 1997). They may learn to help with kitchen chores, to clean up their messes, or to be more considerate. In the single-parent setting, children are encouraged to recognize the work their mother does, assist in household chores, and understand the importance of cooperation (Stewart 2007).

Although most single parents continue to demonstrate love and creativity in the face of adversity, research on their children reveals some negative long-term consequences. In adolescence and young adulthood, children from single-parent families had fewer years of education and were more likely to drop out of high school. They had lower earnings and were more likely to be poor. They were more likely to initiate sex earlier, become pregnant in their teens, and cohabitate but not marry earlier (Furstenberg and Teitler 1994). Furthermore, they were more likely to divorce. These conclusions are consistent for Caucasians, African Americans, Latinos, and Asian Americans. The reviewers note that socioeconomic status accounts for some but not all of the effects. Some effects are attributed to family structure.

Harriette Pipes McAdoo (1988, 1996) attributed the cause to poverty, not to single parenthood. She notes that African American families are able to meet their children's needs in a variety of structures. "The major problem arising from female-headed families is poverty," she writes (McAdoo 1988). "The impoverishment of Black families has been more detrimental than the actual structural arrangement."

Successful Single Parenting

Single parenting is difficult, but for many single parents the problems are manageable. Almost two-thirds of divorced single parents found that single parenting grows easier over time (Richards and Schmiege 1993). Thus, it is important to note that many of the characteristics of successful single parents and their families are shared by all successful families.

Characteristics of Successful Single Parents

In-depth interviews with successful single parents found certain themes running through their lives (Olsen and Haynes 1993):

- *Acceptance of responsibilities and challenges of single parenthood.* Successful single parents saw themselves as primarily responsible for their families; they were determined to do the best they could under varying circumstances.
- *Parenting as first priority.* In balancing family and work roles, their parenting role ranked highest. Romantic relationships were balanced with family needs.
- *Consistent, nonpunitive discipline.* Successful single parents realized that their children's development required discipline. They adopted an authoritative style of discipline that respected their children and helped them develop autonomy.
- *Emphasis on open communication.* They valued and encouraged expression of their children's feelings and ideas. Parents similarly expressed their feelings.
- *Fostering individuality supported by the family.* Children were encouraged to develop their own interests and goals; differences were valued by the family.
- *Recognition of the need for self-nurturance.* Successful single parents realized that they needed time for themselves. They needed to maintain an independent self that they achieved through other activities, such as dating, music, dancing, reading, classes, and trips.
- *Dedication to rituals and traditions.* Successful single parents maintained or developed family rituals and traditions, such as bedtime stories, family prayer or meditation, sit-down family dinners at least once a week, picnics on Sundays, visits to Grandma's, or watching television or going for walks together.

Single-Parent Family Strengths

Although most studies use a **deficit approach** emphasizing the stresses and difficulties faced by single parents and their children, some studies note the potential that single parents have to build strength and confidence. This may be especially true for women (Amato 2000; Coontz 1997).

Qualitative interview research has uncovered ways in which both single mothers and single fathers display and develop strengths that enable them to effectively and successfully raise their children. A study by family researchers Leslie Richards and Cynthia Schmiege (1993) of 60 white single mothers and 11 white single fathers (most of whom were divorced) identified five family strengths associated with successful single parenting:

- *Parenting skills.* Successful single parents have the ability to take on both expressive and instrumental roles and traits. Single mothers may teach their children household repairs or car maintenance; single fathers may become more expressive and involved in their children's daily lives.
- *Personal growth.* Developing a positive attitude toward the changes that have taken place in their lives helps single parents, as does feeling success and pride in overcoming obstacles.
- *Communication.* Through good communication, single parents can develop trust and a sense of honesty with their children as well as an ability to convey their ideas and feelings clearly to their children and friends.
- *Family management.* Successful single parents develop the ability to coordinate family, school, and work activities and to schedule meals, appointments, family time, and alone time.
- *Financial support.* Developing the ability to become financially self-supporting and independent is important to single parents.

Among the single parents in the study, more than 60% identified parenting skills as one of their family strengths. In addition, 40% identified family management as a strength in their families (Richards and Schmiege 1993). About 25% identified personal growth and communication among their family strengths.

Sociologist Barbara Risman's (1986) research on custodial single fathers showed their abilities to be attentive, nurturing caregivers to their children. Rather than relying on paid help or female social supports, men became the nurturers in their children's lives. They were involved in their personal, social, and academic lives and saw to it that their emotional and physical needs were met. To Risman, they affirmatively answer the question in her title, "Can Men Mother?"

Critical Thinking

If you are or have been a member of a single-parent family, what were its strengths and problems? What do you know of the strengths and problems of friends and relatives in single-parent families?

Binuclear Families

One of the most complex and ambiguous relationships in contemporary America is what some researchers call the **binuclear family**—a postdivorce family system with children (Ahrons and Rodgers 1987; Ganong and Coleman 1994). It is the original nuclear family divided in two. The binuclear family consists of two nuclear families—the maternal nuclear family headed by the mother (the ex-wife) and the paternal one headed by the father (the ex-husband). Both single-parent families and stepfamilies are forms of binuclear families.

Divorce ends a marriage but not a family. It dissolves the husband–wife relationship but not necessarily the father–mother, mother–child, or father–child relationship. The family reorganizes itself into a binuclear family. In this new family, ex-husbands and ex-wives may continue to relate to each other and to their children, although in substantially altered ways. The significance of the maternal and paternal components of the binuclear family varies. In families with joint physical custody, the maternal and paternal families may be equally important to their children. In single-parent families headed by women, the paternal family component may be minimal.

Subsystems of the Binuclear Family

To clarify the different relationships, researchers Constance Ahrons and Roy Rodgers (1987) divide the binuclear family into five subsystems: former spouse, remarried couple, parent–child, sibling (step-siblings and half siblings), and mother/stepmother–father/stepfather.

Former Spouse Subsystem

Divorce may sever the marital relationship between two people, but it doesn't bring an end to their parenting responsibilities. Thus, their relationship as coparents endures. As noted in the previous chapter, children tend to benefit when they maintain relationships with both parents. However, achieving this requires that former spouses work through the following:

- Anger and hostility they may feel toward each other as a consequence of their previous marriage, the separation, and whatever postdivorce arrangements have been reached regarding such matters as custody, child support, visitation, and spousal support
- Conflict over different parenting styles, values, and aspirations concerning the children
- Shifting roles and relationships between former spouses when one or both remarries
- The incorporating of others as stepparents, step-siblings and step-grandparents as one or both former spouses remarry

In essence, former spouses must separate their marital and personal issues from their mutual desire to raise their children effectively.

Remarried Couple Subsystems

Remarried life is complicated by the need to manage such issues as the exchange of money, sharing of decision-making power, time, and children with former spouses. Those with physical custody must provide their former spouses with access to the children. As a consequence of these complexities, the remarried couple experiences many typical marital issues that become complicated by the presence or involvement of the former spouse.

Parent–Child Subsystem

Such matters force both biological and stepparents to make adjustments. Former single parents must now make room for the presence and involvement of a second parent in decision making and child-rearing practices. Stepparents who assume that their role will be similar to the parent role have the greatest and hardest adjustment to make. Stepmothers tend to experience greater stress than stepfathers.

Sibling, Step-Sibling, and Half-Sibling Subsystem

A parent's remarriage may introduce "instant" sisters or brothers into one's life. They likely differ, perhaps considerably, in their temperaments. They must compete and contend with and accommodate each other's needs for parental affection, attention, space, and so on. Sharing one's parent with new step-siblings may

Critical Thinking

Are you a part of a binuclear family? If so, in what role? Which subsystems are functional or dysfunctional in your binuclear family? How do you imagine that conflict within the former spouse subsystem would affect children in a binuclear family?

be even harder than sharing one's parent with a new stepparent. Visiting biological children compete with stepchildren who are now living with the visiting children's biological parent. This may affect older children who are out of the house (or on their way) as well as children who must adapt to the feeling of being out of place in their own parent's home.

Mother/Stepmother–Father/Stepfather Subsystems

The relationship between new spouses and former spouses often influences the remarried family. The former spouse can be an intruder in the new marriage and a source of conflict between the remarried couple. Other times, the former spouse is a handy scapegoat for displacing problems. Much of current spouse–former spouse interaction depends on how the ex-spouses feel about each other.

Recoupling: Courtship in Repartnering

Certain norms governing courtship before first marriage are fairly well understood. As courtship progresses, individuals spend more time together; at the same time, their family and friends limit time and energy demands because "they're in love." Courtship norms for second and subsequent marriages, however, are not so clear (Ganong and Coleman 1994; Rodgers and Conrad 1986).

For example, when is it acceptable for formerly married (and presumably sexually experienced) men and women to become sexually involved? What type of commitment validates "premarital" sex among postmarital men and women? When should a parent's new partner be introduced to his or her children? How long should courtship last before a commitment to marriage is made? Should the couple cohabit? Without clear norms, courtship following divorce can be plagued by uncertainty about what to expect.

Remarriage courtships tend to be short unless preceded by cohabitation. Research on how postdivorce cohabitation affects the timing of remarriage shows that postdivorce cohabitation tends to lead to a longer waiting time until remarriage than is experienced by those who don't cohabitate before remarrying (Xu, Hudspeth, and Bartkowski 2006).

As noted earlier, almost one-third of divorced individuals marry within a year of their divorces. This may indicate, however, that they knew their future partners before they were divorced. If neither partner has children, courtship for remarriage may resemble courtship

before the first marriage, with one major exception: the memory of the earlier marriage exists as a model for the second marriage. Courtship may trigger old fears, regrets, habits of relating, wounds, or doubts. At the same time, having experienced the day-to-day living of marriage, the partners may have more realistic expectations. Their courtship may be complicated if one or both are noncustodial parents. In that event, visiting children present an additional element.

Cohabitation

Increases in the rates of cohabitation in the United States include many divorced women and men who cohabit before or instead of remarrying. As great an increase as has occurred in premarital cohabitation, *post-divorce* cohabitation is even more common (Xu et al. 2006). Thus, although remarriage rates have declined in recent years, "recoupling" through cohabitation remains common (Coleman, Ganong, and Fine 2000).

Larry Ganong and Marilyn Coleman (1994) describe cohabitation as "the primary way people prepare for remarriage," making it a major difference between first-time marriages and remarriages. This may reflect the desire to test compatibility in a "trial marriage" to prevent later marital regrets (Buunk and van Driel 1989). However, couples who lived together before remarriage did not discuss stepfamily issues any more than did those who did not cohabit (Ganong and Coleman 1994). Looking at some characteristics of remarriages following divorce and cohabitation reveals the following:

- Remarital happiness is about 28% lower for postdivorce cohabitors than for noncohabitors.
- Remarital instability is around 65% greater for cohabitors than for noncohabitors.
- As of now, it is impossible to determine whether postdivorce cohabitation or the types of individuals who cohabit (the selection effect) are responsible for the effect that cohabitation has on remarriages. This should be familiar; we posed the same question about the effects of cohabitation on first marriages.

Courtship and Children

Courtship before remarriage differs considerably from that preceding a first marriage if one or both members in the dating relationship are custodial parents. Single parents are not often a part of the singles world because such participation requires leisure and money, which single parents generally lack. Children rapidly consume both of these resources.

Although single parents may wish to find a new partner, their children usually remain the central figures in their lives. This creates a number of new problems. First, the single parent's decision to go out at night may lead to guilt feelings about the children. If a single mother works and her children are in day care, for example, should she go out in the evening or stay at home with them? Second, a single parent must look at a potential partner as a potential parent. A person may be a good companion and listener and be fun to be with, but if he or she does not want to assume parental responsibilities, the relationship will often stagnate or be broken off. A single parent's new companion may be interested in assuming parental responsibilities, but the children may regard him or her as an intruder and try to sabotage the new relationship.

A single parent may also have to decide whether to permit a romantic partner to spend the night when children are in the home. This is often an important symbolic act. For one thing, it brings the children into the parent's new relationship. If the couple has no commitment, the parent may fear the consequences of the children's emotional involvement with the romantic partner; if the couple breaks up, the children may be adversely affected.

Remarriage

The eighteenth-century writer Samuel Johnson described **remarriage**—a marriage in which one or both partners have been previously married—as "the triumph of hope over experience." Americans are a hopeful people. Many newly divorced men and women express great wariness about marrying again, yet they are actively searching for mates. Women often view their divorced time as important for their development as individuals, and they tend to have a wider network of friends and, thus, more social support. Men may seek to reenter marriage in search of the emotional and social support they hope marriage will provide (Ganong, Coleman, and Hans 2006).

Rates and Patterns of Remarriage

As Table 14.1 illustrates, based on 2004 Census data, of the 110 million U.S. males age 15 and older, 69% had ever married, with 15% having been married at least twice. Among the 118 million women 15 and older, 74% had ever been married, with 16% marrying at least twice.

Table 14.1 Number of Times Married, Males and Females Age 15 and Older, 2004

	Males	Females
Never married	31%	26%
Married once	54%	58%
Married twice	12%	13%
Married three or more times	3%	3%

SOURCE: U.S. Census Bureau (2004).

Among those who were married at the time of the census survey, 21% of married males and females age 15 and older had been married at least twice. In 36% of marriages that occurred in the 12 months prior to the survey, in 36% at least one of the spouses had been previously married (U.S. Census Bureau 2004).

Remarriage is common among divorced people, especially men, who have higher remarriage rates than women (Coleman et al. 2000). Still, 54% of divorced women remarry within five years, and 75% remarry within 10 years (Bramlett and Mosher 2002). Racially, whites have higher remarriage rates than do African Americans (Kreider 2006).

In recent years, the remarriage rate has declined. The decline may be partly the result of the desire on the part of divorced men and women to avoid the legal responsibilities accompanying marriage. Instead of remarrying, many are choosing to cohabit.

Indeed, cohabitation after divorce and prior to remarriage has become increasingly common. Sociologists Xiaohe Xu, Clark Hudspeth, and John Bartkowski contend that cohabitation after a divorce is more common than cohabitation prior to a first marriage. More than 70% of those who cohabit after a divorce end up marrying their partner within a four year period. In addition, as is true of premarital cohabitation, post-divorce/pre-remarriage cohabitation lowers ones remarital happiness by nearly 30% (Xu et al. 2006).

Research indicates that the likelihood of remarriage is affected by the presence of children (de Graaf and Kalmjin 2003; Goldscheider and Sassler 2006). However, the presence of dependent children appears to be a stronger obstacle for women's chances of remarriage than for men's chances, especially if she has many children or very young children (Stewart, Manning, and Smock 2003). The presence of children especially affects the likelihood of women marrying men without children (Goldscheider and Sassler 2006). Women seem to be less reluctant to marry a man with children than men are to marry a woman with children, especially young children. In fact, research by sociologists Frances Goldscheider and Sharon Sassler (2006) indicates that the presence of children might even positively affect men's establishing new cohabiting or married relationships. Men who themselves have children seem less hesitant to remarry a woman who is also a mother.

Remarriage is more likely among white divorced women and among younger women—women 25 years or younger at the time of divorce. Eighty percent of these younger women remarry within 10 years, compared to 68% of women older than 25 years at the breakup of their marriage. African American women are less likely than Caucasian or Hispanic women to remarry. Within five years after a divorce, approximately 33% of black women, 44% of Hispanic women, and nearly 60% of white women had entered a remarriage (Bramlett and Mosher 2002).

In addition to age and ethnicity, socioeconomic variables such as education may affect remarriage rates, although research that has identified effects is not consistent. Education appears to work differently for women's and men's likelihood of remarriage, raising a man's likelihood of remarriage but reducing a woman's (Coleman et al. 2000).

Gender

There are a number of reasons that more men than women remarry. First, divorced women tend to be older than never-married women. Given the tendency for men to marry women younger than themselves and that older women are seen as less attractive and therefore less desirable as spouses, women face more competition and possess fewer "resources" to bring to a remarriage. They are also more likely to have custody of children, which can reduce both the ease with which they socialize or date and their appeal as potential spouses.

Presence of Children

Children lower the probability of remarriage for both women and men but especially for women (Coleman et al. 2000). The effects are most marked when a woman has three or more children. Most research, however, is 15 to 20 years old, and the increased incidence of single-parent families and stepfamilies may have decreased some of the negative effect of children. Whereas researchers generally speculate that children are a "cost" in remarriage, some point out that some men may regard children as a "benefit" in the form of a ready-made family (Ganong and Coleman 1994).

Some research suggests that the stepparent with no biological children experiences the most negative effect (MacDonald and DeMaris 1995).

Initiator Status

Research suggests that initiators will be more likely to remarry than noninitiators (see Chapter 14). In their decisions about seeking a divorce, initiators may factor in the prospect for reentering marriage. They also may be "better prepared emotionally" than noninitiators to remarry. The advantage initiators have over noninitiators may be temporary because noninitiators lag behind initiators in the process of adjusting to and accepting the ending of their marriages (Sweeney 2002). Indeed, Megan Sweeney found that initiators enter new relationships substantially more quickly than noninitiators.

Need, Attractiveness, or Opportunity?

For women, the highest remarriage rate takes place in the twenties; it declines by 25% in the thirties and by more than 65% in the forties. What's going on that accounts for the changing probabilities? First, older women may have less drive to remarry. Second, the cultural association of youthfulness with attractiveness for women means that as they get older, they may be deemed less desirable to potential partners in a way that won't as strongly limit men's remarriage chances. Third, they are more likely to have children and less likely to desire to have more children. Factors such as these affect their suitability to potential partners. Furthermore, the pool of eligible partners for remarriage continues to grow smaller as women age. More potential partners of their same age will be already married. As a result of these processes, men and women may be willing to "settle for less." They may choose someone they would not have chosen when they were younger (Ganong and Coleman 1994).

The Remarriage Marketplace

There are three main contexts from which divorced women and men might find another partner: in the workplace, through leisure activities, and through

Critical Thinking

If you were seeking a marital partner, would you consider a previously married person? Why or why not? Would it make a difference if he or she already had children?

their social network. Women and men who are employed and who are socially integrated with coworkers are more likely to find a new partner. Employment affects their opportunities to remarry by adding the workplace as a venue in which they are likely to meet potential partners.

Characteristics of Remarriage

Remarriage is different from first marriage in a number of ways. First, the new partners get to know each other during a time of significant changes in life relationships, which include confusion, guilt, stress, and mixed feelings about the past (Keshet 1980). They have great hope that they will not repeat past mistakes, but there is also often some fear that the hurts of the previous marriage will recur (McGoldrick and Carter 1989). The past is still part of the present. A Talmudic scholar once commented, "When a divorced man marries a divorced woman, four go to bed."

Remarriages occur later than first marriages. People are at different stages in their life cycles and may have different goals. Divorced people may have different expectations of their new marriages. A woman who already has had children may enter a second marriage with strong career goals. In her first marriage, raising children may have been more important.

In an early study of second marriages in Pennsylvania, Frank Furstenberg (1980) discovered that three-fourths of the couples had a different conception of love than couples in their first marriages. Two-thirds thought that they were less likely to stay in an unhappy marriage; they had already survived one divorce and knew that they could make it through another. Four out of five believed their ideas of marriage had changed.

Marital Satisfaction and Stability in Remarriage

According to various studies, remarried people are about as satisfied or happy in their second marriages as they were in their first marriages. As in first marriages, marital satisfaction appears to decline with the passage of time (Coleman and Ganong 1991). Yet although marital happiness and satisfaction may be similar in first and second marriages, remarried couples are more likely to divorce. As Coleman et al. (2000) note, "Serial remarriages are increasingly common."

How do we account for this paradox? Researchers have suggested several reasons for the higher divorce rate in remarriage (for a discussion of various models explaining the greater fragility of remarriage, see Ganong and Coleman 1994).

First, people who remarry after divorce often have a different outlook on marital stability and are more likely to use divorce to resolve an unhappy marriage (Booth and Edwards 1992). Frank Furstenberg and Graham Spanier (1987) note that they were continually struck by the willingness of remarried individuals to dissolve unhappy marriages: "Regardless of how unattractive they thought this eventuality, most indicated that after having endured a first marriage to the breaking point they were unwilling to be miserable again simply for the sake of preserving the union."

Second, despite its prevalence, remarriage remains an "incomplete institution" (Cherlin 1981). Society has not evolved norms, customs, and traditions to guide couples in their second marriages. There are no rules, for example, defining a stepfather's responsibility to a child: is he a friend, a father, a sort of uncle, or what? Nor are there rules establishing the relationship between an individual's former spouse and his or her present partner: are they friends, acquaintances, rivals, or strangers? Remarriages don't usually receive the same family and kin support as do first marriages (Goldenberg and Goldenberg 1994).

Remarriages are more likely to be unhappy and/or end in divorce than are first marriages.

Sociologist Constance Ahrons tells the story of a wedding she attended. The bride was her late ex-husband's daughter by his second wife. Looking for ways to describe or explain whose wedding she was flying off to attend, she told some people it was the wedding of her daughters' sister. Confused by this designation (she was really their half sister but they avoided such a potentially condescending qualifier—as far as her daughters were concerned, the bride was their real sister), people were similarly perplexed that Ahrons would travel across the country to attend such a wedding or that she had even be invited.

During the wedding, standing alongside her late ex-husband's widow, she caused further puzzlement when someone from the groom's side asked about her connection to the bride's mother, and they replied, "We used to be married to the same man" (Ahrons 2004, 5). If this was true, the questioner wondered, why was Ahrons at this wedding? How were the two women getting along? This is not the way people assume or expect it to be.

Constance Ahrons has spent more than two decades researching and writing about divorce and the binuclear family. She has followed a sample of divorced families over time, beginning in 1979 with 98 pairs of ex-spouses with children and most recently interviewing 173 of the 204 children from these families. As Ahrons describes her own "extended family rearranged by divorce," she remarks, "My family, and the many others like mine, don't fit the ideal images we have about families . . . they're not tidy. There are extra people and relationships that don't exist in nuclear families and are awkward to describe because we don't have familiar and socially defined kinship terms to do so" (Ahrons 2004, 6).

Yet such families, comprised of people rearranged by divorce and expanded by remarriages, are increasingly common. To the children who grow up in them, they are simply who they call "family": perhaps their stepmom or stepdad, their mom's or dad's boyfriend or girlfriend, their parents who now live apart, their new grandparents or cousins—they are all just family. Ahrons advocates that we reconceptualize what we mean by family; recognize that our nuclear family bias no longer describes the family lives of millions of men, women, and children; and extend the idea of family to include all of those hard-to-describe relationships among those who consider themselves family.

Third, remarriages are subject to stresses that are not present in first marriages. The vulnerability of remarriage to divorce is especially real if children from a prior relationship are in the home (Booth and Edwards 1992). Children can make the formation of the husband–wife relationship more difficult because they compete for their parents' love, energy, and attention. In such families, time together alone becomes a precious and all-too-rare commodity. Furthermore, although children have little influence in selecting their parent's new husband or wife, they have immense power in "deselecting" them by acting in ways that bring out differences in parenting styles and values or create tension between siblings and step-siblings (Ihinger-Tallman and Pasley 1987).

The divorce proneness of remarriages seems to lessen and become more like that of first marriages as people age. People who enter remarriage after turning 40 may face a lower divorce likelihood than that found among first marriages (Coleman et al. 2000). As mentioned earlier in the chapter, those who cohabit before remarrying face greater risk of divorce (Xu et al. 2006).

Blended Families

Remarriages that include children are vastly different from those that do not. The families that emerge from remarriages with children are traditionally known as *stepfamilies*. They are also sometimes called *reconstituted, restructured,* or *remarried families* by social scientists—names that emphasize their structural differences from other families. In popular discourse, we often consider them "blended families," as the new spouses and their children attempt to blend into a single functioning family. Attempting to focus more on the positive aspect of blending (and striving to steer clear of the negative connotations that sometimes accompany "steps," as in "evil stepmother"), some even refer to their new stepchildren or stepparents as "bonus" children or "bonus" parents. A Web site for Bonus Families (www.bonusfamilies.com), a nonprofit organization whose goal is to promote "peaceful coexistence between divorced or separated parents and their new families," suggests that at different phases different terms may be more appropriate or acceptable.

Children in Stepfamilies

The research into effects on children of being in a stepfamily reveal a number of negative outcomes, including academic, psychological, and behavioral outcomes. Academically, attendance in school, school performance, and overall educational attainment tend to be worse among children in stepfamilies than among children in families headed by married biological parents. Psychologically, children in stepfamilies are more likely to have needed and/or received psychological help, to be sad or anxious, and to be less sociable. Behaviorally, they are more likely to smoke cigarettes, drink, use illegal drugs, and engage in early sexual behavior and childbearing. In addition to these, longer-term effects have been identified, including leaving home at an earlier age and maintaining less adult contact with siblings (Stewart 2007).

Of course, comparisons with children in intact two-parent homes are complicated by other differences between stepfamilies and families headed by married biological parents. In part because of some of the economic factors that contribute to divorce, in part because of postdivorce economic consequences, and in part because of the economic context in which non-marital childbearing is most common, stepfamilies are often at a disadvantage. Sociologist Susan Stewart (2007) notes that stepfamilies have less savings and lower incomes and are less likely to be homeowners. These socioeconomic differences could be associated with some of the outcomes that have been reported. For example, the lower earnings and education of stepfathers in comparison to biological fathers is likely among the causes of the lesser academic achievement of children in stepfamilies (Hofferth 2006).

Further complicating assessments of outcomes is the fact that outcomes are not the same for all children or all types of stepfamilies. They vary by such characteristics as the child's age at time of stepfamily formation, the age of the stepparents, the child's gender, whether one or both parents have previous children, and other socioeconomic or demographic characteristics. For example, girls are reported to have more difficulties with stepfamilies than do boys. They are particularly likely to experience poorer relationships with and negative consequences from relationships with stepmothers.

Two other points about effects on children bear mentioning. First, there is evidence suggesting that compared to children in single-parent homes, children with a stepparent are no better off and may be at greater risk for psychological or behavioral problems. Finally, when considering differences between stepchildren and biological children, the differences tend to be small ones (Stewart 2007).

Conflict in Stepfamilies

Conflict takes place in all families: traditional nuclear families, single-parent families, and stepfamilies. If some family members do not like each other, they will bicker, argue, tease, and fight. Sometimes they have no better reason for disruptive behavior than that they are bored or frustrated and want to take it out on someone. These are fundamentally personal conflicts. Other conflicts are about definite issues, such as dating, use of the car, manners, television, or friends. These conflicts can be between partners, between parents and children, or among the children themselves. Certain types of stepfamily conflicts, however, are of a frequency, intensity, or nature that distinguishes them from conflicts in traditional nuclear families. Recent research on how conflict affects children in stepfather households found that parental conflict does not account for children's lower level of well-being (Hanson, McLanahan, and Thomson 1996). These conflicts are about favoritism, divided loyalties, discipline, and money, goods, and services.

Favoritism

Favoritism exists in families of first marriages as well as in stepfamilies. In stepfamilies, however, the favoritism often takes a different form. Whereas a parent may favor a child in a biological family on the basis of age, sex, or personality, in stepfamilies favoritism tends to run along kinship lines. A child is favored by a parent because he or she is the parent's biological child. If a new child is born to the remarried couple, they may favor him or her as a child of their joint love. In American culture, where parents are expected to treat children equally, favoritism based on kinship seems particularly unfair.

Divided Loyalties

"How can you stand that lousy, low-down, sneaky, nasty mother (or father) of yours?" demands a hostile parent. It is one of the most painful questions children can confront because it forces them to take sides against someone they love. One study (Lutz 1983) found that about half the adolescents studied confronted situations in which one divorced parent talked negatively about the other. Almost half the adolescents felt themselves "caught in the middle." Three-quarters found such talk stressful.

Divided loyalties put children in no-win situations, forcing them not only to choose between parents but also to reject new stepparents. Children feel disloyal to one parent for loving the other parent or stepparent. But, as shown in the previous chapter, divided loyalties, like favoritism, can exist in traditional nuclear families as well. This is especially true of conflict-ridden families in which warring parents seek their children as allies.

Discipline

Researchers generally agree that discipline issues are among the most important causes of conflict among remarried families (Ihinger-Tallman and Pasley 1987). Discipline is especially difficult to deal with if the child is not the person's biological child. Disciplining a stepchild often gives rise to conflicting feelings within the stepparent. Stepparents may feel that they are overreacting to the child's behavior, that their feelings are out of control, and that they are being censured by the child's biological parent. Compensating for fears of unfairness, the stepparent may become overly tolerant.

The specific discipline problems vary from family to family, but a common problem is interference by the biological parent with the stepparent (Mills 1984). The biological parent may feel resentful or overreact to the stepparent's disciplining if he or she has been reluctant to give the stepparent authority. As one biological mother who believed that she had a good remarriage stated (Ihinger-Tallman and Pasley 1987),

> Sometimes I feel he is too harsh in disciplining, or he doesn't have the patience to explain why he is punishing and to carry through in a calm manner, which causes me to have to step into the matter (which I probably shouldn't do).... I do realize that it was probably hard for my husband to enter marriage and the responsibility of a family instantly... but this has remained a problem.

As a result of interference, the biological parent implies that the stepparent is wrong and undermines his or her status in the family. Over time, the stepparent may decrease involvement in the family as a parent figure.

Money, Goods, and Services

Problems of allocating money, goods, and services exist in all families, but they can be especially difficult in stepfamilies. In first marriages, husbands and wives form an economic unit in which one or both may produce income for the family, the husband and wife being interdependent. Following divorce, the binuclear family consists of two economic units: the custodial family and the noncustodial family. Both must provide separate housing, dramatically increasing their basic expenses. Despite their separation, the two households may nevertheless continue to be extremely interdependent. The mother in the custodial single-parent family, for example, probably has reduced income. She may be employed but still dependent on child support payments or welfare, such as Temporary Aid for Needy Families. She may have to rely more extensively on child care, draining her resources dramatically. The father in the noncustodial family may make child support payments or contribute to medical or school expenses, depleting his income. Both households may have to deal with financial instability. Custodial parents can't count on always receiving their child support payments, making it difficult to undertake financial planning.

When one or both of the former partners remarry, their financial situation may be altered significantly. On remarriage, the mother receives less income from her former partner or lower welfare benefits. Instead, her new partner becomes an important contributor to the family income. At this point, a major problem in stepfamilies arises. What responsibility does the stepfather have in supporting his stepchildren? Should he or the biological father provide financial support? Because there are no norms, each family must work out its own solution.

Stepfamilies typically have resolved the problem of distributing their economic resources by using a one-pot or two-pot pattern (Fishman 1983). In the *one-pot* pattern, families pool their resources and distribute them according to need rather than biological relationship. It doesn't matter whether the child is a biological child or a stepchild. One-pot families typically have relatively limited resources and consistently fail to receive child support from the noncustodial biological parent. By sharing their resources, one-pot families increase the likelihood of family cohesion.

In *two-pot* families, resources are distributed by biological relationship; need is secondary. These families tend to have a higher income, and one or both parents have former spouses who regularly contribute to the support of their biological children. Expenses relating to children are generally handled separately; usually, there are no shared checking or savings accounts. Two-pot families maintain strong bonds among members of the first family. For these families, a major

What does a stepparent owe a stepchild? What rights and responsibilities do members of stepfamilies have? If the biological parent dies, does a stepparent gain legal custody? Speaking of custody, in a divorce can a stepparent seek custody? What about visitation rights? Must she or he pay child support? Despite legal questions that might arise around stepfamily responsibilities and relationships, as sociologist Susan Stewart notes, when it comes to stepfamilies, there are no clear and consistent legal guidelines. For the most part, stepparents remain legal strangers to their stepchildren. In that vein, they are not guaranteed custody or visitation and may not even be able to include coverage of their stepchildren in health insurance policies (Stewart 2007).

Like other areas of family law, such as divorce law and rights to marry, *where they exist*, laws and policies concerning stepfamilies vary from state to state. Thus, there is no consistency to the answers to questions such as those posed above. In some states, stepparents can seek

and obtain visitation of stepchildren after a divorce or custody of stepchildren after the biological parent dies. In most states, unless one adopts one's stepchild—which requires a biological parent to sign away his or her parental rights—there is no legal standard as to what is expected of one or what one is entitled to as a stepparent. Even the weighty issues of financial support or authority to approve medical treatment are not uniformly treated throughout the United States (Engel 2000).

Thus, what one finds in examining the legal rights and responsibilities of stepfamily members is policy that is inconsistent at best (varying from state to state). As Stewart (2007) describes, inconsistency is not the only problem. In general, there is a deficit of stepfamily law. Given how many men, women, and children are or will be members of stepfamilies, it seems crucial for states to clarify how far stepparent rights extend and how wide stepparent responsibilities range.

problem is achieving cohesion in the stepfamily while maintaining separate checking accounts.

Just as economic resources need to be redistributed following divorce and remarriage, so do goods and services. Whereas a two-bedroom home or apartment may have provided plenty of space for a single-parent family with two children, a stepfamily with additional residing or visiting step-siblings can experience instant overcrowding. Rooms, bicycles, and toys, for example, need to be shared; larger quarters may have to be found. Time becomes a precious commodity for harried parents and stepparents in a stepfamily. When visiting stepchildren arrive, duties are doubled. Stepchildren compete with parents and other children for time and affection.

It may appear that remarried families are confronted with many difficulties, but traditional nuclear families also encounter financial, loyalty, and discipline problems. We need to put these problems in perspective. (After all, half of all current marriages end in divorce, suggesting that first marriages are not problem free.) When all is said and done, the problems that remarried families face may not be any more overwhelming than those faced by traditional nuclear families (Ihinger-Tallman and Pasley 1987).

Strengths of Stepfamilies

Because we have traditionally viewed stepfamilies as less than ideal, we have often ignored their strengths. Instead, we have seen only their problems. We end this chapter by focusing on the strengths of blended families.

Family Functioning

Although traditional nuclear families may be structurally less complicated than stepfamilies, stepfamilies are nevertheless able to fulfill traditional family functions. A binuclear single-parent, custodial, or noncustodial family may provide more companionship, love, and security than the particular traditional nuclear family it replaces. If the nuclear family was ravaged by conflict or violence, for example, the single-parent family or stepfamily that replaces it may be considerably better, and because children now see happy parents, they have positive role models of marriage partners (Rutter 1994). Second families may not have as much emotional closeness as first families, but they generally experience less trauma and crisis (Ihinger-Tallman and Pasley 1987).

New partners may have greater objectivity regarding old problems or relationships. Opportunity presents itself for flexibility and patience. As family boundaries expand, individuals grow and adapt to new personalities and ways of being. In addition, new partners are sometimes able to intervene between former spouses to resolve long-standing disagreements, such as custody or child care arrangements.

Benefits for Children

As illustrated earlier, stepfamilies are often associated with a number of problematic outcomes for children. But potentially, blended families can offer children benefits that can compensate for the negative consequences of divorce and of living with a single parent. Remember the notion of "bonus families" introduced earlier? Here are some ways in which stepfamilies offer children some bonuses:

- Children gain additional role models from which to choose. Instead of having only one mother or father after whom to model themselves, children may have two mothers or fathers: the biological parents and the stepparents.
- Children gain greater flexibility. They may be introduced to new ideas, different values, or alternative politics. For example, biological parents may be unable to encourage certain interests, such as music or model airplanes, whereas a stepparent may play the piano or be a die-hard modeler. In such cases, that stepparent can assist the stepchildren in pursuing their development. In addition, children often have alternative living arrangements that enlarge their perspectives.

- Stepparents may act as a sounding board for their children's concerns. They may be a source of support or information in areas in which the biological parents feel unknowledgeable or uncomfortable.
- Children may gain additional siblings, either as step-siblings or half siblings, and consequently gain more experience in interacting, cooperating, and learning to settle disputes among peers.
- Children gain an additional extended kin network, which may become at least as important and loving as their original kin network.
- A child's economic situation is often improved, especially if a single mother remarried.
- Children may gain parents who are happily married. Most research indicates that children are significantly better adjusted in happily remarried families than in conflict-ridden nuclear families.

If anything, it should be clear by now that the American family is no longer what it was through most of the twentieth century. Remember, though, that families are dynamic and diverse—they change and they differ—and thus the rise of the single-parent family and stepfamily does not imply an end to the family. Rather, these forms provide different paths that contemporary families take as they strive to fulfill the hopes, needs, and desires of their members.

Summary

- Many of today's families depart from the traditional family system based on lifetime marriage and the intact nuclear family.
- Our pluralistic family system consists of types of families other than intact nuclear families, including single-parent families (either never married or formerly married) and stepfamilies.
- Single-parent families tend to be created by divorce or births to unmarried women, are generally headed by women, are predominantly African American or Latino, are usually poor, involve a variety of household types, and are usually a transitional stage.
- Because of gender discrimination and inequality in wages or job opportunities, many female-headed families face economic hardship.
- Single, custodial fathers typically take different paths to single parenthood, obtaining custody because mothers are financially unable to provide adequate care for children, are physically or psychologically unfit, or do not want full-time responsibility for raising children.
- Both mother-only and father-only families are more likely to be poor than are two-parent families.
- Many "single-parent households" actually contain the parent and his or her unmarried partner. Even in the absence of parents' live-in partners, parents' romantic partners, relatives, or family associates may play important roles in children's lives.
- Single parents, especially mothers, often come to rely on a combination of state or federal assistance and *private safety nets*: support from their social networks on which they can fall back in times of economic need.
- Because of a lack of resources (such as money, attention, and guidance) children of single parents are more likely to engage in high-risk, "health-compromising" behaviors and to suffer a variety of educational, economic, and personal costs.

- The inability to marry and/or the inability to adopt the children of one's partner forces many gay or lesbian parents into legal single parenthood while denying others the opportunity to develop and maintain ties with their partners' children.

- There are also positive consequences experienced by many children of single parents and numerous strengths that single mothers and fathers often display.

- The *binuclear family* is a postdivorce family system with children. It consists of two nuclear families: the mother-headed family and the father-headed family.

- The *binuclear family* consists of a number of subsystems—the former spouse subsystem, the remarried couple subsystem, the parent–child subsystem, the mother/stepmother–father/stepfather subsystem, and the sibling subsystem.

- Courtship for second marriage lacks clear norms. Courtship is complicated by the presence of children because remarriage involves the formation of a stepfamily.

- Cohabitation is more common in the "courtship" process leading to remarriages. As with cohabitation in first marriages, cohabitation before remarriages leads to higher rates of marital instability.

- Remarriage rates are lower for those who have children and as adults age. More men than women remarry. Those who initiate the divorce are more likely to remarry within three years than noninitiators.

- Explanations of remarriage focus on factors such as need, attractiveness, and opportunity.

- Remarriage differs from first marriage in several ways.

- Remarried couples are more likely to divorce than couples in their first marriages.

- The stepfamily or blended family differs from the original family because almost all members have lost an important primary relationship, one biological parent lives outside the current family, the relationship between a parent and his or her children pre-dates the new marital relationship, and *stepparent roles* are ill defined.

- Traditionally, researchers viewed stepfamilies from a "deficit" perspective, assuming that stepfamilies are very different from traditional nuclear families. More recently, stepfamilies have been viewed as normal families.

- Research in the United States and a number of other countries reveals some hazards of stepfamilies for children, including academic difficulties, higher risk of physical and mental health problems, earlier onset of sexual activity, and greater risk of dropping out of school, involvement in substance use, and criminal activity. Some research indicates that girls adjust less well than boys to stepfamily life.

- Relations between stepparents and their stepchildren have been characterized as "disengaged."

- Becoming a stepfamily is a process—a series of developmental stages. Each person—the biological parent, the stepparent, and the stepchild (or children)—experiences the process differently. For family members, it involves seven stages. The early stages are fantasy, immersion, and awareness; the middle stages are mobilization and action; and the later stages are contact and resolution.

- Although both often experience difficulty in being integrated into the family, stepmothers tend to experience greater stress in stepfamilies than do stepfathers. The warmer a woman is to her stepchildren, the more hostile they may become to her because they feel that she is trying to replace their "real" mother.

- Men are generally less involved in child rearing; they usually have few "cruel stepparent" myths to counter. A stepfather usually joins an already established single-parent family. The longer a single-parent family has been functioning, the more difficult it usually is to reorganize it.

- Despite the previously mentioned difficulties, many men attempt paternal claiming of stepchildren, embracing them as though they were their own children.

- A key issue for stepfamilies is family solidarity—the feeling of oneness with the family.

- Conflict in stepfamilies is often over favoritism, divided loyalties, discipline, and money, goods, and services. The addition of a new baby into a stepfamily neither solidifies nor divides the family.

- Stepfamily strengths may include improved family functioning and reduced conflict between former spouses. Children may gain multiple role models, more flexibility, concerned stepparents, additional siblings, additional kin, an improved economic situation, and happily married parents.

Key Terms

binuclear family 493	serial monogamy 487
blended families 486	single-parent families 486
deficit approach 492	social capital 490
paternal claiming 504	social father 489
private safety nets 489	stepfamilies 486
remarriage 495	stepparent role 502

RESOURCES ON THE WEB

Book Companion Website

www.cengage.com/sociology/strong

Prepare for quizzes and exams with online resources—including tutorial quizzes, a glossary, interactive flash cards, crossword puzzles, self-assessments, virtual explorations, and more.

Appendix A Sexual Structure and the Sexual Response Cycle

The Female Reproductive System

External Genitalia

The female external genitalia are known collectively as the vulva, which includes the mons veneris, labia, clitoris, urethra, and introitus. The *mons veneris* (literally, "mountain of Venus") is a protuberance formed by the pelvic bone and covered by fatty tissue. The *labia* are the vaginal lips surrounding the entrance to the vagina. The *labia majora* (outer lips) are two large folds of spongy flesh extending from the mons veneris along the midline between the legs. The outer edges of the labia majora are often darkly pigmented and are covered with pubic hair beginning in puberty. Usually the labia majora are close together, giving them a closed appearance. The *labia minora* (inner lips) lie within the fold of the labia majora. The upper portion folds over the clitoris and is called the *clitoral hood*. During sexual excitement, the labia minora become engorged with blood and double or triple in size. The labia minora contain numerous nerve endings that become increasingly sensitive during sexual excitement.

The *clitoris* is the center of erotic arousal in the female. It contains a high concentration of nerve endings and is highly sensitive to erotic stimulation. The clitoris becomes engorged with blood during sexual arousal and may increase greatly in size. Its tip, the clitoral glans, is especially responsive to touch.

Between the folds of the labia minora are the urethral opening and the *introitus*. The introitus is the opening to the vagina; it is often partially covered by a thin perforated membrane called the *hymen*, which may be torn accidentally or intentionally before or during first intercourse. On either side of the introitus is a tiny *Bartholin's gland* that secrets a small amount of moisture during sexual arousal.

Internal Genitalia

The *vagina* is an elastic canal extending from the vulva to the cervix. It envelops the penis during sexual intercourse and is the passage through which a baby is normally delivered. The vagina's first reaction to sexual arousal is "sweating," that is, producing lubrication through the vaginal walls.

Figure A.1 External Female Genitalia

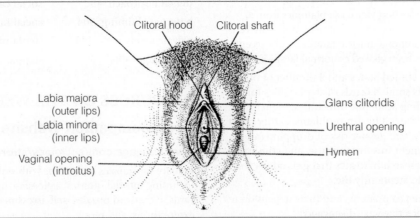

Clitoral hood   Clitoral shaft

Labia majora (outer lips)

Labia minora (inner lips)

Vaginal opening (introitus)

Glans clitoridis

Urethral opening

Hymen

Figure A.2 Cross Section of the Female Reproductive System

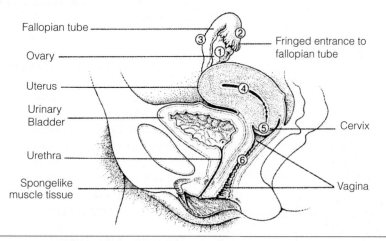

1. A follicle matures in the ovary and releases an ovum. 2. The fimbriae trap the ovum and move it into the fallopian tube. 3. The ovum travels through the fallopian tube to the uterus. 4. If the ovum is fertilized, the resulting blastocyst descends into the uterus. 5. If not fertilized, the ovum is discharged through the cervix into the vagina along with the shed uterine lining during the menstrual flow. 6. The vagina serves as a passageway to the body's exterior.

A few centimeters from the vaginal entrance, on the vagina's anterior (front) wall, there is, according to some researchers, an erotically sensitive area that they have dubbed the "Grafenberg spot" or "G-spot." The spot is associated with female ejaculation, the expulsion of clear fluid from the urethra, which is experienced by a small percentage of women.

A female has two *ovaries*, reproductive glands (gonads) that produce *ova* (eggs) and the female hormones *estrogen* and *progesterone*. At the time a female is born, she already has all the ova she will ever have—more than forty thousand of them. About four hundred will mature during her lifetime and be released during ovulation; ovulation begins in puberty and ends at menopause.

The Path of the Egg

The two *fallopian tubes* extend from the uterus up to, but not touching the ovaries. When an egg is released from an ovary during the monthly *ovulation*, it drifts into a fallopian tube, propelled by waving *fimbriae* (the fingerlike projections at the end of each tube). If it is fertilized by sperm, fertilization usually takes place within the fallopian tube. The fertilized egg will then move into the uterus.

The *uterus* is a hollow, muscular organ within the pelvic cavity. The pear-shape uterus is normally about 3 inches long, 3 inches wide at the top, and 1 inch at the bottom. The narrow, lower part of the uterus projects into the vagina and is called the *cervix*. If an egg is fertilized, it will attach itself to the inner lining

of the uterus, the *endometrium*. Inside the uterus it will develop into an embryo and then into a fetus. If an egg is not fertilized, the endometrial tissue that developed in anticipation of fertilization will be shed during *menstruation*. Both the unfertilized egg and the inner lining of the uterus will be discharged in the menstrual flow.

The Male Reproductive System

The Penis

Both urine and semen pass through the penis. Ordinarily, the penis hangs limp and is used for the elimination of urine because it is connected to the bladder by the urinary duct (urethra). The penis is usually between 2.5 and 4 inches in length. When a man is sexually aroused, it swells to about 5 to 8 inches in length, is hard, and becomes erect (hence, the term *erection*). When the penis is erect, muscle contractions temporarily close off the urinary duct, allowing the ejaculation of semen.

The penis consists of three main parts: the root, the shaft, and the glans penis. The *root* connects the penis to the pelvis. The *shaft*, which is the spongy body of the penis, hangs free. At the end of the shaft is the *glans penis*, the rounded tip of the penis. The opening at the tip of the glans is called the *urethral meatus*. The glans penis is especially important in sexual arousal because it contains a high concentration of nerve endings, making it erotically sensitive. The *frenulum*, a small area of skin on the underside

Figure A.3 External Male Genitalia

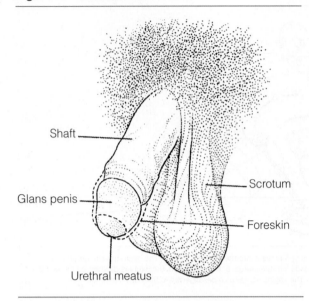

Shaft

Glans penis

Urethral meatus

Scrotum

Foreskin

of the penis where the glans and shaft meet, is especially sensitive. The glans is covered by a thin sleeve of skin called the *foreskin*. Circumcision, the surgical removal of the foreskin, may damage the frenulum.

When the penis is flaccid, blood circulates freely through its veins and arteries, but as it becomes erect, the circulation of blood changes dramatically. The arteries expand and increase the flow of blood into the penis. The spongelike tissue of the shaft becomes engorged and expands, compressing the veins within the penis so that the additional blood cannot leave it easily. As a result, the penis becomes larger, harder, and more erect.

The Testes

Hanging behind the male's penis is his *scrotum*, a pouch of skin holding his two *testes* (singular *testis*; also called *testicles*). The testes are the male reproductive glands (also called *gonads*), which produce both sperm and the male hormone *testosterone*. The testes produce sperm through a process called *spermatogenesis*. Each testis produces between 100 million and 500 million sperm daily. Once the sperm are produced, they move into the *epididymis*, where they are stored prior to ejaculation.

The Path of the Sperm

The epididymis merges into the tubular *vas deferens* (plural *vasa deferentia*). The vasa deferentia can be felt easily within the scrotal sac. Extending into the pelvic cavity, each vas deferens widens into a flask-like area called the *ampulla* (plural *ampullae*). Within the ampullae, the sperm mix with an activating fluid from the *seminal vesicles*. The ampullae connect to the *prostate gland* through the *ejaculatory ducts*. Secretions from the prostate account for most of the milky, gelatinous liquid that makes up the *semen* in which the sperm are suspended. Inside the prostate, the ejaculatory ducts join to the urinary duct from the bladder to form the urethra, which extends to the tip of the penis. The two *Cowper's glands*, located below the

Figure A.4 Cross Section of the Male Reproductive System

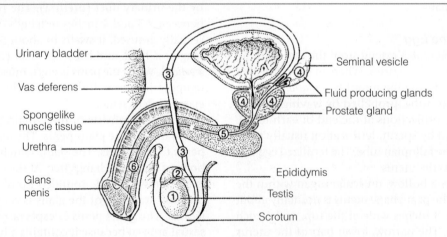

Urinary bladder

Vas deferens

Spongelike muscle tissue

Urethra

Glans penis

Seminal vesicle

Fluid producing glands

Prostate gland

Epididymis

Testis

Scrotum

1. The testis produces sperm. 2. Sperm mature in the epididymis. 3. During ejaculation, sperm travel through the vas deferens. 4. The seminal vesicles and the prostate gland provide fluids. 5. Sperm mix with the fluids, making semen. 6. Semen leaves the penis by way of the urethra.

prostate, secrete a clear, sticky fluid into the urethra that appears as small droplets on the meatus during sexual excitement.

If the erect penis is stimulated sufficiently through friction, an ejaculation usually occurs. *Ejaculation* is the forceful expulsion of semen. The process involves rhythmic contractions of the vasa deferentia, seminal vesicles, prostate, and penis. Altogether, the expulsion of semen may last from three to fifteen seconds. It is also possible to have an orgasm without the expulsion of semen.

The Sexual Response Cycle

Psychological and Physiological Aspects

When we respond sexually, we begin what is known as the *sexual response cycle.* Helen Singer Kaplan (1979) developed a model to describe the sexual response cycle. According to this model, the cycle consists of three phases: the desire phase, the excitement phase, and the orgasmic phase. The desire phase represents the psychological element of the sexual response cycle; the excitement and orgasmic phases represent its physiological aspects.

Sexual Desire

Desire can exist separately from overtly physical sexual responses. It is the psychological component that motivates sexual behavior. We can feel desire but not be physically aroused. It can suffuse our bodies without producing explicit sexual stirrings. We experience sexual desire as erotic sensations or feelings that motivate us to seek sexual experiences. These sensations generally cease after orgasm.

Physiological Responses: Excitement and Orgasm

A person who is sexually excited experiences a number of bodily responses. Most of us are conscious of some of these responses: a rapidly beating heart, an erection or lubrication, and orgasm. Many other responses may take place below the threshold of awareness, such as curling of the toes, the ascent of the testes, the withdrawal of the clitoris beneath the hood, and a flush across the upper body.

The physiological changes that take place during sexual response cycle depend on two processes: vasocongestion and myotonia. *Vasocongestion* occurs when body tissues become engorged with blood. For example, blood fills the genital regions of both males and females, causing the penis and clitoris to enlarge. *Myotonia* refers to increased muscle tension as orgasm approaches. Upon orgasm, the body undergoes involuntary muscle contractions and then relaxes. (The word *orgasm* is derived from the ancient Sanskrit *urja,* meaning "vigor" or "sap.")

Excitement Phase

In women, the vagina becomes lubricated and the clitoris enlarges during the excitement phase. The vaginal barrel expands, and the cervix and uterus elevate, a process called "tenting." The labia majora flatten and rise; the labia minora begin to protrude. The breasts may increase in size, and the nipples may become erect. Vasocongestion causes the outer third of the vagina to swell, narrowing the vaginal opening. This swelling forms the *orgasmic platform;* during sexual intercourse, it increases the friction against the penis. The entire clitoris retracts but remains sensitive to touch.

In men, the penis becomes erect as a result of vasocongestion, and the testes begin to rise. The testes may enlarge to as much as 150 percent of their unaroused size.

Orgasmic Phase

Orgasm is the release of physical tensions after the buildup of sexual excitement; it is usually accompanied by ejaculation of semen in physically mature males. In women, the orgasmic phase is characterized by simultaneous rhythmic contractions of the uterus, orgasmic platform, and rectal sphincter. In men, muscle contractions occur in the vasa deferentia, seminal vesicles, prostate, and the urethral bulb, resulting in the ejaculation of semen; contractions of the rectal sphincter also occur. Ejaculation usually accompanies male orgasm, but ejaculation and orgasm are separate processes.

Following orgasm, one of the most striking differences between male and female sexual response occurs as males experience a *refractory period.* The refractory period denotes the time following orgasm during which male arousal levels return to prearousal or excitement levels. During the refractory period, additional orgasms are impossible. Females do not have any comparable period. As a result, they have greater potential for multiple orgasms—that is, for having a series of orgasms. Although most women have the potential for multiple orgasms, only about

Figure A.5 Stages of Female Sexual Response (internal left; external right)

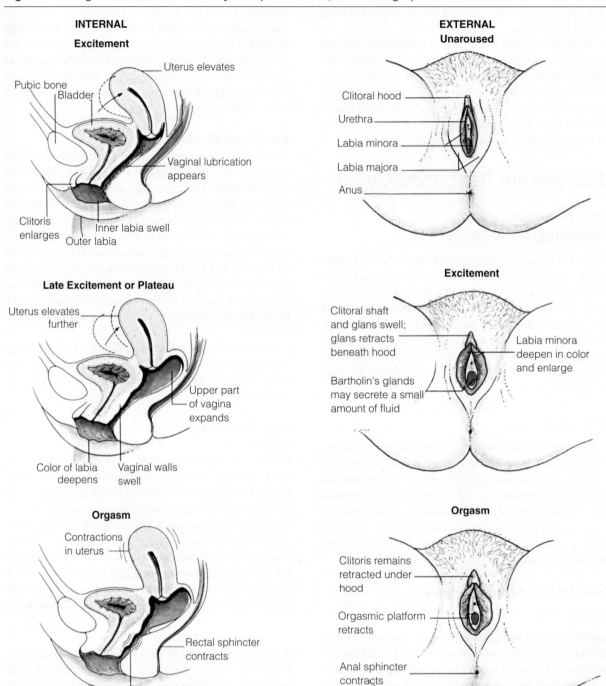

INTERNAL

Excitement

Uterus elevates

Pubic bone
Bladder

Vaginal lubrication appears

Clitoris enlarges
Inner labia swell
Outer labia

Late Excitement or Plateau

Uterus elevates further

Upper part of vagina expands

Color of labia deepens
Vaginal walls swell

Orgasm

Contractions in uterus

Rectal sphincter contracts

Rhythmic contractions in vagina

EXTERNAL

Unaroused

Clitoral hood
Urethra
Labia minora
Labia majora
Anus

Excitement

Clitoral shaft and glans swell; glans retracts beneath hood

Labia minora deepen in color and enlarge

Bartholin's glands may secrete a small amount of fluid

Orgasm

Clitoris remains retracted under hood

Orgasmic platform retracts

Anal sphincter contracts

Appendix C   The Budget Process

A budget is a plan for spending and saving. It requires you to estimate your available income for a particular period of time and decide how to allocate this income toward your expenses. A working budget can help you implement your money management plan. A well-planned budget does several things for you and your household. It can help you do the following:

- Prevent impulse spending
- Decide what you can or cannot afford
- Know where your money goes
- Increase savings
- Decide how to protect against the financial consequences of unemployment, accidents, sickness, aging, and death

A working budget need not be complicated or rigid. However, preparing one takes planning, and following one takes determination. You must do several things to budget successfully.

First, communicate with other members of your household, including older children. Consider each person's needs and wants so that all family members feel they are a part of the plan. Everyone may work harder to make the budget a success and be less inclined to overspend if they realize the consequences. When families fail to communicate about money matters, it is unlikely that a budget will reflect a workable plan.

Second, be prepared to compromise. This is often difficult. Newlyweds, especially, may have problems. Each may have been living on an individual income and not be accustomed to sharing or may have been in school and dependent on parents. If, for example, one wants to save for things and the other prefers buying on credit, the two will need to discuss the pros and cons of both methods and decide on a middle ground that each can accept. A plan cannot succeed unless there is a financial partnership.

Third, exercise willpower. Try not to indulge in unnecessary spending. Once your budget plan is made, opportunities to overspend will occur daily. Each household member needs to encourage the others to stick to the plan

Fourth, develop a good record-keeping system. At first, all members of the household may need to keep records of what they spend. This will show how well they are following the plan and will allow intermediate adjustments in the level of spending. Record keeping is especially important during the first year of a spending plan, when you are trying to find a budget that works best for you. Remember: a good budget is flexible, requires little clerical time, and, most important, works for you.

Choosing a Budget Period

A budget may cover any convenient period of time—one month, three months, or a year. Make sure the period you choose is long enough to cover the bulk of household expenses and income. Remember: Not all bills are due monthly, and every household experiences some seasonal expenses. Most personal budgets are for twelve months. You can begin the twelve-month period at any time during the year. If this is your first budget, you may want to set up a trial plan for a shorter time to see how it works.

After setting up your plan, subdivide it into more manageable operating periods. For a yearly budget, divide income and expenses by 12, 24, 26, or 52, depending on your pay schedule or when your bills are due. Most paychecks are received weekly or every two weeks. Although most bills are due once a month, not all are due at the same time in the month. Try using each paycheck to pay your daily expenses and expenses that will be due within the next week or two. This way, you will be able to pay your bills on time. You may also want to allocate something from each paycheck toward large expenses that will be due soon.

Worksheet 1 Estimating Your Income

Source	January	February	March	April	May	June
Net salary:*						
Household member 1						
Household member 2						
Household member 3						
Household member 4						
Social Security payments						
Pension payments						
Annuity payments						
Veterans' benefits						
Assistance payments						
Unemployment compensation						
Allowances						
Alimony						
Child support						
Gifts						
Interest						
Dividends						
Rents from real estate						
Other						
Monthly Totals						

*Net salary is the amount that comes into the household for spending and saving after taxes, Social Security, and other deductions.

Developing a Successful Budget

Step 1: Estimate Your Income

Total the money you expect to receive during the budget period. Use Worksheet 1 as a guide in estimating your household income. Begin with regular income that you and your family receive—wages, salaries, income earned from a farm or other business, Social Security benefits, pension payments, alimony, child support, veterans' benefits, public assistance payments, unemployment compensation, allowances, and any other income. Include variable income, such as interest from bank accounts and investments, dividends from stock and insurance, rents from property you own, gifts, and money from any other sources.

If your earnings are irregular, it may be difficult to estimate your income. It is better to underestimate than to overestimate income when setting up a budget. Some households have sufficient income, but its receipt does not coincide with the arrival of bills. For these households, planning is very important.

July	August	September	October	November	December

Step 2: Estimate Your Expenses

After you have determined how much your income will be for the planning period, estimate your expenses. You may want to group expenses into one of three categories: fixed, flexible, or set-asides. Fixed expenses are payments that are basically the same amounts each month. Fixed regular expenses include such items as rent or mortgage payments, taxes, and credit installment payments. Fixed irregular expenses are large payments due once or twice a year, such as insurance premiums. Flexible expenses vary from one month to the next, such as amounts spent on food, clothing, utilities, and transportation. Set-asides are variable amounts of money accumulated for special purposes, such as for seasonal expenses, savings and emergency funds, and intermediate and long-term goals.

Use old records, receipts, bills, and canceled checks to estimate future expenses, if you are satisfied with what your dollars have done for you and your family in the past. If you are not satisfied, now is the time for change. Consider which expenses can be cut back and which expenses need to be increased. If you spent a large amount on entertainment, for example, your new budget may reallocate some of this money to a savings account to contribute to some of your future goals.

If you do not have past records of spending, or if this is your first budget, the most accurate way to find out how much you will need to allow for each expense is to keep a record of your household spending. Carry a pocket notebook in which you jot down expenditures during a week or pay period, and total the amounts at the end of each week. You may prefer to keep an account book in a convenient place at home and make entries in it. Kept faithfully for a month or two, the record can help you find out what you spend for categories such as food, housing, utilities, household operation, clothing, transportation, entertainment, and personal items. Use this record to estimate expenses in your plan for future spending. You also need to plan for new situations and changing conditions that increase or decrease expenses. For example, the cost of your utilities may go up.

Total your expenses for a year and divide to determine the amounts that you will have to allocate toward each expense during the budgeting period. Record your estimate for each budgetary expense in the space provided on Worksheet 2. Begin with the regular fixed expenses that you expect to have. Next, enter those fixed expenses that are due once or twice a year. Many households allocate a definite amount each budget period toward these expenses to spread out the cost.

One way to meet major expenses is to set aside money regularly before you start to spend. Keep your set-aside funds separate from other funds so you will not be tempted to spend them impulsively. If possible, put them in an account where they will earn interest. You may also plan at this point to set aside a certain amount toward the long-term and intermediate goals you listed on Worksheet 1. Saving could be almost as enjoyable as spending, once you accept the idea that saving money is not punishment, but instead a systematic way of reaching your goals. You do without some things now in anticipation of buying what will give you greater satisfaction later.

You may want to clear up debts now by doubling up on your installment payments or putting aside an extra amount in your savings fund to be used for this purpose. Also, when you start to budget, consider designating a small amount of money for emergencies. Extras always come up at the most inopportune times. Every household experiences occasional minor crises too small to be covered by insurance but too large to be absorbed into the day-to-day budget. Examples may be a blown-out tire or an appliance that needs replacing. Decide how large a cushion you want for meeting emergencies. As your fund reaches the figure you have allowed for emergencies, you can start saving for something else. Now, record money allocated for occasional major expenses, future goals, savings, emergencies, and any other set-asides in the space provided for them on Worksheet 2.

After you have entered your fixed expenses and your set-asides, you are ready to consider your flexible expenses. Consider including here a personal allowance, or "mad money," for each member of the household. A little spending money that does not have to be accounted for gives everyone a sense of freedom and takes some of the tedium out of budgeting.

Step 3: Balance

Now you are ready for the balancing act. Compare your total expected income with the total of your planned expenses for the budget period. If your planned budget equals your estimated future income, are you satisfied with this outcome? Have you left enough leeway for emergencies and errors? If your expenses add up to more than your income, look again at all parts of the plan. Where can you cut down? Where are you overspending? You may have to decide which things are most important to you and which ones can wait. You may be able to do some trimming on your flexible expenses.

Once you have cut back your flexible expenses, scan your fixed expenses. Maybe you can make some sizable reductions here, too. Rent is a big item in a budget. Some households may want to consider moving to a lower-priced apartment or making different living arrangements. Others may turn in a too-expensive car and seek less expensive transportation. Look back at Worksheet 1. You may need to reallocate some of this income to meet current expenses. Perhaps you may have to consider saving for some of your goals at a later date.

If you have cut back as much as you think you can or are willing to do and your plan still calls for more than you make, consider ways to increase your income. You may want to look for a better-paying job, or a part-time second job may be the answer.

If only one spouse is employed, consider becoming a duel-earner family. The children may be able to earn their school lunch and extra spending money by doing odd jobs in your neighborhood, such as cutting grass or baby-sitting. Older children can work part-time on weekends to help out. Another possibility, especially for short-term problems, is to draw on savings. These are decisions each individual household has to make.

If your income exceeds your estimate of expenses—good! You may decide to satisfy more of your immediate wants or to increase the amount your family is setting aside for future goals.

Carrying Out Your Budget

After your plan is completed, put it to work. This is when your determination must really come into play. Can you and your family resist impulse spending?

Become a Good Consumer

A vital part of carrying out a budget is being a good consumer. Learn to get the most for your money, to recognize quality, to avoid waste, and to realize time costs as well as money costs in making consumer decisions.

Keep Accurate Records

Accurate financial records are necessary to keep track of your household's actual money inflow and outgo. A successful system requires cooperation from everyone in the household. Receipts can be kept and entered at the end of each budget period in a "Monthly Expense Record." It is sometimes a good idea to write on the back of each receipt what the purchase was for, who made it, and the date. Decide which family member will be responsible for paying bills or making purchases, and decide who will keep the record system up to date.

The household business record-keeping system does not need to be complex. The simpler it is, the more likely it will be kept current. Store your records in one spot—a set of folders in a file drawer or other fire-resistant box is a good place. You can assemble a folder for each of several categories, including budget, food, clothing, housing, insurance, investments, taxes, health, transportation, and credit. Use these folders for filing insurance policies, receipts, warranties, cancelled checks, bank statements, purchase contracts, and other important papers. Many households also rent a safe deposit box at the bank for storing deeds, stock certificates, and other valuable items.

Evaluating Your Budget

The information on Worksheet 2 can help you determine whether your actual spending follows your plan. If your first plan did not work in all respects, do not be discouraged. A budget is not something you make once and never touch again. Keep revising until the results satisfy you.

Dealing with Unemployment

Step 1: Take Time to Talk

Come right out and let your family know what is going on. Lay-off? Plant closing? Depressed economy? Business down? Explain what happened. Break down the big words so that everyone understands, especially the kids.

Fill in everyone at a family meeting or on a one-to-one basis. The important thing is not to leave anyone in the dark. If a family meeting seems out of the question, take time to talk when cleaning up after meals, cutting or raking the lawn, or taking trips to the store. Don't sugar coat the facts or tell "fairy tales." Living with less money will force your family to make hard changes. Yet let your kids know that even though there's less money, they can still count on a loving family—maybe more loving than ever.

Step 2: Take Time to Listen

Let everyone have a say about what these changes mean to him or her. Especially now, kids should be seen and heard.

Listen to words and actions. Is someone suddenly having a lot of crying spells, sleeping in late all the time, acting mean, drinking heavily, withdrawing, abusing drugs, complaining of stomach pains?

Step 3: Find Out Who's Hurting

Let everyone say what he or she is really feeling from time to time.

Just repeat whatever you hear, right when it's said. Then look for a nod to see if you heard it right. Is someone feeling helpless, sad, unloved, confused, worried, frightened, angry, like a burden to the family?

Try not to say "You shouldn't feel that way" because someone may be in real pain. The best you can

Worksheet 2 Expense Estimate and Budget
Balancing Sheet, Fixed Expenses
(Prepare for Each Month)

Month:	Amount Estimated	Amount Spent	Difference
Rent			
Mortgage			
Installments:			
Credit card 1			
Credit card 2			
Credit card 3			
Automobile loan			
Personal loan			
Student loan			
Insurance:			
Life			
Health			
Property			
Automobile			
Disability			
Set-asides:			
Emergency fund			
Major expenses			
Goals			
Savings and investments			
Allowances			
Education:			
Tuition			
Books			
Transportation:			
Repairs			
Gas and oil			
Parking and tolls			
Bus and taxi			
Recreation			
Gifts			
Other			
Total Fixed Expenses for Month			

do is let your loved ones have their say and get it off their chests.

Step 4: Let Your Feelings Out, Together or Alone

Give everyone in your family a space and time to let deep feelings out. Don't bottle them up or hide them from yourselves. If you're not comfortable showing others how you feel or fear you may strike someone who's dear to you, consider getting out of the house for a run or a brisk walk; having a good cry, alone; hitting a cushion or pillow; going to your room, shutting the door, and screaming; or all of the above.

Step 5: Solve Problems Together

Every week, look at the changes taking place in your household, and work out ways to deal with them. Working together as a team, your family can do more than survive. It can grow together and come through stronger.

Decide together things like these: what we can't afford now; what things we can do for family fun that will not cost a lot of money; who will do what chores around the house; how we will all get by with less. If your discussions break down, go back to Step 1.

If you have a lot of trouble going through these steps, professional help may be what you need. Call and make an appointment with the family service agency nearest you. Whether or not you have money to pay for the services, the agency will do its best to help your family. Remember: You are not alone.

Glossary

A

abstinence Refraining from sexual intercourse, often on religious or moral grounds.

abuse of power Sexual harassment consisting of unwelcome sexual advances, requests for sexual favors, or other verbal or physical conduct of a sexual nature as a condition of instruction or employment.

accessibility Availability of a parent or caregiver to a child, when the parent is in the same location but not in direct interaction.

acquaintance rape Rape in which the assailant is personally known to the victim, usually in the context of a dating relationship. Also known as *date rape.*

acquired immunodeficiency syndrome (AIDS) An infection caused by the human immunodeficiency virus (HIV), which suppresses and weakens the immune system, leaving it unable to fight opportunistic infections.

adaptation In ecological theory, the processes of responding to the circumstances imposed by the physical, social, cultural, and economic environments within which we live.

adolescence The social and psychological state occurring during puberty.

agents of socialization Individuals, groups, and organizations that shape our self identities, and our ideas about gender differences and our understanding of expected and acceptable behavior for each gender.

affiliated kin Unrelated individuals who are treated as if they were related.

alimony Court-ordered monetary support to a spouse or former spouse following separation or divorce.

ambiguous loss A situation of uncertainty and unclear loss, resulting from confusion about a family's boundaries, and from not knowing who is in or out of a particular family.

anal eroticism Sexual activities involving the anus.

anal intercourse Penetration of the anus by the penis.

androgyny Displaying both masculine and feminine qualities.

anonymity A state or condition requiring that no one, including the researcher can connect particular responses to the individuals who provided them.

anti-gay prejudice Strong dislike, fear, or hatred of gay men and lesbians because of their homosexuality. See also *homophobia.*

applied research The focus of such research is more practical than theoretical. It tends to be less concerned with formulating theories, generating concepts, or testing hypotheses. Data are gathered in an effort to solve problems, evaluate policies or programs, or estimate the outcome of some proposed future change in policy.

assertiveness In conflict situations, assertiveness refers to attempts to satisfy our own concerns.

assisted reproductive technologies (convert to italics)

attachment theory of love A theory maintaining that the degree and quality of an infant's attachment to his or her primary caregiver is reflected in his or her love relationships as an adult.

attributions The ways individuals account for and explain relationship failures. Such attributions influence the level of distress felt after a breakup.

authoritarian child rearing A parenting style characterized by the demand for absolute obedience.

authoritative child rearing A parenting style that recognizes the parent's legitimate power and also stresses the child's feelings, individuality, and need to develop autonomy.

autoeroticism Erotic behavior involving only the self; usually refers to masturbation but also includes erotic dreams and fantasies.

B

basic conflict Pronounced disagreement about funda-mental roles, tasks, and functions. Cf. *nonbasic conflict.*

battered child syndrome Medical term to describe patterns of injuries commonly found in physically abused children.

battering A violent act directed against another, such as hitting, slapping, beating, stabbing, shooting, or threatening with weapons.

bias A personal leaning or inclination.

bifurcation of working time Situation wherein some work longer and longer days and weeks while others work fewer hours than they want or need.

binuclear family A postdivorce family with children, consisting of the original nuclear family divided into two families, one headed by the mother, the other by the father; the two "new" families may be either single-parent or stepfamilies.

bisexuality Sexual involvement with both sexes, usually sequentially rather than during the same time period.

blended family A family in which one or both partners have a child or children from an earlier marriage or relationship; a stepfamily. See also *binuclear family.*

boomerang generation Individuals who, as adults, return to their family home and live with their parents.

bundling A colonial Puritan courtship custom in which a couple slept together with a board separating them.

C

case-study method In clinical research, the in-depth examination of an individual or small group in some form of psychological treatment in order to gather data and formulate hypotheses.

centrists Share aspects of both conservative and liberal positions. Like liberals, they identify wider social changes (e.g., economic or demographic) as major determinants of the changes in family life, but like conservatives, they believe that some familial changes have had negative consequences.

child-free marriage A marriage in which the partners have chosen not to have children.

child maltreatment Instances of the victimization of children from neglect, physical abuse, sexual abuse, and/or emotional abuse.

child sexual abuse Any sexual interaction, including fondling, erotic kissing, oral sex, or genital penetration, that occurs between an adult (or older adolescent) and a prepubertal child.

child support Court-ordered financial support by the noncustodial parent to pay or assist in paying child-rearing expenses incurred by the custodial parent.

clan A group of families related along matrilineal or patrilineal descent lines, regarded as the basic family unit in some cultures.

clinical research The in-depth examination of an individual or small group in clinical treatment in order to gather data and formulate hypotheses. See also *case-study method.*

closed field A setting in which potential partners may meet, characterized by a small number of people who are likely to interact, such as a class, dormitory, or party. Cf. *open field.*

close relationship model of legal marriage View of marriage in which marriage is considered a close, private relationship with emphasis on the emotional, psychological and sexual dimensions.

cognition The mental processes, such as thought and reflection, that occur between the moment we receive a stimulus and the moment we respond to it.

cognitive developmental theory A theory of socialization associated with Swiss psychologist Jean Piaget in which the emphasis was placed on the child's developing abilities to understand and interpret their surroundings.

cohabitation The sharing of living quarters by two heterosexual, gay, or lesbian individuals who are involved in an ongoing emotional and sexual relationship. The couple may or may not be married.

cohabitation effect The cohabitation effect refers to the fact that people who cohabit before marriage appear to have a greater risk of divorce than those couples who don't cohabit.

coitus The insertion of the penis into the vagina and subsequent stimulation; sexual intercourse.

coming out For gay, lesbian, and bisexual individuals, the process of publicly acknowledging one's sexual orientation.

commitment Factors that help maintain a relationship, for better or worse, including love, obligation, and social pressure.

common couple violence Sociologist Michael Johnson's term for the more routine forms of partner violence that results from disputes and disagreements, and for which there is a high degree of gender symmetry.

common law marriage A legal marriage resulting from two people living together as though married, and presenting themselves as married, even though their relationship was never formalized by religious or secular ceremony.

companionate love A form of love emphasizing intimacy and commitment.

companionate marriage A marriage characterized by shared decision making and emotional and sexual expressiveness.

complementary needs theory A theory of mate selection suggesting that we select partners whose needs are different from and/or complement our own needs.

concepts abstract ideas that we use to represent the reality in which we are interested. We use concepts to focus our research and organize our data.

conceptualization The specification and definition of concepts used by the researcher.

conduct of fatherhood Men's actual participation in raising their children.

confidentiality An ethical rule according to which the researcher knows the identities of participants and can connect what was said to who said it but promises not to reveal such information publicly.

conflict-habituated marriage Enduring marriages in which tension and conflict characterize the relationship.

conflict theory A social theory that views individuals and groups as being basically in competition with each other. Power is seen as the decisive factor in interactions.

conjugal family A family consisting of husband, wife, and children. See also *nuclear family.*

conjugal model of legal marriage A view of marriage that stresses the importance of maintaining strong marital ties for the benefit of children.

conjugal relationship A relationship formed by marriage.

consanguineous relationship A relationship formed by common blood ties.

consummate love In Robert Sternberg's theory, a love that includes passion, intimacy and commitment. It is idealized, sought after, yet difficult to find and sustain.

contempt Verbal or nonverbal communication with another that conveys that the recipient is undesirable. Contempt can be displayed verbally through insults, sarcasm, and mockery, or nonverbally through such expressions as rolling one's eyes.

cooperativeness In conflict situations, cooperativeness refers to attempts one makes to satisfy concerns of others.

conservatives tend to believe that cultural values have shifted from individual self-sacrifice for their families toward personal self-fulfillment. Conservatives believe that as a result of such changes, today's families are weaker and less able to meet the needs of children, adults, or the wider society.

coprovider families Families that are dependent on economic activity from both men and women.

crossover A situation in which one's job-related emotional state affects one's partner in the same way.

crude birthrate A statistic reflecting the number of births per thousand people in the population.

crude divorce rate A statistical measure of divorce calculated on the basis of the number of divorces per 1,000 people in the population.

cultural lag the outcome of rapid social change, when part of the culture changes more rapidly than another part.

culture of fatherhood Ralph LaRossa's term for the beliefs we have about the roles, responsibilities, and involvement of fathers in raising their children. LaRossa noted that these beliefs have changed more dramatically than has the conduct of fatherhood.

cunnilingus Oral stimulation for female genitals.

cycle of violence According to Lenore Walker's research, the recurring three-phase battering cycle of (1) tension building, (2) explosion, and (3) reconciliation.

D

date rape Rape in which the assailant is personally known to the victim, usually in the context of a dating relationship. Also known as *acquaintance rape.*

deductive research Research designed to test hypotheses and examine causal relationships between variables.

deep friendship Characteristic of peer marriages, deep friendship is an intense companionship, that produces a profound intimacy and mutual respect.

deficit approach An approach to studying single-parent families which emphasizes the stresses and difficulties faced by parents and children.

deinstitutionalization of marriage The weakening of social norms that define people's behavior in marriage.

Defense of Marriage Act Federal legislation signed into law by President Clinton denying recognition to same-sex couples, should any state legalize same-sex marriage.

demand-withdraw communication A communication pattern in which one person makes an effort to engage the other person in a discussion of some issue of importance. The one raising the issue may criticize, complain, or suggest a need for change in the other's behavior or in the relationship. In response, the other party withdraws by either leaving the discussion, failing to reply, or changing the subject.

dependent variable A variable that is observed or measured in an experiment and may be affected by another variable. See independent variable.

devitalized marriages A marriage type that begins with high levels of emotional intensity that dwindles over time.

distal causes Background characteristics (e.g., age at marriage, prior cohabitation, parental divorce) brought by spouses into marriage that raise the likelihood of later marital problems.

division of labor The interdependence of persons with specialized tasks and abilities. Within the family, labor is traditionally divided along gender lines. See also *complementary marriage model.*

divorce mediation The process in which a mediator (counselor) assists a divorcing couple in resolving personal, legal, and parenting concerns in a cooperative manner.

domestic partners Cohabiting heterosexual or same gender couples can enter legal relationships as domestic partners, wherein they gain many of the legal rights otherwise accorded to married couples.

double standard of aging The devaluation of women in contrast to men in terms of attractiveness as they age.

dual-career families Subcategory of dual-earner families in which both partners or spouses have high achievement orientations, a greater commitment to equality, and a stronger desire to exercise their capabilities through their work.

Duluth Model A multi-stage rehabilitative approach for batterers which emphasizes helping batterers develop critical thinking skills around themes of nonviolence, respect, partnership and negotiation.

duration-of-marriage effect The accumulation over time of various factors, such as poor communication, unresolved conflicts, role overload, heavy work schedules, and child-rearing responsibilities, that negatively affect marital satisfaction.

dyspareunia Painful sexual intercourse.

E

economic distress The stressful aspects of the economic life of individuals or families, including unemployment, poverty, and worrying about money.

egalitarian As used to describe both societies and social groups, instances in which women and men possess similar amounts of power and neither dominates economically or politically.

egocentric fallacy The mistaken belief that one's own personal experience and values are those of others in general.

empty nest The experience of parents when the last grown child has left home. The "empty nest syndrome," in which the mother becomes depressed after the children have gone, is believed to be more of a myth than a reality.

endogamy Marriage within a particular group. Cf. *exogamy*.

engagement A pledge to marry.

environment A central concept in ecological theory, referring to the physical, social, cultural and economic situations and circumstances within which individuals and families live.

environmental influences The wider context and external influences on families that are the focus of family ecological theory.

equity In social exchange theory, the result of exchanges that are fair and balanced.

erectile dysfunction Inability or difficulty in achieving erection.

ethical guidelines Standards agreed upon by professional researchers. These guidelines protect the privacy and safety of individuals who provide information in a research setting.

ethnic group A large group of people distinct from others because of cultural characteristics, such as language, religion, and customs, transmitted from one generation to another. See also *minority group* and *racial group*.

ethnocentric fallacy (also ethnocentrism) The belief that one's own ethnic group, nation, or culture is inherently superior to others. See also *racism*.

exogamy Norm requiring one to marry outside certain groups (e.g., outside one's family).

experimental research A research method involving the isolation of specific factors (variables) under controlled circumstances to determine the effects of each factor.

expressive displays ways of communicating feelings of love, largely through verbal expression.

expressive trait A supportive or emotional personality trait or characteristic.

extended family The family unit of parent(s), child(ren), and other kin, such as grandparents, uncles, aunts, and cousins.

extended household A household composed of several different families.

extrafamilial sexual abuse Child sexual abuse that is perpetrated by nonrelated individuals. Cf. *intrafamilial sexual abuse*.

extramarital sex Sexual activities, especially sexual intercourse, occurring outside the marital relationship.

F

fallacy A fundamental error in reasoning that affects our understanding of a subject.

familism A pattern of social organization in which family loyalty and strong feelings for the family are important.

families of choice Family like relationships, with fluid boundaries crossing multiple households, constructed by gay men and lesbians, consisting of those who can be counted upon to provide emotional, financial and practical support.

family A unit of two or more persons, of which one or more may be children who are related by blood, marriage, or affiliation and who cooperate economically and may share a common dwelling place.

Family and Medical Leave Act of 1993 U.S. policy requiring employers to provide employees 12 weeks of job-protected, unpaid leave from work to meet family or medical needs. Such a leave is required of employers who employ 50 or more employees. To qualify for such a leave, an employee must have been employed for 12 months by the same company and worked a minimum of 1,250 hours.

family development theory A micro-level perspective that emphasizes the patterned changes that occur in families through stages and across time.

family ecology theory A macro-level theory, emphasizing how families are influenced by and in turn influence the wider environment.

family life cycle A developmental approach to studying families, emphasizing the family's changing roles and relationships at various stages, beginning with marriage and ending when both spouses have died.

family of cohabitation The family formed by two people living together whether married or unmarried; may include children or stepchildren.

family of orientation The family in which a person is reared as a child. Cf. *family of procreation*.

family of procreation The family formed by a couple and their child or children. See also *family of cohabitation*.

family policy A set of objectives concerning family well-being and specific measures initiated by government to achieve them.

family systems theory A theory viewing family structure as created by the pattern of interactions between its various subsystems, and individual actions as being strongly influenced by the family context.

fault-based divorce Adversarial process of divorce in which one spouse alleges that the other has committed an act that has caused the marriage to break down or has failed to act in ways that would preserve the marriage.

feedback In communication, an ongoing process in which participants and their messages produce a result and are subsequently modified by the result.

fellatio Oral stimulation of the male genitals.

feminism 1. The principle that women should have equal political, social, and economic rights with men. 2. The social movement to obtain for women political, social, and economic equality with men.

feminist perspectives A variety of perspectives that stress the importance of gender and gender inequality in shaping social and familial experience.

feminization of love The idea that our cultural construction of love is based on mostly expressive qualities, more compatible with women's earlier socialization. More instrumental displays of love tend not to be recognized as love.

feminization of poverty The shift of poverty to females, primarily as a result of high divorce rates and births to unmarried women.

fictive kin ties The extension of kinshiplike attributes to non-blood relationships (such as friends or neighbors) to demonstrate their importance and to symbolize the mutual reciprocity found within them.

filial responsibility Responsibility to see that one's aging parents have support, assistance, company and are well cared for.

flextime A policy allowing workers to choose and change their work schedules to meet family needs.

foreplay Erotic activity prior to coitus, such as kissing, caressing, sex talk, and oral/genital contact; petting.

forgiveness A reduction in negative feelings and an increase in positive feelings toward a "transgressor" after a transgression, an attitude of goodwill toward someone who has done us harm, and showing compassion and forgoing resentment toward someone who has caused us pain.

friendship An attachment between people; the foundation for a strong love relationship.

G

gatekeeping A method of exerting control over and exercising power in such familial situations as caring for children by making the decisions that determine how.

gender The division into male and female, often in a social sense; sex.

gender attribution Procedures through which we come to identify others as unambiguously male or female.

gender identity The psychological sense of whether one is male or female.

gender ideology Arlie Hochschild's term for what individuals believe they ought to do as husbands or wives, and how they believe paid and unpaid work should be divided.

gender-rebellion feminism Versions of feminism that emphasize the interconnectedness between multiple inequalities (race, class, sexual orientation, age, and gender), and see gender inequality as only one aspect of wider social inequality.

gender-reform feminism Versions of feminism that stress how similar women and men are and emphasize the need for equal rights and opportunities for both genders.

gender-resistant feminism Versions of feminism that advocate separatist strategies, wherein women establish women-only social institutions and settings.

gender role The culturally assigned role that a person is expected to perform based on male or female gender.

gender-role stereotype A rigidly held and oversimplified belief that all males and females possess distinctive psychological and behavioral traits as a result of their gender.

gender symmetry The similarity in survey estimates of male-on-female and female-on-male intimate partner violence.

gender theory A theory in which gender is viewed as the basis of hierarchal social relations that justify greater power to males.

generativity A commitment to guiding or nurturing others.

grounded theory Theory that emerges from inductive research and is rooted in repeated observations.

H

halo effect The tendency to infer positive characteristics or traits based on a person's physical attractiveness.

hegemonic models of gender Dominant models of masculinity and femininity.

heterogamy Marriage between those with different social or personal characteristics. Cf. *homogamy*.

heterosexism Actions that deny, denigrate, or stigmatize non-heterosexual behavior, relationships, identity, or community.

heterosexuality Sexual orientation toward members of the opposite sex.

hierarchy of needs Abraham Maslow's theoretical construct in which human needs are ranked from most basic to higher level needs. The most basic needs are physiological (e.g., water, oxygen, food), followed by safety needs. Once these are met, the need for intimacy and belonging is most fundamental.

HIV See *human immunodeficiency virus*.

homemaker role A family role usually allocated to women, in which they are primarily responsible for home management, child rearing, and the maintenance of kin relationships. Traditionally the role is associated with economic dependency and has primacy over other female roles.

homeostasis A social group's tendency to maintain internal stability or balance and to resist change.

homoeroticism Erotic attraction to members of the same sex.

homogamous Relationship characteristic wherein partners or spouses share common demographic or personality characteristics.

homogamy Marriage between those with similar social or personal characteristics. Cf. *heterogamy*.

homophobia Irrational or phobic fear of gay men and lesbians.

homosexuality Sexual orientation toward members of the same sex. See also *gay male* and *lesbian*.

honeymoon effect The tendency of newly married couples to overlook problems, including communication problems.

hostile conflict A pattern of negative interaction wherein couples engage in frequent heated arguments, call each other names and insult each other, display an unwillingness to listen to each other, and lack emotional involvement with each other.

hostile environment An environment created through sexual harassment in which the harassed person's ability to learn or work is negatively influenced by the harasser's actions.

household As defined by the U.S. Census Bureau, one or more people—everyone living in a housing unit makes up a household.

human immunodeficiency virus (HIV) The virus causing AIDS.

hypergamy A marriage in which one's spouse is of a higher social class or rank.

hypogamy A marriage in which one's spouse is from a lower social standing.

hypothesis An unproven theory or proposition tentatively accepted to explain a collection of facts.

I

identity bargaining The process of role adjustment in a relationship, involving identifying with a role, having the role validated by others, and negotiating with the partner to make changes in the role.

impaired fecundity The experience of difficulty conceiving or carrying a pregnancy to term.

incest Sexual intercourse between individuals too closely related to marry, usually interpreted to mean father/daughter, mother/son, or brother/sister. See also *intrafamilial sexual abuse*.

independent variable A variable that may be changed or manipulated in an experiment.

individualized marriage Marriage type where the emphasis is on personal self-fulfillment more than marital commitment.

inductive research Unlike deductive research, inductive research does not begin with hypotheses to test. Research that begins with a topic of interest and some concepts that the researcher explores. As data are collected concepts are refined, patterns identified, and hypotheses are generated.

indulgent child rearing Permissive childrearing.

infant mortality rate The number of deaths for every 1,000 live births.

initiator The spouse or partner whose unhappiness or dissatisfaction leads them to consider divorce or a break-up and sets in motion the process of "uncoupling."

instrumental displays Displaying love more by what one does than what one says; tasks done out of and displaying love.

instrumental trait A practical or task-oriented personality trait or characteristic.

intensive mothering ideology The belief that children need full-time, unconditional attention from mothers to develop into healthy, well-adjusted people.

interaction In communication, a reciprocal act that takes place between at least two people.

intergenerational transmission The increased likelihood that children of divorced parents will themselves later divorce.

intersectionality An approach to studying gender that pays additional attention to race, ethnicity, and class.

serial monogamy A practice in which one person may have several spouses over his or her lifetime although no more than one at any given time.

sex 1. Biologically, the division into male and female. 2. Sexual activities.

sexual double standard The existence of different standards for judging male and female sexual behavior, on such matters as how many lifetime partners are "acceptable" (as opposed to "too many"), who in the relationship should initiate sexual activity, and how openly one should discuss sexual matters.

sexual dysfunction Recurring problems in sexual functioning that cause distress to the individual or partner; may have a physiological or psychological basis.

sexual enhancement Any means of improving a sexual relationship, including developing communication skills, fostering a positive attitude, giving a partner accurate and adequate information, and increasing self-awareness.

sexual harassment Deliberate or repeated unsolicited verbal comments, gestures, or physical contact that is sexual in nature and unwelcomed by the recipient. Two types of sexual harassment involve (1) the abuse of power and (2) the creation of a hostile environment. See also *hostile environment*.

sexual intercourse Coitus; heterosexual penile/vaginal penetration and stimulation.

sexual orientation Sexual identity as heterosexual, gay, lesbian, or bisexual.

sexual script A culturally approved set of expectations as to how one should behave sexually as male or female and as heterosexual, gay, or lesbian.

shift couples Two-earner households in which spouses work different, often nonoverlapping shifts, so that one partner is home while the other is at work.

SIDS See *sudden infant death syndrome*.

single-parent family A family with children, created by divorce or unmarried motherhood, in which only one parent is present. A family consisting of one parent and one or more children.

singlism Prejudice, stereotyping and discrimination experienced by unmarried women and men.

situational couple violence A common form of violence wherein the violence erupts during an argument. It is not part of a wider pattern, is as likely to come from a woman as from a man, rarely escalates, and is less likely to result in serious injury or fatality.

social capital The guidance, attention, and social connections that can enhance the quality of one's life and/or enlarge one's opportunities.

social classes Groupings of people who share a common economic position by virtue of their wealth, income, power, and prestige, and thus have similar social and familial experiences.

social construct An idea or concept created by society. As applied to gender, it refers to the ways in which we define gender and then act on our beliefs about it.

social desirability bias The tendency for participants in research to answer questions with more acceptable answers. Such a bias may be especially prominent in research about sexual behavior.

social exchange theory A theory that emphasizes the process of mutual giving and receiving of rewards, such as love or sexual intimacy, in social relationships, calculated by the equation Reward − Cost = Outcome.

social father: A male relative, family associate or mother's partner who acts like a father to her children.

social feminism A feminist perspective centered around the belief that workplace and family supports are essential if women are to experience a high quality of life.

social institution The organized pattern of statuses and structures, roles, and rules by which a society attempts to meet certain of its most basic needs.

social integration The degree of interaction between individuals and the larger community.

social learning theory A theory of human development that emphasizes the role of cognition (thought processes) in learning.

social mobility A term used to refer to movement up or down the socioeconomic ladder, which can occur within a person's lifetime or between generations.

social role A socially established pattern of behavior that exists independently of any particular person, such as the husband or wife role or the stepparent role.

socialization The shaping of individual behavior to conform to social or cultural norms.

socioeconomic status A term used to refer to the combined effects of income, occupational prestige, wealth, education, and income on a person's lifestyle and opportunities.

sole custody Child custody arrangement in which only one parent has both legal and physical custody of the child. See also *joint custody* and *split custody*.

split custody Custody arrangement when there are two or more children in which custody is divided between the parents, the mother generally receiving the girls and the father receiving the boys.

stations of divorce The multiple experiences people have as their marriages end.

stations of marriage The different dimensions of day-to-day life that change in the process of marrying.

stepfamily A family in which one or both partners have a child or children from an earlier marriage or relationship. Also known as a *blended family*; see also *binuclear family*.

stepparent role The role a stepparent forges for herself or himself within the stepfamily as there is no such role clearly defined by society.

stereotype A rigidly held, simplistic, and overgeneralized view of individuals, groups, or ideas that fails to allow for individual differences and is based on personal opinion and bias rather than critical judgment.

Sternberg's triangular theory of love A theory developed by Robert Sternberg emphasizing the dynamic quality of love as expressed by the interrelationship of three elements: intimacy, passion, and decision/commitment.

stimulus-value-role theory A three-stage theory of romantic development proposed by Bernard Murstein: (1) stimulus brings people together; (2) value refers to the compatibility of basic values; (3) role has to do with each person's expectations of how the other should fulfill his or her roles.

structural functionalism A sociological theory that examines how society is organized and maintained by examining the functions performed by its different structures. In marriage and family studies, structural functionalism examines the functions the family performs for society, the functions the individual performs for the family, and the functions the family performs for its members.

structural mobility When large segments of a population experience upward or downward mobility resulting from changes in the society and economy.

subsystem A system that is part of a larger system, such as family, and religious and economic systems being subsystems of society and the parent/child system being a subsystem of the family.

survey research Research method using questionnaires or interviews to gather information from small, representative groups and to infer conclusions that are valid for larger populations.

suspicious jealousy Jealousy that occurs when there is either no reason for suspicion or only ambiguous evidence that a partner is involved with another. Cf. *reactive jealousy*.

symbolic interaction A theory that focuses on the subjective meanings of acts and how these meanings are communicated through interactions and roles to give shared meaning.

T

theory A set of general principles or concepts used to explain a phenomenon and to make predictions that may be tested and verified experimentally.

time strain Situation in which an individual feels as though they do not have enough time or spend enough time in certain roles or relationships.

total marriages Marriages where the characteristics of vital relationships are present to an even greater degree.

traditional family In popular usage, an intact, married two-parent family with at least one child, which adheres to conservative family values; an idealized family. Popularly used interchangeably with *nuclear family*.

transsexuals Individuals who develop gender identities opposite that of their biological sex.

transgendered A broad category describing individuals who stand outside and apart from societal conventions regarding gender, neither identifying with nor acting within societal conventions for males or females. Some may undergo sex reassignment procedures, while others alter their social but not necessarily their physical characteristics, presenting themselves as of the opposite sex.

transvestites Also known as cross-dressers, transvestites are individuals who wear clothing of the opposite sex.

traumatic sexualization The process of developing inappropriate or dysfunctional sexual attitudes, behaviors, and feelings by a sexually abused child.

trial marriage Cohabitation with the purpose of determining compatibility prior to marriage.

triangulation The use of multiple data collection techniques in a single study.

trust Belief in the reliability and integrity of another.

two-person career An arrangement in which it takes the efforts of two spouses to ensure the career success of one. One spouse, typically the husband, can devote himself or herself fully to career pursuits because of the help and assistance received from his or her spouse. This help and assistance includes taking care of all family and domestic needs, but also often includes unpaid supportive roles (such as entertaining business colleagues).

U

uninvolved parenting Parents who are neither responsive to their children's needs nor demanding of them in their expectations.

unrequited love Love that is not returned.

upper-middle class A socioeconomic class consisting of college-educated, highly paid professionals (for example, lawyers, doctors, engineers) who have annual incomes that may reach into the hundreds of thousands of dollars.

V

value judgment An evaluation based on ethics or morality rather than on objective observation.

value theory The theory that we choose spouses based on similarity of values.

variable In experimental research, a factor, such as a situation or behavior, that may be manipulated. See also *independent variable* and *dependent variable*.

viagra An oral medication for erectile dysfunction that has restored many men's abilities to engage in sexual activity.

violence An act carried out with the intention of causing physical pain or injury to another.

violent resistance Violence used by victims—more often women—as self defense.

virginity The state of not having engaged in sexual intercourse.

vital marriages Marriages that begin and continue with high levels of emotional intensity. Such couples spend much time together and are "bound together" in many ways.

W

wheel theory of love A theory developed by Ira Reiss holding that love consists of four interdependent processes: rapport, self-revelation, mutual dependency, and intimacy fulfillment.

work spillover The effect that employment has on time, energy, activities, and psychological functioning of workers and their families.

working class A socioeconomic class comprised of skilled laborers with high school or vocational educations. The working class lives somewhat precariously, with little savings and few liquid assets should illness or job loss occur.

Bibliography

"A Fatal, Unknowing Dose: With GHB, Line Between a High and Death is Narrow." ABC-News.Go.com., March 2, 2000.

Abbey, A., and R. J. Harnish. "Perception of Sexual Intent: The Role of Gender, Alcohol Consumption, and Rape Supportive Attitudes." *Sex Roles* 32, 5–6 (March 1995): 297–313.

Abelsohn, David. "A 'Good Enough' Separation: Some Characteristic Operations and Tasks." *Family Process* 31, 1 (1992): 61–83.

Abma, J. C., and G. M. Martinez. "Childlessness among Older Women in the United States: Trends and Profiles." *Journal of Marriage and the Family* 68, 4 (2006): 1045–1056.

Abma, J. C., G. M. Martinez, W. D. Mosher, and B. S. Dawson. "Teenagers in the United States: Sexual Activity, Contraceptive Use, and Childbearing, 2002." *Vital Health Statistics* 23, 24 (2004).

Absi-Semaan, Nada, Gail Crombie, and Corinne Freeman. "Masculinity and Femininity in Middle Childhood: Developmental and Factor Analyses." *Sex Roles: A Journal of Research* 28 (1993): 187–207.

"Abuse in Later Life, Centers for Disease Control and Prevention International Violence Against Women." Centers for Disease Control and Prevention.

Acevedo, B., and A. Aron. "Does a Long-Term Relationship Kill Romantic Love?" *Review of General Psychology* 13, 1 (2009): 59–65.

Acker, Joan. "From Sex Roles to Gendered Institutions." *Contemporary Sociology* 21, 5 (1993): 565–569.

Ackerman, Diane. *The Natural History of Love.* New York: Random House, 1994.

Adams, Bert. "The Family Problems and Solutions." *Journal of Marriage and the Family* 47, 3 (August 1985): 525–529.

Adler, Nancy, Susan Hendrick, and Clyde Hendrick. "Male Sexual Preference and Attitudes toward Love and Sexuality." *Journal of Sex Education and Therapy* 12, 2 (September 1996): 27–30.

Adrian, V. and S. Coontz. "Economic Woes-Family Stress." Council on Contemporary Families, July 23, 2008. http://www.contemporaryfamilies.org/economic-issues/woes.html.

Ahlander, N., and K. Bahr, "Beyond Drudgery, Power, and Equity: Toward an Expanded Discourse on the Moral Dimensions of Housework in Families." *Journal of Marriage and the Family* 57, 1 (February 1995): 54–68.

Ahrons, C. *We're Still Family: What Grown Children Have to Say about Their Parents' Divorce.* New York: HarperCollins, 2004.

Ahrons, Constance, and Roy Rodgers. *Divorced Families: A Multidisciplinary View.* New York: Norton, 1987.

Ainsworth, Mary D., M. D. Blehar, E. Waters, and S. Wall. *Patterns of Attachment: A Psychological Study of the Strange Situation.* Hillsdale, NJ: Lawrence Erlbaum, 1978.

Alapack, Richard. "The Adolescent First Kiss." *Humanistic Psychologist* 19, 1 (March 1991): 48–67.

Aldous, Joan. *Family Careers: Developmental Change in Families.* New York: Wiley, 1978.

———. "American Families in the 1980s: Individualism Run Amok?" *Journal of Family Issues* 8, 4 (December 1987): 422–425.

———. "Perspectives on Family Change." *Journal of Marriage and the Family* 52, 3 (August 1990): 571–583.

Aldous, Joan, and Gail M. Mulligan "Fathers' Child Care and Children's Behavior Problems: A Longitudinal Study." *Journal of Family Issues* 23, 5 (July 2002): 624–647.

Ali, L. "True or False: Having Kids Makes You Happy." *Newsweek*, July 14, 2008.

Allen, K. R. "Ambiguous Loss After Lesbian Couples with Children Break Up: A Case for Same-Sex Divorce." *Family Relations* 56 (2007a): 174–182.

Allen, K. R. "The Missing Right to Same-Sex Divorce." *National Council on Family Relations Report: Family Focus on Divorce and Relationship Dissolution*, Issue FF36, F2 (December 2007b): 17.

Allen, Katherine R., and Kristin M. Baber. "Starting a Revolution in Family Life Education: A Feminist Vision." *Family Relations* 41, 4 (October 1992): 378–384.

Allen, Katherine, Rosemary Blieszner, and Karen Roberto. "Families in the Middle and Later Years: A Review and Critique of Research in the 1990's." *Journal of Marriage and the Family* 62, 4 (November 2000): 911–926.

Allen, Katherine R., David H. Demo, and Mark A. Fine, *Handbook of Family Diversity.* New York: Oxford University Press, 2000.

Allen, K., E. Husser, D. Stone, and C. Jordal. *Agency and Error in Young Adults' Stories of Sexual Decision Making. Family Relations* 57, 4 (2008): 517–529.

Altman, I., and M. Chemers. *Culture and Environment.* Belmont, CA: Wadsworth, 1984.

Amato, P. "For the Sake of the Children." Council on Contemporary Families. http://www.contemporary

families.org/subtemplate.php?t=factSheets&ext=fact4, 2006.

Amato, Paul R. "Who Cares for Children in Public Places? Naturalistic Observation of Male and Female Caretakers." *Journal of Marriage and the Family* 51 (November 1989): 981–990.

Amato, P. "Parental Absence During Childhood and Depression in Later Life." *The Sociological Quarterly* 32 (1991): 543–556.

Amato, P. "Children's Adjustment to Divorce: Theories, Hypotheses, and Empirical Support." *Journal of Marriage and the Family* 55 1 (February 1993): 23–32.

Amato, P. "The Consequences of Divorce for Adults and Children," *Journal of Marriage and the Family* 62, 4 (November 2000): 1269–1288.

Amato, P. "Divorce and the Well-Being of Adults and Children." *National Council on Family Relations Report: Family Focus on Divorce and Relationship Dissolution*, Issue FF36, F3–4 (December 2007): 18.

Amato, Paul R., and Tamara D. Afifi. "Feeling Caught between Parents: Adult Children's Relations with Parents and Subjective Well-Being." *Journal of Marriage and Family* 68, 1 (February 2006): 222–235.

Amato, Paul R., and Alan Booth. "A Prospective Study of Divorce and Parent-Child Relationships." *Journal of Marriage and the Family* 58, 2 (May 1996): 356–365.

Amato, P., A. Booth, D. Johnson, and S. Rogers. *Alone Together: How Marriage in America Is Changing*. Cambridge, MA: Harvard University Press, 2007.

Amato, Paul R., and Danelle De Boer. "The Transmission of Marital Instability Across Generations: Relationship Skills or Commitment to Marriage?" *Journal of Marriage and Family* 63 (November 2001): 1038–1051.

Amato, Paul R., and Bruce Keith. "Parental Divorce and the Well-Being of Children: A Meta-Analysis." *Psychological Bulletin* 110 (1991): 26–46.

Amato, Paul R., and Denise Previti. "People's Reasons for Divorcing: Gender, Social Class, the Life Course, and Adjustment." *Journal of Family Issues* 24, 5 (July 2003): 602–626.

Amato, Paul, and Stacy Rogers. "A Longitudinal Study of Marital Problems and Subsequent Divorce." *Journal of Marriage and the Family* 59 (August 1997): 612–624.

American Association of University Women. "Sexual Harassment in the Workplace: Facts and Statistics," http://www.aauw.org/advocacy/laf/lafnetwork/library/workplaceharassmentfacts.cfm (December 1, 2008).

American Federation of State, County and Municipal Employees. "Fact Sheets: The Family and Medical Leave Act (FMLA)," afscme.org, 2006.

American Psychological Association, APA Online. "Elder Abuse and Neglect: In Search of Solutions," Public Interest, Office on Aging. 2009.

American Psychological Association. *Interim Report of the APA Working Group on Investigation of Memories of Childhood Abuse*. Washington, DC: American Psychological Association, 1994.

American Psychological Association, APA Online. "Topic: Sexuality. Answers to Your Questions About Transgender Individuals and Gender Identity," 2009 http://www.apa.org/topics/transgender.html#howprevalent.

American Psychological Association. "Understanding Child Sexual Abuse: Education, Prevention, and Recovery," http://apa.org/releases/sexabuse, 2001.

American Society for Reproductive Medicine. "2003 Assisted Reproductive Technology (ART) Report: Success Rates." In *Assisted Reproductive Technologies: A Guide for Patients, 2003*. Society for Assisted Reproductive Technology, Birmingham, AL, 2003. http://www.sart.org/ARTPatients.html.

American Society of Reproductive Medicine. 2009 http://www.protectyourfertility.org/infertility_stats.html.

American Sociological Association. "24/7 Economy's Work Schedules Are Family Unfriendly and Suggest Needed Policy Changes." ASA News (May 2004): http://www.asanet.org.

———. "Achieving 'Adulthood' Is More Elusive for Today's Youth: Transition to 'Adulthood' Occurring at a Later Age." ASA News (August 2, 2004): http://www.asanet.org.

———. "Data Support Americans' Sense of an Accelerating 'Time Warp'; Balance Between Work and Family Remains Elusive for Many Workers." ASA News (November 2004): http://www.asanet.org.

———. "The Sexual Revolution and Teen Dating Trends is Explored in ASA's Magazine, Contexts," www2.asanet.org/ media/cntrisman.html, March 2002.

"American Time Use Survey—2008 Results." *Bureau of Labor Statistics News*. Washington, DC: U.S. Department of Labor.

American Time Use Survey—2008 Results. USDL 09-0704. http://www.bls.gov/news.release/pdf/atus.pdf.

Ames, B., W. A. Brosi, and K. M. Damiano-Teixeira. "I'm Just Glad My Three Jobs Could Be during the Day: Women and Work in a Rural Community." *Family Relations* 55, 1 (January 2006): 119–131.

Ammons, S., and P. Edgell. "Religious Influences on Work-Family Tradeoffs." *Journal of Family Issues* 28, 6 (June 2007): 794–826.

Ammons, S., and E. Kelly. "Social Class and the Experience of Work-Family Conflict during the Transition to Adulthood." *New Directions for Child and Adolescent Development* 119 (2008): 71–84.

Andersen, J. E., P. A. Andersen, and M. W. Lustig. "Opposite-sex Touch Avoidance: A National Replication and Extension." *Journal of Nonverbal Behavior,* rl (1987): 89–109.

Anderson, K. L. "Is Partner Violence Worse in the Context of Control?" *Journal of Marriage and Family* 70, 5 (2008): 1157–1168.

Andersen, P., L. Guerrero, and S. Jones. "Nonverbal Behavior in Intimate Interactions and Intimate Relationships." in *The Sage Handbook of Nonverbal Communication*, edited by V. Manusov and M. Patterson. Thousand Oaks, CA: Sage, 2006.

Aponte, Robert, with Bruce Beal and Michelle Jiles. "Ethnic Variation in the Family: The Elusive Trend Toward Convergence." In *Handbook of Marriage and the Family,* 2nd ed., edited by M. Sussman, S. Steinmetz, and G. Peterson. New York: Plenum, 1999.

Archer, J. "Sex Differences in Aggression between Heterosexual Partners: A Metaanalytic Review." *Psychological Bulletin* 126, 5 (2000): 651–680.

Arditti, Joyce A . "Noncustodial Fathers: An Overview of Policy and Resources." *Family Relations* 39, 4 (October 1990): 460–465.

———. "Factors Related to Custody, Visitation, and Child Support for Divorced Fathers: An Exploratory Analysis." *Journal of Divorce and Remarriage* 17, 3–4 (1992): 23–42.

Arditti, Joyce A., and Katherine R. Allen. "Understanding Distressed Fathers' Perceptions of Legal and Relational Inequities Postdivorce." *Family and Conciliation Courts Review* 31, 4 (1993): 461–476.

Aries, Phillipe. *Centuries of Childhood.* New York: Vintage, 1962.

Aron, Arthur, and Elaine Aron. "Love and Sexuality." In *Sexuality in Close Relationships,* edited by K. McKinney and S. Sprecher. Hillsdale, NJ: Lawrence Erlbaum, 1991.

Aron, Arthur, et al. "Experiences of Falling in Love." *Journal of Social and Personal Relationships* 6 (1989): 243–257.

Arrighi, Barbara A., and David J. Maume Jr. "Workplace Subordination and Men's Avoidance of Housework." *Journal of Family Issues* 21, 4 (May 2000): 464–487.

"Artificial Insemination," ivf-infertility.com, 2005.

"At a Glance." *OURS: The Magazine of Adoptive Families* 23, 6 (November 1990): 63.

Atkins, D., and D. Kessell. "Religiousness and Infidelity: Attendance, but Not Faith and Prayer, Predict Marital Fidelity." *Journal of Marriage and Family* 70, 2 (May 2008): 407–418.

Atkinson, Maxine P., and Stephen P. Blackwelder. "Fathering in the 20th Century." *Journal of Marriage and the Family* 55, 4 (November 1993): 975–986.

Atwood, J. D., and J. Gagnon. "Masturbatory Behavior in College Youth." *Journal of Sex Education and Therapy* 13 (1987): 35–42.

Avellar, Sarah, and Pamela Smock. "The Economic Consequences of the Dissolution of Cohabiting Unions." *Journal of Marriage and Family* 67, 2 (May 2005): 315–327.

Babbie, E. *The Practice of Social Research,* 11th ed. Belmont, CA: Wadsworth, 2007.

Babbie, Earl. *Basics of Social Research—with SPSS.* Belmont, CA: Wadsworth. 2002.

Babbie, Earl. *The Practice of Social Research,* 6th ed. Belmont, CA: Wadsworth: 1986.

"Baby Blues Family Tree," babyblues.com, May 1999.

Baca Zinn, M. "Feminist Re-Thinking from Racial-Ethnic Families" in *Women of Color in U.S. Society,* edited by M. Baca Zinn and B. T. Dill. Temple University Press, 1994.

Bachu, Amara. "Is Childlessness among American Women on the Rise?" U.S. Census Bureau, Population Division, Fertility and Family Statistics Branch. Washington, DC: U.S. Census Bureau, 1999b.

Bachu, Amara. *Current Population Reports (Series P23, No. 197),* Washington, DC: U.S. Census Bureau, 1999a.

Baggett, Courtney R. "Sexual Orientation: Should It Affect Child Custody Rulings?" *Law and Psychology Review* 16 (1992): 189–200.

Bahroo, B. "Special Issue: Child Protection in the 21st Century: Pedophilia: Psychiatric Insights." *Family Court Review* 41, 4 (October 2003): 497–507.

Baiocchi, F. "The Stink beneath the Ink: How Cartoons Are Animating the Gay and Lesbian Culture Wars." Advocates Forum, School of Social Services Administration, University of Chicago. http://www.ssa.uchicago.edu/publications/advforum/2006.pdf, 2006.

Baker-Sperry, L., and L. Grauerholz. "The Pervasiveness and Persistence of the Feminine Beauty Ideal in Children's Fairy Tales." *Gender and Society* 17, 5 (2003): 711–726.

Bakker, A., M. Westman, and J. Hetty van Emmerik. "Position Paper: Advancements in Crossover Theory." *Journal of Managerial Psychology* 24 (2009): 206–219.

Bakker, A. B., E. Demerouti, E. and R. Burke. "Workaholism and Relationship Quality: A Spillover-Crossover Perspective." *Journal of Occupational Health Psychology,* 14, 1 (2008): 23–33.

Balsam, Kimberly F., and Dawn M. Szymanski. "Relationship Quality and Domestic Violence in Women's Same-Sex Relationships: The Role of Minority Stress." *Psychology of Women Quarterly* 29, 3 (September 2005): 258–269.

Bankston, Carl, and Jacques Henry. "Endogamy Among Louisiana Cajuns: A Social Class Explanation." *Social Forces* 77 (4), 1999: 1317–1338.

Barnett, R., K. Gareis, and R. Brennan. "Wives' Shift Work Schedules and Husbands' and Wives' Well-Being in Dual-Earner Couples with Children: A Within-Couple Analysis." *Journal of Family Issues* 29, 3 (2008): 396–422.

Bartholomew, Kim. "Avoidance of Intimacy: An Attachment Perspective." *Journal of Social and Personal Relationships* 7, 2 (1990): 147–178.

Basow, S. A. Gender: *Stereotyping and Roles.* Pacific Grove, CA: Brooks/Cole, 1992.

Basow, Susan A. *Gender, Stereotypes and Roles,* 4th ed. Pacific Grove, CA. Brooks/Cole: 1993.

Batson, C. D., Z-C. Qian, and D. T. Lichter. "Interracial and Intraracial Patterns of Mate Selection Among America's Diverse Black Populations." *Journal of Marriage and Family* 68 (2006): 658–672.

Baumrind, Diana. "Current Patterns of Parental Authority." *Developmental Psychology Monographs* 4, 1 (1971): 1–102.

Baumrind, Diana. "Rejoinder to Lewis's Reinterpretation of Parental Firm Control Effects: Are Authoritative Families Really Harmonious?" *Psychological Bulletin* 94, 1 (July 1983): 132–142.

Baumrind, Diana. "The Influence of Parenting Style on Adolescent Competence and Substance Use." *Journal of Early Adolescence* 11, 1 (February, 1991): 56–95.

Baxter, Janeen. "To Marry or Not to Marry: Marital Status and the Household Division of Labor." *Journal of Family Issues* 26, 3 (April 2005): 300–321.

BBC. "Map: Parenthood Policies in Europe," news.bbc .co.uk, March 2006.

Beattie, T. *Labor of Love: The Story of One Man's Extraordinary Pregnancy.* Berkeley, CA: Seal Press, 2008.

Becerra, Rosina. "The Mexican American Family." In *Ethnic Families in America: Patterns and Variations,* 3rd ed., edited by C. Mindel et al. New York: Elsevier North Holland, 1988.

Bechhofer, L., and L. Parrot. "What Is Acquaintance Rape?" In *Acquaintance Rape: The Hidden Crime,* edited by A. Parrott and L. Bechhofer. New York: Wiley, 1991.

Beck, Joyce W., and Barbara M. Heinzerling. "Gay Clients Involved in Child Custody Cases: Legal and Counseling Issues." *Psychotherapy in Private Practice* 12, 1 (1993): 29–41.

Beer, William. *American Stepfamilies.* New Brunswick, NJ: Transaction, 1992.

Beier, E. G., and D. P. Sternberg. "Subtle Cues between Newly Weds." *Journal of Communication* 27 (1997): 92–97.

Bell, B., R. Lawton, and H. Dittmar. "The Impact of Thin Models in Music Videos on Adolescent Girls' Body Dissatisfaction." *Body Image* 4, 2 (June 2007): 137–145.

Bellah, Robert, et al. *Habits of the Heart.* Berkeley, CA: University of California Press, 1985.

Bem, S. *The Lenses of Gender: Transforming the Debate on Sexual Inequality.* New Haven, CT: Yale University Press, 1993.

Benenson, Joyce, and Athena Christakos. "The Greater Fragility of Females' Versus Males' Closest Same-Sex Friendships." *Child Development* 74, 4 (July 2003): 1123–1129.

Benokraitis, N. V., and J. R. Feagin. *Modern Sexism: Blatant, Subtle, and Covert Discrimination.* Englewood Cliffs, NJ: Prentice-Hall, 1995.

Benokraitis, Nijole V., ed. *Feuds About Families: Conservative, Centrist, Liberal, and Feminist Perspectives."* Upper Saddle River, NJ: Prentice Hall, 2000.

Berger, C. R. "Planning and Scheming: Strategies for Initiating Relationships." In *Accounting for Relationships: Explanations, Representation and Knowledge,* edited by R. Burnett, P. McChee, and D. Clarke. New York: Methuen, 1987.

Berger, P., and H. Kellner. "Marriage and the Construction of Reality." In *The Family: Its Structure and Functions,* edited by Rose Coser. New York: St. Martin's Press, 1974.

Bernard, Jessie. *The Future of Marriage,* 2nd ed. New York: Columbia University Press, 1982.

Berns, N., and D. Schweingruber. " 'When You're Involved, It's Just Different': Making Sense of Domestic Violence." *Violence Against Women* 13, 3 (2007): 240–261.

Bernstein, R., and T. Edwards. "An Older and More Diverse Nation by Midcentury." *U.S. Census Department News.* Washington, DC: U.S. Department of Commerce, August 14, 2008.

Bersamin, M., M. Todd, D. A. Fisher, et al. "Parenting Practices and Adolescent Sexual Behavior: A Longitudinal Study." *Journal of Marriage and Family* 70 (February 2008): 97–112

Betchen, Stephen. "Male Masturbation as a Vehicle for the Pursuer/Distancer Relationship in Marriage." *Journal of Sex and Marital Therapy* 17, 4 (December 1991): 269–278.

Bevan, Emma, and Daryl Higgins. "Is Domestic Violence Learned? The Contribution of Five Forms of Child Maltreatment to Men's Violence and Adjustment." *Journal of Family Violence* 17, 3 (September 2002): 223–245.

Bialik, C. "Which Is Epidemic—Sexting or Worrying about It? Cyberpolls, Relying on Skewed Samples of Techno-Teens, Aren't Always Worth the Paper They're Not Printed On." *Wall Street Journal* (April 8, 2009): A9.

Bianchi, S., and L. Casper. "American Families." *Population Bulletin* 55, 4 (2000): 1–43.

Bianchi S., J. Robinson, and M. Milkie. *Changing Rhythms of American Family Life.* New York: Russell Sage, 2006.

Bird, Chloe E., and Catherine E. Ross. "Houseworkers and Paid Workers: Qualities of the Work and Effects on Personal Control." *Journal of Marriage and the Family* 55, 4 (November 1993): 913–925.

Bird, Chloe. "Gender Differences in the Social and Economic Burdens of Parenting and Psychological Distress." *Journal of Marriage and the Family* 59(4) 1997: 809–823.

Bisin, A., G. Topa and T. Verdier. "Religious Intermarriage and Socialization in the United States," *Journal of Political Economy,* University of Chicago Press, vol. 112(3), (June, 2004): 615–664.

Bittman, M., P. England, L. Sayer, N. Folbre, and G. Matheson. "When Does Gender Trump Money? Bargaining and Time in Household Work." *American Journal of Sociology* 109, 1 (July 2003): 186–214.

Blackwell, Debra. "Marital Homogamy in the United States: The Influence of Individual and Paternal Education." *Social Science Research* 27 (2) 1998: 159–164.

Blain, G. "Gov. Paterson Signs Nixzmary's Law." *New York Daily News,* October 10, 2009.

Blair, Sampson Lee. "The Sex-Typing of Children's Household Labor: Parental Influence on Daughters' and Sons' Housework." *Youth and Society* 24, 2 (1992): 178–203.

———. "Employment, Family, and Perceptions of Marital Quality among Husbands and Wives." *Journal of Family Issues* 14, 2 (1993): 189–212.

Blair, Sampson, and Daniel Lichter. "Measuring the Division of Household Labor: Gender Segregation and Housework among American Couples." *Journal of Family Issues* 12, 1 (March 1991): 91–113.

Blair-Loy, M. *Competing Devotions: Career and Family among Women Executives.* Cambridge, MA: Harvard University Press, 2003.

Blair-Loy, Mary and Amy Wharton. "Employees Use of Work-Family Policies and the Workplace Social Context." *Social Forces* 80, 3 (March 2002): 813–845.

Blood, Robert, and Donald Wolfe. *Husbands and Wives.* Glencoe, IL: Free Press, 1960.

Blow, Adrian J., and Kelley Hartnett. "Infidelity in Committed Relationships II: A Substantive Review." *Journal of Marital and Family Therapy* 31, 2 (April 2005): 183–216.

Blumenfeld, Warren, and Diane Raymond. *Looking at Gay and Lesbian Life.* Boston: Beacon Press, 1989.

Blumstein, Philip. "Identity Bargaining and Self-Conception." *Social Forces* 53, 3 (1975): 476–485.

Blumstein, Philip, and Pepper Schwartz. *American Couples.* New York: McGraw-Hill, 1983.

Bogle, K. *Hooking Up: Sex, Dating, and Relationships on Campus.* New York: New York University Press, 2008.

———. "The Shift from Dating to Hooking Up: What Scholars Have Missed." Paper presented at the annual meeting of the American Sociological Association, Philadelphia. http://www.allacademic.com/meta/p23315_index.html, 2005.

Bohannan, Paul, ed. *Divorce and After.* New York: Doubleday, 1970.

Bonafide, M. *Tourism Boycott Threatened: Gay Rights Group Angry over Refusal of Partner Benefits.* Asbury Park Press, Asbury Park, NJ: 2006.

Bonus Families. Home page, bonusfamilies.com, 2006.

Book, Cassandra L., et al. *Human Communication: Principles, Contexts, and Skills.* New York: St. Martin's Press, 1980.

Booth, A., and J. N. Edwards. "Starting Over: Why Remarriages Are More Unstable. *Journal of Family Issues* 13, 2 (June 1992): 179–194.

Booth, Alan, and Paul Amato. "Parental Predivorce Relations and Offspring Postdivorce Well-Being." *Journal of Marriage and Family* 63 (February 2001): 197–212.

Booth, Alan, et al. "Divorce and Marital Instability over the Life Course." *Journal of Family Issues* 7 (1986): 421–442.

Borisoff, Deborah, and Lisa Merrill. *The Power to Communicate: Gender Differences as Barriers.* Prospect Heights, IL: Waveland, 1985.

Borland, Dolores. "An Alternative Model of the Wheel Theory." The Family Coordinator (July 1975): 289–292.

Boss, P. "Rethinking Adoptions That Dissolve: What We Need to Know about Research, Practice, Policies and Attitudes." *National Council on Family Relations Report: Family Focus on Adoption,* Issue FF39 (September 2008): F5–7, 19.

Boss, Pauline. "Ambiguous Loss Research, Theory, and Practice: Reflections After 9/11." *Journal of Marriage & Family* 66, 3 (August 2004): 551–566.

Bouchard, Genevieve, "Adult Couples Facing a Planned vs. an Unplanned Pregnancy: Two Realities." *Journal of Family Issues* 26, 5 (July 2005): 619–637.

Boushey, H. "Women Breadwinners, Men Unemployed." Center for American Progress. http://www.american-progress.org/issues/2009/07/breadwin_women.html, 2009.

Bowe, J. "Gay Donor or Gay Dad? Gay Men and Lesbians Are Having Babies and Redefining Fatherhood, Commitment and What a Family Can Be." *New York Times Magazine* (November 19, 2006):66.

Bowen, D. "California General Election: Election Night Results." http://vote.sos.ca.gov/returns/props.

Bowlby, J. *Attachment and Loss. Vol 2: Separation: Anxiety and Anger.* New York: Basic Books, 1973, reissued in 1999.

Bowlby, John. *Attachment and Loss.* 3 vols. New York: Basic Books, 1980.

Bowlby, John. *Attachment and Loss.* New York: Basic Books, 1969.

Bradbury, T. N., and B. R. Karney. "Understanding and Altering the Longitudinal Course of Marriage." *Journal of Marriage and Family* 66, 4 (November 2004): 862–879.

Bradsher, Keith. "3 Guilty of Manslaughter in Slipping Drug to Girl," *New York Times,* March 15, 2000.

Bramlett, M. D., and W. D. Mosher. "Cohabitation, Marriage, Divorce, and Remarriage in the United States." *Vital and Health Statistics* 23, 22 (2002).

Bratter, J., and R. King. "But Will It Last?": Duration of Interracial Unions Compared to Similar Race Relationships." *Family Relations* 57 (2008): 160–171.

Braver, Sanford, et al. "A Social Exchange Model of Nonresidential Parent Involvement." In *Nonresidential Parenting: New Vistas in Family Living,* edited by C. E. Depner and J. H. Bray. Newbury Park, CA: Sage Publications, 1993a.

———. "A Longitudinal Study of Noncustodial Parents: Parents without Children." *Journal of Family Psychology* 7, 1 (June 1993b): 9–23.

Bray, J. H., and S. H. Berger. "Noncustodial Parent and Grandparent Relationships in Stepfamilies." *Family Relations* 39, 4 (1990): 414–419.

Bray, James H., and Charlene Depner. "Nonresidential Parents: Who Are They?" In *Nonresidential Parenting: New Vistas in Family Living,* edited by C. E. Depner and J. H. Bray. Newbury Park, CA: Sage, 1993.

Bray, James. "Family Assessment: Current Issues in Evaluating Families." *Family Relations* 44, 4 (October 1995): 469–477.

Brayfield, April A. "Employment Resources and Housework in Canada." *Journal of Marriage and the Family* 54, 1 (February 1992): 19–30.

Breault, K. D., and Augustine Kposowa. "Explaining Divorce in the United States: A Study of 3,111 Counties, 1980." *Journal of Marriage and the Family* 49, 3 (August 1987): 549–558.

Brennan, Robert, Rosalind Barnett, and Karen Gareis. "When She Earns More Than He Does: A Longitudinal Study of Dual-Earner Couples." *Journal of Marriage & Family* 63, 1 (February 2001): 168–183.

Bretschneider, Judy, and Norma McCoy. "Sexual Interest and Behavior in Healthy 80– to 102–Year-Olds." *Archives of Sexual Behavior* 17, 2 (April 1988): 109–128.

Bretthauer, B., T. Schindler Zillerman, and J. Banning. "A Feminist Analysis of Popular Music: Power over, Objectification of, and Violence against Women." *Journal of Feminist Family Therapy* 18, 4 (2007): 29–51.

Brewster, Karin and Irene Padavic. "Change in Gender-Ideology, 1977–1996: The Contributions of Intracohort Change and Population Turnover. *Journal of Marriage and the Family* 62, 2 (May 2000): 477–488.

Bringle, Robert G., and Glenda J. Bagby. "Self-Esteem and Perceived Quality of Romantic and Family

Relationships in Young Adults." *Journal of Research in Personality* 26, 4 (1992): 340–356.

Bringle, Robert, and Bram Buunk. "Jealousy and Social Behavior: A Review of Person, Relationship, and Situational Determinants." In *Review of Personality and Social Psychology, Vol. 6, Self, Situation, and Social Behavior,* edited by P. Shaver. Newbury Park, CA: Sage, 1985.

———. "Extradyadic Relationships and Sexual Jealousy." In *Sexuality in Close Relationships,* edited by K. McKinney and S. Sprecher. Hillsdale, NJ: Lawrence Erlbaum, 1991.

Brittingham, Angela, and de la Cruz, G. Patricia. "We the People of Arab Ancestry in the United States." Census 2000 Special Reports #21. US Census Bureau, Washington DC, 2005.

Brittingham, Angela, and G. Patricia de la Cruz. "Ancestry 2000: A Census 2000 Brief. C2KBR-35." Washington, DC: United States Census Bureau: 2004.

Britton, D. M. "Homophobia and Homosociality: An Analysis of Boundary Maintenance." *Sociological Quarterly* 31, 3 (September 1990): 423–439.

Bronfenbrenner, Urie. *The Ecology of Human Development.* Cambridge, MA: Harvard University Press, 1979.

Brooke, J. "Home Alone Together." *New York Times,* May 4, 2006.

Brooks, Jane B. *Parenting in the 90s.* Mountain View, CA: Mayfield, 1994.

Brown, H. "U.S. Maternity Leave Benefits Are Still Dismal." *Forbes.com,* May 4, 2009.

Brown, J. D., K. L. L 'Engle, C. J. Pardun, G. Guo, K. Kenneavy, and C. Jackson. "Sexy Media Matter: Exposure to Sexual Content in Music, Movies, Television and Magazines Predicts Black and White Adolescents' Sexual Behavior." *Pediatrics* 117, 4 (2006): 1018–1027.

Brown, Lyn Mikel, and Carol Gilligan. *Meeting at the Crossroads: Women's Psychology and Girl's Development.* Cambridge, MA: Harvard University Press, 1992.

Brown, Susan, and Alan Booth. "Cohabitation Versus Marriage: A Comparison of Relationship Quality," *Journal of Marriage and the Family* 58 (August) 1996: 668–678.

———. "Stress at Home, Peace at Work: A Test of the Time Bind Hypothesis." *Social Science Quarterly* 83, 4 (December 2002a): 905–919.

———. "Bending the Time Bind: Rejoinder to Hochschild and Goodman." *Social Science Quarterly* 83, 4, (December 2002b): 941–946.

Browne, Angela, and David Finkelhor. "Initial and Long-Term Effects: A Review of the Research." In *Sourcebook on Child Sexual Abuse,* edited by D. Finkelhor. Beverly Hills, CA: Sage, 1986.

Brubaker, Timothy H. "Families in Later Life: A Burgeoning Research Area." *Journal of Marriage & Family* 52, 4 (November 1990): 959–981.

Bryant, A. "Changes in Attitudes toward Women's Roles: Predicting Gender Role Traditionalism among College Students." *Sex Roles* 48, 3/4 (February 2003): 131–143.

Bryant, Z. Lois, and Marilyn Coleman. "The Black Family as Portrayed in Introductory Marriage and Family Textbooks." *Family Relations* 37, 3 (July 1988): 255–259.

Budgeon, S. "Couple Culture and the Production of Singleness." *Sexualities* 11, 3 (2008): 301–325.

Budig, M. "Feminism and the Family." In *The Blackwell Companion to the Sociology of Families,* edited by J. Scott Treas and M. Richards. Malden, MA: Blackwell, 2004.

Budig, Michelle, and Paula England. "The Wage Penalty for Motherhood." *American Sociological Review* 66 (April 2001): 204–225.

Buffum J. "Prescription Drugs and Sexual Function." *Psychiatr. Med.* 10 (1992): 181–198.

Bumpass, Larry L., Teresa C. Martin, and James A. Sweet. "The Impact of Family Background and Early Marital Factors on Marital Disruption." *Journal of Family Issues* 12, 1 (1991): 22–42.

Bumpass, Larry, and Hsien-Hen Lu. "Trends in Cohabitation and Implications for Children's Family Contexts in the United States." *Population Studies* 54, 1 (March 2000): 29–41.

Bumpass, Larry, and James Sweet. "Changing Patterns of Remarriage." *Journal of Marriage & Family* 52, 3 (August 1990): 747–756.

Bumpass, Larry, James Sweet, and Teresa Castro Martin. "Changing Patterns of Remarriage." *Journal of Marriage and the Family* 52, 3 (August 1990): 747–756.

Burdette, A., C. G. Ellison, D. E. Sherkat, and K. A. Gore. "Are There Religious Variations in Marital Infidelity?" *Journal of Family Issues* 28, 12 (2007): 1553–1581.

Bureau of Labor Statistics, Employment Situation Summary: March 6, 2009.

Burgess, Ernest. "The Family as a Unity of Interacting Personalities." *The Family* 7, 1 (March 1926): 3–9.

Burgess, M., S. Stermer, and S. Burgess. "Sex, Lies and Video Games: The Portrayal of Male and Female Characters on Video Game Covers." *Sex Roles* 57, 5–6 (September 2007): 419–433.

Burleson, Brant, and Wayne Denton. "The Relationship Between Communication Skill and Marital Satisfaction: Some Moderating Effects." *Journal of Marriage and the Family* 59, 4 (November 1997): 884–902.

Burns, Ailsa, and Cath Scott. *Mother-Headed Families and Why They Have Increased.* Hillsdale, NJ: Erlbaum, 1994.

Butler, Amy C. "Gender Differences in the Prevalence of Same-Sex Sexual Partnering: 1988–2002." Social Forces 84, 1 (September 2005): 421–449.

Buunk, Bram, and Barry van Driel. *Variant Lifestyles and Relationships.* Newbury Park, CA: Sage, 1989.

Buunk, Bram, and Ralph Hupka. "Cross-Cultural Differences in the Elicitation of Sexual Jealousy." *Journal of Sex Research* 23, 1 (February 1987): 12–22.

Byers, E. S., H. A. Sears, and A. D. Weaver. "Parents' Reports of Sexual Communication with Children in Kindergarten to Grade 8." *Journal of Marriage and Family* 70 (2008): 86–96.

Byrne, Michael, Alan Carr, and Marie Clark. "Power in Relationships of Women with Depression." *Journal of Family Therapy* 26, 4 (November 2004): 407–429.

Cabrera, N., J. Fagan, and D. Farrie. "Explaining the Long Reach of Fathers' Prenatal Involvement on Later

Paternal Engagement." *Journal of Marriage and Family* 70, 5 (2008): 1094–1107.

Caetano, R., P. A. C. Vaeth, and S. Ramisetty-Mikler. *Intimate Partner Violence Victim and Perpetrator Characteristics among Couples in the United States.* New York: Springer Science + Business Media, 2008.

Caetano, Raul, John Schafer, Catherine Clark, Carol Cunradi, and Kelly Raspberry. "Intimate Partner Violence, Acculturation, and Alcohol Consumption Among Hispanic Couples in the United States." *Journal of Interpersonal Violence* 15, 1 (January 2000): 30–45.

Cahoon, D. E., M. Edmonds, R. M. Spaulding, and J. C. Dickens. "A Comparison of the Opinions of Black and White Males and Females Concerning the Occurrence of Rape." *Journal of Social Behavior and Personality* 10, 1 (March 1995): 91–100.

Call, V., S. Sprecher, and P. Schwartz. "The Incidence and Frequency of Marital Sex in a National Sample." *Journal of Marriage and Family* 57, 3 (August 1995): 639–652.

Call, Vaughn R. A., and Tim B. Heaton. "Religious Influence on Marital Stability." *Journal for the Scientific Study of Religion* 36, 3 (September 1997): 382–392.

Camarota, Steven. "Immigrants from the Middle East: A Profile of the Foreign-born Population from Pakistan to Morocco." Center for Immigration Studies (August 2002).

Campa, M., and J. J. Eckenrode. "Pathways to Intergenerational Adolescent Childbearing in a High-Risk Sample." *Journal of Marriage and Family* 68, 3 (2006): 558–572.

Campbell, L., R. Martin, and J. Ward. "An Observational Study of Humor Use while Resolving Conflict in Dating Couples." *Personal Relationships* 15 (2008): 41–55.

Cancian, F. M. "Gender Politics: Love and Power in the Private and Public Spheres." In *Family in Transition,* edited by A. S. Skolnick and J. H. Skolnick. Glenview, IL: Scott, Foresman, 1989.

Cancian, Francesca. "Gender Politics: Love and Power in the Private and Public Spheres." In *Gender and the Life Course,* edited by A. Rossi. Hawthorne, NY: Aldine, 1985: 253–262.

———. *Love In America: Gender and Self Development.* New York: Oxford University Press, 1987.

Cannon, H. Brevy. " 'Whose Turn to Pay?' Can Be a Real Deal-Breaker for Cohabiting Couples." UVa Today .http://www.virginia.edu/uvatoday/newsRelease. php?id=8573, 2009.

Canter, Lee, and Marlene Canter. *Assertive Discipline for Parents.* Santa Monica, CA: Canter and Associates, 1985.

Caponera, B. (1998). *The Nature of Sexual Assault in New Mexico I (1995–1996).* New Mexico Clearinghouse on Sexual Abuse and Assault Services, Albuquerque, NM: 1998.

Cargan, Leonard, and Matthew Melko. *Singles: Myths and Realities.* Beverly Hills, CA: Sage, 1982.

Carl, Douglas. "Acquired Immune Deficiency Syndrome: A Preliminary Examination of the Effects on Gay Couples and Coupling." *Journal of Marital and Family Therapy* 12, 3 (July 1986): 241–247.

Caron, Sandra L., and Eilean G. Moskey. "Changes over Time in Teenage Sexual Relationships: Comparing the High School Class of 1950, 1975, and 2000." *Adolescence* 37, 147 (Fall 2002): 515–526.

Carpenter, Laura. "Gender and the Meaning and Experience of Virginity Loss in the Contemporary United States." *Gender and Society* 16, 3 (June 2002): 345–365.

Carpenter, Laura. *Virginity Lost: An Intimate Portrait of First Sexual Experiences.* New York: New York University Press, 2005.

Carr, Deborah. "The Desire to Date and Remarry Among Older Widows and Widowers." *Journal of Marriage and Family* 66, 4 (November 2004): 1051–1068.

Carroll, D. and S. Reid, with K. Moline. *Nanny 911: Expert Advice for All Your Parenting Emergencies.* New York: Harper Collins. 2005.

Carter, Betty, and Monica McGoldrick, eds. *The Changing Family Life Cycle,* 2nd ed. Boston: Allyn and Bacon, 1989.

Casper, L., and L. Sayer. "Cohabitation Transitions: Different Purposes and Goals, Different Paths." Paper presented at the meeting of the Population Association of America, Los Angeles, 2002.

Casper, Lynne M., and Kristin E. Smith. "Dispelling the Myths: Self-care, Class, and Race." *Journal of Family Issues* 23, 6 (2002): 716–727.

Casper, Lynne, and Suzanne Bianchi. *Continuity and Change in the American Family.* Thousand Oaks, CA: Sage Publications, 2002.

Cate, Rodney M., and Sally A. Lloyd. *Courtship.* Newbury Park, CA: Sage, 1992.

Caughlin, J. P., and R. S. Malis. "Demand/Withdraw Communication between Parents and Adolescents as a Correlate of Relational Satisfaction." *Communication Reports,* June 22, 2004: 59–71.

Caughlin, J., and A. Scott. "Toward a Communication Theory of the Demand/Withdraw Pattern of Interaction in Interpersonal Relationships." In *New Directions in Interpersonal Communication Research,* edited by S. Smith and S. Wilson. Thousand Oaks, CA: Sage, 2010.

Cavanagh, S., and K. Sullivan. "Family Instability, Childhood Relationship Skills and Romance in Adolescence." National Center for Family and Marriage, Working Paper Series, WP-09-11. September 2009.

CDC Online Newsroom. "Teen Birth Rate Rises for First Time in 14 Years." Press release, December 5, 2007.

———. "Pregnancy Rate Drops for US Women under Age 25." Press release, April 14, 2008.

———. "CDC Releases New Infant Mortality Data." Press release, October 15, 2008.

CDC. *See* Centers for Disease Control and Prevention.

"Census: 5.4 Million Mothers Are Choosing to Stay at Home." *USA Today,* November 30, 2004.

Centers for Disease Control. Center for Health Statistics. *Vital and Health Statistics* 23, 24 (2004). http://www. edc.gov/nchs/data/series/sr_23/sr23_024.pdf.

Centers for Disease Control and Prevention, National Center for Health Statistics, Health Data Interactive. http://www.cdc.gov/nchs/hdi.htm, 2009.

Fields, J. "Children's Living Arrangements and Characteristics: March 2002." U.S. Census Bureau. Current Population Reports (P20-547).

Fields, J. "America's Families and Living Arrangements: 2003." *Current Population Reports, P20-553*. Washington, DC: U.S. Census Bureau, 2004.

Fields, J. "The Living Arrangements of Children in 2005." http://www.census.gov/population/www/pop-profile/files/dynamic/LivArrChildren.pdf.

Fields, J., and L. Casper. "America's Families and Living Arrangements: March 2000." *Current Population Reports, P20-537*. Washington, DC: U.S. Census Bureau, 2001.

Fields, Jason, and Lynne Casper. "America's Families and Living Arrangements: March 2000." *Current Population Reports* (Series P20, No. 537). Washington, DC: U.S. Census Bureau, 2001.

Fields, Jason. "Living Arrangements of Children: 2001." U.S. Census Bureau. Current Population Reports (P70-104).

Fields, Jason. *America's Families and Living Arrangements: 2003.* Current Population Reports, P20-553. U.S. Census Bureau, Washington, DC, 2003.

Filene, Peter. *Him/Her/Self: Sex Roles in Modern America* (2nd ed.). Baltimore: Johns Hopkins University Press, 1986.

Fincham, F. D., and T. N. Bradbury. "The Impact of Attributions in Marriage: A Longitudinal Analysis." *Journal of Personality and Social Psychology* 53 (1987): 510–517.

Fincham, Frank D., and Steven R. H. Beach. "Conflict in Marriage: Implications for Working with Couples." *Annual Review of Psychology* 50, 1 (1999): 47–77.

Fincham, Frank, and Steven Beach. "Forgiveness in Marriage: Implications for Psychological Aggression and Constructive Communication." *Personal Relationships* 9, 3 (September 2002): 239–251.

Fineman, Martha Albertson, and Roxanne Mykitiuk. *The Public Nature of Private Violence: The Discovery of Domestic Abuse.* New York: Routledge, 1994.

Finkel, Judith A., and Finy J. Hansen. "Correlates of Retrospective Marital Satisfaction in Long-Lived Marriages: A Social Constructivist Perspective." *Family Therapy* 19, 1 (1992): 1–16.

Finkelhor, D. and L. Jones. "Why Have Child Maltreatment and Child Victimization Declined?" *Journal of Social Issues* 62, 4 (2006): 685–716.

Finkelhor, D., and K. Yllo. *License to Rape: Sexual Abuse of Wives.* New York: Holt, Rinehart, and Winston, 1985.

Finkelhor, David, and Angela Browne. "Initial and Long-Term Effects: A Conceptual Framework." In *Sourcebook on Child Sexual Abuse,* edited by D. Finkelhor. Beverly Hills, CA: Sage, 1986.

Finkelhor, David, and Larry Baron. "High Risk Children." In *Sourcebook on Child Sexual Abuse,* edited by D. Finkelhor. Beverly Hills, CA: Sage, 1986.

Finkelhor, David, and Richard Ormrod. "Factors in the Underreporting of Crimes against Juveniles." *Child Maltreatment* 6, 3 (2001): 219–229.

Finkelhor, David, G. Hotaling, I. A. Lewis, and C. Smith. *Missing, Abducted, Runaway, and Throwaway Children in America.* Washington, DC: U.S. Department of Justice, 1990.

Finkelhor, David. *Sexually Victimized Children.* New York: Free Press, 1979.

———. "Common Features of Family Abuse." In *The Dark Side of Families,* edited by D. Finkelhor et al. Beverly Hills, CA: Sage, 1983.

———. *Child Sexual Abuse: New Theory and Research.* New York: Free Press, 1984.

———. "Prevention: A Review of Programs and Research." In *Sourcebook on Child Sexual Abuse,* edited by D. Finkelhor. Beverly Hills, CA: Sage Publications, 1986a.

———. "Prevention Approaches to Child Sexual Abuse." In *Violence in the Home: Interdisciplinary Perspectives,* edited by M. Lystad. New York: Brunner/Mazel, 1986b.

———. "Sexual Abuse of Children." In *Vision 2010: Families and Violence, Abuse and Neglect,* edited by R. J. Gelles. Minneapolis: National Council on Family Relations, 1995.

Fischer, Lucy R. "Mothers and Mothers-in-Law." *Journal of Marriage and the Family* 45, 1 (February 1983): 187–192.

Fisher, C., and R. Lerner. "Sex Differences." In *Encyclopedia of Applied Developmental Science*, vol. 2. Newbury Park, CA: Sage, 2005.

Fisher, J. D., and W. A. Fisher. "Changing AIDS-Risk Behavior." *Psychological Bulletin* 111 (1992): 455–474.

Fishman, Barbara. "The Economic Behavior of Stepfamilies." *Family Relations* 32 (July 1983): 356–366.

Fitting, Melinda, Peter Rabins, M. Jane Lucas, and James Eastham. "Caregivers for Demented Patients: A Comparison of Husbands and Wives." *Gerontologist* 26 (1986): 248–252.

Fitzpatrick, J., S. Liang, D. Feng, D. Crawford, G. Sorell, and B. Morgan-Fleming. "Social Values and Self-Disclosure: A Comparison of Chinese Native, Chinese Resident (in U.S.) and North American Spouses." *Journal of Comparative Family Studies* 37 (2006): 113–127.

Flaks, David K., Ilda Ficher, Frank Masterpasqua, and G. Joseph. "Lesbians Choosing Motherhood: A Comparative Study of Lesbians and Heterosexual Parents and Their Children." *Developmental Psychology* 31, 1 (January 1995): 105–114.

Flynn, Clifton P. "Sex Roles and Women's Response to Courtship Violence." *Journal of Family Violence* 5, 1 (March 1990): 83–94.

Foa, Uriel G., Barbara Anderson, J. Converse, and W. A. Urbansky, "Gender-Related Sexual Attitudes: Some Cross-Cultural Similarities and Differences." *Sex Roles* 16, 19–20 (May 1987): 511–519.

Foley, L., et al. "Date Rape: Effects of Race of Assailant and Victim and Gender of Subjects." *Journal of Black Psychology* 21, 1 (February 1995): 6–18.

Folk, Karen F., and Andrea H. Beller. "Part-Time Work and Child Care Choices for Mothers of Preschool Children." *Journal of Marriage and the Family* 55, 1 (February 1993): 147–157.

Folk, Karen Fox, and Yunae Yi. "Piecing Together Child Care with Multiple Arrangements: Crazy Quilt or

Preferred Pattern for Employed Parents of Preschool Children." *Journal of Marriage and the Family* 56, 3 (August 1994): 669–680.

Follingstad, D. R., E. S. Hause, L. L. Rutledge, and D. S. Polek. "Effects of Battered Women's Early Responses on Later Abuse Patterns." *Violence and Victims* 7 (1992): 109–128.

Follingstad, Diane R., L. L. Rutledge, B. J. Berg, and E. S. Haure. "The Role of Emotional Abuse in Physically Abusive Relationships." *Journal of Family Violence* 5, 2 (June 1990): 107–120.

Foran, H. M., and K. D. O'Leary. "Alcohol and Intimate Partner Violence: A Meta-Analytic Review." *Clinical Psychology Review* 28 (2008): 1222–1234.

Forste, R., and K. Tanfer. "Sexual Exclusivity among Dating, Cohabiting, and Married Women." *Journal of Marriage and Family* 58 (February 1996): 33–47.

Forste, Renata, and Tim Heaton. "Initiation of Sexual Activity among Female Adolescents." *Youth and Society* 19, 3 (March 1988): 250–268.

Fowers, B., K. Montel, and D. Olson. "Predicting Marital Success for Premarital Couple Types Based on PREPARE." *Journal of Marital and Family Therapy* 22, 1 (1996): 103–119.

Fox, Greer L., and Robert F. Kelly. "Determinants of Child Custody Arrangements at Divorce." *Journal of Marriage and the Family* 57, 3 (August 1995): 693–708.

Fox, Greer, and Velma McBride Murry. "Gender and Families: Feminist Perspectives and Family Research." *Journal of Marriage and the Family* 62, 4 (November 2000): 1160–1172.

Fraser, Julie, Eleanor Maticka-Tyndale, and Lisa Smylie. "Sexuality of Canadian Women at Midlife." *Canadian Journal of Human Sexuality* 13, 3–4 (2004): 171–187.

French, J. P., and Bertram Raven. "The Bases of Social Power." In *Studies in Social Power,* edited by L. Cartwright. Ann Arbor: University of Michigan Press, 1959.

Friedan, Betty. *The Feminine Mystique.* New York: Dell, 1963.

Friedman, Rochelle, and Bonnie Gradstein. *Surviving Pregnancy Loss.* Boston: Little, Brown, 1982.

Frieze, Irene H. "Female Violence Against Intimate Partners: An Introduction." *Psychology of Women Quarterly* 29 (2005): 229–237.

Fuchs, Dale. "Spanish Socialists' Proposals Opposed by Church." *New York Times,* May 30, 2004.

Furstenberg, Frank F., Jr. "Good Dads-Bad Dads: Two Faces of Fatherhood" In *The Changing Family,* edited by A. Cherlin. New York: Urban Institute Press, 1988.

———. "Reflections on Remarriage." *Journal of Family Issues* 1, 4 (1980): 443–453.

Furstenberg, Frank F., Jr., and A. J. Cherlin. *Divided Families: What Happens to Children When Parents Part.* Cambridge, MA: Harvard University Press, 1991.

Furstenberg, Frank F., Jr., and J. O. Teitler. "Reconsidering the Effects of Marital Disruption: What Happens to Children of Divorce in Early Adulthood?" *Journal of Family Issues* 15, 2 (June 1994): 173–190.

Furstenberg, Frank F., Jr., and Christine Nord. "Parenting Apart: Patterns in Childrearing after Marital Disruption." *Journal of Marriage and the Family* 47, 4 (November 1985): 893–904.

Furstenberg, Frank F., Jr., and Graham Spanier, eds. *Recycling the Family—Remarriage after Divorce.* Rev. ed. Newbury Park, CA: Sage, 1987.

Furstenberg, Frank F., Jr., Sheela Kennedy, Vonnie C. Mcloyd, Rubén G. Rumbaut, and Richard A. Settersten Jr. "Growing Up Is Harder To Do." *Contexts* 3, 3 (Summer 2004): 33–41.

Fuwa, Makiko. "Macro-level Gender Inequality and the Division of Household Labor in 22 Countries." *American Sociological Review* 69, 6 (December 2004): 751–767.

Gager, C., T. M. Cooney, and K. T. Call. "The Effects of Family Characteristics and Time Use on Teenage Girls' and Boys' Household Labor." *Journal of Marriage and Family* 61 (1999): 982–994.

Gager, Constance T., Teresa M. Cooney, and Kathleen Thiede. "The Effects of Family Characteristics and Time Use on Teenagers' Household Labor." *Journal of Marriage & Family* 61, 4 (November 1999): 982–994.

Gagnon, John, and William Simon. "The Sexual Scripting of Oral Genital Contacts." *Archives of Sexual Behavior* 16, 1 (February 1987): 1–25.

Galinsky, E., et al. *Overwork in America: When the Way We Work Becomes Too Much.* New York, NY: Families and Work Institute, 2005.

Ganong, L. H., and M. Coleman. "Effects of Remarriage on Children: A Review of the Empirical Literature." *Family Relations* 33 (1984): 389–406.

Ganong, L., and M. Coleman. "Stepparent: a Pejorative Term?" *Psychological Reports* 52 (1983): 919–922.

Ganong, L., M. Coleman, and J. Hans. "Divorce as Prelude to Stepfamily Living and the Consequences of Re-Divorce." In *Handbook of Divorce,* edited by M. Fine and J. Harvey. Hillsdale, NJ: Lawrence Erlbaum.

Ganong, Lawrence, and Marilyn Coleman. "Gender Differences in Expectations of Self and Future Partner." *Journal of Family Issues* 13, 1 (March 1992): 55–64.

———. *Remarried Family Relationships.* Newbury Park, CA: Sage Publications, 1994.

Ganong, Lawrence, Marilyn Coleman, and Gregory Kennedy. "The Effects of Using Alternate Labels in Denoting Stepparent or Stepfamily Status." *Journal of Social Behavior and Personality* 5 (1990): 453–463.

Gans, D., and M. Silverstein. "Norms of Filial Responsibility for Aging Parents across Time and Generations." *Journal of Marriage and Family* 68 (2006): 961–986.

Gans, Herbert. "Symbolic Ethnicity: The Future of Ethnic Groups and Cultures in America." In *On the Making of Americans,* edited by H. Gans. Philadelphia: University of Pennsylvania, 1979.

Gao, Ge. "Stability of Romantic Relationships in China and the United States." In *Cross-Cultural Interpersonal Communication,* edited by S. Ting-Toomey and F. Korzenny. Newbury Park, CA: Sage, 1991.

Garbarino, J. *See Jane Hit: Why Girls Are Growing More Violent and What We Can Do about It.* New York: Penguin, 2006.

Garbarino, James. *Children and Families in the Social Environment.* Hawthorne, NY: Aldine De Gruyter, 1982.

Garcia-Moreno, C., H. Jansen, M. Ellsberg, L. Heise, and C. Watts. "Prevalence of Intimate Partner Violence: Findings from the WHO Multi-Country Study on Women's Health and Domestic Violence." *Lancet* 368, 9543 (2006): 1260–1269.

Garfinkel, Irwin, and Sara McLanahan. *Single Mothers and Their Children: A New American Dilemma.* Washington, DC: Urban Institute Press, 1986.

Gariety, B. S., and S. Shaffer. "Wage Differentials Associated with Working at Home." *Monthly Labor Review* (March 2007): 61–67.

Garner, Abigail. *Families Like Mine: Children of Gay Parents Tell It Like It Is.* New York: HarperCollins, 2005.

Garnets, L., et al. "Violence and Victimization of Lesbians and Gay Men: Mental Health Consequences." *Journal of Interpersonal Violence* 5 (1990): 366–383.

Gecas, Viktor, and Monica Seff. "Social Class, Occupational Conditions, and Self-Esteem." *Sociological Perspectives* 32 (1989): 353–364.

———. "Families and Adolescents." In *Contemporary Families: Looking Forward, Looking Back,* edited by A. Booth. Minneapolis: National Council on *Family Relations,* 1991.

Gelles, R., and C. P. Cornell. *Intimate Violence in Families,* 2nd ed. Newbury Park, CA: Sage, 1990.

Gelles, Richard J. "Child Abuse and Violence in Single-Parent Families: Parent Absence and Economic Deprivation." *American Journal of Orthopsychiatry* 59, 4 (October 1989): 492–501.

———. "Through a Sociological Lens: Social Structure and Family Violence." In *Current Controversies in Family Violence,* edited by R. Gelles and D. Loseke. Newbury Park, CA: Sage, 1993.

Gelles, Richard J., and Jon R. Conte. "Domestic Violence and Sexual Abuse of Children: A Review of Research in the Eighties." In *Contemporary Families: Looking Forward, Looking Back,* edited by A. Booth. Minneapolis: National Council on Family Relations, 1991.

Gelles, Richard J., and Claire Pedrick Cornell. *Intimate Violence in Families,* 2nd ed. Newbury Park, CA: Sage, 1990.

Gelles, Richard J., and Claire Pedrick Cornell. *Intimate Violence in Families.* Newbury Park, CA: Sage, 1985.

Gelles, Richard J., and Murray Straus. *Intimate Violence: The Definitive Study of the Causes and Consequences of Abuse in the American Family.* New York: Simon and Schuster, 1988.

General social surveys, 1972–2008[machine-readable data file] /Principal Investigator, James A. Davis; Director and Co-Principal Investigator, Tom W. Smith; Co-Principal Investigator, Peter V. Marsden; Sponsored by National Science Foundation. —NORC ed.— Chicago: National Opinion Research Center [producer]; Storrs, CT: The Roper Center for Public Opinion Research, University of Connecticut [distributor], 2009.

Genevie, Lou, and Eva Margolies. *The Motherhood Report: How Women Feel about Being Mothers.* New York: Macmillan, 1987.

Gentile, D. "Pathological Video Game Use among Youth 8 to 18: A National Study." *Psychological Science,* 2008: 594–602.

Gentile, D. A., and J. R. Gentile. "Violent Video Games as Exemplary Teachers: A Conceptual Analysis." *Journal of Youth and Adolescence* 9 (2008): 127–141.

Gentile, D., and C. Anderson. "Video Games." In *Encyclopedia of Human Development,* Vol. 3, edited by N. J. Salkind. Thousand Oaks, CA: Sage, 2006: 1303–1307.

Gentile, D., and D. Walsch. "A Normative Study of Family Media Habits." *Journal of Applied Developmental Psychology* 23 (2002): 157–178.

Gentile, D., and J. R. Gentile. "Violent Video Games as Exemplary Teachers." Paper presented at the Biennial Meeting of the Society for Research in Child Development, Atlanta, GA, April 9, 2005.

Gentile, D., P. Lynch, J. Linder, and D. Walsch. "The Effects of Violent Video Game Habits on Adolescent Hostility, Aggressive Behaviors, and School Performance." *Journal of Adolescence* 27, 1 (February 2004): 5–22.

Gentleman, A. "Indian Prime Minister Denounces Abortion of Females." *New York Times,* April 28, 2008.

George, Kenneth, and Andrew Behrendt. "Therapy for Male Couples Experiencing Relationship Problems and Sexual Problems." *Journal of Homosexuality* 14, 1–2 (1987): 77–88.

Gerris, Jan, Maja Dekovic, and Jan Janssens. "The Relationship between Social Class and Childrearing Behaviors: Parents' Perspective Taking and Value Orientations." *Journal of Marriage and the Family* 59, 4 (November 1997): 834–847.

Gershoff, E. T. *Report on Physical Punishment in the United States: What Research Tells Us about Its Effects on Children.* Columbus, OH: Center for Effective Discipline, 2008.

Gerson, Kathleen. *Hard Choices: How Women Decide About Work, Career, and Motherhood.* Berkeley: University of California Press, 1985.

———. *No Man's Land: Men's Changing Commitments to Family and Work.* New York: Basic Books, 1993.

Gerstel, N., and N. Sarkisian. "Marriage: The Good, the Bad, and the Greedy." *Contexts* 5, 4 (November 2006): 16–21.

Gibson-Davis, Christina, Kathryn Edin, and Sara McLanahan. "High Hopes But Even Higher Expectations: The Retreat From Marriage Among Low-Income Couples." *Journal of Marriage and Family* 67, 3 (December 2005): 1301–1312.

Gies, Frances, and Joseph Gies. *Marriage and the Family in the Middle Ages.* New York: Harper & Row, Publishers, 1987.

Gilbert, D. *The American Class Structure in an Age of Growing Inequality,* 7th ed. Thousand Oaks, CA: Pine Forge Press, 2008.

Gilbert, Lucia, et al. "Perceptions of Parental Role Responsibilities: Differences between Mothers and Fathers." *Family Relations* 31 (April 1982): 261–269.

Giles, H., and B. LePoire. "Introduction: The Ubiquity and Social Meaningfulness of Nonverbal Communication." In *The Sage Handbook of Nonverbal Communication,* edited by V. Manusov and M. Patterson. Thousand Oaks, CA: Sage, 2006.

Gillen, K., and S. J. Muncher. "Sex Differences in the Perceived Casual Structure of Date Rape: A Preliminary Report." *Aggressive Behavior* 21, 2 (1995): 101–112.

Gillespie, Rosemary. "Childfree and Feminine: Understanding the Gender Identity of Voluntarily Childless Women." *Gender & Society* 17, 1 (February 2003): 122–136.

Gilligan, Carol. *In a Different Voice: Psychological Theory and Women's Development.* Cambridge, MA: Harvard University Press, 1982.

Gilmore, David. *Manhood in the Making.* New Haven, CT: Yale University Press, 1990.

Glazer-Malbin, Nona, ed. *Old Family/New Family.* New York: Van Nostrand, 1975.

Gleick, Elizabeth. "Tower of Psychobabble: Pronouncing on the Differences Between the Sexes Has Made John Gray Master of a Self-Help Universe. But Is He More of a Healer or a Huckster?" *Time,* June 16, 1997.

Glenn, Evelyn N., and Stacey G. H. Yap. "Chinese American Families." In *Minority Families in the United States: A Multicultural Perspective,* edited by R. L. Taylor. Englewood Cliffs, NJ: Prentice Hall, 1994.

Glenn, Norval, and Beth Ann Shelton. "Regional Differences in Divorce in the United States." *Journal of Marriage and the Family* 47, 3 (August 1985): 641–652.

Glenn, Norval, and Michael Supancic. "The Social and Demographic Correlates of Divorce and Separation in the United States: An Update and Reconsideration." *Journal of Marriage and the Family* 46, 3 (August 1984): 563–575.

Glenn, Norval. "Duration of Marriage, Family Composition, and Marital Happiness." *National Journal of Sociology* 3 (1989): 3–24.

———. "The Recent Trend in Marital Success in the United States," *Journal of Marriage and the Family* 53 (2) May 1991: 261–270.

Glick, Paul. "The Family Life Cycle and Social Change." *Family Relations* 38, 2 (April 1989): 123–129.

———. "Fifty Years of Family Demography." *Journal of Marriage and the Family* 50, 4 (November 1988): 861–873.

Gneezy, U., and A. Rustichini. "Gender and Competition at a Young Age." *American Economic Review Papers and Proceedings,* May 2004, 377–381.

Gnezda, Therese. *The Effects of Unemployment on Family Functioning.* Prepared Statement to the Select Committee on Children, Youth and Families, House of Representatives, at Hearings on the New Unemployed, Detroit, March 4, 1984. Washington, DC: Government Printing Office, 1984.

Goeke-Morey, M. C., E. M. Cummings, and L. M. Papp. "Children and Marital Conflict Resolution: Implications for Emotional Security and Adjustment." *Journal of Family Psychology* 21, 4 (2007): 744–753.

Goelman, Hillel, et al. "Family Environment and Family Day Care." *Family Relations* 39, 1 (January 1990): 14–19.

Goetting, Ann. "The Six Stages of Remarriage: Developmental Tasks of Remarriage after Divorce." *Family Relations* 31 (April 1982): 213–222.

———. "Patterns of Support among In-Laws in the United States: A Review of Research." *Journal of Family Issues* 11, 1 (1990): 67–90.

Goetz, Kathryn W., and Cynthia J. Schmiege. "From Marginalized to Mainstreamed: The HEART Project Empowers the Homeless." *Family Relations* 45, 4 (October 1996).

Goldberg, A. E., and A. G. Sayer. Lesbian Couples' Relationship Quality across the Transition to Parenthood. *Journal of Marriage and Family* 68 (2006): 87–100.

Goldenberg, H., and I. Goldenberg. *Counseling Today's Families,* 2nd ed. Pacific Grove, CA: Brooks/Cole, 1994.

Goldman, S., D. Wolcott, and K. Kennedy. "A Coordinated Response to Child Abuse and Neglect: The Foundation for Practice. Chapter Five: What Factors Contribute to Child Abuse and Neglect?" National Clearinghouse on Child Abuse and Neglect Information.http://www .childwelfare.gov/pubs/usermanuals/foundation/ foundatione.cfm#riskfactors, 2009.

Goldscheider, F., and G. Kaufman, "Willingness to Stepparent: Men and Women's Attitudes Toward Marrying Someone with Children" Conference Papers—American Sociological Association, 2003 Annual Meeting, Atlanta, GA, pN.PAG, 0p; (AN 15921331).

Goldscheider, F., and G. Kaufman. "Willingness to Stepparent: Attitudes about Partners Who Already Have Children." *Journal of Family Issues* 27, 10 (2006): 1415–1436.

Goldscheider, F., and S. Sassler. "Creating Stepfamilies: Integrating Children into the Study of Union Formation." *Journal of Marriage and Family* 68 (2006): 275–291.

Goldscheider, Frances and Sharon Sassler. "Creating Stepfamilies: Integrating Children into the Study of Union Formation." *Journal of Marriage and Family* 68, 2 (May 2006): 273–291.

Goldstein, David, and Alan Rosenbaum. "An Evaluation of the Self-Esteem of Maritally Violent Men." *Family Relations* 34, 3 (July 1985): 425–428.

Goleman, Daniel. "Spacing of Siblings Strongly Linked to Success in Life." *New York Times* (May 28, 1985): 17–18.

———. "Gay Parents Called No Disadvantage." *New York Times* (March 11, 1992).

Gondolf, Edward W. "Evaluating Progress for Men Who Batter." *Journal of Family Violence* 2 (1987): 95–108.

———. "The Effect of Batterer Counseling on Shelter Outcome." *Journal of Interpersonal Violence* 3, 3 (September 1988): 275–289.

Gondolf, Edward. "Male Batterers." In *Family Violence: Prevention and Treatment,* edited by R. L. Hampton et al. Newbury Park, CA: Sage, 1993.

Gongla, Patricia, and Edward Thompson, Jr., "Single Parent Families." In *Handbook of Marriage and the Family,* edited by M. Sussman and S. Steinmetz. New York: Plenum Press, 1987.

Gonzalez, N. "Beyond the FMLA: California's Paid Family Leave Is a National First." *National Council on Family Relations Report: Family Focus on Families and Work-Life* 51, 2 (June 2006): 11.

———. "Stuck in the Muddle with You: A Family Life Educator Looks at Divorce Research." *National Council on Family Relations Report: Family Focus on Divorce and Relationship Dissolution*, Issue FF36, December 2007, F7–8.

———. "What Is the Divorce Rate? A Complicated Answer to a Simple Question." *National Council on Family Relations Report: Family Focus on Divorce and Relationship Dissolution*, Issue FF36, December 2007, F1–2.

Goode, William, ed. *The Family.* 2nd ed. Englewood Cliffs, NJ: Prentice Hall, 1982.

Goode, William. "Force and Violence in the Family." *Journal of Marriage and the Family* 33 (November 1971): 624–636.

Goode, William. "Why Men Resist." *Dissent* 27, 2 (1980): 181–193.

Goodman, Lisa, and Deborah Epstein. "Refocusing on Women: A New Direction for Policy and Research on Intimate Partner Violence." *Journal of Interpersonal Violence* 20, 4 (April 2005): 479–487.

Goodman, M. "What Laid-Off Dads Want This Father's Day. Think about What's Best for That Unemployed Dad." http://www.abcnews.go.com, 2009.

Gordon, Thomas. *P.E.T. in Action.* New York: Bantam Books, 1978.

Gornick, Janet C., and Marcia K. Meyers. "Helping America's Working Parents: What We Can Learn from Europe and Canada." Work and Family Program research paper. Washington, DC: New America Foundation, November 2004. www.newamerica.net/publications/policy/helping_americas_working_parents.

Gottman, J. M., R. W. Levenson, J. Gross, B. Fredrickson, K. McCoy, L. Rosenthal, A. Ruel, and D. Yoshimoto. "Correlates of Gay and Lesbian Couples' Relationship Satisfaction and Relationship Dissolution." *Journal of Homosexuality* 45, 1 (2003): 23–43.

Gottman, J., Ryan, K., Carrere, S., and Erley, A., "Toward a scientifically based marital therapy." In *Family psychology: Science-based interventions*, edited by H. Liddle, D. Sanstistban, et al. Washington, DC: American Psychological Association. (2002): 147–174.

Gottman, John M. *What Predicts Divorce? The Relationship between Marital Processes and Marital Outcomes.* Hillsdale, NJ: Lawrence Erlbaum, 1994.

Gottman, John M. *Why Marriages Succeed or Fail and How You Can Make Yours Work.* New York: Simon and Schuster, 1995.

Gottman, John, James Coan, Sybil Carrere, and Catherine Swanson. "Predicting Marital Happiness and Stability from Newlywed Interactions." *Journal of Marriage and the Family* 60, 1 (February 1998): 5–22.

Gottschalk, Lorene. "Same-sex Sexuality and Childhood Gender Non-conformity: A Spurious Connection." *Journal of Gender Studies* 12, 1 (March 2003): 35–51.

Gough, Kathleen. "Is the Family Universal: The Nayer Case." In *A Modern Introduction to the Family*, edited by N. Bell and E. Vogel. New York: Free Press, 1968.

Graham-Kevan, Nicola, and John Archer. "Investigating Three Explanations of Women's Relationship Aggression." *Psychology of Women Quarterly* 29 (2005): 270–277.

Grall, Timothy. "Custodial Mothers and Fathers and Their Child Support: 2001." *Current Population Reports.* Washington, D.C.: U.S. Census Bureau. (October 2003).

Grall, Timothy. "Support Providers: 2002." Current Population Reports (Series P70, No. 99). Washington, DC: U.S. Census Bureau, 2005.

Gray, John. *Men Are From Mars, Women Are From Venus.* New York: Harper Collins. 1992.

Gray, John. *Men Are from Mars, Women Are from Venus.* New York: HarperCollins. 1993.

Greeff, Abraham P., and Tanya deBruyne. "Conflict Man-agement Style and Marital Satisfaction." *Journal of Sex and Marital Therapy* 26, 4 (October 2000): 321–334.

Green, K. M., M. E. Ensminger, J. A. Robertson, and H. Juon. "Impact of Adult Sons' Incarceration on African American Mothers' Psychological Distress." *Journal of Marriage and Family* 68, 2 (May 2006): 430–441.

Green, P. "Your Mother Is Moving In? That's Great." *New York Times*, January 15, 2009, D1.

Greenberger, Ellen, and Robin O'Neil. "Parents' Concern about Their Child's Development: Implications for Father's and Mother's Well-Being and Attitudes toward Work." *Journal of Marriage and the Family* 52, 3 (August 1990): 621–635.

Greenberger, Ellen. "Explaining Role Strain: Intrapersona." *Journal of Marriage and the Family* 52, 1 (February 1994): 115–118.

Greene, K., and S. L. Faulkner. "Gender Belief in the Sexual Double Standard, and Sexual Talk in Heterosexual Dating Relationships." *Sex Roles* 53, 3/4 (August 2005): 239–251.

Greenfield, E. A., and N. F. Marks. "Linked Lives: Adult Children's Problems and Their Parents' Psychological and Relational Well-Being." *Journal of Marriage and Family* 68 (2006): 442–454.

Greenstein, T. N. "Occupation and Divorce." *Journal of Family Issues* 6 (1985): 347–357.

Greenstein, Theodore N. "Husbands' Participation in Domestic Labor: Interactive Effects of Wives' and Husbands' Gender Ideologies." *Journal of Marriage and the Family* 58, 3 (August 1996): 585–595.

Greenstein, Theodore. "Marital Disruption and the Employment of Married Women." *Journal of Marriage and the Family* 52 (1990): 657–676.

Greenwood, G., M. V. Relf, B. Huang, L. M. Pollack, J. A. Canchola, and J. A. Catania. "Battering Victimization among a Probability-Based Sample of Men Who Have Sex with Men." *American Journal of Public Health* 92, 12 (December 2002): 1964–1969.

Greeson, Larry E. "Recognition and Ratings of Television Music Videos: Age, Gender, and Sociocultural Effects." *Journal of Applied Social Psychology* 21, 23 (1991): 1908–1920.

Greif, Geoffrey L., and Joan Kristall. "Common Themes in a Group for Noncustodial Parents." *Families in Society* 74, 4 (1993): 240–245.

Greif, Geoffrey. "Children and Housework in the Single Father Family." *Family Relations* 34, 3 (July 1985): 353–357.

Griffin-Shelley, Eric. "The Internet and Sexuality: A Literature Review—1983–2002." *Sexual & Relationship Therapy* 18, 3 (August 2003): 355–371.

Griswold, Robert. *Fatherhood in America: A History.* New York: Basic Books, 1993.

Gross, J. "Boomerang Parents." *New York Times*, November 18, 2008.

Grossman, C. L., and I. Yoo. "Civil Marriage on Rise across USA." *USA Today*, October 7, 2003, 1A.

Grosswald, Blanche. "The Effects of Shift Work on Family Satisfaction." *Families in Society* 85, 3 (July–September 2004): 413–423.

Grotevant, H. "Open Adoption: What Is It and How Is It Working?" *National Council on Family Relations Report: Family Focus on Adoption*, Issue FF39 (September 2008): F1, 2, 17–19.

Grotevant, H. D., R. G. McCoy, C. Elde, and D. L. Frave. "Adoptive Family System Dynamics: Variations by Level of Openness in the Adoption." *Family Process* 33 (1994): 125–146.

Grotevant, Harold, and Julie Kohler. "Adoptive Families." In *Parenting and Child Development in "Nontraditional" Families,* edited by M. E. Lamb. Mahwah, NJ: Lawrence Erlbaum, 1999.

Groth, Nicholas. *Men Who Rape: The Psychology of the Offender.* New York: Plenum Press, 1980.

Grov, C., D. Bimbi, J. E. Nanin, and J. T. Parsons. "Coming-Out Process among Gay, Lesbian, and Bisexual Individuals." *Journal of Sex Research* 43, 2 (May 2006): 115–121.

Grych, J. "Marital Relationships and Parenting." In *Handbook of Parenting, vol. 4: Social Conditions and Applied Parenting,* edited by Marc Bornstein. Rahwah, NJ: Lawrence Erlbaum, 2008.

Grzywacz, Joseph, and Nadine Marks. "Family, Work, Work-Family Spillover, and Problem Drinking During Midlife." *Journal of Marriage and the Family* 62, 2 (May 2000): 336–348.

Guerrero Pavich, Emma. "A Chicana Perspective on Mexican Culture and Sexuality." In *Human Sexuality, Ethnoculture, and Social Work,* edited by L. Lister. New York: Haworth, 1986.

Guerrero, L. K., and Anderson, P. A. "The waxing and waning of relational intimacy: Touch as a function of relational stage, gender and touch avoidance." *Journal of Social and Personal Relationships* 8 (1991): 147–165.

Guldner, Gregory T., and Clifford H. Swensen. "Time Spent Together and Relationship Quality: Long-Distance Relationships as a Test Case." *Journal of Social and Personal Relationships* 12, 2 (1995): 313–320.

Guldner, Gregory. "Long-Distance Romantic Relationships: Prevalence and Separation-Related Symptoms in College Students." *Journal of College Student Development* 37, 3 (May–June 1996): 289–296.

Gulledge, Andrew, Michelle Gulledge, and Robert Stahmann. "Romantic Physical Affection Types and Relationship Satisfaction." *The American Journal of Family Therapy* 31 (2003): 233–242.

Gullickson, A. "Education and Black-White Interracial Marriage" *Demography,* 43, 4 (November 2006): 673–689.

Gullo, Karen. "Parents Juggling Multiple Care Arrangements for Kids, Study Shows." Associated Press: March 8, 2000.

Gupta, S. "Autonomy, Dependence, or Display? The Relationship between Married Women's Earnings and Housework." *Journal of Marriage and Family* 69 (2007): 399–417.

Guttentag, M., and P. Secord. *Too Many Women.* Newbury Park, CA: Sage, 1983.

Guttman, Herbert. *The Black Family: From Slavery to Freedom.* New York: Pantheon, 1976.

Guttman, Joseph. *Divorce in Psychosocial Perspective: Theory and Research.* Hillsdale, NJ: Lawrence Erlbaum, 1993.

Guzzo, K. B. "Marital Intentions and the Stability of First Cohabitations." *Journal of Family Issues* 20, 2 (2009): 179–205

Haaga, D. A. "Homophobia?" *Journal of Behavior and Personality* 6 (1991): 171–174.

Haberman, M. "Senate OKs Tough 'Nixzmary' Law." *New York Post*, September 12, 2009. http://www.nypost.com

Haferd, Laura. "Paddling Returns to Child Rearing." *San Jose Mercury News* (December 20, 1986): D12.

Hafner, K. "Texting May be Taking a Toll." *New York Times,* May 25, 2009.

Hafstrom, Jeanne, and Vicki Schram. "Chronic Illness in Couples: Selected Characteristics, Including Wife's Satisfaction with and Perception of Marital Relationships." *Family Relations* 33 (1984): 195–203.

Hajek, H. "Teen Pregnancy Rate Is on the Rise," January, 2009. http://www.healthnews.com.

Hall, D. R., and J. Z. Zhao. "Cohabitation and Divorce in Canada: Testing the Selectivity Hypothesis." *Journal of Marriage and the Family* 57 (May 1995): 421–427.

Hall, E. *The Hidden Dimension.* Garden City, NY: Doubleday, 1966

Hamer, J., and K. Marchioro. "Becoming Custodial Dads: Exploring Parenting among Low-Income and Working-Class African American Fathers." *Journal of Marriage and Family* 64 (2002): 116–129.

Hamer, Jennifer, and Kathleen Marchioro. "Becoming Custodial Dads: Exploring Parenting Among Low-Income and Working-Class African American Fathers." *Journal of Marriage & Family* 64, 1 (February 2002): 116–129.

Hamilton, B. E., J. A. Martin, and P. D. Sutton. "Births: Preliminary Data for 2002." National Vital Statistics Report 51 (11). Hyattsville, MD: National Center for Health Statistics (2003).

Hamilton, B. E., J. A. Martin, and S. J. Ventura. "Births: Preliminary Data for 2007." *National Vital Statistics Reports.* Web release, vol. 57, no. 12. Hyattsville, MD: National Center for Health Statistics, 2009.

Hamilton, B., J. Martin, S. Ventura, P. Sutton, and F. Menacker. "Births: Preliminary Data for 2004." *National Vital Statistics Reports* 54, 8. Hyattsville, MD: National Center for Health Statistics, 2004.

Hansen, Christine H., and Ranald D. Hansen. "The Influence of Sex and Violence on the Appeal of Rock Music Videos." *Communication Research* 17, 2 (1990): 212–234.

Hansen, Gary. "Dating Jealousy among College Students." *Sex Roles* 12, 7–8 (April 1985): 713–721.

———. "Extradyadic Relations during Courtship." *Journal of Sex Research* 23, 3 (August 1987): 383–390.

Hansen, Karen. *Not So Nuclear Families: Class Gender and Networks of Care.* New Brunswick, NJ: Rutgers University Press, 2005.

Hanson, B., and C. Knopes. "Prime Time Tuning Out Varied Cultures." *USA Today* (July 6, 1993).

Hanson, Thomas L., Sara S. McLanahan, and Elizabeth Thomson. "Double Jeopardy: Parental Conflict and Stepfamily Outcomes for Children." *Journal of Marriage and the Family,* 58, 1 (February 1996): 141–154.

Hare-Mustin, Rachel T., and Jeanne Marecek. "On Making a Difference." In *Making a Difference: Psychology and the Construction of Gender,* edited by R. T. Hare-Mustin and J. Marecek. New Haven, CT: Yale University Press, 1990.

Harknett, K. "The Relationship between Private Safety Nets and Economic Outcomes among Single Mothers." *Journal of Marriage and Family* 68, 1 (2006): 172–191.

Harknett, Kristen, and Sara McLanahan. "Racial and Ethnic Differences in Marriage After the Birth of a Child." *American Sociological Review* 69 (December 2004): 790–811.

Harknett, Kristen. "The Relationship between Private Safety Nets and Economic Outcomes among Single Mothers." *Journal of Marriage and Family* 68, 1 (February 2006): 172–191.

Harriman, Lynda. "Personal and Marital Changes Accompanying Parenthood." *Family Relations* 32, 3 (July 1983): 387–394.

Harrington, Michael. *The Other America: Poverty in the United States.* New York: Macmillan, 1962.

Harris Interactive. "Technology-Intensive Childbirth is the Norm for Great Majority of Primarily Healthy Women." *Health Care News.* Vol. 7, Issue 1,(January 4, 2007): http://www.harrisinteractive.com.

Harris, C. R. "Sexual and Romantic Jealousy in Heterosexual and Homosexual Adults." *Psychological Science* 13, 1 (January 2002): 7–12.

Harris, David, and Hiromi Ono. "How Many Interracial Marriages Would There Be If All Groups Were of Equal Size in All Places? A New Look at National Estimates of Interracial Marriage." *Social Science Research* 34, 1 (2005): 236–251.

Harris, G. T., N. Z. Hilton, M. E. Rice, and A. W. Eke. "Children Killed by Genetic Parents versus Stepparents." *Evolution and Human Behavior* 28 (2007): 85–95.

Harris, Kathleen Mullan. "The National Longitudinal Study of Adolescent Health (Add Health), Waves I & II, 1994–1996; Wave III, 2001–2002; Wave IV, 2007–2009" [machine-readable data file and documentation]. Chapel Hill, NC: Carolina Population Center, University of North Carolina at Chapel Hill, 2009.

Harry, Joseph. "Decision Making and Age Differences among Gay Male Couples." In *Gay Relationships,* edited by J. DeCecco. New York: Haworth, 1988.

Hart, Betsy. "'Living apart together' relationship ultimately selfish." Deseret News (Salt Lake City), May 14, 2006.

Hart, J., E. Cohen, A. Gingold, and R. Homburg. "Sexual Behavior in Pregnancy: A Study of 219 Women." *Journal of Sex Education and Therapy* 17, 2 (June 1991): 88–90.

Hartwell-Walker, M. "Legal Issues for Cohabiting Couples." http://psychcentral.com/lib/2008/legal-issues-for-cohabiting-couples, 2008.

Hassebrauck, Manfred, and Beverly Fehr. "Dimensions of Relationship Quality." *Personal Relationships* 9 (2002): 253–270.

Hatch, L., and K. Bulcroft. "Does Long-Term Marriage Bring Less Frequent Disagreements? Five Explanatory Frameworks." *Journal of Family Issues* 25, 4 (2004): 465–495.

Hatfield, E., and R. L Rapson. "Love and intimacy." In *Encyclopedia of Human Behavior* 3, edited by V. S. Ramachandran in New York: Academic Press (1994): 1145–1149.

Hatfield, Elaine, and G. William Walster. *A New Look at Love.* Reading, MA: Addison-Wesley, 1981.

Hatfield, Elaine, and R. Rapson. *Love, Sex, and Intimacy: The Psychology, Biology, and History.* New York: Harper-Collins, 1993.

Hatfield, Elaine, and Susan Sprecher. *Mirror, Mirror: The Importance of Looks in Everyday Life.* New York: State University of New York, 1986.

Hawkins, Alan J., and Tomi-Ann Roberts. "Designing a Primary Intervention to Help Dual-Earner Couples Share Housework and Childcare." *Family Relations* 41, 2 (April 1992): 169–177.

Hawkins, Alan J., Tomi-Ann Roberts, Shawn Christiansen, and C. M. Marshall. "An Evaluation of a Program to Help Dual-Earner Couples Share the Second Shift." *Family Relations* 43, 2 (April 1994): 213–220.

Hawkins, Alan, and David Dollahite, eds. *Generative Fathering: Beyond Deficit Perspectives.* Vol. 3, *Current Issues in the Family.* Thousand Oaks, CA: Sage, 1997.

Hawkins, D., and Alan Booth. "'Unhappily Ever After': Effects of Long-Term Low-Quality Marriages on Well Being." *Social Forces* 84, 1 (September 2005): 451–471.

Haynes, Faustina. "Gender and Family Ideals: An Exploratory Study of Black Middle-Class Americans." *Journal of Family Issues* 21, 7 (October 2000): 811–837.

Hays, Dorothea, and Aurele Samuels. "Heterosexual Women's Perceptions of their Marriages to Bisexual or Homosexual Men." *Journal of Homosexuality* 18, 2 (1989): 81–100.

Hays, Sharon. *Cultural Contradictions of Motherhood.* New Haven, CT: Yale University Press, 1996.

Hazan, Cindy, and Philip Shaver. "Romantic Love Conceptualized as an Attachment Process." *Journal of Personality and Social Psychology* 52 (March 1987): 511–524.

Healey, K. M., and C. Smith. "Batterer Programs: What Criminal Justice Agencies Need to Know." National Institute of Justice: Research in Action, U.S. Department of Justice, 1998.

"Health Encyclopedia—Diseases and Conditions: Dyspareunia," healthscout.com, 2001.

Healy, Joseph M., Abigail J. Stewart, and Anne P. Copeland. "The Role of Self-Blame in Children's Adjustment to

Parental Separation." *Personality and Social Psychology Bulletin* 19, 3 (1993): 279–289.

Heaton, T. "Factors Contributing to Increasing Marital Stability in the United States." *Journal of Family Issues* 23, 3 (2002): 392–409.

Hecht, Michael L., Peter J. Marston, and Linda Kathryn Larkey. "Love Ways and Relationship Quality in Heterosexual Relationships." *Journal of Social & Personal Relationships* 11, 1 (February 1994): 25–43.

Hefflinger, M. "Report: Top Markets See Double-Digit Game Sales Growth in '08." *DMW Daily* (February 2, 2009). Digital Media Wire. http://www.dmwmedia.com/news/2009.

Hefner, R., et al. "Development of Sex-Role Transcendence." *Human Development* 18 (1975): 143–158.

Heilbrun, Carolyn. *Toward a Recognition of Androgyny.* New York: Norton, 1982.

Helms-Erikson, Heather. "Marital Quality Ten Years After the Transition to Parenthood: Implications of the Timing of Parenthood and the Division of Housework." *Journal of Marriage and Family* 63 (November 2001): 1099–1110.

Hemstrom, Orjan. "Is Marriage Dissolution Linked to Differences in Morbidity Risks for Men and Women?" *Journal of Marriage and the Family* 58, 2 (May 1996): 366–378.

Henderson, Stephen. "Weddings: Vows—Rakhi Dhanoa and Ranjeet Purewal." *New York Times* (August 18, 2002).

Hendrick, Clyde, and Susan Hendrick. "A Theory and Method of Love." *Journal of Personality and Social Psychology* 50 (February 1986): 392–402.

———. "Attachment Theory and Close Adult Relationships." *Psychological Inquiry* 5, 1 (1994): 38–41.

Hendrick, S. S., and C. Hendrick. "Gender differences and similarities in sex and love." *Personal Relationships* 2 (1995): 55–65.

Hendrick, S., C. Hendrick, and N. Adler. "Romantic relationships: Love, satisfaction, and staying together." *Journal of Personality and Social Psychology,* 54, 6 (June 1988): 980–988.

Hendrick, Susan. "Self-Disclosure and Marital Satisfaction." *Journal of Personality and Social Psychology* 40 (1981): 1150–1159.

Henley, N. *Body Politics: Power, Sex, and Nonverbal Communication.* Englewood Cliffs, NJ: Prentice Hall, 1977.

Henneck, Rachel. "Family Policy in the U.S., Japan, Germany, Italy and France: Parental Leave, Child Benefits/Family Allowances, Child Care, Marriage/Cohabitation, and Divorce." www.contemporaryfamilies.org, May 2003.

Henry J. Kaiser Family Foundation. "HIV Policy Fact Sheet: Black Americans and HIV/AIDS." http://www.kff.org, 2008.

Henslin, J. *Essentials of Sociology: A Down-to-Earth Approach.* Boston: Allyn and Bacon, 2006.

Henslin, James. *Essentials of Sociology: A Down to Earth Approach,* 3rd ed. Needham Heights, MA: Allyn and Bacon, 2000.

Herbert, Bob. "Scapegoat Time." *New York Times* (December 16, 1994): A19.

Herek, G. "Relationship Science, beyond Homophobia: A Weblog about Sexual Orientation, Prejudice, Science, and Policy." 2006. http://www.beyondhomophobia.com/blog/

———. "Hate Crimes and Stigma-Related Experiences among Sexual Minority Adults in the United States: Prevalence Estimates from a National Probability Sample." *Journal of Interpersonal Violence* 24 (2009): 54–74.

Herek, G. M., J. R. Gillis, and J. C. Cogan. "Psychological Sequelae of Hate-Crime Victimization among Lesbian, Gay, and Bisexual Adults." *Journal of Consulting and Clinical Psychology* 67 (1999): 945–951.

Herek, Gregory M. "Heterosexuals' Attitudes Toward Bisexual Men and Women in the United States." *Journal of Sex Research* 39, 4 (November 2002): 264–275.

Hernandez, R. "Law to Ease Disclosures on Child Abuse." *New York Times,* February 13, 1996, B6.

Herring, Cedric, and Karen R. Wilson-Sadberry. "Preference or Necessity? Changing Work Roles of Black and White Women, 1973–1990." *Journal of Marriage and the Family* 55, 2 (May 1993): 314–325.

Herring, Susan, Inna Kouper, Lois Scheidt, and Elijah Wright. "Women and Children Last: The Discursive Construction of Weblogs." In *Into the Blogosphere: Rhetoric, Community, and Culture of Weblogs,* edited by Laura Gurak, Smiljana Antonijevic, Laurie Johnson, Clancy Ratliff, and Jessica Reyman. blog.lib.umn.edu/blogosphere/women_and_children.html, retrieved August 13, 2004.

Heslin, R., and Alper, T. "Touch: A Bonding Issue." In *Nonverbal Communication,* edited by J. M. Weimann and R. P. Harrison. Beverly Hills, CA: Sage, 1983.

Hetherington, E. Mavis, and John Kelly. *For Better or Worse: Divorce Reconsidered.* New York: W.W. Norton, 2002.

Hetherington, E. Mavis, and Margaret Stanley-Hagan. "Stepfamilies." In *Parenting and Child Development in 'Nontraditional' Families* edited by M. Lamb. Mahwah, New Jersey: Lawrence Erlbaum, 1999: 137–160.

Hetherington, E. Mavis. "Intimate Pathways: Changing Patterns in Close Personal Relationships Across Time." *Family Relations* 52, 4 (October 2003): 318–331.

Heuveline, Patrick, and Jeffrey Timberlake. "The Role of Cohabitation in Family Formation: The United States in Comparative Perspective." *Journal of Marriage & Family* 66, 5 (December 2004): 1214–1230.

Hewlett, Sylvia, and Cornel West. *The War Against Parents: What We Can Do for America's Beleaguered Moms and Dads.* New York: Houghton Mifflin, 1998.

Hewlett, Sylvia. *A Lesser Life: The Myth of Women's Liberation in America.* New York: Morrow, 1986.

———. *When the Bough Breaks: The Cost of Neglecting Our Children.* New York: Basic Books, 1991.

Heyl, Barbara. "Homosexuality: A Social Phenomenon." In *Human Sexuality: The Societal and Interpersonal Context,* edited by K. McKinney and S. Sprecher. Norwood, NJ: Ablex, 1989.

Heymann, J., A. Earle, and J. Hayes. "How Does the U.S. Rank in Work Policies for Individuals and Families." Council on Contemporary Families, http://www .con temporaryfamilies.org/subtemplate.php?t= briefingPapers&ext=workpolicies, 2007.

Hibbard, R.A., and G. L. Hartman. "Behavioral Problems in Alleged Sexual Abuse Victims." *Child Abuse and Neglect* 16 (1982): 755–762.

Higginbottom, Susan F., Julian Barling, and E. Kevin Kelloway. "Linking Retirement Experiences and Marital Satisfaction: A Mediational Model." *Psychology and Aging* 8, 4 (1993): 508–516.

Higgins, C., L. Duxbury, and C. Lee. "Impact of Life-cycle Stage and Gender on the Ability to Balance Work and Family Responsibilities." *Family Relations* 43 (1994): 144–150.

Hill, Ivan. *The Bisexual Spouse*. McLean, VA: Barlina Books, 1987.

Hirschl, Thomas A., Joyce Altobelli, and Mark R. Rank. "Does Marriage Increase the Odds of Affluence? Exploring the Life Course Probabilities." *Journal of Marriage and Family* 65, 4 (November 2003): 927–938.

Hirshman, L. "Off to Work She Should Go." *New York Times*, April 25, 2007.

Ho, V., and J. Johnson. "Overview of Florida Alimony." *Florida Bar Journal* 71 (October 2004).

Hochschild, Arlie Russell. *The Time Bind: When Work Becomes Home and Home Becomes Work*. New York: Holt, 1997.

Hochschild, Arlie. *The Second Shift: Working Parents and the Revolution at Home*. New York: Viking Press, 1989.

Hofferth, S. "Residential Father Family Type and Child Well-Being: Investment versus Selection." *Demography* 43 (2006): 53–77.

Hoffman, Kristi L., K. Jill Kiecolt, and John N. Edwards. "Physical Violence between Siblings: A Theoretical and Empirical Analysis." *Journal of Family Issues* 26, 8 (November 2005): 1103–1130.

Hoffman, Lois Wladis. "Maternal Employment: 1979." *American Psychologist* 34 (1979): 859–865.

Holmes, J., and S. Murray. "Conflict in close relation-ships." In *Social psychology: Handbook of basic principles*, edited by E. T. Higgins, A. W. Kruglanski. New York, NY: Guilford Press, 1996: 622–654

Holmes, T., and R. Rahe. "The Social Readjustment Rating Scale." *Journal of Psychosomatic Medicine* 11 (1967): 213–218.

Holtzen, D. W., and A. A. Agresti. "Parental Responses to Gay and Lesbian Children." *Journal of Social and Clinical Psychology* 9, 3 (September 1990): 390–399.

Hook, J. L., and S. Chalasani. "Gendered Expectations? Reconsidering Single Fathers' Childcare Time." Journal of Marriage and Family 70, 4 (November 2008): 835–1091.

Hook, Misty, Lawrence Gerstein, Lacy Detterich, and Betty Gridley. "How Close Are We? Measuring Intimacy and Examining Gender Differences." *Journal of Counseling and Development* 81 (Fall 2003): 462–472.

Hopper, Joseph. "The Symbolic Origins of Conflict in Divorce." *Journal of Marriage and Family* 63 (May 2001): 430–445.

Hort, Barbara E., M. D. Leinbach, and B. I. Fagot. "Is There Coherence among the Cognitive Components of Gender Acquisition?" *Sex Roles* 24, 3–4 (1991): 195–207.

Hort, Barbara, et al. "Are People's Notions of Maleness More Stereotypically Framed Than Their Notions of Femaleness?" *Sex Roles* 23, 3 (February 1990): 197–212.

Horwitz, Alan, and Helene White, "The Relationship of Cohabitation and Mental Health: A Study of a Young Adult Cohort," *Journal of Marriage and the Family* 60 (May 1998): 505–514.

Hotaling, Gerald T., and David B. Sugarman. "A Risk Marker Analysis of Assaulted Wives." *Journal of Family Violence* 5, 1 (March 1990): 1–14.

Hotaling, Gerald T., et al., eds. *Coping with Family Violence: Research and Perspectives*. Newbury Park, CA: Sage, 1988.

———. *Family Abuse and Its Consequences*. Newbury Park, CA: Sage, 1988.

Houran, James, et al. "Do Online Matchmaking Tests Work? An Assessment of Preliminary Evidence for a Publicized 'Predictive Model of Marital Success.'" *North American Journal of Psychology* 6 (2004): 507–526.

Houseknecht, Sharon K. "Voluntary Childlessness." In *Handbook of Marriage and the Family*, edited by M. B. Sussman and S. K. Steinmetz. New York: Plenum Press, 1987.

Houseknecht, Sharon K., Suzanne Vaughan, and Ann Statham. "The Impact of Singlehood on the Career Patterns of Professional Women." *Journal of Marriage and the Family* 49, 2 (May 1987): 353–366.

Houts, Leslie A. "But Was It Wanted? Young Women's First Voluntary Sexual Intercourse." *Journal of Family Issues* 26, 8 (November 2005): 1082–1102.

Howard, Judith. "A Structural Approach to Interracial Patterns in Adolescent Judgments about Sexual Intimacy." *Sociological Perspectives* 31, 1 (January 1988): 88–121.

Howes, Carollee. "Can the Age of Entry into Child Care and the Quality of Child Care Predict Adjustment in Kindergarten?" *Developmental Psychology* 26 (1990): 292–303.

Huang, Chien-Chung, Ronald B. Mincy, and Irwin Garfinkel. "Child Support Obligations and Low-Income Fathers." *Journal of Marriage and Family* 67, 5 (December 2005): 1213–1225.

Huebner, D. M., G. M. Rebchook, and S. M. Kegeles. "Experiences of Harassment, Discrimination, and Physical Violence among Young Gay and Bisexual Men." *American Journal of Public Health* 94 (2004): 1200–1203.

Hughes, Jean O'Gorman, and Bernice R. Sandler. "Friends Raping Friends—Could It Happen to You?" Project on the Status and Education of Women, Association of American Colleges, 1987.

Hughes, M., K. Morrison, and K. J. K. Asada. "What's Love Got to Do with It? Exploring the Impact of Maintenance Rules, Love Attitudes, and Network

Support on Friends with Benefits Relationships."
Western Journal of Communication 69, 1 (January
2005): 49–66.

Humble, A., C. R. Solomon, K. R. Allen, K. R. Blaisure,
and M. P. Johnson. "Feminism and Mentoring of
Graduate Students." *Family Relations* 55 (January
2006): 2–15.

"Humor Page," nokidding.net, 2006.

Huston, Ted L., and Gilbert Geis. "In What Ways Do
Gender-Related Attributes and Beliefs Affect Marriage?"
Journal of Social Issues 49, 3 (1993): 87–106.

Huston, Ted, and Heidi Melz. "The Case for
(Promoting) Marriage: The Devil Is in the Details."
Journal of Marriage and Family 66, 4 (November
2004): 943–958.

Huston, Ted, et al. "From Courtship to Marriage: Mate
Selection as an Interpersonal Process." In *Personal
Relationships 2: Developing Personal Relationships,* edited
by S. Duck and R. Gilmour. London: Academic Press,
1981.

Huston, Ted, S. M. McHale, and A. C. Crouter. "When
the Honeymoon's Over: Changes in the Marriage
Relationship over the First Year." In *The Emerging Field of
Personal Relationships,* edited by S. Duck and R. Gilmour.
Hillsdale, NJ: Lawrence Erlbaum, 1986.

Huston, Ted. "The Social Ecology of Marriage and Other
Intimate Unions." *Journal of Marriage and the Family* 62,
2 (May 2000): 298–321.

Hutchins, Loraine, and Lani Kaahumanu. "Overview." In
Bi Any Other Name: Bisexual People Speak Out, edited
by L. Hutchins and L. Kaahumanu. Boston: Alyson
Publications, 1991.

Hyde, J. S. "The Gender Similarities Hypothesis." *American
Psychologist* 60, 6 (2005): 581–592.

Hynie, Michaela, John E. Lydon, Sylvana Cote, and Seth
Wiener. "Relational Sexual Scripts and Women's
Condom Use: The Importance of Internalized Norms."
Journal of Sex Research 35, 4 (November 1998):
370–380.

Ihinger-Tallman, Marilyn, and Kay Pasley. "Divorce and
Remarriage in the American Family: A Historical
Review." In *Remarriage and Stepparenting: Current
Research and Theory,* edited by K. Pasley and
M. Ihinger-Tallmann. New York: Guilford Press,
1987a.

———. *Remarriage.* Newbury Park, CA: Sage, 1987b.

Impett, Emily, and Letitia Anne Peplau. "Why Some
Women Consent to Unwanted Sex with a Dating
Partner: Insights from Attachment Theory." *Psychology of
Women Quarterly* 26 (2002): 360–370.

Isaacs, J. "Economic Mobility of Black and White
Families." http://www.brookings.edu/papers/2007/11_
blackwhite_isaacs.aspx, 2007.

———. "Economic Mobility of Families across
Generations." 2007. http://www.brookings.edu/
papers/2007/11_generations_isaacs.aspx.

Isensee, Rik. *Love Between Men: Enhancing Intimacy and
Keeping Your Relationship Alive.* New York: Prentice Hall,
1990.

Ishii-Kuntz, Masako. "Racial and Ethnic Diversity
in Marriage: Asian American Perspective." In

Vision 2004: What is the Future of Marriage, edited by
Paul R. Amato and Nancy Gonzalez, Minneapolis,
MN: National Council on Family Relations, 2004:
36–40.

Ishii-Kuntz, Masako. "Japanese American Families." In
Families in Cultural Context, edited by M. K. DeGenova.
Mountain View, CA: Mayfield, 1997.

Jaccard, J., P. J. Dittus, and V. V. Gordon. "Parent-
Adolescent Congruency in Reports of Adolescent
Sexual Behavior and in Communications about Sexual
Behavior." *Child Development* 69, 1 (February 1998):
247–261.

Jaccard, James, Patricia J. Dittus, and Vivian V. Gordan.
"Parent-Adolescent Congruency in Reports of
Adolescent Sexual Behavior and in Communications
about Sexual Behavior." *Child Development* 69, 1
(February 1998): 247–261.

Jackson, Susan M., Fiona Cram, and Fred W. Seymour.
"Violence and Sexual Coercion in High School
Students' Dating Relationships." *Journal of Family
Violence* 15, 1 (March 2000): 23–36.

Jacobs, Jerry, and Kathleen Gerson. The *Time Divide: Work,
Family, and Gender Inequality.* Cambridge, MA: Harvard
University Press, 2004.

Jamieson, P., and D. Romer, eds. *The Changing Portrayal of
Adolescents in the Media since 1950.* New York: Oxford
University Press, 2008.

Jankowiak, W., and E. Fischer. "A Cross-Cultural
Perspective on Romantic Love," *Ethnology,* 31, 1992:
149–155.

Jankowiak, W., M. Sudakov, and B. Wilreker. "Co-Wife
Conflict and Co-Operation." *Ethnology* 44, 1 (Winter
2005): 81–98.

Janus, Samuel, and Cynthia Janus. *The Janus Report.* New
York: Wiley, 1993.

Jayakody, R., and A. Kalil. "Social Fathering in Low-
Income, African American Families with Preschool
Children." *Journal of Marriage and Family* 64, 2 (2002):
504–516.

Jayson, Sharon. "Living Together No Longer 'Playing
House.'" *USA Today,* July 28, 2008.

Jeffrey, T. B., and L. K. Jeffrey. "Psychologic Aspects of
Sexual Abuse in Adolescence." *Current Opinion in
Obstetrics and Gynecology* 3, 6 (December 1991):
825–831.

Jensen, Larry C., and Janet Jensen. "Family Values,
Religiosity, and Gender." *Psychological Reports* 73, 2
(October 1993): 429–430.

Jensen-Scott, Rhonda L. "Counseling to Promote
Retirement Adjustment." *Career Development Quarterly*
1, 3 (1993): 257–267.

Jewkes, R. "Intimate Partner Violence: Causes and
Prevention." *Lancet* 359, 9315: 1423–1429.

Joesch, Jutta. "The Effects of the Price of Child Care on
AFDC Mothers' Paid Work Behavior." *Family Relations*
40, 2 (April 1991): 161–166.

Johnson, Catherine B., Margaret S. Stockdale, and Frank E.
Saal. "Persistence of Men's Misperceptions of Friendly
Cues across a Variety of Interpersonal Encounters."
Psychology of Women Quarterly 15, 3 (September 1991):
463–475.

Johnson, David, Lynn White, John Edwards, and Alan Booth. "Dimensions of Marital Quality: Toward Methodological and Conceptual Refinement." *Journal of Family Issues* 7 (1986): 31–49.

Johnson, I., and R. Sigler (with R. T. Sigler). "Forced Sexual Intercourse among Intimates." *Journal of Family Violence* 15 (March 2000): 95–108.

Johnson, Julia Overturf, and Barbara Downs. "Maternity Leave and Employment Patterns: 1961–2000." *Current Population Reports* (Series P70, No. 103). Washington, DC: U.S. Census Bureau, 2005.

Johnson, K. L., S. Gill, V. Reichman, and L. G. Tassinary. "Swagger, Sway, and Sexuality: Judging Sexual Orientation from Body Motion and Morphology." *Journal of Personality and Social Psychology* 93 (2007): 321–334.

Johnson, Leanor Boulin. "Perspectives on Black Family Empirical Research: 1965–1978. In *Black Families,* edited by H. Pipes McAdoo. Newbury Park, CA: Sage, 1988.

Johnson, M. "An Exploration of Men's Experience and Role at Childbirth" *Journal of Men's Studies* 10, 2 (Winter 2002).

Johnson, M. "Domestic Violence and Child Abuse: What Is the Connection—Do We Know?" *National Council on Family Relations Report: Family Focus on Child Abuse and Neglect,* Issue FF40. December 2008, F17–19.

Johnson, M., J. Caughlin, and T. Huston. "The Tripartite Nature of Marital Commitment: Personal, Moral, and Structural Reasons to Stay Married." *Journal of Marriage and Family* 61 (1999): 160–177.

Johnson, Michael, and Janel Leone. "The Differential Effects of Intimate Terrorism and Situational Couple Violence: Findings from the National Violence Against Women Survey." *Journal of Family Issues* 26, 3 (April 2005): 322–349.

Johnson, Michael, and Kathleen Ferraro. "Research on Domestic Violence in the 1990's: Making Distinctions," *Journal of Marriage and the Family,* 62, 4 (November 2000): 948–963.

Johnson, Michael. "Patriarchal Terrorism and Common Couple Violence: Two Forms of Violence Against Women." *Journal of Marriage and the Family* 57 (2) (May 1995): 283–294.

Johnston, Deirdre, and Debra Swanson. "Invisible Mothers: A Content Analysis of Motherhood Ideologies and Myths in Magazines." *Sex Roles* 49 (July 2003).

Jones J. *Who Adopts? Characteristics of Women and Men Who Have Adopted Children.* NCHS Data Brief, no. 12. Hyattsville, MD: National Center for Health Statistics, 2009.

Jones, C. "Teleworking: The Quiet Revolution (2005 Update)." *Gartner,* September 14, 2005.

Jones, Carl. *Mind Over Labor.* New York: Penguin, 1988.

Jones, T. S., and M. S. Remland. "Sources of Variability in Perceptions of and Responses to Sexual Harassment." *Sex Roles* 27, 3–4 (August 1992): 121–142.

Jorgensen, Stephen, and Russell Adams. "Predicting Mexican-American Family Planning Intentions: An Application and Test of a Social Psychological Model." *Journal of Marriage and the Family* 50, 1 (February 1988): 107–119.

Julian, T. W., P. McKenry, and L. McKelvey. "Cultural Variations in Parenting: Perceptions of Caucasian, African American, Hispanic, and Asian American Parents." *Family Relations* (43), 1994: 30–37.

Julien, D., H. J. Markman, and K. Lindahl. "A Comparison of a Global and Microanalytic Coding System: Implications for Future Trends in Studying Interactions." *Behavioral Assessment* 11 (1989): 81–100.

Juni, Samuel, and Donald W. Grimm. "Marital Satisfaction and Sex-Roles in a New York Metropolitan Sample." *Psychological Reports* 73, 1 (1993): 307–314.

Jurich, Anthony, and Cheryl Polson. "Nonverbal Assessment of Anxiety as a Function of Intimacy of Sexual Attitude Questions." *Psychological Reports* 57 (3, Pt. 2), (December 1985): 1247–1243.

Jurkovic, G. *Lost Childhoods: The Plight of the Parentified Child.* New York: Brunner/Mazell, 1997.

Kübler-Ross, Elisabeth. *On Death and Dying.* New York: Macmillan, 1969.

———. *Working It Through.* New York: Macmillan, 1982.

———. *AIDS: The Ultimate Challenge.* New York: Macmillan, 1987.

Kachadourian, Lorig K., Frank Fincham, and Joanne Davila. "The Tendency to Forgive in Dating and Married Couples: The Role of Attachment and Relationship Satisfaction." *Personal Relationships* 11, 3 (September 2004): 373–393.

Kagan, Jerome, and N. Snidman. "Temperamental Factors in Human Development." *American Psychologist* 46, 8 (1991): 856–862.

Kahlor, L. A., and D. Morrison. "Television Viewing and Rape Myth Acceptance among College Women." *Sex Roles* 56, 11–12 (June 2007): 729–739.

Kaiser Family Foundation. "U.S. Teen Sexual Activity," January 2005

Kalis, Pamela, and Kimberly Neuendorf. "Aggressive Cue-Prominence and Gender Participation in MTV." *Journalism Quarterly* 66, 1 (March 1989): 148–154, 229.

Kalmijn, M. "Educational Inequality and Extended Family Relationships." Paper presented at the Ross Colloquium Series, University of California, Los Angeles (UCLA), April 7, 2004.

Kalmijn, M. "Explaining Cross-National Differences in Marriage, Cohabitation, and Divorce in Europe, 1990–2000." *Population Studies* 61, 3 (2007): 243–263

Kalmijn, Matthijs, and Henk Flap. "Assortative Meeting and Mating: Unintended Consequences of Organized Settings for Partner Choices." *Social Forces* 79, 4 (June 2001): 1289–1312.

Kalmijn, Matthijs. "Intermarriage and Homogamy: Causes, Patterns, Trends." *Annual Review of Sociology* 24 (1998): 395–421.

Kalmijn, Matthijs. "Marriage Rituals as Reinforcers of Role Transitions: An Analysis of Weddings in The Netherlands." *Journal of Marriage & Family* 66, 3 (August 2004): 582–594.

Kantor, J. "A Portrait of Change: Nation's Many Faces in Extended First Family." *New York Times,* January 20, 2009.

Kantrowitz, Barbara. "Gay Families Come Out." *Newsweek* (November 4, 1996): 51–57.

Kapinus, Carolyn. "The Effects of Parents' Attitudes Toward Divorce on Offspring's Attitudes: Gender and Parental Divorce as Mediating Factors." *Journal of Family Issues* 25 1 (January 2004): 112–135.

Kaplan, Helen Singer. *Disorders of Desire.* New York: Simon and Schuster, 1979.

Kaplan, L., C. Hennon, and L. Ade-Ridder. "Splitting Custody between Parents: Impact on the Sibling System." *Families in Society: the Journal of Contemporary Human Services CEU Article* 30 (1993): 131–144.

Karp, D. *The Burden of Sympathy.* New York: Oxford University Press, 2001.

Kassop, Mark. "Salvador Minuchin: A Sociological Analysis of His Family Therapy Theory." *Clinical Sociological Review* 5 (1987): 158–167.

Katchadourian, Herant. *Midlife in Perspective.* San Francisco: Freeman, 1987.

Katz, J. N. " 'Homosexual' and 'Heterosexual': Questioning the Terms" in *Sexualities: Identities, Behaviors and Society,* edited by Michael Kimmel and Rebecca Plante. New York: Oxford University Press, 2004.

Katz, Jennifer, Stephanie Washington Kuffel, and Amy Coblentz. "Are There Gender Differences in Sustaining Dating Violence? An Examination of Frequency, Severity, and Relationship Satisfaction." *Journal of Family Violence* 17, 3 (September 2002): 247–271.

Katz, Lynn Fainsilber, and Erica M. Woodin. "Hostility, Hostile Detachment, and Conflict Engagement in Marriages: Effects on Child and Family Functioning." *Child Development* 73, 2 (March 2002): 636–690.

Katz, Michael B. *The Undeserving Poor: From War on Poverty to War on Welfare.* New York: Pantheon, 1990.

Katzev, A. R., R. L. Warner, and A. C. Acock. "Girls or Boys? Relationship of Child Gender to Marital Instability." *Journal of Marriage and the Family* 56 (February 1994): 89–100.

Kaufman, Joan, and Edward Zigler. "The Intergenerational Transmission of Abuse Is Overstated." In *Current Controversies in Family Violence,* edited by R. Gelles and D. Loseke. Newbury Park, CA: Sage, 1993.

Kaufman, L. "When the Ex Writes a Blog, Dirty Laundry Is Aired." *New York Times,* April 18, 2008.

Kawamoto, D. "Videoconferencing Ties Seniors with Families." CNET News. http://news.cnet.com/Videoconferencing-ties-seniors-with-families/2100-1041_3-6202474.html, 2007.

Kawamoto, Walter T., and Tamara C. Cheshire. "American Indian Families." In *Families in Cultural Context,* edited by M. K. DeGenova. Mountain View, CA: Mayfield, 1997.

Keene-Reid, Jennifer, and John Reynolds. "Gender Differences in the Job Consequences of Work-to-Family Spillover." *Journal of Family Issues* 26, 3 (2005): 275–299.

Keillor, Garrison. "It's Good Old Monogamy That's Really Sexy." *Time* (October 17, 1994): 71.

Keith, Pat, and Robbyn Wacker. "Grandparent Visitation Rights: An Inappropriate Intrusion or Appropriate Protection?" *International Journal of Aging and Human Development* 54, 3 (2002): 191–204.

Keller, David, and Hugh Rosen. "Treating the Gay Couple within the Context of Their Families of Origin." *Family Therapy Collections* 25 (1988): 105–119.

Kellett, J. M. "Sexuality of the Elderly." *Sexual and Marital Therapy* 6, 2 (1991): 147–155.

Kelley, Douglas L., and Judee K. Burgoon. "Understanding Marital Satisfaction and Couple Type as Functions of Relational Expectations." *Human Communication Research* 18, 1 (1991): 40–69.

Kelley, Harold, et al., eds. *Close Relationships.* San Francisco: Freeman, 1983.

Kelley, Harold. "Love and Commitment." In *Close Relationships,* edited by H. Kelley et al. San Francisco: Freeman, 1983.

Kempe, C. H., F. Silverman, B. Steele, W. Droegmueller, and H. Silver. "The Battered Child Syndrome." *Journal of the American Medical Association* 181 (1962): 17–24.

Kendig, S., and S. M. Bianchi. "Single, Cohabitating, and Married Mothers' Time with Children." *Journal of Marriage and Family* 70, 5 (2008): 1228–1240.

Kennedy, Gregory E. "College Students' Expectations of Grandparent and Grandchild Role Behaviors." *Gerontologist* 30, 1 (1990): 43–48.

Kennedy, Gregory E., and C. E. Kennedy. "Grandparents: A Special Resource for Children in Stepfamilies." *Journal of Divorce and Remarriage* 19, 3–4 (1993): 45–68.

Keshet, Jamie. "From Separation to Stepfamily." *Journal of Family Issues* 1, 4 (December 1980): 517–532.

Kiecolt, K. Jill. "Satisfaction with Work and Family Life: No Evidence of a Cultural Reversal." *Journal of Marriage and Family* 65 (February 2003): 23–35.

Kieren, D., T. Maguire, and N. Hurlbut. "A Marker Method to Test a Phasing Hypothesis in Family Problem-Solving Interaction" *Journal of Marriage and the Family,* 58, 2 (1996): 442–455.

Kikumura, Akemi, and Harry Kitano. "The Japanese American Family." In *Ethnic Families in America: Patterns and Variations,* 3rd ed., edited by C. Mindel et al. New York: Elsevier North Holland, 1988.

Kilmartin, Christopher. *The Masculine Self.* New York: Macmillan, 1994.

Kim, Janna L., and L. Monique Ward. "Pleasure Reading: Associations between Young Women's Sexual Attitudes and Their Reading of Contemporary Women's Magazines." *Psychology of Women Quarterly* 28, 1 (March 2004): 48.

Kim, Jungsik, and Elaine Hatfield. "Love Types and Subjective Well-Being: A Cross-Cultural Study." *Social Behavior and Personality* 32, 2 (2004): 173–182.

———. *Manhood in America: A Cultural History.* New York: Free Press, 1994.

———. *The Gendered Society.* New York: Oxford University Press, 2000.

Kimmel, M., and M. Messner. *Men's Lives,* 7th ed. Boston: Pearson/Allyn and Bacon, 2007.

Kimmel, Michael, and Michael Messner, eds. *Men's Lives,* 4th ed. Needham Heights, MA: Allyn and Bacon, 1998.

Kimmel, Michael, and Rebecca Plante. "The Gender of Desire: The Sexual Fantasies of College Women and Men." In *Sexualities: Identities, Behaviors, and Society,* edited by Michael Kimmel and Rebecca Plante. New York: Oxford University Press, 2004.

Kimmel, Michael. "Masculinity as Homophobia: Fear, Shame, and Silence in the Construction of Gender Identity." In *Theorizing Masculinities*, edited by H. Brod and M. Kaufman. Thousand Oaks, CA: Sage, 1995.

———. *Manhood in America: A Cultural History*. Berkeley: University of California Press, 1996.

King, V., K. M. Harris, and H. E. Heard. "Racial and Ethnic Diversity in Nonresident Father Involvement." *Journal of Marriage and Family* 66 (2004): 1–21.

King, Valarie, and Mindy E. Scott. "A Comparison of Cohabiting Relationships among Older and Younger Adults." *Journal of Marriage and Family* 67, 2 (May 2005): 271–285.

King, Valarie, Kathleen Mullan Harris, and Holly E. Heard. "Racial and Ethnic Diversity in Nonresident Father Involvement." *Journal of Marriage and the Family* 66 (2004): 1–21.

Kiselica, M. S., and M. Morrill-Richards. Sibling Maltreatment: The Forgotten Abuse." *Journal of Counseling and Development* 25, 2 (2007): 148.

Kitano, K., and H. Kitano. The Japanese-American Family." In *Ethnic Families in America: Patterns and Variations*, 4th ed., edited by C. Mindel, R. Habenstein, and R. Wright, Jr. Upper Saddle River, NJ: Prentice Hall, 1998: 311–330.

Kitson, Gay. "Marital Discord and Marital Separation: A County Survey." *Journal of Marriage and the Family* 47 (August 1985): 693–700.

Kitson, Gay, and Leslie Morgan. "The Multiple Consequences of Divorce: A Decade Review. In *Contemporary Families: Looking Forward, Looking Back*, edited by A. Booth. Minneapolis: National Council on *Family Relations*, 1991b.

Kitson, Gay, and Leslie Morgan. "Consequences of Divorce." In *Contemporary Families: Looking Forward, Looking Back* edited by A. Booth. Minneapolis: National Council on *Family Relations*, 1991a.

Kitzinger, Sheila. *The Complete Book of Pregnancy and Childbirth*. New York: Knopf, 1989.

Klein, David M., and James M. White. *Family Theories: An Introduction*. Thousand Oaks, CA: Sage, 1996.

Klein, Fritz. "The Need to View Sexual Orientation as a Multivariate Dynamic Process: A Theoretical Perspective." In *Homosexuality/Heterosexuality: Concepts of Sexual Orientation*, edited by David McWhirter, Stephanie Sandovers, and June Machover Reinisch. New York: Oxford University Press, 1990.

Klinetob, Nadya, and David Smith. "Demand-Withdraw Communication in Marital Interaction: Tests of Interspousal Contingency and Gender Role Hypothesis." *Journal of Marriage and the Family*, 58, 4 (November 1996): 945–957.

Kluwer, Esther, Jose Heesink, and Evert Van de Vliert. "Marital Conflict About the Division of Household Labor and Paid Work." *Journal of Marriage and the Family* 58, 4 (November 1996): 958–969.

———. "The Division of Labor across the Transition to Parenthood: A Justice Perspective." Journal of Marriage and Family 64, 4 (November 2002): 930–943.

———. "The Marital Dynamics of Conflict Over the Division of Labor." *Journal of Marriage and the Family* 59, 3 (August 1997): 635–654.

Knafo, D., and Y. Jaffe. "Sexual Fantasizing in Males and Females." *Journal of Research in Personality* 18 (1984): 451–462.

Knoester, Chris, and Alan Booth. "Barriers to Divorce." *Journal of Family Issues* 21, 1 (January 2000): 78–99.

Knox, David, and Caroline Schacht. "Sexual Behaviors of University Students Enrolled in a Human Sexuality Course." *College Student Journal* 26, 1 (March 1992): 38–40.

———. *Choices in Relationships: An Introduction to Marriage and the Family*, 6th Edition. Belmont, CA: Wadsworth Pub Co, 2000.

Knox, David, and K. Wilson. "Dating Behaviors of University Students." *Family Relations* 30 (1981): 83–86.

Knox, David, and Tim Britton. "Age Discrepant Relationships Reported by University Faculty and Their Students." *College Student Journal* 31, 3 (September 1997): 280.

Knox, David, Marty E. Zusman, Vivian Daniels, and Angel Brantley. "Absence Makes the Heart Grow Fonder? Long Distance Dating Relationships Among College Students." *College Student Journal* 36, 3 (2002): 364–366.

Knox, David, Marty Zusman, and Vivian Daniels. "College Students Attitudes Toward Interreligious Marriage." *College Student Journal* 36, 1 (March 2002): 84–87.

Knudson-Martin, Carmen, and Anne Rankin Mahoney. "Language and Processes in the Construction of Equality in New Marriages." *Family Relations* 47, 1 (January 1998): 81–91.

Kohlberg, Lawrence. "The Cognitive-Development Approach to Socialization." In *Handbook of Socialization Theory and Research*, edited by A. Goslin. Chicago: Rand McNally, 1969.

Kohn, M. "Social Structure and Personality: A Quintessentially Sociological Approach to Social Psychology." *Social Forces* 68 (September, 1989): 26–33.

Komarovsky, Mirra. *Blue Collar Marriage*. New York: Vintage, 1962.

———. *Blue-Collar Marriage*. 2nd ed. New Haven, CT: Yale University Press, 1987.

Konner, Melvin. *Childhood*. Boston: Little, Brown, 1991.

Kortenhaus, Carole M., and Jack Demarest. "Gender Role Stereotyping in Children's Literature: An Update." *Sex Roles: A Journal of Research* 28, 3–4 (1993): 219–323.

Kosciw, J. G., and E. M. Diaz. *Involved, Invisible, Ignored: The Experiences of Lesbian, Gay, Bisexual and Transgender Parents and Their Children in Our Nation's K–12 Schools*. New York: Gay, Lesbian and Straight Education Network, 2008.

Koss, Mary P., and Sarah L. Cook. "Facing the Facts: Date and Acquaintance Rape Are Significant Problems for Women." In *Current Controversies in Family Violence*, edited by R. Gelles and D. Loseke. Newbury Park, CA: Sage, 1993.

Koss, Mary. "Hidden Rape: Sexual Aggression and Victimization in a National Sample of Students in

Higher Education." In *Rape and Sexual Assault,* vol. 2, edited by A. W. Burgess. New York: Garland, 1988.

Kotchick, B. A., S. Dorsey, and L. Heller. "Predictors of Parenting among African American Single Mothers: Personal and Contextual Factors." *Journal of Marriage and Family* 67 (2005): 448–460.

Kovach, G. "Texas Seeking Custody of 8 Children from Sect." *New York Times,* August 5, 2008.

Kowal, A. K., and L. Blinn-Pike. "Sibling Influences on Adolescents' Attitudes toward Safe Sex Practices." *Family Relations* 53, 4 (2004): 377–384.

Kposowa, Augustine J. "The Impact of Race on Divorce in the United States." *Journal of Comparative Family Studies* 29, 3 (Autumn 1998): 529–548.

Kranichfeld, Marion. "Rethinking Family Power." *Journal of Family Issues* 8, 1 (March 1987): 42–56.

Kreider, R. "Marital Status in the 2004 American Community Survey." U.S. Census Bureau, Population Division Working Paper No. 83, December 2006a.

———. "Poster 15: Remarriage in the United States." Paper presented at the annual meeting of the American Sociological Association, Montreal. http://www .allacademic.com/meta/p103541_index.html, 2006b.

———. "Living Arrangements of Children: 2004." *Current Population Reports, 70–114.* Washington, DC: U.S. Census Bureau, 2007.

Kreider, Rose M. "Adopted Children and Stepchildren: 2000." *Census 2000 Special Reports.* Washington, DC: U.S. Census Bureau, 2003.

Kreider, Rose M. "Number, Timing, and Duration of Marriages and Divorces: 2001." *Current Population Reports,* P70–97. Washington, DC, U.S. Census Bureau, 2005.

Kruk, Edward. "Promoting Co-operative Parenting after Separation: A Therapeutic/Interventionist Model of Family Mediation." *Journal of Family Therapy* 15, 3 (1993): 235–261.

Kunkel, D., K. Eyal, K. Finnerty, and E. Donnderstein. *Sex on TV5: A Biennial Report to the Kaiser Family Foundation.* Menlo Park, CA: Kaiser Family Foundation, 2005.

Kurdek, L. "The Allocation of Household Labor by Partners in Gay and Lesbian Couples." *Journal of Family Issues* 28, 1 (January 2007): 132–148.

Kurdek, L. A. "Conflict Resolution Styles in Gay, Lesbian, Heterosexual Nonparent, and Heterosexual Parent Couples." *Journal of Marriage and Family* 56, 3 (August 1994): 705–722.

Kurdek, L. A., and M. A. Fine. "The Relation Between Family Structure and Young Adolescents' Appraisals of Family Climate and Parenting Behavior." *Journal of Family Issues* 14 (June 1993): 279–290.

Kurdek, Lawrence, and P.J. Smith. "Partner Homogamy in Married, Cohabiting, Heterosexual, Gay and Lesbian Couples." *Journal of Sex Research* 23 (1987): 212–232.

Kurdek, Lawrence. "Correlates of Negative Attitudes Toward Homosexuals in Heterosexual College Students." *Sex Roles* 18 (1988): 727–738.

Kurz, Demie. "Caring for Teenage Children." *Journal of Family Issues* 23, 6 (September 2002): 748–767.

Kutner, L. "Parent and Child." *New York Times,* January 28, 1988.

Labor Project for Working Families. 2009. http://www .working-families.org.

Lakoff, Robin. *Language and Women's Place.* New York, 1975.

Lamanna, Mary Ann, and Agnes Riedmann. *Marriages and Families; Making Choices in a Diverse Society.* Belmont, CA: Wadsworth, 1997.

Lamb, K. A., G. Lee, and A. DeMaris. "Union Formation and Depression: Selection and Relationship Effects." *Journal of Marriage and the Family* 65, 4 (2003): 953–962.

Lamb, Michael (ed.) *The Father's Role: Applied Perspectives.* New York : J. Wiley, 1986.

Lamb, Michael (ed.) *The Role of the Father in Child Development.* New York : Wiley, 1997.

Lamb, Michael E. "Nonparental Childcare." *Parenting and Child Development in "Nontraditional" Families,* edited by M. Lamb. Mahwah, NJ: Lawrence Erlbaum, 1999.

Lamb, Michael E., ed. *Parenting and Child Development in "Nontraditional" Families.* Mahwah, NJ: Lawrence Erlbaum Associates, 1999.

Lamb, Michael. "Book Review." *Journal of Marriage and the Family* 55, 4 (November 1993): 1047–1049.

Laner, Mary R. "Violence or Its Precipitators: Which Is More Likely to Be Identified as a Dating Problem." *Deviant Behavior* 11, 4 (October 1990): 319–329.

Laner, Mary R., and N. Ventrone. "Dating Scripts Revisited." *Journal of Family Issues* 21, 4 (May 2000): 488–500.

Lang, M. M., and B. Risman. "A Stalled Revolution or a Still Unfolding One? The Continuing Convergence of Men's and Women's Roles." Paper prepared for the 10th Anniversary Conference of the Council on Contemporary Families, May 4–5, Chicago, IL. http://www.contemporaryfamilies.org/docs/ langandrismanbriefingpaper.doc, 2007.

Langman, L. "Social Stratification." In *Handbook of Marriage and the Family,* edited by M. G. Sussman and S. K. Steinmetz. New York: Plenum Press, 1987, pp. 211–246.

Lannutti, Pamela J., and Kenzie A. Cameron. "Beyond the Breakup: Heterosexual and Homosexual Post-Dissolutional Relationships." *Communication Quarterly* 50, 2 (2002), 153–170.

Lantz, Herman. "Family and Kin as Revealed in the Narratives of Ex-Slaves." *Social Science Quarterly* 60, 4 (March 1980): 667–674.

Lareau, Annette. "Invisible Inequality: Social Class and Childrearing in Black Families and White families." *American Sociological Review* 67, 5 (October 2002): 747–776.

Lareau, Annette. *Unequal Childhoods.* Berkeley, CA: University of California Press, 2003.

LaRossa, R. "Mythologizing Fatherhood." *National Council on Family Relations Report: Family Focus on Fatherhood* 54, 1 (Spring 2009): F3–4, 6.

LaRossa, Ralph, and Maureen Mulligan LaRossa. *The Transition to Parenthood: How Infants Change Families.* Beverly Hills, CA: Sage, 1981.

LaRossa, Ralph. "Fatherhood and Social Change." *Family Relations* 37 (1988): 451–458.

Larson, Jeffry H., Stephan M. Wilson, and Rochelle Beley. "The Impact of Job Insecurity on Marital and Family Relationships." *Family Relations* 43, 2 (April 1994): 138–143.

Larson, Reed, and Sally Gillman. "Transmission of Emotions in the Daily Interactions of Single-Mother Families." *Journal of Marriage & Family* 61, 1 (February 1999): 21–373.

LaSala, Michael. "Extradyadic Sex and Gay Male Couples: Comparing Monogamous and Nonmonogamous Relationships." *Families in Society: The Journal of Contemporary Social Services* 85, 3 (2004): 405–412.

Lasch, Christopher. *Haven in a Heartless World.* New York: Basic Books, 1977.

Lauer, Jeanette, and Robert Lauer. *'Til Death Do Us Part: How Couples Stay Together.* New York: Haworth Press, 1986.

Laughlin, L. "Child Care Constraints among America's Families." U.S. Census Bureau, Housing and Household Economic Statistics Division. http://www.census.gov/population/www/socdemo/child/child-care-constraints.pdf.

Laumann, Edward, John Gagnon, Robert Michael, and Stuart Michaels. *The Social Organization of Sexuality: Sexual Practices in the United States.* Chicago: University of Chicago Press, 1994.

Lavee, Y., and D. H. Olson. "Seven Types of Marriage: Empirical Typology Based on ENRICH." *Journal of Marital and Family Therapy* 19 (1993): 325–340.

Layne, Linda. "Breaking the Silence: An Agenda for a Feminist Discourse of Pregnancy Loss." *Feminist Studies* 23, 2 (1997): 289–315.

Lederer, William, and Don Jackson. *Mirages of Marriage.* New York: Norton, 1968.

Lee, John A. *The Color of Love.* Toronto: New Press, 1973.

———. "Love Styles." In *The Psychology of Love*, edited by R. Sternberg and M. Barnes. New Haven, CT: Yale University Press, 1988.

Lee, S. M., and B. Edmonston. New Marriages, New Families: U.S. Racial and Hispanic Intermarriage. *Population Bulletin* 60 (2005): 1–40.

Lee, Thomas, Jay Mancini, and Joseph Maxwell. "Contact Patterns and Motivations for Sibling Relations in Adulthood." *Journal of Marriage and the Family* 52, 2 (May 1990): 431–440.

Lee, Y., and L. Waite. "Husbands and Wives Time Spent on Housework: A Comparison of Measures." *National Council on Family Relations Report: Family Focus on Families and Work-Life* 51, 2 (June 2006): F1.

Leiblum, S. Women, Sex and the Internet. *Sexual and Relationship Therapy* 16, 4 (2001): 389–405.

Leitenberg, H., M. J. Detzer, and D. Srebnik. "Gender Differences in Masturbation and the Relation of Masturbation Experience in Preadolescence and or Early Adolescence to Sexual Behavior and Sexual Adjustment in Young Adulthood." *Archives of Sexual Behavior* 22, 2 (April 1993): 87–98.

Lenhart, A. "Teens and Mobile Phones Over the Past Five Years: Pew Internet Looks Back." Pew Internet & American Life Project. An initiative of the Pew Research Center. http://authoring.pewinternet.org/Reports/2009/14—Teens-and-Mobile-Phones-Data-Memo.aspx.

Lenhart, A., S. Jones, and A. MacGill. "Adults and Video Games." Pew Internet and American Life Project. 2008. http://www.pewinternet.org/Reports/2008/Adults-and-Video-Games.aspx.

Leonard, R., and S. Elias. *Family law dictionary.* Berkeley, CA: Nolo Press, 1990.

Leslie, Leigh, and Bethany Letiecq. "Marital Quality of African American and White Partners in Interracial Couples." *Personal Relationships* 11 (2004): 559–574.

Levaro, L. B. "Living Together or Living Apart Together: New Choices for Old Lovers." *National Council on Family Relations Report: Family Focus on Cohabitation* 54, 2 (Summer 2009): F9–10.

Levenson, Robert W., Laura L. Carstensen, and John M. Gottman. "Long-Term Marriage: Age, Gender, and Satisfaction." *Psychology and Aging* 8, 2 (1993): 301–313.

———. "The Influence of Age and Gender on Affect, Physiology, and Their Interrelations: A Study of Long-Term Marriages." *Journal of Personality & Social Psychology* 67, 1 (July 1994): 56–68.

Levin, Irene. "The Model Monopoly of the Nuclear Family." Paper presented at the National Conference on *Family Relations*, Baltimore, November 1993.

Levine, J., C. Emery, and H. Pollack. "Well-Being of Children Born to Teen Mothers." *Journal of Marriage and Family* 69, 1 (February 2007): 105–122

Levine, James. *Working Fathers: Strategies for Balancing Work and Family.* Reading, MA: Addison Wesley Longman, 1997.

Levitt, H., R. Swanger, and J. Butler. "Male Perpetrators' Perspectives on Intimate Partner Violence, Religion, and Masculinity." *Sex Roles.* 58, 5–6 (March, 2008): 435–448.

Lewin, Tamar. "Financially Set, Grandparents Help Keep Families Afloat, Too." *New York Times* (July 14, 2005): A1.

Lewin, Tamar. "Up From the Holler: Living in Two Worlds, At Home in Neither." *New York Times*, May 19, 2005.

Lewis, Oscar. "The Culture of Poverty." *Scientific American*, 215, 4 (1966): 19–25.

Li, J., and L. L. Wu. "No Trend in the Intergenerational Transmission of Divorce." *Demography* 45 (2008): 875–83.

Lichter, D., and Z. Qian. "Serial Cohabitation and the Marital Life Course." *Journal of Marriage and Family* 70 (2008): 861–878.

Lichter, D. T. "Cohabitation and the Rise in Out-of-Wedlock Childbearing." *National Council on Family Relations Report: Family Focus on Cohabitation* 54, 2 (Summer 2009): F11–13.

Lieberson, Stanley, and Mary Waters. *From Many Strands: Ethnic and Racial Groups in Contemporary America.* New York: Russell Sage Foundation, 1988.

Liebow, Elliot. *Tally's Corner; A Study Of Negro Streetcorner Men.* Boston, Little, Brown, 1967.

Lin, Chin-Yau Cindy, and Victoria Fu. "A Comparison of Child-rearing Practices among Chinese, Immigrant Chinese, and Caucasian-American Parents." *Child Development* 61, 2 (April 1990): 429–434.

Lin, I-Fen, Nora Cate Schaeffer, Judith A. Seltzer, and Kay L. Tuschen "Divorced Parents' Qualitative and Quantitative Reports of Children's Living Arrangements." *Journal of Marriage & Family* 66, 2 (May 2004): 385–397.

Linden, M. "The Cost of Doing Nothing: The Economic Impact of Recession-Induced Child Poverty, First Focus: Making Children and Families the Priority." 2008. http://www.firstfocus.net/Download/CostNothing.pdf.

Lindsey, E. W., Y. Caldera, and L. Tankersley. "Marital Conflict and the Quality of Young Children's Peer Play Behavior: The Mediating and Moderating Role of Parent-Child Emotional Reciprocity and Attachment Security." *Journal of Family Psychology* 23, 2 (2009): 130–145.

Lindsey, Linda. *Gender Roles: A Sociological Perspective,* 3rd ed. Upper Saddle River, NJ: Prentice Hall, 1997.

Lippa, R. A. "Sexual Orientation and Personality." *Annual Review of Sex Research* 16 (2005a): 119–153.

———. "Subdomains of Gender-Related Occupational Interests: Do They Form a Cohesive Bipolar M-F Dimension?" *Journal of Personality* 73 (2005b): 693–730.

Lips, Hilary. *Sex and Gender.* 3rd ed. Mountain View, CA: Mayfield, 1997.

Lloyd, Sally A. "Physical and Sexual Violence During Dating and Courtship." In *Vision 2010: Families and Violence, Abuse and Neglect,* edited by R. J. Gelles. Minneapolis: National Council on *Family Relations,* 1995.

Lloyd, Sally A., and Beth C. Emery. "The Dynamics of Courtship Violence." Paper presented at the annual meeting of the National Council on Family Relations, Seattle, WA, November 1990.

LoPiccolo, J. "Counseling and therapy for sexual problems in the elderly." *Clinics in Geriatric Medicine* 7 (1991): 161–179.

LoPresto, C., M. Sherman, and N. Sherman. "The Effects of a Masturbation Seminar on High School Males' Attitudes, False Beliefs, Guilt, and Behavior." *Journal of Sex Research* 21 (1985): 142–156.

Lorber, Judith. *Paradoxes of Gender.* New Haven, CT: Yale University Press, 1994.

———. *Gender Inequality: Feminist Theories and Politics.* Los Angeles: Roxbury, 1998.

"Louisiana Civil Code Art. 98. Mutual Duties of Married Persons," marriagedebate.com, 1999.

Lowenstein, Ludwig. "Causes and Associated Features of Divorce as Seen by Recent Research." *Journal of Divorce & Remarriage* 42, 3/4 (2005): 153–171.

Lutz, Patricia. "The Stepfamily: An Adolescent Perspective." *Family Relations* 32, 3 (July 1983): 367–375.

MacDermid, S. "Work-Family Research: Learning from the 'Best of the Best.'" *National Council on Family Relations Report: Family Focus on Families and Work-Life* 51, 2 (June 2006): F15–16.

MacDermid, S., J. Jurich, J. Myers-Walls, and A. Pelo. "Feminist Teaching: Effective Education." *Family Relations* 41 (1992): 31–38.

MacDonald, W. L., and A. DeMaris. "Remarriage, Stepchildren, and Marital Conflict. Challenges to the Incomplete Institutionalization Hypothesis." *Journal of Marriage and the Family* 57, 2 (May 1995): 387–398.

MacDorman, M. F., and T. J. Mathews. *Recent Trends in Infant Mortality in the United States.* NCHS Data Brief, no. 9. Hyattsville, MD: National Center for Health Statistics, 2008

Mace, David, and Vera Mace. "Enriching Marriage." In *Family Strengths,* edited by N. Stinnet et al. Lincoln: University of Nebraska Press, 1979.

———. "Enriching Marriages: The Foundation Stone of Family Strength." In *Family Strengths: Positive Models for Family Life,* edited by N. Stinnet et al. Lincoln: University of Nebraska Press, 1980.

Madden, Mary, and Amanda Lenhart. "On-line Dating." Pew Internet and American Life Project, www.pewinter net.org, March 5, 2006.

Mahoney, A. "Religion and Conflict in Marital and Parent-Child Relationships." *Journal of Social Issues* 61, 4 (2005): 689–706.

Mancini, Jay, and Rosemary Bliszner. "Research on Aging Parents and Adult Children." *Journal of Marriage and the Family* 51, 2 (May 1989): 275–290.

———. "Aging Parents and Adult Children: Research Themes in Intergenerational Relations." In *Contemporary Families: Looking Forward, Looking Back,* edited by A. Booth. Minneapolis: National Council on *Family Relations,* 1991.

———. "Social Provisions in Adulthood: Concept and Measurement in Close Relationships." *Journal of Gerontology* 47, 1 (1992): 14–20.

Manning, W., and P. Smock. "Divorce-Proofing Marriage: Young Adults' Views on the Connection between Cohabitation and Marital Longevity." *National Council on Family Relations Report: Family Focus on Cohabitation* 54, 2 (Summer 2009): F13–15.

Manning, W. D., and S. L. Brown. "Children's Economic Well-Being in Married and Cohabiting Parent Families." *Journal of Marriage and Family* 68 (2006): 345–362.

Manning, W. D., P. C. Giordano, and M. A. Longmore. "Hooking Up: The Relationship Contexts of 'Non-Relationship' Sex." *Journal of Adolescent Research* 21, 5 (2006): 459–483.

Manning, W., M. A. Longmore, and P. C. Giordano. "The Changing Institution of Marriage: Adolescents' Expectations to Cohabit and to Marry." *Journal of Marriage and Family* 69 (August 2007): 559–575.

Manning, Wendy D., and Kathleen Lamb. "Adolescent Well-Being in Cohabiting, Married, and Single-Parent Families." *Journal of Marriage & Family* 65, 4 (November 2003): 876–893.

Manning, Wendy D., and Pamela Smock. "First Comes Cohabitation and Then Comes Marriage?" *Journal of Family Issues* 23, 8 (November 2002): 1065–1087.

Manning, Wendy. "Children and the Stability of Cohabiting Couples." *Journal of Marriage & Family* 66, 3 (August 2004): 674–689.

Marecek, Jeanne, et al. "Gender Roles in the Relationships of Lesbians and Gay Men." In *Gay Relationships,* edited by J. DeCecco. New York: Haworth Press, 1988.

Margolin, Gayla, Linda Gorin Sibner, and Lisa Gleberman. "Wife Battering." In *Handbook of Family Violence,* edited by V. B. Van Hasselt et al. New York: Plenum Press, 1988.

Marin, Rick. "At-Home Fathers Step Out to Find They Are Not Alone." *New York Times* (January 2, 2000): 1, 16.

Markman, Howard, et al. "The Prediction and Prevention of Marital Distress: A Longitudinal Investigation." In *Understanding Major Mental Disorders: The Contribution of Family Interaction Research,* edited by K. Hahlweg and M. Goldstein. New York: Family Process Press, 1987.

Markman, Howard. "Prediction of Marital Distress: A Five-Year Followup." *Journal of Consulting and Clinical Psychology* 49 (1981): 760–761.

———. "The Longitudinal Study of Couples' Interactions: Implications for Understanding and Predicting the Development of Marital Distress." In *Marital Interaction,* edited by K. Hahlweg and N. S. Jacobsen. New York: Guilford Press, 1984.

Marks, Nadine, James Lambert, and Heejeong Choi. "Transitions to Caregiving, Gender and Psychological Well-Being: A Prospective U.S. National Study," *Journal of Marriage and the Family* 64, 3 (August 2002): 657–667.

Marks, Nadine. "Flying Solo at Midlife: Gender, Marital Status, and Psychological Well-Being." *Journal of Marriage & Family* 58, 4 (November 1996): 917–932.

Marks, Stephen. "What Is a Pattern of Commitment?" *Journal of Marriage and the Family* 52, 1 (February 1994): 112–115.

Marks, Stephen. *Three Corners: Integrating Marriage & The Self.* Lexington, MA: Lexington Books, 1986.

Marquardt, E. *Between Two Worlds: The Inner Lives of Children of Divorce.* New York: Three Rivers Press, 2006.

"Marriage Rights and Benefits," nolo.com, 2006.

Married Parents' Use of Time Summary, www.2003–2006.bls.gov/tus.

Marsiglio, W. "When Stepfathers Claim Stepchildren: A Conceptual Analysis." *Journal of Marriage and Family* 66 (2004): 22–39.

Marsiglio, William. "Paternal Engagement Activities with Minor Children." *Journal of Marriage & Family* 53, 4 (November 1991): 973–986.

———. *Procreative Man.* New York: New York University, 1998.

———. *Stepdads: Stories Of Love, Hope, And Repair.* Lanham, MD: Rowman and Littlefield, 2004.

Martin, and Ventura, *National Vital Statistics Reports,* 57, 12, (2009), Table 1.

Martin, J. A., B. E. Hamilton, P. D. Sutton, S. J. Ventura, et al. "Births: Final Data for 2006." *National Vital Statistics Reports,* vol. 57, no. 7. Hyattsville, MD: National Center for Health Statistics, 2009.

Martin, Karin A. "Giving Birth Like a Girl," *Gender & Society* 17, 1 (February 2003): 54–72.

Martin, S., and S. Parashar. "Women's Changing Attitudes towards Divorce, 1974–2002: Evidence for an Educational Crossover." *Journal of Marriage and Family* 68, 1 (2006): 29–40.

Martin, S. P. "Growing Evidence for a Divorce Divide? Education and Marital Dissolution Rates in the United States Since the 1970s." Russell Sage Foundation Working Papers: Series on Social Dimensions of Inequality. New York: Russell Sage Foundation, 2004.

Martin, T. C., and L. L. Bumpass. "Recent Trends in Marital Disruption." *Demography* 26 (February 1989): 37–51.

Marwell, G., et al. "Legitimizing Factors in the Initiation of Heterosexual Relationships." Paper presented at the First International Conference on Personal Relationships, Madison, WI, July 1982.

Masheter, Carol. "Postdivorce Relationships Between Ex-Spouses: The Roles of Attachment and Interpersonal Conflict." *Journal of Marriage and the Family* 53 (1) April 1991: 103–110.

Mason, M. A., M. Fine, and S. Carnochan. "Family Law in the New Millennium: For Whose Families?" *Journal of Family Issues* 22, 7 (October 2001): 859–881.

Masood, S. "Buried alive: Islamabad looks at 'honor killings' - Women's plans to wed defied tradition." *International Herald Tribune,* September 4, 2008.

Mass Layoffs in June 2009, http://www.gls.gov/mls.

Masters, William, and Virginia Johnson. *Human Sexual Inadequacy.* Boston: Little, Brown, 1970.

Maticka-Tyndale, Eleanor. "Sexual Scripts and AIDS Prevention: Variations in Adherence to Safer-Sex Guidelines." *Journal of Sex Research* 28, 1 (February 1991): 145–166.

Matsumoto, D. "Culture and Nonverbal Behavior." Edited by Valerie Manusov and Miles Patterson in *The Sage Handbook of Nonverbal Communication,* Thousand Oaks, CA: Sage, 2006.

Matthews, T., and B. Hamilton. "Trend Analysis of the Sex Ratio at Birth in the United States." *National Vital Statistics Reports* 53, 20–37. Hyattsville, MD: National Center for Health Statistics, 2006.

Mattingly, M., and L. C. Sayer. "Under Pressure: Trends and Gender Differences in the Relationship between Free Time and Feeling Rushed." *Journal of Marriage and Family* 68, 1 (2006): 205–221.

Max, S. "Living with the In-Laws: Multigenerational Households, though Still Uncommon, Seem to Be Growing in Popularity." http://money.cnn.com/2004/02/18/pf/yourhome/grannyflats/index.htm, 2004.

Max, Sarah. "Living With the In-Laws: Multigenerational Households, Though Still Uncommon, Seem to Be Growing in Popularity." Money/CNN.com (April 22, 2004).

McAdoo, Harriette Pipes. "Changes in the Formation and Structure of Black Families: The Impact on Black Women." Working paper no. 182, Center for Research on Women, Wellesley College, Wellesley, MA, 1988a.

———, ed. *Black Families,* 2nd ed. Beverly Hills, CA: Sage, 1988b.

McAdoo, H. P. *Black Families,* 3rd ed. Thousand Oaks, CA: Sage, 1996.

McCaghy, Charles, and Timothy Capron. *Deviant Behavior: Crime, Conflict and Interest Groups.* Boston, MA: Allyn & Bacon, 1997.

McCaghy, Charles, Timothy Capron, and J. D. Jamieson. *Deviant Behavior: Crime, Conflict, and Interest Groups,* 5th ed. Needham Heights, MA: Allyn and Bacon, 2000.

McClennen, J. C., A. B. Summers, and J. G. Daley. "The Lesbian Partner Abuse Scale." *Research on Social Work Practice* 12 (2002): 277–292.

McGoldrick, Monica, and B. Carter, eds. *The Changing Family Life Cycle,* 2nd ed. Boston: Allyn and Bacon, 1989.

McHale, Susan, and Ted Huston. "The Effect of the Transition to Parenthood on the Marriage and Family Relationship: A Longitudinal Study." *Journal of Family Issues* 6, 4 (December 1985): 409–433.

McHugh, Maureen, Nicole Livingston, and Amy Ford. "A Postmodern Approach to Women's Use of Violence: Developing Multiple and Complex Conceptualizations." *Psychology of Women Quarterly* 29, 4 (December 2005): 323–336.

McKeever, M., and N. Wolfinger. "Reexamining the Economic Costs of Marital Disruption for Women." *Social Science Quarterly* 82, 1 (2001): 202–217.

McKim, Margaret K. "Transition to What? New Parents' Problems in the First Year." *Family Relations* 36, 1 (January 1987): 22–25.

McKinnon, Jesse, and C. Bennett. "We the People: Blacks in the United States: Census 2000 Special Reports." *U.S. Bureau of the Census,* 2005.

McKinnon, Jesse. "The Black Population in the United States: March 2002." *U.S. Census Bureau Current Population Reports.* Series P20-541. Washington, D.C. (2003).

McLanahan, Sara, and Karen Booth. "Mother-Only Families: Problems, Prospects, and Politics." *Journal of Marriage and the Family* (51) 1989: 557–580.

McLanahan, Sara, and Karen Booth. "Mother-Only Families: Problems, Prospects, and Politics." In *Contemporary Families: Looking Forward, Looking Back,* edited by A. Booth. Minneapolis: National Council on Family Relations, 1991.

McLeer, S. V., et al. "Sexually Abused Children at High Risk for Post-Traumatic Stress Disorder." *Journal of the American Academy of Child and Adolescent Psychiatry* 31, 5 (September 1992): 875–879.

McLennen, J. "Domestic Violence between Same-Gender Partners." *Journal of Interpersonal Violence* 20, 2 (2005): 149–154.

McLoyd, Vonnie, Ana Marie Cauce, David Takeuchi, and Leon Wilson. "Marital Processes and Parental Socialization in Families of Color: A Decade Review of Research." *Journal of Marriage and the Family* 62, 4 (November 2000): 1070–1093.

McLoyd, Vonnie, and Julia Smith. "Physical Discipline and Behavior Problems in African American, European American, and Hispanic Children: Emotional Support as a Moderator," *Journal of Marriage and the Family* 64, 1 (February 2002): 40–53.

McMahon, Kathryn. "The Cosmopolitan Ideology and the Management of Desire." *Journal of Sex Research* 27, 3 (August 1990): 381–396.

McManus, Patricia, and Thomas DiPrete. "Losers and Winners: The Financial Consequences of Separation and Divorce for Men." *American Sociological Review* 66 (April 2001): 246–268.

McMenamin, T. "A time to work: recent trends in shift work and flexible schedules." *Monthly Labor Review* 130, 12 (December 2007): 3–15.

McPherson, M., L. Smith-Lovin, and M. Brashears. "The Ties That Bind Are Fraying." *Contexts* 70, 3 (August 2008): 32–36.

McRoy, R. G., H. D. Grotevant, and S. Ayers-Lopez. *Changing Patterns in Adoption.* Austin, TX: Hogg Foundation for Mental Health, 1994.

Mead, Margaret. *Male and Female.* New York: Morrow, 1975.

Meckel, R. "Childhood and the Historians: A Review Essay." *Journal of Family History* 9, 4 (1984): 415–424.

Medeiros, R. A., and M. A. Straus. *A Review of Research on Gender Differences in Risk Factors for Physical Violence between Partners in Marital and Dating Relationships.* Family Research Laboratory, Durham, NH: 2006.

Mederer, H., and R. J. Gelles. "Compassion or Control: Intervention in Cases of Wife Abuse." *Journal of Interpersonal Violence* 4, 1 (March 1989): 25–43.

Medicine, 7 (1991): 161–179.

Medora, Nilufer P., Jeffry H. Larson, Nuran Hortaçsu, and Parul Dave. "Perceived Attitudes towards Romanticism: A Cross-Cultural Study of American, Asian-Indian, and Turkish Young Adults." *Journal of Comparative Family Studies* 33, 2 (Spring 2002): 155–178.

Melito, Richard. "Adaptation in Family Systems: A Developmental Perspective." *Family Processes* 24 (1985): 89–100.

Menaghan, Elizabeth, and Toby Parcel. "Parental Employment and Family Life Research in the 1980s." In *Contemporary Families: Looking Forward, Looking Back,* edited by A. Booth. Minneapolis: National Council on Family Relations, 1991.

Merriam-Webster Online Dictionary. "Cybersex," m-w.com, 2006.

Mertensmeyer, C., and Coleman, M. " Correlates of Inter-Role Conflict in Young Rural and Urban Parents." *Family Relations* 36 (1976), 425–429. Also in R. Marotz-Baden, C. Hennon, and T. Brubaker (Eds.) *Families in Rural America: Stress, Adaptation, and Revitalization,* Minneapolis: National Council on Family Relations.

Messerschmidt, James. *Masculinities and Crime.* Nanham, MD: Rowman and Littlefield, 1993.

Metts, Sandra, and William Cupach. "The Role of Communication in Human Sexuality." In *Human Sexuality: The Social and Interpersonal Context,* edited by K. McKinney and S. Sprecher. Norwood, NJ: Ablex, 1989.

Metz, M., S. Rosser, and N. Strapko. "Differences in Conflict-Resolution Styles among Heterosexual, Gay, and Lesbian Couples." *Journal of Sex Research* 31, 4 (1994): 293–308.

Metz, Michael E., and Norman Epstein. "Assessing the Role of Relationship Conflict in Sexual Dysfunction." *Journal of Sex and Marital Therapy* 28, 2 (March 2002): 139–164.

Meyer, Daniel R., and Judi Bartfeld. "Compliance with Child Support Orders in Divorce Cases." *Journal*

Palkovitz, Robert. "Fathers' Birth Attendence, Early Contact, and Extended Contact with Their Newborns: A Critical Review." *Child Development* 56, 2 (April 1985): 392–407.

———. "Reconstructing 'Involvement': Expanding Conceptualizations of Men's Caring in Contemporary Families." In *Generative Fathering: Beyond Deficit Perspectives*, Vol. 3, *Current Issues in the Family*, edited by A. Hawkins and D. Dollahite. Thousand Oaks, CA: Sage, 1997: 200–216.

Paludi, M. A. "Sociopsychological and Structural Factors Related to Women's Vocational Development." *Annals of New York Academy of Sciences*, 602 (1990): 157–168.

Papanek, Hannah. "Men, Women, and Work: Reflections on the Two-Person Career." *American Journal of Sociology* 78, 4 (January 1973): 90–110.

Papernow, Patricia L. *Becoming a Stepfamily*. San Francisco: Jossey-Bass, 1993.

Pardun, C. J., K. L. L'Engle, and J. D. Brown. "Linking Exposure to Outcomes: Early Adolescents' Consumption of Sexual Content in Six Media. *Mass Communication & Society* 8, 2 (2005): 75–91.

Parents, Families, and Friends of Lesbians and Gays. "What Is Marriage, Anyway?" pflag.org, 2006.

Park, Kristin. "Stigma Management among the Voluntarily Childless." *Sociological Perspectives* 45, 1 (Spring 2002): 21–45.

Parker-Pope, T. "Gay Unions Shed Light on Gender in Marriage." *New York Times*, June 10, 2008.

Pascoe, C. J. *Dude You're a Fag: Masculinity and Sexuality in High School*. Berkeley: University of California Press, 2007.

Pasley, Kay, and Marilyn Ihinger-Tallman, eds. *Remarriage and Stepparenting: Current Research and Theory*. New York: Guilford Press, 1987.

Pasley, Kay. "Family Boundary Ambiguity: Perceptions of Adult Stepfamily Family Members." In *Remarriage and Stepparenting: Current Research and Theory*, edited by K. Pasley and M. Ihinger-Tallman. New York: Guilford Press, 1987.

Patterson, C J., and L. V. Friel. "Sexual Orientation and Fertility" in *Infertility in the Modern World: Biosocial Perspectives*, edited by G. Bentley and N. Mascie-Taylor. Cambridge: Cambridge University Press, 2000.

Patterson, C. J., and R. W. Chan. "Families Headed by Lesbian and Gay Parents." In *Nontraditional Families: Parenting and Child Development*, 2nd ed., edited by M. E. Lamb. Hillsdale, NJ: Lawrence Erlbaum, 1998.

Patterson, Charlotte. "Lesbian and Gay Parents and Their Children: Summary of Research Findings." In *Lesbian and Gay Parenting American Psychological Association* (2005).

Paul, L., and J. Galloway. "Sexual Jealousy: Gender Differences in Response to Partner and Rival." *Aggressive Behavior* 20, 33 (1994): 203–211.

Paul, P. *Parenting, Inc.* New York: Henry Holt, 2008.

Paulson, Sharon, and Cheryl Somers. "Students' Perceptions of Parent-Adolescent Closeness and Communication about Sexuality: Relations with Sexual Knowledge, Attitudes, and Behaviors." *Journal of Adolescence* 23, 5 (October, 2000): 629–644.

Pawelski, J., E. C. Perrin, J. M. Foy, C. E. Allen, J. E. Crawford, M. Del Monte, M. Kaufman, J. D. Klein, K. Smithi, S. Springer, J. L. Tanner, and D. L. Vickers. "Special Article: The Effects of Marriage, Civil Union, and Domestic Partnership Laws on the Health and Well-Being of Children." *Pediatrics* 118, 1 (July 2006): 349–364.

Pearce, L., and A. Thornton. "Religious Identification and Family Ideologies in the Transition to Adulthood." *Journal of Marriage and Family* 69, 5 (2007): 1227–1243.

Peltola, P., M. Milkie, and S. Presser. "The 'Feminist' Mystique Feminist Identity in Three Generations of Women." *Gender and Society* 18, 1 (February 2004): 122–144.

Peoples, J., and G. Bailey. *Humanity: An Introduction to Cultural Anthropology*, 7th ed. Belmont, CA: Thomson Wadsworth, 2006.

Peoples, James, and Garrick Bailey. *Humanity: An Introduction to Cultural Anthropology*, 7th ed. Belmont, CA: Wadsworth Publishing, 2006.

Peplau, L. A., and A. W. Fingerhut. "The Close Relationships of Lesbians and Gay Men." *Annual Review of Psychology* 58 (2007): 405–424.

Peplau, L. A., R. Veniegas, and S. M. Campbell. Gay and Lesbian Relationships. In Richard Petts and Chris Knoester, "Parents' Religious Heterogamy and Children's Well-Being." *SO: Journal for the Scientific Study of Religion* 46, 3 (2007): 373–389.

Peplau, L., and K. Beales. "Lesbians, Gay Men and Bisexuals in Relationships" in *Encyclopedia of Women and Gender: Sex Similarities and Differences and the Impact of Society on Gender, vol. 1, A–P*, edited by J. Worrell. San Diego, CA: Academic Press, 2002.

Peplau, Letitia Anne, and Steven Gordon. "The Intimate Relationships of Lesbians and Gay Men." In *Gender Roles and Sexual Behavior*, edited by E. Allgeier and N. McCormick. Palo Alto, CA: Mayfield, 1982.

Peplau, Letitia Anne, Charles T. Hill, and Zick Rubin. "Sex Role Attitudes in Dating and Marriage: A 15-Year Follow-Up of the Boston Couples Study." *Journal of Social Issues* 49, 3 (1993): 31–53.

Peplau, Letitia Anne, Rosemary Veniegas, and Susan Miller Campbell. "Gay and Lesbian Relationships," (pp. 200–215). In *Sexualities: Identities, Behaviors, and Society*, edited by Michael Kimmel, Rebecca Plante. New York: Oxford University, 2004.

Peplau, Letitia, and Susan Cochran. "Value Orientations in the Intimate Relationships of Gay Men." *Gay Relationships*, edited by J. DeCecco. New York: Haworth Press, 1988.

Peplau, Letitia. "Research on Homosexual Couples." In *Gay Relationships*, edited by J. DeCecco. New York: Haworth Press, 1988.

Perkins, Kathleen. "Psychosocial Implications of Women and Retirement." *Social Work* 37, 6 (1992): 526–532.

Perlman, Daniel, and Steve Duck, eds. *Intimate Relationships: Development, Dynamics, and Deterioration*. Beverly Hills, CA: Sage, 1987.

Perry-Jenkins, Maureen, and Karen Folk. "Class, Couples, and Conflict: Effects of the Division of Labor on

Assessments of Marriage in Dual-Earner Marriages." *Journal of Marriage and the Family* 56, 1 (February 1994): 165–180.

Perry-Jenkins, Maureen, Rena Repetti, and Ann Crouter. "Work and Family in the 1990's." *Journal of Marriage and the Family*, 62, 4 (November 2000): 981–998.

Peters, Stefanie, et al. "Prevalance." In *Sourcebook on Child Sexual Abuse,* edited by D. Finkelhor. Beverly Hills, CA: Sage, 1986.

Petersen, Larry. "Interfaith Marriage and Religious Commitment among Catholics." *Journal of Marriage and the Family* 48, 4 (November 1986): 725–735.

Petersen, Virginia, and Susan B. Steinman. "Helping Children Succeed after Divorce: A Court-Mandated Educational Program for Divorcing Parents." *Family and Conciliation Courts Review* 32, 1 (1994): 27–39.

Peterson, Gary W., and Boyd C. Rollins. "Parent-Child Socialization." In *Handbook of Marriage and the Family,* edited by M. B. Sussman and S. K. Steinmetz. New York: Plenum Press, 1987.

Peterson, Marie Ferguson. "Racial Socialization of Young Black Children." In *Black Children: Social, Educational, and Parental Environments,* edited by H. Pipes McAdoo and J. McAdoo. Beverly Hills, CA: Sage, 1985.

Peterson, Richard R. "A Reevaluation of the Economic Consequences of Divorce." *American Sociological Review* 61, 3 (June 1996).

Pettigrew, J. "Text Messaging and Connectedness Within Close Interpersonal Relationships." *Marriage & Family Review* 45, 6 (2009): 697–716.

Petts, R., and C. Knoester, "Parents' Religious Heterogamy and Children's Well-Being." *Journal for the Scientific Study of Religion* 46, 3(2007): 373–389.

Peyrot, Mark, et al. "Marital Adjustment to Adult Diabetes: Interpersonal Congruence and Spouse Satisfaction." *Journal of Marriage and the Family* 50, 2 (May 1988): 363–376.

Phillips, J. A., and M. A. Sweeney. "Premarital Cohabitation and Marital Disruption among White, Black, and Mexican American Women." *Journal of Marriage and Family* 67 (2005): 296–314.

Phillips, Julie A., and Megan M. Sweeney. "Premarital Cohabitation and Marital Disruption among White, Black, and Mexican American Women." *Journal of Marriage and Family* 67, 2 (May 2005): 296–314.

———. "Can Differential Exposure to Risk Factors Explain Recent Racial and Ethnic Variation in Marital Disruption?" *Social Science Research* 35, 2 (June 2006): 409–434.

Picard, P. *Research Topline: Tech Abuse in Teen Relationships Study.* Teenage Research Unlimited, Northbrook, IL. 2007.

Pill, Cynthia. "Stepfamilies: Redefining the Family." *Family Relations* 39, 2 (April 1990): 186–193.

Pillemer, K., and D. Finkelhor. "The Prevalence of Elder Abuse: A Random Sample Survey." *Violence Against Women* 4, 5 (1998): 559–571.

Pillemer, K., and J. Suitor. " 'Will I ever escape my child's problems?' Effects of adult children's problems on

elderly parents." *Journal of Marriage and the Family* 53 (1991): 585–594.

Pillemer, Karl, and J. Jill Suitor. "Elder Abuse." In *Handbook of Family Violence,* edited by V. B. Van Hasselt et al. New York: Plenum Press, 1988.

Pistole, M. C., E. M. Clark, and A. L. Tubbs, Jr. "Adult attachment and the investment model." *Journal of Mental Health Counseling* 17 (1995): 199–209.

Pleck, J. H. "Paternal involvement: Levels, origins, and consequences." in *The role of the father in child development,* 3rd ed. edited by M. E. Lamb. New York: Wiley, 1997: (pp. 66–103).

Pleck, Joseph. *Working Wives/Working Husbands.* Beverly Hills, CA: Sage, 1985.

Poehlmann, J., R. Shlafer, E. Maes, and A. Hanneman. "Factors Associated with Young Children's Opportunities for Maintaining Family Relationships during Maternal Incarceration." *Family Relations* 57 (July 2008): 267–280.

Pogarsky, G., T. P. Thornberry, and A. J. Lizotte. "Developmental Outcomes for Children of Young Mothers." *Journal of Marriage and Family* 68 (2006): 332–334.

Polatnik, M. Rivka. "Too Old for Child Care? Too Young for Self Care? Negotiating After-School Arrangements for Middle School." *Journal of Family Issues* 23, 6 (September 2002): 728–747.

Pollack, D., and S. Mason. "Mandatory Visitation: In the Best Interests of the Child." *Family Court Review* 42, 1 (2004): 74–84.

Pollack, William. *Real Boys: Rescuing Our Sons from the Myths of Boyhood.* New York: Random House, 1998.

Pollard, M. and K. M. Harris. "Measuring Cohabitation in Add Health." In *Handbook of Measurement Issues in Family Research,* edited by S. Hofferth and L. Casper. Mahwah, NJ: Lawrence Erlbaum, 2007.

Popenoe, D., and B. Whitehead. *Should We Live Together? What Young Adults Need to Know about Cohabitation Before Marriage.* New Brunswick, NJ: Rutgers University. 1999.

Popenoe, D., and B. D. Whitehead. "Should We Live Together? What Young Adults Need to Know about Cohabitation before Marriage." *A Comprehensive Review of Recent Research,* 2nd ed. The National Marriage Project: The Next Generation Series. New Brunswick, NJ: Rutgers University. 2002.

Popenoe, David. *Life Without Father: Compelling New Evidence That Fatherhood and Marriage Are Indispensable for the Good of Children and Society.* New York: Free Press, 1996.

Population Resource Center. Executive Summary: A Demographic Profile of Hispanics in the U.S. Washington, D.C. (2001).

POV20: People in Families With Related Children Under 6 by Householder's Work Experience and Family Structure: 2006.

Power, C., Rodgers, B., Hope, S. "Heavy alcohol consumption and marital status: disentangling the relationship in a national study of young adults." *Addiction* 94 (1999):1477–1487.

Power, Chris, and Bryan Rodgers. "Heavy Alcohol Consumption and Marital Status: Disentangling the

Relationship in a National Study of Young Adults." *Addiction* 94, 10 (October 1999): 1477–1487.

Presser, Harriet. "Nonstandard Work Schedules and Marital Instability." *Journal of Marriage and the Family,* 62, 1 (February 2000): 93–110.

Presser, Harriet. *Working in a 24/7 Economy: Challenges for American Families.* New York: Russell Sage Foundation, 2003.

Prevent Child Abuse America. 2006 National Child Maltreatment Statistics, April, 2008, 66 (2004): 333–350.

Price, John. "North American Indian Families." In E*thnic Families in America: Patterns and Variations,* 2nd ed., edited by C. Mindel and R. Habenstein. New York: Elsevier-North Holland, 1981.

Priest, Ronnie. "Child Sexual Abuse Histories among African-American College Students: A Preliminary Study." *American Journal of Orthopsychiatry* 62, 3 (July 1992): 475.

Primack, B., E. Douglas, M. Fine, and M. Dalton. "Association of Favorite Music and Sexual Behavior." *American Journal of Preventive Medicine* 36 (2009): 317–323.

Proctor, B., and J. Dalaker. *Poverty in the United States: 2002. Current Population Reports, Series P60-222.* U.S. Bureau of the Census: Washington, D.C. (2003).

Prothrow-Smith, D., and H. Spivak. *Sugar and Spice and No Longer Nice: How We Can Stop Girls' Violence.* San Francisco: Jossey-Bass, 2005.

Putnam, F. "Ten-Year Research Update Review: Child Sexual Abuse." *Journal of the American Academy of Child and Adolescent Psychiatry* 42, 3 (March 2003): 269–278.

Raley, R. K., and M. M. Sweeney. "What Explains Race and Ethnic Variations in Cohabitation, Marriage, Divorce, and Nonmarital Fertility?" California Center for Population Research Online Working Paper Series, University of California, Los Angeles, September 2007.

Rank, Mark R., and Li-Chen Cheng. "Welfare Use Across Generations: How Important Are the Ties That Bind?" *Journal of Marriage and the Family* 57, 3 (August 1995): 673–684.

Raschke, Helen. "Divorce." In *Handbook of Marriage and the Family,* edited by M. Sussman and S. Steinmetz. New York: Plenum Press, 1987.

Reed, J. "Not Crossing the 'Extra Line': How Cohabitors with Children View Their Unions." *Journal of Marriage and Family* 68 (2006): 1117–1131.

Regan, P., E. Kogan, and T. Whitlock. "Ain't Love Grand! A Prototype Analysis of the Concept of Romantic Love." *Journal of Social and Personal Relationships* 15, 3 (1998): 411–420.

Regan, Pamela. *The Mating Game: A Primer on Love, Sex, and Marriage.* Thousand Oaks, CA: Sage, 2003.

Rehman, U. S., and A. Holtzworth-Munroe. "A Cross-Cultural Examination of the Relation of Marital Communication Behavior to Marital Satisfaction." *Journal of Family Psychology* 21, 4 (2007): 759–763.

Reid, A., and N. Purcell, "Pathways to feminist identification." *Sex Roles* 50 11/12 (2004): 759–769.

Reiss, Ira. *Family Systems in America.* 3rd ed. New York: Holt, Rinehart, and Winston, 1980.

Renzetti, C., and D. Curran. *Women, Men and Society,* 4th ed. Boston: Allyn and Bacon, 1999.

———. *Women, Men, and Society,* 5th ed. Needham Heights, MA: Allyn and Bacon, 2003.

Renzetti, Claire, and Daniel Curran. *Living Sociology.* Needham Heights, MA: Allyn and Bacon, 1998.

Renzetti, Claire. "Violence in Gay and Lesbian Relationships." In *Vision 2010: Families and Violence, Abuse and Neglect,* edited by R. J. Gelles. Minneapolis: National Council on *Family Relations,* 1995.

Rettig, M. " 'Outing' of Simpsons Character Consistent with Hollywood Bias." *American Family Association,* http://www.afa.net. 2005.

Rice, F. Phillip, and Kim Gale Dolgin. *The Adolescent: Development, Relationships and Culture,* 10th ed. Boston: Allyn and Bacon, 2002.

Richards, Leslie N., and Cynthia J. Schmiege. "Problems and Strengths of Single-Parent Families: Implications for Practice and Policy." *Family Relations* 42, 3 (July 1993): 277–285.

Ridley, Jane, and Michael Crowe. "The Behavioural-Systems Approach to the Treatment of Couples." *Sexual and Marital Therapy* 7, 2 (1992): 125–140.

Riggs, Janet Morgan. "Impressions of Mothers and Fathers on the Periphery of Childcare." *Psychology of Women Quarterly* 29, 1 (2005): 58–62.

Riseden, Andrea D., and Barbara E. Hort. "A Preliminary Investigation of the Sexual Component of the Male Stereotype." Unpublished paper, 1992.

Risman, Barbara, and Danette Johnson-Sumerford. "Doing It Fairly: A Study of Post Gender Marriages." *Journal of Marriage and the Family* 60 (February 1998): 23–40.

Risman, Barbara. "Can Men Mother? Life as a Single Father." *Family Relations* (35) 1986: 95–102.

Risman, Barbara. "Can Men Mother? Life as a Single Father." In *Gender in Intimate Relationships,* edited by B. Risman and P. Schwartz. Belmont, CA: Wadsworth, 1989.

Risman, Barbara. "Intimate Relationships From a Microstructural Perspective: Men Who Mother." *Gender & Society* 1, 1 (March 1987): 6–32.

———. *Gender Vertigo: American Families in Transition.* New Haven, CT: Yale University Press, 1998.

Roberts, Nicole, and Robert Levenson. "The Remains of the Workday: Impact of Job Stress and Exhaustion on Marital Interaction in Police Couples." *Journal of Marriage and Family* 63, 4 (November 2001): 1052–1067.

Roberts, S. "Two-Parent Black Families Showing Gains." *New York Times,* December 27, 2008.

Rodgers, Kathleen Boyce, and Hilary A. Rose. "Risk and Resiliency Factors among Adolescents Who Experience Marital Transitions." *Journal of Marriage and Family* 64, 4 (November 2002): 1024–1037.

Rodgers, Roy, and L. Conrad. "Courtship for Remarriage: Influences on Family Reorganization after Divorce." *Journal of Marriage and the Family* 48 (1986): 767–775.

Rodgers, Roy. *Family Interaction and Transaction: The Developmental Approach.* Englewood Cliffs, NJ: Prentice Hall, 1973.

Roehling, Patricia V., Lorna Hernandez Jarvis, and Heather Swope. "Variations in Negative Work-Family Spillover Among White, Black, and Hispanic American Men and Women: Does Ethnicity Matter?" *Journal of Family Issues* 26, 6 (September 2005): 840–865.

Roenrich, L., and B. N. Kinder. "Alcohol Expectancies and Male Sexuality: Review and Implications for Sex Therapy." *Journal of Sex and Marital Therapy* 17 (1991): 45–54.

Rogers, Stacy. "Dollars, Dependency, and Divorce: Four Perspectives on the Role of Wives' Income." *Journal of Marriage and the Family* 66, 1 (February 2004): 59–74.

Romance Writers of America. Home page, rwanational.org, 2006.

Romero, A., A. K. Baumle, M. V. L. Badgett, and G. J. Gates. (December 1, 2007) Census Snapshot: United States.

Roper Starch Worldwide. Liz Claiborne Inc. Study of Fortune 1000 Senior Executives. 2002.

Rose, Amanda. "Co-Rumination in the Friendships of Girls and Boys." *Child Development* 73, 6 (December 2002): 1830–1843.

Rosen, L., N. Cheever, C. Cummings, and J. Felt. "The impact of emotionality and self-disclosure on online dating versus traditional dating." *Computers in Human Behavior* 24 (2008): 2124–2157.

Rosen, M. P., et al. "Cigarette Smoking: An Independent Risk Factor for Atherosclerosis in the Hypograstric-Cavernous Arterial Bed of Men with Arteriogenic Impotence." *Journal of Urology* 145, 4 (April 1991): 759–776.

Rosenberg, Joshua D. "In Defense of Mediation." *Family and Conciliation Courts Review* 30, 4 (1992): 422–467.

Rossi, Alice. "Transition to Parenthood." *Journal of Marriage and the Family* 30 (1) February 1968: 26–39.

Roy, R., Weibust, K., and Miller, C. "Effects of stereo-types about feminists on feminist self-identification." *Psychology of Women Quarterly* 31 (2007): 146–156.

Ruane, J. and K. Cerulo. *Second Thoughts: Seeing Conventional Wisdom through the Sociological Eye*, 3rd ed. Thousand Oaks, CA: Pine Forge Press, 2004.

Rubin, Lillian. *Worlds of Pain: Life in the Working Class Family*. New York: Basic Books, 1976.

———. *Intimate Strangers: Men and Women Together*. New York: Perennial, 1983.

———. *Just Friends: The Role of Friendship in Our Lives*. New York: Harper and Row, 1985.

———. *Erotic Wars*. New York: Farrar, Straus and Giroux, 1990.

———. *Families on the Faultline: America's Working Class Speaks about the Family, the Economy, Race, and Ethnicity*. New York: HarperCollins, 1994.

Rudd, J., and S. Herzberger. "Brother-sister Incest—Father-daughter Incest: A Comparison of Characteristics and Consequences." *Child Abuse and Neglect* 23, 9 (September 1999): 915–928.

Rudd, J., and Herzberger, S. "Brother-Sister Incest: Father-Daughter Incest: A Comparison of Characteristics and Consequences." *Child Abuse and Neglect* 23a (1998): 915–928.

Runyan, William. "In Defense of the Case Study Method." *American Journal of Orthopsychiatry* 52, 3 (July 1982): 440–446.

Russell, Diana E. H. *The Secret Trauma: Incest in the Lives of Girls and Women*. New York: Basic Books, 1986.

———. *Rape in Marriage*. Rev. ed. Bloomington, IN: Indiana University Press, 1990.

Russell, Diana. *Sexual Exploitation: Rape, Child Sexual Abuse, and Workplace Harassment*. Beverly Hills, CA: Sage, 1984.

Russell, Graeme. "Problems in Role-Reversed Families." In *Reassessing Fatherhood*, edited by C. Lewis and M. O'Brien. London: Sage, 1987: 161–179.

Rust, Paula. "Two Many and Not Enough: The Meanings of Bisexual Identities." In *Sexualities: Identities, Behaviors, and Society*, edited by Michael Kimmel and Rebecca Plante. New York: Oxford University Press, 2004.

Rutter, V. "Economic Woes = Family Stress." Council on Contemporary Families, July 23, 2008.

Rutter, V. "Lessons from Stepfamilies." *Psychology Today* (May 1994): 30–33, 60.

Saad, L. "Women Slightly More Likely to Prefer Working to Homemaking." Gallup News Service, August 31, 2007.

———. "Americans Evenly Divided on Morality of Homosexuality." Gallup Organization. http://www.gallup.com, 2008.

Sadker, Myra, and David Sadker. *Failing at Fairness: How American Schools Cheat Girls*. New York: C. Scribner's Sons, 1994.

Safilios-Rothschild, Constantina. "The Study of the Family Power Structure." *Journal of Marriage and the Family* 32, 4 (November 1970): 539–543.

———. "Family Sociology or Wives' Sociology? A Cross-Cultural Examination of Decision Making." *Journal of Marriage and the Family* 38 (1976): 355–362.

Sagrestano, Lynda, Christopher Heavey, and Andrew Christensen. "Perceived Power and Physical Violence in Marital Conflict." *Journal of Social Issues* 55, 1 (Spring 1999): 65–79.

Sahadi, J. "Being a Mom Could Be a 6-Figure Job." http://money.cnn.com/2006/05/03/pf/mothers_work, 2006.

San Francisco AIDS Foundation. "United States Statistics." http://www.sfaf.org/aidsinfo/statistics/unitedstates, 2008.

Sanchez, Laura, and Constance T. Gager. "Hard Living, Perceived Entitlement to a Great Marriage, and Marital Dissolution." *Journal of Marriage and Family* 62, 3 (August 2000): 708–722.

Sanchez, Laura, Steven Nock, James D. Wright, and Constance Gager. "Setting the Clock Forward or Back?" *Journal of Family Issues*, 23, 1 (January 2002): 91–120.

Sanchez, Laura. "Feminism and Families." *Journal of Marriage and the Family* 59, 4 (1997): 1031–1032.

Sanchez, Laura. "Feminism, Family Work, and Moral Discourse: A Comment on Rollins Ahlander and Slaugh Bahr's 'Beyond Drudgery, Power, and Equity'." *Journal of Marriage and the Family* 58, 2 (1996): 514–520.

Sanford, Keith. "Problem–Solving Conversations in Marriage: Does It Matter What Topics Couples Discuss?" *Personal Relationships* 10, 1 (March 2003): 97–112.

Sarkisian, N., and N. Gerstel. "Till Marriage Do Us Part: Adult Children's Relationships with Parents." *Journal of Marriage and Family* 70, 2 (May 2008): 360–376.

Sarkisian, N., M. Gerena, and N. Gerstel. "Extended Family Integration among Euro and Mexican Americans:

Ethnicity, Gender, and Class." *Journal of Marriage and Family*, 69, 1 (February 2007): 40–54.

Sassler, Sharon. "The Process of Entering into Cohabiting Unions." *Journal of Marriage and Family* 66, 2 (May 2004): 491–505.

Satir, Virginia. *The New Peoplemaking*. Rev. ed. Mountain View, CA: Science and Behavior Books, 1988.

———. *Peoplemaking*. Palo Alto, CA: Science and Behavior Books, 1972.

Sayer, L. C. "Gender, Time, and Inequality: Trends in Women's and Men's Paid Work, Unpaid Work, and Free Time." *Social Forces* 84, 1 (2005): 285–303.

———. "Economic Aspects of Divorce and Relationship Dissolution." In *Handbook of Divorce and Relationship Dissolution*, edited by M. A. Fine and J. H. Harvey. Mahwah, NJ: Lawrence Erlbaum, 2006.

Sayer, Liana C., Suzanne M. Bianchi, and John P. Robinson. "Are Parents Investing Less in Children? Trends in Mothers' and Fathers' Time with Children." *American Journal of Sociology* 110, 1 (2004): 1–43.

Scanzoni, J. "Social Exchange and Behavioral Interdependence." In *Social Exchange in Developing Relationships*, edited by R. Burgess and T. Huston. New York: Academic Press, 1979.

Scanzoni, J., K. Polonko, J. Teachman, and L. Thompson. *The Sexual Bond: Rethinking Families and Close Relationships*. Newbury Park, CA: Sage, 1989.

Scanzoni, John. *Sexual Bargaining*. 2nd ed. Englewood Cliffs, NJ: Prentice Hall, 1980.

———. "Reconsidering Family Policy: Status Quo or Force for Change?" *Journal of Family Issues* 3, 3 (September 1982): 277–300.

Scelfo, J. "Bad Girls Go Wild: A Rise in Girl-on-Girl Violence Is Making Headlines Nationwide and Prompting Scientists to Ask Why." *Newsweek*, June 13, 2005.

Schaap, Cas, Bram Buunk, and Ada Kerkstra. "Marital Conflict Resolutions." In *Perspectives on Marital Interaction*, edited by P. Noller and M. A. Fitzpatrick. Philadelphia: Multilingual Matters, 1988.

Schoen, Robert, and Yen-Hsin Alice Cheng. "Partner Choice and the Differential Retreat From Marriage." *Journal of Marriage and Family* 68, 1 (February 2006): 1–10.

Schoen, Robert. "Union Disruption in the United States." *International Journal of Sociology* 32, 4 (Winter 2002/2003): 36–50.

Schoenborn, C. A. "Marital Status and Health: United States, 1999–2002." *Vital and Health Statistics*, no. 351. Hyattsville, MD: National Center for Health Statistics, 2004.

Schooler, Carmi. "Psychological Effects of Complex Environments during the Life Span: A Review and Theory." In *Cognitive Functioning and Social Structure over the Life Course*, edited by C. Schooler and K. Warner Schaie. Norwood, NJ: Ablex, 1987.

Schwartz, C., and R. Mare. "Trends in Educational Assortative Marriage from 1940 to 2003." California Center for Population Research. http://computing.ccpr.ucla.edu/ccprwpseries/ccpr_017_05.pdf, 2005.

Schwartz, Pepper. *Peer Marriage: How Love Between Equals Really Works*. New York: Free Press, 1994.

Scott, Janny. "Low Birth Weight's High Cost." *Los Angeles Times* (December 24, 1990): 1.

Scott, Joan. "Gender: A Useful Category of Historical Analysis." *American Historical Review* 91 (1986): 1053–1075.

Seccombe, Karen. "Families in Poverty in the 1990's: Trends, Causes, Consequences, and Lessons Learned." *Journal of Marriage and the Family*, 62, 4 (November 2000): 1094–1113.

Sedikides, C., Oliver, M. B., and Campbell, W. K. "Perceived benefits and costs of romantic relationships for women and men: Implications for exchange theory." *Personal Relationships*, 1 (1994): 5–21.

Sellers, C., and M. Bromley. "Violent Behavior in College Student Dating Relationships." *Journal of Contemporary Criminal Justice* 12, 1 (1996): 1–27.

Seltzer, Judith. "Families Formed Outside of Marriage." *Journal of Marriage and the Family*, 62, 4 (November 2000): 1247–1268.

Sennett, Richard, and Jonathan Cobb. *The Hidden Injuries of Class*. New York: Vintage, 1972.

Serewicz, M. C. M., and E. Gale. "First-Date Scripts: Gender Roles, Context, and Relationship." *Sex Roles* 58 (2008): 149–164.

Sharpsteen, Don J. "Romantic Jealousy as an Emotion Concept: A Prototype Analysis." *Journal of Social and Personal Relationships* 10, 1 (1993): 69–82.

Shaver, Phillip, Cindy Hazan, and D. Bradshaw. "Love as Attachment: The Integration of Three Behavioral Systems." In *The Psychology of Love*, edited by R. Sternberg and M. Barnes. New Haven, CT: Yale University Press, 1988.

Shaw, A. "Psychology Professor Says 'Sexting' Is 'Normal.' " *The York Dispatch*, March 30, 2009.

Shellenbarger, S. "When 20-Somethings Move Back Home, It Isn't All Bad." *Wall Street Journal*, May 21, 2008.

———. "How Growing Up in a Recession Can Shape a Child's Future." *Wall Street Journal*, February 18, 2009, D1.

Shelton, Beth A., and Daphne John. "Does Marital Status Make a Difference? Housework among Married and Cohabiting Men and Women." *Journal of Family Issues* 14, 3 (September 1993): 401–420.

Shibley-Hyde, J., and S. Jaffee. "Becoming a Heterosexual Adult: The Experiences of Young Women." *Journal of Social Issues* 56, 2 (2000): 283–296.

Shields, B. *Down Came the Rain: My Journey through Postpartum Depression*. New York: Hyperion, 2005.

Shostak, Arthur B. "Tomorrow's Family Reforms: Marriage Course, Marriage Test, Incorporated Families, and Sex Selection Mandate." *Journal of Marital and Family Therapy* 7, 4 (October 1981): 521–528.

———. "Singlehood." In *Handbook of Marriage and the Family*, edited by M. Sussman and S. Steinmetz. New York: Plenum Press, 1987.

Silverstein, Meril, Xuan Chen, and Kenneth Heller. "Too Much of a Good Thing? Intergenerational Social Support and the Psychological Well-Being of Older Parents." *Journal of Marriage and the Family* 58, 4 (November 1996): 970–982.

Simenauer, J., and D. Carroll. *Singles: The New Americans*. New York: Simon and Schuster, 1982.

Simon, R. "The Joys of Parenthood, Reconsidered." *Contexts: Understanding people in their social worlds* 7, 2 (Spring 2008): 40–45.

Slater, Alan, Jan A. Shaw, and Joseph Duquesnel. "Client Satisfaction Survey: A Consumer Evaluation of Mediation and Investigative Services: Executive Summary." *Family and Conciliation Courts Review* 30, 2 (1992): 252–259.

"Sleeping on Back Saves 1,500 Babies." *San Mateo County Times* (June 25, 1996): A-4.

Sloan Work and Family Research Network. *Questions and Answers about Spillover: Negative Impacts.* Boston: Boston College Graduate School of Social Work, 2008a.

———. *Questions and Answers about Part-Time Work.* Boston: Boston College Graduate School of Social Work, 2009a.

———. *Questions and Answers about Shift Work.* Boston: Boston College Graduate School of Social Work, 2009b.

———. *Questions and Answers about Telework.* Boston: Boston College Graduate School of Social Work, 2008b.

———. *Questions and Answers about Women in the Workforce.* Boston: Boston College Graduate School of Social Work, 2008c.

Sluzki, Carlos. "The Latin Lover Revisited." In *Ethnicity and Family Therapy,* edited by M. McGoldrick et al. New York: Guilford Press, 1982.

Small, S. "Bridging Research and Practice in the Family and Human Sciences." *Family Relations* 54 (April 2005): 320–334.

Small, S. A., and Riley, D. "Toward a multidimensional assessment of work spillover into family life." *Journal of Marriage and the Family* 52 (1990): 51–61.

Small, Stephen, and Dave Riley. "Toward a Multidimensional Assessment of Work Spillover into Family Life." *Journal of Marriage and the Family* 52, 1 (February 1990): 51–61.

Smalley, S., and R. Kahn. "Violence Raging among Teen Girls." *Boston Globe,* June 20, 2005.

Smith, Carolyn, Marvin Krohn, Rebekah Chu, and Oscar Best. "African American Fathers: Myths and Realities about Their Involvement with Their Firstborn Children." *Journal of Family Issues* 26, 7 (October 2005): 975–1001.

Smith, Craig S. "A Love That Transcends Death Is Blessed by the State." *New York Times,* February 20, 2004.

Smith, D., and M. Hindus. "Premarital Pregnancy in America 1640–1971: An Overview and Interpretation." *Journal of Interdisciplinary History* 4 (Spring 1975): 537–570.

Smith, J. A. *The Daddy Shift: How Stay-at-Home Dads, Breadwinning Moms, and Shared Parenting Are Transforming American Families.* Boston: Beacon Press, 2009.

Smith, T. W. "American Sexual Behavior: Trends, Socio-Demographic Differences, and Risk Behavior." GSS Topical Report Number 25. 2006. http://cloud9.norc.unchicago.edu/dlib/t-25.htm.

Smith, T. W. "Changing Racial Labels: From Negro to Black to African American." *Public Opinion Quarterly* 56, 4 (1992): 496–514.

Smits, Jeroen, Wout Ultee, and Jan Lammers. "Effects of Occupational Status Differences Between Spouses on the Wife's Labor Force Participation and Occupational Achievement: Findings from 12 European Countries." *Journal of Marriage and the Family* 58, 1 (February 1996): 101–115.

Smock, P. J. "The Economic Costs of Marital Disruption for Young Women over the Past Two Decades." *Demography* 30, 3 (August 1993): 353–371.

Smock, P. J., L. M. Casper, and J. Wyse. "Nonmarital Cohabitation: Current Knowledge and Future Directions for Research." Population Studies Center Research Report, Institute for Social Research, University of Michigan, Ann Arbor, MI: July 2008.

Smock, Pamela J. "Cohabitation in the United States: An Appraisal of Research Themes, Findings, and Implications." *Annual Review of Sociology* 26, Summer 2000.

Smock, Pamela J. "The Wax and Wane of Marriage: Prospects for Marriage in the 21st Century." *Journal of Marriage and Family* 66, 4 (November 2004): 966–973.

Snarey, John, et al. "The Role of Parenting in Men's Psychosocial Development." *Developmental Psychology* 23, 4 (July 1987): 593–603.

Snyder, H. *Sexual Assault of Young Children as Reported to Law Enforcement: Victim, Incident, and Offender Characteristics.* Washington, DC: Bureau of Justice Statistics, U.S. Department of Justice, 2000.

Solomon, S. E., E. D. Rothblum, and K. F. Balsam. "Money, Housework, Sex and Conflict: Same-Sex Couples in Civil Unions, Those Not in Civil Unions, and Heterosexual Married Siblings." *Sex Roles* 52 (2005): 561–575.

Solot, Dorian, and Marshall Miller. "Common Law Marriage Fact Sheet." www.unmarried.org/common.html, August 2005.

Sommers-Flanagan, Rita, John Sommers-Flanagan, and Britta Davis. "What's Happening on Music and Television? A Gender Role Content Analysis." *Sex Roles* 28, 11–12 (June 1993): 745–753.

South, Scott, Katherine Trent, and Yang Shen. "Changing Partners: Toward a Macrostructural-Opportunity Theory of Marital Dissolution." *Journal of Marriage and Family* 63, 3 (August 2001): 743–754.

South, Scott. "Time Dependent Effects of Wives' Employment on Marital Dissolution." *American Sociological Review* 66 (April 2001): 226–245.

Spaht, Katherine. "Model Marriage Obligations Statute," marriagedebate.com.

Spanier, Graham B., and Linda Thompson. *Parting: The Aftermath of Separation and Divorce.* Rev. ed. Newbury Park, CA: Sage, 1987.

Spanier, Graham B., and R. L. Margolis. "Marital Separation and Extramarital Sexual Behavior." *Journal of Sex Research* 19 (1983): 23–48.

Spector, I. P., and M. P. Carey. "Incidence and Prevalence of the Sexual Dysfunctions—A Critical Review of the Empirical Literature." *Archives of Sexual Behavior* 19, 4 (August 1990): 389–408.

Speer, S. "The Interactional Organization of the Gender Attribution Process." *Sociology* 39, 1 (2005): 67–87.

Spitalnick, Josh, and Lily McNair. "Couples Therapy with Gay and Lesbian Clients: An Analysis of Important

Clinical Issues." *Journal of Sex & Marital Therapy* 31, 1 (January/February 2005): 43–56.

Spitze, Glenna, and John Logan. "Employment and Filial Relations: Is There a Conflict?" *Sociological Forum* 6 (December 1991): 681–698.

Spock, B., and M. Rothenberg. *Dr. Spock's Baby and Child Care: 40th Anniversary Edition.* New York: Pocket Books, 1985.

Sprecher, S., and D. Felmlee. "The Balance of Power in Romantic Heterosexual Couples over Time from 'His' and 'Her' Perspectives." *Sex Roles* 37, 5/6 (1997): 361–379.

St. Petersburg Times Special Report. "9/11 For the Record." (September 8, 2002). http://www.sptimes.com/2002/09/08/ 911/911__For_the_record.shtml.

Stacey, Judith, and Timothy Biblarz. "(How) Does the Sexual Orientation of Parents Matter?" *American Sociological Review* 66, 2 (2001): 159–183.

Stacey, Judith. "Good Riddance to the Family: A Response to David Popenoe," *Journal of Marriage and the Family* 55 (3) (August 1993): 545–547.

Stack, Carol B. *All Our Kin: Strategies for Survival in a Black Community.* New York: Harper and Row, 1974.

Stanley, S., and G. Rhoades. "'Sliding vs. Deciding': Understanding a Mystery." *National Council on Family Relations Report: Family Focus on Cohabitation* 54, 2 (Summer 2009): F1–4.

Steelman, Lala Carr, and Brian Powell. "The Social and Academic Consequences of Birth Order: Real, Artificial, or Both?" *Journal of Marriage and the Family* 47 (1985): 117–124.

Stein, P. J. "Singlehood: An Alternative to Marriage." *The Family Coordinator* 24, 4 (October 1975): 489–502.

Stein, Peter. "Men and Their Friendships." In *Men In Families,* edited by R. Lewis and R. Salt. Beverly Hills, CA: Sage, 1986: 261–270.

Steinberg, Laurence, and Susan Silverberg. "Marital Satisfaction in Middle Stages of Family Life Cycle." *Journal of Marriage and the Family* 49, 4 (November 1987): 751–760.

Steinhauer, J. "A Cul-de-Sac of Lost Dreams, and New Ones." *New York Times,* August 22, 2009.

Steinmetz, Suzanne, Sylvia Clavan, and K. Stein. *Marriage and Family Realities.* New York: Harper and Row, 1990.

Steinmetz, Suzanne. "Family Violence." In *Handbook of Marriage and the Family,* edited by M. Sussman and S. Steinmetz. New York: Plenum Press, 1987.

Sternberg, R. "Triangulating Love." in *The Altruism Reader,* edited by Thomas Oord. West Conshocken, PA, (2007): 331–347.

Sternberg, Robert, and Michael Barnes, eds. *The Psychology of Love.* New Haven, CT: Yale University Press, 1988.

Stevens, D. P., K. Minnotte, S. Mannon, and G. Kiger. "Examining the 'Neglected Side of the Work-Family Interface': Antecedents of Positive and Negative Family-to-Work Spillover." *Journal of Family Issues* 28, 2 (February 2007): 242–262.

Stevens, Daphne, Gary Kiger, and Pamela J. Riley. "Working Hard and Hardly Working: Domestic Labor and Marital Satisfaction among Dual-Earner Couples." *Journal of Marriage and Family* 63, 2 (May 2001): 514–526.

Stevens, Gillian, and Robert Schoen. "Linguistic Intermarriage in the United States." *Journal of Marriage and the Family* 50, 1 (February 1988): 267–280.

Stevenson, B., and J. Wolfers. "Marriage and Divorce: Changes and Their Driving Forces." *Journal of Economic Perspectives* 21, 2 (Spring 2007): 27–52.

———. "The Paradox of Declining Female Happiness." *American Economic Journal: Economic Policy* 1, 2 (August 2009): 190–225.

Stevenson, M. "Tolerance for Homosexuality and Interest in Sexuality Education." *Journal of Sex Education and Therapy* 16 (1990): 194–197.

Stewart, A. J., and C. McDermott. "Gender in Psychology." *Annual Review of Psychology* 55 (2004): 519–544.

Stewart, S. D. "Boundary Ambiguity in Stepfamilies." *Journal of Family Issues* 26 (2005a): 1002–1029.

———. "How the Birth of a Child Affects Involvement with Stepchildren." *Journal of Marriage and Family* 67 (2005b): 461–473.

———. *Brave New Stepfamilies: Diverse Paths toward Stepfamily Living.* Thousand Oaks, CA: Sage, 2007.

Stewart, S., W. D. Manning, and P. J. Smock. "Union Formation among Men in the U.S.: Does Having Prior Children Matter?" *Journal of Marriage and Family* 65, 1 (February 2003): 90–104.

Stewart, Susan. "Boundary Ambiguity in Stepfamilies." *Journal of Family Issues* 26, 7 (October 2005): 1002–1029.

Stewart, Susan. "How the Birth of a Child Affects Involvement with Stepchildren." *Journal of Marriage & Family* 67, 1 (May 2005): 461–473.

Stimpson, J. P., and F. A. Wilson. "Cholesterol screening by marital status and sex in the United States." *Prev Chronic Dis 2009;* 6(2). http://www.cdc.gov/pcd/issues/2009/.

Stockard, Janice. *Daughters of the Canton Delta: Marriage Patterns and Economic Strategies in South China, 1860–1930.* Stanford, CA: Stanford University Press, 1989.

Stockdale, M. S. "The Role of Sexual Misperceptions of Women's Friendliness in an Emerging Theory of Sexual Harassment." *Journal of Vocational Behavior* 42, 1 (February 1993): 84–101.

Stolzenberg, L., and S. J. D' Alessio. "The Effect of Divorce on Domestic Crime." *Crime and Delinquency* 52, 2 (April 2007): 281–302.

Storaasli, Ragnar D., and Howard J. Markman. "Relationship Problems in the Early Stages of Marriage: A Longitudinal Investigation." *Journal of Family Psychology* 4, 1 (September 1990): 80–98.

Strasburger, V. "Adolescents, Sex and the Media: Ooooo Baby, Baby: A Q and A." *Adolescent Medicine Clinics* 16 (2005): 269–288.

Straus, M. "Prevalence of Violence against Dating Partners by Male and Female University Students Worldwide." *Violence Against Women* 10, 7 (July 2004): 790–811.

———. "Dominance and Symmetry in Partner Violence by Male and Female University Students in 32 Nations." *Children and Youth Services Review* 30 (2008a): 252–275.

———. Ending Spanking Can Make a Major Contribution to Preventing Physical Abuse. *National Council on Family Relations Report: Family Focus on Child Abuse and Neglect,* Issue FF40 (December 2008b): F14–16.

Straus, M., and Ignacio Luis Ramirez. "Gender Symmetry in Prevalence, Severity, and Chronicity of Physical Aggression against Dating Partners by University Students in Mexico and USA. Paper presented at the 15th World Meeting of the International Society for Research on Aggression, Montreal, July 2002.

Straus, Murray A. "Physical Assaults by Wives: A Major Social Problem." In *Current Controversies in Family Violence,* edited by R. Gelles and D. Loseke. Newbury Park, CA: Sage, 1993.

Straus, Murray, and Carolyn Field. "Psychological Aggression by American Parents: National Data on Prevalence, Chronicity, and Severity." *Journal of Marriage and the Family* 65, 4 (November 2003): 795–808.

Straus, Murray, and Carrie Yodanis. "Corporal Punishment in Adolescence and Physical Assaults on Spouses Later in Life: What Accounts for the Link?" *Journal of Marriage and the Family* 58, 4 (November 1996): 825–841.

Straus, Murray, Richard Gelles, and Suzanne Steinmetz. *Behind Closed Doors.* Garden City, NY: Anchor Books, 1980.

Strohschein, Lisa. "Parental Divorce and Child Mental Health Trajectories." *Journal of Marriage and Family* 67, 5 (December 2005): 1286–1300.

Strom, Robert, Pat Collinsworth, Shirley Strom, and D. Griswold. "Strengths and Needs of Black Grandparents." *International Journal of Aging and Human Development* 36, 4 (1992–1993): 255–268.

Strong, Bryan, and Christine DeVault. *Human Sexuality: Diversity in Contemporary America,* 2nd ed. Mountain View, CA: Mayfield, 1997.

"Study Says Telecommunicating May Harm Workers Left Behind in the Office." *Inside Rensselaer* 2, 1 (January 17, 2008). http://www.rpi.edu/about/inside/issue/v2nl/golden.html.

Suitor, J. Jill. "Marital Quality and Satisfaction with Division of Household Labor." *Journal of Marriage and the Family* 53, 1 (February 1991): 221–230.

Sullivan, O., and S. Coltrane. "Men's Changing Contribution to Housework and Childcare: A Discussion Paper on Changing Family Roles." Paper presented at the 11th Annual Conference of the Council on Contemporary Families, University of Illinois, Chicago, April 25–26, 2008.

Sulloway, Frank J. *Born to Rebel; Birth Order, Family Dynamics, and Creative Lives.* New York: David McKay Company, 1996.

Sun, Yong Min. "Family Environment and Adolescents' Well-Being Before and After Parents' Marital Disruption: A Longitudinal Analysis." *Journal of Marriage and Family* 63, 3 (August 2001): 697–713.

Sun, Yongmin. "The Well-Being of Adolescents in Households With No Biological Parents." *Journal of Marriage & Family* 65, 4 (November 2003): 894–909.

Sung, B. L. Mountains of Gold. New York: Macmillan, 1967.

Surra, Catherine, P. Arizzi, and L. L. Asmussen. "The Association between Reasons for Commitment and the Development and Outcome of Marital Relationships." *Journal of Social and Personal Relationships* 5 (1988): 47–63.

Surra, Catherine. "Research and Theory on Mate Selection and Premarital Relationships in the 1980s." In *Contemporary Families: Looking Forward, Looking Back,* edited by A. Booth. Minneapolis: National Council on Family Relations, 1991.

"Surrogacy," ivf-infertility.com, 2005.

Sutfin, E., M. Fulcher, R. P. Bowles, and C. J. Patterson. "How Lesbian and Heterosexual Parents Convey Attitudes about Gender to Their Children: The Role of Gendered Environments." *Sex Roles* 58, 7–8 (April 2008): 501–513.

———. "From 1997 to 2007: Fewer Mothers Prefer Full-Time Work." http://pewresearch.org.

———. "Generation Gap in Values, Behaviors: As Marriage and Parenthood Drift Apart, Public Is Concerned about Social Impact." Pew Research Center. http://pewsocialtrends.org/assets/pdf/Marriage.pdf, 2007.

Swain, Scott. "Covert Intimacy: Closeness in Men's Friendships." In *Gender in Intimate Relationships: A Microstructural Approach,* edited by B. Risman and P. Schwartz. Belmont, CA: Wadsworth, 1989.

Sweeney, Megan M. "Remarriage and the Nature of Divorce: Does It Matter Which Spouse Chose to Leave?" *Journal of Family Issues* 23, 3 (April 2002): 410–440.

Sweeney, Megan, and Julie Phillips. "Understanding Racial Differences in Marital Disruption: Recent Trends and Explanations." *Journal of Marriage & Family* 66, 3 (August 2004): 639–650.

Sweet, James A., and Larry L. Bumpass. "The National Survey of Families and Households—Waves 1, 2, and 3: Data Description and Documentation." Center for Demography and Ecology, University of Wisconsin–Madison. (http://www.ssc.wisc.edu/nsfh/home.htm), 2002.

Sweet, J. A., L. L. Bumpass, and V. R. A. Call. *The Design and Content of the National Survey of Families and Households.* Working Paper NSFH-1. Madison: University of Wisconsin, Center for Demography and Ecology, 1988.

Swift, Carolyn. "Preventing Family Violence: Family-Focused Programs." In *Violence in the Home: Interdisciplinary Perspectives,* edited by M. Lystad. New York: Brunner/Mazel, 1986.

Symons, Donald. *The Evolution of Human Sexuality.* New York: Oxford University Press, 1979.

Szinovacz, Maximiliane, and Adam Davey. "Retirement and Marital Decision Making: Effects on Retirement Satisfaction." *Journal of Marriage & Family* 67, 2 (May 2005): 387–398.

Szinovacz, Maximiliane. "Family Power." In *Handbook of Marriage and the Family,* edited by M. Sussman and S. Steinmetz. New York: Plenum Press, 1987.

Takagi, Diana Y. "Japanese American Families." In *Minority Families in the United States: A Multicultural Perspective,* edited by R. L. Taylor. Englewood Cliffs, NJ: Prentice Hall, 1994.

Tanfer, K., and L. dc Cubbins, "Coital Frequency among Single Women: Normative Constraints and Situational Opportunities." *Journal of Sex Research* 29, 2 (1992): 221–250.

Tannen, Deborah. "Sex, Lies and Conversation: Why Is It So Hard for Men and Women to Talk to Each Other?" *Washington Post* (June 24, 1990).

Tannen, Deborah. *You Just Don't Understand: Women and Men in Conversation.* New York: Morrow, 1990.

Tanner, Litsa, et al. "Images of Couples and Families in Animated Disney Movies." *American Journal of Family Therapy* 31 (2003): 355–374.

Tashiro, Ty, and Patricia Frazier. "'I'll Never Be in a Relationship Like That Again': Personal Growth Following Romantic Relationship Breakups." *Personal Relationships* 10, 1 (March 2003): 113–128.

Taylor, P., C. Funk, P. Craighill, and C. Kennedy. "As Family Forms Change, Bonds Remain Strong: Families Drawn Together by Communication Revolution." Pew Research Center. http://pewresearch.org/assets/social/pdf/FamilyBonds.pdf.

Taylor, Paul, Cary Funk, and Peyton Craighill. "Who's Feeling Rushed? (Hint: Ask a Working Mom)." Pew Research Center. http://pewsocialtrends.org/assets/pdf/Rushed.pdf, 2006.

Taylor, Robert J. "Black American Families." In *Minority Families in the United States: A Multicultural Perspective,* edited by R. L. Taylor. Englewood Cliffs, NJ: Prentice Hall, 1994a.

———. "Minority Families in America." In *Minority Families in the United States: A Multicultural Perspective,* edited by R. L. Taylor. Englewood Cliffs, NJ: Prentice Hall, 1994b.

Taylor, Robert J., Linda M. Chatters, Belinda Tucker, and Edith Lewis. "Developments in Research on Black Families." In *Contemporary Families: Looking Forward, Looking Back,* edited by A. Booth. Minneapolis: National Council on Family Relations, 1991.

Teachman, Jay D., and Karen A. Polonko. "Cohabitation and Marital Stability in the United States." *Social Forces* 69, 1 (September 1990): 207–220.

Teachman, Jay D., R. Vaughn, A. Call, and Karen P. Carver. "Marital Status and Duration of Joblessness Among White Men." *Journal of Marriage and the Family* 56, 2 (May 1994): 415–428.

"Teen Dad Terrell Pough Killed," People.com, 2005.

Tejada-Vera, B., and P. D. Sutton. "Births, Marriages, Divorces, and Deaths: Provisional Data for 2008." *National Vital Statistics Reports,* vol. 57, no. 19. Hyattsville, MD: National Center for Health Statistics, 2009.

Telework Research Network. "The Latest Telecommuting Statistics." August 2, 2009.

Ten Kate, N. "Choosing and Using Contraception." *American Demographics* 20, 6 (June 1998).

Tessina, Tina. *Gay Relationships: For Men and Women. How to Find Them, How to Improve Them, How to Make Them Last.* Los Angeles: Jeremy P. Tarcher, 1989.

Testa, Ronald J., Bill N. Kinder, and G. Ironson. "Heterosexual Bias in the Perception of Loving Relationships of Gay Males and Lesbians." *Journal of Sex Research* 23, 2 (May 1987): 163–172.

Thayer, Leo. *On Communication.* Norwood, NJ: Ablex, 1986.

"The 'Cinderella Effect': Elevated Mistreatment of Stepchildren in Comparison to Those Living with Genetic Parents." http://psych.mcmaster.ca/dalywilson/cinderella%20effect%20facts.pdf, 2009.

The Employment Situation—July 2009, http://www.bls.gov/cps.

Thoits, Peggy A. "Identity Structures and Psychological Well-Being: Gender and Marital Status Comparisons." *Social Psychology Quarterly* 55, 3 (1992): 236–256.

Thompson, Elizabeth, Jane Mosley, Thomas Hanson, and Sara McLanahan. "Remarriage, Cohabitation, and Changes in Mothering Behavior." *Journal of Marriage and Family* 63 (May 2001): 370–380.

Thompson, Anthony. "Emotional and Sexual Components of Extramarital Relations." *Journal of Marriage and the Family* 46, 1 (February 1984): 35–42.

Thompson, Linda, and Alexis J. Walker. "Gender in Families: Women and Men in Marriage, Work, and Parenthood." *Journal of Marriage & Family* 51, 4 (November 1989): 845–871.

Thompson, Linda, and Alexis J. Walker. "The Place of Feminism in Family Studies." *Journal of Marriage & Family* 57, 4 (November 1995): 847–865.

Thompson, Linda. "Family Work: Women's Sense of Fairness," *Journal of Family Issues* 12 (2) June 1991: 181–196.

———. "Conceptualizing Gender in Marriage: The Case of Marital Care." *Journal of Marriage and the Family* 55, 3 (August 1993): 557–569.

Thomson, E., J. Mosley, T. L. Hanson, and S. S. McLanahan. "Remarriage, Cohabitation, and Changes in Mothering Behavior." *Journal of Marriage and Family* 63 (2001): 370–380.

Thornton, Arland. "Changing Attitudes toward Family Issues in the United States." *Journal of Marriage and the Family* 51, 4 (November 1989): 873–893.

Tichenor, V. *Earning More and Getting Less: Why Successful Wives Can't Buy Equality.* New Brunswick, NJ: Rutgers University Press, 2005a.

———. "Maintaining Men's Dominance: Negotiating Identity and Power When She Earns More." *Sex Roles* 53, 3/4 (August 2005b).

Tienda, Marta, and Jennifer Glass. "Household Structure and Labor Force Participation of Black, Hispanic, and White Mothers." *Demography* 22 (1985): 281–394.

Tienda, Marta, and Ronald Angel. "Headship and Household Composition among Blacks, Hispanics, and Other Whites." *Social Forces* 61 (1982): 508–531.

Ting-Toomey, Stella. "An Analysis of Verbal Communication Patterns in High and Low Marital Adjustment Groups." *Human Communications Research* 9, 4 (June 1983): 306–319.

Toews, M. L., B. S. Catlett, and P. C. McKenry. "Women's Use of Aggression during Marital Separation." *Journal of Divorce and Remarriage* 42 (2005): 1–14.

Toews, Michelle, Beth Catlett, and Patrick McKenry. "Women's Use of Aggression Toward Their Former Spouses During Marital Separation." *Journal of Divorce and Remarriage* 42, (2005): 1–14.

Tolman, Richard M. "Treatment Program for Men Who Batter." In *Vision 2010: Families and Violence, Abuse and Neglect,* edited by R. J. Gelles. Minneapolis: National Council on Family Relations, 1995.

Topham, Glade L., Jeffry H. Larson, and Thomas B. Holman. "Family-of-Origin Predictors of Hostile

Conflict in Early Marriage." *Contemporary Family Therapy* 27, 1 (March 2005): 101–121.

Torpey, E. M. "Flexible Work Adjusting the When and Where of Your Job." *Occupational Outlook Quarterly*, Summer 2007.

Toufexis, Anastasia. "Older—But Coming on Strong." *Time* (February 22, 1988): 76–79.

Tracy, Kathleen. *The Secret Story of Polygamy.* Naperville, IL: Sourcebooks, 2002.

Tran, Than Van. "The Vietnamese American Family." In *Ethnic Families in America: Patterns and Variations*, 3rd ed., edited by C. H. Mindel et al. New York: Elsevier, 1988.

Treas, Judith, and Deirdre Giesen, "Sexual Infidelity Among Married and Cohabiting Americans." *Journal of Marriage and the Family* 62, 1 (2000): 48–60.

Treas, Judith, and Vern L. Bengtson. "The Family in Later Years." In *Handbook of Marriage and the Family*, edited by M. Sussman and S. Steinmetz. New York: Plenum Press, 1987.

Treichler, Paula. "Feminism, Medicine, and the Meaning of Childbirth." In *Body/Politics: Women and the Discourses of Science*, edited by Mary Jacobus, Evelyn Fox Keller, and Sally Shuttleworth. New York: Routledge, 1990: 113–138.

Trejos, N. "Back Home to Roost: Some Make Do in the Recession by Moving in with Mom and Dad." *Washington Post*, April 26, 2009.

Tremblay, D-G., R. Paquet, and E. Najem. "Telework: A Way to Balance Work and Family or an Increase in Work-Family Conflict?" *Canadian Journal of Communication* 31, 3 (2006).

Troiden, Richard. *Gay and Lesbian Identity: A Sociological Analysis.* New York: General Hall, 1988.

Troll, L. E. "Family Embedded vs. Family-Deprived Oldest-Old: A Study of Contrasts." *International Journal of Aging and Human Development* 38 (1994), 51–63.

Troll, Lillian. "The Contingencies of Grandparenting." In *Grandparenthood*, edited by V. Bengtson, and J. Robertson. Beverly Hills, CA: Sage 1985.

Tucker, Judith Stadtman. "The Mother and the Magazine." *The Mothers Movement Online*, mothersmovement.org, 2003.

Tucker, Raymond K., M. G. Marvin, and B. Vivian. "What Constitutes a Romantic Act." *Psychological Reports* 89, 2 (October 1991): 651–654.

Turner, R. Jay, and William R. Avison. "Assessing Risk Factors for Problem Parenting: The Significance of Social Support." *Journal of Marriage and the Family* 47, 4 (November 1985): 881–892.

Turner, Robert L., and M. E. Fakouri. "Androgyny and Differences in Fantasy Patterns." *Psychological Reports* 73, 3 (1993): 1164–1166.

Turnto23.com. "Bakersfield News: Family Goes Homeless Due to Recession." http://www.turnto23.com/news/18788458/detail.html.

Twohey, Denise, and Micheal Ewing. "The Male Voice of Emotional Intimacy." *Journal of Mental Health Counseling* 17, 1 (January 1995): 54–63.

Tyre, P. "Daddy's Home, and a Bit Lost." *New York Times*, January 9, 2009.

Tyre, Peg. "The Trouble with Boys." *Newsweek* (January 10, 2006).

Tyre, Peg. "The Trouble With Boys." *Newsweek* (January 30, 2006): 44–51.

U.S. Bureau of Justice Statistics, *Violence by Intimates: Analysis of Data on Crimes by Current or Former Spouses, Boyfriends, and Girlfriends.* Washington, DC: Department of Justice, 1998.

U.S. Bureau of Labor Statistics. A *Profile of the Working Poor: 2003.* Report 983. Washington, D.C. (March 2005).

U.S. Census Bureau. Table 13.2 Total Money Income in 2003 of Families by Type, and by Hispanic Origin Type of Householder: 2004. Washington, D.C. : U.S. Census Bureau, Population Division, Ethnic & Hispanic Statistics Branch Washington, DC: U.S. Government Printing Office, 160–171, 1989.

———. "2005 American Community Survey," www.census.gov, 2005a.

———. "Annual Social and Economic Supplement," Current Population Reports, Series P20-553. Washington, DC: U.S. Government Printing Office, 2003a.

———. "Annual Social and Economic Supplement." Current Population Reports. Washington, DC: U.S. Government Printing Office, 2004a.

———. "Census of Population and Housing Characteristics of American Indians and Alaska Natives by Tribe and Language: 2000, PHC-5. Washington, DC: U.S. Government Printing Office, 2003b.

———. "Facts for Features: Unmarried and Single Americans Week," www.census.gov, 2005b.

———. "Hispanic Population in the United States: 2004 March CPS," www.census.gov, 2005c.

———. Income, Poverty, and Health Insurance Coverage in the United States: 2004. Report P60, n. 229. Washington, DC: U.S. Government Printing Office, Table B-2, pp. 52–57, 2004b.

———. Statistical Abstract of the United States. Washington, DC: U.S. Government Printing Office, 2006.

U.S. Census Bureau News. "Families and Living Arrangements: Americans Marrying Older, Living Alone More, See Households Shrinking, Census Bureau Report (2006) http://www.census.gov/Press-Release/www/releases/ archives/families_households/006840.html

———. "Race and Hispanic or Latino." U.S. Census Bureau, Census 2000 Summary File 1, Matrices P3, P4, PCT4, PCT5, PCT8, and PCT11, Washington, DC: U.S. Government Printing Office 2000a.

———. *Infant Mortality.* International Data Base, 1996a.

———. *Statistical Abstract of the United States.* Washington, DC: U.S. Government Printing Office, 1996b.

———. *Statistical Abstract of the United States.* Washington, DC: U.S. Government Printing Office, 2000b.

———. *Statistical Abstract of the United States.* Washington, DC: U.S. Government Printing Office, 2001b.

———. *Statistical Abstract of the United States.* Washington, DC: U.S. Government Printing Office, 2002.

———. *Statistical Abstract of the United States.* Washington, DC: U.S. Government Printing Office, 2003c.

U.S. Census Bureau, *Current Population Report,* 1998. www.census.gov.

U.S. Census Bureau, Annual Social and Economic (ASEC) Supplement, U.S. Department of Agriculture Center for Nutrition Policy and Promotion. "Expenditures on Children by Families," Miscellaneous Publication Number 1528, 2007.

U.S. Census Bureau, Historical Abstracts of the United States, Colonial Times to 1970, Series A. "Marital Status of the Population 15 Years and Over, by Sex and Race: 1950 to Present," MS-1. Washington, DC: U.S. Government Printing Office, 2001a.

U.S. Census Bureau, Survey of Income and Program Participation. "Children in Self-Care, by Age of Child, Employment Status of Mother, and Selected Characteristics of Children Living with Mother," 2004.

U.S. Census Bureau. "American Community Survey, SO201 3–88," 2007.

U.S. Census Bureau. "American Community Survey, SO201," 2007.

U.S. Census Bureau. "American Community Survey," 2005.

U.S. Census Bureau. "Current Population Survey Reports, America's Families and Living Arrangements," 2008.

U.S. Census Bureau. "Current Population Survey, June 1976–2006," table 1.

U.S. Census Bureau. "Facts for Features: Asian Pacific-American Heritage Month," 2009.

U.S. Census Bureau. "Population Profile: American Indian," 2007.

U.S. Census Bureau. "Survey of Income and Program Participation," 2004.

U.S. Census Bureau. "Who's Minding the Kids? Child Care Arrangements: Summer 2006 Detailed Tables—2006."

U.S. Census Bureau. *Statistical Abstract of the United States: 2009,* 128th ed. http://www.census.gov/statab/www.

U.S. Census Bureau. http://census.gov/pop/socdemo/hh-fam/cps2008/tabA1-all.xls.

U.S. Census Bureau. Population Division, Table 4, "Projections of the Population by Sex, Race and Hispanic Origin for the United States: 2010–2050," NP2008-T4.

U.S. Census Bureau. Table MS-2. http://www.census.gov/population/socdemo/hh-fam/ms2.xls.

U.S. Census Bureau. Women's History Month: March 2008, January 2, 2008.

U.S. Department of Agriculture, Center for Nutrition Policy and Promotion. http://www.cnpp.usda.gov, 2008.

U.S. Department of Commerce, Economics and Statistics Administration. "Living Arrangements of Children: 2004," February 2008.

U.S. Department of Education. Institute of Education Sciences, National Center for Education Statistics.

U.S. Department of Health and Human Services, Administration for Children and Families. "Child Maltreatment 2002."http://www.acf.hhs.gov/programs/cb/pubs/cm02/index.htm, 2002

U.S. Department of Justice Office of Justice Programs, Office of Juvenile Justice and Delinquency Prevention. "Girls Study Group Understanding and Responding to Girls' Delinquency," May 2008.

U.S. Department of Justice. "Sexual Offenses and Offenders." Washington, DC: Bureau of Justice Statistics, 1997.

U.S. Department of Labor, Bureau of Labor Statistics. "A Profile of the Working Poor, 2006." Report 1006, August 2008.

———. "Employment Situation Summary," March 6, 2009.

U.S. Department of Labor, Employment Standards Administration, Wage and Hour Division. "Fact Sheet #28: The Family and Medical Leave Act of 1993," January 2009.

U.S. Department of Labor. "The Family and Medical Leave Act of 1993," Public Law 103-3, www.dol.gov/esa/regs/statutes/whd/fmla.htm, 1993.

U.S. Supreme Court, *Loving v. Virginia* 388 U.S. 1 (1967).

Udry, J. Richard. *The Social Context of Marriage.* Philadelphia: Lippincott, 1974.

Uecker, J. E., and C. E. Stokes. "Early Marriage in the United States." *Journal of Marriage and Family* 70 (2008): 835–846.

Ulman, A., and M. Straus. "Violence by Children against Mothers in Relation to Violence between Parents and Corporal Punishment by Parents." *Journal of Comparative Family Studies* 34 (2003): 41–60.

UNAIDS. "2008 Report on the Global AIDS Epidemic." http://www.unaids.org, 2008.

United Nations. "Demographic Yearbook 2006, Series R, No. 37. Table 25." http://unstats.un.org/unsd/pubs, 2006.

United Nations. Demographic Yearbook, 2002. New York: United Nations, 2005.

"University Students Worldwide." *Violence Against Women* 10, 7 (July 2004): 790–811.

Vaillant, Caroline O., and George E. Vaillant. "Is the U-Curve of Marital Satisfaction an Illusion? A 40–Year Study of Marriage." *Journal of Marriage and the Family* 55, 1 (February 1993): 230–240.

Valentine, Deborah. "The Experience of Pregnancy: A Developmental Process." *Family Relations* 31, 2 (April1982): 243–248.

Van Buskirk. "Soap Opera Sex: Tuning In, Tuning Out." *US* (August 1992): 64–67.

Van Horn, K. Roger, Angela Arnone, Kelly Nesbitt, Laura Desilets, Tanya Sears, Michelle Giffin, and Rebecca Brudi "Physical Distance and Interpersonal Characteristics in College Students' Romantic Relationships." *Personal Relationships* 4, 1 (1997): 25–34.

Van Laningham, J., D. Johnson, and P. Amato. "Marital Happiness, Marital Duration, and the U-Shaped Curve: Evidence from a Five-Wave Panel Study." *Social Forces* 79, 4 (June 2001): 1313–1341.

Vande Berg, Leah R., and Diane Streckfuss. "Prime-Time Television's Portrayal of Women and the World of Work: A Demographic Profile." *Journal of Broadcasting and Electronic Media* (March 1992): 195–207.

Vaquera, Elizabeth, and Grace Kao. "Private and Public Displays of Affection among Interracial and Intra-Racial Adolescent Couples." *Social Science Quarterly* 86, 2 (June 2005): 484–508.

Vaselle-Augenstein, Renata, and Annette Ehrlich. "Male Batterers: Evidence for Psychopathology." In *Intimate Violence: Interdisciplinary Perspectives,* edited by E. C. Viano. Washington, DC: Hemisphere, 1992.

Vasquez-Nuttall, E., et al. "Sex Roles and Perceptions of Femininity and Masculinity of Hispanic Women: A Review of the Literature." *Psychology of Women Quarterly* 11 (1987): 409–426.

Vasta, Ross. "Physical Child Abuse: A Dual-Component Analysis." *Developmental Review* 2, 2 (June 1992): 125–149.

Vaughan, Diane. *Uncoupling: Turning Points In Intimate Relationships.* New York: Vintage Reprint Edition, 1990.

Vaughan, Diane. *Uncoupling: Turning Points in Intimate Relationships.* New York: University Press, 1986.

Veevers, Jean. *Childless by Choice.* Toronto: Butterworth, 1980.

Vega, William. "Hispanic Families." In *Contemporary Families: Looking Forward, Looking Back,* edited by A. Booth. Minneapolis: National Council on Family Relations, 1991.

Vemer, Elizabeth, et al. "Marital Satisfaction in Remarriage: A Meta-analysis." *Journal of Marriage and the Family* 53, 3 (August 1989): 713–726.

Ventura, Jacqueline N. "The Stresses of Parenthood Reexamined." *Family Relations* 36, 1 (January 1987): 26–29.

Ventura, S. "Changing Patterns of Nonmarital Childbearing in the United States." NCHS Data Brief, No. 18, May 2009.

Ventura, S., J. Abma, W. Mosher, and S. K. Henshaw. "Estimated Pregnancy Rates by Outcome for the United States, 1990–2004." *National Vital Statistics Reports,* vol. 56, no. 15. Hyattsville, MD: National Center for Health Statistics, 2008.

Vermont Guide to Civil Unions. http://www.sec.state.vt.us/otherprg/civilunions/civilunions.html, 2005.

Vermont Secretary of State. "The Vermont Guide to Civil Unions," sec.state.vt.us/otherprg/civilunions/civilunions .html, August 2006.

Visher, Emily B., and John S. Visher. *Stepfamilies: A Guide to Working with Stepparents and Stepchildren.* New York: Brunner/Mazel, 1979.

———. *How to Win as a Stepfamily.* New York: Brunner/Mazel, 1991.

Voeller, Bruce. "Society and the Gay Movement." In *Homosexual Behavior,* edited by J. Marmor. New York: Basic Books, 1980.

Vogel, D., A., M. A. Lake, and S. Evans. "Children's and Adults' Sex-Stereotyped Perceptions of Infants." *Sex Roles* 24 (1991): 605–616.

Vogel, D., and B. Karney. "Demands and Withdrawals in Newlyweds: Elaborating on the Social Structure Hypothesis." *Journal of Social and Personal Relationships* 19, 5 (October 2002): 685–701.

Vogel, David L., Stephen R. Wester, and Martin Heesacker. "Dating Relationships and the Demand–Withdraw Pattern of Communication." *Sex Roles* 41, 3–4 (1999): 297–306.

Voydanoff, Patricia, and Brenda Donnelly. "Work and Family Roles and Psychological Distress." *Journal of Marriage and the Family* 51, 4 (November 1989): 933–941.

———. "Economic Distress and Family Relations: A Review of the Eighties." In *Contemporary Families: Looking Forward, Looking Back,* edited by A. Booth. Minneapolis: National Council on *Family Relations,* 1991.

Voydanoff, Patricia. *Work and Family Life.* Newbury Park, CA: Sage, 1987.

Vredevelt, P. *Empty Arms: Emotional Support for Those Who Have Suffered Miscarriage or Stillbirth.* Sisters, OR: Questar, 1994.

Wade, Terrance J., and David J. Pevalin. "Marital Transitions and Mental Health." *Journal of Health and Social Behavior* 45, 2 (June 2004): 155–170.

Wagner, Cynthia G. "Homosexual Relationships." *Futurist* 40, 3 (May/June 2006): 6.

Waite, Linda, and Maggie Gallagher. *The Case for Marriage: Why Married People are Happier, Healthier, and Better off Financially.* New York: Doubleday, 2000; Broadway Books, 2001.

Walker, Karen. "Men, Women, and Friendship: What They Say, What They Do," *Gender and Society* 8, 2 (June 1994).

Walker, Lenore. *The Battered Woman Syndrome.* New York: Harper Colophon, 1979.

Walker, L. "The Battered Woman Syndrome Is a Psychological Consequence of Abuse." In *Current Controversies in Family Violence,* edited by R. Gelles and D. Loseke. Newbury Park, CA: Sage Publications, 1993.

Waller, Willard, and Reuben Hill. *The Family: A Dynamic Interpretation.* New York: Dryden Press, 1951.

Wallerstein, J. Children of Divorce: The Psychological Tasks of the Child." *American Journal of Orthopsychiatry* 53, 2 (April 1983): 230–243.

Wallerstein, Judith (with Julia Lewis, Sandra Blakeslee). *The Unexpected Legacy Of Divorce: A 25 Year Landmark Study.* New York : Hyperion, 2000.

Wallerstein, Judith, and Joan Kelly. "Effects of Divorce on the Visiting Father-Child Relationship." *American Journal of Psychiatry* 137, 12 (December 1980a): 1534–1539.

———. *Surviving the Breakup: How Children and Parents Cope with Divorce.* New York: Basic Books, 1980b.

Wallerstein, Judith, and Sandra Blakeslee. *Second Chances: Men, Women, and Children a Decade after Divorce.* New York: Ticknor & Fields, 1989.

———. *Second Chances: Men, Women, and Children a Decade after Divorce* (paperback, revised). New York: Houghton Mifflin, 1995.

Wallerstein, Judith, Julia Lewis, and Sandra Blakeslee. *The Unexpected Legacy of Divorce: A Twenty-Five Year Landmark Study.* New York: Hyperion Press, 2000.

Walsh, J., and M. Ward. "Adolescent Gender Role Portrayals in the Media: 1950 to the Present." in *The Changing Portrayal of Adolescents in the Media since 1950*, edited by P. Jamieson and D. Romer. New York: Oxford University Press, 2008.

Walzer, Susan, and Thomas P. Oles. "Accounting for Divorce: Gender and Uncoupling Narratives." *Qualitative Sociology* 26, 3 (Fall 2003): 331–349.

Walzer, Susan. *Thinking About the Baby: Gender and Transitions into Parenthood*. Philadelphia: Temple University, 1998.

Wang, C., and J. Holton. *Economic Impact Study: Total Estimated Cost of Child Abuse and Neglect in the United States*. Prevent Child Abuse America, Chicago, IL. 2007.

Wang, H., and P. Amato. "Predictors of Divorce Adjustment: Stressors, Resources, and Definitions." *Journal of Marriage and the Family* 62 (August 2000): 655–668.

Wang, T., W. Parish, E. Laumann, and Y. Luo. "Partner Violence and Sexual Jealousy in China: A Population-Based Survey." *Violence Against Women* 15, 7 (2009): 774–798.

Ward, M., and K. Friedman. "Using TV as a Guide: Associations between Television Viewing and Adolescents' Sexual Attitudes and Behavior." *Journal of Research on Adolescence* 16, 1 (March 2006): 105–131.

Wareham, J., D. P. Boots, and J. Chavez. "A Test of Social Learning and Intergenerational Transmission among Batterers." *Journal of Criminal Justice* 37, 2 (March–April 2009): 163–173.

Warren, Jennifer A., and Phyllis J. Johnson. "The Impact of Workplace Support on Work-Family Role Strain." *Family Relations* 44, 2 (April 1995): 163–169.

Watkins, William G., and Arnon Bentovim. "The Sexual Abuse of Male Children and Adolescents: A Review of Current Research." *Journal of Child Psychology and Psychiatry and Allied Disciplines* 33, 1 (January 1992): 197–248.

Weeks, Jeffrey. *Sexuality and Its Discontents*. London: Routledge, 1985.

Wei, G. S., and J. E. Herbers. "Reporting Elder Abuse: A Medical, Legal, and Ethical Overview." *Journal of the American Medical Women's Association* 59, 4 (2004): 248–254.

Weinberg, M. S., and C. J. Williams. *Male Homosexuals: Their Problems and Adaptations*. New York: Penguin, 1974.

Weinberg, Martin S., C. J. Williams, and Douglas W. Pryor. *Dual Attraction: Understanding Bisexuality*. New York: Oxford University Press, 1994.

Weiss, David. "Open Marriage and Multilateral Relationships: The Emergence of Nonexclusive Models of the Marital Relationship." In *Contemporary Families and Alternative Lifestyles*, edited by E. Macklin and R. Rubin. Beverly Hills, CA: Sage, 1983.

Weiss, Robert. *Marital Separation*. New York: Basic Books, 1975.

Weitzman, Lenore. *The Marriage Contract: Spouses, Lovers, and the Law*. New York: Macmillan, 1981.

———. *The Divorce Revolution: The Unexpected Social and Economic Consequences for Women and Children in America*. New York: Free Press, 1985.

Weitzman, Susan. *Not to People Like Us: Hidden Abuse in Upscale Marriages*. New York: Basic Books, 2000.

Weizman, R., and J. Hart. "Sexual Behavior in Healthy Married Elderly Men." *Archives of Sexual Behavior* 16, 1 (February 1987): 39–44.

Wellman, B., A. Smith, A. Wells, and T. Kennedy. "Networked Families. Pew Internet and American Life Project." (October 2008). http://www.pewinternet.org/Reports/2008/Networked-Families.aspx.

Werner, Carol M., Barbara B. Brown, Irwin Altman, and Brenda Staples. "Close Relationships in Their Physical and Social Contexts: A Transactional Perspective." *Journal of Social and Personal Relationships* 9, 3 (1992): 411–431.

West, Candace, and Don Zimmerman. "Doing Gender." *Gender and Society* 1 (1987): 125–151.

Westberg, Heather, Thorana S. Nelson, and Kathleen W. Piercy. "Disclosure of Divorce Plans to Children: What the Children Have To Say." *Contemporary Family Therapy* 24, 4 (December 2002): 525–542.

Westman, M., D. Etzion, and S. Horovitz. "The Toll of Unemployment Does Not Stop with the Unemployed." *Human Relations* 57, 7 (July 2004): 823–844.

Whealin, Julia. "Men and Sexual Trauma: A National Center for PTSD Fact Sheet," ncptsd.va.gov, 2006.

Whisman, Mark, Amy Dixon, and Benjamin Johnson. "Therapists' Perspectives of Couple Problems and Treatment Issues in Couple Therapy." *Journal of Family Psychology* 11, 3 (September 1997): 361–366.

Whitbourne, Susan, and Joyce Ebmeyer. *Identity and Intimacy in Marriage: A Study of Couples*. New York: Springer-Verlag, 1990.

White, G., and P. Mullen. *Jealousy: Theory Research and Clinical Strategies*. New York: Guilford, 1989.

White, Jacquelyn W. "Feminist Contributions to Social Psychology." *Contemporary Social Psychology* 17, 3 (September 1993): 74–78.

White, James, and D. Klein. *Family Theories*, 2nd ed. Thousand Oaks, CA: Sage, 2002.

White, Joseph, and Thomas Parham. *The Psychology of Blacks: An African-American Perspective*, 2nd ed. Englewood Cliffs, NJ: Prentice Hall, 1990.

White, Lynn, and Alan Booth. "The Transition to Parenthood and Marital Quality." *Journal of Family Issues* 6 (1985): 435–449.

———. "Divorce over the Life Course: The Role of Marital Happiness." *Journal of Family Issues* 12, 1 (March 1991): 5–22.

White, Lynn, and Bruce Keith. "The Effect of Shift Work on the Quality and Stability of Marital Relations." *Journal of Marriage and the Family* 52, 2 (May 1990): 453–462.

White, Lynn. "Determinants of Divorce." In *Contemporary Families: Looking Forward, Looking Back*, edited by A. Booth. Minneapolis: National Council on Family Relations, 1991.

Whitehead, B., and D. Popenoe, "The State of Our Unions: 2001." New Brunswick, NJ: National Marriage Project at Rutgers University, June 27, 2001.

Whitehead, Barbara Dafoe, and David Popenoe. "Social Indicators of Marital Health and Well Being." The State of Our Unions, 2004: The Social Health of Marriage in

America. The National Marriage Project. New Brunswick, NJ: Rutgers University, 2004.

Whitehead, Barbara Dafoe, and David Popenoe. "Social Indicators of Marital Health and Well Being." The State of Our Unions, 2005: The Social Health of Marriage in America. The National Marriage Project. New Brunswick, NJ: Rutgers University, 2005.

Whitehead, Barbara Dafoe. "Dan Quayle Was Right," *Atlantic Monthly.* April 1993: 47–84.

———. *The Divorce Culture.* New York: Knopf, 1997.

Whitehead, Barbara Dafoe. *The Divorce Culture: Rethinking Poor Commitments to Marriage and the Family.* New York: Knopf Publishing Group, 1996.

Whiteman, S., S. McHale, and A. Crouter. "Longitudinal Changes in Marital Relationships: The Role of Offspring's Pubertal Development." *Journal of Marriage and Family* 69 (2007): 1005–1020.

"Who's Who in the Family Wars: A Characterization of the Major Ideological Factions," In *Feuds About Families: Conservative, Centrist, Liberal, and Feminist Perspectives,* edited by N. Benokraitis. Upper Saddle River, NJ: Prentice Hall, 2000: 2–13.

Widmer, Eric D., Judith Treas, and Robert Newcomb. "Attitudes toward Nonmarital Sex in 24 Countries." *Journal of Sex Research* 35 (November 1998): 349–358.

Wilcox, Allen, et al. "Incidents of Early Loss of Pregnancy." *New England Journal of Medicine* 319, 4 (July 28, 1988): 189–194.

Wildsmith, E., and R. K. Raley. "Race-ethnic differences in nonmarital fertility: A focus on Mexican American women." *Journal of Marriage and Family* 68, 2 (2006): 491–508.

Wilkie, Colleen F., and Elinor W. Ames. "The Relationship of Infant Crying to Parental Stress in the Transition to Parenthood." *Journal of Marriage and the Family* 48, 3 (August 1986): 545–550.

Wilkinson, Doris Y. "American Families of African Descent." In *Families in Cultural Context,* edited by M. K. DeGenova. Mountain View, CA: Mayfield, 1997.

Wilkinson, Doris, et al., eds. "Transforming Social Knowledge: The Interlocking of Race, Class, and Gender." *Gender and Society* [Special issue] (September 1992).

Willetts, Marion. "An Exploratory Investigation of Heterosexual Licensed Domestic Partners." *Journal of Marriage and the Family* 65, 4 (November 2003): 939–952.

Williams, John, and Arthur Jacoby. "The Effects of Premarital Heterosexual and Homosexual Experience on Dating and Marriage Desirability." *Journal of Marriage and the Family* 51 (May 1989): 489–497.

Willing, Richard. "Research Downplays Risk of Cousin Marriages." *USA Today* (April 4, 2002).

Wills, J. B. and B. J. Risman. "The Visibility of Feminist Thought in Family Studies." *Journal of Marriage and the Family.* 68, 3 (2006): 690–700.

Wilson, J. Q. *The Marriage Problem: How Our Culture Has Weakened Families.* New York: HarperCollins, 2002.

Wilson, K. J. *When Violence Begins at Home: A Comprehensive Guide to Understanding and Ending Abuse.* Salt Lake City, UT: Publishers Press, 1997.

Wilson, L. "Sexting and Minors: Child Pornography, Not Child's Play." 2008. http://blog.lawinfo.com/2008/10/08/sexting-and-minors-child-pornography-not-childs-play.

Wilson, Pamela. "Black Culture and Sexuality." *Journal of Social Work and Human Sexuality* 4, 3 (March 1986): 29–46.

Wilson, S. M., and N. P. Medora. "Gender Comparisons of College Students' Attitudes toward Sexual Behavior." *Adolescence* 25, 99 (September 1990): 615–627.

Wilson, William Julius. *The Truly Disadvantaged: The Inner City, the Underclass, and Public Policy.* Chicago: University of Chicago Press, 1987.

Wineberg, H. "Marital Reconciliation in the United States: Which Couples Are Successful?" *Journal of Marriage and the Family* 56, 1 (February 1994): 80–88.

Wineberg, H. "The Timing of Remarriage among Women Who Have a Failed Marital Reconciliation in the First Marriage." *Journal of Divorce and Remarriage* 30, 3/4 (June 1999): 57–69.

Wineberg, H., and J. McCarthy. "Separation and Reconciliation in American Marriages." *Journal of Divorce & Remarriage* 20, 1–2 (1993): 21–42.

Winslow, Sarah. "Work–Family Conflict, Gender, and Parenthood, 1977–1997." *Journal of Family Issues* 26, 6 (2005): 727–755.

Winslow-Bowe, S. "The Persistence of Wives' Income Advantage." *Journal of Marriage and Family* 68 (November, 2006): 824–842.

Winton, Chester. *Frameworks for Studying Families.* Guilford, CT: Dushkin, 1995.

———. *Children as Caregivers: Parental and Parentified Children.* Boston: Pearson Allyn and Bacon, 2003.

Wisdom, S., and J. Green. *Stepcoupling: Creating and Sustaining a Strong Marriage in Today's Blended Family.* New York: Crown Publishing, 2002.

Wise, T. N., S. Epstein, and R. Ross. "Sexual Issues in the Medically Ill and Aging." *Psychiatric Medicine* 10 (1992): 169–180.

Wohl, M., and A. McGrath. "The Perception of Time Heals All Wounds: Temporal Distance Affects Willingness to Forgive Following an Interpersonal Transgression." *Personality and Social Psychology Bulletin* 33, 7 (2007): 1023–1035.

Wolf, Michelle, and Alfred Kielwasser. "Introduction: The Body Electric: Human Sexuality and the Mass Media." *Journal of Homosexuality* 21, 1/2 (1991): 7–18.

Wolf, Rosalie S. "Abuse and Neglect of the Elderly." In *Vision 2010: Families and Violence, Abuse and Neglect,* edited by R. J. Gelles. Minneapolis: National Council on Family Relations, 1995.

Wolfinger, N. *Understanding the Divorce Cycle: The Children of Divorce in Their Own Marriages.* New York: Cambridge University Press, 2005.

———. "Hello and Goodbye to Divorce Reform." *National Council on Family Relations Report: Family Focus on Divorce and Relationship Dissolution,* Issue FF36 (December 2007): F15–16.

Wolfinger, Nicholas H. "Parental Divorce and Offspring Marriage: Early or Late?" *Social Forces* 82, 1 (September 2003): 337–353.

Subject Index

Homogamy, 140, **272–276**
  age, 276
  defined, **153**
  marital history and, 276–277
  religious, 275
  residential propinquity and, 277
Homophobia, **193**
Homosexuals, **187**. *See also* Gays; Lesbians
  popular culture and, 22, 23
  public attitudes toward, 191
Honeymoon effect, **229**, 284
Honeymoon tradition, 286
Honor killings, 106
Hooking up, 159–160
Hooking Up: Sex, Dating, and relationships on Campus (Bogle), 159
Hostile conflict, **246**
Hostile environment, **403**
Household, **11**
  African American, 92
  extended, **93**
  multigenerational, 17
Household work
  changes in gender-role attitudes and, 123–124
  by children, gender roles and, 117
  by cohabiting *vs.* married couples, 318
  conflict over, 251–252
  with dual earners, 392–393
  fairness in distribution of, 400
  familial division of labor and, 388–389
  gender inequality and, 107
  impact on women, 123
  marital roles and, 288
  social class and, 83
Househusbands, 389
HPV, 211, 212
Human immunodeficiency virus (HIV).
  See HIV
Human papilloma virus (HPV), 214
Hungary, crude divorce rate in, 457
Hurricane Katrina, 38
Hypotheses, **36**

I

Iceland, crude divorce rate in, 457
Identity bargaining, **289**
Identity, postdivorce, 464–466
Ideology of intensive mothering, 356–357, 359
Immersion stage, in stepfamilies, 501
Immigrants/immigration, nineteenth-century, 67–68
Impaired fecundity, **337**
Incest, **445**
Incest taboo, 445
Income
  African American, 92
  Asian Americans, 96
  divorce and, 458–459
  of dual-earner families, 392
  familial division of labor and, 387, 388
  family violence and, 427–428
  fertility rates and, 335

gender and, 107, 110
  gender disparities in, 403
  Hispanics, 94
Independent variable, **36**, 54
India
  arranged marriages in, 267
  divorce and seeking remarriage in, 458
  selective abortion in, 106
Indirect crossover, 386
Individualism, 78
Individualistic cultural values, divorce and, 457
Individualistic explanation of family violence, 422
Individualized marriage, 268
Indulgent child rearing, **363–364**
Industrialization, 65, 387
Inequality, gender and, 105–108
Infant attachment, 151
Infant mortality, **347–348**, 348
Infatuation, 150
Infertility, 337
Infidelity, 205–207
  Hite report on, 181–182
  jealousy and, 162–163
Initiator of separation, **463**
  remarriage and, 496
In-laws, marital satisfaction and, 289–290
Instrumental displays, **142**
Instrumental traits, 40, **104**, 111
Intensive mothering ideology, **126**
Interaction, **43**
Intercourse
  anal, **200**, 201
  premarital, during colonial era, 63–64
  sexual, **200**
Interdependence, romantic partners and, 39
Intergenerational transmission, **460–461**
Intermarriages, 91, 278
International Dating Violence Study, 55
Internet
  changes in families/marriage and, 75
  divorce and, 465
  mate selection and, 157
  personal ads on, 153
  sex on the, 180–181, 188, 189
Interpersonal cognitive complexity, 234
Interpersonal sexuality, 198–201
Interracial relationships, displays of affection and, 146
Intersectionality, **127**
Intersexed, **108**
Intervening variable, **36**
Interviews (survey research), 51–52
Intimacy
  communication, conflict and, 222
  conflict and, 241–243
  defined, **139**
  displays of, 146
  family violence and, 424
  of friendship, 139–140
  function of marriage/family and, 12–13
  gender and, 142–146
  of love, 139–140
  need for, 138–142
  over time, 168

power and, 238
  sexual dysfunctions and, 209
  Sternberg's triangular theory of, 150–151
  in wheel theory of love, 161
Intimate and family violence. *See* Violence
Intimate partner abuse, **419**, 443
Intimate partner violence, **419**
Intimate Portrait of First Sexual Experiences, An (Carpenter), 186
Intimate terrorism, **419**
Intimate violence and abuse. *See* Violence
Intimate zone, **225**
Intrafamilial sexual abuse, **445**
Involuntarily childless, 337
Involuntary celibate relationships, 46–47
Involuntary sex, 185
Involved grandparents, 371
Iowa, 272, 322
Iraqw of Tanzania, 8
Issei (immigrant generation), 97
Italy, 227, 457

J

Japanese-American families, 11
Jealousy, **161–163**
  gender differences in, 162–163
  managing, 163
  reactive, 161
  suspicious, 161
  test of love and, 161
Jews, marriage and, 267
Job sharing, 412
Joint custody, **477**
Joint legal custody, **477**
Joint physical custody, **477**
Journal of Marriage and the Family (journal), 262

K

Kansas, 322
Katrina disaster, 38
Kin relationships, 98
Kinship system, **18–19**
  among immigrants, 68
  Hispanic families, 95
Kissing, 199
Korean immigrants, marital conflict in, 244

L

Labor force, women in. *See* Working women
Language, diversity and, 90
Later-life marriage, 294–298
Latinos/Latinas. *See also* Hispanics
  family and, 11
  gender roles, 125–126
  HIV/AIDS among, 213
  love style and, 150
  machismo, 127
  on motherhood, 126
  poverty and, 81
  proximity and, 226

rape, **428–429**
reasons for remaining in relationships with, 432–433
responding to, 442–444
sibling, 438–440
types of intimate, 419–420
in video games, 125
women as perpetrators of, 426–427
Violent resistance, **420**
Virginity, 178, 185–**186**
Virginity loss, 178
different meanings of, 186–187
gender and, 187
Virtual Family Dinner, 18
Vital marriages, 298, 299
Voice-Male (Chethik), 210
Vows, wedding, 262

W

Wages. *See* Income
Washington (state), 272, 323
Washington D.C., 272
The Way We Really Are: Coming to Terms with America's Changing Families (Coontz), 475
"We are Family" video, 22
Weblog (blog), 180–191
Weddings, 285–286
Wedding vows, 262
Wheel theory of love, 160–161
White ethnicity, 99–100. *See also* Caucasians
adolescent sexual behavior, 181, 185, 186
anal eroticism, 201
child abuse and, 435
cohabitation, 313
divorce and, 454, 459
family violence and, 428
oral-genital sex among, 201
pregnancies among, 341, 342
racial intermarriage and, 272–274
remarriage and, 496
sexual experiences, 202
sexually transmitted infection rate among, 213
shift work by, 398
single-parent families and, 488
unmarried Americans, 305

Who Wants to Marry a Multi-Millionaire (television show), 153
Widowhood, 296–298
Wife Swap (television program), 290
Wisconsin, 323
Wishfuls, **308**
Women. *See also* Gender differences; Mothers/motherhood; Working women
alimony paid to, 468
changes in sexual responsiveness in, 196
changing opportunities for, 76–77
colonial era, 64
conflict and, 242
contribution to household economy, 13
dating and, 158
economic impact of divorce on, 466–467
family division of labor and, 387–388
family-to-work spillover and, 385
fertility/childbearing rates, 334–336
involuntarily childless, 337
later-life marriage and, 294–295
life pattern of, 487
middle-class, 83
nineteenth-century, 66
as perpetrators of violence, 426–427
postpartum blues in, 353
in poverty, 82–83
remaining in violent relationships, 432–433
roles in families and work, 124–126
sexual double standard and, 176
sexual script of, 175
in stepfamilies, 502–503
time spent on housework, 392–393
twentieth-century, 69
upper-class, 83
as victims of violence, 425–426
virginity and, 186–187
voluntarily childless, 337–338
widowed, 297
Women and Love (Hite), 181–182
Women's movement(s), 77, 129–132
Work. *See also* Employment; Household work; Working women
crossover and, **386**
discrimination against women, 403
Family and Medical Leave Act (FMLA), **413**

family life and, 380–381
family policy, **411–413**
home life affected by, 383–384
overwork, 382–383
time spent on, 381–382
unemployment, 408–411
work and family spillover, **383–386**
Working class, **80**
childrearing, 84, 85
gender roles and, 123
marriage relationships, 83
Working poor, the, 81–82
Working women, 69. *See also* Dual-earner marriages/families
African American, 93
at-home fathers and, 401–402
changes in marriage/families and, 74
employment patterns, 391–392
gender disparities in earnings and, 403
gender-role attitudes and, 123, 126
during the Great Depression and World Wars, 70–71
historical development, 74, 389
rates of maternal labor force participation, 389–390
reasons for increase in, 390–391
self-criticism and, 128–129
sexual harassment and, **403**
Workplace, gender learning in, 123
Work spillover, **383–385**
Work-to-family spillover, 383–384
World War I, 70
World War II, 69, 70

Y

Yemen, 106
Young adulthood. *See also* Adolescence/adolescents; Teenagers
media influence, 182, 183
patterns for women and men, 187
sex on the Internet and, 180–181
sexual development tasks, 182–184
sexuality in, 182–187
unwanted and forced sex, 185
virginity loss, 185–187, **186**

Z

Zits (comic strip), 76